Handbook of Nanomaterials for Cancer Theranostics

Handbook of Nanomaterials for Cancer Theranostics

Edited by

João Conde

ELSEVIER

Elsevier
Radarweg 29, PO Box 211, 1000 AE Amsterdam, Netherlands
The Boulevard, Langford Lane, Kidlington, Oxford OX5 1GB, United Kingdom
50 Hampshire Street, 5th Floor, Cambridge, MA 02139, United States

Library of Congress Cataloging-in-Publication Data
A catalog record for this book is available from the Library of Congress

British Library Cataloguing-in-Publication Data
A catalogue record for this book is available from the British Library

ISBN: 978-0-12-813339-2

Publisher: Matthew Deans
Acquisition Editor: Simon Holt
Editorial Project Manager: Joshua Mearns
Production Project Manager: Kamesh Ramajogi
Cover Designer: Miles Hitchen

Typeset by SPi Global, India

Contents

Contributors

Marta Alonso-Nocelo Autónoma University of Madrid (UAM); "Alberto Sols" Biomedical Research Institute, CSIC-UAM; Ramón y Cajal Health Research Institute (IRYCIS), Madrid, Spain

Ramon Alvarez-Puebla Department of Physical and Inorganic Chemistry - EMaS, Universitat Rovira i Virgili, Tarragona; ICREA, Barcelona, Spain

Tomás Bauleth-Ramos University of Helsinki, Helsinki, Finland; University of Porto, Porto, Portugal

Mattias Björnmalm Imperial College London, London, United Kingdom

Belén López Bouzo Health Research Institute of Santiago de Compostela (IDIS), SERGAS, CIBERONC, Santiago de Compostela, Spain

Michael J. Burgum Swansea University Medical School, Institute of Life Sciences, Swansea, Wales, United Kingdom

Semih Calamak Stanford University, Palo Alto, CA, United States; Hacettepe University, Ankara, Turkey

Gizem Calibasi Kocal Stanford University, Palo Alto, CA, United States; Dokuz Eylul University, Izmir, Turkey

Ai-Zheng Chen Huaqiao University, Xiamen, PR China

Ciro Chiappini Centre for Craniofacial and Regenerative Biology, King's College London, London, United Kingdom

Martin J.D. Clift Swansea University Medical School, Institute of Life Sciences, Swansea, Wales, United Kingdom

Ana F. Cruz University of Coimbra, Coimbra, Portugal

Daxiang Cui Shanghai Jiao Tong University, Shanghai, PR China

Jiwei Cui Shandong University, Jinan, China

María de la Fuente Health Research Institute of Santiago de Compostela (IDIS), SERGAS, CIBERONC, Santiago de Compostela, Spain

Utkan Demirci Stanford University, Palo Alto, CA, United States

Shareen H. Doak Swansea University Medical School, Institute of Life Sciences, Swansea, Wales, United Kingdom

Naside G. Durmus Stanford University, Palo Alto, CA, United States

Menekse Ermis Middle East Technical University, Ankara, Turkey; Stanford University, Palo Alto, CA, United States

Stephen J. Evans Swansea University Medical School, Institute of Life Sciences, Swansea, Wales, United Kingdom

Kelong Fan Institute of Biophysics, Chinese Academy of Sciences, Beijing, China

Michelle E. Farkas University of Massachusetts Amherst, Amherst, MA, United States

Patrícia Figueiredo University of Helsinki, Helsinki, Finland

Nuno A. Fonseca University of Coimbra; TREAT U, SA, Coimbra, Portugal

Ana C. Gregório University of Coimbra, Coimbra, Portugal

Luca Guerrini Department of Physical and Inorganic Chemistry - EMaS, Universitat Rovira i Virgili; Centre Tecnològic de la Química de Catalunya, Tarragona, Spain

Sinan Guven Stanford University, Palo Alto, CA, United States; Dokuz Eylul University, Izmir, Turkey

Kimberly Hamad-Schifferli University of Massachusetts, Boston, MA, United States

Jingcheng Hao Shandong University, Jinan, China

Joseph Hardie University of Massachusetts Amherst, Amherst, MA, United States

Tayyaba Hasan Harvard Medical School, Boston; Harvard University and Massachusetts Institute of Technology, Cambridge, MA, United States

Nesrin Hasirci Middle East Technical University, Ankara, Turkey

Vasif Hasirci Middle East Technical University, Ankara, Turkey

Ting He Shenzhen University, Shenzhen, China

Jouni Hirvonen University of Helsinki, Helsinki, Finland

Chih-Chia Huang National Cheng Kung University, Tainan, Taiwan

Yan Huang Shenzhen University, Shenzhen, China

Peng Huang Shenzhen University, Shenzhen, China

Shuaidong Huo National Center for Nanoscience and Technology, Chinese Academy of Sciences (CAS), Beijing, China

Roshni Iyer University of Texas at Arlington, Arlington, TX, United States

Gareth J. Jenkins Swansea University Medical School, Institute of Life Sciences, Swansea, Wales, United Kingdom

Ying Jiang University of Massachusetts Amherst, Amherst, MA, United States

Ranjith K. Kankala Huaqiao University, Xiamen, PR China

Chien-Wei Lee National Cheng Kung University, Tainan, Taiwan

Yue Li University of Macau, Macao SAR, China

Fan Li Shenzhen University, Shenzhen, China

Xing-Jie Liang National Center for Nanoscience and Technology, Chinese Academy of Sciences (CAS), Beijing, China

Eliana M. Lima Federal University of Goiás, Goiânia, Brazil

Jing Lin Shenzhen University, Shenzhen, China

Tzu-Ming Liu University of Macau, Macao SAR, China

Gang Liu Xiamen University, Xiamen, China

Rui M. Lopes University of Coimbra, Coimbra, Portugal

Davide A. Martella Centre for Craniofacial and Regenerative Biology, King's College London, London, United Kingdom

Livia P. Mendes Northeastern University, Boston, MA, United States; CAPES Foundation, Ministry of Education of Brazil, Brasilia, Brazil

João Nuno Moreira University of Coimbra, Coimbra, Portugal

Vera Moura University of Coimbra; TREAT U, SA, Coimbra, Portugal

Kytai T. Nguyen University of Texas at Arlington, Arlington; University of Texas Southwestern Medical Center, Dallas, TX, United States

Mary Okesola Centre for Craniofacial and Regenerative Biology, King's College London, London, United Kingdom

Dona Padanilam University of Texas at Arlington, Arlington, TX, United States

Adriele Prina-Mello Department of Clinical Medicine; Centre for Research on Adaptive Nanostructures and Nanodevices (CRANN) Institute and AMBER Centre; Trinity Translational Medicine Institute, Trinity College Dublin, Dublin, Ireland

Eduardo G. Rico Fundacion de Investigacion HM Hospitales, Madrid; Centro Integral Oncologico Clara Campal (CIOCC), Madrid; Hospital Universitario HM Torrelodones, Torrelodones; San Pablo CEU, Madrid, Spain

Imran Rizvi Harvard Medical School, Boston, MA, United States

Cristina Rodriguez-Quijada University of Massachusetts, Boston, MA, United States

Vincent M. Rotello University of Massachusetts Amherst, Amherst, MA, United States

Juan Ruiz-Bañobre Health Research Institute of Santiago de Compostela (IDIS), SERGAS, CIBERONC, Santiago de Compostela, Spain

Hélder A. Santos University of Helsinki; Helsinki Institute of Life Science, HILIFE, Helsinki, Finland

Bruno Sarmento University of Porto, Porto, Portugal; CESPU, Gandra, Portugal

Sérgio Simões University of Coimbra, Coimbra, Portugal

Tuanwei Sun Shenzhen University, Shenzhen, China

Vladimir P. Torchilin Northeastern University, Boston, MA, United States

Ângela Valério-Fernandes University of Coimbra, Coimbra, Portugal

Abi Judit Vázquez-Ríos Health Research Institute of Santiago de Compostela (IDIS), SERGAS, CIBERONC, Santiago de Compostela, Spain

Shi-Bin Wang Huaqiao University, Xiamen, PR China

Yu-Chi Wang National Cheng Kung University, Tainan, Taiwan

Ling Wang Shandong University, Jinan, China

Zhe Wang The Chinese University of Hong Kong, Hong Kong SAR, China

Junqing Wang The Chinese University of Hong Kong, Hong Kong SAR; Xiamen University, Xiamen, China

Joseph Wolf University of Texas at Arlington, Arlington, TX, United States

Xiyun Yan Institute of Biophysics, Chinese Academy of Sciences; University of Chinese Academy of Sciences, Beijing, China

Yu S. Zhang Harvard Medical School, Cambridge, MA, United States

Daria Zhukova University of Texas at Arlington, Arlington, TX, United States

Preface

As a disease, cancer is especially suitable for a theranostic approach. The use of nanomedicine for cancer is based on the enhanced permeability and retention (EPR) effect (Matsumura & Maeda, 1986). Recent reports have established the existence of the EPR effect in humans (Clark et al., 2016; Harrington et al., 2001). The leaky vasculature of tumors allows selective accumulation and retention of intravenously administered nanoparticles. However, the degree of the EPR effect is highly variable among different individuals and cancer types. It is imperative to ascertain that the medicine carried by the nanoparticles can get to the tumors. Thus, a theranostic nanoparticle will report if the tumors in the patients can be accessed by the nanoparticles. If the EPR effect in the patient is low, there is no point in injecting the nanoparticles. Thus, nanomedicine for cancer should be an individualized medicine. Theranostic nanomedicine promises to fulfill this need.

Nanoparticles can be functionalized for prolonged circulation, target specificity, responsiveness to the tumor microenvironment, and controlled release of the cargo in the tumor. These features have been extensively studied over the last two decades. The nanomedicines approved for clinical use are mostly liposomal anticancer drugs. Many of these drugs are actively loaded into the liposomes with a gradient of ammonium sulfate, such as in the case of Doxil™. Lee et al. used the residual ammonium sulfate gradient to load ^{64}Cu into a Doxil-like liposome formulation and showed a positive EPR effect in metastatic breast cancer patients using position emission tomography. This is an excellent example for how cancer theranostics can benefit patients (Lee et al., 2015).

I am very enthusiastic about cancer nanomedicine, especially the theranostic approach.

— Professor Leaf Huang

University of North Carolina at Chapel Hill, Chapel Hill, NC, United States

References

Clark, A. J., Wiley, D. T., Zuckerman, J. E., Webster, P., Chao, J., Lin, J., et al. (2016). CRLX101 nanoparticles localize in human tumors and not in adjacent, nonneoplastic tissue after intravenous dosing. *Proceedings of the National Academy of Sciences, 113*(14), 3850–3854. https://doi.org/10.1073/pnas.1603018113.

Harrington, K. J., Mohammadtaghi, S., Uster, P. S., Glass, D., Peters, A. M., Vile, R. G., et al. (2001). Effective targeting of solid tumors in patients with locally advanced cancers by radiolabeled pegylated liposomes effective targeting of solid tumors in patients with locally advanced cancers by radiolabeled pegylated liposomes. *Clinical Cancer Research*, *7*(February), 243–254.

Lee, H., Zheng, J., Gaddy, D., Orcutt, K. D., Leonard, S., Geretti, E., et al. (2015). A gradient-loadable 64Cu-chelator for quantifying tumor deposition kinetics of nanoliposomal therapeutics by positron emission tomography. *Nanomedicine: Nanotechnology, Biology and Medicine*, *11*(1), 155–165. https://doi.org/10.1016/j.nano.2014.08.011.

Matsumura, Y., & Maeda, H. (1986). A new concept for macromolecular therapeutics in cnacer chemotherapy: Mechanism of tumoritropic accumulatio of proteins and the antitumor agents Smancs. *Cancer Research*, *46*(12 Pt 1), 6387–6392. https://doi.org/10.1021/bc100070g.

CHAPTER 1

The Emerging Role of Multifunctional Theranostic Materials in Cancer Nanomedicine

Patrícia Figueiredo[*], Tomás Bauleth-Ramos[*,†], Jouni Hirvonen[*], Bruno Sarmento[†,‡], Hélder A. Santos[*,§]

[*]*University of Helsinki, Helsinki, Finland* [†]*University of Porto, Porto, Portugal* [‡]*CESPU, Gandra, Portugal* [§]*Helsinki Institute of Life Science, HILIFE, Helsinki, Finland*

1 Introduction

Cancer has one of the highest incidence and mortality rates of all diseases worldwide. According to the World Health Organization (WHO), there were 14.1 million new cases and 8.2 million deaths in 2012 (Ferlay et al., 2015). However, this is expected to increase in the future to 20 million new cancer cases per year by 2025 (Ferlay et al., 2015). Cancer causes cells to divide and grow in an uncontrolled manner. The tumor cells experience several transformations (caused by their genetic instability and multiple molecular alterations) which are responsible for inexhaustible replication and ability to invade the surrounding tissue and spread to other organs. Ultimately, the failure to control or prevent such a spread of cancerous cells often leads to patient death (Mitra et al., 2015). The most commonly diagnosed cancers were lung, breast, colorectal, and prostate, whereas lung, liver, and stomach cancers were responsible for the most deaths (Ferlay et al., 2015).

Currently, cancer treatment can include initial chemotherapy to shrink the tumor, followed by surgical intervention to remove the tumor (if possible), and then more chemotherapy combined with radiation (Brannon-Peppas & Blanchette, 2012). The main goal of chemotherapeutic drugs and radiation is to kill tumor cells, because their ability to divide and grow faster than healthy cells makes them more susceptible to these treatments (Brannon-Peppas & Blanchette, 2012). However, conventional chemotherapeutic drugs are far from being ideal mainly due to their toxicity, tumor resistance, and ability to kill healthy cells, all of which can cause severe side effects and ultimately lead to patient death. Other issues associated with treatment safety and efficacy include short circulation

Handbook of Nanomaterials for Cancer Theranostics. https://doi.org/10.1016/B978-0-12-813339-2.00001-3

time and difficulty in targeting these drugs to the tumor sites (Chakraborty & Rahman, 2012). Taking into account the intrinsic limitations of conventional cancer therapy, cancer nanomedicine has emerged as a promising tool to fight cancer (Misra, Acharya, & Sahoo, 2010). The development of nanotechnology opens up an opportunity to achieve sophisticated multifunctional and targeting strategies, allowing the simultaneous detection, diagnosis, and treatment of cancer. Nanomedicines present the ability to improve the pharmacokinetic and pharmacodynamic profiles of conventional therapeutics and optimize the efficacy of these compounds (Wicki, Witzigmann, Balasubramanian, & Huwyler, 2015). Furthermore, the growing potential utility of nanoparticles plays a critical role in the development of theranostic platforms in cancer, combining both diagnostic and therapeutic abilities (Ahmed, Fessi, & Elaissari, 2012).

In this chapter, we present an overview of the development of multifunctional nanomedicine platforms for theranostic applications, highlighting some of the most advanced drug delivery systems, targeting strategies, and nanotools used for cancer imaging and diagnostics (Fig. 1).

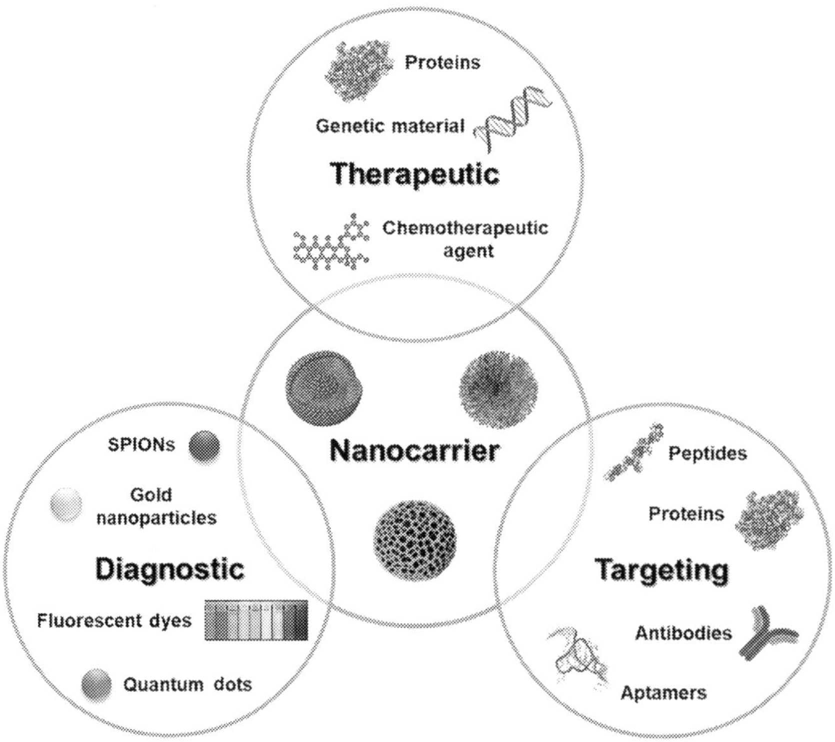

Fig. 1
General overview of the multifunctional materials addressed in this chapter for cancer theranostic purposes.

2 Rational Design of Nanocarriers for Cancer

Nanotechnology has been emerging as an important tool in the oncology field over the past few decades due to nanosystems' unique physicochemical properties, such as nanoscale size, high surface-to-volume ratio, different chemical composition, and the presence of targeting ligands that can be attached to modify the nanoparticles' surfaces (Fig. 2;

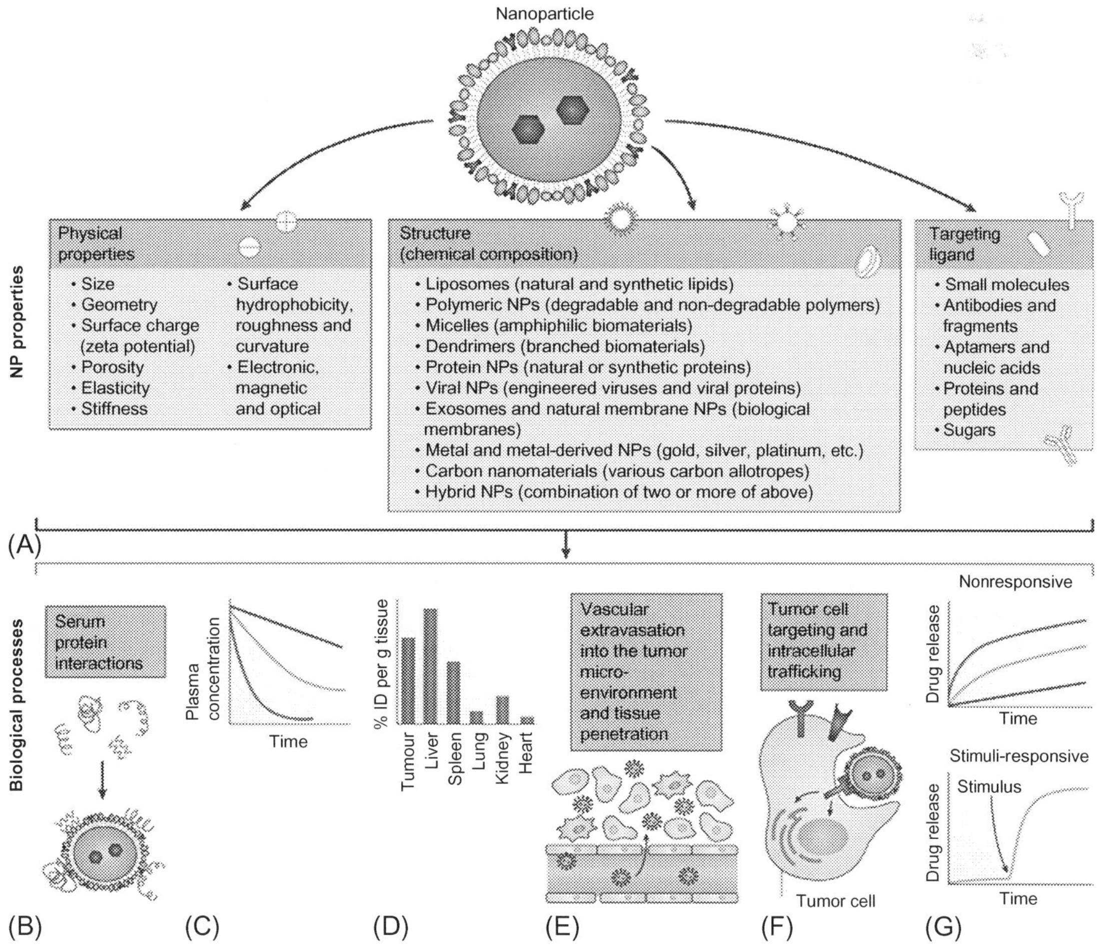

Fig. 2

Physicochemical characterization and chemical composition of nanoparticles and targeting ligands that can be attached to modify the nanoparticles' surfaces (A). These properties can affect different biological processes, such as interaction with serum proteins (B), blood circulation (C), biodistrubution evaluated by percentage of injected dose (ID) per gram of tissue (D), vascular extravasation to the tumor microenvironment through the leaky tumor vessels and penetration within the tumor tissue (E), tumor cell targeting and cellular uptake (F), and controlling the release profiles of the cargo (G). *Reproduced with permission from Shi, J., Kantoff, P. W., Wooster, R., & Farokhzad, O. C. (2017). Cancer nanomedicine: progress, challenges and opportunities. Nature Reviews Cancer, 17, 20–37, with Nature Publishing Group, copyright 2017.*

Shi, Kantoff, Wooster, & Farokhzad, 2017; Wicki et al., 2015). Consequently, they affect the biological response by modulating both the pharmacokinetic and pharmacodynamic profiles of drugs, enhancing their therapeutic index. Drug loading into nanocarriers can increase in vivo stability and extend the drug's blood circulation time in a controlled manner (Balasubramanian, Liu, Hirvonen, & Santos, 2017). Nanomedicines can also modify drug biodistribution by preferentially accumulating the drug at the tumor site, because of the enhanced permeability and retention (EPR) effect (Shi et al., 2017). Moreover, targeting moieties attached to the nanosystems allow the active targeting of nanoparticles, enhancing the cellular uptake by increasing the endocytosis of the nanoparticles and consequently the delivery of the cargo (Byrne, Betancourt, & Brannon-Peppas, 2008). In addition, stimuli-responsive nanosystems can act and release their cargo in response to specific stimuli that can be exogenous (e.g., variations in temperature, magnetic field, ultrasound intensity, light, and electric pulses) or endogenous (e.g., changes in pH, enzyme concentration, or redox gradients) (Mura, Nicolas, & Couvreur, 2013). Finally, advanced multifunctional nanosystems with theranostic ability have been developed to provide both diagnosis and treatment in the same nanoformulation (Ahmed et al., 2012).

In the following subsections, general strategies and advanced functionalities of the nanosystems to improve the drug's therapeutic index and targeting of nanocarriers towards to the tumor are discussed, including passive targeting, active targeting, and stimuli-responsive systems, as well as the development of multifunctional systems for theranostic purposes.

2.1 Passive Targeting

General features of tumors and their microenvironment have been exploited for passive targeting, especially the EPR effect that allows the accumulation of nanocarriers in the tumor (Fig. 3). Along with nanocarriers whose surface is tailored for prolonged blood circulation times, this concept is referred to as passive targeting. This effect happens due to the presence of leaky intratumoral blood vessels, whose endothelia display large fenestrations from 100 to 780 nm (Hirsjarvi, Passirani, & Benoit, 2011; Hobbs et al., 1998). The tumor neovasculature, which is characterized by the formation of new blood vessels, increases the permeability of the blood vessels in tumors and leads to the extravasation of nanoparticles from the bloodstream to the tumor site. Moreover, the compromised lymphatic drainage in tumors retains the accumulated nanocarriers and allows them to release their payloads into the vicinity of the tumor cells (Peer et al., 2007). In this way, several studies using liposomes with different average sizes showed that the threshold particle size for extravasation into tumors is ~400 nm, whereas other studies suggested that particles with average sizes of <200 nm are more effective (Deshpande, Biswas, & Torchilin, 2013).

However, passive targeting approaches are not sufficient to control the side effects of cytotoxic drugs, and they present several limitations. First, targeting cells within a tumor is

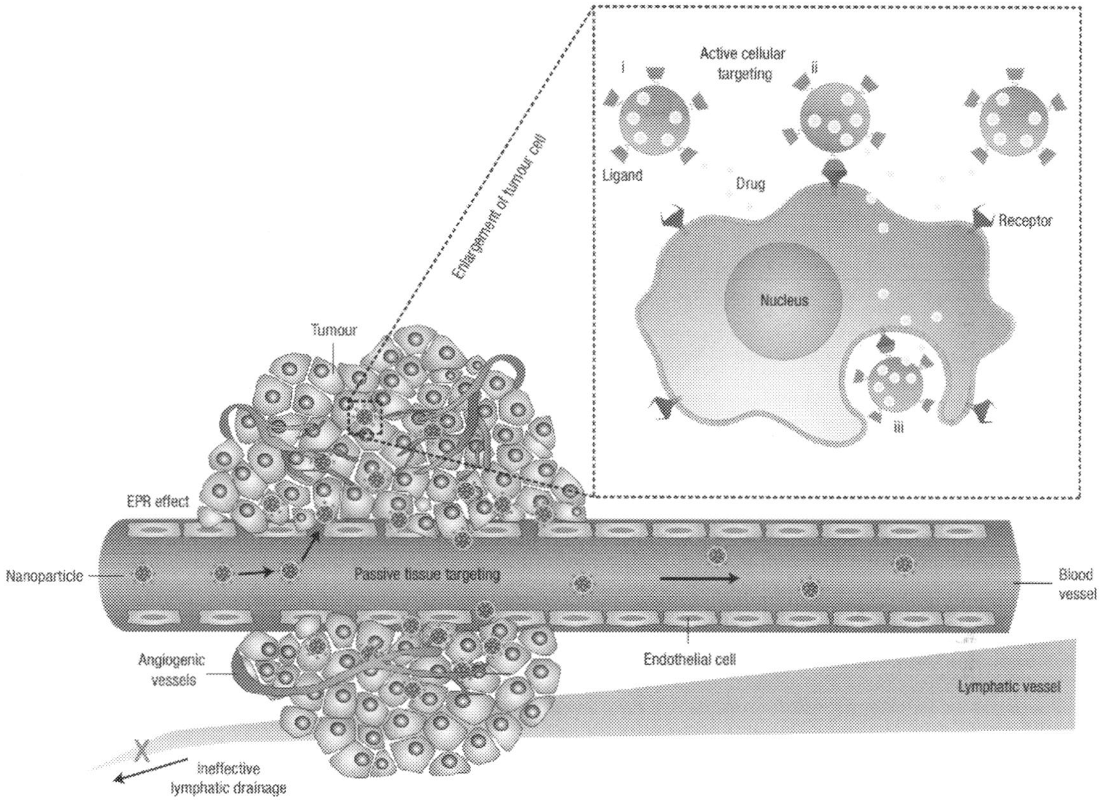

Fig. 3

Schematic representation of passive targeting, accomplished by extravasation of nanoparticles through increased permeability of the tumor vasculature and defective lymphatic drainage (EPR effect), and active targeting, achieved by functionalizing the nanoparticles' surfaces with ligands that promote cellular interaction. The nanoparticles can (i) release their cargo in close proximity to the target cells; (ii) attach to the membrane of the cell and act as an extracellular sustained-release drug depot; or (iii) internalize into the cell. *Reproduced with permission from Peer, D., Karp, J. M., Hong, S., Farokhzad, O. C., Margalit, R., & Langer, R. (2007). Nanocarriers as an emerging platform for cancer therapy. Nature Nanotechnology, 2, 751–760, with Nature Publishing Group, copyright 2007.*

not ubiquitously achievable because the random nature of this approach makes it difficult to efficiently control drug diffusion. Consequently, this defective control can lead to multiple-drug resistance (MDR), in which treatment with chemotherapeutic drugs fail due to cancer cells' resistance to those drugs (Nakamura, Mochida, Choyke, & Kobayashi, 2016; Peer et al., 2007). The MDR phenomenon can be due to the overexpression of transporter proteins on the surfaces of cancer cells, which expels the drugs from the cells, lowering their therapeutic effect, and ultimately leads to development of resistance to a variety of drugs (Peer et al., 2007). Second, certain tumors do not exhibit the EPR effect, and the endothelial pore sizes are different in primary and metastatic tumors, as well as within the same tumor type. Consequently, the permeability of vessels may not be the same throughout a tumor (Sriraman, Aryasomayajula, & Torchilin, 2014). For example, the extracellular matrix of certain tumors, such as pancreatic cancer, limits drug diffusion into the tumor (Provenzano et al., 2012). Finally, passive targeting does not prevent accumulation of nanocarriers in other organs with fenestrated endothelia, for example, the liver and spleen, causing side effects in these tissues (Gaumet, Vargas, Gurny, & Delie, 2008). Therefore, development of next-generation nanosystems with an active-targeting ligand or stimuli-responsive properties is needed (Wicki et al., 2015).

2.2 Active Targeting

The second generation of nanomedicines with an active targeting moiety has been developed in order to overcome the limitations given by the passive targeting discussed above. In this way, these nanocarriers can complement the EPR effect by actively binding to specific cells after extravasation, increasing the efficacy of the payloads and minimizing side effects (Fig. 3) (Peer et al., 2007). The decoration of nanoparticles can be achieved by attaching targeting ligands to the surface of the nanocarrier using different conjugation chemistries. The most commonly used targeting agents are classified as proteins (mainly antibodies and their fragments), nucleic acids (aptamers), or other receptor ligands (peptides, vitamins, and carbohydrates) (Friedman, Claypool, & Liu, 2013). The ligand must be chosen to allow the binding to the target cells and minimize the binding to healthy cells. Thus, the nanocarriers bind to the target cells through ligand-receptor interactions, and they are internalized before the drug is released inside the cell. Here, the targeting specificity and the delivering capacity are two significant issues to take into consideration when evaluating the efficiency of an active targeting system. Targeting specificity depends upon the biodistribution of the ligand-functionalized nanoparticles and how they interact with cells and off-target molecules, whereas the delivering capacity is determined by the nanoparticle material and structure (Bertrand, Wu, Xu, Kamaly, & Farokhzad, 2014). Generally, the targeting agent used to deliver nanosystems to cancer cells needs to bind with high selectivity to the molecules expressed specifically on the cell surface. In addition, the surface molecule should be overexpressed on the target cells, as compared to the normal cells, in order to maximize specificity (Peer et al., 2007). Another aspect to consider is the nanocarriers' stability, to avoid premature drug release and its degradation in the bloodstream. Therefore, the density of the targeting ligand should be optimized in order to maintain stealth properties and avoid rapid recognition and clearance by the reticule-endothelial system (Wicki et al., 2015).

2.3 Stimuli-Responsive Nanosystems

Stimuli-responsive nanocarriers have been developed to improve the biological specificity and therapeutic efficacy of the payloads. The advantage of using stimuli-responsive nanosystems is that the drug is released in response to specific stimuli, minimizing systemic exposure to the compound (Qian et al., 2017). The triggers for the drug release can be exogenous (e.g., variations in temperature, magnetic field, ultrasound intensity, light and electric pulses) or endogenous (e.g., changes in pH, enzyme concentration, and redox gradients) (Mura et al., 2013).

In the case of endogenous stimuli, solid tumors often have more acidic microenvironments (pH 6.5–7.2) than the physiological pH of 7.4 in blood and normal tissues; this can be exploited for tumor specific delivery, either by developing pH-sensitive nanocarriers that release their cargo at the tumor site, or by using pH-sensitive chemical bonds that are stable at neutral pH but are cleaved under acidic conditions (Du, Lane, & Nie, 2015). Hypoxia is characterized by low oxygen pressure and poor nutrient levels in tumor tissues. Consequently, the redox potential in cancer cells differs from normal tissue, and this difference can be exploited strategically for selectively killing tumor cells (Wilson & Hay, 2011). The tumor environment is rich in reductive agents that can act as a trigger for redox-responsive nanocarriers. For example, the presence of higher glutathione (GSH) levels in cancer cells compared to normal cells can work as a trigger to release the drug (Estrela, Ortega, & Obrador, 2006). This can be achieved using reductively degradable micelles from self-assembled, amphiphilic copolymers containing disulfide bonds, or containing GSH-sensitive crosslinking agents, as the disulfide bonds are prone to rapid cleavage by glutathione (Mura et al., 2013). Enzymes such as proteases, phospoholipases, or glycosidases present an altered expression profile in cancer, which can be used to achieve enzyme-mediated drug release. Some reports suggested the use of short peptide sequences that are cleavable by matrix metalloproteinases, leading to the release and accumulation of the cargo at the tumor site (Harris et al., 2008).

On the other hand, the exogenous stimuli-responsive nanocarriers allow a precise drug release in a spatial, temporal, and dose-controlled manner. Thermo-responsive drug delivery systems take advantage of the local hyperthermia that increases vascular permeability in the tumor, the variation in the surrounding temperature inducing the release of drugs from the nanocarriers. Ideally, thermosensitive nanocarriers should retain their cargo at body temperature (~37°C), and deliver the drug in the tumor (~39–42°C) (Needham, Anyarambhatla, Kong, & Dewhirst, 2000). Additionally, magnetically responsive nanosystems enable transportation of the drug to a specific target using a magnetic field; this process relies on the different forms that magnetic responses can take: (1) magnetic guidance under a permanent magnetic field; (2) application in an alternating magnetic field; and (3) both, alternatively used. Iron oxide nanoparticles have been extensively studied for this purpose, and they can present different chemical compositions, for example, magnetite (Fe_3O_4) or maghemite (γ-Fe_2O_3). Furthermore, iron oxide nanoparticles can be encapsulated into different colloidal systems that become sensitive to an external magnetic field, releasing the encapsulated drug in a controlled way. This allows magnetic resonance imaging (MRI) and combines diagnostics and therapy into a single system for theranostic purposes (Estelrich, Escribano, Queralt, & Busquets, 2015; Mura et al., 2013). Ultrasounds are also an effective and noninvasive strategy to spatiotemporally control the release of drugs and contrast agents at a specific site. Photo-responsive systems have been developed using light as an external stimulus to remotely control drug release in response to different wavelengths in the ultraviolet, visible, or near-infrared (NIR) regions (Mura et al., 2013). Finally, electro-responsive systems use electric fields to achieve a pulsed or sustained drug release. For example, conductive polypyrrole nanoparticles can tailor the release profile with an excellent spatial, temporal, and dosage control (Ge, Neofytou, Cahill, Beygui, & Zare, 2012).

Overall, the combination of nanocarriers with a stimuli-responsive strategy is a promising opportunity to integrate imaging and therapeutic ability into a single nanosystem for theranostic purposes, leading to real-time monitoring of the drugs and assessment of in vivo therapeutic efficacy.

3 Nanomedicines as Platforms for Drug Delivery, Diagnosis, and Imaging

Over the past decade, research into theranostic application of nanoparticles has been increasing exponentially. Analyzing publications about theranostic nanoparticles in the PubMed database, we see that more than 90% of the research papers were published in the last five years (https://www.ncbi.nlm.nih.gov/pubmed/?term=theranostic+nanoparticles).

By combining diagnosis and treatment in the same platform, theranostic agents are able to monitor the accumulation of nanomedicine compounds at a specific site, visualize their biodistribution, trigger drug release, and evaluate therapeutic efficacy (Wicki et al., 2015). Ideal theranostic nanoparticles should be able to: (1) selectively and rapidly accumulate in target tissue, (2) report biochemical and morphological characteristics of the target area, (3) effectively deliver the therapeutic agent without damaging healthy organs, and (4) be safe and biodegradable with nontoxic byproducts (Jokerst & Gambhir, 2011).

Theranostic nanoparticles can be engineered by conjugating or loading therapeutic compounds onto the imaging nanoparticles, including quantum dots (QDs), magnetic nanoparticles, and gold nanoparticles. Another strategy includes the encapsulation/conjugation of the therapeutic compounds and imaging agents (e.g., fluorescent dyes, QDs, superparamagnetic iron oxide nanoparticles (SPIONs), and gold nanoparticles) into biocompatible nanoplatforms, such as liposomes, polymeric nanoparticles, dendrimers, and porous silicon nanoparticles (Chen, Ehlerding, & Cai, 2014; Muthu, Leong, Mei, & Feng, 2014). In addition, several surface modifications with poly(ethylene glycol) (PEG) and/or different targeting ligands can be performed in order to increase the blood circulation time and provide an active targeting towards the tumor tissue (Chen et al., 2014). The characteristic features of the most commonly used nanoparticles for theranostic applications are described in Table 1. In addition, several imaging techniques can take advantage of the physicochemical properties and stimuli-responsive behavior of the nanomaterials, offering new opportunities for application of nanosystems in theranostics (Fig. 4). These techniques

Table 1: Overview of the main nanoplatforms used for theranostic applications

Type of Nanosystems	Schematic Representation	Characteristic Features for Theranostic Applications	References
Magnetic nanoparticles (< 50 nm)		■ Monodisperse ■ Superparamagnetic behavior ■ High saturation magnetization ■ *Applications:* contrast agent for MRI induce therapeutic response via hyperthermia	Ho, Sun, and Sun (2011), Mahmoudi, Sant, Wang, Laurent, and Sen (2011), and Singh and Sahoo (2014)
Gold nanoparticles (1–100 nm)		■ Different sizes and shapes (e.g., nanospheres, nanoshells and nanorods) ■ Intense light absorption ■ High photo-thermal conversion rate ■ Colloidal stability ■ Biocompatibility ■ *Applications*: light scattering imaging, two-photon fluorescence imaging, photoacoustic imaging/tomography, computed tomography, surface plasmon resonance and MRI	Akhter, Ahmad, Ahmad, Storm, and Kok (2012) and Cabral and Baptista (2014)
Quantum dots (2–10 nm)		■ High photoluminescence quantum yield ■ Broad absorption with narrow and symmetric emission spectra ■ Resistant to metabolic degradation ■ Photostable ■ Cytotoxicity ■ *Applications*: infrared fluorescence, positron emission tomography, computed tomography and photodynamic therapy	Iga, Robertson, Winslet, and Seifalian (2007), Tan et al. (2011), and Yaghini, Seifalian, and Macrobert (2009)
Liposomes (50–400 nm)	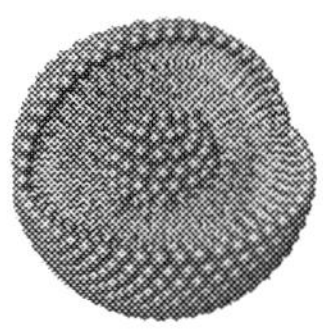	■ Ability to entrap hydrophobic and hydrophilic chemotherapeutic and imaging agents ■ High loading efficiency ■ Good stability in biological environments ■ Drug release in a controlled manner ■ Biocompatibility ■ Surface functionalization	Bozzuto and Molinari (2015), Dai and Yue (2017), and Xing, Hwang, and Lu (2016)

(Continued)

Table 1: Overview of the main nanoplatforms used for theranostic applications—cont'd

Type of Nanosystems	Schematic Representation	Characteristic Features for Theranostic Applications	References
Polymeric nanoparticles (20–600 nm)	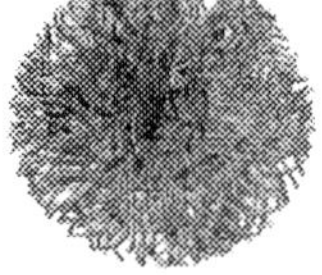	■ Colloidal stability ■ Efficient loading chemotherapeutic drugs and imaging agents ■ Surface functionalization ■ Excellent cellular uptake ■ Stimuli-responsive behavior according to the chosen polymer ■ Biodegradability ■ Biocompatibility ■ Nontoxic	Masood (2016), Prabhu, Patravale, and Joshi (2015), and Rao and Geckeler (2011)
Porous silicon nanoparticles (≈ 200 nm)	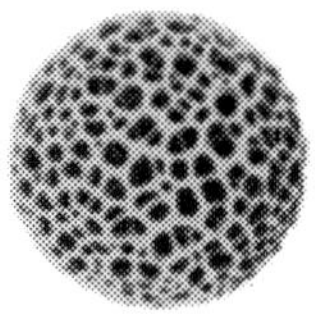	■ Large surface area ■ High loading efficiency of drugs and imaging agents ■ Drug release in a controlled manner ■ Enable the surface modification with targeting ligands ■ Biocompatibility and biodegradability	Bimbo et al. (2010), Liu et al. (2013), Santos et al. (2011), Santos, Mäkilä, Airaksinen, Bimbo, and Hirvonen (2014), and Shahbazi, Herranz, and Santos (2012)

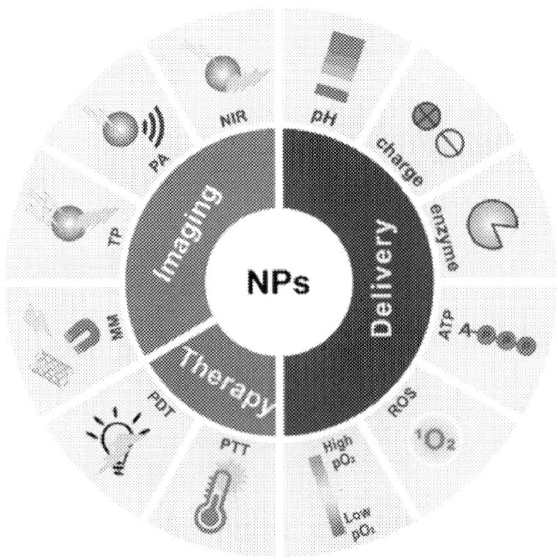

Fig. 4

Schematic representation of the rationale of using nanoparticles for theranostic applications. *Modified from Qian, C.-G., Chen, Y.-L., Feng, P.-J., Xiao, X.-Z., Dong, M., Yu, J.-C., et al. (2017). Conjugated polymer nanomaterials for theranostics. Acta Pharmacologica Sinica, 38, 764–781, with permission from Nature Publishing Group, copyright 2017.*

include near-infrared (NIR) imaging, two-photon (TP) imaging, photoacoustic (PA) imaging, and multimodal (MM) imaging, which can provide deep tissue penetration and high spatial resolution. Furthermore, photodynamic therapy (PDT) and photothermal therapy (PTT) can also be used in combination with drug delivery systems to achieve an enhanced therapeutic efficacy (Qian et al., 2017).

Inorganic nanoparticles present unique physicochemical characteristics that make them ideal functional probes for simultaneous imaging and therapeutic purposes. In this regard, magnetic nanoparticles have been widely used as drug delivery vehicles and imaging probes. Among magnetic nanoparticles, SPIONs are especially attractive for theranostic applications, due to their chemical/magnetic stability and low cytotoxicity. SPIONs have cores made of iron oxides such as magnetite (Fe_3O_4) and ferrite (Fe_2O_4) that can be directed to a specific area through use of an external magnetic field (Ho et al., 2011; Mahmoudi et al., 2011). In general, SPIONs should be small (hydrodynamic size $<50\,nm$) and monodisperse, with superparamagnetic behavior, high field irreversibility, and high saturation magnetization. Because of these properties, the particles no longer show magnetic interaction after the external magnetic field is removed (Mahmoudi et al., 2011). In addition, SPIONs also present a functional surface that allows an active targeting towards tumor tissue. Taking all of this into account, they can induce a therapeutic response via hyperthermia, and they are also efficient contrast agents for MRI (Singh & Sahoo, 2014).

Gold nanoparticles are other example of inorganic nanoparticles that can be used for theranostic applications, because of their properties such as easy synthesis that confers different sizes and shapes (e.g., nanospheres, nanoshells and nanorods) with specific optical properties, intense light absorption, high photothermal conversion rate and photostability, colloidal stability and biocompatibility, and an increased molecule loading per particle. Furthermore, gold nanoparticles can load therapeutic compounds and imaging agents simultaneously, and their surface can be functionalized with multiple targeting ligands. The main imaging modalities that include gold nanoparticles are light scattering imaging, two-photon fluorescence (TPF) imaging, photoacoustic imaging/tomography (PAT), X-ray computed tomography (CT), and surface plasmon resonance (Akhter et al., 2012; Cabral & Baptista, 2014).

Finally, quantum dots (QDs) have also emerged as contrast agents to be included in imaging platforms. QDs are semiconductor nanocrystals with hydrodynamic diameters ranging between 2 and 10 nm depending upon their fluorescence spectrum (i.e., larger QDs fluoresce in red, whereas smaller QDs fluoresce in blue) (Tan et al., 2011). According to their chemical composition, QDs can be generally divided into two groups: (1) II–VI, in which the QDs are made of elements from groups II–VI of the periodic table (e.g., cadmium selenide [CdSe] and cadmium telluride [CdTe]); and (2) III–V, where the QDs are made of elements from groups III to V (e.g., InP and P). The second category of QDs are the most commonly used because of their higher quantum yield, as compared to QDs from the first category (Iga et al., 2007). In general, QDs consist of an inorganic core, an inorganic shell, and an aqueous organic coating that allows their functionalization with targeting ligands or conjugation with chemotherapeutic drugs (Tan et al., 2011). Compared to conventional fluorophores, QDs display superior fluorescent properties. They present high photoluminescence quantum yield and broad absorption with narrow and symmetric emission spectra, resulting in a reduced cellular autofluorescence and increased sensitivity, compared to conventional fluorophores. They are also more photostable than conventional flurorphores and resistant to metabolic degradation. However, QDs present inherent cytotoxicity, which must be overcome by coating with polymers or proteins. Therefore, QDs can act as multimodal contrast imaging agents that can be employed in infrared fluorescence, positron emission tomography, CT, and photodynamic therapy (Yaghini et al., 2009).

Different nanomaterials have also been investigated as a platform for loading therapeutic compounds and imaging agents. For example, liposomes are spherical vesicles formed from lipid bilayers that can be used for encapsulating chemotherapeutics and imaging nanoparticles, due to their unique structure that can entrap hydrophobic agents within the lipid bilayers and hydrophilic agents inside the aqueous compartment, protecting the drugs from degradation (Dai & Yue, 2017). The hydrodynamic diameter of liposomes can range from a few nanometers to several micrometers. However, when used for biomedical applications, the sizes of liposomes range between 50 and 450 nm (Bozzuto & Molinari, 2015). Other intrinsic advantages of liposomes include high loading efficiency and stability in biological environments, controllable release kinetics, and biocompatibility. The multimodality imaging functionalization of liposomes makes them a good platform for monitoring in vivo cancer targeting and pharmacokinetics of the loaded drugs, predicting their therapeutic efficacy (Dai & Yue, 2017; Xing et al., 2016). For example, core-shell giant liposomes were produced by water-in-oil (w/o/w) emulsion templates, in which hydrophilic doxorubicin and porous silicon nanoparticles (loaded with hydrophobic erlotinib) were encapsulated in the core, and a hydrophobic drug 17-AAG in the bilayer membrane. In addition, magnetic nanoparticles and gold nanorods were also encapsulated in the core, providing magnetic and photo-thermal responsive behaviors (Kong et al., 2015).

Polymeric nanoparticles have also been widely explored for theranostic applications. Depending on the morphology and composition in the core and periphery, polymeric nanoparticles can be solid polymeric nanoparticles, polymeric micelle, polymer conjugate, dendrimer, polymersome, polyplex, and polymer-lipid hybrid system (Prabhu et al., 2015). The wide range of polymers that can be used to prepare these nanocarriers can have either natural or synthetic origin, and they enable the synthesis of polymeric nanoparticles with desired properties (Rao & Geckeler, 2011). In general, these polymeric nanoparticles present an excellent colloidal stability, and they can efficiently load a wide range of chemotherapeutic drugs and imaging agents and sustain their release in a controlled manner at a specific cancer tissue, minimizing systemic side effects. Depending on the polymers employed in their preparation, polymeric nanoparticles can also be stimuli-responsive, biodegradable, biocompatible, and nontoxic, and can protect the cargo during systemic circulation and rapid clearance (Masood, 2016). Additionally, different polymeric nanoparticles can also carry imaging probes for theranostic purposes, such as iron oxide nanoparticles (Figueiredo et al., 2017; Herranz-Blanco et al., 2016), gold nanoparticles (Almeida et al., 2017; Kong et al., 2016), and fluorescent dyes (Figueiredo et al., 2016; Herranz-Blanco et al., 2015; Liu et al., 2015; Liu, Bernuz, et al., 2017; Shahbazi, Almeida, Mäkilä, Correia, et al., 2014; Shahbazi, Almeida, Makila, Kaasalainen, et al., 2014; Shahbazi et al., 2017; Xu et al., 2015).

Porous silicon (PSi) nanoparticles have also been investigated as a platform for carrying therapeutic compounds and imaging agents. Due to their large surface area and internal volume and easily functionalized surface properties, PSi nanoparticles can efficiently load different therapeutic agents and release them in a controlled manner. Furthermore, surface functionalization (with, e.g., antibodies and peptides) can enable effective targeting of the payloads. Other outstanding properties of PSi nanoparticles include excellent biosafety and in vivo biocompatibility and biodegradability (Bimbo et al., 2010; Liu et al., 2013; Santos et al., 2011; 2014; Shahbazi et al., 2012). Several strategies using PSi nanoparticles to simultaneous carry chemotherapeutic drugs and the imaging agents described above will be described in detail in the following section.

4 Multifunctional Silicon-Based Nanomaterials for Theranostic Applications

Considering the myriad of possible approaches for theranostics applications using nanoplatforms, this subsection will discuss different strategies using inorganic nanoparticles, such as PSi, as an example material, which are summarized in Table 2.

4.1 Intrinsic Luminescence

PSi-based nanomaterials are considered remarkable platforms for biomedical purposes. Among the vast possibility of medical applications, these nanomaterials can be exploited for theranostic goals, combining both diagnosis and anticancer therapy. Because of their biodegradability, low toxicity, and intrinsic luminescence (Anglin, Cheng, Freeman, & Sailor, 2008; Hua et al., 2005), PSi materials can be used for in vitro and in vivo imaging (Gu et al., 2013; Joo et al., 2015; Park et al., 2009), overcoming the existent liabilities of cytotoxic or nonbiodegradable fluorescent probes currently in use. Taking advantages of these properties, and the ability to functionalize PSi particles with different targeting moieties (Secret et al., 2013; Shahbazi et al., 2015; Wang et al., 2014; Wang, Sarparanta, et al., 2015), PSi particles were functionalized with iRGD tumor-targeting-peptide and their fate successfully monitored

Table 2: Summary of the main silicon nanoparticles used for theranostics applications

Type of Nanoparticles	Payloads	Surface Modification	Chemistry Used	In Vitro Studies	In Vivo Tumor Model	Outcome	References
Si NPs	–	Organic molecules	Photoinitiated hydrosilylation	–	–	The functional groups provided by the different organic molecules can allow the use of these particles as reactive probes in bioimaging, in sensors, or for assembling nanoscale semiconductor devices	Hua, Swihart, and Ruckenstein (2005)
PSi NPs	Doxorubicin	Dextran	Physical absorption	Cytotoxicity using MDA-MB-435 cell line	Nude mouse bearing MDA-MB-435 breast tumor cells	The intrinsic near-infrared photoluminescence enabled to follow both accumulation and degradation in vivo, with a low-toxicity degradation pathway The tumor imaging using dextran-coated PSi NPs was demonstrated in vivo	Park et al. (2009)
PSi NPs	–	PEG-silane	Physical absorption	–	Nude mice bearing a SKOV3 ovarian tumor cells	Improvement of signal to background contrast ratio by >20-fold in vivo when imaging PSi NPs. Time-gated imaging of porous silicon nanoparticles accumulated in a human ovarian cancer xenograft in a live mouse	Gu et al. (2013)
PSi NPs	–	iRGD	Coupling via free cysteine residue	–	BALB/c mice bearing 4T1 breast tumor cells	The time-gated imaging of low cytotoxic PSi NPs provided a method with superior sensitivity to eliminate tissue autofluorescence and to distinguish the nanoparticle from exogenous fluorescent probes	Joo et al. (2015)

UnTHCPSi NPs	Sorafenib	CD326-FITC	EDC/NHS chemistry	MCF-7 and MDA-MB-231 cells	–	High cellular interaction and internalization of the conjugated NPs into the cells expressing CD326 antigen (MCF-7 cells) via CD326 targeting, inducing antibody dependent cellular cytotoxicity and cytokine secretion by effector immune cells	Shahbazi et al. (2015)
TCPSi NPs	Sorafenib	FITC, RGDS and iRGD	Copper-free azide-alkyne cycloaddition	EA.hy926 cells	–	The peptides modification enhanced the cell uptake efficiency of PSi NPs in EA.hy926 cells. The NPs covalently labeled with FITC allowed the visualization of the NPs by confocal fluorescence microscopy	Wang et al. (2014)
UnTHCPSi NPs	Sorafenib	DBCO-PEG4-amine and Alexa Fluor 488 DOTA and iRGD ^{111}In-radiolabeling	EDC/NHS chemistry Strain-promoted azide-alkyne cycloaddition	PC3-MM2 cells	Male Hsd:NMRI-*Foxnl*$^{nu/nu}$ nude mice bearing PC3-MM2 prostate tumor cells	Radiolabeling of the NPs with ^{111}In allowed the monitoring of their in vivo biodistribution by single photon emission computed tomography (SPECT), and their fluorescent labeling with Alexa Fluor 488 was used to determine the long-term biodistribution of the NPs by immunofluorescence at the tissue level ex vivo, with an enhanced tumor uptake of the iRGD-conjugated NPs after i.v. administration	Wang, Sarparanta, et al. (2015)

(Continued)

Table 2: Summary of the main silicon nanoparticles used for theranostics applications—cont'd

Type of Nanoparticles	Payloads	Surface Modification	Chemistry Used	In Vitro Studies	In Vivo Tumor Model	Outcome	References
THCPSi NPs	TRITC	Tumor homing peptide (CooP)	–	MDA-MB-231 and MDGI-c expressing MDA-MB-231 cells	Nude mice bearing MDGI-expressing tumors	Functionalized PSi NPs showed higher uptake by MDGI-expressing cancer cells measured by confocal microscopy and flow cytometry. After i.v. injections into nude mice bearing MDGI-expressing tumors, the functionalized PSi NPs showed ~9-fold higher accumulation in the tumor site compared to the control NPs	Kinnari et al. (2013)
THCPSi NPs	–	^{18}F-radiolabeling	–	Caco-2 and RAW 264.7 cells	Male Wistar Han rats	Biodistribution study demonstrated that the particles passed intact through the gastrointestinal tract after oral administration and were not absorbed from a subcutaneous deposit	Bimbo et al. (2010)
THCPSi NPs	–	HFBII and ^{18}F ^{125}I-HFBII coating	–	CRL-1739 human gastric adenocarcinoma cell line	Male Wistar Han rats	The HFBII-coated NPs were noncytotoxic and showed mucoadhesive in the tested cells. The passage of HFBII-NPs in the rat GI tract was significantly slower than that of uncoated NPs, being retained in stomach by gastric mucoadhesion up to 3 h after oral administration. Upon entry to the small intestine, the mucoadhesive properties were lost, resulting in the rapid transit of the nanoparticles through the remainder of the GI tract	Sarparanta et al. (2012)

THCPSi NPs	–	SLNC, ^{18}F	Emulsion solvent evaporation	MDA-MB-231 cells	Female Crl:CD1 mice bearing orthotopic human breast carcinoma	The solid lipid nanocoating improved the dispersibility of the NPs, resulting in a lower uptake of the NPs in the lung. ^{18}F-labeled SLNCs-NPs showed higher tumor uptake than the ^{18}F-labeled bare NPs at 7 weeks after tumor inoculation	Kallinen et al. (2014)
UnTHCPSi NPs	Doxorubicin	PEG, VD11-4-2 (fluorescent compound)	EDC/NHS chemistry	MCF-7 cells	–	VD11-4-2, a novel inhibitor for carbonic anhydrase IX, was anchored on NPs and these NPs showed a strong affinity towards hypoxic human breast adenocarcinoma (MCF-7) cells. Thus, an improved killing efficiency for the hypoxia induced drug resistance cancer cell was observed	Janoniene et al. (2017)
UnTHCPSi NPs	Doxorubicin	AmQu, Alexa-Fluor 488	EDC/NHS chemistry	MCF-7 and doxorubicin-resistant MCF-7 cells	–	The resulting nanosystem presented several advanced properties, such as promoted anticancer efficiency, multiple drug resistance reversing, stimuli-responsive drug release, and enhanced particle-cell interactions	Liu, Balasubramanian, et al. (2017)

(Continued)

Table 2: Summary of the main silicon nanoparticles used for theranostics applications—cont'd

Type of Nanoparticles	Payloads	Surface Modification	Chemistry Used	In Vitro Studies	In Vivo Tumor Model	Outcome	References
PSi NPs	Fe_3O_4	Cy7 fluorophore	NHS-conjugation	Rat hepatocytes and HepG2	Sprague-Dawley rats bearing liver cancer	No cytotoxicity of the resulting formulations was observed in vitro or in vivo, and the NPs were degraded and eliminated after 4–8 h in vivo. The ability to tailor the magnetic properties of such materials may be useful for in vivo imaging, magnetic hyperthermia, and drug-delivery applications	Kinsella et al. (2011)
APTES-modified PSi MPs	SPIONs	LPS OVA	–	BMDCs	Female nu/nu mice	An increase in negative contrast in the spleen was observed after i.v. injection of the MPs. Incubation of BMDCs with the MPs containing SPIONs, OVA and LPS stimulated the T cell and BMDCs interaction and IFN-γ production by $CD8^+$ T cells, enabling the presentation of a payload that was traceable	Lundquist et al. (2014)
APTES-modified PSi MPs	Amine-PEG SPIONs	pH sensitive dye	–	HMVECs	–	The loading of SPIONs into PSi MPs allowed to follow the intracellular trafficking of the particles	Ferrati et al. (2010)

Luminescent PSi MPs	Fe_3O_4, doxorubicin	–	–	HeLa cells	–	The luminescent drug-containing particles were delivered to HeLa cells in vitro, under the guidance of a magnetic field, leading to a high concentration of particles in the proximity of the magnetic field that results in a high concentration of drug released in that region of the Petri dish	Gu, Park, Duong, Ruoslahti, and Sailor (2010)
UnTHCPSi NPs	Sorafenib	CPP nanocomplex, Alexa-Fluor 488	Ionotropic gelation technique	MDA-MB-231 cells	–	The resulting nanocomplexes significantly enhanced their cellular internalization and enabled their endosomal escape, resulting in the efficient cytoplasmic delivery of the sorafenib-loaded nanosystems, leading to a potent in vitro antiproliferative effect on breast cancer cells	Almeida et al. (2017)
UnTHCPSi NPs	Sorafenib, calcein	Cystine, 9-aminoacridine and DNA, FITC, CPP polyplex	EDC/NHS chemistry, click chemistry, ionotropic gelation technique	MDA-MB-231 cells	–	The combined nanocomposite containing an imaging agent showed a high loading capacity, low cytotoxicity and hemotoxicity, glutathione responsive on-command drug release, and superior cytosolic delivery	Shahbazi et al. (2017)

Abbreviations: *AmQu*, 3-aminopropoxy-linked quercetin; *APTES*, 3-aminopropyl-triethoxysilane; *CPP*, cysteine-polyethylenimine-poly(methyl vinyl ether-alt-maleic acid); *DBCO-PEG4-amine*, dibenzocyclooctyne-PEG4-amine; *DOTA*, tetraazacyclododecane-1,4,7-tris acetic acid; *EDC*, 1-ethyl-3-[3-dimethylaminopropyl] carbodiimide hydrochloride; *FITC*, fluorescein isothiocyanate; *HFBII*, 86-amino acid class II hydrophobin; *i.v.*, intravenous; *LPS*, lipopolysaccharide; *MPs*, microparticles; *NHS*, *N*-hydroxysuccinimide; *NPs*, nanoparticles; *PEG*, poly(ethylene glycol); *PSi*, porous silicon; *Si*, silicon; *SLNC*, solid lipid nanocoating; *SPIONs*, superparamagnetic iron oxide nanoparticles; *TCPSi*, thermally carbonized porous silicon; *UnTHCPSi*, undecylenic acid modified thermally hydrocarbonized porous silicon.

in vivo, using a time gating technique (Joo et al., 2015). The results showed an enhanced sensitivity, with >100 × contrast improvement relative to steady state imaging, providing a promising strategy for in vivo imaging.

4.2 Labeling

Since not all modified PSi-based nanomaterials exhibit luminescence, there is the need to find suitable alternative approaches. On this matter, fluorescent-labeled PSi nanoparticles have demonstrated the ability on multiple occasions to act as diagnostic systems, making it possible to follow their fate in vitro (Kinnari et al., 2013; Santos et al., 2011; Shahbazi et al., 2015; Wang et al., 2014; Wang, Makila, et al., 2015). The in vivo biodistribution can be monitored, for example, by labeling the nanoparticles with radioisotopes (Bimbo et al., 2010; Kallinen et al., 2014; Sarparanta et al., 2012). In a study performed by Wang, Sarparanta, et al. (2015), both techniques were utilized. In this study, a multifunctional nanosystem for cancer theranostics was developed by dual labeling PSi nanoparticles with Alexa Fluor® 488 to determine their long-term distribution ex vivo, and with the radioisotope ^{111}In to monitor the in vivo biodistribution of the nanocarrier in a prostate cancer xenograft mice model by single photon emission computed tomography (SPECT) technique. Additionally, the nanoparticles were loaded with sorafenib anticancer agent, and further functionalized with iRGD, producing a nanoplatform combining diagnostic imaging modalities with therapeutic payload. To evaluate biodistribution and anticancer effect, the nanoplatform was injected both intravenously (i.v.) and intratumorally (i.t.). After i.v. administration, the particles accumulated mainly in the liver and spleen, while after i.t. administration, they were mainly in the injection site (Fig. 5A). Moreover, iRGD functionalization lead to an increased tumor accumulation of the nanoparticles, both after i.v. and i.t. administration, displaying the targeting properties of the peptide (Fig. 5B). Finally, the in vivo tumoricidal effect of the system was analyzed by comparing, over time, the tumor growth before and after treatment. The treatment was performed by dual injection, 24 h apart, of sorafenib-loaded PSi particles both nonfunctionalized and functionalized with iRGD (amount of nanoparticles corresponding to 3 mg/kg of sorafenib) and free sorafenib, through i.v. and i.t. When administering i.v., the loaded nanoparticles exerted effects similar to free drug (Fig. 5C). However, after i.t injection, the nanoparticles had a significantly higher antiproliferative effect than the free drug. This result further indicated the retention of the PSi nanoparticles in the tumor, meaning the drug released in a sustained manner.

4.3 Multifunctional Ligands

Another possible strategy to achieve theranostic goals is to modify the surface of the nanoparticles with ligands that can achieve multiple purposes. These multifunctional moieties can act as targeting and imaging agents, bringing additional advantages to the nanosystems,

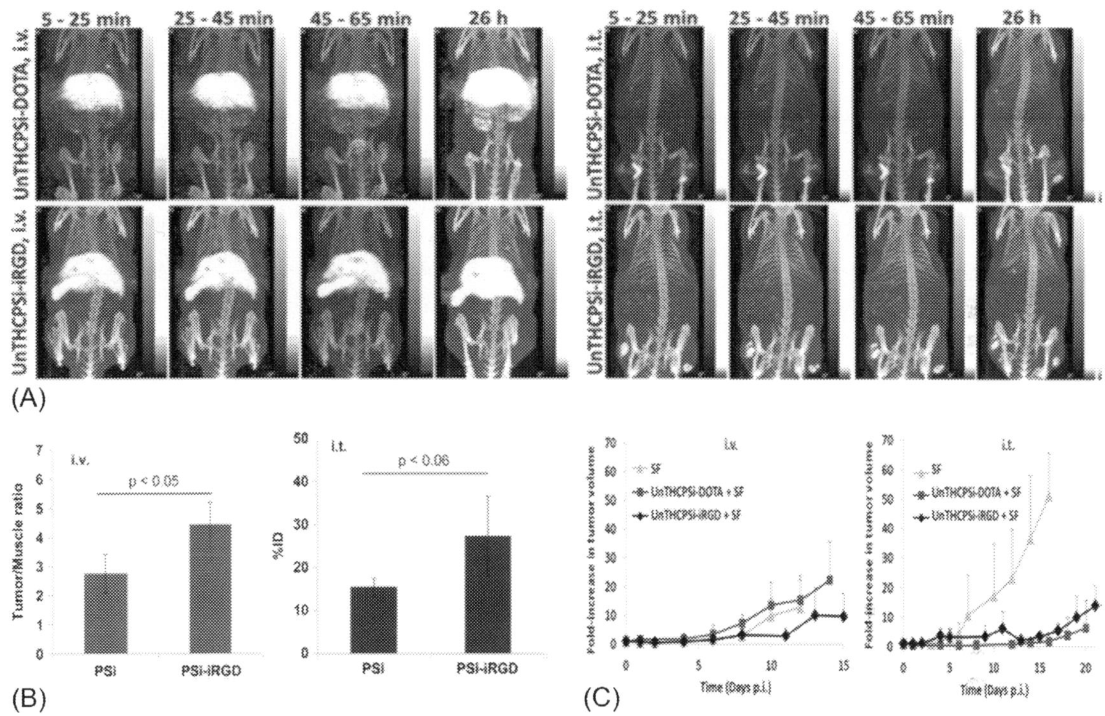

Fig. 5

(A) Nano-SPECT/CT-fused images taking in dynamic mode during 1 h after 5 min postinjection (p.i.) and at 26 h p.i. of the whole mouse. Images were taken after i.v. and i.t. administration of dual labeled PSi particles with (PSi-iRGD) and without (PSi) iRGD modification. (B) Tumor accumulation and retention of PSi and PSi-iRGD nanoparticles, after i.v. and i.t. administration. (C) Tumor growth inhibition of PC3-MM2 xenografts in male nude mice by the treatment with sorafenib (SF) loaded nanoparticles. The data is represented as mean ± SD (n = 4–6). The statistical differences are UnTHCPSi-iRGD + SF vs UnTHCPSi-Dota + SF vs SF, P = .3238 (i.v. administration route, Kruskall-Wallis test) and P = .0107 (i.t. administration route, Kruskall-Wallis test). *Modified from Wang, C. F., Sarparanta, M. P., Makila, E. M., Hyvonen, M. L., Laakkonen, P. M., Salonen, J. J., et al. (2015). Multifunctional porous silicon nanoparticles for cancer theranostics. Biomaterials, 48, 108–118, with permission from Elsevier, copyright 2015.*

such as allowing control and monitoring of the fate of both particles and payloads. In this context, a PSi nanoplatform for theranostic applications was designed by anchoring VD11-4-2, an inhibitor for carbonic anhydrase IX (CA IX) to doxorubicin (DOX; anticancer drug) loaded PSi nanoparticles (Janoniene et al., 2017). CA IX is a transmembrane protein highly expressed in cells under hypoxia in tumors, when compared to normal tissues (Mahon, Pinard, & Mckenna, 2015), and VD11-4-2 is a recently synthesized compound with intrinsic fluorescence and high affinity and selectivity for this protein (Dudutiene et al., 2015). Thereby, VD11-4-2 was introduced to the surface of the drug-loaded nanoparticles to induce cell targeting and to control the fate of the DOX. In this study, the functionalized nanoparticles were tested in MCF-7 breast cancer cells, under hypoxia and normoxia conditions.

Cells under hypoxia showed increase expression of CA IX when compared to cells in normaxia conditions, leading to an enhanced targeted effect of the functionalized nanoparticles and a consequent increased antiproliferative effect. Moreover, it was possible to observe a Forster resonance energy transfer (FRET) between VD11-4-2 and DOX, being the fluorescence of VD11-4-2 quenched as a result of DOX absorbance. This phenomenon allowed the monitorization of the release of DOX through confocal microscopy and flow cytometry (Fig. 6). After 3h incubation of the functionalized nanoparticles with hypoxia

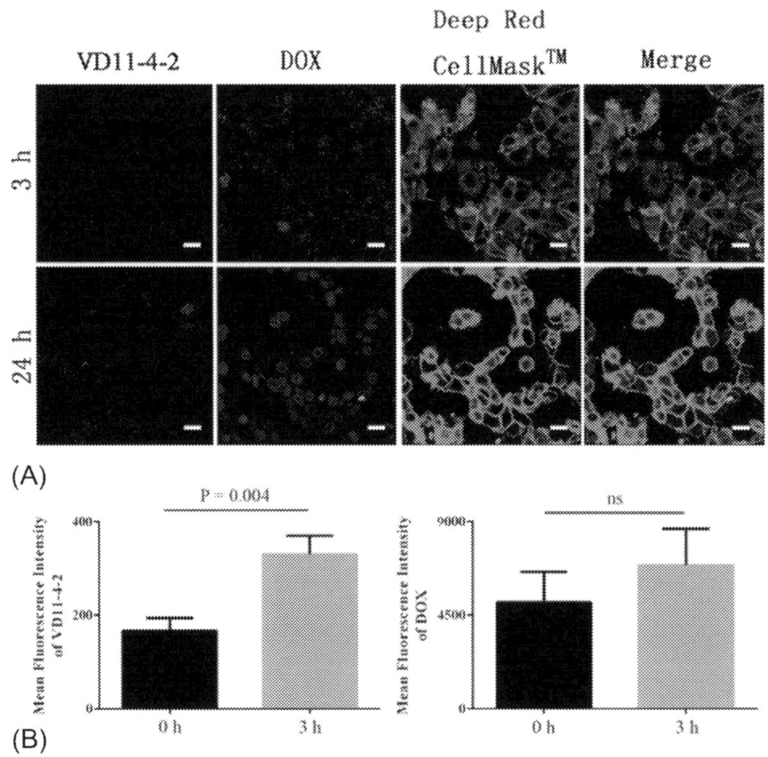

Fig. 6

Nanoparticles cellular interaction. (A) Confocal microscope images of DOX@VD-PSi after 3- and 24-h incubation with MCF-7 breast cancer cells under hypoxic conditions. *Blue*: VD-PSi; *green*: DOX; *red*: cell membranes stained with CellMask DeepRed. All the scale bars indicate 20 μm. (B) Flow cytometry analysis of the mean fluorescence intensity of both VD11-4-2 and DOX. Error bars represent mean ± SD (n = 3) (For interpretation of the references to color in this figure legend, the reader is referred to the web version of this article). *Reproduced with permission from Janoniene, A., Liu, Z. H., Baranauskiene, L., Makila, E., Ma, M., Salonen, J., et al. (2017). A versatile carbonic anhydrase IX targeting ligand-functionalized porous silicon nanoplatform for dual hypoxia cancer therapy and imaging. ACS Applied Materials & Interfaces, 9, 13976–13987, with American Chemical Society, copyright 2017.*

cancer cells (Fig. 6A), and despite the existence of cellular uptake, there was no observed signal from VD11-4-2, meaning DOX was still loaded in the particles. However, after 24 h, not only was DOX signal accumulated in the nucleus, but there was also a clear signal from VD11-4-2, indicating the release of the drug. The results were further confirmed with flow cytometry (Fig. 6B), as after DOX release, there was an increment of VD11-4-2 signal while the DOX signal remained unchanged.

Similarly, in a study by Liu, Balasubramanian, et al. (2017), a novel flavonoid derivative, 3-aminopropoxy-linked quercetin (AmQu), was synthesized and conjugated to the surface of DOX-loaded PSi nanoparticles. The properties of the DOX-loaded AmQu-functionalized PSi nanoparticles were tested by incubation of the particles with MCF-7 breast cancer cells and with DOX-resistant MCF-7 cells. The results showed that this newly synthesized multifunctional compound had the ability to enhance cellular interaction, control DOX release and at the same time, by quenching the fluorescence of the loaded DOX, allowed real-time imaging monitoring release of DOX. Moreover, it improved the cytotoxicity toward MCF-7 and DOX-resistant MCF-7 cells.

4.4 Synergism Between Nanoparticles

The ability of combining PSi-based materials with secondary nanoparticles allows the production of multifunctional systems which not only inherit the different properties of the constituent particles but can also improve them. These combinations can be useful for the enhancement of drug loading and release profiles, overcoming biological barriers, modulation of imaging properties, and even controlling activation via external stimuli.

For the abovementioned purposes, several studies have reported the combination of SPIONs with PSi (Ferrati et al., 2010; Kinsella et al., 2011; Lundquist et al., 2014). By loading SPIONs into PSi particles, it is possible to enhance their MRI contrast, creating traceable platforms which are useful for monitoring the particles trafficking in vivo. Furthermore, SPIONs can be used for hyperthermia due to their magnetic properties, and be applied for remote-guided target drug delivery. In this respect, both DOX and SPIONs were co-loaded into PSi particles, forming a complex that was simultaneously fluorescent, magnetic and loaded with an anticancer agent (Gu et al., 2010). To evaluate the magnetic guidance properties, these particles were incubated with a cell culture, in vitro, in the presence or absence of a magnetic field in the edge of a Petri dish. The presence of the magnet led to accumulation of the particles in the vicinity of the magnetic field, resulting in higher cell death in that area compared to other regions of the plate. At same time, in the absence of the magnetic field, there was a uniformly dispersed lower density of dead cells. This studied showed that it was possible to manipulate the guidance of the particles, in vitro, using magnetic field and localize the delivery of the payload.

PSi nanomaterials can both encapsulate different payloads and also be encapsulated themselves into nano-in-nano complexes. The use of polyplexes allows the simultaneous encapsulation of different type of nanoparticles. For example, Almeida et al. (2017) co-encapsulated sorafenib-loaded PSi nanoparticles with gold nanoparticles into CPP polymeric nanocomplex, enhancing cellular association and internalization, promoting endosomal escape and controlling the release of sorafenib. Additionally, this nanosystem prevented the in vitro proliferation of MDA-MB-231 breast cancer cells using low concentrations of drug. The successful encapsulation of gold nanoparticles can create opportunities for different imaging approaches based in their characteristics described in Section 3, which can be one valuable proof-of-concept for cancer theranostic applications.

In other study, DNA-coated PSi nanoparticles were encapsulated in a polyplex for simultaneous co-delivery of hydrophobic or hydrophilic compounds with an imaging agent to achieve theranostic goals (Shahbazi et al., 2017). In this study, first, either sorafenib (hydrophobic molecule) or calcein (hydrophilic molecule) was loaded into PSi nanoparticles. Afterward, the loaded nanoparticles' surfaces were modified with DNA, cystine (redox responsive inducer), and 9-aminoacridine (DNA intercalator), for gate capping the porous nanoparticles and to create a redox and DNase responsive carrier. Furthermore, the particles were encapsulated within cysteine-polyethyleneimine (PEI)-poly(methyl vinyl ether-*alt*-maleic acid) (PMVE-MA) polymeric nanoshell (CPP), which was further conjugated with FITC (as model imaging agent) (Fig. 7).

These drug-loaded DNA-gated PSi nanoparticles, both with and without CPP encapsulation, were tested in release medium, in the presence and absence of either glutathione (GSH) or DNase, to evaluate their ability to release their payloads upon stimuli. Samples were compared to a control of unmodified PSi nanoparticles. DNA capping could successfully trigger stimuli controlled release as in the absence of stimuli agents, it prevented the release of payloads, while in their presence, there was sustained release over time (Fig. 8A). To study cellular interaction, endosomal escape and imaging properties, the particles, with or without CPP encapsulation, were incubated with MDA-MB-231 breast cancer cells and evaluated through flow cytometry and confocal microscopy techniques. The measurement of the granularity of the cells by side scatter (SSC) analysis (Fig. 8B), showed that encapsulation of the particles with CPP promoted the internalization of the nanoparticles, as there was higher localization within the intracellular compartments (higher % of cells in higher SSC values). Moreover, due to the fluorescent properties of the particles, it was possible to follow their intracellular fate by confocal microscopy (Fig. 8C), and observe that CPP led to the rupture of the membrane of the acidic vesicles in which the particles were confined, enabling their release into the cytoplasm.

Altogether, the ability to co-deliver both hydrophobic/hydrophilic molecules with imaging agents, improve cellular uptake, induce endosomal escape, and control the release by DNA cleavage, makes PSi encapsulation by polyplexes a promising strategy for theranostics.

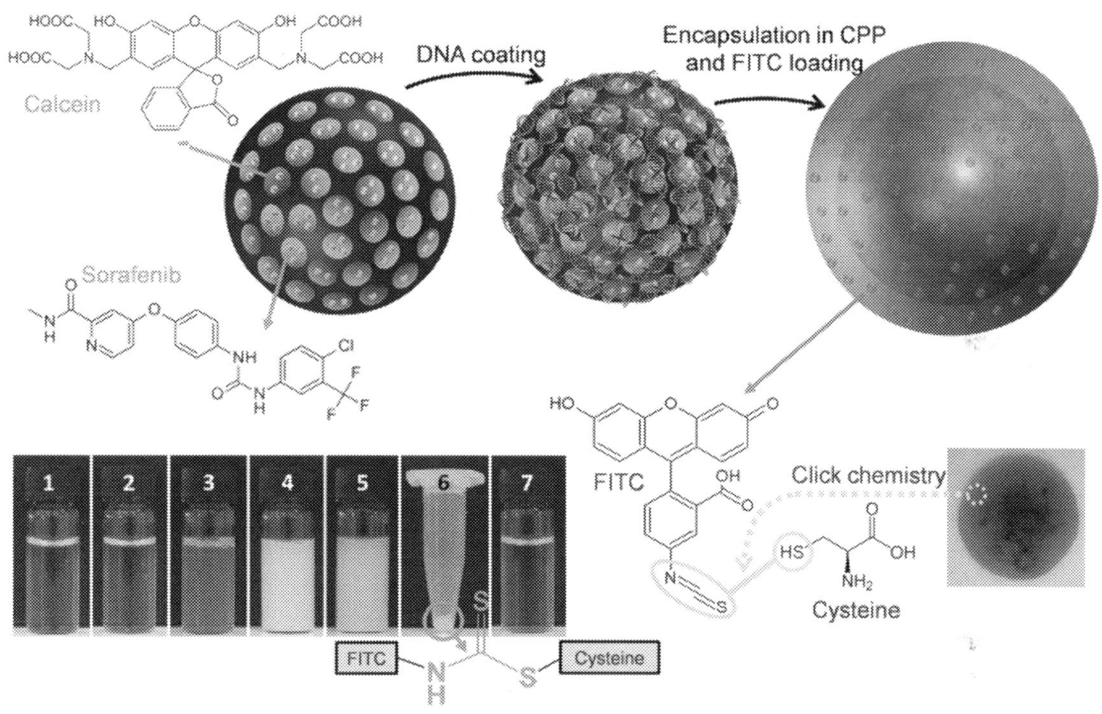

Fig. 7

Schematic illustration of the formation of the nanocomplex. Bottles numbered from 1 to 7 represent cysteine, PEI, FITC and PMVE-MA, CPP polyplexes without FITC, CPP polyplexes loaded with FITC, the FTIC-loaded CPP pellet after centrifugation, and the supernatant of the CPP pellet, respectively. *Reproduced with permission from Shahbazi, M.-A., Almeida, P. V., Correia, A., Herranz-Blanco, B., Shrestha, N., Mäkilä, E., et al. (2017). Intracellular responsive dual delivery by endosomolytic polyplexes carrying DNA anchored porous silicon nanoparticles. Journal of Controlled Release, 249, 111–122, with Elsevier, copyright 2017.*

5 Conclusion

Cancer nanomedicine has undergone tremendous advances in the past decade. The expansion of the materials repertoire, their versatility, and ability to modify their properties, such as size, shape, and charge, provides a unique opportunity for the production of multifunctional nanosystems. Nanoparticles hold several advantages for cancer therapy and diagnosis with the ability to load different payloads, increasing their half-life and solubility. These particles might also possess intrinsic properties such as luminescence or magnetism, opening the possibility for different therapies modalities. Moreover, not only do they have an inherent ability to accumulate in the tumor site through the EPR effect, but they can also be modified and functionalized with different moieties to target specific cells, prolong their circulation times, minimize the clearance by the immune system, and control and monitor the release

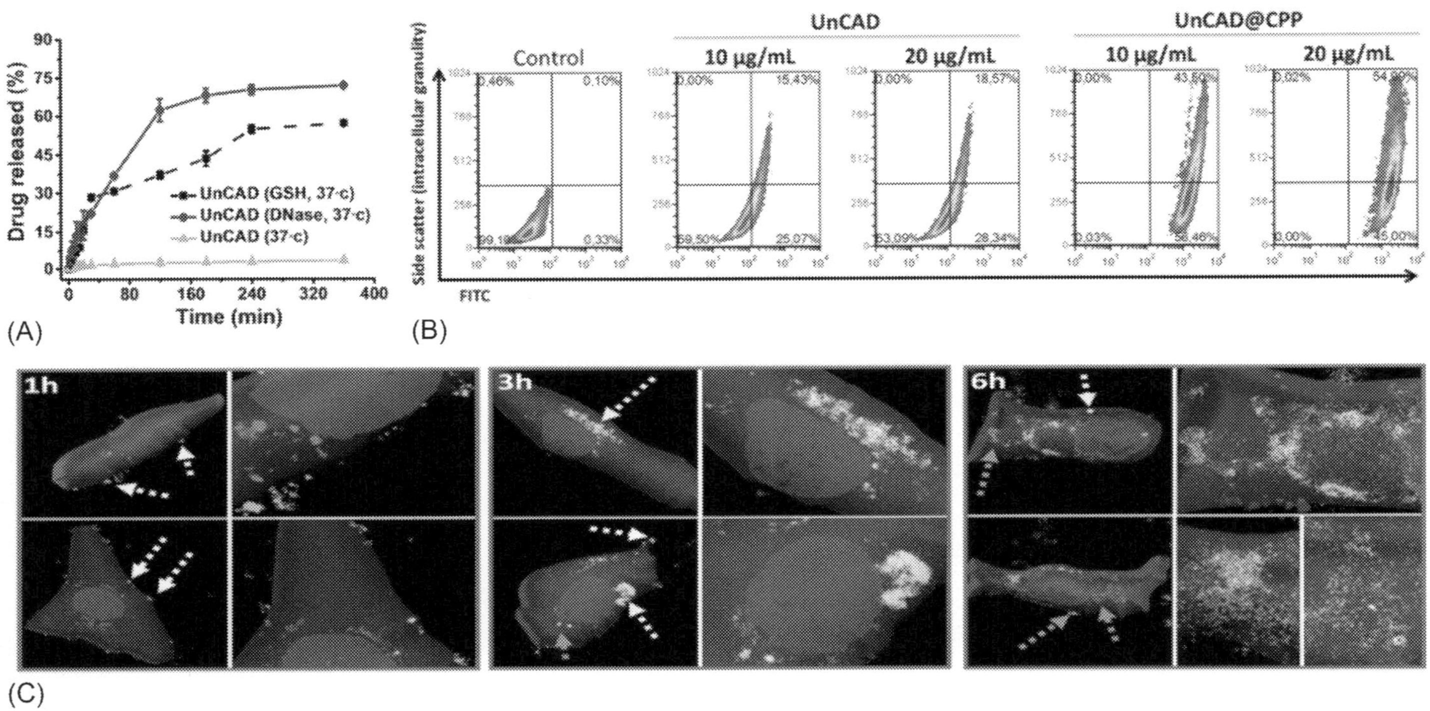

Fig. 8

(A) Release of sorafenib from DNA-capped PSi nanoparticles (UnCAD) and UnCAD encapsulated in CPP (UnCAD@CPP) in the absence of stimuli or presence of either GSH or DNase. Error bars represent mean ± SD ($n \geq 3$). (B) Analysis of intracellular localization of UnCAD and UnCAD@CPP nanoparticles by flow cytometry, after 3-h incubation at 37°C of the nanoparticles with MDA-MB-231 cells. Increased SSC concurrent with enhanced FITC intensity represents increased intracellular granularity. (C) Confocal images of MDA-MB-231 cells incubated with 20 μg/mL of nanoparticles for 1, 3, and 6 h at 37°C. Nuclei of the cells were stained with DAPI (*blue color*) and the acidic compartments with LysoTracker Red DND-99 (*red color*). Nanoparticles when at the surface of the cells or in the cytosol can be detected in *green*. Nanoparticles co-localized in the lysosomes are shown in *yellow color* (For interpretation of the references to color in this figure legend, the reader is referred to the web version of this article). *Arrows* show the nanoparticles on the surface, within acidic organelles, or in the cytoplasmic compartments. *Reproduced with permission from Shahbazi, M.-A., Almeida, P. V., Correia, A., Herranz-Blanco, B., Shrestha, N., Mäkilä, E., et al. (2017). Intracellular responsive dual delivery by endosomolytic polyplexes carrying DNA anchored porous silicon nanoparticles. Journal of Controlled Release, 249, 111–122, with Elsevier, copyright 2017.*

of therapeutic agents. Thus, nanoplatforms can be designed for simultaneous detection, treatment, and monitoring of tumors. This theranostic approach has the goal of maximizing diagnostic information and diminishing the frequency of interventions to treat the disease, enhancing the quality of the patients' lives.

Despite the fact that nanomaterials have revolutionized the theranostic field, many challenges still need to be addressed before translation into clinical use. Focus should be directed toward deeper understanding of the interaction between nanoparticles and tumors, assessment of long-term toxicity of nanoplatforms, scaled-up synthesis, and translation from in vivo to clinical trials. Only afterwards will it be possible to deliver efficient drugs for targeted and personalized therapy.

Acknowledgments

T. Bauleth-Ramos acknowledges financial support from Fundação para a Ciência e a Tecnologia (SFRH/BD/110859/2015). Prof. H. A. Santos acknowledges financial support obtained from the University of Helsinki Research Funds, the Helsinki Institute of Life Science, the Sigrid Jusélius Foundation (Decision No. 4704580), and the European Research Council under the European Union's Seventh Framework Programme (FP/2007-2013, Grant No. 310892).

References

Ahmed, N., Fessi, H., & Elaissari, A. (2012). Theranostic applications of nanoparticles in cancer. *Drug Discovery Today*, *17*, 928–934.

Akhter, S., Ahmad, M. Z., Ahmad, F. J., Storm, G., & Kok, R. J. (2012). Gold nanoparticles in theranostic oncology: current state-of-the-art. *Expert Opinion on Drug Delivery*, *9*, 1225–1243.

Almeida, P. V., Shahbazi, M. A., Correia, A., Makila, E., Kemell, M., Salonen, J., et al. (2017). A multifunctional nanocomplex for enhanced cell uptake, endosomal escape and improved cancer therapeutic effect. *Nanomedicine (London, England)*, *12*, 1401–1420.

Anglin, E. J., Cheng, L., Freeman, W. R., & Sailor, M. J. (2008). Porous silicon in drug delivery devices and materials. *Advanced Drug Delivery Reviews*, *60*, 1266–1277.

Balasubramanian, V., Liu, Z., Hirvonen, J., & Santos, H. A. (2017). Bridging the knowledge of different worlds to understand the big picture of cancer nanomedicines. *Advanced Healthcare Materials*, 1700432.

Bertrand, N., Wu, J., Xu, X., Kamaly, N., & Farokhzad, O. C. (2014). Cancer nanotechnology: the impact of passive and active targeting in the era of modern cancer biology. *Advanced Drug Delivery Reviews*, *66*, 2–25.

Bimbo, L. M., Sarparanta, M., Santos, H. A., Airaksinen, A. J., Makila, E., Laaksonen, T., et al. (2010). Biocompatibility of thermally hydrocarbonized porous silicon nanoparticles and their biodistribution in rats. *ACS Nano*, *4*, 3023–3032.

Bozzuto, G., & Molinari, A. (2015). Liposomes as nanomedical devices. *International Journal of Nanomedicine*, *10*, 975–999.

Brannon-Peppas, L., & Blanchette, J. O. (2012). Nanoparticle and targeted systems for cancer therapy. *Advanced Drug Delivery Reviews*, *64*(Supplement), 206–212.

Byrne, J. D., Betancourt, T., & Brannon-Peppas, L. (2008). Active targeting schemes for nanoparticle systems in cancer therapeutics. *Advanced Drug Delivery Reviews*, *60*, 1615–1626.

Cabral, R. M., & Baptista, P. V. (2014). Anti-cancer precision theranostics: a focus on multifunctional gold nanoparticles. *Expert Review of Molecular Diagnostics*, *14*, 1041–1052.

Chakraborty, S., & Rahman, T. (2012). The difficulties in cancer treatment. *Ecancermedicalscience*, *6*, ed16.

Chen, F., Ehlerding, E. B., & Cai, W. (2014). Theranostic nanoparticles. *Journal of Nuclear Medicine*, *55*, 1919–1922.

Dai, Z., & Yue, X. (2017). Liposomal nanotechnology for cancer theranostics. *Current Medicinal Chemistry*.

Deshpande, P. P., Biswas, S., & Torchilin, V. P. (2013). Current trends in the use of liposomes for tumor targeting. *Nanomedicine (London, England)*, *8*, 1509–1528. https://doi.org/10.2217/nnm.13.118.

Du, J., Lane, L. A., & Nie, S. (2015). Stimuli-responsive nanoparticles for targeting the tumor microenvironment. *Journal of Controlled Release: Official Journal of the Controlled Release Society*, *219*, 205–214.

Dudutiene, V., Zubriene, A., Smirnov, A., Timm, D. D., Smirnoviene, J., Kazokaite, J., et al. (2015). Functionalization of fluorinated benzenesulfonamides and their inhibitory properties toward carbonic anhydrases. *ChemMedChem*, *10*, 662–687.

Estelrich, J., Escribano, E., Queralt, J., & Busquets, M. A. (2015). Iron oxide nanoparticles for magnetically-guided and magnetically-responsive drug delivery. *International Journal of Molecular Sciences*, *16*, 8070–8101.

Estrela, J. M., Ortega, A., & Obrador, E. (2006). Glutathione in cancer biology and therapy. *Critical Reviews in Clinical Laboratory Sciences*, *43*, 143–181.

Ferlay, J., Soerjomataram, I., Dikshit, R., Eser, S., Mathers, C., Rebelo, M., et al. (2015). Cancer incidence and mortality worldwide: sources, methods and major patterns in GLOBOCAN 2012. *International Journal of Cancer*, *136*, E359–86.

Ferrati, S., Mack, A., Chiappini, C., Liu, X., Bean, A. J., Ferrari, M., et al. (2010). Intracellular trafficking of silicon particles and logic-embedded vectors. *Nanoscale*, *2*, 1512–1520.

Figueiredo, P., Balasubramanian, V., Shahbazi, M.-A., Correia, A., Wu, D., Palivan, C. G., et al. (2016). Angiopep2-functionalized polymersomes for targeted doxorubicin delivery to glioblastoma cells. *International Journal of Pharmaceutics*, *511*, 794–803.

Figueiredo, P., Lintinen, K., Kiriazis, A., Hynninen, V., Liu, Z., Bauleth-Ramos, T., et al. (2017). In vitro evaluation of biodegradable lignin-based nanoparticles for drug delivery and enhanced antiproliferation effect in cancer cells. *Biomaterials*, *121*, 97–108.

Friedman, A. D., Claypool, S. E., & Liu, R. (2013). The smart targeting of nanoparticles. *Current Pharmaceutical Design*, *19*, 6315–6329.

Gaumet, M., Vargas, A., Gurny, R., & Delie, F. (2008). Nanoparticles for drug delivery: the need for precision in reporting particle size parameters. *European Journal of Pharmaceutics and Biopharmaceutics*, *69*, 1–9.

Ge, J., Neofytou, E., Cahill, T. J., Beygui, R. E., & Zare, R. N. (2012). Drug release from electric-field-responsive nanoparticles. *ACS Nano*, *6*, 227–233.

Gu, L., Hall, D. J., Qin, Z. T., Anglin, E., Joo, J., Mooney, D. J., et al. (2013). In vivo time-gated fluorescence imaging with biodegradable luminescent porous silicon nanoparticles. *Nature Communications*, *4*. Article number 2326.

Gu, L., Park, J. H., Duong, K. H., Ruoslahti, E., & Sailor, M. J. (2010). Magnetic luminescent porous silicon microparticles for localized delivery of molecular drug payloads. *Small*, *6*, 2546–2552.

Harris, T. J., von Maltzahn, G., Lord, M. E., Park, J.-H., Agrawal, A., Min, D.-H., et al. (2008). Protease-triggered unveiling of bioactive nanoparticles. *Small*, *4*, 1307–1312.

Herranz-Blanco, B., Liu, D., Mäkilä, E., Shahbazi, M.-A., Ginestar, E., Zhang, H., et al. (2015). Drug delivery: on-chip self-assembly of a smart hybrid nanocomposite for antitumoral applications. *Advanced Functional Materials*, *25*, 1612.

Herranz-Blanco, B., Shahbazi, M. A., Correia, A. R., Balasubramanian, V., Kohout, T., Hirvonen, J., et al. (2016). pH-switch nanoprecipitation of polymeric nanoparticles for multimodal cancer targeting and intracellular triggered delivery of doxorubicin. *Advanced Healthcare Materials*, *5*, 1904–1916.

Hirsjarvi, S., Passirani, C., & Benoit, J. P. (2011). Passive and active tumour targeting with nanocarriers. *Current Drug Discovery Technologies*, *8*, 188–196.

Ho, D., Sun, X., & Sun, S. (2011). Monodisperse magnetic nanoparticles for theranostic applications. *Accounts of Chemical Research*, *44*, 875–882.

Hobbs, S. K., Monsky, W. L., Yuan, F., Roberts, W. G., Griffith, L., Torchilin, V. P., et al. (1998). Regulation of transport pathways in tumor vessels: role of tumor type and microenvironment. *Proceedings of the National Academy of Sciences of the United States of America*, *95*, 4607–4612.

Hua, F., Swihart, M. T., & Ruckenstein, E. (2005). Efficient surface grafting of luminescent silicon quantum dots by photoinitiated hydrosilylation. *Langmuir*, *21*, 6054–6062.

Iga, A. M., Robertson, J. H. P., Winslet, M. C., & Seifalian, A. M. (2007). Clinical potential of quantum dots. *Journal of Biomedicine and Biotechnology*, *2007*, 76087.

Janoniene, A., Liu, Z. H., Baranauskiene, L., Makila, E., Ma, M., Salonen, J., et al. (2017). A versatile carbonic anhydrase IX targeting ligand-functionalized porous silicon nanoplatform for dual hypoxia cancer therapy and imaging. *ACS Applied Materials & Interfaces*, *9*, 13976–13987.

Jokerst, J. V., & Gambhir, S. S. (2011). Molecular imaging with theranostic nanoparticles. *Accounts of Chemical Research*, *44*, 1050–1060.

Joo, J., Liu, X., Kotamraju, V. R., Ruoslahti, E., Nam, Y., & Sailor, M. J. (2015). Gated luminescence imaging of silicon nanoparticles. *ACS Nano*, *9*, 6233–6241.

Kallinen, A. M., Sarparanta, M. P., Liu, D., Makila, E. M., Salonen, J. J., Hirvonen, J. T., et al. (2014). In vivo evaluation of porous silicon and porous silicon solid lipid nanocomposites for passive targeting and imaging. *Molecular Pharmaceutics*, *11*, 2876–2886.

Kinnari, P. J., Hyvonen, M. L., Makila, E. M., Kaasalainen, M. H., Rivinoja, A., Salonen, J. J., et al. (2013). Tumour homing peptide-functionalized porous silicon nanovectors for cancer therapy. *Biomaterials*, *34*, 9134–9141.

Kinsella, J. M., Ananda, S., Andrew, J. S., Grondek, J. F., Chien, M. P., Scadeng, M., et al. (2011). Enhanced magnetic resonance contrast of Fe3O4 nanoparticles trapped in a porous silicon nanoparticle host. *Advanced Materials*, *23*, H248.

Kong, F., Zhang, H., Zhang, X., Liu, D., Chen, D., Zhang, W., et al. (2016). Biodegradable photothermal and pH responsive calcium carbonate@phospholipid@acetalated dextran hybrid platform for advancing biomedical applications. *Advanced Functional Materials*, *26*, 6158–6169.

Kong, F., Zhang, X., Zhang, H., Qu, X., Chen, D., Servos, M., et al. (2015). Inhibition of multidrug resistance of cancer cells by co-delivery of DNA nanostructures and drugs using porous silicon nanoparticles@giant liposomes. *Advanced Functional Materials*, *25*, 3330–3340.

Liu, Z., Balasubramanian, V., Bhat, C., Vahermo, M., Makila, E., Kemell, M., et al. (2017). Quercetin-based modified porous silicon nanoparticles for enhanced inhibition of doxorubicin-resistant cancer cells. *Advanced Healthcare Materials*, *6*. 1601009.

Liu, D., Bernuz, C. R., Fan, J., Li, W., Correia, A., Hirvonen, J., et al. (2017). A nano-in-nano vector: merging the best of polymeric nanoparticles and drug nanocrystals. *Advanced Functional Materials*, *27*, 1604508.

Liu, D., Mäkilä, E., Zhang, H., Herranz, B., Kaasalainen, M., Kinnari, P., et al. (2013). Nanostructured porous silicon-solid lipid nanocomposite: towards enhanced cytocompatibility and stability, reduced cellular association, and prolonged drug release. *Advanced Functional Materials*, *23*, 1893–1902.

Liu, D., Zhang, H., Makila, E., Fan, J., Herranz-Blanco, B., Wang, C. F., et al. (2015). Microfluidic assisted one-step fabrication of porous silicon@acetalated dextran nanocomposites for precisely controlled combination chemotherapy. *Biomaterials*, *39*, 249–259.

Lundquist, C. M., Loo, C., Meraz, I. M., Cerda, J. D., Liu, X., & Serda, R. E. (2014). Characterization of free and porous silicon-encapsulated superparamagnetic iron oxide nanoparticles as platforms for the development of theranostic vaccines. *Medical Science*, *2*, 51–69.

Mahmoudi, M., Sant, S., Wang, B., Laurent, S., & Sen, T. (2011). Superparamagnetic iron oxide nanoparticles (SPIONs): development, surface modification and applications in chemotherapy. *Advanced Drug Delivery Reviews*, *63*, 24–46.

Mahon, B. P., Pinard, M. A., & Mckenna, R. (2015). Targeting carbonic anhydrase IX activity and expression. *Molecules*, *20*, 2323–2348.

Masood, F. (2016). Polymeric nanoparticles for targeted drug delivery system for cancer therapy. *Materials Science and Engineering: C*, *60*, 569–578.

Misra, R., Acharya, S., & Sahoo, S. K. (2010). Cancer nanotechnology: application of nanotechnology in cancer therapy. *Drug Discovery Today*, *15*, 842–850.

Mitra, A. K., Agrahari, V., Mandal, A., Cholkar, K., Natarajan, C., SHAH, S., et al. (2015). Novel delivery approaches for cancer therapeutics. *Journal of Controlled Release: Official Journal of the Controlled Release Society*, *219*, 248–268.

Mura, S., Nicolas, J., & Couvreur, P. (2013). Stimuli-responsive nanocarriers for drug delivery. *Nature Materials*, *12*, 991–1003.

Muthu, M. S., Leong, D. T., Mei, L., & Feng, S.-S. (2014). Nanotheranostics—application and further development of nanomedicine strategies for advanced theranostics. *Theranostics*, *4*, 660–677.

Nakamura, Y., Mochida, A., Choyke, P. L., & Kobayashi, H. (2016). Nanodrug delivery: is the enhanced permeability and retention effect sufficient for curing cancer? *Bioconjugate Chemistry*, *27*, 2225–2238.

Needham, D., Anyarambhatla, G., Kong, G., & Dewhirst, M. W. (2000). A new temperature-sensitive liposome for use with mild hyperthermia: characterization and testing in a human tumor xenograft model. *Cancer Research*, *60*, 1197–1201.

Park, J. H., Gu, L., von Maltzahn, G., Ruoslahti, E., Bhatia, S. N., & Sailor, M. J. (2009). Biodegradable luminescent porous silicon nanoparticles for in vivo applications. *Nature Materials*, *8*, 331–336.

Peer, D., Karp, J. M., Hong, S., Farokhzad, O. C., Margalit, R., & Langer, R. (2007). Nanocarriers as an emerging platform for cancer therapy. *Nature Nanotechnology*, *2*, 751–760.

Prabhu, R. H., Patravale, V. B., & Joshi, M. D. (2015). Polymeric nanoparticles for targeted treatment in oncology: current insights. *International Journal of Nanomedicine*, *10*, 1001–1018.

Provenzano, P. P., Cuevas, C., Chang, A. E., Goel, V. K., von Hoff, D. D., & Hingorani, S. R. (2012). Enzymatic targeting of the stroma ablates physical barriers to treatment of pancreatic ductal adenocarcinoma. *Cancer Cell*, *21*, 418–429.

Qian, C.-G., Chen, Y.-L., Feng, P.-J., Xiao, X.-Z., Dong, M., Yu, J.-C., et al. (2017). Conjugated polymer nanomaterials for theranostics. *Acta Pharmacologica Sinica*, *38*, 764–781.

Rao, J. P., & Geckeler, K. E. (2011). Polymer nanoparticles: preparation techniques and size-control parameters. *Progress in Polymer Science*, *36*, 887–913.

Santos, H. A., Bimbo, L. M., Lehto, V. P., Airaksinen, A. J., Salonen, J., & Hirvonen, J. (2011). Multifunctional porous silicon for therapeutic drug delivery and imaging. *Current Drug Discovery Technologies*, *8*, 228–249.

Santos, H. A., Mäkilä, E., Airaksinen, A. J., Bimbo, L. M., & Hirvonen, J. (2014). Porous silicon nanoparticles for nanomedicine: preparation and biomedical applications. *Nanomedicine*, *9*, 535–554.

Sarparanta, M. P., Bimbo, L. M., Makila, E. M., Salonen, J. J., Laaksonen, P. H., Helariutta, A. M., et al. (2012). The mucoadhesive and gastroretentive properties of hydrophobin-coated porous silicon nanoparticle oral drug delivery systems. *Biomaterials*, *33*, 3353–3362.

Secret, E., Smith, K., Dubljevic, V., Moore, E., Macardle, P., Delalat, B., et al. (2013). Antibody-functionalized porous silicon nanoparticles for vectorization of hydrophobic drugs. *Advanced Healthcare Materials*, *2*, 718–727.

Shahbazi, M.-A., Almeida, P. V., Correia, A., Herranz-Blanco, B., Shrestha, N., Mäkilä, E., et al. (2017). Intracellular responsive dual delivery by endosomolytic polyplexes carrying DNA anchored porous silicon nanoparticles. *Journal of Controlled Release*, *249*, 111–122.

Shahbazi, M.-A., Almeida, P. V., Mäkilä, E., Correia, A., Ferreira, M. P. A., Kaasalainen, M., et al. (2014). Poly(methyl vinyl ether-alt-maleic acid)-functionalized porous silicon nanoparticles for enhanced stability and cellular internalization. *Macromolecular Rapid Communications*, *35*, 624–629.

Shahbazi, M. A., Almeida, P. V., Makila, E. M., Kaasalainen, M. H., Salonen, J. J., Hirvonen, J. T., et al. (2014). Augmented cellular trafficking and endosomal escape of porous silicon nanoparticles via zwitterionic bilayer polymer surface engineering. *Biomaterials*, *35*, 7488–7500.

Shahbazi, M. A., Herranz, B., & Santos, H. A. (2012). Nanostructured porous Si-based nanoparticles for targeted drug delivery. *Biomatter*, *2*, 296–312.

Shahbazi, M. A., Shrestha, N., Makila, E., Araujo, F., Correia, A., Ramos, T., et al. (2015). A prospective cancer chemo-immunotherapy approach mediated by synergistic CD326 targeted porous silicon nanovectors. *Nano Research*, *8*, 1505–1521.

Shi, J., Kantoff, P. W., Wooster, R., & Farokhzad, O. C. (2017). Cancer nanomedicine: progress, challenges and opportunities. *Nature Reviews Cancer*, *17*, 20–37.

Singh, A., & Sahoo, S. K. (2014). Magnetic nanoparticles: a novel platform for cancer theranostics. *Drug Discovery Today*, *19*, 474–481.

Sriraman, S. K., Aryasomayajula, B., & Torchilin, V. P. (2014). Barriers to drug delivery in solid tumors. *Tissue Barriers*, *2*, e29528.

Tan, A., Yildirimer, L., Rajadas, J., de la Pena, H., Pastorin, G., & Seifalian, A. (2011). Quantum dots and carbon nanotubes in oncology: a review on emerging theranostic applications in nanomedicine. *Nanomedicine (London, England)*, *6*, 1101–1114.

Wang, C. F., Makila, E. M., Bonduelle, C., Rytkonen, J., Raula, J., Almeida, S., et al. (2015). Functionalization of alkyne-terminated thermally hydrocarbonized porous silicon nanoparticles with targeting peptides and antifouling polymers: effect on the human plasma protein adsorption. *ACS Applied Materials & Interfaces*, *7*, 2006–2015.

Wang, C. F., Makila, E. M., Kaasalainen, M. H., Liu, D., Sarparanta, M. P., Airaksinen, A. J., et al. (2014). Copper-free azide-alkyne cycloaddition of targeting peptides to porous silicon nanoparticles for intracellular drug uptake. *Biomaterials*, *35*, 1257–1266.

Wang, C. F., Sarparanta, M. P., Makila, E. M., Hyvonen, M. L., Laakkonen, P. M., Salonen, J. J., et al. (2015). Multifunctional porous silicon nanoparticles for cancer theranostics. *Biomaterials*, *48*, 108–118.

Wicki, A., Witzigmann, D., Balasubramanian, V., & Huwyler, J. (2015). Nanomedicine in cancer therapy: challenges, opportunities, and clinical applications. *Journal of Controlled Release*, *200*, 138–157.

Wilson, W. R., & Hay, M. P. (2011). Targeting hypoxia in cancer therapy. *Nature Reviews Cancer*, *11*, 393–410.

Xing, H., Hwang, K., & Lu, Y. (2016). Recent developments of liposomes as nanocarriers for theranostic applications. *Theranostics*, *6*, 1336–1352.

Xu, W., Thapa, R., Liu, D., Nissinen, T., Granroth, S., Närvänen, A., et al. (2015). Smart porous silicon nanoparticles with polymeric coatings for sequential combination therapy. *Molecular Pharmaceutics*, *12*, 4038–4047.

Yaghini, E., Seifalian, A. M., & Macrobert, A. J. (2009). Quantum dots and their potential biomedical applications in photosensitization for photodynamic therapy. *Nanomedicine*, *4*, 353–363.

CHAPTER 2

Self-Assembled Nanogels: From Particles to Scaffolds and Membranes

Ranjith K. Kankala*, Shi-Bin Wang*, Ai-Zheng Chen*, Yu S. Zhang†
**Huaqiao University, Xiamen, PR China †Harvard Medical School, Cambridge, MA, United States*

1 Introduction

Cancer is one of the leading causes of death, accounting for millions of mortalities annually (Mantyh, 2006; Sun et al., 2014). Chemotherapy is the primarily practiced treatment regimen, along with surgery and/or radiation therapy, which can simultaneously improve the survival rate of patients with cancers (Kankala et al., 2017; Mantyh, 2006). However, it can result in severe adverse effects due to a lack of specificity. To overcome this limitation, a carrier immobilized/encapsulated with chemotherapeutic agents is preferred for its efficient conveyance to and effective action at the tumor site for a prolonged period through controlled delivery (Kankala, Zhang, Wang, Lee, & Chen, 2017; Sun et al., 2014).

In recent years, the design of smart materials in pharmaceutical applications using nanotechnology has gained increasing attention for efficient delivery of therapeutic moieties, as part of a multidisciplinary effort involving various fields of science including, but not limited to, chemistry, medicine, and materials (Raemdonck, Demeester, & De Smedt, 2009). The tunable/tailorable physical and chemical properties (such as electronic [interplay between charge transfer], mechanical [potential superplasticity, high strength, toughness, and ductility], magnetic, and optical properties [refractive indices, absorption, and photo-active effects]) of the nano-sized materials are noticeably different from their structural units or the bulk matter of similar compositions (Farokhzad & Langer, 2006; Huczko, 2000; Sun et al., 2014). These differences are related to their spatial structure, shape, inherent energetics, electronic architecture, phase changes, and assemblies (Hoffman, 2002; Sun et al., 2014).

Numerous techniques such as synthetic (chemical and physical), biological, and hybrid methods have been reported to prepare nano-sized particles (Kulkarni, 2015). Commonly used strategies for the synthesis of different types of nanoconstructs include electrospinning, phase separation, template-based synthesis, self-assembly, and sonochemical process, among others (Fleischer & Dvir, 2013; Huczko, 2000; Okitsu, Ashokkumar, & Grieser, 2005;

Handbook of Nanomaterials for Cancer Theranostics. https://doi.org/10.1016/B978-0-12-813339-2.00002-5

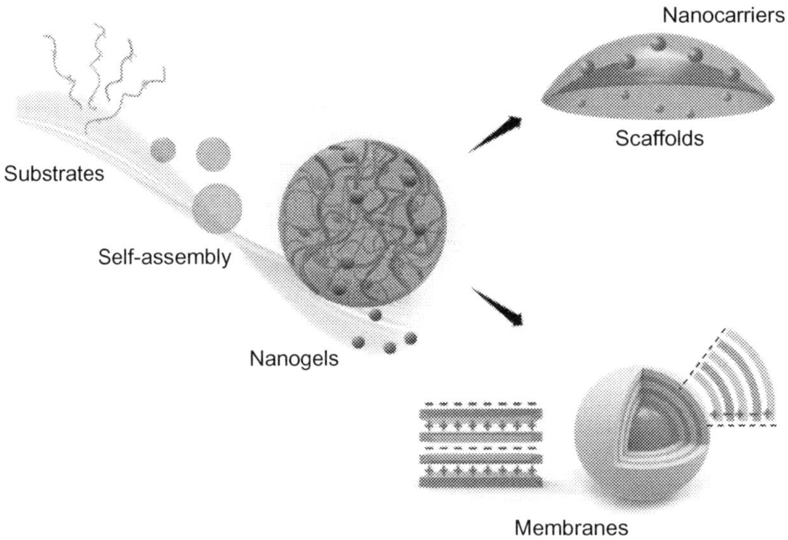

Fig. 1
Schematic illustration showing the self-assembling process of nano-sized hydrogels and their advancements in the design of polymeric scaffolds and membranes.

Takai, Tamura, & Fuji, 2014; Yang et al., 2017). Out of various methods used in the preparation of nanomaterials, the self-assembly process is of high relevance in different fields, and has been predominantly applied in electronics (Hirst, Escuder, Miravet, & Smith, 2008) and biomedical applications such as drug/gene/protein delivery and imaging (Akiyoshi, Ueminami, Kurumada, & Nomura, 2000; Ayame, Morimoto, & Akiyoshi, 2008; Chacko, Ventura, Zhuang, & Thayumanavan, 2012; Guarrotxena & Quijada-Garrido, 2016; Hasegawa, Nomura, Kaul, Hirano, & Akiyoshi, 2005; Hoffman, 2002; Nomura, Ikeda, Yamaguchi, Aoyama, & Akiyoshi, 2003; Ryan & Nilsson, 2012).

In this chapter, we give an overview of self-assembled nanomaterials (Fig. 1) by focusing on distinctive features of substrates and interactions involved during the self-assembly process. We then elaborate on self-assembled hydrogels and their important properties such as ease of synthesis, controllable swelling, viscoelasticity, high loading efficiency, and desired release characteristics, among others, for their utilization as controlled or stimuli-responsive release carriers in cancer therapy. In addition, we highlight the pros and cons of nanogel technology relevant to its biomedical applications. Finally, we extend our discussions to the advancement of nanogels for the preparation of polymeric scaffolds and membranes for cancer theranostics.

2 Self-Assembled Nanoparticles

Self-assembly is a process adopted by the disordered molecules of preexisting components for their spontaneous rearrangement into a distinct pattern or structure, which can be caused by a range of physical mechanisms (Grzelczak, Vermant, Furst, & Liz-Marzán, 2010; Whitelam, 2015; Whitesides & Grzybowski, 2002). Assembly of molecular building blocks can be driven by specific noncovalent interactions between them, which might be direct between the preexisting components (molecule-molecule forces and molecule-surface forces) or indirect through the influence of environment such as ions in the solution favoring the assembly (Grzelczak et al., 2010). This process enables exquisite control at the molecular scale, and creates a definite pattern among building blocks that remain too small to be manipulated individually (Pastuszka & Mackay, 2016). From a thermodynamic point of view, self-assembly is an equilibrium process in which the system exists in the lowest free-energy state, resulting in weaker intermolecular forces between the components and guiding them to arrange spontaneously (Whitesides, Kriebel, & Love, 2005).

Self-assembly is a significant phenomenon that is often observed in natural processes, for instance, most of the complex processes in our body, such as the formation of lipid membranes, folding of proteins, protein aggregation, and many others (Whitesides & Boncheva, 2002). In addition, it provides a synthetic route for the formation of molecular crystals, liquid crystals, and other regular structures of materials, making it one of the most generalized bottom-up strategies for the generation of three-dimensional (3D) nano-sized materials (Schwiebert, Chin, Macdonald, & Whitesides, 1996). Various types of self-assembled nanoparticles include micelles, polymersomes, and layer-by-layer nanoparticles have been reported (Tahara & Akiyoshi, 2015). Typically, the self-assembly process is of two types, static and dynamic, based on the arrangement of the building units (Timonen, Latikka, Leibler, Ras, & Ikkala, 2013). The static type of self-assembly involves the formation of particles in the system that are at an equilibrium state and do not dissipate energy. The dynamic-type process involves the organization of components in a specific pattern by specific local interactions during the assembly process, and it happens only if the system is dissipating energy (Grzelczak et al., 2010; Timonen et al., 2013; Whitesides & Grzybowski, 2002).

2.1 Features of Self-Assembly

In general, a few distinctive features inherent to the self-assembly process include an order of arrangement/adjustability, interactions between the components, and the preexisting substituents in the system and their physicochemical properties (Whitelam, 2015; Whitesides & Grzybowski, 2002). To generate ordered structures during self-assembly, structural formation must allow the components to adjust their position in the complex, and the strength of interaction formation should be competitive with their disruption, resulting in the stable self-assembled nanoaggregate (Whitelam, 2015). Eventually, the components in the system equilibrate and tend to adjust their positions during aggregation. The other characteristic feature that drives the self-assembly process is the molecular interactions associated with the components (Whitesides & Boncheva, 2002). Various interactions are involved in the formation of self-assembled aggregates such as van der Waals force, hydrogen bonding, and ionic and metallic interactions (Grzelczak et al., 2010; Whitesides & Grzybowski, 2002). In a few cases, weak covalent interactions may also form during the self-assembly process (Olenyuk, Whiteford, Fechtenkotter, & Stang, 1999). These interactions play a crucial role in the determination of physical properties of aggregates in various environments, especially in biological fluids during the drug delivery process (Demortiere et al., 2014). The third distinctive feature of the self-assembly process lies in the components of aggregates and their properties, which play crucial roles and guide the process of assembly. Vast research has shown that the self-assembly process involves a broad range of components and different surfaces for arrangement, which results in diverse structures that are in equilibrium or kinetically trapped, with altered morphological and other physicochemical properties, functionalities, and chemical composition, among others (Grzelczak et al., 2010).

2.2 Molecular Interactions

The self-assembly resulting in supramolecular architectures involves distinctive molecular interactions between the molecular building blocks, which drive them to associate and adjust position during the aggregation process (Whitesides et al., 2005). Various molecular interactions such as intermolecular forces and Hamaker interaction forces are advantageous to the components and result in desired self-assembled nanoaggregates (Demortiere et al., 2014; Grzelczak et al., 2010; Whitesides & Grzybowski, 2002). The direct intermolecular forces include van der Waals force, hydrogen bonding, ionic and metallic interactions, and charge-transfer interactions, as well as other supramolecular weak forces and internal forces such as hydrophobic interactions (Grzelczak et al., 2010).

In addition to intermolecular forces, externally driven forces such as the electric and magnetic fields can also influence the self-assembly process. These external forces at a balanced state enable interactions by overcoming the typical Brownian motion of components and result in the enhancement of their alignment (Kralj & Makovec, 2015). Application of these external forces adds the advantage of polarizability of the nanoparticles and their functional groups (Grzelczak et al., 2010). Other forces for self-assembly include a combination of fields (electromagnetic field) as well as capillary and entropic interactions, a combination which influences the outcome and results in the flexibility of the self-assembled design (Kralj & Makovec, 2015).

3 Hydrogels

Hydrogels are water-absorbed, 3D, swollen materials formed by stabilization of hydrophilic polymeric chains in the aqueous environment (Peppas, Hilt, Khademhosseini, & Langer, 2006; Zhang & Khademhosseini, 2017). The gelation of hydrogels can be achieved through a variety of mechanisms such as electrostatic interactions, spanning physical entanglement of polymer chains, and covalent chemical crosslinking (Raemdonck et al., 2009).

These water-rich constructs at an arbitrary gauge ranging from macro- to micro- to nanoscale, with varied internal structure, are utilized in various fields such as drug delivery, tissue engineering, and soft electronics, due to their broadly tunable physical and chemical properties (Hoffman, 2002). Further advancements have been made to improve the physicochemical properties of hydrogels to make them suitable for additional specific applications (Zhang & Khademhosseini, 2017).

Although extensive research has been focused on macroscopic hydrogels, there is an increasing interest in their confinement to the micro- and nanoscale dimensions. Moreover, confined hydrogels possess additional advantages in drug delivery applications such as ease of conveyance when administered as injectable, and are ideal candidates for cellular internalization of sensitive therapeutic agents such as nucleotides and peptides, which are of particular interest to transport therapeutic moieties in vivo (Peppas et al., 2006; Raemdonck et al., 2009).

3.1 Nanotechnology and Nanogel Technology

In general, chemotherapeutic agents (drugs, peptides, or enzymes) administered through oral or other routes exhibit poor pharmacokinetic behavior due to poor solubility and absorption, rapid metabolism and excretion, as well as high plasma protein binding, among others (Kai et al., 2015). In addition, most anticancer agents have a narrow therapeutic window, which requires thorough monitoring to achieve effective levels while simultaneously avoiding toxic effects. To overcome these obstacles, nanotechnology-based approaches offer new strategies for potential therapeutic and diagnostic applications, which provide beneficial contributions to the healthcare field (Jabir et al., 2012). Various types of formulations have been reported using several available nanotechnological tools for treating different types of cancers by drugs/small molecules, nucleotides, proteins, and other biomolecules (Farokhzad & Langer, 2006; Hassan et al., 2017; Kankala, Kuthati, Liu, Mou, & Lee, 2015; Sun et al., 2014). The delivery can be achieved either by passive targeting or through various active-targeting approaches (Sun et al., 2014). However, nanoparticle-based carriers suffer from certain limitations in vivo such as poor colloidal stability, low aqueous solubility, and short systemic circulation times due to rapid elimination by the mononuclear phagocytic system (Hassan et al., 2017; Soni, Desale, & Bronich, 2016; Yallapu, Jaggi, & Chauhan, 2011).

To this end, nanogels have attracted enormous attention in recent years. These nanogels, termed as confined hydrogels, are similar in internal structure but vary in size and responsiveness to hydrogels. Their characteristics such as size, shape, charge, softness, porosity, hydrophilicity, and degradability can be fine-tuned by altering the chemical composition of the nanogels (Zhang, Malhotra, Molina, & Haag, 2015). In particular, nanogels possessing hydrophilic groups such as hydroxyl, carboxyl, amino, amide, and sulfonic groups are pursued for biomedical applications due to their advantageous properties such as high biocompatibility with an extensive drug loading capability and their small sizes, which are approximately confined to 200 nm (Molina et al., 2015; Yallapu et al., 2011). Additional features of nanogels for drug delivery include protection of drug cargo from degradation and elimination as well as their controlled release (Tahara & Akiyoshi, 2015). The architecture of nanogels is versatile, as it allows to encapsulate various guest molecules ranging from small therapeutic molecules to biomacromolecules such as proteins and nucleic acids with suitable modifications in their construction, without compromising their gel-like behavior (Soni et al., 2016). Drug encapsulation and its release can be guided by various interactions with the host, that is, the polymeric network, such as complexation, hydrogen bonding, and hydrophobic interactions, among others (Tahara & Akiyoshi, 2015). Based on the type of polymeric chain in the formulation, nanogels are diverse, and to a substantial degree, drug release can be designed to assume a desired pattern for cancer therapy, for example, stimuli-responsive polymeric units for regulated drug release at the tumor site (Molina et al., 2015).

3.2 Preparation Methods

Nanogels are predominantly prepared using various crosslinking reactions of polymeric network chains to yield 3D water-rich environments. Various crosslinking methods include ionic crosslinking, radiation crosslinking, crosslinking polymerization, functional group crosslinking, self-assembly, crystallization, and precipitation polymerization (Chacko et al., 2012; Hoffman, 2002; Raemdonck et al., 2009). Broadly speaking, the preparation strategies can be classified into two categories, chemical crosslinking and physical crosslinking, based on the type of interactions between the polymeric chains. For successful preparation of nano-sized hydrogels, three aspects play crucial roles: the nature and chemical composition of the polymeric chain, control over the size of the particles, and crosslinking strategy (Asadian-Birjand, Sousa-Herves, Steinhilber, Cuggino, & Calderon, 2012). In addition, rigorous control over the crosslinking of the chains is essential to substantially confine the size of the particles to the nano-scale range. Typically, this can be achieved by performing gelation in highly diluted solutions of polymers, which reduces the risk of intermolecular crosslinking and subsequently increases the intramolecular crosslinking that prevents the macroscopic gelation (Asadian-Birjand et al., 2012).

The chemical crosslinking method involves the formation of irreversible covalent bonds, leading to the formation of an insoluble polymeric framework (Chacko et al., 2012). Among various preparation strategies available for chemically crosslinked hydrogels, free-radical crosslinking copolymerization of monomers and co-monomers remains the simplest and most efficient method (Chacko et al., 2012; Hoffman, 2002; Raemdonck et al., 2009). However, uniform-sized, chemically crosslinked nanogels can be predominantly obtained by emulsion-based polymerization and precipitation polymerization (Pelton, 2000; Raemdonck et al., 2009). Various substituents such as *N,N′*-methylenebisacrylamide (MBA), diallyl phthalate, divinylbenzene, and poly(ethylene glycol) diacrylate (PEGDA), have been commonly used as crosslinkers in the preparation of nanogels (Pelton, 2000). It has been demonstrated that the chemical crosslinkers play a crucial role in tailoring the properties of hydrogels such as size, swelling, and architecture, and are also responsible for the establishment of the predetermined release kinetics of entrapped drug molecules (Eichenbaum et al., 2005; Raemdonck et al., 2009).

On the other hand, nanogels self-assembled through hydrophobic interactions between the substrates in water have been reported (Hoffman, 2002; Raemdonck et al., 2009). This hierarchical self-assembly process results in the controlled formation of nano-sized aggregates through supramolecular self-assembly with respect to polymeric concentration and other factors such as temperature, pH, and enzymes (Akiyama et al., 2007; Kuroda, Fujimoto, Sunamoto, & Akiyoshi, 2002; Morimoto, Winnik, & Akiyoshi, 2007; Wang et al., 2010). These hydrogels can be assembled into different shapes such as spherical (Cheng, Liang, Liao, Huang, & Lee, 2017; Lin, Deng, Xu, & Dong, 2012), botryoidal (Morimoto et al., 2007), janus (Wei et al., 2013), and leaf-like (Wang et al., 2010), among others. Akiyoshi et al. synthesized self-assembled nanogels for the first time by modifying the polymer with hydrophobic groups such as cholesterol, which instantaneously resulted in the formation of monodispersed nanogels by intermolecular self-aggregation in water as tectonic materials (Akiyoshi, Deguchi, Moriguchi, Yamaguchi, & Sunamoto, 1993). Furthermore, they developed self-assembled nanogels using pullulan polymer and its modified derivative with alkyl chains or others, for the entrapment of various drugs and biomolecules (Akiyama et al., 2007; Akiyoshi et al., 2000; Ayame et al., 2008; Chan, lux, Nishimura, Akiyoshi, & Almutairi, 2015; Fujii et al., 2014; Kageyama et al., 2008; Kawasaki, Sasaki, & Akiyoshi, 2017; Kuroda et al., 2002; Morimoto et al., 2007; Nomura et al., 2003; Shimizu et al., 2008; Tahara & Akiyoshi, 2015; Takahashi, Tahara, Sawada, & Akiyoshi, 2013). Similarly, various polymers such as poly(*N*-isopropylacryamide) (PNIPAAm), cycloamylose, chitosan, heparin, hyaluronic acid (HA), dextran, cluster dextrin, glycogen, mannan, and poly(amino acids) have been used to synthesize various self-assembled nanogels entrapping drugs, proteins, small molecules, and other biomolecules for biomedical applications (Akiyoshi et al., 1993, 2000; Hoffman, 2002; Nomura et al., 2003; Shimizu et al., 2008; Tahara & Akiyoshi, 2015).

Another interesting method for the preparation of nanogels is the gelation of polymers in liposomal templates, where self-assembled liposomal templates composed of phospholipids entrap the solvent containing gelation initiator along with monomers as the core, which subsequently results in the formation of nanogels (Asadian-Birjand et al., 2012). The process is advantageous over others (such as the emulsion polymerization technique) because it can be performed without the need of organic solvents.

3.3 Properties of Nanogels

Nanogel technology is a rapidly growing field that has gained significant interest of researchers due to the unique physicochemical properties of nanogels such as tunable morphology including size, shape, and molecular architecture, distinctive swelling behavior, elasticity and

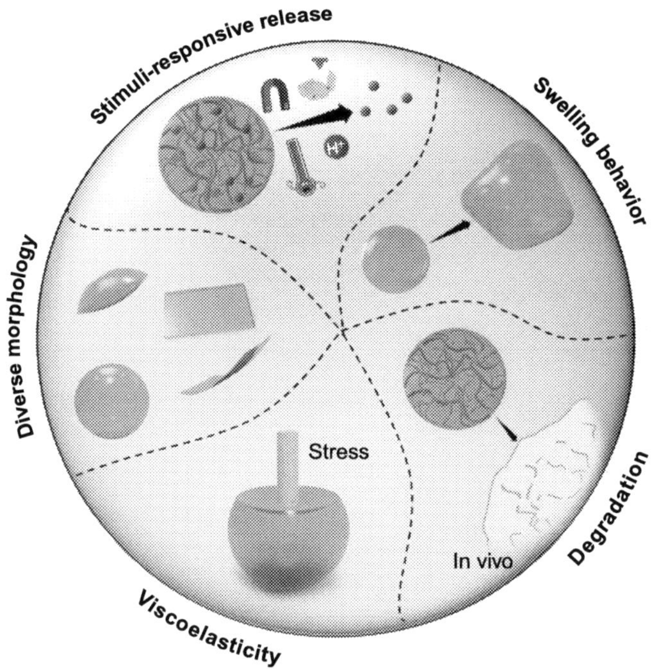

Fig. 2
Schematic representation illustrating the summarized properties of nanogels that have been explored as carriers for cancer therapy.

considerable mechanical integrity, and stimuli-responsive behavior, among others (Fig. 2) (Raemdonck et al., 2009; Soni et al., 2016; Yallapu et al., 2011). The ability to control the processing of self-assembled nanogels to obtain desired features makes it easier to tailor their nanoscale properties and functionality in designing drug delivery systems. Tailored properties increase the therapeutic efficiency of the encapsulated drug and reduce its nonspecific toxicity through the enhanced permeation and retention (EPR) effect. In addition, specific targeting ligands can be linked to the self-assembling substrates to direct the delivery systems toward the desired sites (Kim et al., 2008; Shu, Sabi-Mouka, Wang, & Ding, 2017; Tan & Liu, 2011).

In general, an ideal nanogel-based drug delivery carrier should possess characteristics such as a smaller particle size, proper biodegradability and/or biocompatibility, prolonged systemic circulation time, higher amount of drug or protein loading, and protection of molecules from the immune system of the body (Asadian-Birjand et al., 2012; Eckmann, Composto, Tsourkas, & Muzykantov, 2014). Recently, many new advancements have been made to improve the molecular architecture and the delivery pattern during the formulation of nanogels, which is essential to optimize them for in vivo applications (Soni et al., 2016; Vinogradov, Zeman, Batrakova, & Kabanov, 2005). Herein, we discuss a few properties of self-assembled nanogels highlighting their utilization for cancer theranostics.

3.3.1 Morphology

Rigorous control over the size of nanogels is critical for their biomedical applications. The nanogels are designed such that their optimum size should be in a range of 10–200 nm, which results in large surface areas and feasibility for multivalent conjugations. Unlike micro-sized hydrogels, the injectable nanogels in the specified size range may be easily applied for intravenous administration, which can extravasate through the capillary endothelium to reach the target tissue and deliver the drug cargo. For example, Gao et al. prepared ultrafine hydrophilic polyacrylamide-based nanogels for therapeutic applications, evading reticuloendothelial system (RES) uptake and reducing the risk of accumulation in the cell (Gao, Xu, Philbert, & Kopelman, 2007).

Typically, nanogels are spherical; however, current advancements in synthetic strategies have also resulted in the fabrication of nanogels in different forms (Soni et al., 2016). More often, micromolding, a unique synthetic technique, is utilized to prepare nanogels with control over desired manifold forms, sizes, and compositions, which does not require harsh experimental conditions such as high shear, high temperature, or presence of organic solvents (Asadian-Birjand et al., 2012; Dunn et al., 2012).

3.3.2 Swelling behavior

Swelling behavior is an intrinsic property of hydrogels, where the nanogels enlarge due to solvent penetration into the void space between the polymeric chain network. In some cases, swelling behavior can also be influenced by the external triggers such as pH, ionic strength, and temperature of the environment (Guarrotxena & Quijada-Garrido, 2016; Yadav et al., 2011). Nanogels respond faster to exhibit swelling capacity compared to conventional hydrogels (micro- and macro-sized hydrogels), due to the large surface areas of the nanogels, which create a high possibility of fluid exchange with the environment (Chacko et al., 2012). In addition, the pH-dependent release behavior of encapsulated drug cargo can be correlated to the higher swelling rate of nanogels in the acidic environment due to a considerable increase in their volume, which may accelerate endosome disruption and enhances the cytosolic levels of the drug, known as the "proton sponge" effect (Chiang et al., 2012; Kim, Kim, & Kim, 2011; Vinogradov et al., 2005).

3.3.3 Biocompatibility and degradability

Biocompatibility and degradability are the nanogel properties most essential for their utilization in biomedical applications. In general, nanogels are highly biocompatible and free from toxicity, so they do not provoke/induce any undesirable adverse biological responses at the molecular, cellular, or organ level, largely due to their high water content and the hydrophilic functional groups in the polymeric network (Eckmann et al., 2014; Urakami et al., 2013).

Most nanogel-based drug delivery systems prepared via self-assembly are degradable because the polymeric substrates are arranged through various weak interactions. However, there are a few polymers such as PNIPAAm whose degradability in physiological fluids is low. To overcome this limitation, the nanogel network has been prepared by crosslinking the polymers with degradable substrates such as albumin, HA, and acid-labile orthoester-based crosslinkers to enhance their degradation kinetics during drug delivery (Tada, Tanabe, Tachibana, & Yamauchi, 2005; Yallapu et al., 2011; Zhang et al., 2015). In addition, polymeric substrates from natural and synthetic sources, which possess sensitive functional groups such as ester, anhydride, amide, and phosphazene and phosphate esters in the polymeric chains, could augment the formulations' degradability (Yallapu et al., 2011). However, it should be noted that chemical functionalization of self-assembling polymers for immobilizing drugs or various targeting moieties can alter the pattern or rate of degradation or even reduce their degradation ability. More often, degradation of nanogels occurs through solubilization, hydrolysis, and enzymatic mechanisms, among others (Yallapu et al., 2011).

3.3.4 Viscoelasticity

Due to their highly solvated and water-rich environment, nanogels exhibit both solid as well as liquid-like behaviors and are highly viscoelastic in nature (Jahren, Butler, Adams, & Cameron, 2010; Raemdonck et al., 2009; Zhang et al., 2015). These viscoelastic nanogel particles are flexible enough that they can deform and navigate in the presence of flow in the extracellular environment, pass cell barriers, and behave the same intracellularly. Various rheological methods can readily characterize bulk gels; however, there is still no specific technique for demonstrating the rheological behavior of nanogels (Eckmann et al., 2014).

3.3.5 Drug loading capacity and drug release

The optimum amount of drug encapsulation is an important feature of any delivery system, and it needs to be addressed during the delivery vehicle design. Furthermore, encapsulation stability should also be addressed to avoid unwanted drug leakage before reaching the target site. Generalized strategies used for drug loading into nanogels include self-assembling encapsulation either during the synthesis of nanogels or following the process through electrostatic or other interactions (Cheng et al., 2017). High drug loading in nanogels can be achieved through prior conjugation to the polymeric chains that undergo the self-assembly process (Chacko et al., 2012). However, the loading efficiency resulting in stable encapsulation utterly depends on the selection of polymeric substrates (Chacko et al., 2012; Soni et al., 2016; Tahara & Akiyoshi, 2015).

Drugs encapsulated within the biocompatible polymeric networks are released through two main ways of diffusional release and burst release (Kankala, Zhang, et al., 2017; Tahara & Akiyoshi, 2015). In a diffusional release phenomenon, high-affinity interactions between the self-assembled layers as well as crosslinks with the drug are weakened and subsequently become porous, eventually releasing the drug (Eckmann et al., 2014; Schwiebert et al., 1996; Tahara & Akiyoshi, 2015). The other method is stimuli-responsive release, where the host nanogel matrix is sensitive to the external environment and degrades in response to environmental cues such as biological or external triggers (explained under "stimuli-responsive drug release"). The drug release mechanism predominantly depends on the type of polymer used in the nanogel preparation, the type of crosslinking in the formulation, and other factors such as environmental pH and biological triggers (Tahara & Akiyoshi, 2015).

4 Nanogels in Cancer Theranostics

Considerable effort has been expended to design smart delivery systems for cancer theranostics. Nanogel technology has been at the forefront of the development of versatile drug delivery systems with desired features, as they possess advantageous properties such as softness, high water content, biodegradability, high drug payload, and tunable size range, among others, which have opened up a new paradigm for their utilization in biomedical applications, especially in cancer theranostics (Asadian-Birjand et al., 2012). More often, the nanogel-based delivery systems carrying a high payload of therapeutic moieties such as drugs, proteins, genes, radioactive substances, and other biomolecules, facilitate their delivery at the desired site in a controlled mode (Kawasaki et al., 2017; Lalitha et al., 2014; Takahashi et al., 2013). In addition, nanogel systems can be tailor-made to respond to various stimuli in the tumor environment. Common stimuli for triggered delivery include pH, enzymes, redox potential, and temperature, which occur naturally in vivo, and magnetic field and light irradiation applied externally (Molina et al., 2015; Motornov, Roiter, Tokarev, & Minko, 2010).

4.1 Controlled Release Nanogels

A suitable carrier for controlled release of anti-cancer drugs should be able to carry a high payload of drugs, sensitive to the environment at the target site, and should control release of the loaded therapeutic cargo. Self-assembled nanogels are one such delivery system which can deliver therapeutic cargos at the tumor sites (Kawasaki et al., 2017; Lalitha et al., 2014; Takahashi et al., 2013). Incorporating active moieties within biocompatible and/or biodegradable polymeric nanogels has been shown to increase their solubility in physiological fluids and feature site-specific accumulation, which reduces the systemic toxicity and undesirable drug waste (Ruiz, Gantner, & Talevi, 2014). The materials for preparing self-assembled nanogel delivery systems include various polymers that usually form stable and soft supramolecular hydrogels in the aqueous environment via various interactions such as π-π interactions and multiple hydrogen bonding (Wei et al., 2013). In some cases, polymeric substrates are modified with hydrophobic moieties to enhance their interactions, promote self-aggregation and drug encapsulation efficiency, and eventually result in sustained drug release (Tan & Liu, 2009). As depicted in Table 1, many active pharmaceutical

Table 1: Drug release from self-assembled nanogels for cancer theranostics

Drug	Polymer	Size (nm)	Outcome	References
5-Fluorouracil (5-FU)	Fmoc-FF, TPGS	~100	Peptide-based nanogels resulted in controlled drug release	Ischakov, Adler-Abramovich, Buzhansky, Shekhter, and Gazit (2013)
	Lysozyme, Sodium CMC	241	The release of drug was more in intestinal fluids compared to gastric fluids	Zhu et al. (2013)
Adriamycin	Chitosan, Linoleic acid	~400	Self-aggregated nanogels resulted in high drug loading and encapsulation efficiencies	Tan and Liu (2009)
Boron	Carborane, Pullulan	150	Nanogels for boron neutron captured therapy localized in tumor through EPR	Kawasaki et al. (2017)
Cisplatin	PAE, PEG	100–200	Grafted carboxylic groups enhanced cisplatin loading through complexation	Jin et al. (2007)
	PEG, PLA	20–30	Thermogel of prodrug showed the sustained release effect	Shen et al. (2015)
Chlorin e6 (Ce6)	Chondroitin sulfate	120–160	Acetylated-nanogels upon self-aggregation enhanced the photodynamic effects of the drug	Li and Na (2011)
Coumarin	Pyrene	~200	Nanogels inhibited prostate tumor effectively	Lalitha et al. (2014)
Doxorubicin	EG-grafted 1,8-naphthalimide	100–200	Efficient drug release from nanogels resulted in enhanced anticancer effect	Xu et al. (2015)
	PLGA-PEG-PLGA	50–150	Highly stable janus nanogels inhibited breast tumor efficiently	Wei et al. (2013)
	Fmoc-FF, TPGS	~100	Peptide-based nanogels resulted in controlled drug release	Ischakov et al. (2013)
	Chondroitin sulfate	~100	Stable nanogels were internalized to deliver drug via endocytic mechanism exploited by sugar receptors	Park, Park, and Na (2010)
	Low-density lipoproteins, Sodium CMC	~90	Green-step assembly of nanogels for facile loading and pH-dependent release of drug	He et al. (2015)
	Cytosine, PPG	~80	Thermoresponsive release of drug significantly enhanced the cytotoxic effects and reduced the dosage	Cheng et al. (2017)
Fludarabine	PEI, PEG	20–100	Polyplex nanogels enhanced cytotoxicity significantly compared to the pure form of the drug	Vinogradov et al. (2005)
Indocarbocyanine	PNIPAAm-co-AAc, Iron oxide	~100	Thermosensitive self-assembled nanogels for effective bioimaging	Park, Yang, Woo, Jeon, and Park (2013)

Table 1: Drug release from self-assembled nanogels for cancer theranostics—cont'd

Drug	Polymer	Size (nm)	Outcome	References
Interleukin-12 (IL-12)	Cholesterol, Pullulan	20–30	Subcutaneously injected nanogels resulted in prolonged elevation of IL-12 levels	Shimizu et al. (2008)
Methotrexate (MTX)	Lysozyme, CMC	123	Nanogels resulted in controlled drug delivery with concomitant bioavailability enhancement of the drug	Li, Xu, Zhang, Chen, and Li (2015)
pDNA	PNIPAAm-co-AAc, Iron oxide	~100	Thermosensitive self-assembled nanogels delivered gene efficiently	Park et al. (2013)
Porphyrin	PNIPAAm	~66	Porphyrin incorporation resulted in stable thermoresponsive and viscoelastic properties of nanogels	Belali, Karimi, and Hadizadeh (2017)
RNA	Cholesterol, Cycloamylose	~30	Intracellularly delivered siRNA/ nanogels resulted in efficient knockdown of VEGF	Fujii et al. (2014)
Taxol	PLGA-PEG-PLGA	80–200	Highly stable Janus nanogels for efficient inhibition of breast tumor	Wei et al. (2013)
	FA, Gly-Tyr-Lys	50	FA-based nanogel as a targeted delivery system	Wang et al. (2011)
Tretinoin	Curdlan derivatives	20–50	Designed nanogels exhibited potential immunological enhancement	Na, Park, Kim, and Bae (2000)

Abbreviations: CMC, carboxymethyl cellulose; CD, cyclodextrin; TPGS, D-α-tocopheryl, polyethylene glycol 1000 succinate; EPR, enhanced permeation and retention; EG, ethylene glycol; FA, folic acid; Gly-Tyr-Lys, glycine-tyrosine-lysine; Fmoc-FF, *N*-fluorenylmethoxycarbonyl-diphenylalanine; pDNA, plasmid deoxyribose nucleic acid; PAE, poly(β-amino ester; PLA, poly(D,L-lactide); PLGA, poly(lactide-*co*-glycolide); PEG, poly(ethylene glycol); PEI, polyethylenimine; PNIPAAm, poly(*N*-isopropylacrylamide); PNIPAAm-*co*-AAc, poly(*N*-isopropylacrylamide-*co*-acrylic acid); PPG, polypropylene glycol; siRNA, small interfering ribose nucleic acid; VEGF, vascular endothelial growth factor.

ingredients have been encapsulated in biodegradable polymeric substrates to prepare self-assembled nanogels. Here we emphasize a few reports relevant to the controlled release of drugs from self-assembled nanogels for cancer theranostics.

Typically, self-assembly of nanogels is observed at an optimum critical gel concentration (CGC) through various interactions. In some cases, the nanogel formation happens to be favorable at a low CGC by applying ultrasound treatment. In an example, Zhao et al. synthesized stable nano-sized gels of hydrophilic ethylene glycol-decorated 1,8-naphthalimide with a low CGC via π-π interactions and multiple hydrogen bonds (Xu et al., 2015). Initially, the synthesized derivative of naphthalimide was subjected to ultrasound treatment for several minutes with a low CGC (0.36 wt%) resulting in the formation of long and uniform nanofibers that were entangled to form stable hydrogel networks. Doxorubicin encapsulation in these nanogels resulted in high loading efficiency, and the subsequent cellular internalization process demonstrated controlled drug release and eventually, strong anticancer effect.

4.2 Stimuli-Responsive Nanogels

Despite their significant advantages and excellent physicochemical properties, nanogels still possess a few disadvantages such as insufficient drug-entrapment stability over extended periods of time and lack of well-controlled drug stability under specific microenvironment conditions, resulting in damage to normal tissues and therapeutic inefficiency (Cheng et al., 2017). To overcome these curbs, a highly stable nanogel formulation with the ability to release the drug at the desired site is preferred. Several attempts have been made to design intelligent drug delivery systems which undergo physical or chemical modification in response to external stimuli and release the drugs. These external stimuli include typical biological triggers such as temperature change, pH of the environment, functional groups/ linkers in the polymeric chain network sensitive to specific biomolecules, and external triggers such as the application of a magnetic field or light irradiation (Table 2) (Motornov et al., 2010). The advantages of stimuli-responsive delivery systems over conventional systems include improved drug loading/release, minimized premature release under physiological conditions, and reduced undesirable side effects, among others (Arunraj et al., 2014; Fleige, Quadir, & Haag, 2012; Molina et al., 2015; Motornov et al., 2010; Raemdonck et al., 2009; Soni et al., 2016; Yallapu et al., 2011).

Table 2: Representative responsiveness of self-assembled nanogels for cancer theranostics

Responsive Modality	Components With Specific Functionality	Application	References
Light	Imino and diazo-functionalities	Release favored when exposed to near-infrared light	Asadian-Birjand et al. (2012) and Chen et al. (2016)
Magnetic	Iron oxide in a nanogel core	Agents for imaging such as MRI	Park et al. (2013) and Zhang and Song (2016)
Molecular	Sensitive linker for a molecule Acetals-pH, disulfide-glutathione, esters-enzymes	Controlled release by degradation of respective linker in the nanogel	Bae, Mok, and Park (2008), Dunn et al. (2012), Lee, Mok, Lee, Oh, and Park (2007), and Shi et al. (2012)
pH	Polymer chains containing ionizable repeating groups (weakly acidic/basic) are suitable	Delivery favored at a site with lower pH, specifically acidic (4.5–6.0) in cancer	Arunraj, Sanoj Rejinold, Ashwin Kumar, and Jayakumar (2014), He et al. (2015), Na & Bae, 2002, Na, Lee, and Bae (2007), and Zhang et al. (2013)
Thermoresponsive	PNIPAAm, PCL derivatives, and others	Release/gelation favored by temperature trigger	Cheng et al. (2017), Gong et al. (2012), Lin et al. (2012), and Shen et al. (2015)
Multiresponsive	Combination of two or more responsiveness	Above-discussed responsive behaviors act simultaneously	Asadian-Birjand et al. (2012), Chen et al. (2016), Khatun et al. (2015), Park et al. (2013), and Shi, Zhang, Qiao, Bi, and Dai (2014)

Abbreviations: PCL, poly(ε-caprolactone); PNIPAAm, poly(*N*-isopropylacrylamide); MRI, magnetic resonance imaging.

4.2.1 Biological triggers

Both intracellular and extracellular environments of most of the solid tumors possess lower pH (i.e., 4.5–6.0) compared to physiological body fluids (i.e., 7.4) surrounding it (Matsumoto, Buchman, Rome, & Maynard, 2015; Van Dyke, 1996). This different condition can be utilized as one of the biological triggers that favor drug release (Kocak, Tuncer, & Butun, 2017). Polymers that are pH-responsive and possess ionizable groups such as weakly acidic or basic groups have been utilized; these respond to the changes in the environment pH and alter themselves physically, resulting in the release of drug in the tumor environment (Na & Bae, 2002). Polymers from both natural (e.g., chitosan and dextran) and synthetic origins (polymers with carboxylic and amine functional groups) have been chosen for the design of pH-responsive nanogels (Kocak et al., 2017; Na et al., 2007; Na & Bae, 2002).

Among various stimuli, environmental temperature is a promising stimulus that can influence the release of therapeutic cargo. This can be attained by increasing the temperature of the tumor site (a process known as hyperthermia) to trigger drug release. More often, polymeric substrates containing thermoresponsive units are utilized to construct the molecular architecture, which reversibly respond to the temperature (Fleige et al., 2012). These materials can be readily tailored during nanogel synthesis to obtain the desired phase-transition temperature and environmental temperature-induced release of drugs. For example, aqueous solutions of PNIPAAm exhibit a de-mixing phase transition with a lower critical solution temperature (LCST) at approximately 32°C, which is triggered by changes in hydrophobic and hydrogen bond interactions (Das, Mardyani, Chan, & Kumacheva, 2006). In another approach, Li et al. adjusted the LCST of nanogels prepared using copolymers to a level between normal body temperature (37°C) and hyperthermia temperature (43°C) as stimuli (Li, Pan, Zhang, & Du, 2009). The nanogels loaded with paclitaxel via self-assembly were found to possess long circulation times under physiological conditions and exhibited the thermoresponsive release behavior necessary for anticancer therapy. In addition, various supramolecular substrates such as cytosine have been linked to polymeric substrates to establish multiple hydrogen-bonded systems for the thermoresponsive behavior of self-assembled nanogels (Cheng et al., 2017). Thus, these thermoresponsive nanogels can be used in combination with the hyperthermia technique to achieve efficient anticancer therapy (Cheng et al., 2017).

Thermogelling polymers have been used for localized controlled delivery of drugs (Shen et al., 2015). It should be noted that this study demonstrated thermoresponsive gelation but not the release of the therapeutic moiety. The unique feature of these polymers is that the gelation of the polymer occurs at elevated temperatures (Koppolu et al., 2012; Pelton, 2000). For example, poly(ethylene glycol) (PEG) in combination with co-block polymers exhibit this property at body equivalent temperature (Shen et al., 2015). Interestingly, Yu and Shen et al. prepared biocompatible and biodegradable diblock copolymer methoxyl PEG-b-poly(D,L-lactide) (mPEG-PLA) and a succinic acid-modified platinum prodrug substrates, which upon

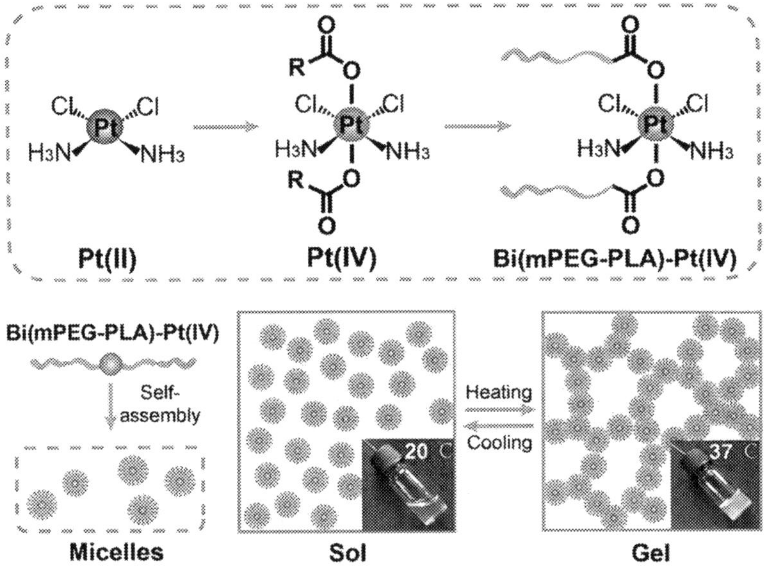

Fig. 3

Molecular design of the unique polymer-Pt(IV) conjugate and corresponding thermogel. The resultant conjugates could self-assemble into micelles in water. The sol was a suspension of micelles at room temperature and spontaneously turned into a semisolid gel with percolated micelle network upon heating. *Reproduced from reference Shen, W., Luan, J., Cao, L., Sun, J., Yu, L., & Ding, J. (2015). Thermogelling polymer-platinum(IV) conjugates for long-term delivery of cisplatin. Biomacromolecules, 16, 105–115 with permission from American Chemical Society, 2015.*

self-assembly yielded the micellar structures in water and the concentrated polymeric solution underwent reversible sol-gel transition with increasing temperature (Fig. 3) (Shen et al., 2015). Moreover, the prodrug conjugates in the formulation had a significant influence on the thermogelling property of the polymer, and the in vitro release of platinum from the hydrogel lasted for a couple of months and led to enhanced cytotoxic effect.

Smart nanogels possessing sensitive crosslinkers/functional groups in their polymeric chains respond to the available biomolecules in the environment such as enzymes and others, facilitating the delivery of encapsulated drugs in the hydrogel framework. More often, polymeric chains possessing or linked with disulfide bonds or other reducible functional groups are utilized to design molecular-responsive nanogels, which can be reduced by the intracellular glutathione, because the intracellular environment of the tumor is more reducing compared to the extracellular environment (Bae et al., 2008; Dunn et al., 2012; Shi et al., 2012). Similarly, another classic example of molecular recognition is enzyme-dependent release. More often, enzyme-responsive nanogels rely on the cleavage of its ester functional group or short peptide sequences by intracellular esterases or proteases (Fleige et al., 2012). For example, HA is degraded by its specific enzyme hyaluronidase, whose levels are high in the tumor (Khatun et al., 2015) resulting in the degradation-dependent cytotoxicity

(Lee et al., 2007; Park, Kim, Bae, Kim, & Na, 2010). In addition, nanogels have been prepared by utilizing different polymeric substrates with ester or peptide bonds in the hydrogel matrix via self-assembly that are sensitive to various tumor-associated molecular recognition enzymes such as matrix metalloproteinase (MMP)-2, MMP-9, esterase, and others, for on-demand release of therapeutic moieties (Fleige et al., 2012). These nanogels were found to disassemble and release the encapsulated drug cargo in response to the intracellular enzymes (Fleige et al., 2012). In a case, Kim et al., developed a bioresponsive system via self-assembling peptide amphiphiles possessing the cell-adhesive MMP-2-sensitive GTAGLIGQRGDS peptide and a fatty acid for the delivery of cisplatin (Kim, Anderson, Jun, Repka, & Jo, 2009). These molecular recognition substrates in nanogels provide a spatially and temporally controlled delivery system for efficient delivery of antitumor drugs.

4.2.2 External triggers

In addition to intrinsic biological triggers in tumor cells, other external triggers such as light and magnetic field can also influence the release of drugs in the tumor. Light irradiation at a specific wavelength, particularly near-infrared (NIR) range, is a desirable stimulus for drug delivery due to the ease with which it penetrates tissues at a depth larger than 1 mm, and the fact it is inexpensive and easily controlled (Yun & Kwok, 2017). Applying light induces the release of therapeutic moieties by increasing the temperature of the nanogels. Light-sensitive molecule-immobilized polymeric substrates absorb light and subsequently increase the nanogels' temperature, which in turn alters the swelling behavior and releases the drugs (Suzuki & Tanaka, 1990). In addition, light triggers the cleavage of sensitive links such as urethane bonds within the polymeric substrates, which enhances the release of drugs specifically at the tumor site (Hang, Zou, Zhong, Zhong, & Meng, 2017).

The magnetic field is another commonly used external trigger for the delivery of various therapeutic agents due to its noninvasive nature, biocompatibility, and minimal adverse effects (Koppolu et al., 2012). Nanogels containing superparamagnetic substances have been a promising alternative, and they can be used for multipurpose applications such as combinations of magnetic resonance imaging (MRI), magnetically targeted delivery, and others (Park et al., 2013). An alternating magnetic field can be applied to control the time and dose of the release of therapeutic cargos from nanogels. In addition, these incorporated magnetic substances in nanogels may enhance the temperature inside the tumor cells for MRI-guided magnetic hyperthermia therapy (Zhang & Song, 2016).

4.3 Multiresponsive Nanogels

In recent years, multiresponsive nanogels have emerged as a promising strategy for biomedical applications. Combining various stimuli-responsive triggers in a single formulation has been utilized to fulfill therapeutic needs that eventually augment drug levels at the desired target site to produce persistent therapeutic efficacy for a prolonged duration.

Self-assembled nanogels with dual triggers such as light and hyaluronidase (Chen et al., 2016), pH and glutathione (Shi et al., 2014), pH and temperature (Belali et al., 2017), and other combinations (Ayame et al., 2008) have been predominantly utilized for efficacious cancer theranostics. In some cases, multiple triggers in a single nanogel formulation have been applied for enhancing the releasing ability of nanogels and subsequently intracellular therapeutic cargo amounts. In one case, Khatun and Lee et al. prepared a self-assembled nanogel formulation by encapsulating graphene-drug conjugate with HA through physical deposition that exerted multiple-responsive delivery patterns involving pH-, light-, and glutathione-responsiveness (Fig. 4) (Khatun et al., 2015). These multiresponsive nanogels demonstrated enhanced intracellular amounts of drug and substantially resulted in synergistic therapeutic effect. These nanogels were effective against a human lung cancer cell line (A549 cells) with limited toxicity in the normal cell line (MDCK cells).

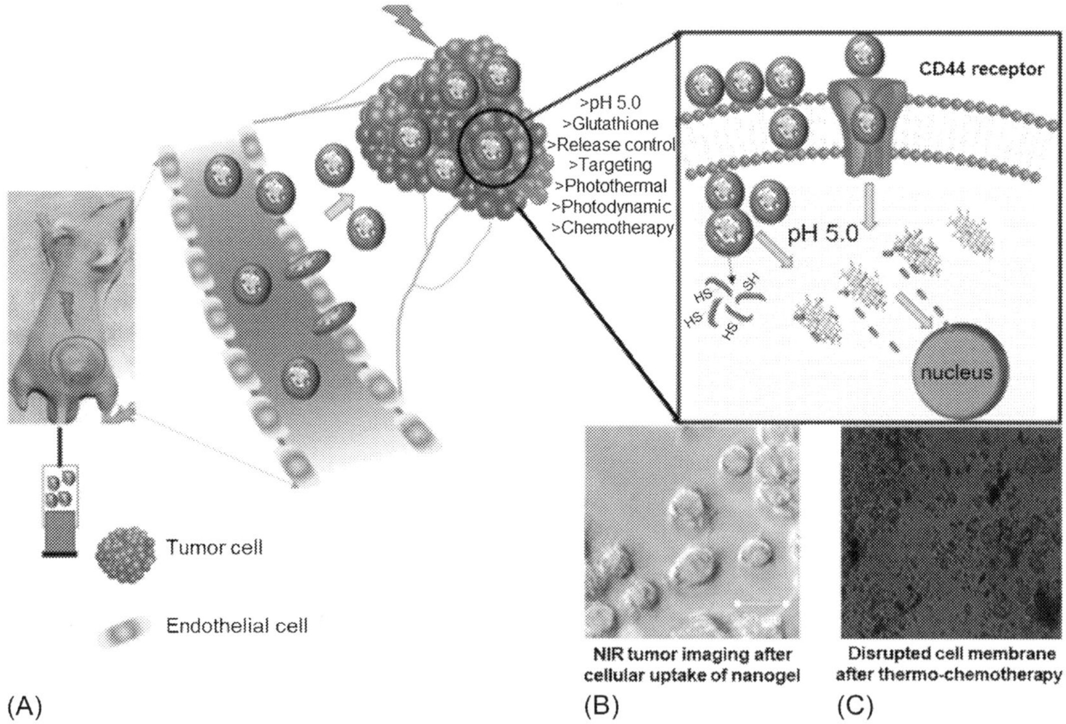

Fig. 4

Multiresponsive self-assembled nanogels. (A) Schematic illustration representing the accumulation of the nanogels around tumor cells and receptor-mediated endocytosis of the nanogels. (B) Optical tumor imaging after treatment with the nanogels. (C) Disruption of tumor cells by thermochemotherapy. *Reproduced from reference Khatun, Z., Nurunnabi, M., Nafiujjaman, M., Reeck, G. R., Khan, H. A., Cho, K. J., & Lee, Y. K. (2015). A hyaluronic acid nanogel for photo-chemo theranostics of lung cancer with simultaneous light-responsive controlled release of doxorubicin. Nanoscale, 7, 10680–10689 with permission from Royal Society of Chemistry, 2015.*

4.4 Targeted Drug Delivery

Nanogels offer enormous potential as targeted drug delivery vehicles due to numerous advantages such as ability to reduce off-target effects, extend circulation time, carry high payload of drugs and other bioactive molecules to the desired organ, and target the specific cell population via conjugating with high-affinity ligands (Kim et al., 2011; Shu et al., 2017). These versatile nanogels not only protect the drug cargo from rapid degradation but also facilitate crossing of multiple tissue barriers (Eckmann et al., 2014). Other unique features of targeted self-assembled nanogels include enhanced stability and deformability to augment binding and retention within the targeted tissue. The classical examples of targeting moieties immobilized on self-assembled nanogels include folate (Kim et al., 2008; Shu et al., 2017; Tan & Liu, 2011), RGD peptide (Shu et al., 2017; Xu et al., 2016), and cell surface peptide (Dickerson et al., 2010), among others, which have the ability to target the desired tumor tissues and convey drug cargos with minimal adverse effects.

4.5 Tumor Imaging

Imaging techniques are critical to diagnosing and monitoring various diseases by enabling noninvasive assessment of the extent of disease, which also facilitates treatment feedback (Weissleder, 2006). Cancer can be analyzed by various imaging techniques such as MRI, a method offering numerous advantages like high anatomical resolution with unlimited tissue depth and soft tissue contrast, among others (Caravan et al., 2007). However, a suitable contrast agent is necessary to deliver a correct dose of the imaging agent to the appropriate location for successful diagnosis (Peppas et al., 2006). Several United States Food and Drug Administration (FDA)-approved MRI contrast agents have been encapsulated in self-assembled nanogels for effective tumor imaging due to their versatility in encapsulation, crosslinked polymer network, and high water content (Jiang, Chen, Deng, Suuronen, & Zhong, 2014). In addition, the unique advantage of using nanogels is their hydrophilic nature, which increases the water exchange rate and consequently enhances relaxivity during imaging (Chan et al., 2015). For example, Hasegawa et al. prepared a novel carrier based on the cholesterol-bearing pullulan modified with amino groups via self-assembly for intracellular labeling (Hasegawa et al., 2005). These stable nanogels were highly effective in carrying quantum dots and have a great potential for use in long-term live cell imaging over other carriers such as liposomes. Furthermore, the encapsulation of sensitive imaging agents in nanogels can overcome certain limitations for their utilization in the biomedical field such as fast elimination, lack of stability, and others (Hasegawa et al., 2005).

4.6 Nanogel Architecture for In Vivo Application and Their Clinical Translation

Ideally, nanogel delivery vehicles should convey their drug cargo safely to the desired target tissue by surpassing numerous extracellular barriers during the transportation

(Fleige et al., 2012; Molina et al., 2015). To achieve this, considerable amounts of effort have been expended utilizing various approaches such as nanogels shielded with inert and hydrophilic polymer coatings like PEG, which would significantly improve in vivo delivery of therapeutic cargos. Shielding nanocarriers by utilizing PEG offers numerous advantages such as prevention of uptake by mononuclear phagocytes, improvement of plasma circulation time, and avoidance of the nonspecific interactions with the serum proteins (Tahara & Akiyoshi, 2015). Another interesting approach is the immobilization of targeting ligands, which enhances the uptake efficiency of the nanogels specifically in cancer cells, and in turn, minimizes the adverse effects in normal cells. An additional valuable approach for in vivo application of nanogels is the utilization of stimuli responsive polymeric substrates, which are sensitive to the microenvironments and effectively deliver the drug cargo in tumors (Molina et al., 2015).

Despite their success in the development of synthetic methodologies, enormous progress and advancement of formulations, and obvious usefulness in many cases, nanogels have not yet gained ubiquity in clinical practice. However, a few self-assembled nanoparticles have been translated into clinical trials (Costantino & Boraschi, 2012). Among them, cholesteryl pullulan-based self-assembled nanogels comprising 146HER2 vaccine were analyzed for humoral responses in patients that elicited 146HER2-specific CD8+ and CD4+ T-cell immune responses in patients with HER2-expressing tumors (Aoki et al., 2009; Kageyama et al., 2008). The vaccine-immobilized nanogels via self-assembly have evidenced promising results with no significant adverse effects, suggesting that these can be utilized to treat multiple tumors (Aoki et al., 2009).

5 Hydrogel-Based Scaffolds in Cancer

Among various nano-sized carriers, self-assembling nanogels are one of the most promising delivery systems for cancer theranostics. However, substantial improvement in their physicochemical properties is necessary to obtain desired therapeutic effects. This can be achieved by modifying the rational design at the molecular level and controlling their multiscale architecture (Zhang & Khademhosseini, 2017). The mechanical properties of self-assembled collagen hydrogels can be improved by introducing covalently attached crosslinkers (Lee, Singla, & Lee, 2001). However, these introduced crosslinkers led to their irreversibility and unacceptable toxicity in a few cases, which limited their use as injectable hydrogels (Xing et al., 2016). In addition, mechanical properties can also be improved by integrating the nanotechnology platform with the hydrogels (Hosoya et al., 2016). Moreover, a few reports suggested that the fabrication of hydrogel-based scaffolds produced excellent physicochemical properties such as shear-thinning and others (Hosoya et al., 2016). Recently, Yan et al. fabricated biomineralization-triggered injectable scaffolds based on a self-assembling collagen-gold hybrid hydrogel with adjustable mechanical properties yielding shear-thinning and self-healing ability, and excellent biocompatibility as well as biodegradability (Xing et al., 2016). The design consisted of the complexation of photoactive and tunable gold ionic clusters with the positively charged collagen resulting in the gold-collagen hybrid scaffold. Furthermore, the self-assembled hydrogels encapsulated with water-soluble anticancer moiety led to its sustained release and substantially synergistic effects against cancer. In another study, the injectable preparations of thermosensitive, nanoparticle-self-aggregated, hydrogel-based scaffolds demonstrated a precise spatiotemporal control over the conveyance of the encapsulated drug cargo payloads (Huang et al., 2016). These preparations are highly efficient in delivering the bioactive species during the postoperative lesions that may prevent the recurrence and complete suppression of tumor at the lesion site. Other significant contributions toward preparing hydrogel-based scaffolds for cancer theranostics were discussed.

Hydrogel formation via self-assembling peptides are highly suitable for cancer therapy and have been shown to form functional nanofibrillar structures (Li et al., 2016). Facile control over the formation of nanostructures makes them excellent candidates as multicomponent adjuvant scaffolds for the delivery of various therapeutic moieties such as proteins, drugs, and other bioactive molecules (Li et al., 2016). In one case, Li et al. prepared co-assembled nanostructures of bioscaffold based on the fluorenylmethoxycarbonyl (Fmoc)-FRGDF peptide and fucoidan polymer, which reduced the expression of pro-inflammatory cytokines to induce apoptosis in cancerous epithelial cells (Li et al., 2016). This potent self-assembled design demonstrated that it provided a nontoxic, biocompatible scaffold for treating pro-inflammatory cancers.

Hydrogel-based scaffolds have also been used as efficient delivery vehicles to realize the potential of RNAs in abrogating cancer by successfully releasing them in tumor cells. Interestingly, Conde et al. developed a RNA-triple-helix structure and conjugated them to dendrimers yielding stable triplex nanoparticles, which formed adhesive scaffold upon interaction with dextran (Fig. 5) (Conde, Oliva, Atilano, Song, & Artzi, 2016). The resultant self-assembled RNA-triple-helix conjugates modulated the expression of endogenous RNAs in cancer in vivo, leading to a nearly 90% level of breast tumor shrinkage at 2 weeks post-gel implantation.

6 Hydrogel-Based Membranes in Cancer Therapy

Organic thin films with remarkable properties are of increasing interest in nanotechnology, and they can be used in building bioactive membranous structures for diverse biomedical applications (Dai et al., 2009). Nano-sized thin films can be assembled in a variety of ways. However, layer-by-layer (LbL) self-assembly, one of the latest techniques first demonstrated by Decher in 1991 (Decher, MacLennan, Reibel, & Sohling, 1991), has been widely used for organizing the conjugates/films/monolayers of polymeric substrates into layered membranes (Shaik, Shaikh Mohammed, Mcshane, & Mills, 2013). The self-assembled hydrogels

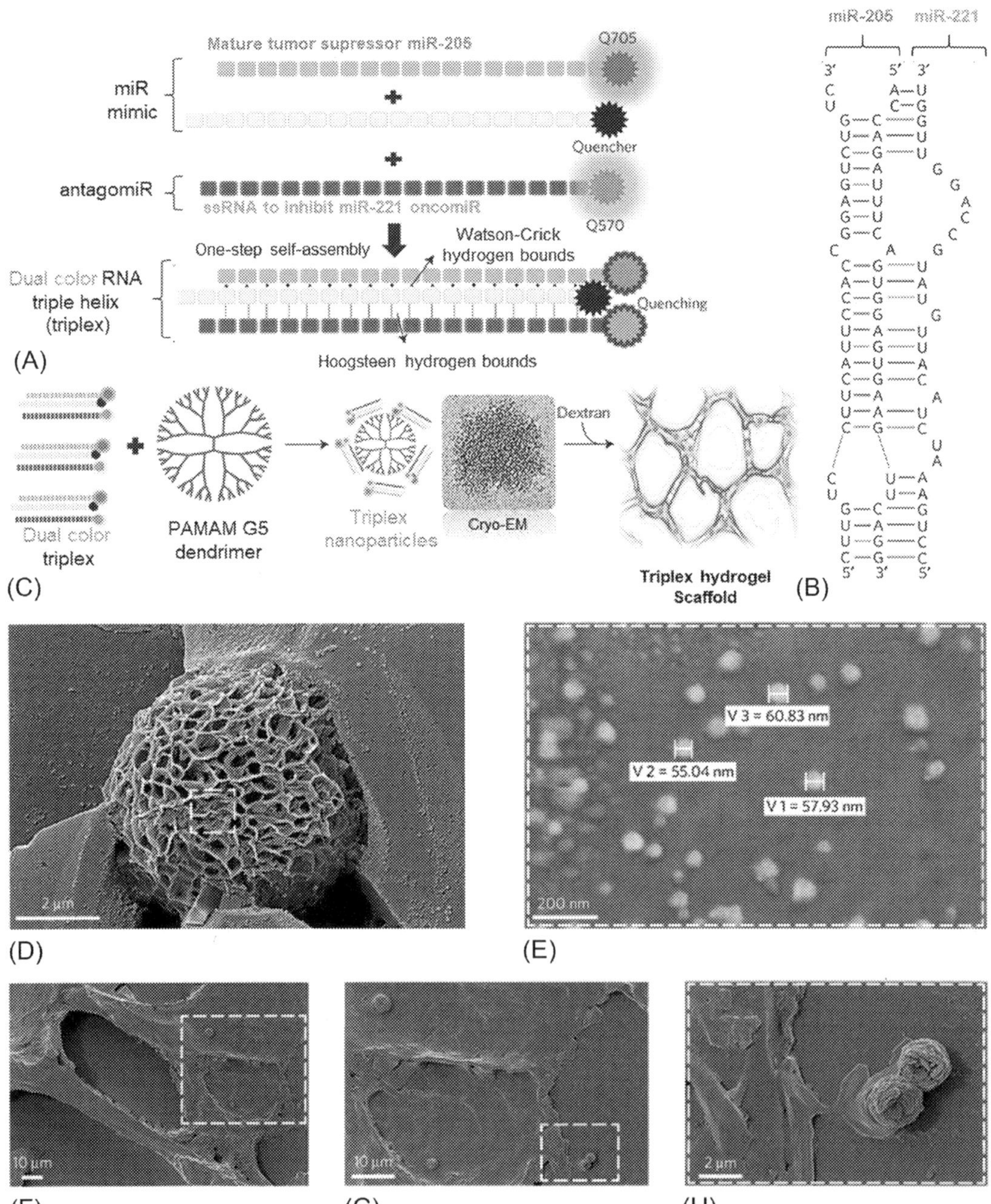

See the legend on next page.

are arranged into membranes by the noncovalent interactions such as hydrogen bonding and π-π stacking between layers (Johnson, Adams, & Cameron, 2010). Self-assembled membranes on a certain surface could manipulate their physicochemical characteristics toward biological membranes (such as regulating coagulation in blood) and improve the cell attachment and subsequently proliferation (Choi et al., 2016). In a case, onion structure-like multimembranous hydrogels were fabricated through a template gel-core to shells by a dynamic self-assembly method (Dai et al., 2009). The advantages of this facile method included rapid formation of organic, solvent-free membranes, assembly of the multilayered membrane with controllable thickness, as well as interlayer space and desired bio-active molecules of interest that could be loaded into a target layer during membrane formation, facilitating multistaged release patterns.

The ability to prepare thin films/membranes of hydrogels would introduce a range of innovative applications in nanomedicine (Dai et al., 2009; Zarket & Raghavan, 2017). For example, peptide amphiphiles with polymeric substrates such as HA were hierarchically self-assembled to create robust membranous structures with a molecular order at 1-μm scale yielding spherical- to fibrous-shaped constructs for cancer theranostics (Capito, Azevedo, Velichko, Mata, & Stupp, 2008; Zha, Sur, & Stupp, 2013). Membrane properties can be tuned by varying the polymeric substrate concentration (Zha et al., 2013). The mechanism underlying the membrane growth/assembly is the dynamic synergy between the osmotic pressure of ions and static self-assembly of nanostructured fibers (Capito et al., 2008). In another study, Hoare et al. synthesized magnetically triggered PNIPAAm hydrogels-based composite membranes for the on-demand release of antitumor drugs (Hoare et al., 2009). The switchable release of drugs was successfully demonstrated by applying the oscillating magnetic field over multiple "on-off" cycles, and the release of drug was proportional to the duration of the "on" pulse. These biocompatible membranes with excellent physicochemical

Fig. 5

Self-assembled RNA-triple-helix hydrogel nanoconjugates and scaffold for microRNA (miR) delivery. (A) Schematic illustration showing the self-assembly process of three RNA strands to form a dual-color RNA triple helix. The RNA triplex nanoparticles consist of stable two-pair fluorescence resonance energy transfer (FRET) donor/quencher RNA oligonucleotides used for in vivo miRNA inhibition (miR-221 antagomiR) strategy and miRNA replacement therapy (miR-205 mimic). (B) The secondary structure of an RNA triple helix bearing both miR-205 mimic and miR-221 antagomiR (structure produced/adapted using M-fold software). (C) Formation of the RNA-triple-helix hydrogel scaffolds. (D) High-resolution scanning electron microscopy (SEM) image of RNA microstructures (3.4 ± 1.1 μm) composed of (E) small RNA-triple-helix-dendrimer nanoparticles (56.6 ± 3.9 nm). (F–H) High-resolution SEM images of RNA macrostructures embedded in the dextran-dendrimer hydrogel scaffolds. *Reproduced from reference Conde, J., Oliva, N., Atilano, M., Song, H. S., & Artzi, N. (2016). Self-assembled RNA-triple-helix hydrogel scaffold for microRNA modulation in the tumour microenvironment. Nature Materials, 15, 353–363 with permission from Nature Publishing Group, 2016.*

properties retained their switchable flux properties for more than 45 days after subcutaneous implantation, demonstrating that this versatile long-term delivery system has great potential for various biomedical applications.

7 Conclusions and Future Trends

In summary, this chapter has highlighted insights on self-assembled nanogels for cancer theranostics. These nanogels are advantageous over other nanoparticle-assisted drug delivery systems due to their improvable physicochemical properties and ability to respond to the external stimuli by undergoing the volume phase transitions, which are usually manifested in the form of swelling of their polymeric chain networks. These properties have enabled them to find their way into various biomedical applications such as drug delivery, diagnosis, and imaging, among others.

Despite significant progress in the development of nano-sized gels and their advancement to polymeric scaffolds and membranes, there is still a long way to go before they enter clinics for cancer therapy. The complexity of the architecture of nanogels and intricate structural organization demands careful engineering for improving the desired therapeutic effect. In addition, several challenges associated with the manufacturing of nanogels need to be addressed for promoting scalability and batch-to-batch reproducibility. Progress in the design along with the thorough investigations in vivo and by addressing the challenges in scalability will eventually bring nanogels from the bench to clinics.

References

Akiyama, E., Morimoto, N., Kujawa, P., Ozawa, Y., Winnik, F. M., & Akiyoshi, K. (2007). Self-assembled Nanogels of cholesteryl-modified polysaccharides: Effect of the polysaccharide structure on their association characteristics in the dilute and semidilute regimes. *Biomacromolecules*, *8*, 2366–2373.

Akiyoshi, K., Deguchi, S., Moriguchi, N., Yamaguchi, S., & Sunamoto, J. (1993). Self-aggregates of hydrophobized polysaccharides in water. Formation and characteristics of nanoparticles. *Macromolecules*, *26*, 3062–3068.

Akiyoshi, K., Ueminami, A., Kurumada, S., & Nomura, Y. (2000). Self-association of cholesteryl-bearing poly(L-lysine) in water and control of its secondary structure by host-guest interaction with cyclodextrin. *Macromolecules*, *33*, 6752–6756.

Aoki, M., Ueda, S., Nishikawa, H., Kitano, S., Hirayama, M., Ikeda, H., et al. (2009). Antibody responses against NY-ESO-1 and HER2 antigens in patients vaccinated with combinations of cholesteryl pullulan (CHP)-NY-ESO-1 and CHP-HER2 with OK-432. *Vaccine*, *27*, 6854–6861.

Arunraj, T. R., Sanoj Rejinold, N., Ashwin Kumar, N., & Jayakumar, R. (2014). Bio-responsive chitin-poly(L-lactic acid) composite nanogels for liver cancer. *Colloids and Surfaces B: Biointerfaces*, *113*, 394–402.

Asadian-Birjand, M., Sousa-Herves, A., Steinhilber, D., Cuggino, J. C., & Calderon, M. (2012). Functional nanogels for biomedical applications. *Current Medicinal Chemistry*, *19*, 5029–5043.

Ayame, H., Morimoto, N., & Akiyoshi, K. (2008). Self-assembled cationic nanogels for intracellular protein delivery. *Bioconjugate Chemistry*, *19*, 882–890.

Bae, K. H., Mok, H., & Park, T. G. (2008). Synthesis, characterization, and intracellular delivery of reducible heparin nanogels for apoptotic cell death. *Biomaterials*, *29*, 3376–3383.

Belali, S., Karimi, A. R., & Hadizadeh, M. (2017). Novel nanostructured smart, photodynamic hydrogels based on poly(N-isopropylacrylamide) bearing porphyrin units in their crosslink chains: A potential sensitizer system in cancer therapy. *Polymer*, *109*, 93–105.

Capito, R. M., Azevedo, H. S., Velichko, Y. S., Mata, A., & Stupp, S. I. (2008). Self-assembly of large and small molecules into hierarchically ordered sacs and membranes. *Science*, *319*, 1812–1816.

Caravan, P., Das, B., Dumas, S., Epstein, F. H., Helm, P. A., Jacques, V., et al. (2007). Collagen-targeted MRI contrast agent for molecular imaging of fibrosis. *Angewandte Chemie*, *119*, 8319–8321.

Chacko, R. T., Ventura, J., Zhuang, J., & Thayumanavan, S. (2012). Polymer nanogels: A versatile nanoscopic drug delivery platform. *Advanced Drug Delivery Reviews*, *64*, 836–851.

Chan, M., lux, J., Nishimura, T., Akiyoshi, K., & Almutairi, A. (2015). Long-lasting and efficient tumor imaging using a high relaxivity polysaccharide nanogel magnetic resonance imaging contrast agent. *Biomacromolecules*, *16*, 2964–2971.

Chen, X., Liu, Z., Parker, S. G., Zhang, X., Gooding, J. J., Ru, Y., et al. (2016). Light-induced hydrogel based on tumor-targeting mesoporous silica nanoparticles as a theranostic platform for sustained cancer treatment. *ACS Applied Materials & Interfaces*, *8*, 15857–15863.

Cheng, C. C., Liang, M. C., Liao, Z. S., Huang, J. J., & Lee, D. J. (2017). Self-assembled supramolecular nanogels as a safe and effective drug delivery vector for cancer therapy. *Macromolecular Bioscience*, *17*.

Chiang, W.-H., Ho, V. T., Huang, W.-C., Huang, Y.-F., Chern, C.-S., & Chiu, H.-C. (2012). Dual stimuli-responsive polymeric hollow nanogels designed as carriers for intracellular triggered drug release. *Langmuir*, *28*, 15056–15064.

Choi, D., Heo, J., Park, J. H., Jo, Y., Jeong, H., Chang, M., et al. (2016). Nano-film coatings onto collagen hydrogels with desired drug release. *Journal of Industrial and Engineering Chemistry*, *36*, 326–333.

Conde, J., Oliva, N., Atilano, M., Song, H. S., & Artzi, N. (2016). Self-assembled RNA-triple-helix hydrogel scaffold for microRNA modulation in the tumour microenvironment. *Nature Materials*, *15*, 353–363.

Costantino, L., & Boraschi, D. (2012). Is there a clinical future for polymeric nanoparticles as brain-targeting drug delivery agents? *Drug Discovery Today*, *17*, 367–378.

Dai, H., Li, X., Long, Y., Wu, J., Liang, S., Zhang, X., et al. (2009). Multi-membrane hydrogel fabricated by facile dynamic self-assembly. *Soft Matter*, *5*, 1987–1989.

Das, M., Mardyani, S., Chan, W. C. W., & Kumacheva, E. (2006). Biofunctionalized pH-responsive microgels for cancer cell targeting: Rational design. *Advanced Materials*, *18*, 80–83.

Decher, G., MacLennan, J., Reibel, J., & Sohling, U. (1991). Highly-ordered ultrathin LC multilayer films on solid substrates. *Advanced Materials*, *3*, 617–619.

Demortiere, A., Snezhko, A., Sapozhnikov, M. V., Becker, N., Proslier, T., & Aranson, I. S. (2014). Self-assembled tunable networks of sticky colloidal particles. *Nature Communications*, *5*, 3117.

Dickerson, E. B., Blackburn, W. H., Smith, M. H., Kapa, L. B., Lyon, L. A., & McDonald, J. F. (2010). Chemosensitization of cancer cells by siRNA using targeted nanogel delivery. *BMC Cancer*, *10*, 10.

Dunn, S. S., Tian, S., Blake, S., Wang, J., Galloway, A. L., Murphy, A., et al. (2012). Reductively responsive siRNA-conjugated hydrogel nanoparticles for gene silencing. *Journal of the American Chemical Society*, *134*, 7423–7430.

Eckmann, D. M., Composto, R. J., Tsourkas, A., & Muzykantov, V. R. (2014). Nanogel carrier design for targeted drug delivery. *Journal of Materials Chemistry B*, *2*, 8085–8097.

Eichenbaum, K. D., Thomas, A. A., Eichenbaum, G. M., Gibney, B. R., Needham, D., & Kiser, P. F. (2005). Oligo-α-hydroxy ester cross-linkers: Impact of cross-linker structure on biodegradable hydrogel networks. *Macromolecules*, *38*, 10757–10762.

Farokhzad, O. C., & Langer, R. (2006). Nanomedicine: Developing smarter therapeutic and diagnostic modalities. *Advanced Drug Delivery Reviews*, *58*, 1456–1459.

Fleige, E., Quadir, M. A., & Haag, R. (2012). Stimuli-responsive polymeric nanocarriers for the controlled transport of active compounds: Concepts and applications. *Advanced Drug Delivery Reviews*, *64*, 866–884.

Fleischer, S., & Dvir, T. (2013). Tissue engineering on the nanoscale: Lessons from the heart. *Current Opinion in Biotechnology*, *24*, 664–671.

Fujii, H., Shin-Ya, M., Takeda, S., Hashimoto, Y., Mukai, S. A., Sawada, S., et al. (2014). Cycloamylose-nanogel drug delivery system-mediated intratumor silencing of the vascular endothelial growth factor regulates neovascularization in tumor microenvironment. *Cancer Science*, *105*, 1616–1625.

Gao, D., Xu, H., Philbert, M. A., & Kopelman, R. (2007). Ultrafine hydrogel nanoparticles: Synthetic approach and therapeutic application in living cells. *Angewandte Chemie International Edition*, *46*, 2224–2227.

Gong, C., Wang, C., Wang, Y., Wu, Q., Zhang, D., Luo, F., et al. (2012). Efficient inhibition of colorectal peritoneal carcinomatosis by drug loaded micelles in thermosensitive hydrogel composites. *Nanoscale*, *4*, 3095–3104.

Grzelczak, M., Vermant, J., Furst, E. M., & Liz-Marzán, L. M. (2010). Directed self-assembly of nanoparticles. *ACS Nano*, *4*, 3591–3605.

Guarrotxena, N., & Quijada-Garrido, I. (2016). Optical and swelling stimuli-response of functional hybrid nanogels: Feasible route to achieve tunable smart Core@Shell Plasmonic@polymer nanomaterials. *Chemistry of Materials*, *28*, 1402–1412.

Hang, C., Zou, Y., Zhong, Y., Zhong, Z., & Meng, F. (2017). NIR and UV-responsive degradable hyaluronic acid nanogels for CD44-targeted and remotely triggered intracellular doxorubicin delivery. *Colloids and Surfaces B: Biointerfaces*, *158*, 547–555.

Hasegawa, U., Nomura, S. M., Kaul, S. C., Hirano, T., & Akiyoshi, K. (2005). Nanogel-quantum dot hybrid nanoparticles for live cell imaging. *Biochemical and Biophysical Research Communications*, *331*, 917–921.

Hassan, S., Prakash, G., Bal Ozturk, A., Saghazadeh, S., Farhan Sohail, M., Seo, J., et al. (2017). Evolution and clinical translation of drug delivery nanomaterials. *Nano Today*, *15*, 91–106.

He, L., Liang, H., Lin, L., Shah, B. R., Li, Y., Chen, Y., et al. (2015). Green-step assembly of low density lipoprotein/sodium carboxymethyl cellulose nanogels for facile loading and pH-dependent release of doxorubicin. *Colloids and Surfaces B: Biointerfaces*, *126*, 288–296.

Hirst, A. R., Escuder, B., Miravet, J. F., & Smith, D. K. (2008). High-tech applications of self-assembling supramolecular nanostructured gel-phase materials: From regenerative medicine to electronic devices. *Angewandte Chemie International Edition*, *47*, 8002–8018.

Hoare, T., Santamaria, J., Goya, G. F., Irusta, S., Lin, D., Lau, S., et al. (2009). A magnetically triggered composite membrane for on-demand drug delivery. *Nano Letters*, *9*, 3651–3657.

Hoffman, A. S. (2002). Hydrogels for biomedical applications. *Advanced Drug Delivery Reviews*, *54*, 3–12.

Hosoya, H., Dobroff, A. S., Driessen, W. H., Cristini, V., Brinker, L. M., Staquicini, F. I., et al. (2016). Integrated nanotechnology platform for tumor-targeted multimodal imaging and therapeutic cargo release. *Proceedings of the National Academy of Sciences of the United States of America*, *113*, 1877–1882.

Huang, P., Song, H., Zhang, Y., Liu, J., Zhang, J., Wang, W., et al. (2016). Bridging the gap between macroscale drug delivery systems and nanomedicines: A nanoparticle-assembled thermosensitive hydrogel for Peritumoral chemotherapy. *ACS Applied Materials & Interfaces*, *8*, 29323–29333.

Huczko, A. (2000). Template-based synthesis of nanomaterials. *Applied Physics A*, *70*, 365–376.

Ischakov, R., Adler-Abramovich, L., Buzhansky, L., Shekhter, T., & Gazit, E. (2013). Peptide-based hydrogel nanoparticles as effective drug delivery agents. *Bioorganic & Medicinal Chemistry*, *21*, 3517–3522.

Jabir, N. R., Tabrez, S., Ashraf, G. M., Shakil, S., Damanhouri, G. A., & Kamal, M. A. (2012). Nanotechnology-based approaches in anticancer research. *International Journal of Nanomedicine*, *7*, 4391–4408.

Jahren, S. L., Butler, M. F., Adams, S., & Cameron, R. E. (2010). Swelling and viscoelastic characterisation of pH-responsive chitosan hydrogels for targeted drug delivery. *Macromolecular Chemistry and Physics*, *211*, 644–650.

Jiang, Y., Chen, J., Deng, C., Suuronen, E. J., & Zhong, Z. (2014). Click hydrogels, microgels and nanogels: Emerging platforms for drug delivery and tissue engineering. *Biomaterials*, *35*, 4969–4985.

Jin, W., Xu, P., Zhan, Y., Shen, Y., Van Kirk, E. A., Alexander, B., et al. (2007). Degradable cisplatin-releasing core-shell nanogels from zwitterionic poly(beta-aminoester)-graft-PEG for cancer chemotherapy. *Drug Delivery*, *14*, 279–286.

Johnson, E. K., Adams, D. J., & Cameron, P. J. (2010). Directed self-assembly of dipeptides to form ultrathin hydrogel membranes. *Journal of the American Chemical Society*, *132*, 5130–5136.

Kageyama, S., Kitano, S., Hirayama, M., Nagata, Y., Imai, H., Shiraishi, T., et al. (2008). Humoral immune responses in patients vaccinated with 1–146 HER2 protein complexed with cholesteryl pullulan nanogel. *Cancer Science*, *99*, 601–607.

Kai, M. P., Keeler, A. W., Perry, J. L., Reuter, K. G., Luft, J. C., O'neal, S. K., et al. (2015). Evaluation of drug loading, pharmacokinetic behavior, and toxicity of a cisplatin-containing hydrogel nanoparticle. *Journal of Controlled Release*, *204*, 70–77.

Kankala, R. K., Kuthati, Y., Liu, C.-L., Mou, C.-Y., & Lee, C.-H. (2015). Killing cancer cells by delivering a nanoreactor for inhibition of catalase and catalytically enhancing intracellular levels of ROS. *RSC Advances*, *5*, 86072–86081.

Kankala, R. K., Liu, C.-G., Chen, A.-Z., Wang, S.-B., Xu, P.-Y., Mende, L. K., et al. (2017). Overcoming multidrug resistance through the synergistic effects of hierarchical pH-sensitive, ROS-generating nanoreactors. *ACS Biomaterials Science & Engineering*, *3*, 2431–2442.

Kankala, R. K., Zhang, Y. S., Wang, S.-B., Lee, C.-H., & Chen, A.-Z. (2017). Supercritical fluid technology: An emphasis on drug delivery and related biomedical applications. *Advanced Healthcare Materials*, *6*, 1700433.

Kawasaki, R., Sasaki, Y., & Akiyoshi, K. (2017). Intracellular delivery and passive tumor targeting of a self-assembled nanogel containing carborane clusters for boron neutron capture therapy. *Biochemical and Biophysical Research Communications*, *483*, 147–152.

Khatun, Z., Nurunnabi, M., Nafiujjaman, M., Reeck, G. R., Khan, H. A., Cho, K. J., et al. (2015). A hyaluronic acid nanogel for photo-chemo theranostics of lung cancer with simultaneous light-responsive controlled release of doxorubicin. *Nanoscale*, *7*, 10680–10689.

Kim, J. K., Anderson, J., Jun, H. W., Repka, M. A., & Jo, S. (2009). Self-assembling peptide amphiphile-based nanofiber gel for bioresponsive cisplatin delivery. *Molecular Pharmaceutics*, *6*, 978–985.

Kim, S., Kim, J.-H., & Kim, D. (2011). pH sensitive swelling and releasing behavior of nano-gels based on polyaspartamide graft copolymers. *Journal of Colloid and Interface Science*, *356*, 100–106.

Kim, S., Park, K. M., Ko, J. Y., Kwon, I. C., Cho, H. G., Kang, D., et al. (2008). Minimalism in fabrication of self-organized nanogels holding both anti-cancer drug and targeting moiety. *Colloids and Surfaces B: Biointerfaces*, *63*, 55–63.

Kocak, G., Tuncer, C., & Butun, V. (2017). pH-responsive polymers. *Polymer Chemistry*, *8*, 144–176.

Koppolu, B., Bhavsar, Z., Wadajkar, A. S., Nattama, S., Rahimi, M., Nwariaku, F., et al. (2012). Temperature-sensitive polymer-coated magnetic nanoparticles as a potential drug delivery system for targeted therapy of thyroid cancer. *Journal of Biomedical Nanotechnology*, *8*, 983–990.

Kralj, S., & Makovec, D. (2015). Magnetic assembly of superparamagnetic iron oxide nanoparticle clusters into nanochains and nanobundles. *ACS Nano*, *9*, 9700–9707.

Kulkarni, S. K. (2015). Synthesis of nanomaterials—I (physical methods). In *Nanotechnology: Principles and practices*. Cham: Springer International Publishing.

Kuroda, K., Fujimoto, K., Sunamoto, J., & Akiyoshi, K. (2002). Hierarchical self-assembly of hydrophobically modified pullulan in water: Gelation by networks of nanoparticles. *Langmuir*, *18*, 3780–3786.

Lalitha, K., Jenifer, P., Prasad, Y. S., Muthusamy, K., John, G., & Nagarajan, S. (2014). A self-assembled [small pi]-conjugated system as an anti-proliferative agent in prostate cancer cells and a probe for intra-cellular imaging. *RSC Advances*, *4*, 48433–48437.

Lee, C. H., Singla, A., & Lee, Y. (2001). Biomedical applications of collagen. *International Journal of Pharmaceutics*, *221*, 1–22.

Lee, H., Mok, H., Lee, S., Oh, Y.-K., & Park, T. G. (2007). Target-specific intracellular delivery of siRNA using degradable hyaluronic acid nanogels. *Journal of Controlled Release*, *119*, 245–252.

Li, F., & Na, K. (2011). Self-assembled chlorin e6 conjugated chondroitin sulfate nanodrug for photodynamic therapy. *Biomacromolecules*, *12*, 1724–1730.

Li, R., Pavuluri, S., Bruggeman, K., Long, B. M., Parnell, A. J., Martel, A., et al. (2016). Coassembled nanostructured bioscaffold reduces the expression of proinflammatory cytokines to induce apoptosis in epithelial cancer cells. *Nanomedicine*, *12*, 1397–1407.

Li, Y., Pan, S., Zhang, W., & Du, Z. (2009). Novel thermo-sensitive core-shell nanoparticles for targeted paclitaxel delivery. *Nanotechnology*, *20*, 065104.

Li, Z., Xu, W., Zhang, C., Chen, Y., & Li, B. (2015). Self-assembled lysozyme/carboxymethylcellulose nanogels for delivery of methotrexate. *International Journal of Biological Macromolecules*, *75*, 166–172.

Lin, X., Deng, L., Xu, Y., & Dong, A. (2012). Thermosensitive in situ hydrogel of paclitaxel conjugated poly(ε-caprolactone)-poly(ethylene glycol)-poly(ε-caprolactone). *Soft Matter*, *8*, 3470–3477.

Mantyh, P. W. (2006). Cancer pain and its impact on diagnosis, survival and quality of life. *Nature Reviews Neuroscience*, *7*, 797–809.

Matsumoto, N. M., Buchman, G. W., Rome, L. H., & Maynard, H. D. (2015). Dual pH- and temperature-responsive protein nanoparticles. *European Polymer Journal*, *69*, 532–539.

Molina, M., Asadian-Birjand, M., Balach, J., Bergueiro, J., Miceli, E., & Calderon, M. (2015). Stimuli-responsive nanogel composites and their application in nanomedicine. *Chemical Society Reviews*, *44*, 6161–6186.

Morimoto, N., Winnik, F. M., & Akiyoshi, K. (2007). Botryoidal assembly of cholesteryl-pullulan/poly(N-isopropylacrylamide) nanogels. *Langmuir*, *23*, 217–223.

Motornov, M., Roiter, Y., Tokarev, I., & Minko, S. (2010). Stimuli-responsive nanoparticles, nanogels and capsules for integrated multifunctional intelligent systems. *Progress in Polymer Science*, *35*, 174–211.

Na, K., & Bae, Y. H. (2002). Self-assembled hydrogel nanoparticles responsive to tumor extracellular pH from pullulan derivative/sulfonamide conjugate: Characterization, aggregation, and adriamycin release in vitro. *Pharmaceutical Research*, *19*, 681–688.

Na, K., Lee, E. S., & Bae, Y. H. (2007). Self-organized nanogels responding to tumor extracellular pH: pH-dependent drug release and in vitro cytotoxicity against MCF-7 cells. *Bioconjugate Chemistry*, *18*, 1568–1574.

Na, K., Park, K.-H., Kim, S. W., & Bae, Y. H. (2000). Self-assembled hydrogel nanoparticles from curdlan derivatives: Characterization, anti-cancer drug release and interaction with a hepatoma cell line (HepG2). *Journal of Controlled Release*, *69*, 225–236.

Nomura, Y., Ikeda, M., Yamaguchi, N., Aoyama, Y., & Akiyoshi, K. (2003). Protein refolding assisted by self-assembled nanogels as novel artificial molecular chaperone. *FEBS Letters*, *553*, 271–276.

Okitsu, K., Ashokkumar, M., & Grieser, F. (2005). Sonochemical synthesis of gold nanoparticles: Effects of ultrasound frequency. *The Journal of Physical Chemistry B*, *109*, 20673–20675.

Olenyuk, B., Whiteford, J. A., Fechtenkotter, A., & Stang, P. J. (1999). Self-assembly of nanoscale cuboctahedra by coordination chemistry. *Nature*, *398*, 796–799.

Park, J. S., Yang, H. N., Woo, D. G., Jeon, S. Y., & Park, K.-H. (2013). Poly(N-isopropylacrylamide-co-acrylic acid) nanogels for tracing and delivering genes to human mesenchymal stem cells. *Biomaterials*, *34*, 8819–8834.

Park, W., Kim, K. S., Bae, B. C., Kim, Y. H., & Na, K. (2010). Cancer cell specific targeting of nanogels from acetylated hyaluronic acid with low molecular weight. *European Journal of Pharmaceutical Sciences*, *40*, 367–375.

Park, W., Park, S.-J., & Na, K. (2010). Potential of self-organizing nanogel with acetylated chondroitin sulfate as an anti-cancer drug carrier. *Colloids and Surfaces B: Biointerfaces*, *79*, 501–508.

Pastuszka, M. K., & Mackay, J. A. (2016). Engineering structure and function using thermoresponsive biopolymers. *Wiley Interdisciplinary Reviews: Nanomedicine and Nanobiotechnology*, *8*, 123–138.

Pelton, R. (2000). Temperature-sensitive aqueous microgels. *Advances in Colloid and Interface Science*, *85*, 1–33.

Peppas, N. A., Hilt, J. Z., Khademhosseini, A., & Langer, R. (2006). Hydrogels in biology and medicine: From molecular principles to bionanotechnology. *Advanced Materials*, *18*, 1345–1360.

Raemdonck, K., Demeester, J., & De Smedt, S. (2009). Advanced nanogel engineering for drug delivery. *Soft Matter*, *5*, 707–715.

Ruiz, M. E., Gantner, M. E., & Talevi, A. (2014). Applications of nanosystems to anticancer drug therapy. Part II. Dendrimers, micelles, lipid-based nanosystems. *Recent Patents on Anti-Cancer Drug Discovery*, *9*, 99–128.

Ryan, D. M., & Nilsson, B. L. (2012). Self-assembled amino acids and dipeptides as noncovalent hydrogels for tissue engineering. *Polymer Chemistry*, *3*, 18–33.

Schwiebert, K. E., Chin, D. N., Macdonald, J. C., & Whitesides, G. M. (1996). Engineering the solid state with 2-benzimidazolones. *Journal of the American Chemical Society*, *118*, 4018–4029.

Shaik, J., Shaikh Mohammed, J., Mcshane, M. J., & Mills, D. K. (2013). In vitro evaluation of chondrosarcoma cells and canine chondrocytes on layer-by-layer (LbL) self-assembled multilayer nanofilms. *Biofabrication*, *5*, 015004.

Shen, W., Luan, J., Cao, L., Sun, J., Yu, L., & Ding, J. (2015). Thermogelling polymer-platinum(IV) conjugates for long-term delivery of cisplatin. *Biomacromolecules*, *16*, 105–115.

Shi, B., Zhang, H., Qiao, S. Z., Bi, J., & Dai, S. (2014). Intracellular microenvironment-responsive label-free autofluorescent nanogels for traceable gene delivery. *Advanced Healthcare Materials*, *3*, 1839–1848.

Shi, F., Ding, J., Xiao, C., Zhuang, X., He, C., Chen, L., et al. (2012). Intracellular microenvironment responsive PEGylated polypeptide nanogels with ionizable cores for efficient doxorubicin loading and triggered release. *Journal of Materials Chemistry*, *22*, 14168–14179.

Shimizu, T., Kishida, T., Hasegawa, U., Ueda, Y., Imanishi, J., Yamagishi, H., et al. (2008). Nanogel DDS enables sustained release of IL-12 for tumor immunotherapy. *Biochemical and Biophysical Research Communications*, *367*, 330–335.

Shu, C., Sabi-Mouka, E. M. B., Wang, X., & Ding, L. (2017). Self-assembly hydrogels as multifunctional drug delivery of paclitaxel for synergistic tumour-targeting and biocompatibility in vitro and in vivo. *Journal of Pharmacy and Pharmacology*, *69*, 967–977.

Soni, K. S., Desale, S. S., & Bronich, T. K. (2016). Nanogels: An overview of properties, biomedical applications and obstacles to clinical translation. *Journal of Controlled Release*, *240*, 109–126.

Sun, T., Zhang, Y. S., Pang, B., Hyun, D. C., Yang, M., & Xia, Y. (2014). Engineered nanoparticles for drug delivery in cancer therapy. *Angewandte Chemie International Edition*, *53*, 12320–12364.

Suzuki, A., & Tanaka, T. (1990). Phase transition in polymer gels induced by visible light. *Nature*, *346*, 345–347.

Tada, D., Tanabe, T., Tachibana, A., & Yamauchi, K. (2005). Drug release from hydrogel containing albumin as crosslinker. *Journal of Bioscience and Bioengineering*, *100*(5), 551.

Tahara, Y., & Akiyoshi, K. (2015). Current advances in self-assembled nanogel delivery systems for immunotherapy. *Advanced Drug Delivery Reviews*, *95*, 65–76.

Takahashi, H., Tahara, Y., Sawada, S.-I., & Akiyoshi, K. (2013). Cationic amphiphilic polysaccharide nanoballs: Protein stabilization and intracellular delivery by nano-encapsulation. *Biomaterials Science*, *1*, 842–849.

Takai, C., Tamura, A., & Fuji, M. (2014). Synthesis of a silver nanoparticle network induced by a micro-phase separation. *Advanced Powder Technology*, *25*, 621–624.

Tan, Y.-L., & Liu, C.-G. (2009). Self-aggregated nanoparticles from linoleic acid modified carboxymethyl chitosan: Synthesis, characterization and application in vitro. *Colloids and Surfaces B: Biointerfaces*, *69*, 178–182.

Tan, Y.-L., & Liu, C.-G. (2011). Preparation and characterization of self-assemblied nanoparticles based on folic acid modified carboxymethyl chitosan. *Journal of Materials Science: Materials in Medicine*, *22*, 1213–1220.

Timonen, J. V., Latikka, M., Leibler, L., Ras, R. H., & Ikkala, O. (2013). Switchable static and dynamic self-assembly of magnetic droplets on superhydrophobic surfaces. *Science*, *341*, 253–257.

Urakami, H., Hentschel, J., Seetho, K., Zeng, H., Chawla, K., & Guan, Z. (2013). Surfactant-free synthesis of biodegradable, biocompatible, and stimuli-responsive cationic nanogel particles. *Biomacromolecules*, *14*, 3682–3688.

Van Dyke, R. W. (1996). Acidification of lysosomes and endosomes. *Sub-Cellular Biochemistry*, *27*, 331–360.

Vinogradov, S. V., Zeman, A. D., Batrakova, E. V., & Kabanov, A. V. (2005). Polyplex nanogel formulations for drug delivery of cytotoxic nucleoside analogs. *Journal of Controlled Release*, *107*, 143–157.

Wang, H., Ren, C., Song, Z., Wang, L., Chen, X., & Yang, Z. (2010). Enzyme-triggered self-assembly of a small molecule: A supramolecular hydrogel with leaf-like structures and an ultra-low minimum gelation concentration. *Nanotechnology*, *21*, 225606.

Wang, H., Yang, C., Wang, L., Kong, D., Zhang, Y., & Yang, Z. (2011). Self-assembled nanospheres as a novel delivery system for taxol: A molecular hydrogel with nanosphere morphology. *Chemical Communications*, *47*, 4439–4441.

Wei, J., Wang, H., Zhu, M., Ding, D., Li, D., Yin, Z., et al. (2013). Janus nanogels of PEGylated Taxol and PLGA-PEG-PLGA copolymer for cancer therapy. *Nanoscale*, *5*, 9902–9907.

Weissleder, R. (2006). Molecular imaging in cancer. *Science*, *312*, 1168–1171.

Whitelam, S. (2015). Examples of molecular self-assembly at surfaces. *Advanced Materials*, *27*, 5720–5725.

Whitesides, G. M., & Boncheva, M. (2002). Beyond molecules: Self-assembly of mesoscopic and macroscopic components. *Proceedings of the National Academy of Sciences of the United States of America*, *99*, 4769–4774.

Whitesides, G. M., & Grzybowski, B. (2002). Self-assembly at all scales. *Science*, *295*, 2418–2421.
Whitesides, G. M., Kriebel, J. K., & Love, J. C. (2005). Molecular engineering of surfaces using self-assembled monolayers. *Science Progress*, *88*, 17–48.
Xing, R., Liu, K., Jiao, T., Zhang, N., Ma, K., Zhang, R., et al. (2016). An injectable self-assembling collagen-gold hybrid hydrogel for combinatorial antitumor photothermal/photodynamic therapy. *Advanced Materials*, *28*, 3669–3676.
Xu, F., Liu, J., Tian, J., Gao, L., Cheng, X., Pan, Y., et al. (2016). Supramolecular self-assemblies with nanoscale RGD clusters promote cell growth and intracellular drug delivery. *ACS Applied Materials & Interfaces*, *8*, 29906–29914.
Xu, X. D., Li, X., Chen, H., Qu, Q., Zhao, L., Agren, H., et al. (2015). Host-guest interaction-mediated construction of hydrogels and nanovesicles for drug delivery. *Small*, *11*, 5901–5906.
Yadav, K. S., Jacob, S., Sachdeva, G., Chuttani, K., Mishra, A. K., & Sawant, K. K. (2011). Long circulating PEGylated PLGA nanoparticles of cytarabine for targeting leukemia. *Journal of Microencapsulation*, *28*, 729–742.
Yallapu, M. M., Jaggi, M., & Chauhan, S. C. (2011). Design and engineering of nanogels for cancer treatment. *Drug Discovery Today*, *16*, 457–463.
Yang, S., Yan, Y., Huang, J., Petukhov, A. V., Kroon-Batenburg, L. M. J., Drechsler, M., et al. (2017). Giant capsids from lattice self-assembly of cyclodextrin complexes. *Nature Communications*, *8*, 15856.
Yun, S. H., & Kwok, S. J. J. (2017). Light in diagnosis, therapy and surgery. *Nature Biomedical Engineering*, *1*, 0008.
Zarket, B. C., & Raghavan, S. R. (2017). Onion-like multilayered polymer capsules synthesized by a bioinspired inside-out technique. *Nature Communications*, *8*, 193.
Zha, R. H., Sur, S., & Stupp, S. I. (2013). Self-assembly of cytotoxic peptide amphiphiles into supramolecular membranes for cancer therapy. *Advanced Healthcare Materials*, *2*, 126–133.
Zhang, J., Lin, X., Liu, J., Zhao, J., Dong, H., Deng, L., et al. (2013). Sequential thermo-induced self-gelation and acid-triggered self-release process of drug-conjugated nanoparticles: A strategy for the sustained and controlled drug delivery to tumors. *Journal of Materials Chemistry B*, *1*, 4667–4677.
Zhang, X., Malhotra, S., Molina, M., & Haag, R. (2015). Micro- and nanogels with labile crosslinks—From synthesis to biomedical applications. *Chemical Society Reviews*, *44*, 1948–1973.
Zhang, Y. S., & Khademhosseini, A. (2017). Advances in engineering hydrogels. *Science*, *356*.
Zhang, Z. Q., & Song, S. C. (2016). Thermosensitive/superparamagnetic iron oxide nanoparticle-loaded nanocapsule hydrogels for multiple cancer hyperthermia. *Biomaterials*, *106*, 13–23.
Zhu, K., Ye, T., Liu, J., Peng, Z., Xu, S., Lei, J., et al. (2013). Nanogels fabricated by lysozyme and sodium carboxymethyl cellulose for 5-fluorouracil controlled release. *International Journal of Pharmaceutics*, *441*, 721–727.

CHAPTER 3

Single and Multiphoton Responsive Nanomaterials for the Investigation of Cancer Microenvironment

Yue Li*, Chien-Wei Lee†, Yu-Chi Wang†, Chih-Chia Huang†, Tzu-Ming Liu*

**University of Macau, Macao SAR, China †National Cheng Kung University, Tainan, Taiwan*

Abbreviations

^{18}F-FDG	fluorine-18 fluorodeoxyglucose
1-NBDF	1-(7-nitro-1,2,3-benzadiazole)-fructose
2-DG-750	2-deoxyglucosone 750
2-NBDG	2-(N-(7-nitrobenz-2-oxa-1,3-diazol-4-yl)amino)-2-deoxyglucose)
ACPPs	activatable cell-penetrating peptides
Apo-NP	apoptosis nanoprobe
ApoPep-1	apoptosis-targeting peptide-1
ATP	adenosine triphosphate
CA IX	carbonic anhydrase IX
CD206	macrophage mannose receptor
CPPs	cell penetrating peptides
CQRPPR	amino acid sequence
CS-ATC	chondroitin sulfate-anthocyanin
CTC	circulating tumor cell
Cy5.5	cyanine 5.5
DID	a kind of long-chain dialkylcarbocyanines
DNA	deoxyribonucleic acid
DOX	doxorubicin
ECM	extracellular matrix
EGF	epidermal growth factor
EGFR	epidermal growth factor receptor
ELP	elastin-like polypeptide
EPR effect	enhanced permeability and retention effect
FAD	flavin adenine dinucleotide
Glut1	glucose transporter 1
GPU-167	a novel near-infrared fluorescent probe
HA	hyaluronic acid
HIF	hypoxia-inducible factor
HPHDs	HIF prolyl hydroxylases
HS680	a NIR fluorescent derivative of acetazolamide
mAb	monoclonal antibody

Handbook of Nanomaterials for Cancer Theranostics. https://doi.org/10.1016/B978-0-12-813339-2.00003-7

MMPs	matrix metalloproteinases
MRI	magnetic resonance imaging
NADH	reduced form of nicotinamide-adenine dinucleotide
NIR	near infrared
ODD domain	oxygen-dependent degradation domain
PD-1 receptor	programmed cell death-1 receptor
PD-L1	programmed death ligand 1
PEG	polyethylene glycol
PEI	polyethylenimine
PEI/ARAD	polyethylenimine/ATP-responsive aptamer duplex
PET	positron emission tomography
PFFP	PEG-grafted conjugated polymer
pH	potential of hydrogen
PLGA	PEGylated poly(lactic-*co*-glycolic acid)
pmeLUC	plasma membrane extracellular luciferase
POH-N	a near-infrared fluorescently labeled recombinant protein
PPADT	poly-(1,4-phenyleneacetone dimethylenethioketal)
PS	phosphatidylserine
PSP1 peptide	plasmatocyte spreading peptide
QDs	quantum dots
Q-MSNs	mesoporous silica nanoparticles
ROS	reactive oxygen species
SHG microscopy	second harmonic generation microscopy
ssDNA	single-stranded DNA
TCAD	D-α-tocopheryl polyethylene glycol 1000 succinate (TPGS) conjugated to acid-sensitive cis-aconitic anhydride-modified doxorubicin (CAD)
THG microscopy	third harmonic generation microscopy
TRITC	tetramethylrhodamine isothiocyanate
UCNP	upconversion nanoparticles
UPSe	ultra-pH sensitive
VEGF	vascular endothelial growth factor
VEGFR	vascular endothelial growth factor receptor

1 Introduction

Cancer is a disease threatening human health worldwide. With decades of research effort, cancer prevention, diagnosis, and therapy have made great progress. The latest data provided by the American Cancer Society indicated that all cancer mortality has been continuously declining in the United States since 1991. However, the incidence and mortality of special cancers, such as breast cancer, liver cancer, melanoma, and thyroid cancer, are remarkably increasing (Siegel, Miller, & Jemal, 2017). Furthermore, cancer-related death is still the leading cause of death in developing countries due to poor cancer diagnosis and therapy (Chen et al., 2016). Cancer research will increasingly improve our knowledge on tumor biology and help to develop new means to treat cancer.

Like a tree in the forest, the cancer microenvironment is the soil feeding tumor development and progression. The cancer microenvironment is a highly heterogeneous matrix including tumor cells, stromal cells, a microcirculation system, and an extracellular matrix (ECM) (Liotta & Kohn, 2001). Normalizing the tumor microenvironment could potentially treat cancer. Fortunately, research on tumor microenvironment has been one of the most successfully translational cases in clinical cancer therapy. For example, Bevacizumab, the first approved drug targeting angiogenesis, was designed to treat human cancer by destroying tumor blood supply (Hurwitz et al., 2004). Now, Bevacizumab has become standard therapy for several cancers (Ferrara & Adamis, 2016). Moreover, the popular immune therapy is also designed to activate the function of immune cells in the tumor microenvironment (Topalian, Drake, & Pardoll, 2015).

To monitor the tumor microenvironment and evaluate treatment efficacy at a high temporal resolution and subcellular spatial resolution, optical (light) imaging method is recommended. Compared with other imaging tools, such as magnetic resonance imaging (MRI), positron emission tomography (PET), and ultrasound imaging, optical-based detection methods combined with microscopy techniques enables versatile cellular level imaging in the sensitive detection of molecules at low concentration levels. To date, there are several studies which have successfully monitored the vascular permeation, diffusion, docking, and cellular internalization of nanomaterials in cellular level imaging or evaluated the practical function of nanoagents in an animal model. For the latter, an excitation light source operated at the near-infrared (NIR) wavelength and nanoprobes with NIR emission were considered to increase light penetration across the skin tissue in the front of the targeted area but prevent local tissue heating after imaging.

Among different optical analysis processes, such as absorption and scattering, the fluorescence signal is the most sensitive and specific for being responsive to various physiological parameters in different biological reactions. Single-photon transition fluorescent probe involves the excitation of ground-state electrons and the transitions of electrons to conduction band within the molecule, followed by recombination of electrons and holes in the generation of relaxation process (singlet-singlet fluorescence) for subsequently emitting photons. Generally, to achieve sensitive and robust fluorescent contrast in cellular microenvironment, it is necessary to encapsulate or conjugate organic dyes into the interior or onto the surface of nanoparticles. This nanoformulation method not only avoids the concern of intrinsic toxicity but also enhances inherent fluorescent contrast by blocking the H_2O molecule quencher (McGuire & Feldman, 1973). Moreover, nanoformulation in the imaging probe has the unique advantage of internalization through different delivery pathways in vivo, for example, active transports of the energy-dependent process and receptor-mediated endocytosis, and enhanced delivery via the enhanced permeability and retention (EPR) effect (Wang et al., 2014). Another critical subject is nanodelivery conjugated with a targeting ligand molecule that can promote active transport to increase accumulation and thus improve detection sensitivity.

To reduce absorption and scattering loss at short wavelengths, the femtosecond laser excitation at a wavelength longer than 700 nm excites the multiphoton fluorescence of nanoprobes and mitigates the autofluorescence interference from pigments and ECM. This up-converted fluorescence also offers deep tissue penetration and improves spatial resolution as well as imaging sectioning capability, in contrast to a single-photon process. However, high-excitation dosage might cause photo damage of fluorescent nanoprobes and nanoreporters, which limits its application. It is worth mention that several reports demonstrated minimal photobleaching for those dyes/nanoparticles, as the combination of dyes was protected by the nanostructure.

Much of our knowledge about tumor microenvironment comes from traditional histopathology inspection and biochemical assays; however, these techniques can only capture evidence at a certain moment of dynamics, and cannot recapitulate the living microenvironment of the tumor. With the development of imaging modalities and nanoprobes, imaging platforms provide a feasible approach to directly visualizing aspects of the tumor microenvironment in vivo, such as cell metabolism, molecular distribution, and cell-cell interaction. Upon the nanoprobes' exposure to abnormal expression in the intracellular system, fluorescence was usually activated (fluorescence signal recovery) either by dissociating self-quenching dyes, or through chemical species-inducement during biological sensing. Imaging physiological parameters in the tumor microenvironment will provide new opportunities for cancer diagnosis, therapeutic intervention, and therapeutic assessment. In this chapter, we will review the application of single and multiphoton responsive nanomaterials in different contexts of the cancer microenvironment, including metabolic reprogramming, neovasculature, immune evasion, and therapeutic responses (Fig. 1). The advances and development of these nanoprobes will be indicated (Table 1).

2 Metabolism Reprogramming

Metabolism reprogramming occurs concurrently with tumor development and has been recognized as a new hallmark of cancer (Hanahan & Weinberg, 2011). Tumor cells require sufficient energy supply and active macromolecular biosynthesis to sustain their proliferative ability (Deberardinis, Lum, Hatzivassiliou, & Thompson, 2008). In general, normal cells obtain energy through glycolysis in the cytoplasm followed by oxidation of pyruvate in mitochondria under aerobic conditions. However, most tumor cells prefer to obtain energy through glycolysis followed by lactic acid fermentation in the cytoplasm under aerobic conditions. This interesting phenomenon is termed the *Warburg effect*, a theory accounting for tumor metabolism reprogramming (Vander Heiden, Cantley, & Thompson, 2009). Based on the difference between metabolic activities in tumors and adjacent tissues, various molecular probes have been designed to image tumor metabolism and deliver therapeutics selectively in vivo (Fig. 2A).

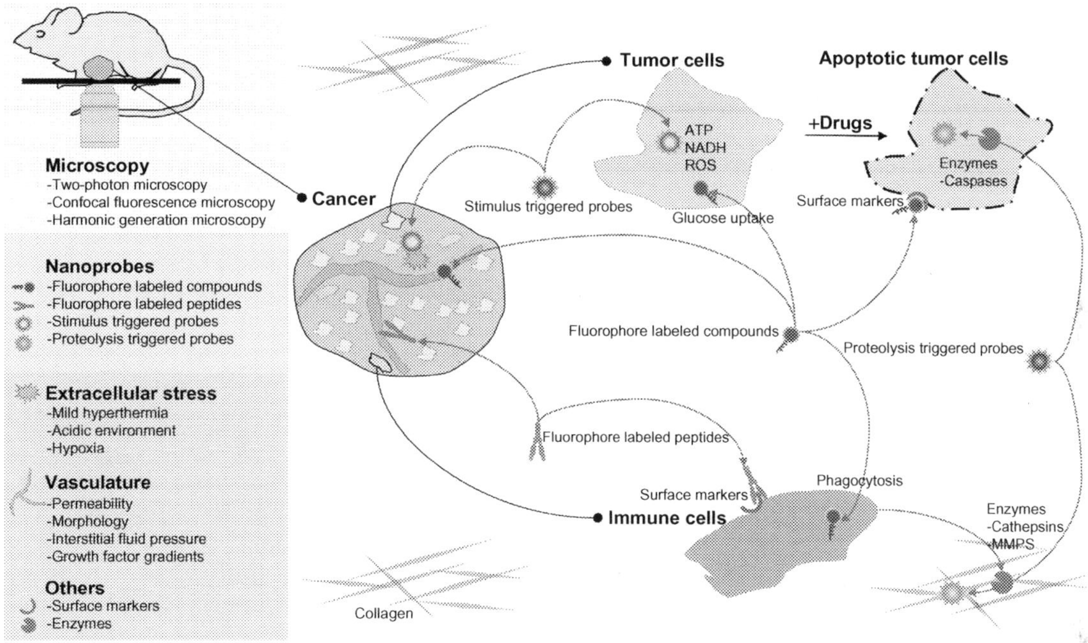

Fig. 1

Photon-responsive nanoprobes for the investigation of the cancer microenvironment. This chapter mainly reviews the application of single and multiphoton-responsive nanomaterials in different contexts of the cancer microenvironment, including metabolic reprogramming, neovasculature, immune evasion, and therapeutic responses.

2.1 Glucose

Aerobic glycolysis is an inefficient pathway for energy production, compared with the chain glycolysis-tricarboxylic acid cycle-oxidative phosphorylation processes (4 ATPs vs. 36 ATPs per glucose) (Vander Heiden et al., 2009). To meet the high demands at low production, tumor cells tend to alter their metabolic activities by reprogramming metabolic genes. For example, tumor cells increase glucose intake by upregulating glucose transporter 1 (Glut1) expression and enhance glycolysis by upregulating glycolytic enzyme expression relative to normal cells (Kroemer & Pouyssegur, 2008). Now, various molecular probes for tumor functional imaging exploit the high glucose intake of tumor cells. These nanoprobes are mainly designed on glucose analogues, which are candidate substrates for glucose transporter. A classic case is fluorine-18 fluorodeoxyglucose (^{18}F-FDG) with PET-CT, which has become a favored imaging technique for cancer diagnosis, tumor staging, and metastases detection in clinical oncology (Almuhaideb, Papathanasiou, & Bomanji, 2011; Fletcher et al., 2008).

Moreover, fluorescent nanoprobes are also used for imaging glucose metabolism in the tumor microenvironment. For instance, 2-NBDG, a fluorescent analog of deoxy-glucose, is

Table 1: Photon-responsive nanoprobes for the investigation of the cancer microenvironment

Parameters	Nanoprobes	Targets	Carriers	Mechanism	References
Tumor metabolism					
Glucose	2-NBDG, Cy5.5-2DG, ICG-Der-02-2DG, Cypate-2DG	Glut1	None	Fluorophore-labeled deoxyglucose derivatives	Cheng et al. (2006), Frees et al. (2014), Guo et al. (2012), Rajaram et al. (2013), and Sheth, Josephson, and Mahmood (2009)
	1-Cy5.5-DF, 1-NBDF	Glut5	None	Fluorophore-labeled fructose derivatives	Levi et al. (2007)
Redox-potential	Intrinsic coenzymes	NADH, FAD	None	Protein-binding-induced fluorescence intensity and lifetime change	Shah, Heaster, and Skala (2017), Skala et al. (2007), and Walsh et al. (2013)
	Genetically fluorescent sensors	NADH	None	Protein-binding conformational change-induced fluorescence intensity change	Zhao et al. (2011)
ROS	CNP/DOX,Cy5.5-labeled CPT-loaded micelles	ROS	ROS-sensitive micelles	Carrier degradation-induced fluorescence recovery from self-quenching	Chiang, Yen, and Lo (2015) and Hu et al. (2017)
	CS-ATC-DOX	ROS	CS-ATC nanocomplex	Carrier degradation-induced fluorescence recovery from self-quenching	Jeong et al. (2016)
	UCNP@mSiO2/Ce6-DOX-TK	ROS	ROS-sensitive mesoporous silica shell	Carrier degradation-induced fluorescence recovery from self-quenching	Zhang et al. (2016)
	(UCN/SiO2-RB + DOX)@PPADT	ROS	PPADT	Carrier degradation-induced fluorescence recovery from self-quenching	Zhou et al. (2016)

Energy state	DOX/Cy5.5-NG	ATP	HA crosslinked gel shell	DNA-binding-induced fluorescence recovery from self-quenching	Mo et al. (2014)
	DOX/PFFP NPs	ATP	PFFP	DNA-binding-induced fluorescence recovery from self-quenching	Qian et al. (2016)
	Oxy-A/D-ONC	ATP	Cationic lipid shell	DNA-binding-induced fluorescence recovery from self-quenching	Zhao et al. (2016)
	PEI/ARAD/DOX	ATP	PEI/ARAD	DNA-binding-induced fluorescence recovery from self-quenching	Wang et al. (2016)
	DOX/DNA-GC	ATP	Graphene oxide	DNA-binding-induced fluorescence recovery from self-quenching	Lai, Shah, Zhang, Yang, and Lee (2015)
	DOX/TDPA-Zn2þ-UCNP@ MSNs	ATP	UCNP@MSN	TDPA-Zn^{2+}-binding-induced fluorescence recovery from self-quenching	Mo, Jiang, Sun, and Gu (2015)
	pmeLUC	Pericellular ATP	HEK293 cells	ATP-driven luciferase/luciferin photoemission	Falzoni et al. (2013)
Temperature	DOX-encapsulated TTSL-Ab, DOX-encapsulating RELs, DITSL, iRGD-LTSL-DOX, LTSLs-DOX	Mild hyperthermia	Thermosensitive liposomes	Carrier degradation-induced fluorescence recovery from self-quenching	Al-Ahmady, Chaloin, and Kostarelos (2014), Deng et al. (2016), Kim et al. (2014), Senavirathna et al. (2013), and Yan et al. (2016)
	ICG-loaded PF-127 micelle	Mild hyperthermia	Thermosensitive micelles	Carrier degradation-induced fluorescence recovery from self-quenching	Kim et al. (2010)
	YSAP UCNPs	Mild hyperthermia	Thermosensitive mesoporous silica	Carrier degradation-induced fluorescence recovery from self-quenching	Lv et al. (2015)
	GNR-FNDs	Hypothermia to hyperthermia	Nanodiamond	Temperature-induced fluorescence spectra shift	Tsai et al. (2017)

Continued

Table 1: Photon-responsive nanoprobes for the investigation of the cancer microenvironment—cont'd

Parameters	Nanoprobes	Targets	Carriers	Mechanism	References
pH value	DiEtN-BDP-labeled trastuzumab	Acidic intracellular environment	Cancer-targeting macromolecule	Chemical reaction-induced fluorescence activation	Urano et al. (2009)
	DmNP Cy5-siRNA	Acidic extracellular environment	Dm-NPsiRNA	Chemical reaction-mediated dye accumulation-induced fluorescence intensity change	Sun et al. (2015)
	DOX-BTTPF	Acidic environment	pH-sensitive dextran/ pendant acetals	Carrier degradation-induced FRET efficiency change	Yu et al. (2014)
	Calcein-loaded TW20-GLY	Acidic intracellular environment	pH-sensitive niosomes	Carrier degradation-induced fluorescence intensity change	Masotti et al. (2010)
	DOX@ MSN-NH-N=C-PADs	Acidic intracellular environment	pH-sensitive dextran/ mesoporous silica nanoparticles	Carrier degradation-induced fluorescence recovery from self-quenching	Zhang et al. (2017)
	ZnPc-HDH, Triblock Copolymer Conjugated with DOX and NIR Probe, UPSe and UPSi	Acidic extracellular and intracellular environment	pH-sensitive micelle	Carrier degradation-induced fluorescence recovery from self-quenching	Debele, Mekuria, and Tsai (2017), Fu, Sun, and Yan (2015), and Wang, Chen, et al. (2014)
	PEG-PTTMA-PAA	Acidic intracellular environment	pH-sensitive polymersomes	Carrier degradation-induced fluorescence recovery from self-quenching	Du, Chen, Zheng, Meng, and Zhong (2012)
	PLGA DOX	Acidic intracellular environment	PLGA hollow microsphere	Carrier degradation-induced fluorescence recovery from self-quenching	Ke et al. (2011)
	DOPE:CHEMS:DSPE-PEG2000:PR_b	Acidic intracellular environment	pH-sensitive liposomes	Carrier degradation-induced fluorescence recovery from self-quenching	Garg and Kokkoli (2011)
	TCAD@Ce6 NPs	Acidic extracellular and intracellular environment	D-α-tocopheryl polyethylene glycol 1000 succinate	Carrier degradation-induced fluorescence recovery from self-quenching	Hou et al. (2016)

Vasculature system					
Hypoxia (oxygen level)	POH-N	HIF prolylhydroxylases	Fusion protein	Proteolysis-induced fluorescence intensity change	Kuchimaru et al. (2010)
	HS680	Carbonic Anhydrase IX	None	Enzyme-binding accumulation-induced fluorescence intensity change	Bao et al. (2012) and Dubois et al. (2009)
	Hemoglobin	Oxygen level	None	Hyperspectral imaging to map hemoglobin saturation	(Sorg, Moeller, Donovan, Cao, and Dewhirst (2005)
	Metal porphyrin complexes	Oxygen level	None	Bioreductive activation-induced phosphorescence quenching	Babilas et al. (2005), Dewhirst et al. (1999), Helmlinger, Yuan, Dellian, and Jain (1997), and Jain, Munn, and Fukumura (2013b)
	GPU-167, Pimonidazole hydrochloride	Oxygen level	None	Bioreductive activation-induced fluorescence intensity change	Okuda et al. (2012) and Sobhanifar, Aquino-Parsons, Stanbridge, and Olive (2005)
	$[Ru(dpp)_3]^{2+}Cl_2$-UCNPs	Oxygen level	UCNP core/ hollow mesoporous silica shell	Bioreductive activation-induced fluorescence intensity change	Liu, Liu, et al. (2014)
	QCys	Oxygen level	None	Bioreductive activation-induced FRET efficiency change	Kiyose et al. (2010)
Growth factor gradients	EGF-IRDye® 800CW, EGF-Cy5.5, EGF-QD, EGF-Cy5.5-BHQ3	EGFR	EGF	Fluorophore-labeled peptide	Adams et al. (2007), Diagaradjane et al. (2008), and Ryu et al. (2013)
	VEGF-QDs, GO-IR800-VEGF	VEGFR	VEGF	Fluorophore-labeled peptide	Hamada et al. (2011) and Sun et al. (2013)
	AlexaFluor 750-bevacizumab, QD-bevacizumab, Bevacizumab-800CW, Bevacizumab-6S-ICG, Bevacizumab-IRDye800CW, Bevacizumab-AF680	VEGF	Bevacizumab	Fluorophore-labeled peptide	Cilliers et al. (2017), Gazouli et al. (2014), Lamberts et al. (2017), Meyer et al. (2016), Paudyal et al. (2014), and Terwisscha Van Scheltinga et al. (2011)
Permeability	TRITC-dextran, Rhodamine-dextran	Pore size	Low molecular dextran	Fluorophore-labeled dextran	He et al. (2016) and Nakasone et al. (2012)
	Angiosense^{750} NIR probe	Pore size	Macromolecule	Fluorophore-labeled molecule	Xing and Zhang (2012)

Continued

Table 1: Photon-responsive nanoprobes for the investigation of the cancer microenvironment—cont'd

Parameters	Nanoprobes	Targets	Carriers	Mechanism	References
Morphology	FITC-Dextran, Texas red-dextran	Retention	High molecular dextran	Fluorophore-labeled dextran	Jain, Munn, and Fukumura (2013a) and Wyckoff et al. (2007)
	Quantum dot	Retention	QDs	Fluorescence intensity depends on the size	Kirkpatrick et al. (2012)
	DID-labeled red blood cell	Retention	Blood cells	Fluorophore-labeled cell	Kamoun et al. (2010)
Interstitial fluid pressure	FITC-dextran	Liquid pressure	High molecular dextran	Fluorophore-labeled dextran	Leunig et al. (1992)
Immune system					
Macrophage	Texas red-dextran	Phagocytosis	Dextran	Fluorophore-labeled dextran	Lin et al. (2006) and Wyckoff et al. (2007)
	Quantum dots	Phagocytosis	Quantum dots	Fluorescence intensity depends on sizes	Stroh et al. (2005)
	Dye-anti-CD206	Macrophage mannose receptor	CD206 antibody	Fluorophore-labeled peptide	Sun, Shen, et al. (2015)
	Mannose-NIR probes	Macrophage mannose receptor	Mannose	Fluorophore-labeled mannose	Baker et al. (2014)
	Folate-FITC, Folate-NIR probes	Folate receptor alpha	Folate	Fluorophore-labeled folate	Baker et al. (2014) and Sun et al. (2014)
Check Point	Cy5.5-DPPA-1	PD-L1	Antagonists targeting PD-L1	Fluorophore-labeled peptide	Chang et al. (2015)
	NIR-PD-L1-mAb	PD-L1	PD-L1 mAb	Fluorophore-labeled peptide	Chatterjee et al. (2016)

Networks					
Collagen	SHG	Collagen	None	Nonlinear optical characteristics	Provenzano et al. (2008), Soon, Braet, and Condeelis (2007), and Wu et al. (2015)
Elastin	THG	Elastin	None	Nonlinear optical characteristics	Yu et al. (2007)
Cleaver enzyme	DiR-labeled T2K	MMP-2 and -9	Micelles	Carrier degradation-induced FRET efficiency change	Zhang et al. (2016)
	MMPSense 680	MMP-2, -3, -9 and -13	None	Proteolysis-triggered fluorescence dequenching	Lohela et al. (2014)
	Cy5-labeled ACPPs	MMP-2	CPPs	Proteolysis-triggered fluorescence dequenching	Jiang et al. (2004) and Nguyen et al. (2010)
	BMV083, ProSense, Lipidated probe	Cathepsin B, L, and S	None	Proteolysis-triggered fluorescence dequenching	Hu et al. (2014), Onda, Kemmochi, Morita, Ishihara, and Shibutani (2013), and Verdoes et al. (2013)
Therapeutic response					
Apoptosis	Annexin V, Cy-labeled annexin V, Annexin-Vivo 750, Zn^{2+}-DPA	Phosphatidylserine on outer membrane	None	Labeled fluorophore	Kim et al. (2015), Ntziachristos et al. (2004), Smith et al. (2011), and Zhao et al. (2015)
	FITC-ApoPep-1	Histone H1	CQRPPR peptide	Fluorophore-labeled peptide	Wang et al. (2010)
	Cy7.5-labeled PSP1 peptide	Phosphatidylserine on outer membrane	PSP1 peptide	Fluorophore-labeled peptide	Kim et al. (2015)
	Z-DEVD-aminoluciferin	Caspase-3	None	Enzyme-triggered luminescent signal	Hickson et al. (2010)
	Quenched C8 and Quenched C3	Caspases	None	Proteolysis-triggered fluorescence dequenching	Zhu et al. (2012)
	TcapQ647	Caspases	CPPs	Proteolysis-triggered fluorescence dequenching	Barnett, Zhang, Maxwell, Chang, and Piwnica-Worms (2009) and Bullok and Piwnica-Worms (2005)
	Apo-NP	Caspase-3	Polymeric nanoparticles	Proteolysis-triggered fluorescence dequenching	Lee et al. (2011)
	Q-MSNs	Caspases	Mesoporous silica nanoparticles	Proteolysis-triggered fluorescence dequenching	Huang et al. (2012)

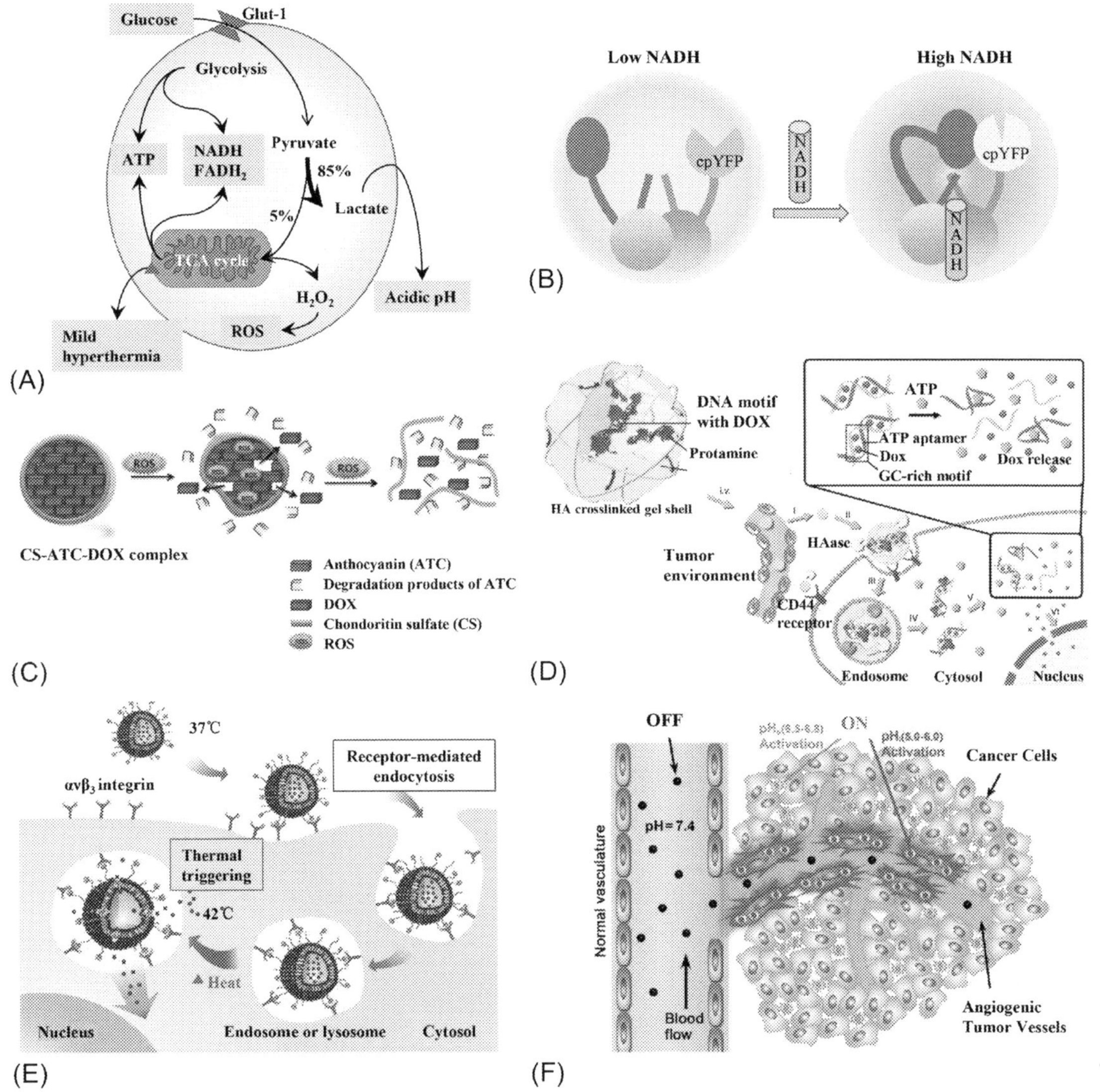

Fig. 2

Nanoprobes for imaging tumor metabolism. (A) Tumor cells prefer to obtain energy through glycolysis even in an aerobic context, which provides the physiological differences necessary to design imaging nanoprobes. (B) A fluorescent sensor of fusion proteins to monitor NADH dynamics in live cells. The sensor activates fluorescence once NADH binds to the sensor (old graphical abstract in previous paper) (Zhao et al., 2011). (C) Schematic illustration of ROS-sensitive imaging nanoprobes. ROS triggers nanocarrier degradation to release DOX, which induces fluorescence recovery from self-quenching (old Scheme 1 in previous paper) (Jeong, Bae, Park, & Na, 2016). (D) Schematic design of ATP-sensitive imaging nanoprobes. Mechanism of ATP-triggered release based on the structural change of duplex-to-aptamer (old Fig. 1 in previous paper) (Mo, Jiang, Disanto, Tai, & Gu, 2014). (E) Schematic illustration of temperature-sensitive imaging nanoprobes (old Scheme 2 in previous paper) (Kim et al., 2014). (F) Schematic illustration of pH sensitive imaging nanoprobes. The sensors stay "OFF" at pH 7.4 during blood circulation, whereas the sensors turn "on" under extracellular acidic pH_e (6.5–6.8) in the tumor microenvironment, or under intracellular acidic pH_i (5.0–6.0) in endocytic organelles of tumor cells (old Fig. 1 in previous paper) (Wang et al., 2014).

widely used to monitor tumor metabolism by imaging glucose uptake (Chitcholtan, Sykes, & Evans, 2012; Rajaram et al., 2013). Recently, Frees et al. developed a delivery-corrected imaging method to reflect glucose (2-NBDG) demand of tumors, and found that delivery-corrected 2-NBDG uptake significantly increased as vascular oxygenation decreased in tumors. These findings indicate that the combination of 2-NBDG uptake and vascular oxygenation imaging may diagnose the metabolic phenotypes more specifically in vivo (Frees et al., 2014).

Moreover, modification on 2-NBDG has been conducted to improve the penetration depth in fluorescent imaging. For example, 2-DG-750 (2-deoxyglucosone 750), a NIR fluorescent nanoprobe, not only can be used to detect a tumor in vivo but also serve as a tracer for the fluorescence imaging of circulating tumor cell (CTC) (Cai & Peng, 2013). Glut1, a glucose transporter, is widely overexpressed in various tumor cells, whereas Glut5, a fructose transporter, is exclusively expressed in breast cancer cells (Godoy et al., 2006). Levi et al. developed two fluorophores labeled fructose derivatives (1-NBDF and 1-Cy5.5-DF) for the imaging of breast cancer cell. Fructose derivatives uptake was comparable to that of 2-NBDG in three breast cancer cell lines. This study indicated that fructose derivatives may be a selective molecular probe for the imaging of breast cancer metabolism (Levi et al., 2007).

2.2 FAD/NADH

As mentioned above, cancer cells may use different metabolic pathways to obtain energy from glucose. There are remarkable differences in the molecular status of bioenergetic intermediates, such as NADH, NADPH, $FADH_2$, and FAD, between aerobic glycolysis and the chain reactions of glycolysis-tricarboxylic acid cycle-oxidative phosphorylation (Huang, Heikal, & Webb, 2002). Because NADH and FAD are endogenous fluorophores, imaging NADH and FAD pool can reflect the level of redox states (NADH and $FADH_2$) in live cells (Bartolome & Abramov, 2015). Importantly, NADH and FAD have two molecular statuses: binding to the mitochondrial membrane or not. Their fluorescence lifetimes are sensitive to binding status, which provides an opportunity to identify metabolic patterns. Multiphoton imaging of NADH/FAD has been used to differentiate between precancerous cells and normal cells in precancerous tissues (Skala et al., 2007). Moreover, optical metabolic imaging of NADH/FAD has been widely used to monitor treatment response in vivo and in vitro (Walsh et al., 2013, 2014; Walsh, Cook, Rexer, Arteaga, & Skala, 2012). Recently, Zhao et al. developed a genetically encoded fluorescent sensor to monitor NADH dynamics in live cells. The fluorescent sensor is a fusion of Rex protein, yellow fluorescent protein, and the NADH binding domain. NADH binds to the sensor and thereby induces a conformational sensor change, which activates the sensor's fluorescence capability (Fig. 2B). A change in the sensor's fluorescence intensity can reflect the glucose metabolism and redox status in cells or subcellular organelles (Zhao et al., 2011).

2.3 ROS

Reactive oxygen species (ROS) is a byproduct of normal aerobic metabolism. Tumor cells generate more ROS than surrounding normal cells do, mainly due to mitochondria dysfunction or chronic inflammation in the tumor. Accumulating evidence shows that ROS is associated with tumor development and progression (Pelicano, Carney, & Huang, 2004). The concentration of ROS in tumor cells is 100 times higher than that of normal cells (Giorgio, Trinei, Migliaccio, & Pelicci, 2007). Now various ROS-responsive nanomaterials, including micelles and mesoporous silica shells, are widely used to evaluate site-specific delivery of the drug. The design of ROS imaging nanoprobes is mainly based on dyes uncaging induced fluorescence recovery. Due to self-quenching, the concentration of fluorescent dyes in nanocarriers is too high to emit fluorescence. For instance, Jeong et al. developed an ROS-responsive probe by loading DOX (doxorubicin) in the chondroitin sulfate (CS)-anthocyanin (ATC)-based nanocomplex via intermolecular stacking interaction. CS-ATC nanocomplex is an efficient ROS-scavenging nanocarrier for DOX. A chemotherapy DOX release and fluorescence recovery were induced by ROS-mediated nanocarrier degradation (Fig. 2C). This study indicated that ROS-sensitive imaging nanoprobes can not only be used to detect a tumor but also to monitor anticancer drug delivery (Jeong et al., 2016). Another ROS-sensitive degradable material is poly-(1,4-phenyleneacetone dimethylenethioketal) (PPADT), which can be utilized for the controlled release of carried DOX in mesoporous silica, allowing the treatment of cancer cells (Zhou et al., 2016). The generation of ROS was stimulated by NIR-induced singlet oxygen species from the upconversion-based nanoparticles.

2.4 ATP

Generally, adenosine triphosphate (ATP) is the ubiquitous energy currency in biological energy transfer. However, accumulating evidence indicates that ATP is also a pleiotropic extracellular messenger of intercellular communication (Burnstock, 2007). Although ATP is present in live cells and the extracellular environment, there is sharp concentration contrast between the extracellular (<0.4 mM) and intracellular (1–10 mM) environment (Gribble et al., 2000). Now, there are various ATP responsive nanomaterials for the evaluation of anticancer drug delivery and the tumor metabolic imaging (Sun & Gu, 2016). The design strategy is hybridization of the fluorescent dye/drug probe into the inorganic or organic matrix, leading first to nanoprecipitation. Differing from small molecule diffusion, nanosized particles favor endocytosis delivery, resulting in high accumulation in cells. Endogenous ATP will disturb the stability of the dye/drug-loading nanocarrier and thus increase emission intensity immediately due to high escape efficiency from the endosomal vehicles.

So far, ATP-sensitive nanoprobes can be divided into two categories based on imaging mechanism. The first kind is imaging probes designed for the specific recognition of ATP by single-stranded DNA (ssDNA) aptamers (Huizenga & Szostak, 1995). In a nanocarrier, the ssDNA aptamer forms a DNA duplex by hybridizing with its complementary nucleotides, which serve as a caging or linker module for loading fluorescent molecules, such as DOX and NIR dyes. DNA duplex is rather stable in the absence of ATP, whereas DNA duplex is dissociating due to competitive binding of ATP under high concentrations of ATP. DNA duplex degradation induces fluorescent recovery from self-quenching. For instance, Mo et al. developed an ATP-sensitive probe comprised of DNA, protein, and polysaccharide. CD44 and RHAMM proteins were crosslinked on the surface of the nanoparticle to target tumor cells. Once nanoparticles were taken up by tumor cells, ATP binding to DNA motif triggered the release of DOX (Fig. 2D) (Mo et al., 2014). The second kind of nanoprobes is based on an ATP-driven enzyme reaction. Falzoni et al. developed a method to measure extracellular ATP by using luciferin/luciferase bioluminescence system. They designed a novel luciferase probe, pmeLUC, to image the pericellular ATP in the tumor microenvironment (Falzoni, Donvito, & Di Virgilio, 2013). Now, the ATP-responsive drug delivery or imaging systems are mostly in the proof-of-principle stages.

2.5 Temperature

Heat release, a by-product of metabolism, plays an important role in body temperature maintenance. Mitochondria is >48°C when the respiratory chain is fully functional. Dysregulated energy metabolism leads to mild hyperthermia (40–42°C) in tumors compared to surrounding tissues, which provides a chance to conduct drug release and tumor imaging. New temperature-responsive strategies have been widely explored in tumor diagnosis and treatment (Mura, Nicolas, & Couvreur, 2013). Mainly, these strategies revolve around the design of temperature-responsive nanocarriers, such as liposome, micelles, and mesoporous silica shells. Similarly, the imaging mechanism of temperature-responsive nanoparticles is uncaging-induced fluorescence recovery from high concentration-mediated self-quenching. Temperature-responsive nanocarriers are expected to retain their load at body temperature, and rapidly deliver the fluorescent dyes or drug within a tumor (40–42°C). Furthermore, local heat in the tumor also provides an external thermal stimulus for temperature-sensitive nanoparticles. For instance, Kim et al. developed a monoclonal antibody-targeted, temperature-sensitive nanoparticle for drug delivery and tumor imaging. Elastin-like polypeptide (ELP), a thermally responsive phase transition peptide, was incorporated into the liposome surface. DOX fluorescence was quenched by encapsulating them in ELP-modified liposomes. The release of DOX was triggered by external heat stimulus, leading to fluorescence recovery from self-quenching (Fig. 2E) (Kim et al., 2014). Recently, precise temperature measurement within living cells has been realized using the fluorescent spectra of a nanodiamond (Tsai et al., 2017). The spectral shift can be used to evaluate temperature change. These temperature probes can be used to study thermal properties and energy dissipation in the tumor microenvironment.

2.6 pH Value

Tumors tend to produce lactate under upregulated glycolysis, even in the presence of sufficient oxygen. Tumor cells drain lactic acid into the extracellular area to equilibrate the intracellular pH, leading to an acidic tumor microenvironment. Previous studies have suggested a correlation between acid microenvironment and tumor progression. Acidic microenvironment not only helps tumor cells evade immune surveillance, but also promotes tumor angiogenesis and cellular motility (Huber et al., 2017; Romero-Garcia, Moreno-Altamirano, Prado-Garcia, & Sanchez-Garcia, 2016). Moreover, a microenvironment rich in lactic acid contributes to tumor drug-resistance (Wojtkowiak, Verduzco, Schramm, & Gillies, 2011). From another point of view, an acidic tumor microenvironment also provides a platform to design imaging probes and drug delivery systems. Besides the extracellular acidic environment, lysosomes are intracellular acidic compartments in tumor cells. The lysosome is a membrane-bound organelle found in almost all animal cells. The pH value in the lysosome is 4.5–5.0, an optimal microenvironment for hydrolysis, which also makes a feasible strategy for intracellular drug delivery system (Mindell, 2012).

Now the pH-sensitive material system has been widely used to design pH-sensitive nanocarriers for drug delivery in cancer therapy (Liu et al., 2014). These imaging nanoprobes can be divided into two broad categories based on the imaging mechanism. First, pH-responsive degradation of nanocarriers induces fluorescent recovery efficiency change (FRET). Commonly used pH-responsive nanocarriers include micelles, noisomes, hollow microspheres, liposomes, and mesoporous silica. For example, Wang developed pH-sensitive fluorescent nanoprobes for tumor imaging. Fluorescent dyes were caged in the micelle, leading to fluorescence silencing by self-quenching. Micelle dissociation in the acidic area resulted in fluorescence signal recovery (Fig. 2F) (Wang, Chen, et al., 2014). Second, acidic pH-mediated chemical reaction induces fluorescence activation. For instance, Urano et al. developed a pH-activatable fluorescence probe for the selective imaging of tumor cells. This probe is based on the boron-dipyrromethene fluorophore and then conjugated to a cancer-targeting monoclonal antibody. The probe is nonfluorescent when outside the tumor cells. Once the probe is internalized through endocytosis, the probe is accumulated in lysosomes, where the acidic pH activates the probe (Urano et al., 2009).

3 Neovasculature

Blood vessels not only promote tumor growth by supplying nutrients and oxygen but also assist tumor metastasis by providing a hematogenous dissemination pathway (Folkman, 2002). In general, new blood vessel formation (angiogenesis) falls behind tumor growth. In the initiation phase, tumor cells mainly get nutrients and oxygen from the surrounding normal tissues. When tumor growth reaches certain critical volume, tumor cells in the central area can't get diffusing nutrients and oxygen from adjacent tissues. In this hypoxic context, tumor cells secrete various pro-angiogenic factors, such as vascular endothelial growth factor (VEGF) and endothelial growth factor (EGF), to stimulate the sprouting of adjacent blood vessels and initiate angiogenesis. Today, anti-angiogenesis agents, such as monoclonal antibodies (Bevacizumab, Aflibercept, Ranibizumab) and tyrosine kinase inhibitors (Sorafenib, Sunitinib), have become standard therapy for several cancers (Ye, 2016). In most cases, the surface of fluorescent nanoparticles was functionalized with anti-EGFR or anti-Her2 antibodies (Cilliers, Nessler, Christodolu, & Thurber, 2017; Lamberts et al., 2017; Meyer et al., 2016) and fluorescence imaging showed very strong emission brightening around the cell membrane in a short time, indicating that those nanoparticles targeted the corresponding protein receptor in the membrane structure. This surface modification approach provides a more sophisticated sensing method, revealing the efficiency of targeted drug delivery.

However, the survival benefits of anti-angiogenesis therapy are relatively modest, and there are individual differences in the clinical response. Alarmingly, anti-angiogenesis therapy can even promote tumor progression, invasion, and metastatic formation in some cancer cases (Carbone et al., 2011; Paez-Ribes et al., 2009). Tumor microenvironment and vasculature heterogeneity are the factors possibly accounting for the failure of anti-angiogenesis therapy. Tracking tumor angiogenesis may enrich our knowledge of cancer progression and design of therapy.

3.1 Hypoxia

Hypoxia is a common characteristic of solid tumors. Emerging evidence indicates that hypoxia promotes tumor progression by reconstructing a series of proteomic and genomic changes to activate angiogenesis, anaerobic metabolism, and therapeutic resistance (Harris, 2002). Therefore, tumor hypoxia detection and imaging are necessary to guide subsequent treatments. Moreover, hypoxia levels can directly reflect the efficacy of anti-angiogenesis therapy; therefore, imaging hypoxia conditions can evaluate the therapeutic outcome. Hypoxia-sensitive molecular imaging is crucial to the study of the underlying biology of hypoxic tumors and the development and monitoring of novel anti-cancer therapies. Various hypoxia-sensitive fluorescence molecules have been developed in recent years (Kizaka-Kondoh & Konse-Nagasawa, 2009; Krohn, Link, & Mason, 2008).

Tumor hypoxia can be measured directly using oxygen levels and indirectly using hypoxia-related proteins. First, direct tumor pO_2 measurement is detected by phosphorescence quenching, fluorescent intensity change (such as pimonidazole and 2-nitroimidazole), and fluorescence resonance energy transfer (FRET) efficiency change. For instance, Okuda developed a fluorescent nanoprobe for tumor hypoxia imaging. This probe consisted of two 2-nitroimidazole moieties as hypoxia markers, which were activated by bioreductive reaction in a hypoxic context, leading to fluorescent intensity change (Okuda et al., 2012). A popular tris(4,7-diphenyl-1,10-phenanthroline) ruthenium (II) dichloride (designated as $[Ru(dpp)_3]^{2+}Cl_2$) is another oxygen indicator which emitted weak fluorescence under the oxygen-rich conditions, while yielding stronger fluorescence under the hypoxic environment. Liu et al. designed the $Ru(dpp)_3]^{2+}$-loaded mesoporous silica nanoshell around UCNPs, allowing 980 nm-NIR light excitation and then converted to generate blue light for further exciting the out-shell $Ru(dpp)_3]^{2+}$ molecules. The authors demonstrated that this sensing nanoprobe could be in situ monitors of living zebrafish embryos under oxygen-deficient conditions (Liu et al., 2014). The other case is the encapsulation of $[Ru(dpp)_3]^{2+}$ into the hydrophobic compartment of polyethylenimine (PEI) as a polycationic polymer vehicle. The added QDs act as O_2 insensitive fluorophores which can emit at 470 nm, which allows excitation of the O_2 sensitive phosphor $[Ru(dpp)_3]^{2+}$ molecules. The emission at 620 nm of $[Ru(dpp)_3]^{2+}$ molecules would be quenched upon a reaction to O_2 molecules. Importantly, the surface of the PEI polymer vehicle could be further conjugated with hyaluronic acids as well for targeted labeling (Park et al., 2015).

Another method is based on detecting endogenous hypoxia-associated proteins, such as hypoxia-inducible factor (HIF) and carbonic anhydrase IX (CA IX). HIF is a heterodimer composed of an alpha subunit (HIF-1α) and a beta subunit (HIF-1β). HIF-1α is relatively stable under hypoxic context and translocates into the nucleus to initiate the expression of hypoxia-related genes. In an aerobic context, HIF-1α is generally degraded through an enzyme catalytic reaction. HIF prolyl hydroxylases (HPHDs) are the main enzymes to control the fate of HIF-1α through its oxygen-dependent degradation (ODD) domain (Schofield & Ratcliffe, 2004). Nanoprobes targeting HIF-1α activity have been used to monitor tumor hypoxia indirectly (Brown & Wilson, 2004). For example, Kuchimaru et al. developed a recombinant protein PTD-ODD-HaloTag (POH) that is under the same ODD regulation as HIF-1α. Intravital imaging showed that NIR fluorescent dye labeled POH (POH-N) successfully monitored HIF-1α active regions in mouse models of the tumor (Fig. 3A) (Kuchimaru et al., 2010). Moreover, CA IX is a transmembrane protein that has been shown to be greatly upregulated in various tumors in a hypoxia context. Bao et al. developed a novel method to detect tumor hypoxia by imaging CA IX activity. This molecular probe (HS680) is based on a NIR fluorescent derivative of the CA IX inhibitor acetazolamide (AZ). HS680 is a candidate noninvasive nanoprobe to quantify CA IX expression in vivo (Bao et al., 2012). A current review has detailed discussion of state-of-the-art design of functionalized probes for imaging and treating tumor hypoxia (Liu, Bu, & Shi, 2017).

3.2 Growth Factor and Receptor

Angiogenesis begins with vessel sprouting and is followed by vessel fusion. First, endothelial cell-lined blood vessels are activated to proliferate through pro-angiogenic factors (such as VEGF and EGF), then migrate into tumor tissues following the pro-angiogenic factor

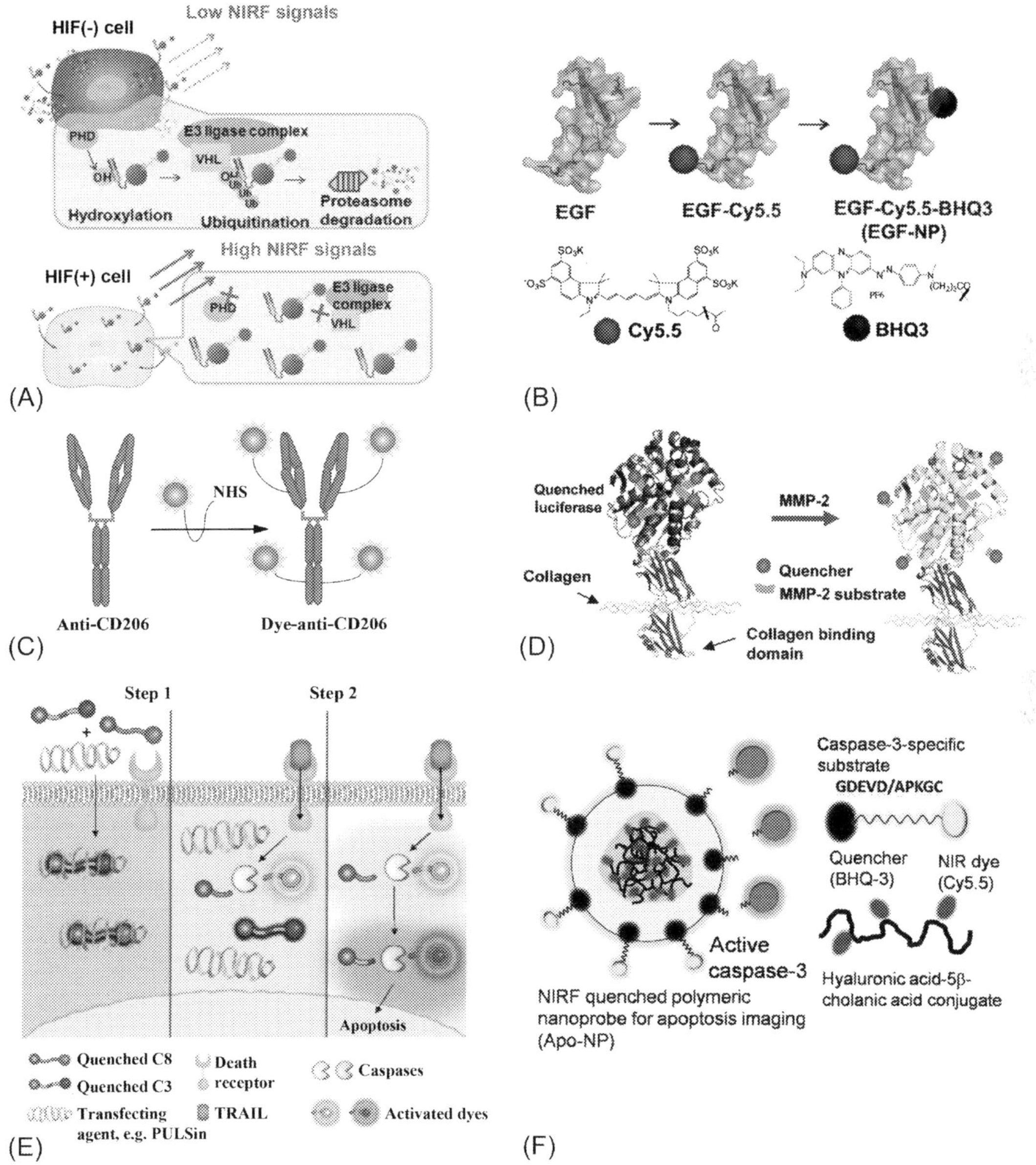

Fig. 3

Nanoprobes for imaging tumor vasculature, immune cells, and therapeutic response. (A) Schematic illustration of HIF-1α imaging nanoprobes. The sensors are degraded through HPHD-mediated ODD in live cells without HIF-1α. Under hypoxic context, the sensors are more stable and release fluorescence (old Fig. 1C in previous paper) (Kuchimaru et al., 2010). (B) Schematic preparation of EGFR-targeted fluorescence nanoprobes. Illustration of chemical reactions used to conjugate Cy5.5 and BHQ3 onto EGF (old Scheme 1 in previous paper) (Ryu et al., 2013). (C) Schematic illustration of Dye-anti-CD206 synthesis (old Fig. 1A in previous paper) (Sun et al., 2015). (D) Schematic illustration of bioluminescent reporter CB-qLuc for imaging MMP-2/9 activity (old Fig. 3A in previous paper) (Xia et al., 2011). (E) Schematic diagram of nanoprobe for imaging the caspase cascade. Once the sensors are delivered into the cells, the nanoprobes are sequentially activated by caspases (old Fig. 1A in previous paper) (Zhu et al., 2012). (F) Schematic illustration of nanoprobe preparation for imaging caspase-3 activity (old Fig. 1A in previous paper) (Lee et al., 2011).

gradients. Second, the adjacent endothelial cells located in the same vessel tightly connect with each other through protein-protein interaction. Developing a nanoprobe for imaging the discrimination of growth factor and receptor in tumors from surrounding tissues has an important role in cancer diagnosis. Moreover, it is still a major challenging problem to visualize pharmacodynamic properties of anti-angiogenesis drugs in vivo.

Various nanoprobes have been designed to image growth factor gradients and its receptor distribution in vivo. These nanoprobes are mainly based on fluorophore-labeled peptides (Fig. 3B). Commonly used peptides include commercial monoclonal antibody drugs (such as Bevacizumab) and targeting antibodies. For instance, Lamberts et al. developed a nanoprobe targeting VEGF-A by using bevacizumab conjugated to the NIRF dye IRDye800CW. They found that Bevacizumab-IRDye800CW could assess tumors and tumor margins in patients with breast cancer (Lamberts et al., 2017). In summary, these nanoprobes not only aim at in vivo tumor diagnosis but also provide a novel drug assessment tool to provide a detailed picture of drug distribution in the tumor microenvironment.

3.3 Vessel Permeability and Interstitial Fluid Pressure

Newborn blood vessels are characterized by high permeability due to their immature structure in a normal context. However, tumors commonly develop an abnormal vascular network characterized by variable vascular leakage and deformity (Pink, Schulte, Parseghian, Zijlstra, & Lewis, 2012). This porous vasculature provides a convenient channel for tumor metastasis. Moreover, tumor cells secrete VEGF to change blood vessels' permeability to aid cell intravasation (Harney et al., 2015). Imaging permeability not only provides a feasible platform for evaluating tumor angiogenesis and cancer targeting therapy, but also gives direct evidence for tumor metastasis (Xing & Zhang, 2012). Importantly, quantifying the pore size of tumor vessels also helps to design nanodrugs with sizes suitable for anti-cancer therapy (Yuan et al., 1995). Nanoprobes commonly used for tumor permeability evaluation are fluorophore-labeled dyes, such as low molecular weight FITC-dextran (Nakasone et al., 2012). Furthermore, these nanoprobes with high molecular weight can be introduced to visualize the morphology of tumor vessels, which provides a tool to quantify microvessel density and a fluorescent background to observe tumor cell intravasation or extravasation (Wyckoff et al., 2007).

As mentioned above, tumor vessels are highly leaky and allow an influx of circulating blood components. Accumulation of cellular and blood proteins in interstitial space leads to a high interstitial fluid pressure in the tumor, which inhibits the diffusion of small molecular drugs. Therefore, grasping the features of interstitial fluid pressure in tumors plays a crucial role in targeting drug delivery. Similarly, fluorophore-labeled dyes are also used to measure liquid pressure in tumor tissues (Leunig et al., 1992).

4 Immune Evasion

The immune system has two primary roles in the prevention of tumors. First, immune cells can eliminate potential risk factors conducive to tumorigenesis, such as virus-induced carcinogenic infections and pathogen-induced chronic inflammation. Second, immune cells can identify tumor cells and/or precancerous cells based on tumor-specific antigens and eliminate them (Swann & Smyth, 2007). But in most cases, tumors can secrete cytokine or express surface receptors to recruit macrophages and regulate local immune dynamics. This tumor survival strategy has been uncovered recently and has been used to design new therapy (Gordon et al., 2017).

4.1 Macrophage

The immune cells in tumors mainly include lymphocytes, macrophages, neutrophils, and natural killer cells. Pioneering studies showed that macrophages are able to eliminate tumor cells through phagocytosis and present antigens to cytotoxic T-cells for cell killing (Fidler & Schroit, 1988).

However, accumulating evidence indicates that macrophages do not suppress tumor development in various tumors. Moreover, macrophages are involved in tumor angiogenesis, invasion, and metastasis, and are immune evading (Qian & Pollard, 2010). Imaging macrophages can not only be used to detect a tumor and predict the prognosis, but can also be used to monitor macrophage targeting therapy. Although there are many nanoprobes which have been applied to imaging macrophages in vivo and in vitro, the design of nanoprobes for macrophage imaging remains a great challenge. To specifically label macrophages, the robust stability of the nanoprobes is necessary to overcome photo-oxidation and enzyme-related digestion, to make them suitable for long-term tracking. The motif in recognition is mainly based on its biological characteristics, such as fluorescence-labeled dextran for endocytosis activity and fluorophore-labeled substrate for specific receptors (Weissleder, Nahrendorf, & Pittet, 2014). For instance, Sun et al. developed an imaging probe, Dye-anti-CD206, targeting macrophage mannose receptor (CD206) expressing in macrophage (Fig. 3C). Dye-anti-CD206 sensitively detected the dynamics of macrophages in the tumor, and provided a nanocarrier for targeted drug delivery (Sun, Shen, et al., 2015).

4.2 Immune Checkpoints

In addition to recruitment of suppressive immune cells, tumor cells also evade immune surveillance by exploiting immune checkpoints through the expression of immunosuppressive molecules. Clinical trials indicate antibodies targeting immune checkpoints efficiently inhibited tumor progression and improved patient survival

(Topalian et al., 2015). PD-1 (programmed cell death-1) receptor is expressed on the surface of activated T cells, whereas programmed death ligand-1 (PD-L1), the primary ligand of PD-1, is commonly expressed on the surface of dendritic cells or macrophages. When PD-L1 binds to PD-1, an inhibitory signal is transmitted into the T cell, which can suppress T cell proliferation and immune activity. However, PD-L1 is commonly overexpressed on tumor cells. PD-L1 expressed on the tumor cells binds to PD-1 receptors on the activated T cells, which inhibits the activity of cytotoxic T cells and blunts tumor cell killing (Okazaki & Honjo, 2007). Recently, monoclonal antibody therapies against PD-1 and PD-L1 have shown promising results in early clinical trials. Intravital imaging of PD-L1 or PD-1 provides a platform for clinical translation of antibody derivatives for imaging and monitors the outcome of antibody delivery. For instance, Chatterjee et al. developed a humanized antibody for imaging PD-L1 expression in tumors. They conjugated NIR dye to a PD-L1-mAb. NIR-PD-L1-mAb can detect the expression of PD-L1 in a mouse model of tumor in vivo (Chatterjee et al., 2016).

5 Network and Cleaver Enzyme

Traditionally, ECM is primarily regarded as a scaffold for tumor construction. However, recent studies have shown that ECM is also involved tumor cell invasion and migration, immune cell infiltration and angiogenesis. Collagen is the most abundant constituent of ECM and also accounts for the major function of ECM. Abnormal deposition of collagen can be associated with increased malignancy (Fang, Yuan, Peng, & Li, 2014). Nonlinear optical techniques, such as second harmonic generation (SHG) microscopy, have been applied to image the collagen in vivo (Provenzano et al., 2008; Wu, Hsieh, Tsai, & Liu, 2015). Collagen imaging provides a visible background for track tumor cells motility in vivo. Similarly, elastin also has nonlinear optical characteristics, and has been imaged using third harmonic generation (THG) microscopy (Yu et al., 2007).

In general, ECM degradation is under tight regulation in normal tissues, whereas this process is typically deregulated in the tumor (Page-McCaw, Ewald, & Werb, 2007). Matrix metalloproteinases (MMPs) and cathepsins are main proteolytic enzymes for ECM proteolysis, which main substrates are collagen and gelatin. Now various nanoprobes have been developed to visualize cleaver enzyme activity in the tumor (Fig. 3D). Enzyme activity based nanoprobes are mainly designed on substrate-conjugated quenched fluorescent groups. The imaging mechanism can be divided into proteolysis-triggered fluorescence dequenching and nanocarrier degradation-induced FRET efficiency change. For instance, a commercial probe MMPSense 680 has been widely used to monitor MMPs enzymatic activity in vivo. An anticancer therapy with anti-CSF1R antibody dramatically depleted MMPSense 680-labeled macrophages and inhibited tumor growth. This study supported that MMPs activity-based probes with intravital imaging techniques may be a platform to evaluate the effect of novel anticancer therapies (Lohela et al., 2014).

6 Therapeutic Apoptosis

Apoptosis is a process of programmed cell death in developing and adult animal tissues. Moreover, induced apoptosis is also an intentional strategy to treat cancer. Previous studies indicate that radiotherapy and chemotherapy cause rapid apoptosis in the tumor after therapy initiation (Eisenhauer et al., 2009; Gerl & Vaux, 2005). Thus, in vivo imaging of apoptosis plays a crucial role in therapeutic assessment.

Phosphatidylserine (PS) exposure on the outer surface of the cell membrane is an early event in apoptosis cascade, which is a potential target for imaging apoptotic cells. One of the most successful and widely applied PS indicators is Annexin V, an endogenous protein that binds to PS (Boersma et al., 2005). Moreover, various other molecular probes with high affinity to PS also be applied to imaging apoptotic cells. For example, Kim et al. developed a peptide-based PS indicator, PSP1, for molecular imaging of tumor apoptosis. The PSP1 peptides specifically targeted apoptotic cells in the tumor (Kim et al., 2015).

Another approach is to monitor specific apoptosis-signaling molecules such as caspases. Caspases are the direct and crucial mediators of apoptosis. Recently, significant efforts have focused on the development of optical nanoprobes targeting caspases. These molecular probes are mainly designed around proteolysis-triggered fluorescence dequenching. For instance, Zhu et al. developed a conventional, dark-quenched fluorescent probe to monitor caspase cascade activation in living cells. Once the nanoprobe was delivered into the cells, quenched probes were sequentially activated by caspases proteolysis (Fig. 3F) (Zhu et al., 2012). Lee and colleagues loaded NIR fluorescence quencher into Cy5.5 conjugated polymer nanoparticles and linked a specific peptide substrate recognition by the caspase-3 (Lee et al., 2011). This hybrid integration into the formation of nanopolymers would increase cell-permeable delivery because the polymer matrix is comprised of hyaluronic acid, which provides active transport ability to recognize membrane proteins of cancer cells. After internalization and cleavage of the substrate by caspase-3, pronounced NIR fluorescence signal recovery from Cy5.5 molecules was observed (Fig. 3E). This biologically stimulated fluorescence turn-on phenomenon is able to identify the caspase-related apoptosis in pathologies. Replacing the polymer nanocapsule with mesoporous silica by the same groups, multiple fluorescent dyes could be co-bound to the interior of the silica channels (Lee et al., 2011). This design not only presented multiplex imaging to detect intracellular proteolytic cascade, but also provided a new platform for monitoring therapeutic efficacy of apoptotic drugs in the future cancer treatment development.

7 Conclusion and Outlook

The cancer hallmarks were updated from six to eight biological capabilities in 2011, which has directed cancer therapy and research in the last decade. These hallmarks mainly include reprogramming energy metabolism, favorable survival (sustaining proliferative signaling, evading growth suppressors, resisting cell death, and enabling replicative immortality), inducing angiogenesis, evading immune destruction, and activating invasion and metastasis (Hanahan & Weinberg, 2011). However, it is important to note that the tumor microenvironment provides the soil for the growth of these cancer hallmarks acquired during tumor development and progression. In this chapter, we have reviewed photon-responsive nanomaterials for the investigation of the cancer microenvironment.

First, the reprogramming of tumor metabolism provides many physiological parameters for tumor imaging and therapy. For instance, fluorophore-labeled analogs are used to image high glucose uptake. Like the radioactive contrast in the clinic (^{18}F-FDG), it is expected that researchers will develop more alternative fluorescent probes for visualizing intraoperative tumor margins and screening suspected tumors during endoscopy examination. Moreover, stimulus-sensitive biomaterials are the nanocarriers most commonly used to image abnormal redox potential (NADH and FAD), up-regulated intracellular energy status (ATP), a higher level of reactive species (ROS and NO), bulk lactic acid production (pH), and intra-tumoral mild hyperthermia. Specific proteins, targeting the overexpressing receptors on the tumor cells, are embedded into the surface of nanocarriers to improve the accumulation of nanoprobes at the tumor sites. Because the nanocarriers are macromolecules, potential concerns regarding cytotoxicity and immunogenicity need to be addressed before clinical translation.

Second, common hypoxia and sustained angiogenesis are a pair of contradictory events during the multistep development of tumors. Hypoxia-related proteins (HIF-1α and carbonic anhydrase IX) and oxygen-sensitive compounds are used to image tumor hypoxia. Tumor vasculature is characterized by sustained sprouting and deformed structure. Angiogenic growth factors (VEGF) and receptors (VEGFR and EGFR) are mainly tracked by fluorophore-labeled antibodies; however, this requires a large quantity of fluorophore-conjugated antibodies to conduct in vivo imaging. Therefore, alternative techniques need to be explored to decrease antibody dosages, such as local delivery or folate-mediated delivery. Vessel permeability and interstitial fluid pressure are commonly tested using fluorophore-labeled molecules.

Third, tumor cells mainly evade immune surveillance by expressing suppressive receptors and training immune cells. Fluorophore-labeled peptides are used to image immune checkpoints, whereas fluorophore-labeled molecules are used to image macrophages. However, there are limited imaging techniques to identify and distinguish macrophages from other phagocytes (such as neutrophils). ECM is the scaffold for tumor construction, but also a roadblock for tumor cell invasion and metastasis. ECM degradation and turnover are mediated by cleaver enzymes, which are mainly imaged by proteolysis-triggered fluorescence dequenching. Last, imaging of apoptosis plays a crucial role in therapeutic assessing. Fluorophore-labeled peptides and proteolysis triggered fluorescence recovery has been used to image apoptosis-related proteins and enzymes in vivo, respectively. Therefore, noninvasively imaging apoptotic cells may become a clinical application after toxicological and clinical trials. Taken together, with continuing progress in optical techniques and nanotechnology, we expect that photon-responsive nanomaterials will provide a favored probe to dissect the tumor microenvironment in the future.

Acknowledgments

This review work was supported by the Faculty of Health Sciences, University of Macau, and Science and Technology Development Fund (FDCT) of Macao SAR under grant number 122/2016/A3 and the Ministry of Science and Technology (grant no. MOST 105-2738-M-006-003) of Taiwan.

References

Adams, K. E., Ke, S., Kwon, S., Liang, F., Fan, Z., Lu, Y., et al. (2007). Comparison of visible and near-infrared wavelength-excitable fluorescent dyes for molecular imaging of cancer. *Journal of Biomedical Optics*, *12*, 024017.

Al-Ahmady, Z. S., Chaloin, O., & Kostarelos, K. (2014). Monoclonal antibody-targeted, temperature-sensitive liposomes: In vivo tumor chemotherapeutics in combination with mild hyperthermia. *Journal of Controlled Release*, *196*, 332–343.

Almuhaideb, A., Papathanasiou, N., & Bomanji, J. (2011). ^{18}F-FDG PET/CT imaging in oncology. *Annals of Saudi Medicine*, *31*, 3–13.

Babilas, P., Liebsch, G., Schacht, V., Klimant, I., Wolfbeis, O. S., Szeimies, R. M., et al. (2005). In vivo phosphorescence imaging of pO_2 using planar oxygen sensors. *Microcirculation*, *12*, 477–487.

Baker, D. W., Zhou, J., Tsai, Y. T., Patty, K. M., Weng, H., Tang, E. N., et al. (2014). Development of optical probes for in vivo imaging of polarized macrophages during foreign body reactions. *Acta Biomaterialia*, *10*, 2945–2955.

Bao, B., Groves, K., Zhang, J., Handy, E., Kennedy, P., Cuneo, G., et al. (2012). In vivo imaging and quantification of carbonic anhydrase IX expression as an endogenous biomarker of tumor hypoxia. *PLoS ONE*, *7*, e50860.

Barnett, E. M., Zhang, X., Maxwell, D., Chang, Q., & Piwnica-Worms, D. (2009). Single-cell imaging of retinal ganglion cell apoptosis with a cell-penetrating, activatable peptide probe in an in vivo glaucoma model. *Proceedings of the National Academy of Sciences of the United States of America*, *106*, 9391–9396.

Bartolome, F., & Abramov, A. Y. (2015). Measurement of mitochondrial NADH and FAD autofluorescence in live cells. *Methods in Molecular Biology*, *1264*, 263–270.

Boersma, H. H., Kietselaer, B. L., Stolk, L. M., Bennaghmouch, A., Hofstra, L., Narula, J., et al. (2005). Past, present, and future of annexin A5: From protein discovery to clinical applications. *Journal of Nuclear Medicine*, *46*, 2035–2050.

Brown, J. M., & Wilson, W. R. (2004). Exploiting tumour hypoxia in cancer treatment. *Nature Reviews Cancer*, *4*, 437–447.

Bullok, K., & Piwnica-Worms, D. (2005). Synthesis and characterization of a small, membrane-permeant, caspase-activatable far-red fluorescent peptide for imaging apoptosis. *Journal of Medicinal Chemistry*, *48*, 5404–5407.

Burnstock, G. (2007). Physiology and pathophysiology of purinergic neurotransmission. *Physiological Reviews*, *87*, 659–797.

Cai, H., & Peng, F. (2013). 2-NBDG fluorescence imaging of hypermetabolic circulating tumor cells in mouse xenograft model of breast cancer. *Journal of Fluorescence*, *23*, 213–220.

Carbone, C., Moccia, T., Zhu, C. H., Paradiso, G., Budillon, A., Chiao, P. J., et al. (2011). Anti-VEGF treatment-resistant pancreatic cancers secrete proinflammatory factors that contribute to malignant progression by inducing an EMT cell phenotype. *Clinical Cancer Research*, *17*, 5822–5832.

Chang, H. N., Liu, B. Y., Qi, Y. K., Zhou, Y., Chen, Y. P., Pan, K. M., et al. (2015). Blocking of the PD-1/PD-L1 interaction by a D-peptide antagonist for cancer immunotherapy. *Angewandte Chemie (International Edition in English)*, *54*, 11760–11764.

Chatterjee, S., Lesniak, W. G., Gabrielson, M., Lisok, A., Wharram, B., Sysa-Shah, P., et al. (2016). A humanized antibody for imaging immune checkpoint ligand PD-L1 expression in tumors. *Oncotarget*, *7*, 10215–10227.

Chen, W., Zheng, R., Baade, P. D., Zhang, S., Zeng, H., Bray, F., et al. (2016). Cancer statistics in China, 2015. *CA: A Cancer Journal for Clinicians*, *66*, 115–132.

Cheng, Z., Levi, J., Xiong, Z., Gheysens, O., Keren, S., Chen, X., et al. (2006). Near-infrared fluorescent deoxyglucose analogue for tumor optical imaging in cell culture and living mice. *Bioconjugate Chemistry*, *17*, 662–669.

Chiang, Y. T., Yen, Y. W., & Lo, C. L. (2015). Reactive oxygen species and glutathione dual redox-responsive micelles for selective cytotoxicity of cancer. *Biomaterials*, *61*, 150–161.

Chitcholtan, K., Sykes, P. H., & Evans, J. J. (2012). The resistance of intracellular mediators to doxorubicin and cisplatin are distinct in 3D and 2D endometrial cancer. *Journal of Translational Medicine*, *10*, 38.

Cilliers, C., Nessler, I., Christodolu, N., & Thurber, G. M. (2017). Tracking antibody distribution with near-infrared fluorescent dyes: Impact of dye structure and degree of labeling on plasma clearance. *Molecular Pharmaceutics*, *14*, 1623–1633.

Debele, T. A., Mekuria, S. L., & Tsai, H. C. (2017). A pH-sensitive micelle composed of heparin, phospholipids, and histidine as the carrier of photosensitizers: Application to enhance photodynamic therapy of cancer. *International Journal of Biological Macromolecules*, *98*, 125–138.

Deberardinis, R. J., Lum, J. J., Hatzivassiliou, G., & Thompson, C. B. (2008). The biology of cancer: Metabolic reprogramming fuels cell growth and proliferation. *Cell Metabolism*, *7*, 11–20.

Deng, Z., Xiao, Y., Pan, M., Li, F., Duan, W., Meng, L., et al. (2016). Hyperthermia-triggered drug delivery from iRGD-modified temperature-sensitive liposomes enhances the anti-tumor efficacy using high intensity focused ultrasound. *Journal of Controlled Release*, *243*, 333–341.

Dewhirst, M. W., Ong, E. T., Braun, R. D., Smith, B., Klitzman, B., Evans, S. M., et al. (1999). Quantification of longitudinal tissue pO_2 gradients in window chamber tumours: Impact on tumour hypoxia. *British Journal of Cancer*, *79*, 1717–1722.

Diagaradjane, P., Orenstein-Cardona, J. M., Colon-Casasnovas, N. E., Deorukhkar, A., Shentu, S., Kuno, N., et al. (2008). Imaging epidermal growth factor receptor expression in vivo: Pharmacokinetic and biodistribution characterization of a bioconjugated quantum dot nanoprobe. *Clinical Cancer Research*, *14*, 731–741.

Du, Y., Chen, W., Zheng, M., Meng, F., & Zhong, Z. (2012). pH-sensitive degradable chimaeric polymersomes for the intracellular release of doxorubicin hydrochloride. *Biomaterials*, *33*, 7291–7299.

Dubois, L., Lieuwes, N. G., Maresca, A., Thiry, A., Supuran, C. T., Scozzafava, A., et al. (2009). Imaging of CA IX with fluorescent labelled sulfonamides distinguishes hypoxic and (re)-oxygenated cells in a xenograft tumour model. *Radiotherapy and Oncology*, *92*, 423–428.

Eisenhauer, E. A., Therasse, P., Bogaerts, J., Schwartz, L. H., Sargent, D., Ford, R., et al. (2009). New response evaluation criteria in solid tumours: Revised RECIST guideline (version 1.1). *European Journal of Cancer*, *45*, 228–247.

Falzoni, S., Donvito, G., & Di Virgilio, F. (2013). Detecting adenosine triphosphate in the pericellular space. *Interface Focus*, *3*, 20120101.

Fang, M., Yuan, J. P., Peng, C. W., & Li, Y. (2014). Collagen as a double-edged sword in tumor progression. *Tumor Biology*, *35*, 2871–2882.

Ferrara, N., & Adamis, A. P. (2016). Ten years of anti-vascular endothelial growth factor therapy. *Nature Reviews Drug Discovery*, *15*, 385–403.

Fidler, I. J., & Schroit, A. J. (1988). Recognition and destruction of neoplastic cells by activated macrophages: Discrimination of altered self. *Biochimica et Biophysica Acta*, *948*, 151–173.

Fletcher, J. W., Djulbegovic, B., Soares, H. P., Siegel, B. A., Lowe, V. J., Lyman, G. H., et al. (2008). Recommendations on the use of ^{18}F-FDG PET in oncology. *Journal of Nuclear Medicine*, *49*, 480–508.

Folkman, J. (2002). Role of angiogenesis in tumor growth and metastasis. *Seminars in Oncology*, *29*, 15–18.

Frees, A. E., Rajaram, N., Mccachren, S. S., 3rd, Fontanella, A. N., Dewhirst, M. W., & Ramanujam, N. (2014). Delivery-corrected imaging of fluorescently-labeled glucose reveals distinct metabolic phenotypes in murine breast cancer. *PLoS ONE*, *9*, e115529.

Fu, L., Sun, C., & Yan, L. (2015). Galactose targeted pH-responsive copolymer conjugated with near infrared fluorescence probe for imaging of intelligent drug delivery. *ACS Applied Materials & Interfaces*, *7*, 2104–2115.

Garg, A., & Kokkoli, E. (2011). pH-sensitive PEGylated liposomes functionalized with a fibronectin-mimetic peptide show enhanced intracellular delivery to colon cancer cells. *Current Pharmaceutical Biotechnology*, *12*, 1135–1143.

Gazouli, M., Bouziotis, P., Lyberopoulou, A., Ikonomopoulos, J., Papalois, A., Anagnou, N. P., et al. (2014). Quantum dots-bevacizumab complexes for in vivo imaging of tumors. *In Vivo*, *28*, 1091–1095.

Gerl, R., & Vaux, D. L. (2005). Apoptosis in the development and treatment of cancer. *Carcinogenesis*, *26*, 263–270.

Giorgio, M., Trinei, M., Migliaccio, E., & Pelicci, P. G. (2007). Hydrogen peroxide: A metabolic by-product or a common mediator of ageing signals? *Nature Reviews Molecular Cell Biology*, *8*, 722–728.

Godoy, A., Ulloa, V., Rodriguez, F., Reinicke, K., Yanez, A. J., Garcia Mde, L., et al. (2006). Differential subcellular distribution of glucose transporters GLUT1-6 and GLUT9 in human cancer: Ultrastructural localization of GLUT1 and GLUT5 in breast tumor tissues. *Journal of Cellular Physiology*, *207*, 614–627.

Gordon, S. R., Maute, R. L., Dulken, B. W., Hutter, G., George, B. M., McCracken, M. N., et al. (2017). PD-1 expression by tumour-associated macrophages inhibits phagocytosis and tumour immunity. *Nature*, *545*, 495–499.

Gribble, F. M., Loussouarn, G., Tucker, S. J., Zhao, C., Nichols, C. G., & Ashcroft, F. M. (2000). A novel method for measurement of submembrane ATP concentration. *The Journal of Biological Chemistry*, *275*, 30046–30049.

Guo, J., Du, C., Shan, L., Zhu, H., Xue, B., Qian, Z., et al. (2012). Comparison of near-infrared fluorescent deoxyglucose probes with different dyes for tumor diagnosis in vivo. *Contrast Media & Molecular Imaging*, *7*, 289–301.

Hamada, Y., Gonda, K., Takeda, M., Sato, A., Watanabe, M., Yambe, T., et al. (2011). In vivo imaging of the molecular distribution of the VEGF receptor during angiogenesis in a mouse model of ischemia. *Blood*, *118*, e93–e100.

Hanahan, D., & Weinberg, R. A. (2011). Hallmarks of cancer: The next generation. *Cell*, *144*, 646–674.

Harney, A. S., Arwert, E. N., Entenberg, D., Wang, Y., Guo, P., Qian, B. Z., et al. (2015). Real-time imaging reveals local, transient vascular permeability, and tumor cell intravasation stimulated by TIE2hi macrophage-derived VEGFA. *Cancer Discovery*, *5*, 932–943.

Harris, A. L. (2002). Hypoxia—A key regulatory factor in tumour growth. *Nature Reviews Cancer*, *2*, 38–47.

He, H., Mack, J. J., Guc, E., Warren, C. M., Squadrito, M. L., Kilarski, W. W., et al. (2016). Perivascular macrophages limit permeability. *Arteriosclerosis, Thrombosis, and Vascular Biology*, *36*, 2203–2212.

Helmlinger, G., Yuan, F., Dellian, M., & Jain, R. K. (1997). Interstitial pH and pO2 gradients in solid tumors in vivo: High-resolution measurements reveal a lack of correlation. *Nature Medicine*, *3*, 177–182.

Hickson, J., Ackler, S., Klaubert, D., Bouska, J., Ellis, P., Foster, K., et al. (2010). Noninvasive molecular imaging of apoptosis in vivo using a modified firefly luciferase substrate, Z-DEVD-aminoluciferin. *Cell Death and Differentiation*, *17*, 1003–1010.

Hou, W., Zhao, X., Qian, X., Pan, F., Zhang, C., Yang, Y., et al. (2016). pH-sensitive self-assembling nanoparticles for tumor near-infrared fluorescence imaging and chemo-photodynamic combination therapy. *Nanoscale*, *8*, 104–116.

Hu, Y., Deng, M., Yang, H., Chen, L., Xiao, C., Zhuang, X., et al. (2017). Multi-responsive core-crosslinked poly (thiolether ester) micelles for smart drug delivery. *Polymer*, *110*, 235–241.

Hu, H. Y., Vats, D., Vizovisek, M., Kramer, L., Germanier, C., Wendt, K. U., et al. (2014). In vivo imaging of mouse tumors by a lipidated cathepsin S substrate. *Angewandte Chemie (International Edition in English)*, *53*, 7669–7673.

Huang, S., Heikal, A. A., & Webb, W. W. (2002). Two-photon fluorescence spectroscopy and microscopy of NAD(P)H and flavoprotein. *Biophysical Journal*, *82*, 2811–2825.

Huang, X., Swierczewska, M., Choi, K. Y., Zhu, L., Bhirde, A., Park, J., et al. (2012). Multiplex imaging of an intracellular proteolytic cascade by using a broad-spectrum nanoquencher. *Angewandte Chemie (International Edition in English)*, *51*, 1625–1630.

Huber, V., Camisaschi, C., Berzi, A., Ferro, S., Lugini, L., Triulzi, T., et al. (2017). Cancer acidity: An ultimate frontier of tumor immune escape and a novel target of immunomodulation. *Seminars in Cancer Biology*, *43*, 74–89.

Huizenga, D. E., & Szostak, J. W. (1995). A DNA Aptamer that binds adenosine and ATP. *Biochemistry*, *34*, 656–665.

Hurwitz, H., Fehrenbacher, L., Novotny, W., Cartwright, T., Hainsworth, J., Heim, W., et al. (2004). Bevacizumab plus irinotecan, fluorouracil, and leucovorin for metastatic colorectal cancer. *The New England Journal of Medicine*, *350*, 2335–2342.

Jain, R. K., Munn, L. L., & Fukumura, D. (2013a). Measuring angiogenesis and hemodynamics in mice. *Cold Spring Harbor Protocols*, *2013*, 354–358.

Jain, R. K., Munn, L. L., & Fukumura, D. (2013b). Measuring interstitial pH and pO_2 in mouse tumors. *Cold Spring Harbor Protocols*, *2013*, 675–677.

Jeong, D., Bae, B. C., Park, S. J., & Na, K. (2016). Reactive oxygen species responsive drug releasing nanoparticle based on chondroitin sulfate-anthocyanin nanocomplex for efficient tumor therapy. *Journal of Controlled Release*, *222*, 78–85.

Jiang, T., Olson, E. S., Nguyen, Q. T., Roy, M., Jennings, P. A., & Tsien, R. Y. (2004). Tumor imaging by means of proteolytic activation of cell-penetrating peptides. *Proceedings of the National Academy of Sciences of the United States of America*, *101*, 17867–17872.

Kamoun, W. S., Chae, S. S., Lacorre, D. A., Tyrrell, J. A., Mitre, M., Gillissen, M. A., et al. (2010). Simultaneous measurement of RBC velocity, flux, hematocrit and shear rate in vascular networks. *Nature Methods*, *7*, 655–660.

Ke, C. J., Su, T. Y., Chen, H. L., Liu, H. L., Chiang, W. L., Chu, P. C., et al. (2011). Smart multifunctional hollow microspheres for the quick release of drugs in intracellular lysosomal compartments. *Angewandte Chemie (International Edition in English)*, *50*, 8086–8089.

Kim, S., Bae, S. M., Seo, J., Cha, K., Piao, M., Kim, S. J., et al. (2015). Advantages of the phosphatidylserine-recognizing peptide PSP1 for molecular imaging of tumor apoptosis compared with annexin V. *PLoS ONE*, *10*, e0121171.

Kim, T. H., Chen, Y., Mount, C. W., Gombotz, W. R., Li, X., & Pun, S. H. (2010). Evaluation of temperature-sensitive, indocyanine green-encapsulating micelles for noninvasive near-infrared tumor imaging. *Pharmaceutical Research*, *27*, 1900–1913.

Kim, M. S., Lee, D. W., Park, K., Park, S. J., Choi, E. J., Park, E. S., et al. (2014). Temperature-triggered tumor-specific delivery of anticancer agents by cRGD-conjugated thermosensitive liposomes. *Colloids and Surfaces B, Biointerfaces*, *116*, 17–25.

Kirkpatrick, N. D., Chung, E., Cook, D. C., Han, X., Gruionu, G., Liao, S., et al. (2012). Video-rate resonant scanning multiphoton microscopy: An emerging technique for intravital imaging of the tumor microenvironment. *Intravital*, *1*.

Kiyose, K., Hanaoka, K., Oushiki, D., Nakamura, T., Kajimura, M., Suematsu, M., et al. (2010). Hypoxia-sensitive fluorescent probes for in vivo real-time fluorescence imaging of acute ischemia. *Journal of the American Chemical Society*, *132*, 15846–15848.

Kizaka-Kondoh, S., & Konse-Nagasawa, H. (2009). Significance of nitroimidazole compounds and hypoxia-inducible factor-1 for imaging tumor hypoxia. *Cancer Science*, *100*, 1366–1373.

Kroemer, G., & Pouyssegur, J. (2008). Tumor cell metabolism: Cancer's Achilles' heel. *Cancer Cell*, *13*, 472–482.

Krohn, K. A., Link, J. M., & Mason, R. P. (2008). Molecular imaging of hypoxia. *Journal of Nuclear Medicine*, *49*(Suppl 2), 129S–148S.

Kuchimaru, T., Kadonosono, T., Tanaka, S., Ushiki, T., Hiraoka, M., & Kizaka-Kondoh, S. In vivo imaging of HIF-active tumors by an oxygen-dependent degradation protein probe with an interchangeable labeling system. *PLoS ONE*, *5*, (2010). e15736.

Lai, J. P., Shah, B. R., Zhang, Y. X., Yang, L. T., & Lee, K. B. (2015). Real-time monitoring of ATP-responsive drug release using mesoporous-silica-coated multicolor upconversion nanoparticles. *ACS Nano*, *9*, 5234–5245.

Lamberts, L. E., Koch, M., De Jong, J. S., Adams, A. L. L., Glatz, J., Kranendonk, M. E. G., et al. (2017). Tumor-specific uptake of fluorescent bevacizumab-IRDye800CW microdosing in patients with primary breast cancer: A phase I feasibility study. *Clinical Cancer Research*, *23*, 2730–2741.

Lee, S., Choi, K. Y., Chung, H., Ryu, J. H., Lee, A., Koo, H., et al. (2011). Real time, high resolution video imaging of apoptosis in single cells with a polymeric nanoprobe. *Bioconjugate Chemistry*, *22*, 125–131.

Leunig, M., Yuan, F., Menger, M. D., Boucher, Y., Goetz, A. E., Messmer, K., et al. (1992). Angiogenesis, microvascular architecture, microhemodynamics, and interstitial fluid pressure during early growth of human adenocarcinoma LS174T in SCID mice. *Cancer Research*, *52*, 6553–6560.

Levi, J., Cheng, Z., Gheysens, O., Patel, M., Chan, C. T., Wang, Y., et al. (2007). Fluorescent fructose derivatives for imaging breast cancer cells. *Bioconjugate Chemistry*, *18*, 628–634.

Lin, E. Y., Li, J. F., Gnatovskiy, L., Deng, Y., Zhu, L., Grzesik, D. A., et al. (2006). Macrophages regulate the angiogenic switch in a mouse model of breast cancer. *Cancer Research*, *66*, 11238–11246.

Liotta, L. A., & Kohn, E. C. (2001). The microenvironment of the tumour-host interface. *Nature*, *411*, 375–379.

Liu, J. N., Bu, W., & Shi, J. (2017). Chemical design and synthesis of functionalized probes for imaging and treating tumor hypoxia. *Chemical Reviews*, *117*, 6160–6224.

Liu, J., Huang, Y., Kumar, A., Tan, A., Jin, S., Mozhi, A., et al. (2014). pH-sensitive nano-systems for drug delivery in cancer therapy. *Biotechnology Advances*, *32*, 693–710.

Liu, J., Liu, Y., Bu, W., Bu, J., Sun, Y., Du, J., et al. (2014). Ultrasensitive nanosensors based on upconversion nanoparticles for selective hypoxia imaging in vivo upon near-infrared excitation. *Journal of the American Chemical Society*, *136*, 9701–9709.

Lohela, M., Casbon, A. J., Olow, A., Bonham, L., Branstetter, D., Weng, N., et al. (2014). Intravital imaging reveals distinct responses of depleting dynamic tumor-associated macrophage and dendritic cell subpopulations. *Proceedings of the National Academy of Sciences of the United States of America*, *111*, E5086–95.

Lv, R., Yang, P., He, F., Gai, S., Yang, G., Dai, Y., et al. An imaging-guided platform for synergistic photodynamic/photothermal/chemo-therapy with pH/temperature-responsive drug release. *Biomaterials*, *63*, (2015). 115–127.

Masotti, A., Vicennati, P., Alisi, A., Marianecci, C., Rinaldi, F., Carafa, M., et al. (2010). Novel Tween 20 derivatives enable the formation of efficient pH-sensitive drug delivery vehicles for human hepatoblastoma. *Bioorganic & Medicinal Chemistry Letters*, *20*, 3021–3025.

McGuire, R., & Feldman, I. (1973). The quenching of tyrosine and tryptophan fluorescence by H20 and D20. *Photochemistry and Photobiology*, *18*, 119–124.

Meyer, J. H., Cunea, A., Licha, K., Welker, P., Sonntag-Bensch, D., Wafula, P., et al. (2016). In vivo imaging of fluorescent probes linked to antibodies against human and rat vascular endothelial growth factor. *Investigative Ophthalmology & Visual Science*, *57*, 759–770.

Mindell, J. A. (2012). Lysosomal acidification mechanisms. *Annual Review of Physiology*, *74*, 69–86.

Mo, R., Jiang, T., Disanto, R., Tai, W., & Gu, Z. (2014). ATP-triggered anticancer drug delivery. *Nature Communications*, *5*, 3364.

Mo, R., Jiang, T., Sun, W., & Gu, Z. (2015). ATP-responsive DNA-graphene hybrid nanoaggregates for anticancer drug delivery. *Biomaterials*, *50*, 67–74.

Mura, S., Nicolas, J., & Couvreur, P. (2013). Stimuli-responsive nanocarriers for drug delivery. *Nature Materials*, *12*, 991–1003.

Nakasone, E. S., Askautrud, H. A., Kees, T., Park, J. H., Plaks, V., Ewald, A. J., et al. (2012). Imaging tumor-stroma interactions during chemotherapy reveals contributions of the microenvironment to resistance. *Cancer Cell*, *21*, 488–503.

Nguyen, Q. T., Olson, E. S., Aguilera, T. A., Jiang, T., Scadeng, M., Ellies, L. G., et al. (2010). Surgery with molecular fluorescence imaging using activatable cell-penetrating peptides decreases residual cancer and improves survival. *Proceedings of the National Academy of Sciences of the United States of America*, *107*, 4317–4322.

Ntziachristos, V., Schellenberger, E. A., Ripoll, J., Yessayan, D., Graves, E., Bogdanov, A., Jr., et al. (2004). Visualization of antitumor treatment by means of fluorescence molecular tomography with an annexin V-Cy5.5 conjugate. *Proceedings of the National Academy of Sciences of the United States of America*, *101*, 12294–12299.

Okazaki, T., & Honjo, T. (2007). PD-1 and PD-1 ligands: From discovery to clinical application. *International Immunology*, *19*, 813–824.

Okuda, K., Okabe, Y., Kadonosono, T., Ueno, T., Youssif, B. G., Kizaka-Kondoh, S., et al. (2012). 2-Nitroimidazole-tricarbocyanine conjugate as a near-infrared fluorescent probe for in vivo imaging of tumor hypoxia. *Bioconjugate Chemistry*, *23*, 324–329.

Onda, N., Kemmochi, S., Morita, R., Ishihara, Y., & Shibutani, M. (2013). In vivo imaging of tissue-remodeling activity involving infiltration of macrophages by a systemically administered protease-activatable probe in colon cancer tissues. *Translational Oncology*, *6*, 628–637.

Paez-Ribes, M., Allen, E., Hudock, J., Takeda, T., Okuyama, H., Vinals, F., et al. (2009). Antiangiogenic therapy elicits malignant progression of tumors to increased local invasion and distant metastasis. *Cancer Cell*, *15*, 220–231.

Page-McCaw, A., Ewald, A. J., & Werb, Z. (2007). Matrix metalloproteinases and the regulation of tissue remodelling. *Nature Reviews Molecular Cell Biology*, *8*, 221–233.

Park, J., Lee, J., Kwag, J., Baek, Y., Kim, B., Yoon, C. J., et al. (2015). Quantum dots in an amphiphilic polyethyleneimine derivative platform for cellular labeling, targeting, gene delivery, and ratiometric oxygen sensing. *ACS Nano*, *9*, 6511–6521.

Paudyal, B., Paudyal, P., Shah, D., Tominaga, H., Tsushima, Y., & Endo, K. (2014). Detection of vascular endothelial growth factor in colon cancer xenografts using bevacizumab based near infrared fluorophore conjugate. *Journal of Biomedical Science*, *21*, .

Pelicano, H., Carney, D., & Huang, P. (2004). ROS stress in cancer cells and therapeutic implications. *Drug Resistance Updates*, *7*, 97–110.

Pink, D. B., Schulte, W., Parseghian, M. H., Zijlstra, A., & Lewis, J. D. (2012). Real-time visualization and quantitation of vascular permeability in vivo: Implications for drug delivery. *PLoS ONE*, *7*, e33760.

Provenzano, P. P., Inman, D. R., Eliceiri, K. W., Knittel, J. G., Yan, L., Rueden, C. T., et al. (2008). Collagen density promotes mammary tumor initiation and progression. *BMC Medicine*, *6*, 11.

Qian, C., Chen, Y., Zhu, S., Yu, J., Zhang, L., Feng, P., et al. (2016). ATP-responsive and near-infrared-emissive nanocarriers for anticancer drug delivery and real-time imaging. *Theranostics*, *6*, 1053–1064.

Qian, B. Z., & Pollard, J. W. (2010). Macrophage diversity enhances tumor progression and metastasis. *Cell*, *141*, 39–51.

Rajaram, N., Frees, A. E., Fontanella, A. N., Zhong, J., Hansen, K., Dewhirst, M. W., et al. Delivery rate affects uptake of a fluorescent glucose analog in murine metastatic breast cancer. *PLoS ONE*, *8*, (2013). e76524.

Romero-Garcia, S., Moreno-Altamirano, M. M., Prado-Garcia, H., & Sanchez-Garcia, F. J. (2016). Lactate contribution to the tumor microenvironment: Mechanisms, effects on immune cells and therapeutic relevance. *Frontiers in Immunology*, *7*, 52.

Ryu, J. H., Shin, M., Kim, S. A., Lee, S., Kim, H., Koo, H., et al. (2013). In vivo fluorescence imaging for cancer diagnosis using receptor-targeted epidermal growth factor-based nanoprobe. *Biomaterials*, *34*, 9149–9159.

Schofield, C. J., & Ratcliffe, P. J. (2004). Oxygen sensing by HIF hydroxylases. *Nature Reviews Molecular Cell Biology*, *5*, 343–354.

Senavirathna, L. K., Fernando, R., Maples, D., Zheng, Y., Polf, J. C., & Ranjan, A. (2013). Tumor spheroids as an in vitro model for determining the therapeutic response to proton beam radiotherapy and thermally sensitive nanocarriers. *Theranostics*, *3*, 687–691.

Shah, A. T., Heaster, T. M., & Skala, M. C. (2017). Metabolic imaging of head and neck cancer Organoids. *PLoS ONE*, *12*, e0170415.

Sheth, R. A., Josephson, L., & Mahmood, U. (2009). Evaluation and clinically relevant applications of a fluorescent imaging analog to fluorodeoxyglucose positron emission tomography. *Journal of Biomedical Optics*, *14*, 064014.

Siegel, R. L., Miller, K. D., & Jemal, A. (2017). Cancer statistics, 2017. *CA: A Cancer Journal for Clinicians*, *67*, 7–30.

Skala, M. C., Riching, K. M., Gendron-Fitzpatrick, A., Eickhoff, J., Eliceiri, K. W., White, J. G., et al. (2007). In vivo multiphoton microscopy of NADH and FAD redox states, fluorescence lifetimes, and cellular morphology in precancerous epithelia. *Proceedings of the National Academy of Sciences of the United States of America*, *104*, 19494–19499.

Smith, B. A., Gammon, S. T., Xiao, S., Wang, W., Chapman, S., McDermott, R., et al. (2011). In vivo optical imaging of acute cell death using a near-infrared fluorescent zinc-dipicolylamine probe. *Molecular Pharmaceutics*, *8*, 583–590.

Sobhanifar, S., Aquino-Parsons, C., Stanbridge, E. J., & Olive, P. (2005). Reduced expression of hypoxia-inducible factor-1alpha in perinecrotic regions of solid tumors. *Cancer Research*, *65*, 7259–7266.

Soon, L., Braet, F., & Condeelis, J. (2007). Moving in the right direction—nanoimaging in cancer cell motility and metastasis. *Microscopy Research and Technique*, *70*, 252–257.

Sorg, B. S., Moeller, B. J., Donovan, O., Cao, Y., & Dewhirst, M. W. (2005). Hyperspectral imaging of hemoglobin saturation in tumor microvasculature and tumor hypoxia development. *Journal of Biomedical Optics*, *10*, 44004.

Stroh, M., Zimmer, J. P., Duda, D. G., Levchenko, T. S., Cohen, K. S., Brown, E. B., et al. (2005). Quantum dots spectrally distinguish multiple species within the tumor milieu in vivo. *Nature Medicine*, *11*, 678–682.

Sun, X., Gao, D., Gao, L., Zhang, C., Yu, X., Jia, B., et al. (2015). Molecular imaging of tumor-infiltrating macrophages in a preclinical mouse model of breast cancer. *Theranostics*, *5*, 597–608.

Sun, W., & Gu, Z. (2016). ATP-responsive drug delivery systems. *Expert Opinion on Drug Delivery*, *13*, 311–314.

Sun, Z., Huang, P., Tong, G., Lin, J., Jin, A., Rong, P., et al. (2013). VEGF-loaded graphene oxide as theranostics for multi-modality imaging-monitored targeting therapeutic angiogenesis of ischemic muscle. *Nanoscale*, *5*, 6857–6866.

Sun, J. Y., Shen, J., Thibodeaux, J., Huang, G., Wang, Y., Gao, J., et al. (2014). In vivo optical imaging of folate receptor-beta in head and neck squamous cell carcinoma. *Laryngoscope*, *124*, E312–9.

Sun, C. Y., Shen, S., Xu, C. F., Li, H. J., Liu, Y., Cao, Z. T., et al. (2015). Tumor acidity-sensitive polymeric vector for active targeted siRNA delivery. *Journal of the American Chemical Society*, *137*, 15217–15224.

Swann, J. B., & Smyth, M. J. (2007). Immune surveillance of tumors. *The Journal of Clinical Investigation*, *117*, 1137–1146.

Terwisscha Van Scheltinga, A. G., Van Dam, G. M., Nagengast, W. B., Ntziachristos, V., Hollema, H., Herek, J. L., et al. (2011). Intraoperative near-infrared fluorescence tumor imaging with vascular endothelial growth factor and human epidermal growth factor receptor 2 targeting antibodies. *Journal of Nuclear Medicine*, *52*, 1778–1785.

Topalian, S. L., Drake, C. G., & Pardoll, D. M. (2015). Immune checkpoint blockade: A common denominator approach to cancer therapy. *Cancer Cell*, *27*, 450–461.

Tsai, P. C., Epperla, C. P., Huang, J. S., Chen, O. Y., Wu, C. C., & Chang, H. C. (2017). Measuring nanoscale thermostability of cell membranes with single gold-diamond nanohybrids. *Angewandte Chemie (International Edition in English)*, *56*, 3025–3030.

Urano, Y., Asanuma, D., Hama, Y., Koyama, Y., Barrett, T., Kamiya, M., et al. (2009). Selective molecular imaging of viable cancer cells with pH-activatable fluorescence probes. *Nature Medicine*, *15*, 104–109.

Vander Heiden, M. G., Cantley, L. C., & Thompson, C. B. (2009). Understanding the Warburg effect: The metabolic requirements of cell proliferation. *Science*, *324*, 1029–1033.

Verdoes, M., Oresic Bender, K., Segal, E., Van Der Linden, W. A., Syed, S., Withana, N. P., et al. (2013). Improved quenched fluorescent probe for imaging of cysteine cathepsin activity. *Journal of the American Chemical Society*, *135*, 14726–14730.

Walsh, A. J., Cook, R. S., Manning, H. C., Hicks, D. J., Lafontant, A., Arteaga, C. L., et al. (2013). Optical metabolic imaging identifies glycolytic levels, subtypes, and early-treatment response in breast cancer. *Cancer Research*, *73*, 6164–6174.

Walsh, A., Cook, R. S., Rexer, B., Arteaga, C. L., & Skala, M. C. (2012). Optical imaging of metabolism in HER2 overexpressing breast cancer cells. *Biomedical Optics Express*, *3*, 75–85.

Walsh, A. J., Cook, R. S., Sanders, M. E., Aurisicchio, L., Ciliberto, G., Arteaga, C. L., et al. (2014). Quantitative optical imaging of primary tumor organoid metabolism predicts drug response in breast cancer. *Cancer Research*, *74*, 5184–5194.

Wang, Y. H., Chen, S. P., Liao, A. H., Yang, Y. C., Lee, C. R., Wu, C. H., et al. (2014). Synergistic delivery of gold nanorods using multifunctional microbubbles for enhanced plasmonic photothermal therapy. *Scientific Reports*, *4*, 5685.

Wang, G.-H., Huang, G.-L., Zhao, Y., Pu, X.-X., Li, T., Deng, J.-J., et al. (2016). Atp triggered drug release and DNA co-delivery systems based on ATP responsive aptamers and polyethylenimine complexes. *Journal of Materials Chemistry B*, *4*, 3832–3841.

Wang, K., Purushotham, S., Lee, J. Y., Na, M. H., Park, H., Oh, S. J., et al. (2010). In vivo imaging of tumor apoptosis using histone H1-targeting peptide. *Journal of Controlled Release*, *148*, 283–291.

Wang, Y., Zhou, K., Huang, G., Hensley, C., Huang, X., Ma, X., et al. (2014). A nanoparticle-based strategy for the imaging of a broad range of tumours by nonlinear amplification of microenvironment signals. *Nature Materials*, *13*, 204–212.

Weissleder, R., Nahrendorf, M., & Pittet, M. J. (2014). Imaging macrophages with nanoparticles. *Nature Materials*, *13*, 125–138.

Wojtkowiak, J. W., Verduzco, D., Schramm, K. J., & Gillies, R. J. (2011). Drug resistance and cellular adaptation to tumor acidic pH microenvironment. *Molecular Pharmaceutics*, *8*, 2032–2038.

Wu, P. C., Hsieh, T. Y., Tsai, Z. U., & Liu, T. M. (2015). In vivo quantification of the structural changes of collagens in a melanoma microenvironment with second and third harmonic generation microscopy. *Scientific Reports*, *5*.

Wyckoff, J. B., Wang, Y., Lin, E. Y., Li, J. F., Goswami, S., Stanley, E. R., et al. (2007). Direct visualization of macrophage-assisted tumor cell intravasation in mammary tumors. *Cancer Research*, *67*, 2649–2656.

Xia, Z., Xing, Y., Jeon, J., Kim, Y. P., Gall, J., Dragulescu-Andrasi, A., et al. (2011). Immobilizing reporters for molecular imaging of the extracellular microenvironment in living animals. *ACS Chemical Biology*, *6*, 1117–1126.

Xing, H. R., & Zhang, Q. (2012). Real-time visualization and characterization of tumor angiogenesis and vascular response to anticancer therapies. *Methods in Molecular Biology*, *872*, 115–127.

Yan, F., Duan, W., Li, Y., Wu, H., Zhou, Y., Pan, M., et al. (2016). NIR-laser-controlled drug release from DOX/IR-780-loaded temperature-sensitive-liposomes for chemo-photothermal synergistic tumor therapy. *Theranostics*, *6*, 2337–2351.

Ye, W. (2016). The complexity of translating anti-angiogenesis therapy from basic science to the clinic. *Developmental Cell*, *37*, 114–125.

Yu, J. C., Chen, Y. L., Zhang, Y. Q., Yao, X. K., Qian, C. G., Huang, J., et al. (2014). pH-responsive and near-infrared-emissive polymer nanoparticles for simultaneous delivery, release, and fluorescence tracking of doxorubicin in vivo. *Chemical Communications (Cambridge, England)*, *50*, 4699–4702.

Yu, C. H., Tai, S. P., Kung, C. T., Wang, I. J., Yu, H. C., Huang, H. J., et al. (2007). In vivo and ex vivo imaging of intra-tissue elastic fibers using third-harmonic-generation microscopy. *Optics Express*, *15*, 11167–11177.

Yuan, F., Dellian, M., Fukumura, D., Leunig, M., Berk, D. A., Torchilin, V. P., et al. (1995). Vascular permeability in a human tumor xenograft: Molecular size dependence and cutoff size. *Cancer Research*, *55*, 3752–3756.

Zhang, T., Lin, H., Cui, L., An, N., Tong, R., Chen, Y., et al. (2016). Near infrared light triggered reactive oxygen species responsive upconversion nanoplatform for drug delivery and photodynamic therapy. *European Journal of Inorganic Chemistry*, *2016*, 1206–1213.

Zhang, M., Liu, J., Kuang, Y., Li, Q., Zheng, D. W., Song, Q., et al. (2017). Ingenious pH-sensitive dextran/mesoporous silica nanoparticles based drug delivery systems for controlled intracellular drug release. *International Journal of Biological Macromolecules*, *98*, 691–700.

Zhang, X., Wang, X., Zhong, W., Ren, X., Sha, X., & Fang, X. (2016). Matrix metalloproteinases-2/9-sensitive peptide-conjugated polymer micelles for site-specific release of drugs and enhancing tumor accumulation: Preparation and in vitro and in vivo evaluation. *International Journal of Nanomedicine*, *11*, 1643–1661.

Zhao, Y., Jin, J., Hu, Q., Zhou, H. M., Yi, J., Yu, Z., et al. (2011). Genetically encoded fluorescent sensors for intracellular NADH detection. *Cell Metabolism, 14*, 555–566.

Zhao, P., Zheng, M., Luo, Z., Fan, X., Sheng, Z., Gong, P., et al. (2016). Oxygen nanocarrier for combined cancer therapy: Oxygen-boosted ATP-responsive chemotherapy with amplified ROS lethality. *Advanced Healthcare Materials, 5*, 2161–2167.

Zhao, P., Zheng, M., Luo, Z., Gong, P., Gao, G., Sheng, Z., et al. (2015). NIR-driven smart theranostic nanomedicine for on-demand drug release and synergistic antitumour therapy. *Scientific Reports, 5*, 14258.

Zhou, F., Zheng, B., Zhang, Y., Wu, Y., Wang, H., & Chang, J. (2016). Construction of near-infrared light-triggered reactive oxygen species-sensitive (UCN/SiO2-RB + DOX)@PPADT nanoparticles for simultaneous chemotherapy and photodynamic therapy. *Nanotechnology, 27*, 235601.

Zhu, L., Huang, X., Choi, K. Y., Ma, Y., Zhang, F., Liu, G., et al. (2012). Real-time monitoring of caspase cascade activation in living cells. *Journal of Controlled Release, 163*, 55–62.

CHAPTER 4

Graphene as 2D Nano-Theranostic Materials for Cancer

Ting He, Fan Li, Yan Huang, Tuanwei Sun, Jing Lin, Peng Huang
Shenzhen University, Shenzhen, China

1 Introduction

In recent years, theranostics (the combination of diagnostics and therapeutics) has been proposed to enable personalized medicine. With the rapid advances in nanotechnology, numerous classes of theranostic nanomaterials have been developed for cancer diagnosis and therapy (Mura & Couvreur, 2012). Compared with the traditional nanotheranostic materials, two-dimensional (2D) nanomaterials and their nanocomposites exhibited remarkable advantages in cancer diagnosis and therapy due to their unique physical and chemical properties (Chen, Wu, Sun, Liu, & Liu, 2017; Huang, Qi, Boey, & Zhang, 2012; Yang, Feng, Shi, & Liu, 2013). Among these 2D nanomaterials, graphene and its derivatives, including graphene oxide (GO), reduced graphene oxide (rGO), and their nanocomposites, have attracted tremendous attention in cancer theranostics (Lin, Chen, & Huang, 2016; Yang et al., 2013; Zhou, Jing, & Chen, 2017).

As a unique monolayer graphite, graphene has an extended 2D honeycomb network constructed of single-layered carbon atoms. Since Geim and colleagues first discovered this ultrathin 2D structure in 2004 (Geim & Novoselov, 2007), graphene has been extensively explored for applications in chemistry, materials, physics, catalysis, and so on (Geim, 2009; Song, Qu, Zhao, Ren, & Qu, 2010; Wang, Li, Tang, Lu, & Li, 2009; Zhang, Li, Cui, & Lin, 2010). In recent years, graphene-based nanomaterials have attracted significant interest in various biomedical applications such as drug delivery, gene delivery, bioimaging, biosensors, cancer therapy, and so on (Lin et al., 2016; Rahman et al., 2015). Because of its high specific surface area and delocalized π electrons, graphene and its derivatives can bind with unsaturated and aromatic biomolecules through π-π conjugation, and show great potential in cancer treatment by combining with different approaches including drug delivery, gene transfection, etc. (Georgakilas et al., 2012; Yang et al., 2013; Zhu et al., 2010). Moreover, graphene and its derivatives have high optical absorption in the near-infrared (NIR) region,

Handbook of Nanomaterials for Cancer Theranostics. https://doi.org/10.1016/B978-0-12-813339-2.00004-9

which endows them with great advantages in multimodal imaging guided photothermal therapy (PTT) (Nurunnabi, Khatun, Reeck, Lee, & Lee, 2014; Zhou et al., 2017). In addition, graphene and its derivatives can be functionalized with different kinds of metal and metal oxide nanoparticles such as Au, Ag, Pt, MnO_2, Fe_3O_4, etc., for a variety of biological imaging modalities including photoacoustic imaging (PAI), fluorescence imaging (FLI), magnetic resonance imaging (MRI), computed tomography (CT), etc. (Chen, Wen, Zhen, & Zheng, 2015; Jin, Wang, Ke, Wang, & Dai, 2013; Yang et al., 2013; Zheng, Chen, Cui, Wang, & Zhang, 2014). The combination of ultimate thinness, NIR absorption, delocalized electrons, ultra-high surface area, and versatile surface functionalization make graphene and its derivatives ideal materials for cancer diagnosis and therapy.

In this chapter, we will discuss the synthesis and surface functionalization of graphene and its derivatives, the biomedical applications of graphene-based nanomaterials for cancer theranostics from cellular levels to small animals and the current challenges and future prospects in the use of graphene-based nanomaterials for cancer theranostics (Scheme 1).

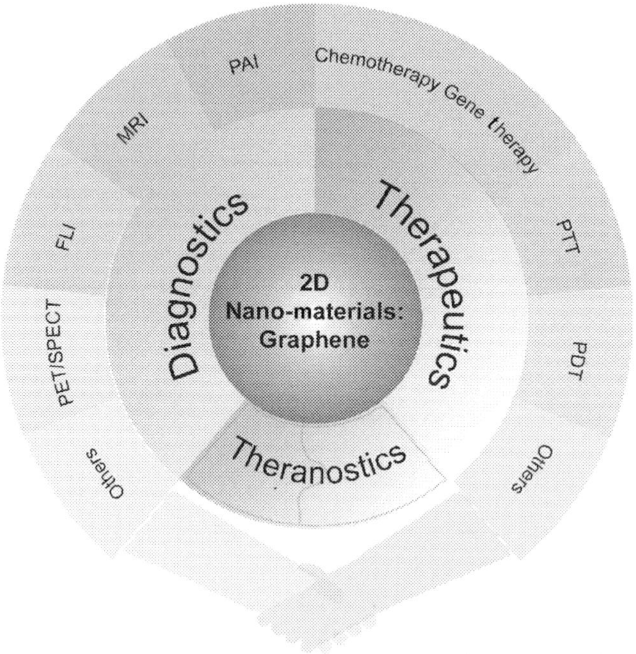

Scheme 1
The summary of graphene as 2D nano-theranostic materials.

2 Synthesis of Graphene-Based 2D Nano-Theranostic Materials

2.1 Synthesis of Graphene, GO, and rGO

Graphene can be obtained by either bottom-up or top-down strategy. In the earlier stage, graphene was prepared by mechanical exfoliation, using adhesive tape to generate single or few layers of pristine graphene (Novoselov et al., 2004). But this "scotch tape" method was only suitable for small-scale laboratory research purposes. Thus, many methods have been proposed to solve this problem; for example, epitaxial growth has been used to produce graphene at high temperature by graphitization on insulator or metal substrates such as SiC or Ni (111) (Emtsev et al., 2009; Meyer et al., 2007). Chemical vapor deposition (CVD) is an alternative method for large-scale fabrication of high quality graphene and has been employed to generate graphene by a catalytic reaction between catalytic metal substrates and hydrocarbon gases (Chen et al., 2017). This technology still suffers from rigorous growing conditions, such as high temperature and critical vacuum. Those as-prepared graphene would not be suitable for biomedical applications. Currently, the mainstream way to synthesize graphene for biomedical applications is the liquid-phase method (Coleman et al., 2011; Hernandez et al., 2008; Stankovich et al., 2007; Tyagi, Attia, & Geckeler, 2017).

For liquid-phase synthesis, GO is typically prepared by the Hummer's method through the oxidation of graphite using $KMnO_4/H_2SO_4$ and subsequent exfoliation in an aqueous solution (Hummers & Offeman, 1958), which endows its surface with large numbers of residual epoxides (–O–), hydroxides (–OH), and carboxylic acid (–COOH) groups. Those oxygen-containing hydrophilic groups further promote the potential applications of GO in cancer theranostics (Chen et al., 2014; Li et al., 2015; Sun et al., 2008). The aforementioned methods for fabrication of graphene are summarized in Fig. 1. rGO can be synthesized by reducing GO through different reduction strategies such as thermal reduction, chemical reduction, electrochemical reduction, solvothermal reduction, etc. (Pei & Cheng, 2012; Zhang et al., 2010).

2.2 Noncovalent Functionalization of Graphene and Its Derivatives

Depending on the design, various surface modification strategies have been used to synthesize graphene-based nanocomposites for cancer theranostics (Lin et al., 2016). The two main methods include non-covalent functionalization based on the adsorption of aromatic-like compounds via π-π and Van der Waals interactions with carbon framework, and covalent functionalization via the grafting of molecules onto the sp^2 carbon.

Noncovalent functionalization by π-π stacking, electrostatics, or Van der Waals force can easily decorate graphene with biomolecules to achieve the combination of imaging and therapy (Kim, Park, Miao, Lee, & Oh, 2015; Sheng et al., 2013). For example, Ling and

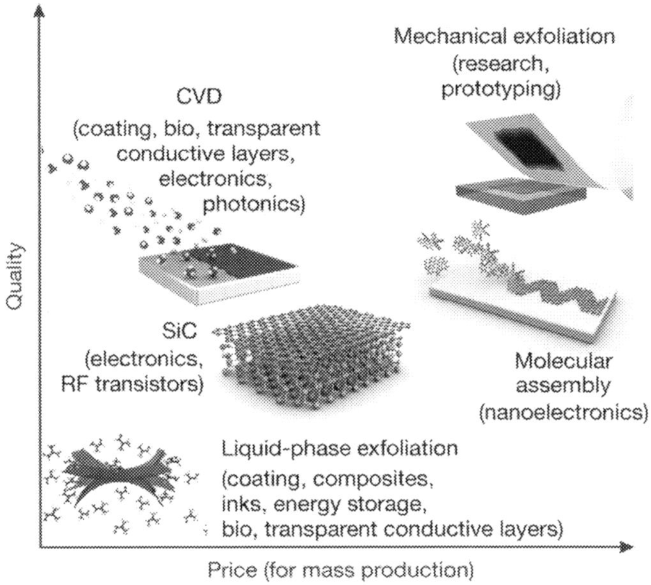

Fig. 1

Methods for the mass-production of graphene, which allow a wide choice depending on size, quality, and price for any particular application. *Adapted and reproduced with permission from reference Novoselov, K. S., Fal'ko, V. I., Colombo, L., Gellert, P. R., Schwab, M. G., & Kim, K. (2012). A roadmap for graphene. Nature, 490, 192–200, copyright 2012, Nature.*

colleagues synthesized a theranostic nanocarrier by immobilizing hydrophobic silicon napthalocyanine bis (trihexylsilyloxide) ($SiNc_4$) photosensitizer onto water-dispersible magnetic and fluorescent graphene (MFG) via π-π stacking (Gollavelli & Ling, 2014). Moreover, the single-stranded DNA(ssDNA)-functionalized graphene sensors, which show enhanced sensitivity and selectivity to NH_3 and H_2S vapors of exhaled breath gas at high humidity, have been synthesized via noncovalent π-π stacking interactions by Kim and colleagues (Jung et al., 2017). The interaction between DNA and graphene involves some theoretical calculations which provide an in-depth understanding of the interaction. Lei et al. clarified the dynamic process of ssDNA adsorbed on GO surface through hydrogen bonding and π-π stacking interactions (Xu et al., 2017).

In addition, electrostatic interaction is also a good approach for graphene functionalization. Polyethyeneimine (PEI), as a positively charged polymer, can be adsorbed on the surface of GO via electrostatic interactions. Liu group improved the physiological stability of GO by noncovalently coating with PEI (Feng, Zhang, & Liu, 2011). Based on the electrostatic interactions between negatively charged GO and positively charged mesoporous silica nanoparticles (MSN), the GO-wrapped MSN (MSN@GO) were easily synthesized (Tang et al., 2015). Mokdad et al. adopted a self-assembly process to construct a hybrid GO system by coupling together electrostatic and π-π interactions between a mononuclear iron complex Fe(II)-Pz_2Py and GO (Mokdad et al., 2015). Moreover, the electrostatic and π-π interactions are also used as driving forces to merge the organometallic complex with GO through an assembly process (Wang et al., 2014).

2.3 Covalent Functionalization of Graphene and Its Derivatives

Most covalent functionalization utilized–COOH, epoxy and hydroxyl groups of GO and rGO. Polyethylene glycol (PEG), as one of the most widely used biomaterials, could be applied to modify GO and rGO via covalent coupling, which could improve their biocompatibility and reduce nonspecific binding to biological molecules and cells (Akhavan, Ghaderi, & Emamy, 2012; Chen, Wen, et al., 2015; Zhang, Lu, Tao, Wan, & Zhao, 2016). Liu group studied in detail the effects of surface coatings on the cytotoxicity and extracellular biodegradation behaviors of graphene and its derivatives (Li et al., 2014). The toxicity of GO and rGO could be greatly attenuated by covalent functionalization with PEG or bovine serum albumin (BSA).

Besides PEG, many other hydrophilic polymers such as amine modified dextran (DEX), chitosan, poly acrylic acid (PAA), etc., have been used to improve the stability and biocompatibility of graphene and its derivatives by covalent conjugation (Gollavelli & Ling, 2012; Siddique et al., 2017; Yang et al., 2013; Zhang, Yang, Feng, & Liu, 2011). Moreover, the covalent functionalization strategy has been used to synthesize graphene-based hydrogels. The GO-PAA nanocomposite hydrogels with good self-healing properties and superior toughness have been developed through a one-pot, in situ, free radical polymerization strategy by using Fe^{3+} as a cross-linker (Zhong, Liu, & Xie, 2015). Liu and colleagues covalently functionalized GO and carbon nanotubes to obtain an electrically conductive hydrogel for enhanced nerve cell responses (Liu et al., 2016).

2.4 Decoration of Nanoparticles on Graphene and Its Derivatives

Graphene and its derivatives have been functionalized with numerous nanostructures such as metals, metal oxides, and upconversion nanoparticles for multimodality imaging and enhanced cancer therapy (Choi et al., 2017; Yang et al., 2013). In consideration of their unique magnetic property, magnetic iron oxide nanoparticles have been incorporated with graphene and its derivatives for biomedical applications. Chen group decorated PEG-functionalized GO with Fe_3O_4 nanoparticles for efficient MRI and drug delivery (Chen, Wen, et al., 2015). Ma and colleagues synthesized a GO-iron oxide nanocomposite loaded with a chemotherapy drug of doxorubicin (DOX) for MRI, magnetically targeted drug delivery, and PTT (Ma et al., 2012). Moreover, GO-$BaGdF_5$ nanocomposites have been synthesized for effective dual-modality MR and X-ray/CT imaging and PTT in vivo (Zhang et al., 2015). Gold nanoclusters have been incorporated with rGO to construct nanotheranostics for drug delivery and PAI of cancer cells (Wang et al., 2011).

3 Graphene-Based Nanomaterials for Cancer Diagnostics

Early detection of malignant neoplasms have a significant effect of the cure of cancer (Wulfkuhle, Liotta, & Petricoin, 2003). Graphene-based nanocomposites can integrate FLI, PAI, MRI, positron emission tomography (PET) or other imaging modes into one platform to effectively diagnose tumors (Chen et al., 2015; Rong et al., 2016; Zhang et al., 2017; Zheng et al., 2014).

3.1 Fluorescence Imaging

FL imaging is a sensitive technique which can detect the photons emitted from fluorescent probes (Bruchez, Moronne, Gin, Weiss, & Aliualatos, 1998). Graphene itself can also be a fluorescent probe. For instance, Han et al. synthesized uniform-sized graphene quantum dots (GQDs) and efficiently labeled stem cells with a strong yellow emission with 14% quantum yield (Zhang et al., 2012). However, short band emission always has strong background interference, which limits its in vivo application in FL imaging. Therefore, the conjugation of NIR dyes, such as Cy5.5, Cy7, etc., onto GO has become the research hotspot (Akhavan & Ghaderi, 2013; Rong et al., 2014; Tang et al., 2015; Yan et al., 2015). For instance, Chen et al. assembled Cy5.5-AS1411 aptamer conjugate onto the surface of graphene oxide-wrapped, DOX-loaded mesoporous silica nanoparticles (MSN-Dox@GO-Apt) for light-mediated drug release and aptamer-targeted cancer therapy (Fig. 2A) (Tang et al., 2015). For the MSN-Dox@GO-Apt, there were two "off-on" switches controlled by aptamer targeting and light triggering. Cy5.5-AS1411 ligand provided the targeting ability against nucleolin (NCL) with "off-on" Cy5.5 fluorescence recovery. FL emission spectra of Cy5.5-Apt quenched by GO and fluorescence signal recovered as NCL's concentration increased (Fig. 2B). This coincided with FL images of 293T cells, which had low expression of NCL, and MCF-7 cells which had a high expression of NCL (Fig. 2C and D). MCF-7 cells showed triggered receptor-mediated endocytosis with MSN-Dox@GO-Apt, and Cy5.5 fluorescence were recovered. This could suggest a suitable time to start laser irradiation, which can turn "on" the other switch to release DOX for chemotherapy.

Because the signal of fluorescent dyes modified on the graphene will be quenching, many studies use this characteristic to achieve signal changes of fluorescent. Stobiecka et al. utilized the strong FL quenching caused by the interactions of GO nanosheets (GONs) with fluorophores to construct a model ternary system: GONs-FITC-ATP (Ratajczak & Stobiecka, 2017). Fluorescein isothiocyanate (FITC) is a model fluorophore, and adenosine-5′-triphosphate (ATP) is a common biomolecule. They found that FITC's fluorescence

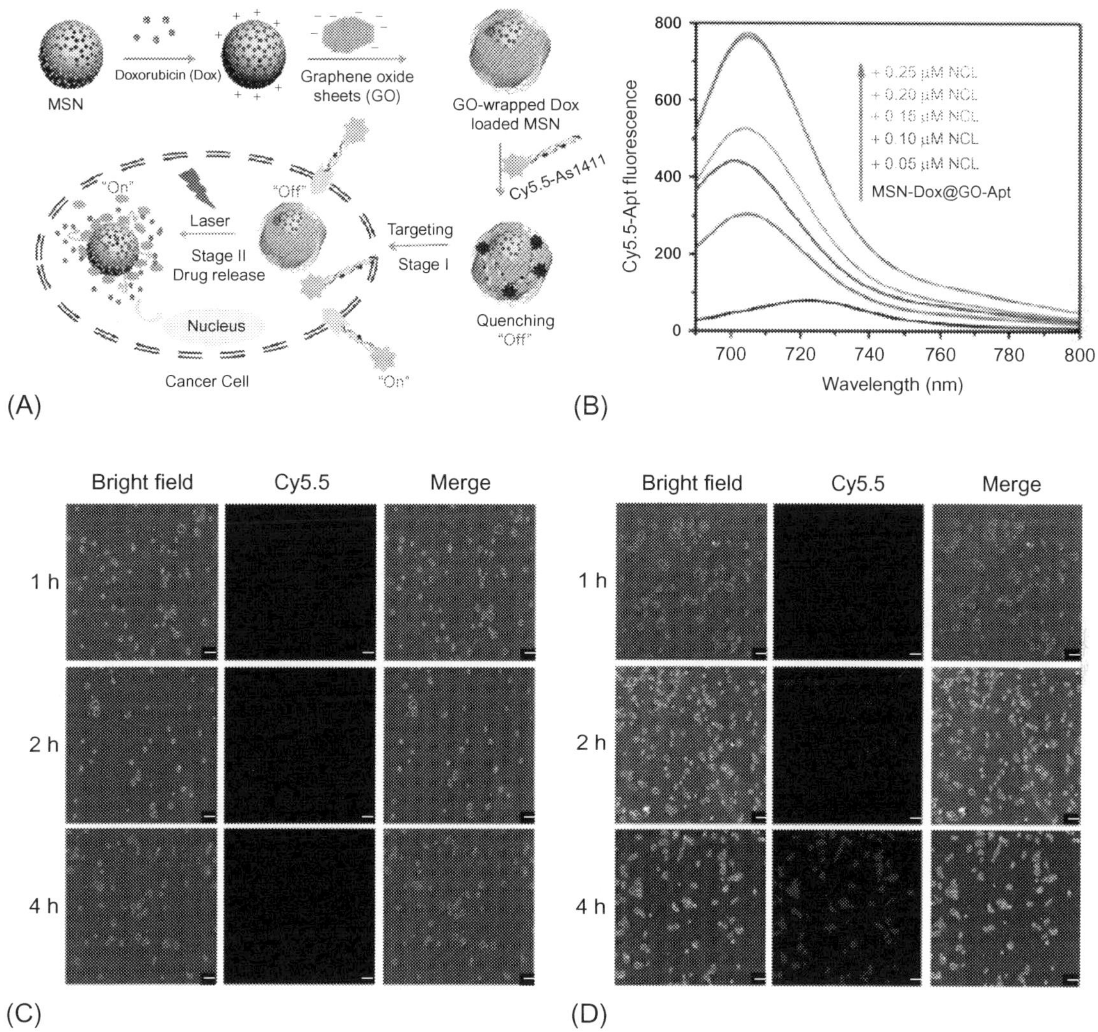

Fig. 2

(A) Schematic illustration of GO-wrapped Dox-loaded MSN-NH_2 bound with Cy5.5-labeled AS1411 aptamer and the corresponding NIR light-controlled intracellular drug release. (B) The fluorescence response of MSN-Dox@GO-Apt toward nucleolin (NCL). Fluorescence emission spectra of 0.05 μM Cy5.5-Apt quenched with GO (bottom line) and fluorescence recovery by adding NCL with concentration range of 0.05–0.25 μM (from bottom to top). FL images of 293T cells (C) and MCF-7 cells (D) incubated with MSN-Dox@GO-Apt at different time points. *Adapted and reproduced with permission from reference Tang, Y. X., Hu, H., Zhang, M. G., Song, J. B., Nie, L. M., Wang, S. J., et al. (2015). An aptamer-targeting photoresponsive drug delivery system using "off-on" graphene oxide wrapped mesoporous silica nanoparticles. Nanoscale, 7, 6304–6310, copyright 2012, The Royal Society of Chemistry.*

was considerably quenched by GONs ($K_{SV} = 598 \pm 29\,M^{-1}$) and associated with the formation of supramolecular assemblies due to the hydrogen bonding and π-π stacking interactions. Furthermore, pH and temperature can also stimulate "off-on" fluorescence on GO nanocarriers. Park et al. reported a pH- and thermoresponsive fluorescent nanomaterial of functionalized rGO with cross-linked polymer hydrophobic dye boron dipyrromethane (BODIPY), which exhibited intense fluorescence at acidic pH and low fluorescence at physiological pH (Khoerunnisa et al., 2017).

3.2 Magnetic Resonance Imaging

Because of its noninvasiveness and high spatial resolution, MRI has played an important role in molecular imaging and clinical diagnosis (Henkelman, Stanisz, & Graham, 2001; Shen et al., 2014; Wang, Ma, Wei, & Qi, 2016; Yang et al., 2016). However, the difference of relaxation time between normal and abnormal soft tissues is little. Therefore, contrast agent is necessary for more accurate diagnosis. The most common contrast agent is superparamagnetic iron oxide nanoparticle (SPIO) that is introduced to a living body to enhance T_2-weighted MR imaging quality (Johnson, Stoupis, Torres, Rosenberg, & Ros, 1996). Graphene can carry SPIO as a MRI contrast agent. For example, Kim's group developed a multifunctional nanocomposite consisting of GO-SPIO-DOX as a theranostic cancer platform (Sasikala et al., 2016). The smart magnetic nanoplatform acted as a chemotherapeutic agent with a pH-dependent drug release to administer a synergistic anticancer treatment with an enhanced T_2 contrast for MRI. Graphene can not only act as a nanocarrier, but can also be an armor of contrast agent. For instance, Chen et al. reported a stable magnetic graphitic nanocapsules (MGNs) for in situ targeted MRI detection of *Helicobacter pylori* (Fig. 3) (Li, Hu, et al., 2017). *H. pylori* is a Gram-negative bacterium typically found in the stomach which can be detected by boronic acid modified molecular binded to peptidoglycan of bacteria (Fig. 3B). The scheme illustrated synthesis of boronic acid modified MGN@B-PEG (Fig. 3A). An obvious T_2-contrast MRI signal was observed with MGN@B-PEG, while the MR signal brightened when using MGNs as a contrast agent with or without the hydrochloric acid pretreatments (Fig. 3C and D). The T_2-weighted images of gastric mucosa infected by *H. pylori* with MGN@B-PEG treatments showed the stable T_2-weighted MR signal after 7 days (Fig. 3E). This indicated that several layers of graphene as the shell can effectively protect the magnetic core from corrosion while retaining the superior contrast effect for MRI in the very harsh conditions. Besides T_2-weighted MRI, T_1 contrast can enhance the brightness of soft organs, which may give more information about the organism. Zhang and colleagues also prepared the GO/$BaGdF_5$/PEG nanocomposites by a solvothermal method in the presence of PEG (Zhang et al., 2015). The results of GO/$BaGdF_5$/PEG shown low cytotoxicity and positive MRI contrast effect.

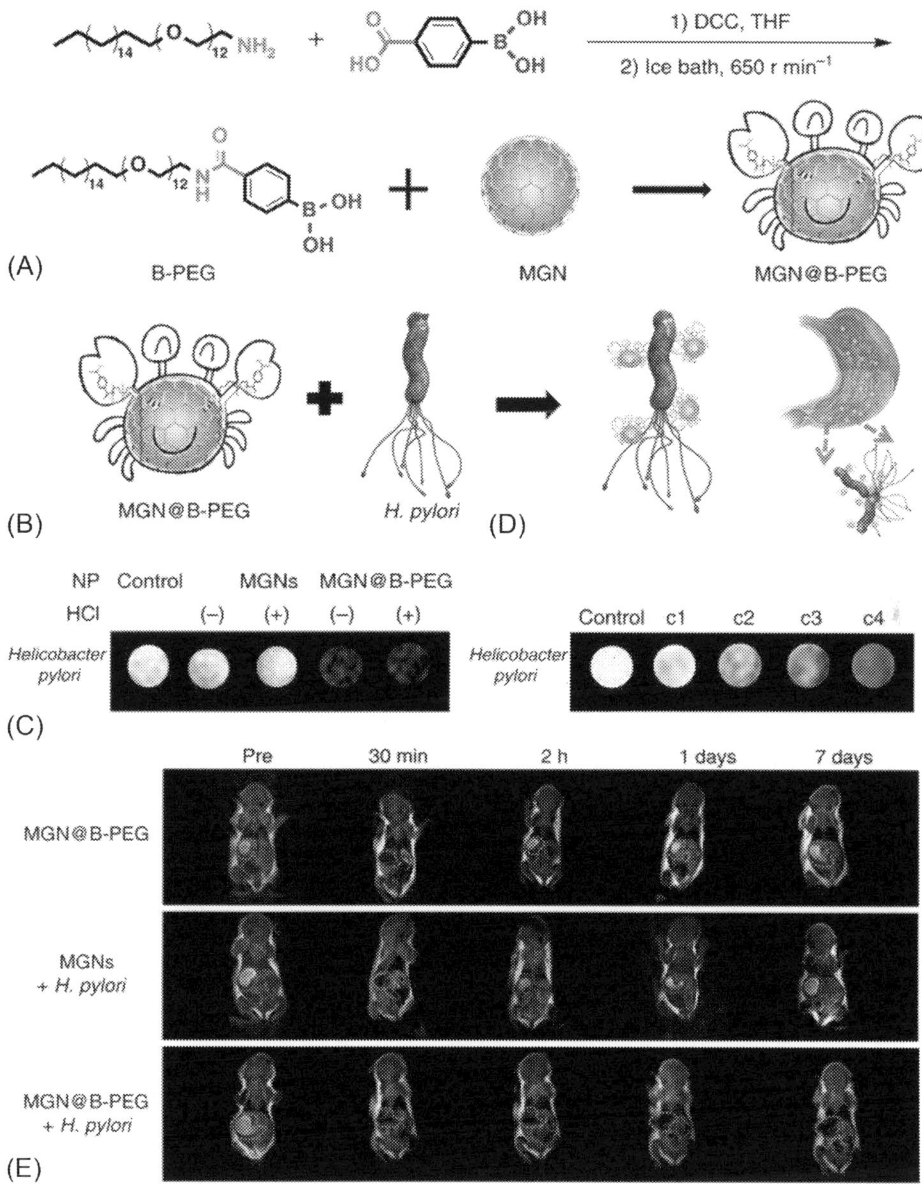

Fig. 3

(A) Schematic illustration of B-PEG synthesis and MGN functionalization. (B) Schematic illustration of *H. pylori* detection with MGN@B-PEG. (C) MR images of *H. pylori* with MGN or MGN@B-PEG after different treatments. (D) MR images of different concentrations of *H. pylori* with MGN@B-PEG. (E) T_2-weighted images at different times with or without MGN@B-PEG treatments. *Adapted and reproduced with permission from reference Li, Y. J., Hu, X. X., Ding, D., Zou, Y. X., Xu, Y. T., Wang, X. W., et al. (2017). In situ targeted MRI detection of Helicobacter pylori with stable magnetic graphitic nanocapsules. Nature Communications, 8, 15653–15663, copyright 2017, Nature.*

3.3 Photoacoustic Imaging

The motivation for PAI is to combine the high contrast of optical imaging with the high penetration depth of ultrasound imaging (Toumia et al., 2016; Xu & Wang, 2006; Zhang, Maslov, Stoica, & Wang, 2006). Because of its high NIR absorption, graphene is often used as a PAI contrast agent to enhance the photoacoustic performance (Sasikala et al., 2016). For example, Cai et al. have synthesized hybrid nanoclusters for PAI and PTT using the natural active protein BSA as a reductant and stabilizer of rGO (Sheng et al., 2013). After intravenous injection of nano-rGO into tumor-bearing mice, the tumor region showed enhanced photoacoustic signals. It indicated that rGO is excellent for passive targeting and PAI.

Moreover, Wang's group specialized investigated graphene nanoribbons and nano-platelets as contrast agents for photoacoustic tomography (PAT). Their results shown that multiwalled GO exhibits approximately 5- to 10-fold signal enhancement for PAT in comparison to blood at the wavelength of 755 nm, and approximately 10%–28% signal enhancement for PAT in comparison to deionized water at 3 GHz (Lalwani, Cai, Nie, Wang, & Sitharaman, 2013).

Recently Paradossi et al. developed hybrid constructs with new electrical, optical, and mechanical properties by coupling graphene with a soft polymer surface, poly(vinyl alcohol)-based microbubbles (PVA MBs) (Fig. 4A) (Toumia et al., 2016). Tethering pristine graphene sheets to the surfaces (PVA MBs) was accomplished by using suitable reactive functional groups of the surfactant scaffold. In vivo PA images of subcutaneous injection of G/PVA MBs into a normal mouse hind limb showed an obvious increase of photoacoustic signal (Fig. 4B and C), including the deep tissues. Despite a very low dose, 30 μL G/PVA (3×10^7 MBs/mL, graphene content 5% w/w), the PA signal could increase from 0.6 to 6.6 a.u. of the ROI (Fig. 4D and E). This impressive performance could be attributed to those major advantages of the hybrid graphene/PVA MBs construct: (i) the preservation of graphene properties, (ii) biocompatibility which improved after anchoring of pristine graphene to the bioinert surface of the PVA bubble, and (iii) a very good enhancement in a NIR spectral region of the PA signal, which does not overlap with the signals of PA active endogenous molecules such as hemoglobin. A similar report from Cai group, they developed an agent of indocyanine green loaded polydopamine-reduced GO nanocomposites (ICG-PDA-rGO) which can also amplify PA singles by PDA layer coating on the surface of rGO (Hu et al., 2016).

3.4 Radionuclide-Based Imaging

Radionuclide imaging would be able to accurately track the labeled substances in vivo in a quantitative manner with excellent sensitivity ($\sim 10^{-11}$–10^{-12} mol/L) and limitless penetration depth (Bao et al., 2003; Gambhir, 2002; Phelps, 2000). The radionuclide-based imaging mainly contains PET and single photon emission computed tomography (SPECT) (Janib, Moses, & MacKay, 2010). PET and SPECT images are acquired over a nominally low background

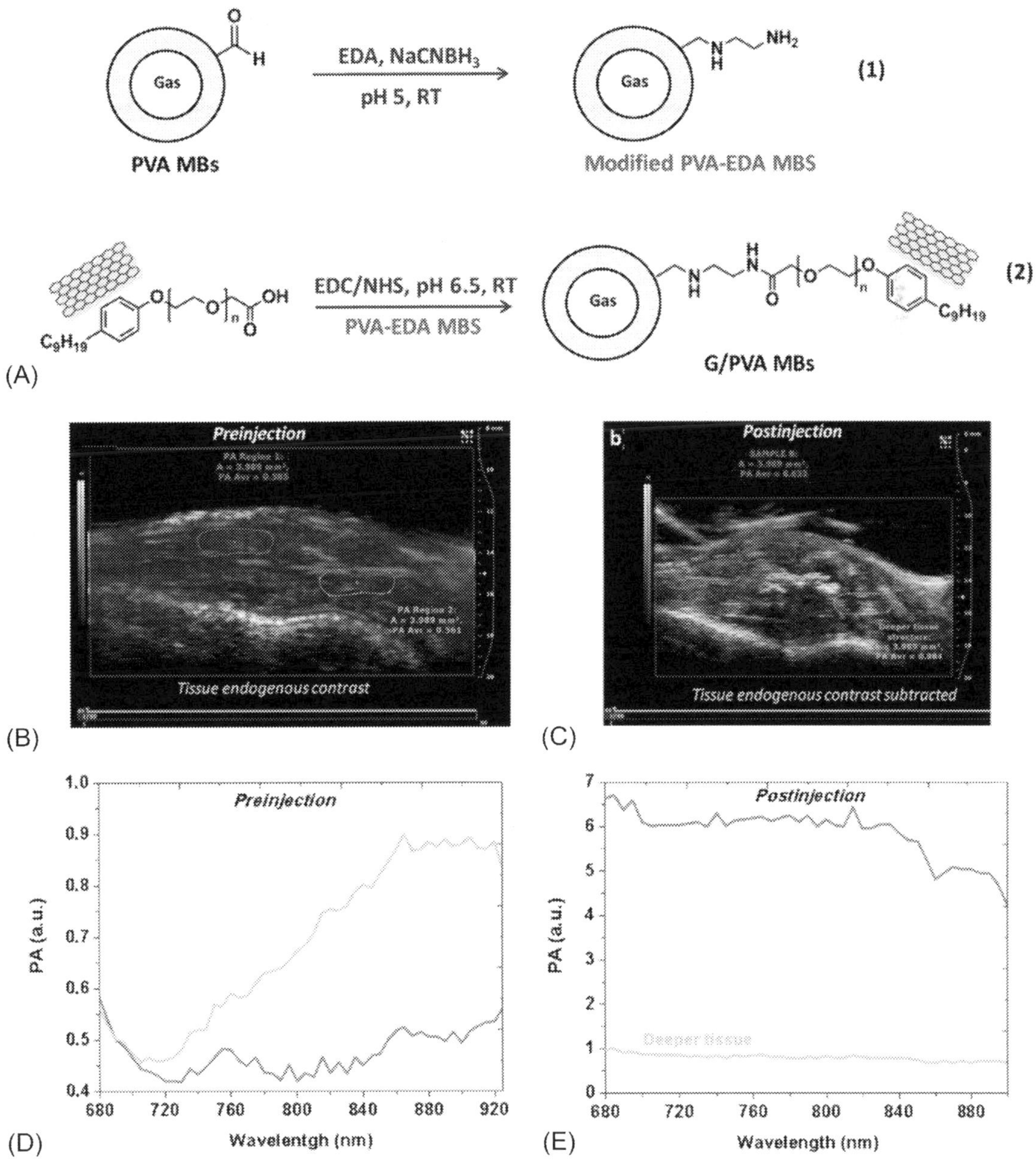

Fig. 4

(A) Scheme of graphene/PVA microbubbles functionalization reaction. Real time in vivo PA images of subcutaneous injection of G/PVA MBs into normal mouse hind limb. (B) Pre-injection tissue endogenous contrast. (C) Post-injection tissue endogenous contrast subtracted. (D) Pre-injection PA spectra. (E) Post-injection PA spectra. *Adapted and reproduced with permission from reference Toumia, Y., Domenici, F., Orlanducci, S., Mura, F., Grishenkov, D., Trochet, P., et al. (2016). Graphene meets microbubbles: A superior contrast agent for photoacoustic imaging. ACS Applied Materials & Interfaces, 8, 16465–16475, copyright 2016, American Chemical Society.*

signal and require little signal amplification. Graphene-based nanomaterials as promising nanoplatforms are playing an important role in PET/SPECT imaging (Lin et al., 2016).

After being labeled with a radioactive element, graphene can be used as a contrast agent for PET. For instance, Cai group labeled rGO and GO with ^{64}Cu by transition metal-π electron interaction based on the electron transfer between $^{64}Cu^{2+}$ cation and π bond on the surface of graphene (Shi et al., 2017). Two subtypes of graphene nanomaterials, rGO and GO, were labeled with ^{64}Cu nanosheets without the chelator conjugation. Due to the abundant π bonds, rGO exhibited significantly higher labeling efficiency in comparison with GO, and, excellent PET imaging in vivo.

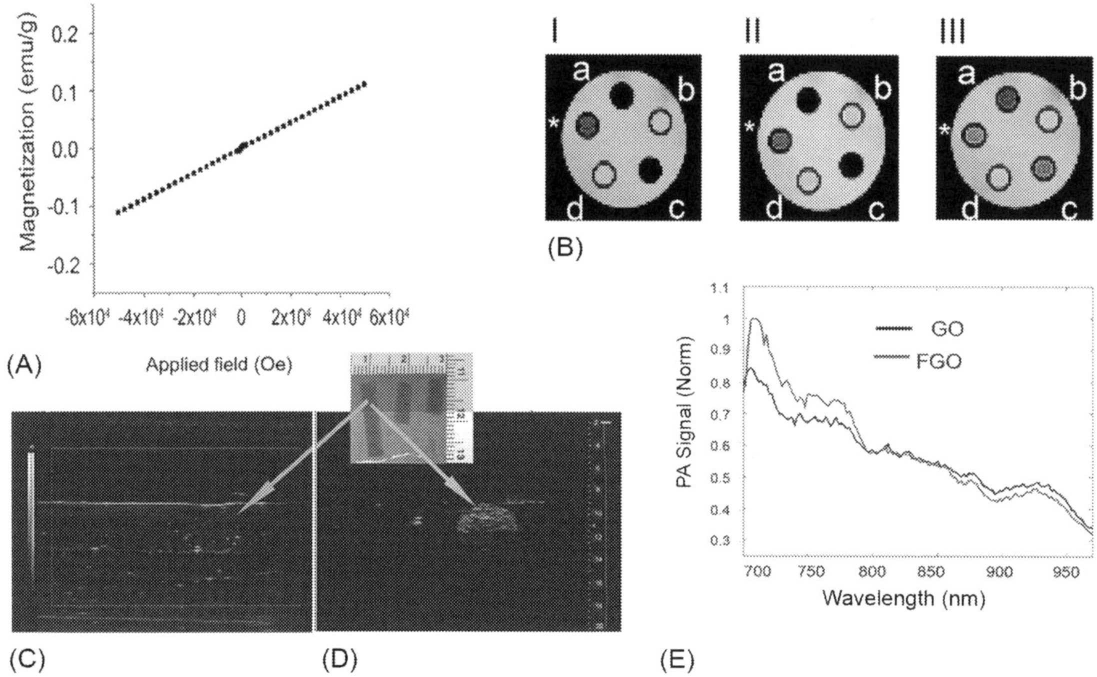

Fig. 5

(A) Super-conducting quantum interference device (SQUID) characterization at 300 K. (B) Spin-spin (T_2) relaxation measurements acquired on a 4.7T MRI of FGO (I: a = 625, c = 500; II: a = 313, c = 250; III: a = 156, c = 125 μg mL^{-1}) and GO (I: b = 625, d = 500; II: b = 313, d = 250; III: b = 156, d = 125 μg mL^{-1}). The positive control (*) consists of diluted magnevist (0.5 mg mL^{-1}). Ultrasound alone (C) and PA imaging (D) of FGO gelatin phantom under a 21 MHz ultrasound equipment. (E) PA normalized signal spectrum of FGO (top line at 710 nm) and GO (bottom line at 710 nm). *Adapted and reproduced with permission from reference Romero-Aburto, R., Narayanan, T. N., Nagaoka, Y., Hasumura, T., Mitcham, T. M., Fukuda, T., et al. (2013). Fluorinated graphene oxide; a new multimodal material for biological applications. Advanced Materials, 25, 5632–5637, copyright 2013, Wiley.*

Graphene can also improve the cellular uptake. Yang et al. developed ^{131}I labeled, PEG coated rGO (^{131}I-rGO-PEG) for nuclear imaging guided combined radiotherapy and PTT of cancer (Chen, Zhong, et al., 2015). Compared with free ^{131}I, ^{131}I-rGO-PEG exhibited enhanced cellular uptake and thus improved radiotherapeutic efficacy against cancer cells. As revealed by gamma imaging, efficient tumor accumulation of ^{131}I-rGO-PEG was observed after its intravenous injection.

3.5 Multimodal Imaging

Compared to the individual imaging modality, multimodal imaging offers higher diagnostic efficiency and accuracy, because it combines the advantages of individual imaging modality and collects all information from a variety of imaging modalities. Graphene, as an ideal nanoplatform, can successfully achieve multimodal imaging because it can be loaded with some functional imaging molecules. For example, Wang and colleagues constructed $RGI_{1.8k}$-ICG which could be used as T_2-weighted MR and infrared thermal imaging agents (Du et al., 2017). Fluorinated GO (FGO) can serve both as a MRI and PAI contrast agent under preclinical settings (Fig. 5) (Romero-Aburto et al., 2013). Superconducting quantum interference device (SQUID) characterization at 300 K showed a linear dependency to the magnetic field, proving that FGO exhibited paramagnetic behavior magnetization curve (Fig. 5A). FGO exhibited the ability to serve as an MRI contrast agent in T_2 mode (Fig. 5B), while GO did not. Spectroscopic imaging of FGO and GO generated similar photoacoustic signal spectrum (Fig. 5E), which was well visualized with PA imaging (Fig. 5D) FGO was poorly visualized when relying on ultrasound alone (Fig. 5C).

4 Therapeutics

4.1 Drug and Gene Delivery

Targeted delivery of drugs in a specific manner represents a potential powerful technology in cancer therapy (Cole, Yang, & David, 2011; Goenka, Sant, & Sant, 2014; Orecchioni, Cabizza, Bianco, & Delogu, 2015). Compared to free drugs, targeted delivery systems exhibit more effective for therapy. For instance, Shen and colleagues prepared a multifunctional targeted delivery system comprised of a molecular biotargeting ligand and SPIO on the surface of GO (Song et al., 2017). Superparamagnetic Fe_3O_4 nanoparticles was loaded on the surface of GO via chemical precipitation method to form GO@Fe_3O_4 nanocomposites. Lactoferrin (Lf), an iron-transporting serum glycoprotein, was chosen as the targeted ligand to construct the targeted delivery system Lf@GO@Fe_3O_4 due to binding with receptors overexpressed at the surface of glioma cells and vascular endothelial cell of the blood brain barrier. The nanodelivery system had a high loading capacity and exhibited a pH-dependent release behavior. Compared with free DOX and DOX@GO@Fe_3O_4, Lf@GO@Fe_3O_4@DOX displayed greater intracellular delivery efficiency and stronger cytotoxicity against C6 glioma cells. Geckeler et al. developed graphene poly vinylpyrrolidone nanoparticle composite (GRP-PVP-NP) in a dual role function as anti-cancer agent and an efficient pH-sensitive drug carrier for the doxorubicin (GRP-PVP-DOX) (Fig. 6) (Tyagi et al., 2017). Graphene sheets were wrapped onto PVP nanoparticles with an average size of 42 nm. The GRP-PVP-NP showed cytotoxicity to HCT-116, HeLa, SCC-9, but was biologically safe to normal cells (NIH-3T3 and HEK-293). This was attributed to that GRP-PVP-NP can affect the reactive oxygen species (ROS), which was increased drastically at all the concentrations (5–200 $\mu g\,mL^{-1}$). Because of the high surface area for drug binding on both sides of the single sheet, GRP-PVP-DOX can also serve as a pH-sensitive drug carrier and can be released into the tumor microenvironment. Other similar strategy was reported by Szunerits and colleagues. They first prepared GO modified with 2-nitrodopamine-coated magnetic particle (GO-MP_{dop}) matrices loaded with insulin (Turcheniuk et al., 2014). The GO-based matrix can ensure the stability of insulin at low pH. *Xenopus laevis* oocytes model showed that the meiotic resumption rate of GO-MP_{dop}

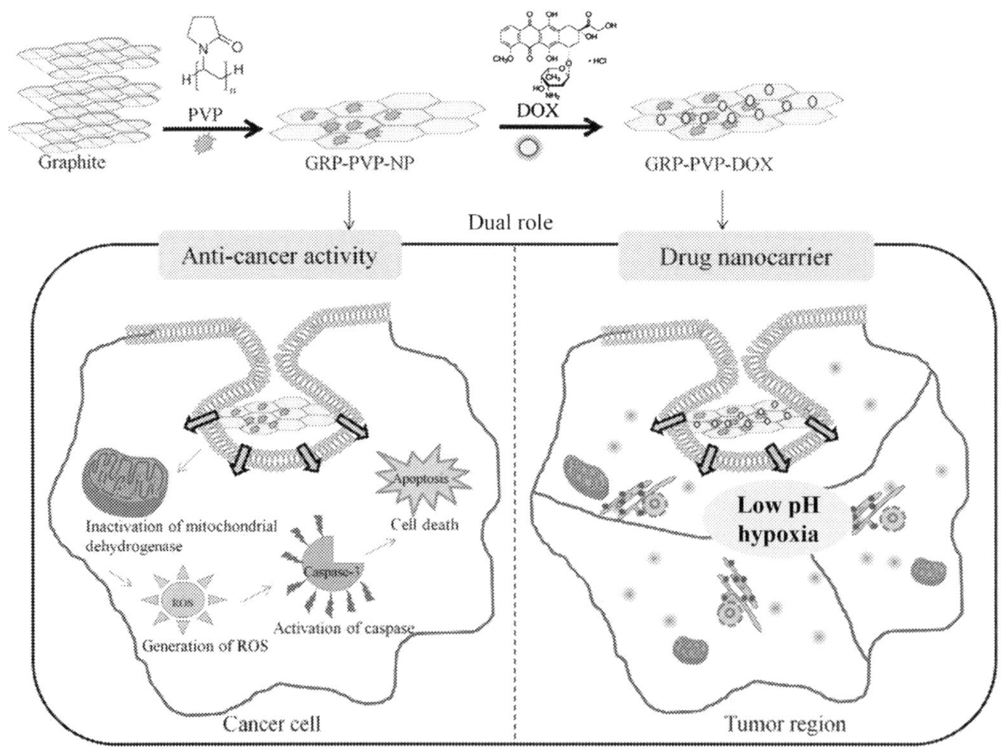

Fig. 6

The scheme of GRP-PVP-NP as a pH-sensitive drug carrier and its anticancer activity. *Adapted and reproduced with permission from reference Tyagi, N., Attia, N. F., Geckeler, K. E. (2017). Exfoliated graphene nanosheets: pH-sensitive drug carrier and anti-cancer activity. Journal of Colloid and Interface Science, 498, 364–377, copyright 2017, Elsevier.*

remained unaltered when pre-treated in acidic conditions, while preincubated free insulin almost entirely lost its maturation effect.

Moreover, Xu et al. developed a glioma-targeted drug delivery system that based on targeted peptide chlorotoxin-conjugated GO (CTX-GO) sheets (Wang, Gu, Xiao, Ye, & Xu, 2014). DOX was loaded onto CTX-GO (CTX-GO/DOX) with high efficiency (570 mg DOX per gram CTX-GO) and showed pH-dependent sustained-release properties. The evaluations experiments demonstrated that CTX-GO/DOX mediated the highest rate of death of glioma cells.

In addition to its application in drug delivery, GO can also be used in gene therapy for cancer treatments. For example, Li et al. utilized multifunctionalized monolayer GO as a gene delivery system to efficiently co-deliver HDAC1 and K-Ras siRNAs (small interfering RNAs targeting the HDAC1 gene and the G12C mutant K-Ras gene, respectively) to specifically target pancreatic cancer cells MIA PaCa-2 (Fig. 7) (Yin et al., 2017). FA/GOs were prepared by conjugating the amine-functionalized FA/PEG-NH_2 which conjugated FA with NH_2-mPEG-NH_2 to increase water solubility and biocompatibility, and then formed positively charged GO/PEG/PAH or GO/PEG/FA/PAH for siRNA delivery by electrostatic interaction (Fig. 7A). After being treated with FA/GO/HDAC1 siRNA, FA/GO/K-Ras siRNA or FA/GO/(HDAC1 + K-Ras) siRNAs, MIA PaCa-2 cells showed remarkable inhibition of HDAC1 and K-Ras gene expression (Fig. 7B and C). Compared to untreated cells, a total of 1344 downregulated and 766 upregulated genes were detected and clustered in MIA PaCa-2 cells with inactivation of HDAC1 and K-Ras, which affected more genes than the inactivation of either HDAC1 or K-Ras alone (Fig. 7D). The results elucidated that dual gene silencing affects the inactivation of both the HDAC1 and the K-Ras gene, thereby causing apoptosis, proliferation inhibition, and cell cycle arrest in treated MIA PaCa-2 cells. The synergistic combination of gene silencing and NIR light thermotherapy showed significant anticancer efficacy, inhibiting in vivo tumor volume growth by >80% without obvious side effects (Fig. 7E). This result shows the promising potential of functionalized GO as a vehicle for gene therapy delivery with low toxicity for the treatment of pancreatic adenocarcinoma.

GO can help small RNA to cross the cell membrane and protect it from deactivation caused by enzymolysis. For instance, Ge group used GO-PLL-SDGR (poly-L-lysine and Arg-Gly-Asp-Ser functionalized GO) as a gene carrier (Ren, Zhang, Cui, Bi, & Ge, 2017). The GO-PLL-SDGR/VEGF-siRNA showed an efficiency of in vitro gene silencing that the expressions of VEGF-mRNA and VEGF protein were down-regulated by 40.86% and 51.71%, respectively. The Cy3 labeled VEGF-siRNA was observed to assemble in tumor tissues, and the tumor inhibitory rate of GO-PLL-SDGR/VEGF-siRNA was 51.74%.

Moreover, Liu et al. explored that PEG and PEI are covalently conjugated to GO via amide bonds, obtaining a physiologically stable dual-polymer-functionalized nano-GO conjugate (NGO-PEG-PEI) with ultra-small size (Feng et al., 2013). Compared with free PEI and the GO-PEI conjugate without PEGylation, NGO-PEG-PEI shows superior gene transfection

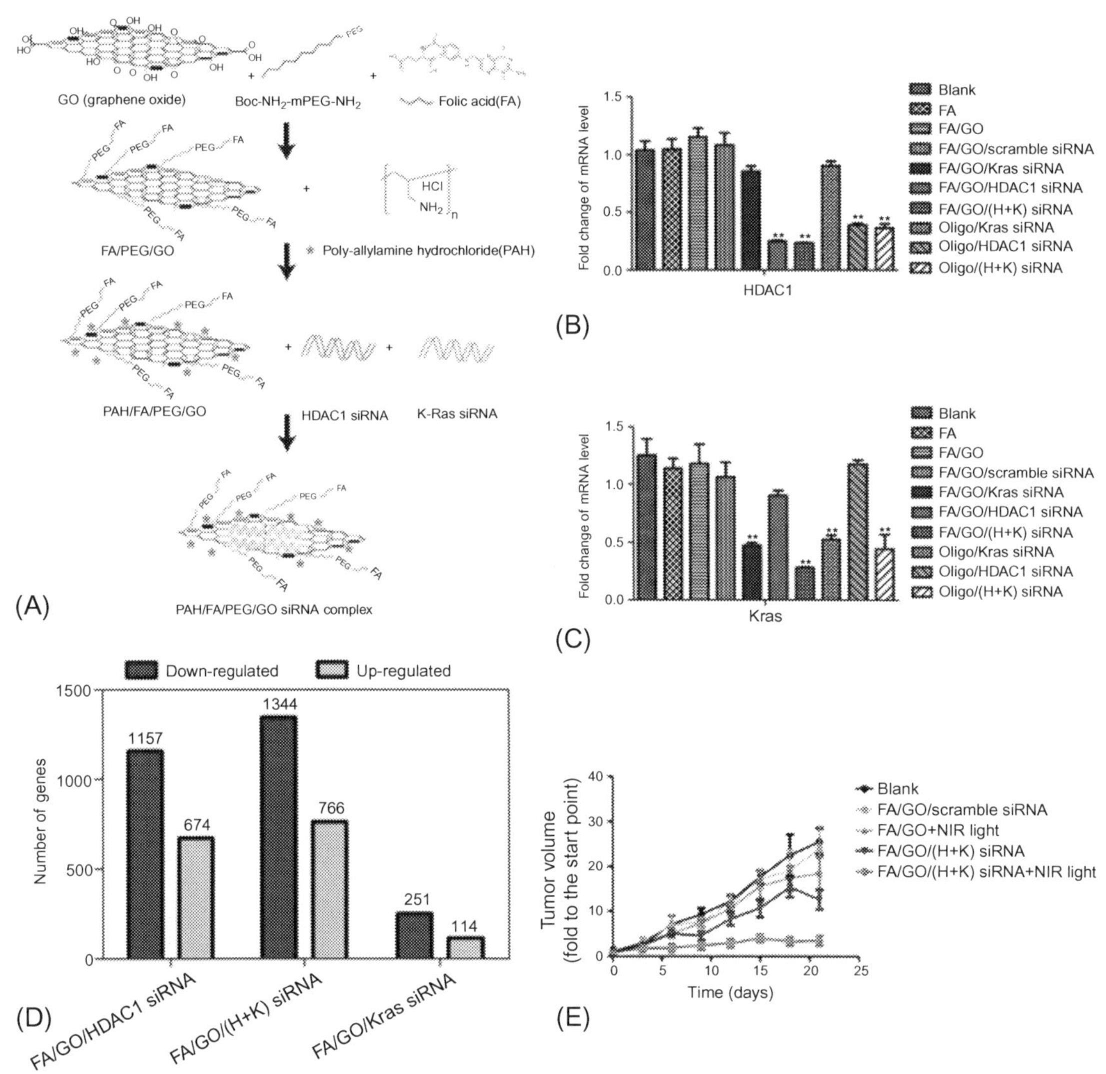

Fig. 7

(A) Schematic overview of the FA/PEG/GO synthesis and gene loading process using engineered GO-based nanocarriers. (B and C) mRNA levels of MIA PaCa-2 cells after treated with PBS, FA, FA/GO, FA/GO/scramble siRNA, FA/GO/K-Ras siRNA, FA/GO/HDAC1 siRNA, FA/GO/(H + K) siRNA, Oligo/K-Ras siRNA, Oligo/HDAC1 siRNA and Oligo/(H + K) siRNA for 4 h (D) Microarray analysis revealed a total of 1344 downregulated and 766 upregulated genes in MIA PaCa-2 cells with combined inactivation of HDAC1 and K-Ras ($n = 3$; fold change >2.0; $P < 0.05$). Gene ontology analysis revealed the different pathways involving these genes. (E) Relative changes in tumor volume over time. *Adapted and reproduced with permission from reference Yin, F., Hu, K., Chen, Y. Z., Yu, M. Y., Wang, D. Y., Wang, Q., et al. (2017). SiRNA delivery with PEGylated graphene oxide nanosheets for combined photothermal and genetherapy for pancreatic cancer. Theranostics, 7, 1133–1148, copyright 2017, Ivyspring International Publisher.*

efficiency without serum interference, as well as reduced cytotoxicity. Utilizing the NIR optical absorbance of NGO, the cellular uptake of NGO-PEG-PEI was shown to be enhanced under a low power NIR laser irradiation, because of mild photothermal heating that increased the cell membrane permeability without significantly damaging cells. As a result, remarkably enhanced plasmid DNA transfection efficiencies induced by the NIR laser are achieved using NGO-PEG-PEI as the light-responsive gene carrier. Intriguing results indicated that NGO-PEG-PEI could efficiently transport the loaded plasmid DNA into cancer cells at ~43°C induced by 808 nm laser irradiation, resulting in greatly enhanced gene transfection efficiency.

4.2 Photothermal/Photodynamic Therapy

PDT and PTT, due to their specific spatiotemporal selectivity and minimal invasiveness, have been widely investigated as alternative treatments of malignant diseases (Chen et al., 2017; Lin et al., 2016; Yan et al., 2015; Yang et al., 2013). For example, Wang and colleagues reported resveratrol (Res) loaded folate-terminated PEG-phospholipid coated rGO nanoassembly (FA-PEG-Lip@rGO/Res) by simply sonicating Res and rGO in FA-PEG linked liposome (FA-PEG-liposome) suspensions (Hai et al., 2017). Yang et al. fabricated GO/Bi_2Se_3/PVP (Bi_2Se_3 deposited on GO with polyvinylpyrrolidone) using a one-pot solvothermal method (Zhang et al., 2017). GO/Bi_2Se_3/PVP nanocomposites showed low cytotoxicity in vitro and significant photothermal cytotoxicity to cancer cells under 808 nm laser irradiation. The prepared FA-PEG-Lip@rGO/Res exhibited highly efficient combined chemotherapy and PTT to eradicate xenografted tumor with a single dose intratumoral injection under NIR laser irradiation. Graphene and its derivatives have been used as carriers to deliver photosensitizers and photothermal conversion agents for combining PDT and PTT. Chen et al. strategically designed and produced a novel phototheranostic platform based on DVDMS photosensitizer-loaded PEGylated GO (GO-PEG-DVDMS) for FL/PA imaging guiding PDT and PTT (Fig. 8A) (Yan, Hu, et al., 2015). In vivo FL and PA dual-modality imaging showed increased tumor signals after intravenous injection with GO-PEG-DVDMS (Fig. 8B and C), which indicated that the GO-PEG-DVDMS had a highly efficient tumor passive targeting ability and long retention time in the tumor tissue. DVDMS significantly enhanced the NIR absorption of GO to improve PTT, as shown in Fig. 8D–G. GO-PEG-DVDMS exhibited significantly higher therapeutic efficacy compared with either individual PDT or PTT alone (Fig. 8F).

Magnetic graphene-based nanostructures can also be used for PTT. For example, Mahmoudi et al. synthesized superparamagnetic zinc ferrite spinel-graphene nanostructures as therapeutic agents for the magnetic targeted PTT of cancer (Hajipour, Akhavan, Meidanchi, Laurent, & Mahmoudi, 2014). The corona-coated spinel-graphene nanomaterials induced a higher intracellular concentration of ROS and exhibited a significantly higher cellular uptake efficacy under laser irradiation compared to that without laser irradiation.

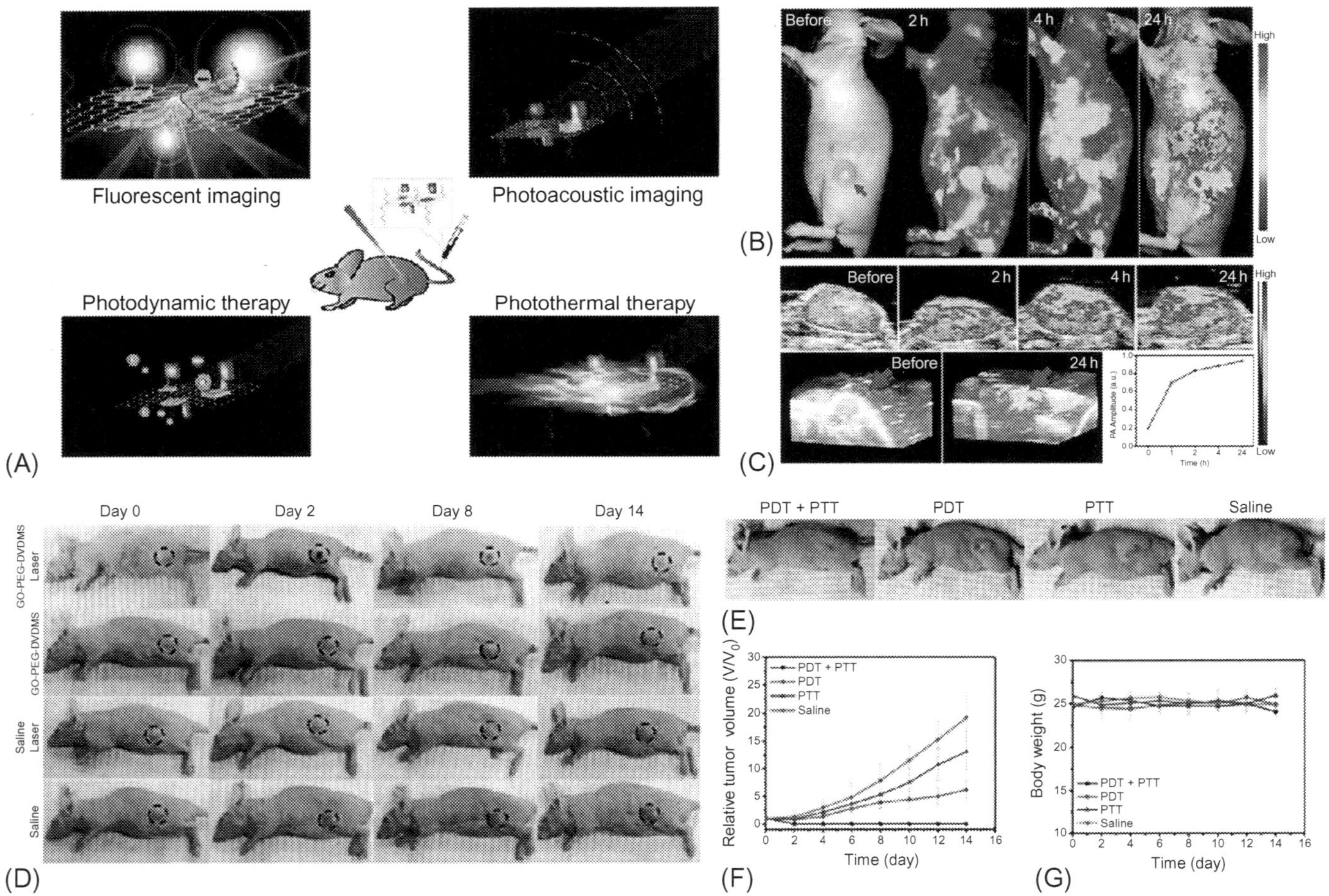

Fig. 8

(A) Schematic illustration of the multimodality imaging guided synergistic photodynamic/photothermal therapies based on PEG-GO carrier and DVDMS photosensitizer. (B) In vivo FL images and (C) ultrasound/photoacoustic images and quantitative analysis of PA signal of PC9 bearing mice treated with GO-PEG-DVDMS. (D) Representative photos of PC9 tumor-bearing mice before and after PTT treatment from different groups. (E) Representative photos of PC9 tumor-bearing mice from different group after treatment of PDT, PTT or their synergism on day 14. (F) Tumor growth curves of different groups of tumor-bearing mice after different treatments. (G) Body weight curves of PC9 tumor mice after different treatments. *Reprinted with permission from reference Yan, X. F., Hu, H., Lin, J., Jin, A. J., Niu, G., Zhang, S. L., et al. (2015). Optical and photoacoustic dual-modality imaging guided synergistic photodynamic/photothermal therapies. Nanoscale, 7, 2520–2526, copyright 2015, Royal Society of Chemistry.*

5 Imaging-Guided Therapy of Graphene-Based Nanocomposites

As a theranostic nanoplatform, imaging-guided therapy is the final goal of graphene-based nanocomposites. Fortunately, some of these materials have achieved this goal. For example, Oh and colleagues reported the image-guided synergistic photothermal antitumor effects of photo-responsive NIR imaging agent constructed by hyaluronic acid-anchored rGO (HArGO) loaded with ICG (Miao et al., 2015) (Fig. 9A). ICG was loaded onto rGO or HArGO via π-π interaction with as high as 100.4% and 92.3% loading efficiency, respectively (Fig. 9B). The temperature can be observed in the tumor tissues as high as 49.4°C after 3 min NIR laser irradiation (Fig. 9C). rGO can also increase the accumulation of ICG in KB tumor tissues. But free ICG mainly located at the liver (Fig. 9D–F). This result reflected by the sufficient tumor ablation after PTT (Fig. 9G), and a visualized shrink of tumor (Fig. 9I and J). rGO with ICG loaded could enhance photothermal activity and guide identification of the tumor location for PTT.

Another example, Liu et al. synthesized a multifunctional rGO-based nanocomposite by functionalization of a nanoscale rGO-IONP complex with PEG. The rGO-based nanocomposite showed excellent physiological stability, strong NIR optical absorbance, and superparamagnetic properties (Yang et al., 2012). In vivo FL, PA, and MR tri-modal imaging were carried out, and the high passive tumor targeting was used for effective photothermal ablation of tumors in mice.

6 Prospects and Challenges

Theranostic agents based on graphene and its derivatives have been developed for decades since their discovery in 2004, and some of them showed great potential in bio-application (Miao et al., 2015; Yan, Hu, et al., 2015; Yang et al., 2012). However, their potential as theranostic agents remain to be further developed, and there is still a long way to go for clinical diagnosis. Graphene involved in theranostic should be considered on its own characteristics: (i) strong absorption in the NIR region that can be used for PAI and PTT (Miao et al., 2015; Song et al., 2015; Yang et al., 2012); (ii) a large specific surface area that can be used as a drug carrier and controlled release upon stimulations (Ji et al., 2015; Turcheniuk et al., 2014; Wang, Gu, et al., 2014); (iii) delocalized electrons that can be used to adsorb dye molecules and temporarily quench its fluorescence; (iv) size that can be adjusted to increase endocytosis and targeting (Rong et al., 2014). Ways to comprehensively utilize the above characteristics of graphene to form a powerful theranostic agent, to achieve multimodality imaging and combination therapy, to improve the accuracy of tumor diagnosis, and to improve the efficiency of tumor treatment need to be seriously considered. For example, how can we modify the surface of graphene to improve its fluorescence quantum yield? How can we improve the targeting of graphene carriers and increase drug accumulation in tumors? How can we improve its biocompatibility through surface modification? Can we build biodegradable graphene materials, and so on? In short, graphene-based theranostic

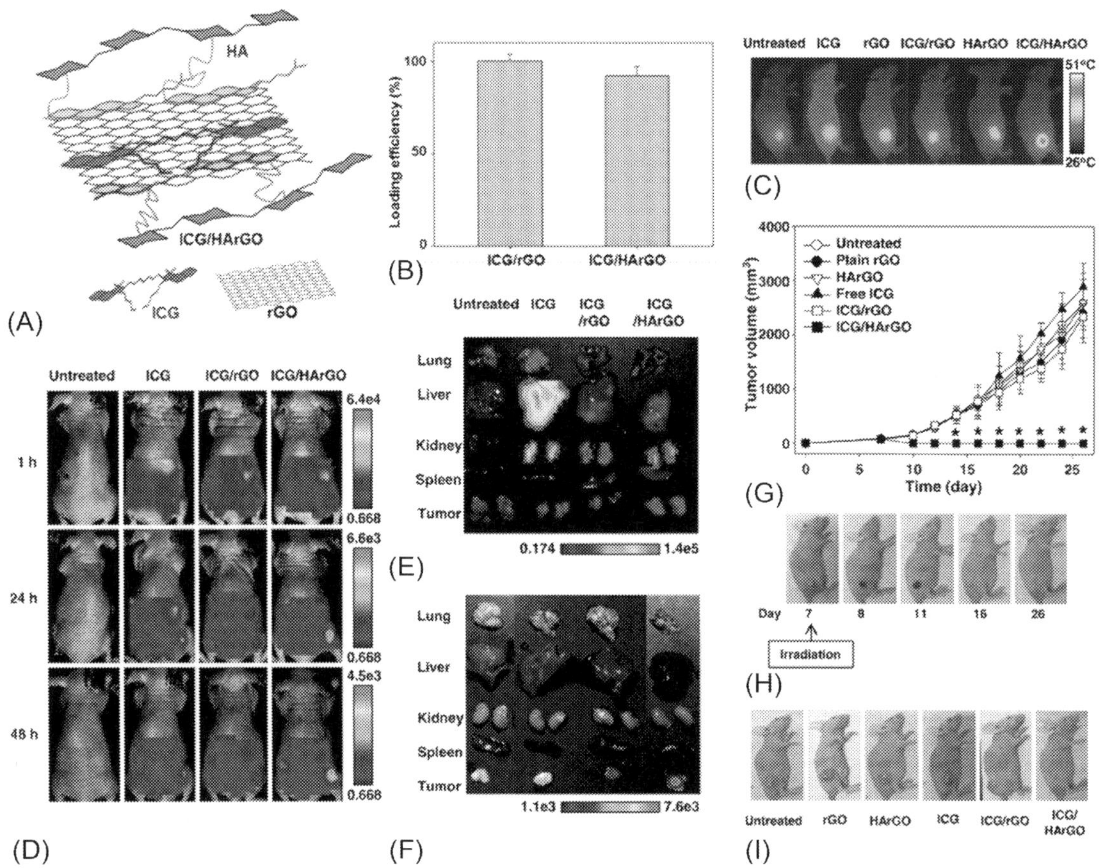

Fig. 9

(A) Scheme of loading ICG on HArGO. (B) The loading efficiency of ICG on rGO or HArGO was measured using UV/vis spectrophotometry. (C) Photothermal effect of the ICG/HArGO after image-guided irradiation. KB-bearing mice were injected intravenously with free ICG, rGO, ICG/rGO, HArGO, or ICG/HArGO. Immediately after irradiation with an 808 nm laser, real-time temperature changes at the tumor site were visualized by real-time IR thermal imaging. (D) Molecular images of ICG in tumor-bearing mice. (E and F) *Ex vivo* images (lung, liver, kidney, spleen, and tumor tissues). (G) KB-bearing mice were intravenously treated with free ICG, rGO, ICG/rGO, HArGO, or ICG/HArGO by administering a single injection on day 6. (H) The appearances of tumor sites in ICG/HArGO-treated mice, monitored starting on day 7. (I) The appearances of tumor sites in all groups, observed on day 26. *Reprinted with permission from reference Miao, W., Shim, G., Kim, G., Lee, S., Lee, H. J., Kim, Y. B., et al. (2015). Image-guided synergistic photothermal therapy using photoresponsive imaging agent-loaded graphene-based nanosheets. Journal of Controlled Release, 211, 28–36, copyright 2015, Elsevier.*

agents have shown tremendous potential for biomedical applications, but their clinical transformation remains a huge challenge.

7 Conclusions

In this chapter, we discussed the synthesis and surface functionalization of graphene and its derivatives, and introduced the progress of graphene-based nanomaterials for cancer theranostics in recent years. Graphene-based nanotheranostic materials as imaging agents show great potential in cancer imaging diagnosis. They have been used for molecular imaging diagnosis by FLI, MRI, PAI, PET and SPECT, etc. (summarized in Table 1). With respect to therapy, graphene-based nanotheranostic materials have been mainly used for drug and gene delivery, PDT, PTT, and synergistic therapy. As the best candidates integrated with diagnostic and therapy, graphene-based, smart nanoprobes have been designed to achieve imaging guided therapy. In addition, the practical application of graphene also faces many challenges, such as biosafety (Seabra, Paula, de Lima, Alves, & Duran, 2014). Although the literature cited above shows that graphene has no obvious toxicity, its physiological toxicity in the long term still needs to be examined.

Conflict of Interest

The authors declare no competing financial interest.

Table 1: Graphene and its directives for cancer theranostics

Materials	Modification	Tissues	Functions	Reference
GO	Poly(vinyl alcohol)	Hind limb	PAI	Toumia et al. (2016)
rGO	Gold nanorod vesicle, ^{64}Cu, DOX	U87MG tumor	PAI, PTT, PET	Song et al. (2015)
GO	Fluorinated polymer	MCF-7 cells, glioma cells	MRI, PAI, ultrasound imaging, PTT	Romero-Aburto et al. (2013)
rGO	Iron oxide nanoparticles, PEG, ^{125}I, Cy5	4T1 murine breast tumor	FLI, PAI, MRI, PTT	Yang et al. (2012)
GO	Folic acid, Pt nanoparticles	MCF-7 cells	Detection of cancer cells	Zhang et al. (2014)
GO, rGO	^{64}Cu, PEG	4T1 tumor	PET	Shi et al. (2017)

Table 1 Graphene and its directives for cancer theranostics—cont'd

Materials	Modification	Tissues	Functions	Reference
rGO	^{131}I, PEG	4T1 tumor	Photothermal imaging, radiotherapy, PTT	Chen, Zhong, et al. (2015)
GO	Hyaluronic acid, Cy3-PNA21	MDA-MB-231 tumor	FLI, genetherapy	Hwang et al. (2017)
rGO	BSA	MCF-7 tumor	Photothermal imaging, PAI, PTT	Sheng et al. (2013)
GO	$BaGdF_5$ NPs, PEG	Hela tumor	CT, MRI, PTT	Zhang et al. (2015)
GO-COOH	Chlorotoxin, DOX	C6 (glioma) cells	Drug delivery	Wang, Gu, et al. (2014)
Graphite	PVP, DOX	NIH-3T3, HCT-116, HeLa, SCC-9 and HEK-293 cells	Drug delivery, ROS-mediated cell apoptosis	Tyagi et al. (2017)
rGO	Hyaluronic acid, ICG	KB tumor	Photothermal imaging, FLI, PTT	Miao et al. (2015)
GO	PEG, Cy5.5, ^{125}I	4T1	FLI, SPECT/CT	Cao et al. (2016)
rGO	FA-PEG-liposome, resveratrol	MCF-7 tumor	Chemotherapy, PTT	Hai et al. (2017)
GO	DK1, CPT	Liver cancer	FLI, drug delivery	Ji et al. (2015)
GO	Bi_2Se_3, PVP	Hela	PAI, CT, PTT	Zhang et al. (2017)
rGO	$ZnFe_2O_4$	U87MG tumor	Magneto-PTT	Akhavan et al. (2014)
GO	Anti-HER2/c-erb-2, S6 aptamer	SKBR3 cells	Raman	Antwi-Boasiako et al. (2017)
GO	Cross-linked polymer	MDA-MB-231 cells	FLI	Khoerunnisa et al. (2017)
GO	Fe_3O_4, lactoferrin, DOX	C6 gliomas	FLI, chemotherapy	Song et al. (2017)
GO	Fe_3O_4, PEI, PEG, ICG	MCF-7 tumor	MRI, PTT	Du et al. (2017)
GO	Peptide-ligand-based VAR probe	Hela cells	FLI	Li et al. (2017)
Graphitic nanocapsule	Boronic-polyethylene	*Helicobacter pylori*	MRI	Li, Hu, et al. (2017)
GO	PEG, CysCOOH	4T1 tumor	PTT, PAI, FLI	Rong et al. (2016)
GO	$Fe\text{-}Pz_2Py$	Fibroblast L929	Drug delivery	Mokdad et al. (2015)
GO	PLL, RGDS, VEGF-siRNA	S189 tumor	Gene delivery	Ren et al. (2017)
GO	Insulin, 2-nitrodopamine-coated magnetic particle	HEK cells	Drug delivery	Turcheniuk et al. (2014)
GO	paclitaxel, ICG, PEG	MG-63 tumor	FLI, chemotherapy	Zhang et al. (2016)
GO	Fe_3O_4, PEG, DOX	CT26 cells	MRI, thermo-chemosensitization	Sasikala et al. (2016)
rGO	ICG, PDA	4T1 tumor	PAI, FLI, PTT	Hu et al. (2016)
GO	FA, PEG, PAH, siRNA	Pancreatic cancer	PTT, gene therapy	Yin et al. (2017)

Acknowledgments

This work was supported by the startup fund from the Shenzhen University, the National Natural Science Foundation of China (31771036, 51703132, 51573096, 81401465), and the Basic Research Program of Shenzhen (JCYJ20170412111100742, JCYJ20160422091238319), Fok Ying-Tong Education Foundation for Young Teachers in the Higher Education Institutions of China (161032).

References

Akhavan, O., & Ghaderi, E. (2013). Graphene nanomesh promises extremely efficient in vivo photothermal therapy. *Small*, *9*, 3593–3601.

Akhavan, O., Ghaderi, E., & Emamy, H. (2012). Nontoxic concentrations of PEGylated graphene nanoribbons for selective cancer cell imaging and photothermal therapy. *Journal of Materials Chemistry*, *22*, 20626–20633.

Akhavan, O., Meidanchi, A., Ghaderi, E., & Khoei, S. (2014). Zinc ferrite spinel-graphene in magneto-photothermal therapy of cancer. *Journal of Materials Chemistry B*, *2*, 3306–3314.

Antwi-Boasiako, A. A., Dunn, D., Dasary, S. S. R., Jones, Y. K., Barnes, S. L., & Singh, A. K. (2017). Bioconjugated graphene oxide-based Raman probe for selective identification of SKBR3 breast cancer cells. *Journal of Raman Specroscopy*, *48*, 1056–1064.

Bao, A., Goins, B., Klipper, R., Negrete, G., Mahindaratne, M., & Phillips, W. T. (2003). A novel liposome radiolabeling method using ^{99m}Tc-"SNS/S" complexes: In vitro and in vivo evaluation. *Journal of Pharmaceutical Sciences*, *92*, 1893–1904.

Bruchez, M., Moronne, M., Gin, P., Weiss, S., & Aliualatos, A. P. (1998). Semiconductor nanocrystals as fluorescent biological labels. *Science*, *281*, 2013–2016.

Cao, T., You, P., Zhou, X., Luo, J., Xu, X., Zhou, Z., et al. (2016). Visualization of size-dependent tumour retention of PEGylated nanographene oxide via SPECT imaging. *Journal of Materials Chemistry B*, *4*, 6446–6453.

Chen, W. H., Wen, X., Zhen, G. R., & Zheng, X. W. (2015). Assembly of Fe_3O_4 nanoparticles on PEG-functionalized graphene oxide for efficient magnetic imaging and drug delivery. *RSC Advances*, *5*, 69307–69311.

Chen, Y. J., Wu, Y. K., Sun, B. B., Liu, S. J., & Liu, H. Y. (2017). Two-dimensional nanomaterials for cancer nanotheranostics. *Small*, *13*, 1603446–1603460.

Chen, Y., Xu, P. F., Shu, Z., Wu, M. Y., Wang, L. Z., Zhang, S. J., et al. (2014). Multifunctional graphene oxide-based triple stimuli-responsive nanotheranostics. *Advanced Functional Materials*, *24*, 4386–4396.

Chen, L., Zhong, X. Y., Yi, X., Huang, M., Ning, P., Liu, T., et al. (2015). Radionuclide ^{131}I labeled reduced graphene oxide for nuclear imaging guided combined radio- and photothermal therapy of cancer. *Biomaterials*, *66*, 21–28.

Choi, S. Y., Baek, S. H., Chang, S. J., Song, Y., Rafique, R., Lee, K. T., et al. (2017). Synthesis of upconversion nanoparticles conjugated with graphene oxide quantum dots and their use against cancer cell imaging and photodynamic therapy. *Biosensors & Bioelectronics*, *93*, 267–273.

Cole, A. J., Yang, V. C., & David, A. E. (2011). Cancer theranostics: The rise of targeted magnetic nanoparticles. *Trends in Biotechnology*, *29*, 323–332.

Coleman, J. N., Lotya, M., O'Neill, A., Bergin, S. D., King, P. J., Khan, U., et al. (2011). Two-dimensional nanosheets produced by liquid exfoliation of layered materials. *Science*, *331*, 568–571.

Du, B. J., Liu, J. H., Ding, G. Y., Han, X., Li, D., Wang, E. K., et al. (2017). Positively charged graphene/Fe_3O_4/polyethylenimine with enhanced drug loading and cellular uptake for magnetic resonance imaging and magnet-responsive cancer therapy. *Nano Research*, *10*, 2280–2295.

Emtsev, K. V., Bostwick, A., Horn, K., Jobst, J., Kellogg, G. L., Ley, L., et al. (2009). Towards wafer-size graphene layers by atmospheric pressure graphitization of silicon carbide. *Nature Materials*, *8*, 203–207.

Feng, L. Z., Yang, X. Z., Shi, X. Z., Tan, X. F., Peng, R., Wang, J., et al. (2013). Polyethylene glycol and polyethylenimine dual-functionalized nano-graphene oxide for photothermally enhanced gene delivery. *Small*, *9*, 1989–1997.

Feng, L. Z., Zhang, S. A., & Liu, Z. A. (2011). Graphene based gene transfection. *Nanoscale*, *3*, 1252–1257.

Gambhir, S. S. (2002). Molecular imaging of cancer with positron emission tomography. *Nature Reviews Cancer*, *2*, 683–693.

Geim, A. K. (2009). Graphene: Status and prospects. *Science*, *324*, 1530–1534.

Geim, A. K., & Novoselov, K. S. (2007). The rise of graphene. *Nature Materials*, *6*, 183–191.

Georgakilas, V., Otyepka, M., Bourlinos, A. B., Chandra, V., Kim, N., Kemp, K. C., et al. (2012). Functionalization of graphene: Covalent and non-covalent approaches, derivatives and applications. *Chemical Reviews*, *112*, 6156–6214.

Goenka, S., Sant, V., & Sant, S. (2014). Graphene-based nanomaterials for drug delivery and tissue engineering. *Journal of Controlled Release*, *173*, 75–88.

Gollavelli, G., & Ling, Y. C. (2012). Multi-functional graphene as an in vitro and in vivo imaging probe. *Biomaterials*, *33*, 2532–2545.

Gollavelli, G., & Ling, Y. C. (2014). Magnetic and fluorescent graphene for dual modal imaging and single light induced photothermal and photodynamic therapy of cancer cells. *Biomaterials*, *35*, 4499–4507.

Hai, L., He, D. G., He, X. X., Wang, K. M., Yang, X., Liu, J. Q., et al. (2017). Facile fabrication of a resveratrol loaded phospholipid@reduced graphene oxide nanoassembly for targeted and near-infrared laser-triggered chemo/photothermal synergistic therapy of cancer in vivo. *Journal of Materials Chemistry B*, *5*, 5783–5792.

Hajipour, M. J., Akhavan, O., Meidanchi, A., Laurent, S., & Mahmoudi, M. (2014). Hyperthermia-induced protein corona improves the therapeutic effects of zinc ferrite spinel-graphene sheets against cancer. *RSC Advances*, *4*, 62557–62565.

Henkelman, R. M., Stanisz, G. J., & Graham, S. J. (2001). Magnetization transfer in MRI: A review. *NMR in Biomedicine*, *14*, 57–64.

Hernandez, Y., Nicolosi, V., Lotya, M., Blighe, F. M., Sun, Z. Y., De, S., et al. (2008). High-yield production of graphene by liquid-phase exfoliation of graphite. *Nature Nanotechnology*, *3*, 563–568.

Hu, D. H., Zhang, J. N., Gao, G. H., Sheng, Z. H., Cui, H. D., & Cai, L. T. (2016). Indocyanine green-loaded polydopamine-reduced graphene oxide nanocomposites with amplifying photoacoustic and photothermal effects for cancer theranostics. *Theranostics*, *6*, 1043–1052.

Huang, X., Qi, X. Y., Boey, F., & Zhang, H. (2012). Graphene-based composites. *Chemical Society Reviews*, *41*, 666–686.

Hummers, W. S., & Offeman, R. E. (1958). Preparation of graphitic oxide. *Journal of the American Chemical Society*, *80*, 1339.

Hwang, D. W., Kim, H. Y., Li, F., Park, J. Y., Kim, D., Park, J. H., et al. (2017). In vivo visualization of endogenous miR-21 using hyaluronic acid-coated graphene oxide for targeted cancer therapy. *Biomaterials*, *121*, 144–154.

Janib, S. M., Moses, A. S., & MacKay, J. A. (2010). Imaging and drug delivery using theranostic nanoparticles. *Advanced Drug Delivery Reviews*, *62*, 1052–1063.

Ji, D.-K., Zhang, Y., Zang, Y., Liu, W., Zhang, X., Li, J., et al. (2015). Receptor-targeting fluorescence imaging and theranostics using a graphene oxide based supramolecular glycocomposite. *Journal of Materials Chemistry B*, *3*, 9182–9185.

Jin, Y. S., Wang, J. R., Ke, H. T., Wang, S. M., & Dai, Z. F. (2013). Graphene oxide modified PLA microcapsules containing gold nanoparticles for ultrasonic/CT bimodal imaging guided photothermal tumor therapy. *Biomaterials*, *34*, 4794–4802.

Johnson, W. K., Stoupis, C., Torres, G. M., Rosenberg, E. B., & Ros, P. R. (1996). Superparamagnetic iron oxide (SPIO) as an oral contrast agent in gastrointestinal (GI) magnetic resonance imaging (MRI): Comparison with state-of-the-art computed tomography (CT). *Magnetic Resonance Imaging*, *14*, 43–49.

Jung, Y., Moon, H. G., Lim, C., Choi, K., Song, H. S., Bae, S., et al. (2017). Humidity-tolerant single-stranded DNA-functionalized graphene probe for medical applications of exhaled breath analysis. *Advanced Functional Materials*, *27*.

Khoerunnisa, Kang, E. B., Mazrad, Z. A. I., Lee, G., In, I., & Park, S. Y. (2017). Preparation of dual-responsive hybrid fluorescent nano probe based on graphene oxide and boronic acid/BODIPY-conjugated polymer for cell imaging. *Materials Science and Engineering: C*, *71*, 1064–1071.

Kim, M. G., Park, J. Y., Miao, W., Lee, J., & Oh, Y. K. (2015). Polyaptamer DNA nanothread-anchored, reduced graphene oxide nanosheets for targeted delivery. *Biomaterials*, *48*, 129–136.

Lalwani, G., Cai, X., Nie, L. M., Wang, L. V., & Sitharaman, B. (2013). Graphene-based contrast agents for photoacoustic and thermoacoustic tomography. *Photoacoustics*, *1*, 62–67.

Li, Y. J., Feng, L. Z., Shi, X. Z., Wang, X. J., Yang, Y. L., Yang, K., et al. (2014). Surface coating-dependent cytotoxicity and degradation of graphene derivatives: Towards the design of non-toxic, degradable nano-graphene. *Small*, *10*, 1544–1554.

Li, Y. J., Hu, X. X., Ding, D., Zou, Y. X., Xu, Y. T., Wang, X. W., et al. (2017). In situ targeted MRI detection of *Helicobacter pylori* with stable magnetic graphitic nanocapsules. *Nature Communications*, *8*, 15653–15663.

Li, Q. R., Jiao, J. B., Li, L. L., He, X. P., Zang, Y., James, T. D., et al. (2017). Graphene oxide-enhanced cytoskeleton imaging and mitosis tracking. *Chemical Communications*, *53*, 3373–3376.

Li, Z. J., Li, C., Zheng, M. G., Pan, J. D., Zhang, L. M., & Deng, Y. F. (2015). Functionalized nano-graphene oxide particles for targeted fluorescence imaging and photothermy of glioma U251 cells. *International Journal of Clinical and Experimental Medicine*, *8*, 1844–1852.

Lin, J., Chen, X. Y., & Huang, P. (2016). Graphene-based nanomaterials for bioimaging. *Advanced Drug Delivery Reviews*, *105*, 242–254.

Liu, X. F., Miller, A. L., Park, S., Waletzki, B. E., Terzic, A., Yaszemski, M. J., et al. (2016). Covalent crosslinking of graphene oxide and carbon nanotube into hydrogels enhances nerve cell responses. *Journal of Materials Chemistry B*, *4*, 6930–6941.

Ma, X. X., Tao, H. Q., Yang, K., Feng, L. Z., Cheng, L., Shi, X. Z., et al. (2012). A functionalized graphene oxide-iron oxide nanocomposite for magnetically targeted drug delivery, photothermal therapy, and magnetic resonance imaging. *Nano Research*, *5*, 199–212.

Meyer, J. C., Geim, A. K., Katsnelson, M. I., Novoselov, K. S., Booth, T. J., & Roth, S. (2007). The structure of suspended graphene sheets. *Nature*, *446*, 60–63.

Miao, W., Shim, G., Kim, G., Lee, S., Lee, H. J., Kim, Y. B., et al. (2015). Image-guided synergistic photothermal therapy using photoresponsive imaging agent-loaded graphene-based nanosheets. *Journal of Controlled Release*, *211*, 28–36.

Mokdad, A., Dimos, K., Zoppellaro, G., Tucek, J., Perman, J. A., Malina, O., et al. (2015). The non-innocent nature of graphene oxide as a theranostic platform for biomedical applications and its reactivity towards metal-based anticancer drugs. *RSC Advances*, *5*, 76556–76566.

Mura, S., & Couvreur, P. (2012). Nanotheranostics for personalized medicine. *Advanced Drug Delivery Reviews*, *64*, 1394–1416.

Novoselov, K. S., Geim, A. K., Morozov, S. V., Jiang, D., Zhang, Y., Dubonos, S. V., et al. (2004). Electric field effect in atomically thin carbon films. *Science*, *306*, 666–669.

Nurunnabi, M., Khatun, Z., Reeck, G. R., Lee, D. Y., & Lee, Y. K. (2014). Photoluminescent graphene nanoparticles for cancer phototherapy and imaging. *ACS Applied Materials & Interfaces*, *6*, 12413–12421.

Orecchioni, M., Cabizza, R., Bianco, A., & Delogu, L. G. (2015). Graphene as cancer theranostic tool: Progress and future challenges. *Theranostics*, *5*, 710–723.

Pei, S. F., & Cheng, H. M. (2012). The reduction of graphene oxide. *Carbon*, *50*, 3210–3228.

Phelps, M. E. (2000). Positron emission tomography provides molecular imaging of biological processes. *Proceedings of the National Academy of Sciences of the United States of America*, *97*, 9226–9233.

Rahman, M., Ahmad, M. Z., Ahmad, J., Firdous, J., Ahmad, F. J., Mushtaq, G., et al. (2015). Role of graphene nano-composites in cancer therapy: Theranostic applications, metabolic fate and toxicity issues. *Current Drug Metabolism*, *16*, 397–409.

Ratajczak, K., & Stobiecka, M. (2017). Ternary interactions and energy transfer between fluorescein isothiocyanate, adenosine triphosphate, and graphene oxide nanocarriers. *The Journal of Physical Chemistry: B*, *121*, 6822–6830.

Ren, L. L., Zhang, Y. F., Cui, C. Y., Bi, Y. Z., & Ge, X. (2017). Functionalized graphene oxide for anti-VEGF siRNA delivery: Preparation, characterization and evaluation in vitro and in vivo. *RSC Advances*, *7*, 20553–20566.

Romero-Aburto, R., Narayanan, T. N., Nagaoka, Y., Hasumura, T., Mitcham, T. M., Fukuda, T., et al. (2013). Fluorinated graphene oxide; a new multimodal material for biological applications. *Advanced Materials*, *25*, 5632–5637.

Rong, P. F., Wu, J. Z., Liu, Z. G., Ma, X. Q., Yu, L., Zhou, K. C., et al. (2016). Fluorescence dye loaded nano-graphene for multimodal imaging guided photothermal therapy. *RSC Advances*, *6*, 1894–1901.

Rong, P., Yang, K., Srivastan, A., Kiesewetter, D. O., Yue, X., Wang, F., et al. (2014). Photosensitizer loaded nano-graphene for multimodality imaging guided tumor photodynamic therapy. *Theranostics*, *4*, 229–239.

Sasikala, A. R. K., Thomas, R. G., Unnithan, A. R., Saravanakumar, B., Jeong, Y. Y., Park, C. H., et al. (2016). Multifunctional nanocarpets for cancer theranostics: Remotely controlled graphene nanoheaters for thermo-chemosensitisation and magnetic resonance imaging. *Scientific Reports*, *6*, 20543–20557.

Seabra, A. B., Paula, A. J., de Lima, R., Alves, O. L., & Duran, N. (2014). Nanotoxicity of graphene and graphene oxide. *Chemical Research in Toxicology*, *27*, 159–168.

Shen, J. M., Huang, G., Zhou, X., Zou, J., Yang, Y., Chen, Y. F., et al. (2014). Safety evaluation of graphene oxide-based magnetic nanocomposites as MRI contrast agents and drug delivery vehicles. *RSC Advances*, *4*, 50464–50477.

Sheng, Z. H., Song, L., Zheng, J. X., Hu, D. H., He, M., Zheng, M. B., et al. (2013). Protein-assisted fabrication of nano-reduced graphene oxide for combined in vivo photoacoustic imaging and photothermal therapy. *Biomaterials*, *34*, 5236–5243.

Shi, S. X., Xu, C., Yang, K., Goel, S., Valdovinos, H. F., Luo, H. M., et al. (2017). Chelator-free radiolabeling of nanographene: Breaking the stereotype of chelation. *Angewandte Chemie International Edition*, *56*, 2889–2892.

Siddique, A., An, J. W., Kim, H. J., Park, H., Lee, G. C., Lee, J. Y., et al. (2017). Chemical modification of graphene oxide through poly(ethylene oxide)-conjugations. *Macromolecular Research*, *25*, 452–460.

Song, Y. J., Qu, K. G., Zhao, C., Ren, J. S., & Qu, X. G. (2010). Graphene oxide: Intrinsic peroxidase catalytic activity and its application to glucose detection. *Advanced Materials*, *22*, 2206–2210.

Song, M. M., Xu, H. L., Liang, J. X., Xiang, H. H., Liu, R., & Shen, Y. X. (2017). Lactoferrin modified graphene oxide iron oxide nanocomposite for glioma-targeted drug delivery. *Materials Science and Engineering: C*, *77*, 904–911.

Song, J., Yang, X., Jacobson, O., Lin, L., Huang, P., Niu, G., et al. (2015). Sequential drug release and enhanced photothermal and photoacoustic effect of hybrid reduced graphene oxide-loaded ultrasmall gold nanorod vesicles for cancer therapy. *ACS Nano*, *9*, 9199–9209.

Stankovich, S., Dikin, D. A., Piner, R. D., Kohlhaas, K. A., Kleinhammes, A., Jia, Y., et al. (2007). Synthesis of graphene-based nanosheets via chemical reduction of exfoliated graphite oxide. *Carbon*, *45*, 1558–1565.

Sun, X. M., Liu, Z., Welsher, K., Robinson, J. T., Goodwin, A., Zaric, S., et al. (2008). Nano-graphene oxide for cellular imaging and drug delivery. *Nano Research*, *1*, 203–212.

Tang, Y. X., Hu, H., Zhang, M. G., Song, J. B., Nie, L. M., Wang, S. J., et al. (2015). An aptamer-targeting photoresponsive drug delivery system using "off-on" graphene oxide wrapped mesoporous silica nanoparticles. *Nanoscale*, *7*, 6304–6310.

Toumia, Y., Domenici, F., Orlanducci, S., Mura, F., Grishenkov, D., Trochet, P., et al. (2016). Graphene meets microbubbles: A superior contrast agent for photoacoustic imaging. *ACS Applied Materials & Interfaces*, *8*, 16465–16475.

Turcheniuk, K., Khanal, M., Motorina, A., Subramanian, P., Barras, A., Zaitsev, V., et al. (2014). Insulin loaded iron magnetic nanoparticle-graphene oxide composites: Synthesis, characterization and application for in vivo delivery of insulin. *RSC Advances*, *4*, 865–875.

Tyagi, N., Attia, N. F., & Geckeler, K. E. (2017). Exfoliated graphene nanosheets: pH-sensitive drug carrier and anti-cancer activity. *Journal of Colloid and Interface Science*, *498*, 364–377.

Wang, H., Gu, W., Xiao, N., Ye, L., & Xu, Q. Y. (2014). Chlorotoxin-conjugated graphene oxide for targeted delivery of an anticancer drug. *International Journal of Nanomedicine*, *9*, 1433–1442.

Wang, C. S., Li, J. Y., Amatore, C., Chen, Y., Jiang, H., & Wang, X. M. (2011). Gold nanoclusters and graphene nanocomposites for drug delivery and imaging of cancer cells. *Angewandte Chemie, International Edition*, *50*, 11644–11648.

Wang, Y., Li, Y. M., Tang, L. H., Lu, J., & Li, J. H. (2009). Application of graphene-modified electrode for selective detection of dopamine. *Electrochemistry Communications*, *11*, 889–892.

Wang, G. S., Ma, Y. Y., Wei, Z. Y., & Qi, M. (2016). Development of multifunctional cobalt ferrite/graphene oxide nanocomposites for magnetic resonance imaging and controlled drug delivery. *Chemical Engineering Journal*, *289*, 150–160.

Wang, X., Wang, Q. X., Wang, Q. H., Gao, F., Gao, F., Yang, Y. Z., et al. (2014). Highly dispersible and stable copper terephthalate metal-organic framework-graphene oxide nanocomposite for an electrochemical sensing application. *ACS Applied Materials & Interfaces*, *6*, 11573–11580.

Wulfkuhle, J. D., Liotta, L. A., & Petricoin, E. F. (2003). Proteomic applications for the early detection of cancer. *Nature Reviews Cancer*, *3*, 267–275.

Xu, Z., Lei, X. L., Tu, Y. S., Tan, Z. J., Song, B., & Fang, H. P. (2017). Dynamic cooperation of hydrogen binding and π stacking in ssDNA adsorption on graphene oxide. *Chemistry—A European Journal*, *23*, 13100–13104.

Xu, M. H., & Wang, L. H. V. (2006). Photoacoustic imaging in biomedicine. *The Review of Scientific Instruments*, *77*, 041101–041122.

Yan, X. F., Hu, H., Lin, J., Jin, A. J., Niu, G., Zhang, S. L., et al. (2015). Optical and photoacoustic dual-modality imaging guided synergistic photodynamic/photothermal therapies. *Nanoscale*, *7*, 2520–2526.

Yan, X. F., Niu, G., Lin, J., Jin, A. J., Hu, H., Tang, Y. X., et al. (2015). Enhanced fluorescence imaging guided photodynamic therapy of sinoporphyrin sodium loaded graphene oxide. *Biomaterials*, *42*, 94–102.

Yang, K., Feng, L. Z., Shi, X. Z., & Liu, Z. (2013). Nano-graphene in biomedicine: Theranostic applications. *Chemical Society Reviews*, *42*, 530–547.

Yang, K., Hu, L. L., Ma, X. X., Ye, S. Q., Cheng, L., Shi, X. Z., et al. (2012). Multimodal imaging guided photothermal therapy using functionalized graphene nanosheets anchored with magnetic nanoparticles. *Advanced Materials*, *24*, 1868–1872.

Yang, Y., Shi, H. L., Wang, Y. P., Shi, B. Z., Guo, L. L., Wu, D. M., et al. (2016). Graphene oxide/manganese ferrite nanohybrids for magnetic resonance imaging, photothermal therapy and drug delivery. *Journal of Biomaterials Applications*, *30*, 810–822.

Yin, F., Hu, K., Chen, Y. Z., Yu, M. Y., Wang, D. Y., Wang, Q. Q., et al. (2017). SiRNA delivery with PEGylated graphene oxide nanosheets for combined photothermal and genetherapy for pancreatic cancer. *Theranostics*, *7*, 1133–1148.

Zhang, M., Bai, L. L., Shang, W. H., Xie, W. J., Ma, H., Fu, Y. Y., et al. (2012). Facile synthesis of water-soluble, highly fluorescent graphene quantum dots as a robust biological label for stem cells. *Journal of Materials Chemistry*, *22*, 7461–7467.

Zhang, L. N., Deng, H. H., Lin, F. L., Xu, X. W., Weng, S. H., Liu, A. L., et al. (2014). In situ growth of porous platinum nanoparticles on graphene oxide for colorimetric detection of cancer cells. *Analytical Chemistry*, *86*, 2711–2718.

Zhang, X. Y., Li, H. P., Cui, X. L., & Lin, Y. H. (2010). Graphene/TiO_2 nanocomposites: Synthesis, characterization and application in hydrogen evolution from water photocatalytic splitting. *Journal of Materials Chemistry*, *20*, 2801–2806.

Zhang, C., Lu, T., Tao, J. G., Wan, G., & Zhao, H. X. (2016). Co-delivery of paclitaxel and indocyanine green by PEGylated graphene oxide: A potential integrated nanoplatform for tumor theranostics. *RSC Advances*, *6*, 15460–15468.

Zhang, H. F., Maslov, K., Stoica, G., & Wang, L. H. V. (2006). Functional photoacoustic microscopy for high-resolution and noninvasive in vivo imaging. *Nature Biotechnology*, *24*, 848–851.

Zhang, H., Wu, H. X., Wang, J., Yang, Y., Wu, D. M., Zhang, Y. J., et al. (2015). Graphene oxide-$BaGdF_5$ nanocomposites for multi-modal imaging and photothermal therapy. *Biomaterials*, *42*, 66–77.

Zhang, S. A., Yang, K., Feng, L. Z., & Liu, Z. (2011). In vitro and in vivo behaviors of dextran functionalized graphene. *Carbon*, *49*, 4040–4049.

Zhang, J. L., Yang, H. J., Shen, G. X., Cheng, P., Zhang, J. Y., & Guo, S. W. (2010). Reduction of graphene oxide via L-ascorbic acid. *Chemical Communications*, *46*, 1112–1114.

Zhang, Y. X., Zhang, H., Wang, Y. K., Wu, H. X., Zeng, B., Zhang, Y. J., et al. (2017). Hydrophilic graphene oxide/bismuth selenide nanocomposites for CT imaging, photoacoustic imaging, and photothermal therapy. *Journal of Materials Chemistry B*, *5*, 1846–1855.

Zheng, X. W., Chen, W. H., Cui, P., Wang, Z. M., & Zhang, W. (2014). Design of multifunctional FePt/GO nanocomposites for targeting, dual-modal imaging diagnostic and in situ therapeutic potential theranostic platform. *RSC Advances*, *4*, 58489–58494.

Zhong, M., Liu, Y. T., & Xie, X. M. (2015). Self-healable, super tough graphene oxide-poly(acrylic acid) nanocomposite hydrogels facilitated by dual cross-linking effects through dynamic ionic interactions. *Journal of Materials Chemistry B*, *3*, 4001–4008.

Zhou, Y., Jing, X. X., & Chen, Y. (2017). Material chemistry of graphene oxide-based nanocomposites for theranostic nanomedicine. *Journal of Materials Chemistry B*, *5*, 6451–6470.

Zhu, Y. W., Murali, S., Cai, W. W., Li, X. S., Suk, J. W., Potts, J. R., et al. (2010). Graphene and graphene oxide: Synthesis, properties, and applications. *Advanced Materials*, *22*, 3906–3924.

Further Reading

Novoselov, K. S., Fal'ko, V. I., Colombo, L., Gellert, P. R., Schwab, M. G., & Kim, K. (2012). A roadmap for graphene. *Nature*, *490*, 192–200.

CHAPTER 5

Applications of Plasmonic Nanomaterials for Phototriggered Theranostics

Cristina Rodriguez-Quijada, Kimberly Hamad-Schifferli
University of Massachusetts, Boston, MA, United States

1 Introduction

Nanotechnology has held great promise for cancer therapy and diagnosis. In particular, plasmonic nanomaterials such as gold and silver nanoparticles (NPs) have been attractive for cancer applications because of their unique optical properties (Lee & Lee, 2010). The ability to externally address NPs has effectively given us a nanoscale "handle" by which we can control biological function. This phenomenon has allowed spatial and temporal control over release from a NP. Since it was discovered that laser excitation of gold NPs can trigger release of conjugated species or generate heat, there has been exciting progress in developing different ways to exploit this process for cancer therapy. We have seen numerous novel implementations, such as triggered release of siRNA or drugs from the gold NPs themselves or from liposomes, polymeric and porous silica NPs (Agarwal, Mackey, El-Sayed, & Bellamkonda, 2011; Grobmyer & Moudgil, 2010; Liu et al., 2017; Yamashita et al., 2011; Zhang et al., 2012). NPs can be used to target a site and control the rate of release of cargo molecules from the particles, and thus, be used to improve the efficacy of drugs by reducing toxicity to healthy cells and tissues. In addition, NP laser excitation has been used to specifically heat the NPs, inducing damage to a tissue at the site of the photoexcitation (Pissuwan, Valenzuela, & Cortie, 2006). Photoexcitation of gold NPs has also led to novel photothermal and photodynamic therapies (PTT and PDT) and tumor ablation strategies (Fig. 1).

Since its inception, the field of phototriggered excitation has reached a level of technical maturity and thus, has emerged from a wide array of initial challenges. Many of these challenges were in the area of interfacing the nanomaterial to the biological molecule or system. However, our knowledge of nano-bio interfaces and how to control them effectively has met great progress. We will discuss the different ways that the field has overcome challenges in using gold NPs in biological environments. Nevertheless, this field is still an emerging area, with new innovations being developed. Therefore, the purpose of this chapter is to provide a review of the current status of state-of-the-art in the field, and point to where

Handbook of Nanomaterials for Cancer Theranostics. https://doi.org/10.1016/B978-0-12-813339-2.00005-0

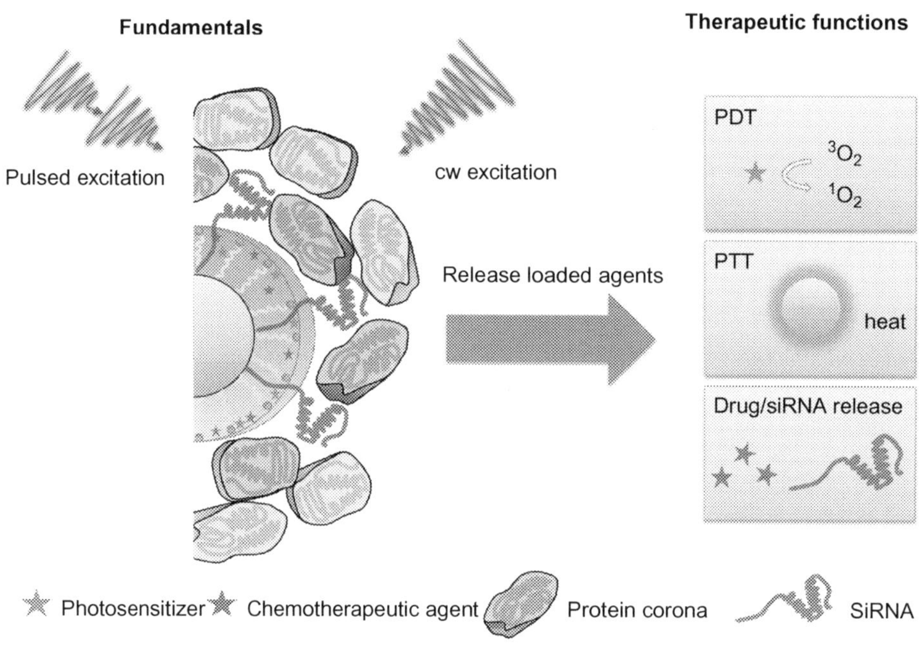

Fig. 1
Overview of the aspects of phototriggered release to be discussed.

it is going. In addition, the goal of this work is to give guidelines for using NPs in triggered release and/or triggered excitation applications. Moving forward, best practices will help to determine ways to optimally utilize NPs for cancer nanomedicine and improve its prospects for the future.

2 Mechanism of Photoexcitation of Plasmonic NPs

Plasmonic nanomaterials such as gold or silver NPs possess a surface plasmon resonance (SPR), which is due to the oscillation of the electrons in the NP. The electronic oscillation gives rise to a strong optical absorption which typically occurs in the visible spectrum. Consequently, NPs possess a strong visible color, which has made them useful in a variety of colorimetric applications (Dreaden, Austin, Mackey, & El-Sayed, 2012; Fig. 2A and B).

Furthermore, the peak wavelength of the SPR is dependent on the NP size and shape. Therefore, by simply changing NP shape and size via chemical synthesis conditions, their optical properties can be varied, resulting in dramatic color changes in the samples (Fig. 2C). This has been useful for shifting the excitation to coincide with the tissue window in the near infrared (NIR), where tissue absorbs minimally, 800–1100 nm. By doing so, through-tissue applications such as triggered release in vivo and photoacoustic imaging in tissue are possible (Eustis & El-Sayed, 2006; Tam et al., 2010). Simulations based on Mie theory have been able to predict plasmon shifts with NP shape, which has enabled the design

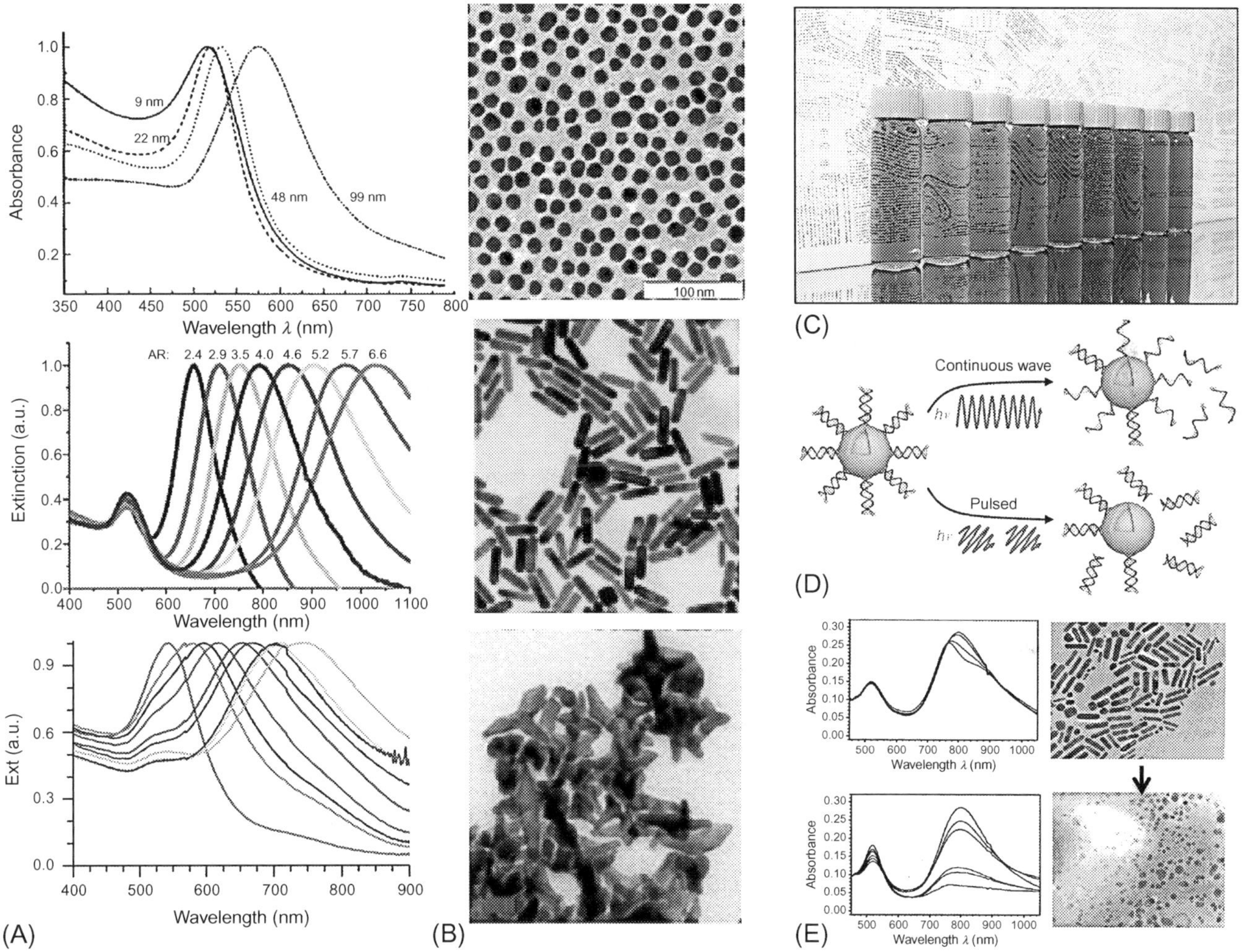

Fig. 2

Types of gold NPs used in phototriggered release. (A) Optical absorption spectra of spherical gold NPs, nanorods, and nanostars can be tuned via tuning the particle size and shape. From Link, S. & El-Sayed, M. A. (2000). Spectral properties and relaxation dynamics of surface plasmon electronic oscillations in gold and silver nanodots and nanorods. *Journal of Physical Chemistry B, 103* (40), 8410–8426; Huang, X. & El-Sayed, M. A. (2010). Gold nanoparticles: optical properties and implementations in cancer diagnosis and photothermal therapy. *Journal of Advanced Research 1* (1), 13–28; de Puig, H., Tam, J. O., Yen, C. W., Gehrke, L., & Hamad-Schifferli, K. (2015). Extinction coefficient of gold nanostars. *Journal of Physical Chemistry. C, Nanomaterials and Interfaces, 119* (30), 17408–17415. (B) Different shapes of soluble gold nanostructures that can be synthesized. From Chen, C.-K., Huang, C.-C., & Chang, H.-T. (2010). Label-free colorimetric detection of picomolar thrombin in blood plasma using a gold nanoparticle-based assay. *Biosensors and Bioelectronics, 25* (8), 1922–1927; Huang, X. & El-Sayed, M. A. (2010). Gold nanoparticles: optical properties and implementations in cancer diagnosis and photothermal therapy. *Journal of Advanced Research 1* (1), 13–28; de Puig, H., Tam, J. O., Yen, C. W., Gehrke, L., & Hamad-Schifferli, K. (2015). Extinction coefficient of gold nanostars. *Journal of Physical Chemistry. C, Nanomaterials and Interfaces, 119* (30), 17408–17415. (C) The solution of gold NPs can be made to have strong colors that span the visible. Image courtesy of Carre-Camps and de Puig. (D) Mechanism of heating for pulsed versus cw laser heating of gold nanoshells coated with DNA. From Goodman, A. M., Hogan, N. J., Gottheim, S., Li, C., Clare, S. E., & Halas, N. J. (2017). Understanding resonant light-triggered DNA release from plasmonic nanoparticles. *ACS Nano, 11* (1), 171–179. (E) Gold nanorod melting upon laser excitation. From Link, S., Burda, C., Nikoobakht, B., & El-Sayed, M. A. (2000). Laser-induced shape changes of colloidal gold nanorods using femtosecond and nanosecond laser pulses. *Journal of Physical Chemistry B, 104* (26), 6152–6163.

of NPs with optimal absorption peaks. Several different types of plasmonic NPs have been used for phototriggered applications, some of which are shown in Table 1.

For nanomedicine applications, NPs of a variety of shapes have been studied, ranging from spheres, rods, stars, cubes, hollow cages, bones, rice, and numerous others. Advances in gold NP synthesis techniques has enabled tuning of many shapes, sometimes with high precision. In addition, for these biocompatible for biological and biomedical applications, NPs must be stable in biological buffers, solutions, and fluids such as blood, serum, and cell media. Furthermore, they must be able to be conjugated to biomolecules such as proteins or DNA, either by covalent or non-covalent attachments (Aubin-Tam & Hamad-Schifferli, 2008).

The most studied NPs outside of spherical gold NPs include gold nanorods (NRs), rod-shaped particles (Alkilany, Thompson, Boulos, Sisco, & Murphy, 2012). NRs have both a transverse and longitudinal SPR, the latter of which is sensitive to the NR aspect ratio (AR). Tuning the AR from 1 to 6 or greater can shift the SPR from ~520 nm to >1000 nm, which is a broad dynamic range spanning the visible to the NIR spectrum (Gou & Murphy, 2005; Lee & El-Sayed, 2005; Sau & Murphy, 2004; Wijaya & Hamad-Schifferli, 2008).

Gold nanoshells (Barhoumi, Huschka, Bardhan, Knight, & Halas, 2009; Huschka et al., 2011), thin hollow Au shells or templated onto SiO_2 NPs, also have been interesting for cancer therapy as PTT agents. Their SPR is tunable from 650 to 950 nm by changing the dimensions of the shell relative to the SiO_2 core. Moreover, this shift can be predicted by theory, facilitating nanoshell design and optimization (Prodan, Radloff, Halas, & Nordlander, 2003).

Gold nanostars, or star-shaped gold NPs, are also attractive for phototriggered release as they also have tunable absorption spectra in the visible and NIR, but their synthesis is much easier than gold NRs (Barbosa et al., 2010; Dam, Culver, & Odom, 2014; Xie, Zhang, Lee, & Wang, 2008). One attractive synthesis is a one-pot approach using gold salts ($HAuCl_4$) and a Goods buffer. Changing the Au:buffer ratio changes the arm length of the nanostar, which results in a tunable SPR. Resulting gold nanostars are water soluble and stable in buffer, which greatly facilitates their incorporation into biological environments and conjugation to biomolecules.

The ability to tune the SPR with the NP shape and size has enabled selective excitation (de Puig, Cifuentes Rius, Flemister, Baxamusa, & Hamad-Schifferli, 2013; Lee et al., 2012; Lee, Liu, Kim, & Lee, 2009; Wijaya, Schaffer, Pallares, & Hamad-Schifferli, 2009), where different NPs can be excited by lasers with different wavelengths. This has been used for selective release of DNA, where two different NRs that absorb at different SPR wavelengths can be excited by different lasers (Wijaya et al., 2009).

There have been many studies of the mechanism of triggered NP melting, triggered release, and photothermal heating of gold NPs under laser excitation. Excitation of the gold NP at the SPR has been achieved by both continuous wave (cw) lasers and also pulsed laser excitation in the femtosecond to nanosecond timescale. Each type of excitation results in a distinct release mechanism (Fig. 2D).

Table 1: Summary of plasmonic nanoparticles used for phototriggered release for different applications

NP Type	Surface Modification/ Chemistry Used	Target	Disease	References	Results/Observations
Gold nanorods	dsDNA coating obtained by a multiple layer process	Tumors from mouse rectum carcinoma colon-26 cells injected in male BALB/c mice	Rectum carcinoma	Yamashita et al. (2011)	Photothermal release of ssDNA from dsDNA modified gold nanorods is achieved in tumors in mice
	PEGylated gold nanorods	Mouse tumor model of human glioblastoma U87	Glioblastoma	Agarwal et al. (2011)	PEGlyated nanorods trigger the photothermal release of doxorubicin encapsulated in thermosensitive liposomes
	Mesoporous silica-coated gold nanorods	In vitro studies with human lung adenocarcinoma epithelial cell line A549	Lung adenocarcinoma	Zhang et al. (2012)	Subcellular localization of nanoparticles was achieved with two-photon imaging. NRI laser irradiation at low intensity triggered the release of DOX and at higher intensities triggered the hyperthermia of AuNR
	Protein corona loaded with thrombin binding aptamer TBA aptamer or antidote	N/A	N/A	de Puig et al. (2013)	Selective light-triggered release of TBA or the antidote loaded in two nanorods with different SPR wavelengths to switch on and off the blood clotting
	IkB siRNA and p65 siRNA coated in two different gold nanorods	Human cervical carcinoma cell line HeLa	Cervical carcinoma	Lee et al. (2012)	Selective light-triggered release of two strands of siRNA loaded in two NR with different SPR peaks to manipulate the gene circuits
Gold nanostars	$CaCO_3$ coating loaded with indocyanine green (ICG)	(in vitro and in vivo) Human gastric carcinoma MGC-803 cell line injected subcutaneously in female or male BALB/c mice	Gastric carcinoma	Liu et al. (2017)	Combination of PDT (ICG) and PTT(Nstar) was superior than solely PDT or PTT. The platform also pH-triggered the drug release due to the $CaCO_3$ coating which allows the tumor imaging and bio-distribution of the nanomaterial
	Thiolated anti-nucleolin aptamer AS1411 covalently attached to gold nanostars	In vitro studies with 12 cancer cell lines	Multiple cancer cell lines	Dam et al. (2014)	Light-triggered release of the aptamer loaded in AuNStar enhanced the in vitro anticancer effect in all cancer cell lines
Nanocages	Coating of nanocages loaded with three effectors with the thermosensitive polymer pNIPAAm	In vitro studies with human breast cancer cell line SK-BR-3	Breast adenocarcinoma	Yavuz et al. (2009)	Photothermal release of precursors driven by the conformational change of the thermosensitive polymer coating
Gold nanoshells	Thiolated dsDNA covalently attached to the gold surface	N/A	N/A	Barhoumi et al. (2009)	Thermal and plasmon resonant light-triggered dehybridization of dsDNA was characterized

Cw laser excitation of the NP at its SPR is strongly absorbed, and the absorbed energy is converted into heat. This results in a temperature increase of the gold NP, which is then transferred to its surroundings. Heating is a collective effect, so the temperature rise is dependent on the NP concentration. Generally, temperature increases of 42°C are desirable for hyperthermia. Because cw excitation also heats the surroundings, selective excitation of two NPs may seem limited. However, there have been several demonstrations of successful selective release of two different species (Lee et al., 2012), most likely because the NPs are not in proximity. Cw lasers are small, inexpensive, and compatible with fiber optics, making them amenable for portable uses.

The other type of excitation used for gold NPs is ultrafast, short-pulsed excitation at the SPR. Pulsed excitation induces melting of the NR, resulting in rapid heating and then melting of the NP depending on the laser power, as was first observed by Link, Burda, Nikoobakht, and El-Sayed (2000). First, the electron-electron energy transfer occurs on the fastest timescale, then the electron-phonon transfer, and finally the heat is transferred to the NP surroundings, which occurs on the timescale of picoseconds. Because this is faster than the timescale of the excitation, this results in thermal confinement, where the heating is localized to the vicinity of the NPs. Thus, the temperature of the NP can increase high enough to induce NP melting (Huettmann, Radt, Serbin, & Birngruber, 2003). For NRs, this results in a shape change from rods to spheres and a change in their absorption spectrum (Fig. 2E; Horiguchi, Honda, Kato, Nakashima, & Niidome, 2008). The excitation can result in a hot electron, which can break the Au-thiol bond for molecules that are conjugated to the NP surface, resulting in triggered release. This strategy has been used for triggered release of thiolated DNA on gold NPs. Ultrafast excitation can also release molecules that are adsorbed onto the surface by accelerating exchange off the particle surface (Alper, Crespo, & Hamad-Schifferli, 2009). In contrast to cw excitation, the ultrafast pulsed excitation is a single particle effect and thus independent of particle concentration. For the most part, it does not result in a temperature increase of the surroundings. This property can facilitate selective excitation of multiple NPs, as cross-talk would be minimized. Unfortunately, the equipment for ultrafast pulsed lasers is much larger than cw lasers, and at this time not portable or easily delivered via a fiber optic. Because of this, their utility in fieldable applications is somewhat limited.

There have been studies of the two mechanisms of release of DNA from gold nanoshells, with discussion of the advantages and disadvantages of each. Goodman et al. (Barhoumi et al., 2009; Huschka et al., 2011) make recommendations for based on these respective mechanisms (Goodman et al., 2017). Because cw excitation results in heating of the surrounding tissue, it could result in undesirable side effects, such as damage to healthy tissue. Consequently, ultrafast pulsed excitation offers distinct advantages over cw laser excitation due to its selectivity.

3 Nano-Bio Interfaces

In addition to the physical phenomenon of heating, the gold NPs must also be linked to or introduced into a biological environment for the aforementioned purposes. The major challenge of utilizing nanomaterials in biological applications is the nano-bio interface (Nel et al., 2009; Park & Hamad-Schifferli, 2010). Complicating issues including protein adsorption, protein corona formation, and protein denaturation on the NP surface have all been difficult to overcome. Initially, the nano-bio interface issues presented major challenges for phototriggered release or PTTs that use NPs. However, nano-biological interactions have been the subject of numerous studies, and our understanding of and control over the nano-bio interface has advanced considerably in recent years. In this section, we will describe the basic fundamentals of these interface effects and how they impact the use of nanomaterials for medical applications.

3.1 Protein Corona Formation Around NPs

Nanomaterial targeting to the tumor is desirable; thus, strategies for nanomedicine applications have typically focused on the decoration of the nanomaterial surface with tumor binding peptides, antibodies, or other targeting species. Nevertheless, studies have shown that the nanomaterial surface changes when introduced into the biological fluids (Tenzer et al., 2013). Macromolecules present at high concentration in the biological milieu, such as proteins and lipids, adsorb onto the NP forming a cloud known as the "protein corona." The species covalently attached to the NP surface can be obscured by the protein corona, and thus cannot be "seen" by the targeted cell or tumor tissue, causing nonspecific uptake or biodistribution (Fig. 3A and B; Ritz et al., 2015; Walczyk, Bombelli, Monopoli, Lynch, & Dawson, 2010). This can decrease the therapeutic efficacy of the nanomaterial and/or lead to negative safety and toxicological implications (Mirshafiee, Mahmoudi, Lou, Cheng, & Kraft, 2013).

The composition and behavior of the layers in the protein corona are difficult to predict. Blood, serum, and biological media are incredibly complex, with thousands of different protein species and small molecules, so the resulting corona is also complex. Because the biocorona is held together by weak noncovalent bonds, the proteins in the corona are constantly exchanging with proteins in the surroundings (Cedervall et al., 2007). Hence, its composition is continuously evolving due to this dynamic exchange (Capriotti et al., 2014). Also, when the NP transitions from one environment to another, such as moving from outside of a cell to inside one, the corona composition also changes, but still retains a "fingerprint" of its previous environment (Casals, Pfaller, Duschl, Oostingh, & Puntes, 2010; Lundqvist et al., 2011). All of these factors have contributed to a lack of understanding of how protein coronas behave.

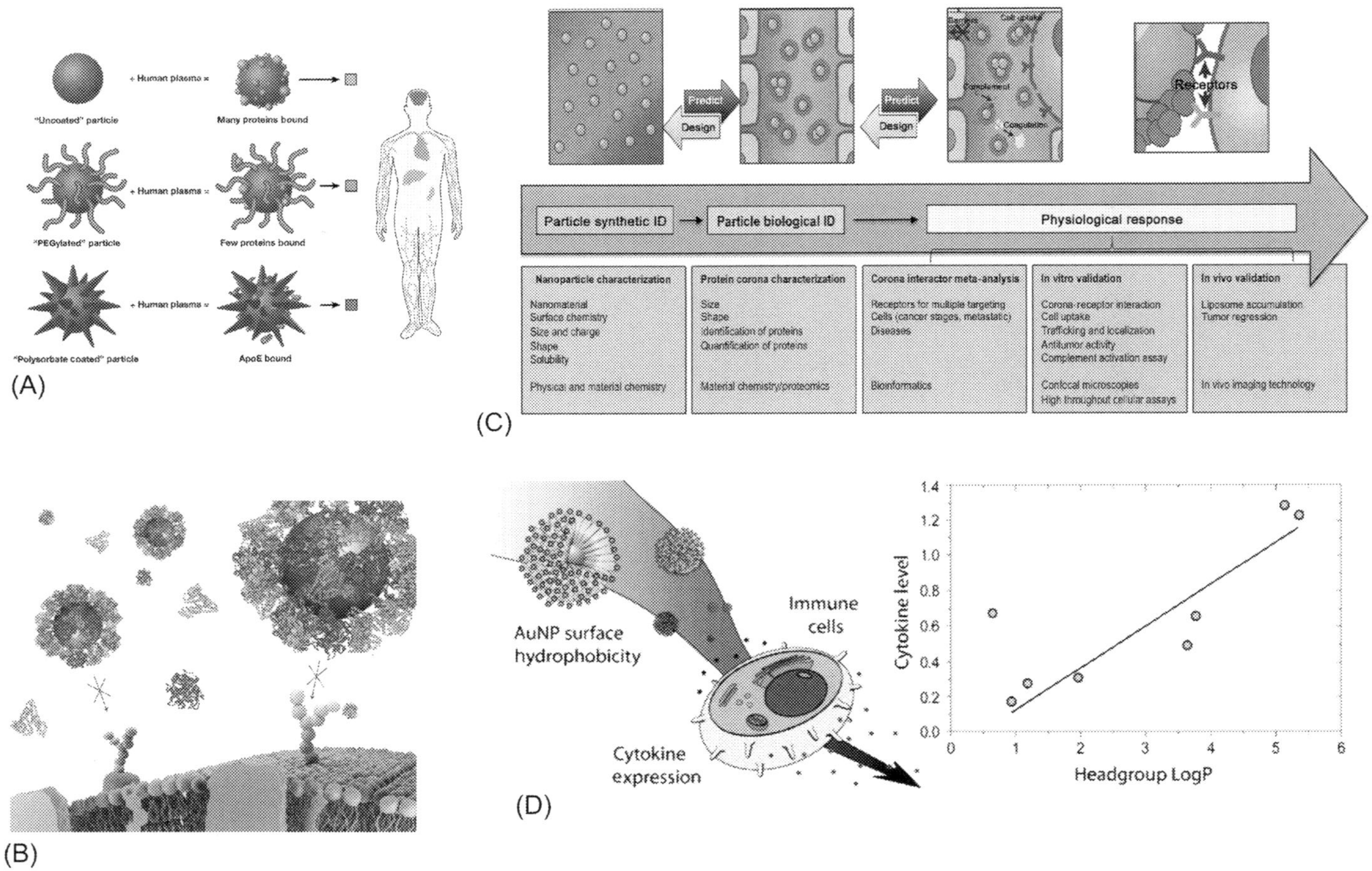

Fig. 3

Challenges of the nano-bio interface. (A) Protein coronas can form around NPs, the composition of which changes for NPs with different coatings. This can impact their biodistribution. From Aggarwal, P., Hall, J. B., McLeland, C. B., Dobrovolskaia, M. A., & McNeil, S. E. (2009). Nanoparticle interaction with plasma proteins as it relates to particle biodistribution, biocompatibility and therapeutic efficacy. *Advanced Drug Delivery Reviews, 61*(6), 428–437. (B) Protein coronas can decrease the specificity of ligand-receptor interactions on the cell surface. From Gaspar, R. (2013). Nanoparticles: pushed off target with proteins. *Nature Nanotechnology, 8* (2), 79–80. (C) NPs have a chemical, or "synthetic" identification, as well as a "biological" identity. From Caracciolo, G. (2015). Liposome-protein corona in a physiological environment: challenges and opportunities for targeted delivery of nanomedicines. *Nanomedicine, 11* (3), 543–557. (D) The immune response is impacted by NP surface chemistry. From Moyano, D. F., Goldsmith, M., Solfiell, D. J., Landesman-Milo, D., Miranda, O. R., Peer, D., et al. (2012). Nanoparticle hydrophobicity dictates immune response. *Journal of the American Chemical Society, 134* (9), 3965–3967.

Nevertheless, the field has made important findings about the fundamental properties of protein coronas. The abundance of certain proteins in a corona depends on their affinity for the NP surface, which impacts their adsorption lifetime (Docter et al., 2014). Two types of protein coronas can be distinguished based on protein affinity: hard corona proteins versus soft corona proteins. The hard corona proteins are thought to adsorb more strongly and thus, reside for longer times on the surface of the NP. The other type of proteins, known as the soft corona, have weaker binding affinities, so they exchange more rapidly with free macromolecules in the surrounding biological milieu. It is thought that the soft corona proteins tend to interact with the hard corona instead of the NP surface itself. Because of these exchange effects, the composition of the protein corona changes over time, which can result in a changing and dynamic biological performance of the nanomaterial (Aggarwal, Hall, McLeland, Dobrovolskaia, & McNeil, 2009; Darabi Sahneh, Scoglio, & Riviere, 2013).

The resulting composition of the protein corona not only depends on the protein species found in a specific environment, but also the physicochemical properties of the nanomaterial. Recent studies have shown that protein corona formation depends on the material, size and shape of the carrier (Lundqvist et al., 2008). In vitro studies showed that protein coronas formed around silica and sulfonate polystyrene NP had different corona compositions regardless of the human serum concentration (Monopoli et al., 2011). Going a step further, Candace et al. demonstrated that the surface chemistries affect also the secondary structure of proteins adsorbed (Fleischer & Payne, 2014). They determined that the epitope exposure was altered, and thus the binding with membrane receptors led to different cell uptakes. Therefore, it is expected that the coating strategies used in PDT for targeting, drug encapsulation, and the use of photosensitizers will also be affected by the protein corona composition and epitope exposure.

The protein corona impacts any nanomaterial introduced into a biological fluid, including plasmonic nanomaterials for phototriggered release, which rely on cellular uptake or tumor targeting of gold NPs. Evidence is increasingly pointing towards the notion that it is the protein corona which ultimately defines the biological performance of the NP (Ritz et al., 2015). Therefore, it is essential to investigate the mechanisms of protein corona interactions with cells to better address the biological behavior of the nanomaterial and its toxicology implications.

3.2 Protein Corona Defines the Biological Identity of a Nanomaterial

Many factors affected by the presence of this spontaneous coating will define the biological identity of the nanomaterial in vivo. Protein coronas can affect the biodistribution, cell uptake and immunological responses that will define the fate of the NP. Recent studies have explored the implications of this protein cloud, which may elucidate better ways to engineer this spontaneous formation to handle the biological performance of the nanomaterial (Fig. 3C).

Recent studies now establish the protein corona as the critical factor that determines the fate of NPs in cell association and uptake in vitro, as determined by a thorough analysis of the protein corona "fingerprint" by mass spectrometry (Salvati et al., 2013; Walkey et al., 2014). These differences in cell uptake influence the biodistribution of a nanomaterial (Wilhelm et al., 2016). Lesniak et al. (2012) showed that different compositions of protein coronas formed on silica-based nanomaterials by incubation with two cell culture medias led to distinct cell uptakes. Also, the presence of apolipoproteins in the protein corona have been shown to give rise to longer residence times in the blood stream as they prevent triggering of the mononuclear phagocytotic system. This was also observed in sulfonate or amino-functionalized NPs that enhance the apolipoprotein adsorption leading to a lower uptake in cells (Ritz et al., 2015). Therefore, the biological performance of the nanomaterial is dictated by the proteins adsorbed to their surface much more than the bare NP itself. Attempts to generate mathematical models that predict the biological performance of the nanomaterials have been previously reported (Dell'Orco, Lundqvist, Oslakovic, Cedervall, & Linse, 2010; Walkey et al., 2014). In vitro experiments have also explored the protein corona impact on drug delivery NPs, and have shown a diminished binding of the nanocarrier to endothelial cells in the presence of protein corona (Sobczynski et al., 2014). Proteins in the corona can also trigger the immunological cascades of the host, what dictates the fate of the nanomaterial. This phenomenon is due to the adsorption of certain proteins and products that will activate the complement cascades and alert the immune system about the presence of the nanomaterial (Dobrovolskaia, Aggarwal, Hall, & McNeil, 2008; Dobrovolskaia et al., 2009). The coating strategies used in the NP for PDT affect the immunological response of the host. This can be exemplified with the different gene expressions of cytokines that were reported depending on the hydrophobicity of the nanomaterial by Moyano et al. (2012) (Fig. 3D).

Taking the aforementioned aspects into consideration, the synthesis of nanomaterials for PDT should explore the chemical surfaces that lead to a protein corona that do not compromise its targeting and efficacy. Recently, there have been several ways to exploit the use of the protein corona that may open a whole different understanding towards the use of NPs in the clinical field (Hamad-Schifferli, 2015; Yallapu et al., 2015).

4 Examples of Phototriggered Release Applications in Nanomedicine

Gold NPs have been used in numerous different ways for phototriggered therapies. This has been partly facilitated by the fact that their surface chemistry is the most versatile compared to other inorganic materials, and thus, their interface issues are the easiest to address. The basic principle behind phototriggered therapy is to use laser light at the NP SPR to trigger release of a drug, or generate heat, providing temporal and spatial control over therapy. Embodiments include triggered release from smart polymers on gold nanocages (Yavuz et al., 2009), release of drugs from liposomes loaded with both the drug and the gold NRs

(Agarwal et al., 2011), or release of nucleic acids (DNA, siRNA) for control of gene expression (Barhoumi et al., 2009; Braun et al., 2009). Combination of PTT with optical imaging of gold nanoshells has also been explored for cancer therapy (Gobin et al., 2007). Protein coronas can even be exploited for applications in triggered release, where coronas are purposely formed to act as a sponge for a drug that can be released from the corona upon light excitation (Cifuentes-Rius, de Puig, Kah, Borros, & Hamad-Schifferli, 2013; de Puig et al., 2013; Kah, Chen, Zubieta, & Hamad-Schifferli, 2012).

There have been some innovative approaches to using phototriggered release for different application which either utilize the interface, the material properties of the gold nanomaterial, or combine a phototriggered response into a combination therapy. This is not a comprehensive list, but is intended only to give examples that illustrate some of the numerous ways that phototriggered responses have been utilized in therapies.

4.1 Using the Protein Corona Around Gold NRs to Improve Combined PDT and PTT

Yeo et al. (2017) utilize the protein corona made of human serum proteins that form around gold NRs and take advantage of the corona proteins to load it with the hydrophobic photosensitizer chlorin e6 (Ce6). They strategically use NRs with a longitudinal surface plasmon resonance (LSPR) that matches the excitation of Ce6 Q band at 665 nm, a strategy used for combined PDT and PTT (Fig. 4A). The properties of the corona and NR work synergistically to kill cancer cells, where laser excitation of the NR is used for photothermal ablation, while release of Ce6 increases cell kill through oxidative stress. NR-human serum (HS) coronas have a high payload capacity due to their high surface area, and can load more Ce6 than the HS alone. They find that with oral squamous cell carcinoma cells, the NR-corona also results in an increased cell uptake of Ce6. The NR-HS-Ce6 can enhance ROS production due to the field enhancement from the excitation of the NRs. Thus, the author has determined a way to utilize the properties of protein coronas to enhance combined PTT and PDT.

4.2 Dynamic Control of Gene Expression

Another key result in using triggered release from gold NPs is where Lee et al. (2012) use gold NR "nanoantennas" to dynamically control gene circuits (Fig. 4B). They exploit the properties of gold NRs for precise spatiotemporal manipulation of gene expression. They utilize the fact that NRs can be made to have different SPR wavelengths and thus can be excited by different lasers, where this was used initially as a proof of principle experiment for selective release of two different strands of DNA on the NRs (Wijaya et al., 2009). In addition, selective release was utilized for controlling blood clotting, where one NR was decorated with a DNA aptamer that can inhibit thrombin and thus the blood clotting cascade, and then the other was decorated with its complement, which prevents aptamer binding to thrombin (de Puig et al., 2013). Thus, the two NRs can be used to switch blood clotting on and then off, where release from a single NR would not make this possible.

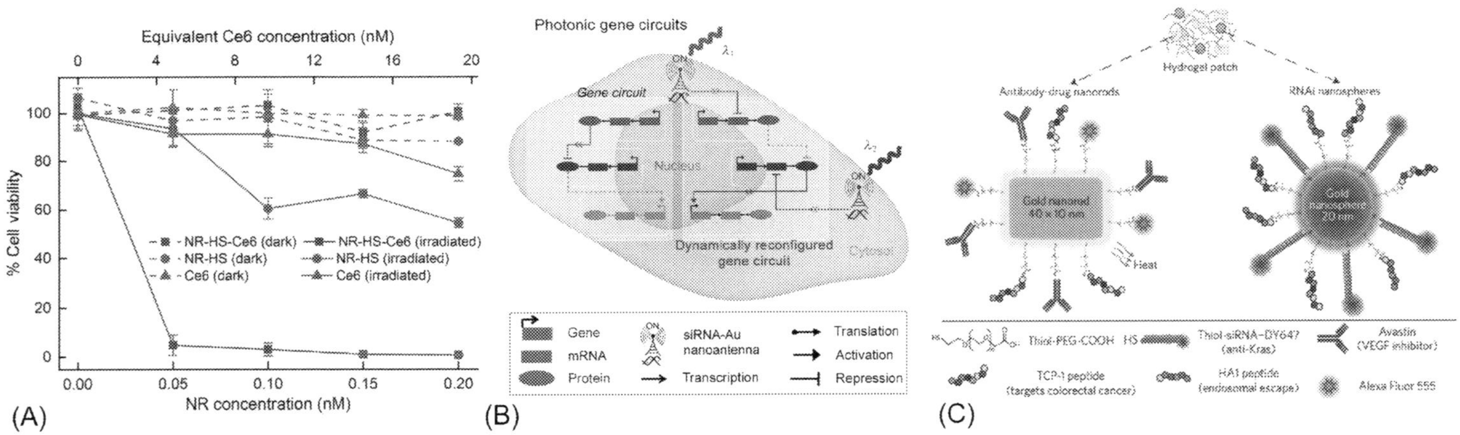

Fig. 4

Different modes of use of phototriggered release. (A) Using protein coronas for combined photothermal therapy and release. From Yeo, E. L. L., Cheah, J. U. J., Neo, D. J. H., Goh, W. I., Kanchanawong, P., Soo, K. C., et al. (2017). Exploiting the protein corona around gold nanorods for low-dose combined photothermal and photodynamic therapy. *Journal of Materials Chemistry B, 5* (2), 254–268. (B) Two-nanoantenna release for dynamic control of gene circuits. From Lee, S. E., Sasaki, D. Y., Park, Y., Xu, R., Brennan, J. S., Bissell, M. J., et al. (2012). Photonic gene circuits by optically addressable siRNA-Au nanoantennas. *ACS Nano, 6* (9), 7770–7780. (C) Triple therapy of gene, drug, and photothermal therapy using hydrogels loaded with gold nanorods decorated with drugs, and gold nanospheres decorated with siRNA. From Conde, J., Oliva, N., Zhang, Y., & Artzi, N. (2016). Local triple-combination therapy results in tumour regression and prevents recurrence in a colon cancer model. Nature Materials, *15* (10), 1128–1138.

These selective release experiments demonstrated that control of multiple species in phototriggered release can manipulation of enable more complex biological processes that control of the release of one species could not. Furthermore, they build upon this ability to trigger siRNA release from NRs by photothermal excitation (Braun et al., 2009) to use selective excitation of two different NRs to enable independent excitation of each nanoantenna. A key innovation is that they use the released siRNA to not just silence expression, but to reconfigure existing gene circuits, thus enabling engineering new gene circuits via optical modulation. Part of their study modeled the response of the NR and the resulting siRNA concentration profile so as to strategically design the multicomponent circuit. By doing so, they demonstrate dynamic control of circuits in living cells (HeLa) with the two proteins IP-10 and RANTES. Even though they use cw laser excitation, which raises the temperature around the NR, they use the fact that the thermal dissipation drops off rapidly with distance from the NR (within 100 nm) to enable selective excitation. This work is a novel implementation of selective release from gold NRs that is outside the typical use of phototriggered release, such as cancer therapy or drug delivery.

4.3 Triple Therapy

The use of phototriggered release in combination therapy has been explored, (Jokerst & Gambhir, 2011) where the majority of uses involve two different types of therapy, such as photothermal heating and drug release. However, the use of more than two modes has been relatively unexplored. Conde, Oliva, Zhang, and Artzi (2016) use triple therapy specifically for colorectal cancer (CRC), which has a high recurrence rate post-surgery and which drug treatments are not specific (Fig. 4C). They construct a system that can perform gene delivery, chemotherapeutic drug delivery, and PTT. To do so, they use a hydrogel scaffold made of a polyamidoamine dendrimer/dextran scaffold and load the hydrogel with multiple species for local drug and gene delivery. One species that is loaded is a gold NR conjugated to the antibody drug bevacuzimab (Avastin), which is an anti-vascular endothelial growth factor (anti-VEGF) antibody that inhibits VEGF-induced cell proliferation. In addition, the NRs are decorated with targeting peptides (TCP-1) to target the tumor vasculature. The NRs also provide the PTT when irradiated at their SPR in the NIR.

They load the hydrogel with a second nanocarrier, gold nanospheres with siRNA for Kras, which is known to be an effector of signal transduction in CRC, and to enhance cell uptake for gene silencing, they functionalize the gold nanospheres with the peptide HA1 for enhanced endosomal escape. They perform an impressive study demonstrating in mice that the all three treatments have a synergistic effect on therapeutic efficiency, where the triple therapy is able to induce tumor regression and eliminate recurrence. They also perform a comprehensive gene expression analysis to determine the response to the triple therapy and each of the individual therapies. From this, they are able to identify biomarkers that respond to the therapies, which could enable further optimization of treatments.

This work shows that strategic use of PTT can work synergistically with other therapies, and that a multimodal approach can minimize side effects because of the local delivery of the therapies.

5 Conclusions and Outlook

In summary, the properties of plasmonic nanomaterials has given rise to many exciting new applications in nanomedicine. Despite initial challenges of the interface, the use of phototriggered therapies has reached a stage where new innovations are constantly being discovered, bringing us closer to realizing the immense potential of nanomedicine. Our understanding of the mechanism of photothermal heating and phototriggered release as well as the nano-bio interface has reached a sufficient level where the design of these nanomaterials for therapeutic applications is possible, and we are now realizing their potential that was originally envisioned. Furthermore, the combination of phototriggered therapies with other functions such as imaging have led to exciting progress. This has been achieved by exploiting the properties of the plasmonic nanomaterial for photoacoustic imaging, or by combining the system with species that can be imaged via other methods such as magnetic resonance imaging.

However, many challenges still remain. In particular, nanomaterial safety is still a relatively open question, where the toxicity and clearance of nanomaterials is still being studied (Tsoi et al., 2016; Wilhelm et al., 2016). There are some studies of long-term toxicity and residence of gold NPs (Ali et al., 2017), though much more work still needs to be done in this area if we are to have a broad understanding of the long term-effects of different classes of NPs.

Acknowledgments

The authors acknowledge support for CRQ from the Rafael del Pino Fellowship and the University of Massachusetts Boston Goranson Award.

References

Agarwal, A., Mackey, M. A., El-Sayed, M. A., & Bellamkonda, R. V. (2011). Remote triggered release of doxorubicin in tumors by synergistic application of thermosensitive liposomes and gold Nanorods. *ACS Nano*, *5*(6), 4919–4926.

Aggarwal, P., Hall, J. B., McLeland, C. B., Dobrovolskaia, M. A., & McNeil, S. E. (2009). Nanoparticle interaction with plasma proteins as it relates to particle biodistribution, biocompatibility and therapeutic efficacy. *Advanced Drug Delivery Reviews*, *61*(6), 428–437.

Ali, M. R. K., Rahman, M. A., Wu, Y., Han, T., Peng, X., Mackey, M. A., et al. (2017). Efficacy, long-term toxicity, and mechanistic studies of gold nanorods photothermal therapy of cancer in xenograft mice. *Proceedings of the National Academy of Sciences*, *114*(15), E3110–E3118.

Alkilany, A. M., Thompson, L. B., Boulos, S. P., Sisco, P. N., & Murphy, C. J. (2012). Gold nanorods: their potential for photothermal therapeutics and drug delivery, tempered by the complexity of their biological interactions. *Advanced Drug Delivery Reviews*, *64*(2), 190–199.

Alper, J., Crespo, M., & Hamad-Schifferli, K. (2009). Release mechanism of octadecyl rhodamine B chloride from Au nanorods by ultrafast laser pulses. *Journal of Physical Chemistry C*, *113*(15), 5967–5973.

Aubin-Tam, M.-E., & Hamad-Schifferli, K. (2008). Structure and function of nanoparticle–protein conjugates. *Biomedical Materials*, *3*(3), 034001.

Barbosa, S., Agrawal, A., Rodriguez-Lorenzo, L., Pastoriza-Santos, I., Alvarez-Puebla, R. A., Kornowski, A., et al. (2010). Tuning size and sensing properties in colloidal gold nanostars. *Langmuir*, *26*(18), 14943–14950.

Barhoumi, A., Huschka, R., Bardhan, R., Knight, M. W., & Halas, N. J. (2009). Light-induced release of DNA from plasmon-resonant nanoparticles: towards light-controlled gene therapy. *Chemical Physics Letters*, *482*(4–6), 171–179.

Braun, G. B., Pallaoro, A., Wu, G., Missirlis, D., Zasadzinski, J. A., Tirrell, M., et al. (2009). Laser-activated gene silencing via gold nanoshell–siRNA conjugates. *ACS Nano*, *3*(7), 2007–2015.

Capriotti, A., Caracciolo, G., Cavaliere, C., Colapicchioni, V., Piovesana, S., Pozzi, D., et al. (2014). Analytical methods for characterizing the nanoparticle–protein corona. *Chromatographia*, *77*(11–12), 755–769.

Casals, E., Pfaller, T., Duschl, A., Oostingh, G. J., & Puntes, V. (2010). Time evolution of the nanoparticle protein corona. *ACS Nano*, *4*(7), 3623–3632.

Cedervall, T., Lynch, I., Lindman, S., Berggård, T., Thulin, E., Nilsson, H., et al. (2007). Understanding the nanoparticle–protein corona using methods to quantify exchange rates and affinities of proteins for nanoparticles. *Proceedings of the National Academy of Science*, *104*, 2050–2055.

Cifuentes-Rius, A., de Puig, H., Kah, J. C. Y., Borros, S., & Hamad-Schifferli, K. (2013). Optimizing the properties of the protein corona surrounding nanoparticles for tuning payload release. *ACS Nano*, *7*(11), 10066–10074.

Conde, J., Oliva, N., Zhang, Y., & Artzi, N. (2016). Local triple-combination therapy results in tumour regression and prevents recurrence in a colon cancer model. *Nature Materials*, *15*(10), 1128–1138.

Dam, D. H. M., Culver, K. S. B., & Odom, T. W. (2014). Grafting aptamers onto gold nanostars increases in vitro efficacy in a wide range of cancer cell types. *Molecular Pharmaceutics*, *11*(2), 580–587.

Darabi Sahneh, F., Scoglio, C., & Riviere, J. (2013). Dynamics of nanoparticle-protein corona complex formation: analytical results from population balance equations. *PLoS One*, *8*(5), e64690.

de Puig, H., Cifuentes Rius, A., Flemister, D., Baxamusa, S. H., & Hamad-Schifferli, K. (2013). Selective light-triggered release of DNA from gold nanorods switches blood clotting on and off. *PLoS One*, *8*(7), e68511.

Dell'Orco, D., Lundqvist, M., Oslakovic, C., Cedervall, T., & Linse, S. (2010). Modeling the time evolution of the nanoparticle-protein corona in a body fluid. *PLoS One*, *5*(6), e10949.

Dobrovolskaia, M. A., Aggarwal, P., Hall, J. B., & McNeil, S. E. (2008). Preclinical studies to understand nanoparticle interaction with the immune system and its potential effects on nanoparticle biodistribution. *Molecular Pharmaceutics*, *5*(4), 487–495.

Dobrovolskaia, M. A., Patri, A. K., Zheng, J., Clogston, J. D., Ayub, N., Aggarwal, P., et al. (2009). Interaction of colloidal gold nanoparticles with human blood: effects on particle size and analysis of plasma protein binding profiles. *Nanomedicine: Nanotechnology, Biology and Medicine*, *5*(2), 106–117.

Docter, D., Distler, U., Storck, W., Kuharev, J., Wünsch, D., Hahlbrock, A., et al. (2014). Quantitative profiling of the protein coronas that form around nanoparticles. *Nature Protocols*, *9*(9), 2030–2044.

Dreaden, E. C., Austin, L. A., Mackey, M. A., & El-Sayed, M. A. (2012). Size matters: gold nanoparticles in targeted cancer drug delivery. *Therapeutic Delivery*, *3*(4), 457–478.

Eustis, S., & El-Sayed, M. A. (2006). Why gold nanoparticles are more precious than pretty gold: noble metal surface plasmon resonance and its enhancement of the radiative and nonradiative properties of nanocrystals of different shapes. *Chemical Society Reviews*, *35*(3), 209–217.

Fleischer, C. C., & Payne, C. K. (2014). Nanoparticle–cell interactions: molecular structure of the protein corona and cellular outcomes. *Accounts of Chemical Research*, *47*(8), 2651–2659.

Gobin, A. M., Lee, M. H., Halas, N. J., James, W. D., Drezek, R. A., & West, J. L. (2007). Near-infrared resonant nanoshells for combined optical imaging and photothermal cancer therapy. *Nano Letters*, *7*(7), 1929–1934.

Goodman, A. M., Hogan, N. J., Gottheim, S., Li, C., Clare, S. E., & Halas, N. J. (2017). Understanding resonant light-triggered DNA release from plasmonic nanoparticles. *ACS Nano*, *11*(1), 171–179.

Gou, L., & Murphy, C. J. (2005). Fine-tuning the shape of gold nanorods. *Chemistry of Materials*, 3668–3672.

Grobmyer, S. R., & Moudgil, B. M. (2010). *Cancer nanotechnology*. Vol. 624. (p. 366). New York: Springer.

Hamad-Schifferli, K. (2015). Exploiting the novel properties of protein coronas: emerging applications in nanomedicine. *Nanomedicine (London, England)*, *10*(10), 1663–1674.

Horiguchi, Y., Honda, K., Kato, Y., Nakashima, N., & Niidome, Y. (2008). Photothermal reshaping of gold nanorods depends on the passivating layers of the nanorod surfaces. *Langmuir*, *24*(20), 12026–12031.

Huettmann, G., Radt, B., Serbin, J., & Birngruber, R. (2003). Inactivation of proteins by irradiation of gold nanoparticles with nano- and picosecond laser pulses. In R. Steiner (Ed.), *Therapeutic laser applications and laser-tissue interactions, Vol. 5142 of Proc. SPIE*. Optical Society of America, paper 5142_88.

Huschka, R., Zuloaga, J., Knight, M. W., Brown, L. V., Nordlander, P., & Halas, N. J. (2011). Light-induced release of DNA from gold nanoparticles: nanoshells and nanorods. *Journal of the American Chemical Society*, *133*(31), 12247–12255.

Jokerst, J. V., & Gambhir, S. S. (2011). Molecular imaging with theranostic nanoparticles. *Accounts of Chemical Research*, *44*(10), 1050–1060.

Kah, J. C. Y., Chen, J., Zubieta, A., & Hamad-Schifferli, K. (2012). Exploiting the protein corona around gold nanorods for loading and triggered release. *ACS Nano*, *6*(8), 6730–6740.

Lee, K. S., & El-Sayed, M. A. (2005). Dependence of the enhanced optical scattering efficiency relative to that of absorption for gold metal nanorods on aspect ratio, size, end-cap shape, and medium refractive index. *Journal of Physical Chemistry B*, *109*, 20331–20338.

Lee, S. E., & Lee, L. P. (2010). Biomolecular plasmonics for quantitative biology and nanomedicine. *Current Opinion in Biotechnology*, *21*(4), 489–497.

Lee, S. E., Liu, G. L., Kim, F., & Lee, L. P. (2009). Remote optical switch for localized and selective control of gene interference. *Nano Letters*, *9*(2), 562–570.

Lee, S. E., Sasaki, D. Y., Park, Y., Xu, R., Brennan, J. S., Bissell, M. J., et al. (2012). Photonic gene circuits by optically addressable siRNA-Au nanoantennas. *ACS Nano*, *6*(9), 7770–7780.

Lesniak, A., Fenaroli, F., Monopoli, M. P., Åberg, C., Dawson, K. A., & Salvati, A. (2012). Effects of the presence or absence of a protein corona on silica nanoparticle uptake and impact on cells. *ACS Nano*, *6*(7), 5845–5857.

Link, S., Burda, C., Nikoobakht, B., & El-Sayed, M. A. (2000). Laser-induced shape changes of colloidal gold nanorods using femtosecond and nanosecond laser pulses. *Journal of Physical Chemistry B*, *104*(26), 6152–6163.

Liu, Y., Zhi, X., Yang, M., Zhang, J., Lin, L., Zhao, X., et al. (2017). Tumor-triggered drug release from calcium carbonate-encapsulated gold nanostars for near-infrared photodynamic/photothermal combination antitumor therapy. *Theranostics*, *7*(6), 1650–1662.

Lundqvist, M., Stigler, J., Cedervall, T., Berggård, T., Flanagan, M. B., Lynch, I., et al. (2011). The evolution of the protein corona around nanoparticles: a test study. *ACS Nano*, *5*(9), 7503–7509.

Lundqvist, M., Stigler, J., Elia, G., Lynch, I., Cedervall, T., & Dawson, K. A. (2008). Nanoparticle size and surface properties determine the protein corona with possible implications for biological impacts. *Proceedings of the National Academy of Sciences*, *105*(38), 14265–14270.

Mirshafiee, V., Mahmoudi, M., Lou, K., Cheng, J., & Kraft, M. L. (2013). Protein corona significantly reduces active targeting yield. *Chemical Communications*, *49*(25), 2557–2559.

Monopoli, M. P., Walczyk, D., Campbell, A., Elia, G., Lynch, I., Baldelli Bombelli, F., et al. (2011). Physical–chemical aspects of protein corona: relevance to in vitro and in vivo biological impacts of nanoparticles. *Journal of the American Chemical Society*, *133*(8), 2525–2534.

Moyano, D. F., Goldsmith, M., Solfiell, D. J., Landesman-Milo, D., Miranda, O. R., Peer, D., et al. (2012). Nanoparticle hydrophobicity dictates immune response. *Journal of the American Chemical Society*, *134*(9), 3965–3967.

Nel, A. E., Madler, L., Velegol, D., Xia, T., Hoek, E. M. V., Somasundaran, P., et al. (2009). Understanding biophysicochemical interactions at the nano-bio interface. *Nature Materials*, *8*, 543–557.

Park, S., & Hamad-Schifferli, K. (2010). Nanoscale interfaces to biology. *Current Opinion in Chemical Biology*, *14*(5), 616–622.

Pissuwan, D., Valenzuela, S. M., & Cortie, M. B. (2006). Therapeutic possibilities of plasmonically heated gold nanoparticles. *Trends in Biotechnology*, *24*(2), 62–67.

Prodan, E., Radloff, C., Halas, N. J., & Nordlander, P. (2003). A hybridization model for the plasmon response of complex nanostructures. *Science*, *302*(5644), 419.

Ritz, S., Schöttler, S., Kotman, N., Baier, G., Musyanovych, A., Kuharev, J., et al. (2015). Protein corona of nanoparticles: distinct proteins regulate the cellular uptake. *Biomacromolecules*, *16*(4), 1311–1321.

Salvati, A., Pitek, A. S., Monopoli, M. P., Prapainop, K., Bombelli, F. B., Hristov, D. R., et al. (2013). Transferrin-functionalized nanoparticles lose their targeting capabilities when a biomolecule corona adsorbs on the surface. *Nature Nanotechnology*, *8*(2), 137–143.

Sau, T. K., & Murphy, C. J. (2004). Seeded high yield synthesis of short Au nanorods in aqueous solution. *Langmuir*, *20*(15), 6414–6420.

Sobczynski, D. J., Charoenphol, P., Heslinga, M. J., Onyskiw, P. J., Namdee, K., Thompson, A. J., et al. (2014). Plasma protein corona modulates the vascular wall interaction of drug carriers in a material and donor specific manner. *PLoS One*, *9*(9), e107408.

Tam, J. M., Tam, J. O., Murthy, A., Ingram, D. R., Ma, L. L., Travis, K., et al. (2010). Controlled assembly of biodegradable plasmonic nanoclusters for near-infrared imaging and therapeutic applications. *ACS Nano*, *4*(4), 2178–2184.

Tenzer, S., Docter, D., Kuharev, J., Musyanovych, A., Fetz, V., Hecht, R., et al. (2013). Rapid formation of plasma protein corona critically affects nanoparticle pathophysiology. *Nature Nanotechnology*, *8*(10), 772–781.

Tsoi, K. M., MacParland, S. A., Ma, X.-Z., Spetzler, V. N., Echeverri, J., Ouyang, B., et al. (2016). Mechanism of hard-nanomaterial clearance by the liver. *Nature Materials*, *15*(11), 1212–1221.

Walczyk, D., Bombelli, F. B., Monopoli, M. P., Lynch, I., & Dawson, K. A. (2010). What the cell "sees" in bionanoscience. *Journal of the American Chemical Society*, *132*(16), 5761–5768.

Walkey, C. D., Olsen, J. B., Song, F., Liu, R., Guo, H., Olsen, D. W. H., et al. (2014). Protein corona fingerprinting predicts the cellular interaction of gold and silver nanoparticles. *ACS Nano*, *8*(3), 2439–2455.

Wijaya, A., & Hamad-Schifferli, K. (2008). Ligand customization and DNA functionalization of gold nanorods via round-trip phase transfer ligand exchange. *Langmuir*, *24*(18), 9966–9969.

Wijaya, A., Schaffer, S., Pallares, I., & Hamad-Schifferli, K. (2009). Selective release of multiple DNA oligonucleotides from gold nanorods. *ACS Nano*, *3*(1), 80–86.

Wilhelm, S., Tavares, A. J., Dai, Q., Ohta, S., Audet, J., Dvorak, H. F., et al. (2016). Analysis of nanoparticle delivery to tumours. *Nature Reviews Materials*, *1*, 16014.

Xie, J., Zhang, Q., Lee, J. Y., & Wang, D. I. C. (2008). The synthesis of SERS-active gold nanoflower tags for in vivo applications. *ACS Nano*, *2*(12), 2473–2480.

Yallapu, M. M., Chauhan, N., Othman, S. F., Khalilzad-Sharghi, V., Ebeling, M. C., Khan, S., et al. (2015). Implications of protein corona on physico-chemical and biological properties of magnetic nanoparticles. *Biomaterials*, *46*, 1–12.

Yamashita, S., Fukushima, H., Akiyama, Y., Niidome, Y., Mori, T., Katayama, Y., et al. (2011). Controlled-release system of single-stranded DNA triggered by the photothermal effect of gold nanorods and its in vivo application. *Bioorganic & Medicinal Chemistry*, *19*(7), 2130–2135.

Yavuz, M. S., Cheng, Y., Chen, J., Cobley, C. M., Zhang, Q., Rycenga, M., et al. (2009). Gold nanocages covered by smart polymers for controlled release with near-infrared light. *Nature Materials*, *8*(12), 935–939.

Yeo, E. L. L., Cheah, J. U. J., Neo, D. J. H., Goh, W. I., Kanchanawong, P., Soo, K. C., et al. (2017). Exploiting the protein corona around gold nanorods for low-dose combined photothermal and photodynamic therapy. *Journal of Materials Chemistry B*, *5*(2), 254–268.

Zhang, Z., Wang, L., Wang, J., Jiang, X., Li, X., Hu, Z., et al. (2012). Mesoporous silica-coated gold nanorods as a light-mediated multifunctional theranostic platform for cancer treatment. *Advanced Materials*, *24*(11), 1418–1423.

Further Reading

Caracciolo, G. (2015). Liposome-protein corona in a physiological environment: challenges and opportunities for targeted delivery of nanomedicines. *Nanomedicine*, *11*(3), 543–557.

Chen, C.-K., Huang, C.-C., & Chang, H.-T. (2010). Label-free colorimetric detection of picomolar thrombin in blood plasma using a gold nanoparticle-based assay. *Biosensors and Bioelectronics*, *25*(8), 1922–1927.

de Puig, H., Tam, J. O., Yen, C. W., Gehrke, L., & Hamad-Schifferli, K. (2015). Extinction coefficient of gold nanostars. *Journal of Physical Chemistry C, Nanomaterials and Interfaces*, *119*(30), 17408–17415.

Gaspar, R. (2013). Nanoparticles: pushed off target with proteins. *Nature Nanotechnology*, *8*(2), 79–80.

Huang, X., & El-Sayed, M. A. (2010). Gold nanoparticles: optical properties and implementations in cancer diagnosis and photothermal therapy. *Journal of Advanced Research*, *1*(1), 13–28.

Link, S., & El-Sayed, M. A. (2000). Spectral properties and relaxation dynamics of surface plasmon electronic oscillations in gold and silver nanodots and nanorods. *Journal of Physical Chemistry B*, *103*(40), 8410–8426.

CHAPTER 6

Bioengineered Ferritin Nanoprobes for Cancer Theranostics

Kelong Fan[*], **Xiyun Yan**[*,†]

[*]*Institute of Biophysics, Chinese Academy of Sciences, Beijing, China* [†]*University of Chinese Academy of Sciences, Beijing, China*

1 Introduction

The ideal nanoprobes for imaging or treating clinical tumors should possess the ability to efficiently load imaging agents or drugs, exhibit excellent biocompatibility and be able to specifically target tumor sites (Chauhan & Jain, 2013; Lim et al., 2015; Strebhardt & Ullrich, 2008; Tosi et al., 2016). Numerous efforts have been made to develop active tumor targeting nanoprobes in the past few decades, while a handful of these systems have been evaluated in clinical trials but none was approved for clinical use (Chen et al., 2017a; Venditto & Szoka, 2013; Wang, Billone, & Mullett, 2013). It is hard to translate chemically prepared nanoprobes from bench to bedside because their synthesis procedures are usually complicated, and they often have unknown long-term toxicity (Tosi et al., 2016). Under these circumstances, delivery of imaging agents or nanodrugs to the disease sites using biological vehicles maybe more advantageous. Endogenous self-assembling proteins are promising candidates considering they are nontoxic for the organism and therefore, safe (Karimi et al., 2017).

Because of their unique protein architecture and physicochemical properties, bioengineered ferritin nanoprobes have been extensively used as versatile nanocarriers for tumor theranostic applications (Fan, Gao, & Yan, 2013; He & Marles-Wright, 2015; Heger et al., 2014; Truffi et al., 2016; Wang et al., 2017). In recent years, the identification of the specific receptors of ferritins has not only helped us understand the physiological function of ferritins, but also prompted the use of the intrinsic tumor targeting property of ferritin nanoprobes. Taken together, these favorable properties make ferritin an ideal base of engineered nanoprobes for tumor imaging and therapy.

Handbook of Nanomaterials for Cancer Theranostics. https://doi.org/10.1016/B978-0-12-813339-2.00006-2

2 Ferritin: Its Structure, Function, and Membrane Receptors

2.1 Structure and Function of Ferritin

Ferritin is a natural iron storage protein which plays an important role in iron homeostasis. Ferritin was first isolated from horse spleen by Laufberger in 1937, and was subsequently found in many other organisms, including humans and other mammals, plants, fungi and bacteria (Arosio, Ingrassia, & Cavadini, 2009; Theil, 1987; Wang et al., 2010). In spite of large variations in amino acid sequences from bacteria to humans, ferritin proteins essentially have the same architecture (Harrison & Arosio, 1996). The typical structure of ferritin is a 24-subunit spherical protein cage encapsulating an iron oxide core (Harrison & Arosio, 1996). Mammalian ferritins are mainly present intracellularly in the cytosol, as well as in the nuclei and mitochondria. Extracellular ferritins are found in serum as well as synovial and cerebrospinal fluids (Meyron-Holtz, Moshe-Belizowski, & Cohen, 2011). The cytosolic ferritins play important roles in iron storage and detoxification, while the physiological function of secreted ferritin is still unclear. It has been shown that elevated serum ferritin levels are linked to inflammation, angiogenesis and tumors (Knovich et al., 2009; Meyron-Holtz et al., 2011; Wang et al., 2010), and therefore, are considered a marker for these conditions.

The ferritin superfamily are spherical proteins typically composed of 24 subunits, with an outer diameter of 12 nm and an interior cavity of 8 nm diameter (Fig. 1). The interior cavity can accommodate up to 4500 Fe(III) atoms as an iron mineral core, traditionally described as ferrihydrite (Harrison & Arosio, 1996). Apoferritin refers to the iron-free form of the protein, while the iron-containing form is termed holoferritin, or simply ferritin. The apoferritin is normally composed of a mixture of ferritin H-chain and L-chain subunits, arranged with fourfold, threefold, and twofold symmetry axes, of which the threefold axes form hydrophilic channels that allow transport of metal in and out of the protein cage (Harrison & Arosio, 1996;

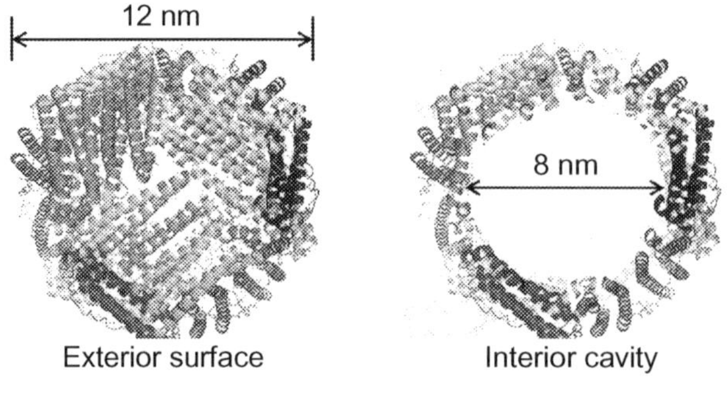

Fig. 1
Cartoon diagrams of exterior surface view of threefold symmetry axes and interior cavity of human H-ferritin (2FHA).

Theil, 1987). The H-chain is named for its initial isolation from heart, whereas the L-chain was isolated from liver (Torti & Torti, 2002). In humans, the H-chain is also heavier, with a molecular mass of 21 kDa, whereas the L-chain mass is 19 kDa (Harrison & Arosio, 1996; Theil, 1987). Therefore, the ferritin subunits are sometimes referred to as heavy (H) and light (L) ferritins, respectively.

Physiologically, H-ferritin and L-ferritin perform different functions. The H-chain with a ferroxidase center is essential for Fe(II) oxidation, while the L-chain assists in iron core formation, and lacks the catalytic center. This is the reason why recombinant L-ferritin, unlike H-ferritin, is typically iron poor. Moreover, the ratio of these two subunits in ferritin varies widely depending on tissue type, with H-chains predominant in the heart and L-chains predominant in the liver (Torti & Torti, 2002). In addition, the H/L ratio changes in inflammation and other pathological conditions.

Natural ferritin exists in both intracellular and extracellular compartments. In most tissues, ferritins are mainly present in the cytosol, nucleus, and mitochondria, and play a role in iron storage as well as iron homeostasis. Iron, a major trace element, is an integral component of many proteins and both potentially toxic and essential for life. In particular, free Fe^{2+} ions are lethal because they catalyze the formation of reactive hydroxyl radicals in oxygenated tissues by the Fenton reaction, leading to damage to DNA, lipids, and proteins (Harrison & Arosio, 1996; Torti & Torti, 2002). The balance between iron storage and utilization is maintained by the regulation of intestinal absorption of the metal from diet, along with the expression of iron transport and storage proteins, including ferritin, transferrin and the transferrin receptor. In addition to iron storage, ferritins also play an important role in iron detoxification by capturing and sequestering the intracellular labile iron pool. This protective ability of ferritin is based on the ferroxidase activity of the H-chain to catalyze the transformation of highly toxic Fe(II) to less toxic Fe(III) (Harrison & Arosio, 1996).

Ferritins in the nucleus (nuclear ferritin (Alkhateeb & Connor, 2010)) and mitochondria (mitochondrial ferritin (Drysdale et al., 2002)) protect DNA or mitochondria from iron toxicity and oxidative damage (Arosio et al., 2009). The nuclear translocation of ferritin, especially H-ferritin, is a selective energy-dependent process that does not seem to require a consensus nuclear localization signal (Thompson et al., 2002). Furthermore, the phosphorylation and *O*-glycosylation of H-ferritin seem to be involved in its nuclear translocation (Alkhateeb & Connor, 2010). Mitochondrial ferritin is an H-ferritin, with a precursor that is targeted to mitochondria by a leader sequence (Drysdale et al., 2002). Thus, H-ferritin is important for iron detoxification and cell protection. Recently, some studies have shown that cytosolic ferritins are elevated in malignant tissues; for example, cytosolic ferritin is expressed in mammary carcinomas at levels up to 10-fold higher than in benign breast tissues (Elliott et al., 1993; Weinstein et al., 1989; Werneke, Elliott, & Ledford, 1990), indicating that cytosolic ferritin may be involved in tumor progression.

Serum ferritins are predominantly composed of L chains, which have low iron content. Although several studies have reported that serum ferritin might arise from the secretion of hepatocytes, macrophages and Kupfer cells (Meyron-Holtz et al., 2011; Wang et al., 2010), the source of serum ferritins still needs to be further investigated. In addition, the physiological functions of serum ferritin as well as the identity of the L-ferritin receptor are not yet clear. Nevertheless, a significant increase in serum ferritin levels has been confirmed to be related to pathological processes, including inflammation, angiogenesis and tumor formation, implying that serum ferritins are a potential biomarker for clinical diagnosis (discussed in following sections) (Harrison & Arosio, 1996; Meyron-Holtz et al., 2011; Wang et al., 2010).

Besides their distinct architectural features, ferritins possess unique physical and chemical properties. Unlike most other proteins, which are sensitive to temperature and pH outside of the physiological range, ferritin is able to bear high temperatures of up to 75°C for 10 min and is stable in various denaturants, including urea or guanidinium chloride. These unique features are due to the large numbers of salt bridges and hydrogen bonds formed between subunits of ferritin (Harrison & Arosio, 1996; Santambrogio et al., 1992). Interestingly, the assembly of ferritin, despite its rigidness under physiological conditions, is pH-dependent (Kang et al., 2008). The ferritin architecture breaks down to subunits in extreme acidic (pH 2–3) or basic (pH 10–12) environments and reconstructs, almost completely, when pH returns back to the physiological range (Belletti et al., 2017; Kang et al., 2008; Santambrogio et al., 1992; Truffi et al., 2016; Uchida et al., 2010). These unique properties make ferritin an ideal and powerful nanoplatform to construct multifunctional nanoprobes for imaging and drug delivery.

2.2 Ferritins and Their Membrane Receptors

Several research groups had already reported that human ferritin is selectively taken up by tumor cells as early as in the 1960s (Caulfield, 1963; Easty, Yarnell, & Andrews, 1964; Ryser, Caulfield, & Aub, 1962). Fargion et al. then reported in the 1980s that human H-ferritin specifically binds to a protein of ~100 kDa molecular weight on tumor cells (Fargion et al., 1988). However, it was not until 2010 that transferrin receptor 1 (TfR1) was identified as the human H-ferritin receptor by Seaman's group using expression cloning (Li et al., 2010), following their earlier identification in 2005 of the mouse T cell immunoglobulin-domain and mucin domain-2 (TIM-2) as the H-ferritin specific receptor (Table 1) (Chen et al., 2005). They found that TfR1 binds specifically to human H-ferritin with little or no binding to L-ferritin. Furthermore, after binding to TfR1 on the cell surface, human H-ferritin enters both endosomes and lysosomes. The finding that TfR1 can bind to H-ferritin as well as transferrin suggests that this dual receptor function may coordinate the use of iron through these iron-binding molecules. Soon after the identification of the H-ferritin receptor, our group reported that human H-ferritins target TfR1 on both tumor cells and tumor tissues from clinical samples (Fan et al., 2012).

Table 1: Ferritins and their membrane receptors

Ferritins	Membrane Receptors	Years	References
Mouse H-ferritin	T cell immunoglobulin-domain and mucin domain-2 (Mouse TIM2)	2005	Chen et al. (2005)
Mouse L-ferritin	Scavenger receptor class A member 5 (Mouse Scara5)	2009	Li et al. (2009)
Human H-ferritin	Transferrin receptor 1 (Human TfR1)	2010	Li et al. (2010)
Human L-ferritin	-	-	-

Because serum ferritin changes with physiological or pathological process, it is suggested that the assumed receptors in liver for both H-ferritin and L-ferritin may serve to regulate the levels of serum ferritin (Moss et al., 1992). A putative ferritin receptor was purified from human liver, which showed higher affinity for L-ferritin and lacked the specificity for human H-ferritin (Adams, Powell, & Halliday, 1988; Moss et al., 1992). Although the molecular weight of this putative receptor is consistent with the glycosylated TIM receptors, a human ortholog for mouse TIM-2 (receptor for mouse H-ferritin) has not been identified yet, and this receptor still remains putative.

The scavenger receptor Class A member 5 (Scara5) has already been identified as the receptor for mouse L-ferritin on mouse embryo kidney cells (Table 1) (Li et al., 2009). While human orthologs of Scara5 exist, no direct evidence has demonstrated that human L-ferritin binds to human Scara5. Finding the receptors for L-ferritin is fundamental for us to better understand the relationships between serum ferritin (iron homeostasis) and physical conditions or pathological processes. Because no receptors for human L-ferritin have been found, this area is well worth further study.

2.3 Cross-Interaction of Ferritins and Their Membrane Receptors From Different Species

Horse spleen ferritin (HosFn) has been widely used as a nanocarrier in nanomedicine due to its commercial availability and reasonable cost. Thus, cross-interaction studies of ferritins and ferritin membrane receptors in different species mainly concentrate on HosFn, while the human ferritin nanoprobes do not have such an issue in pharmaceutical use.

HosFn is nearly 92% (22L/2H) composed of L subunits (Harrison, 1986; Sun et al., 2011a), thus, it is typically regarded as L-ferritin by many researchers (Conti et al., 2016; Geninatti Crich et al., 2015; Mendes-Jorge et al., 2014; Sun et al., 2011a). Li et al. mentioned in their research that they observed the direct interaction of HosFn with human TfR1 (the receptor of human H-ferritin) (Table 2), but they did not present the detailed data (Li et al., 2010). Mendes-Jorge et al. reported that HosFn binds to mouse endothelial cells via Scara5 (mouse L-ferritin receptor) by observation of the colocalization of HosFn and Scara5 on mouse cells (Table 2) (Mendes-Jorge et al., 2014). Little else is known about the endogenous receptors of HosFn in humans.

Table 2: Cross-interaction of HosFn and ferritin membrane receptors from different species

Ferritins	Interaction with Receptors	Years	References
Horse spleen ferritin (22L/2H)	Human TfR1	2010	Li et al. (2010)
	Mouse Scara5	2014	Mendes-Jorge et al. (2014)
		2016	Conti et al. (2016)
	Human Scara5?	2015	Geninatti Crich et al. (2015)

Interestingly, Crich and colleagues reported that horse spleen ferritin can be effectively internalized by both mouse (Conti et al., 2016) and human breast cancer cells (Geninatti Crich et al., 2015). Crich and colleagues further demonstrated that HosFn is specifically taken up by mouse breast cancer cells via binding Scara5, which is confirmed by the siRNA knock-down of mouse Scara5 on mouse breast cancer cells reducing the uptake of HosFn. This finding is consistent with the report of Mendes-Jorge et al. (2014). Together, these evidences indicate that HosFn (22 L of 24 subunits) shares (cross-interacts with) the same receptor as mouse L-ferritin.

For human breast cancer cells, Crich and colleagues deduced that HosFn binds to tumor cells through Scara 5, based on the fact that the human breast cancer cells also express Scara5 (Geninatti Crich et al., 2015), the human ortholog protein of mouse Scara5. So far, this theory still lacks evidence that directly proves the interaction of HosFn and human Scara5 protein. Thus, as shown in Table 2, this interaction is still to be confirmed.

In summary, HosFn can cross interact with the receptor of human H-ferritin—human TfR1, and the receptor of mouse L-ferritin—mouse Scara5 (Fan, Zhou, & Yan, 2017). Other interactions still need to be explored.

3 Natural Human Ferritins as Prognostic Tumor Markers

3.1 Secretory Ferritins

Although the physiological function and source of secretory ferritins are still unknown, numerous evidence indicate that secretory ferritin is involved in pathological processes. For instance, a low concentration of serum ferritin indicates iron deficiency (e.g., anemia) and a high serum ferritin indicates iron overload (e.g., hemochromatosis) (Knovich et al., 2009; Wang et al., 2010). In addition, elevated serum ferritin is also found in inflammation, infection, and liver diseases (Arosio et al., 2009; Knovich et al., 2009; Wang et al., 2010). There is increasing evidence that serum ferritin levels are elevated in many malignancies, which attracted widespread attention considering that serum ferritin can be used as a tumor biomarker. For instance, serum ferritin is elevated in breast cancer patients (Jezequel et al., 2012; Knovich et al., 2009; Wang et al., 2010). Serum ferritin is also used as a biomarker for relapse of malignant diseases. Matzner et al. reported that serum ferritin was markedly increased in all relapsed cases of acute leukemia (Matzner, Konijn, & Hershko, 1980). In all cases, remission was associated with the normalization of serum ferritin levels. These correlations suggest that serum ferritin may be useful in the initial clinical evaluation and in the assessment of response to therapy in patients with acute leukemia and malignant lymphoma (Matzner et al., 1980). Jain et al. recently reported that elevated serum ferritin is predictive of inferior survival in patients with acute leukemia and may be an early marker of an underlying systemic pathological inflammation (Jain et al., 2016). Szymendera et al. and Volpino et al. found that measuring serum ferritin levels is a useful clinical indicator in patients with testicular germ-cell tumors (Szymendera et al., 1985) and lung cancer (Volpino et al., 1984), respectively. Moreover, it has been found that melanoma cells can secrete ferritin, which contributes to the progression of melanoma (Gray, Arosio, & Hersey, 2003). The H/L ratio of serum ferritin also varies in pathological conditions (Cazzola et al., 1985). Under normal physiological conditions, serum ferritin is predominantly composed of L-chains. However, in many malignant conditions, the ratio of H/L in serum ferritin is increased (Arosio et al., 2009; Knovich et al., 2009; Meyron-Holtz et al., 2011; Wang et al., 2010).

Besides serum levels, secretory ferritin levels in cerebrospinal fluid (CSF) have also been found to increase during malignant infiltration of the central nervous system, and have been studied for their ability to serve as a biomarker for the diagnosis of brain malignancies (de Almeida et al., 2008; Meyron-Holtz et al., 2011). Intriguingly, exhaled ferritin has been reported as a potential biomarker for lung cancer (Carpagnano et al., 2012).

3.2 Tissue Ferritins

A number of studies have shown that tissue ferritins are highly expressed in tumorigenic cell lines as well as in some malignant tissues from patients. Although mechanisms underlying these changes are still unclear, these studies reveal that tissue ferritin, especially H-ferritin, may play an important role in malignancy (Arosio et al., 2009; Knovich et al., 2009; Meyron-Holtz et al., 2011; Wang et al., 2010). Importantly, recent reports demonstrated that H-ferritin exerts a negative control on the formation of ovarian cancer stem cells. Moreover, the change of expression levels of H-ferritin may play a role in regulating the amount of cancer stem cells generated, and therefore the formation of metastasis and relapse after chemotherapy (Lobello et al., 2016). Consistent with this report, Rosager et al. found that low levels of H-ferritin-positive tumor cells were associated with poor survival in astrocytic brain tumors (Rosager et al., 2017). Thus, H-ferritin may behave as a prognostic factor for different types of cancers, and it also could be used as a potential cancer therapy target.

Intriguingly, several recent studies also demonstrated that human L-ferritin is a potential prognostic marker in different kinds of cancers. In breast cancer, Jezequel et al. demonstrated that L-ferritin expression is a potential node-negative breast cancer prognostic marker (Jezequel et al., 2012). In gastric cancer, Zhang et al. found that L-ferritin protein was significantly higher in gastric tissues than in adjacent tissues. In addition, the associations between L-ferritin expression and depth of tumor invasion, differentiation grade, lymph node metastasis and TNM (Tumor, Nodes, Metastasis) stage were significant (Zhang, Chen, & Xu, 2017). This work demonstrated that L-ferritin might be used as a prognostic molecular biomarker in gastric cancer patients and may be a potential novel therapeutic target for gastric cancer. Similarly with the report of Zhang et al., Rosager et al. found that high amounts of L-ferritin positive microglia/macrophages had a negative prognostic value for astrocytic brain tumors (Rosager et al., 2017). These reports indicate that, besides assisting with iron formation, L-ferritin may play an important role in the tumorigenesis and progress of cancers. This novel function of L-ferritin in tumors is worth further exploration.

In conclusion, human ferritins themselves in tissues or secretions can act as prognostic markers for tumor theranostics.

4 Tumor Targeting Strategies of Ferritin Nanoprobes

In general, tumor targeting strategies of nanoprobes can be divided into passive and active targeting. Passive targeting of nanoprobes mainly depend on the enhanced permeability and retention (EPR) effects due to the disordered blood and lymph vessels systems in tumor tissues (Greish, 2010). The EPR effects of nanoprobes rely on their size distribution (Ngoune et al., 2016). Because the sizes of the different bioengineered ferritins are typically around 12–20 nm, bioengineered ferritin nanoprobes should all have similar passive tumor targeting abilities. Thus, here we focus on active targeting strategies using ferritin nanoprobes.

4.1 Indirect Targeting of Bioengineered Ferritin Nanoprobes Against Tumors

Engineered nanoparticles have been extensively used to develop nanoprobes that provide diagnostic, therapeutic, and prognostic information about the status of tumors (Cheng et al., 2012; Xie, Lee, & Chen, 2010). Nanoprobes developed for these purposes are typically modified with targeting ligands, such as antibodies, peptides, or small molecules, to enhance their tumor targeting capability (Shi et al., 2016; Wilhelm et al., 2016). Before the cell membrane receptors of ferritin were identified, the tumor targeting strategies of ferritin nanoprobes were bioengineering and chemical modification. Due to ferritin nanoprobes being proteins encoded by genes, it is convenient to manipulate the functionalization of ferritins via gene engineering (Fan et al., 2013). In addition, both lysine and cysteine residues exposed on the surface of ferritin nanocages can be used to effectively conjugate with chemical groups via cross-linking with *N*-hydroxysuccinimide (NHS) ester or maleimide groups (Truffi et al., 2016). Thus, ferritin nanocages can be either genetically or chemically modified to construct active tumor targeting nanoprobes (Fig. 2) (Wang et al., 2017).

Most of the bioengineered ferritin nanoprobes choose human H-ferritin as the structural motif of the nanoprobe (Table 3). Several studies have genetically functionalized the H-ferritin

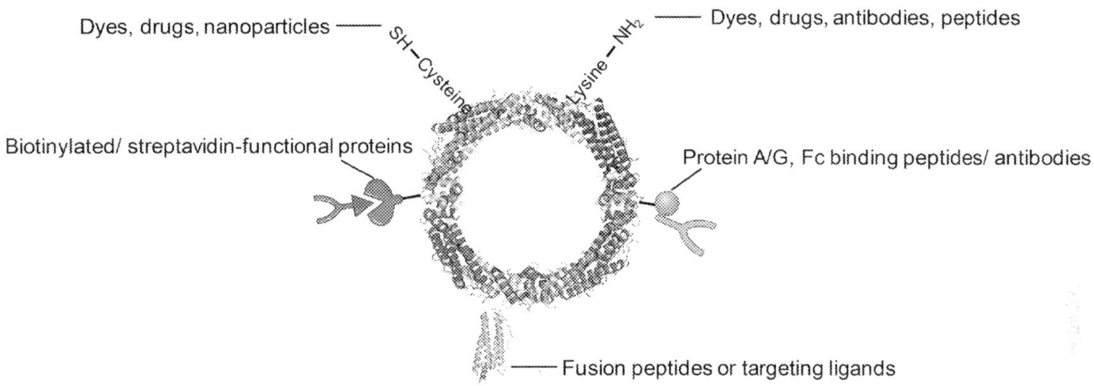

Fig. 2
Illustration of strategies to provide surface functionalization of ferritin nanoprobes by chemical or genetic modification.

Table 3: Tumor targeting moieties of engineered ferritin nanoprobes

Ferritin Types in Engineered Nanoprobes	Targeting Moieties	Targeting Receptors	Tumor Cell Types	References
Human H-ferritin	RGD-4C (CDCRGDCFC) peptide	$\alpha_v\beta_3$ integrins	Human melanoma cell, C32; Human glioma cell, U87MG; Human epithelial lung cancer cell, A549; mouse breast cancer cell, 4T1	Huang et al. (2017a, 2014), Li et al. (2012b), Lin et al. (2011a), Uchida et al. (2006), and Zhen et al. (2013a, 2013b, 2014)
Human H-ferritin	EGF	EGFR	Human breast cancer cell, MDA-MB-231 and MCF-7	Li et al. (2012a)
Human H-ferritin	EGF-derived peptide	EGFR	Human breast cancer cell, MDA-MB-468	Lee et al. (2015)
Human H-ferritin	VNTANST peptide	Vimentin	Mouse melanoma cell, B16F10	Lee et al. (2015)
Human H-ferritin	α-Melanocyte-stimulating hormone peptide (MSH)	Melanocortin receptor	Mouse melanoma cell, B16F10	Fantechi et al. (2014) and Vannucci et al. (2012)
Human H-ferritin	Tumor targeting proapoptotic peptide $CGKRK(KLAKLAK)_2$	p32-mediated internalization	Human breast cancer cell, MDA-MB-231, and human promyelocytic leukemia cell, HL60	Kim et al. (2016a)
Human H-ferritin	Human IgG-specific repebody + Cetuximab/ Trastuzumab	EGRF/HER2	Human breast cancer cell, MDA-MB-468, and SK-BR-3	Kim et al. (2017)

Continued

Table 3: Tumor targeting moieties of engineered ferritin nanoprobes—cont'd

Ferritin Types in Engineered Nanoprobes	Targeting Moieties	Targeting Receptors	Tumor Cell Types	References
Human H-ferritin	Mouse anti CSPG4 Ab	Melanoma-specific antigen CSPG4	Human melanoma cell, Colo 38	Falvo et al. (2013)
Human H-ferritin	Protein G + mouse antihuman-CLDN4	Claudin-4	Human pancreatic cancer cell, Capan-1	Hwang et al. (2013)
Human H-ferritin	FAP-specific single chain variable fragment (scFv)	Carcinoma-associated fibroblasts	Mouse breast cancer cell, 4T1	Zhen et al. (2016)
Human L-ferritin	RGD (GRGDSP) peptide	$\alpha_v\beta_3$ integrins	Human glioma cell, U87MG, and fibrosarcoma cell, HT1080	Kim et al. (2016b)
Horse spleen ferritin	DRGETGPAC peptide	Carcinoma-associated fibroblasts	Human prostate cancer cell, PC-3, and prostate cancer-associated fibroblasts	Ji et al. (2013)
Horse spleen ferritin	Anti-PSMA antibodies	Prostate specific membrane antigen (PSMA)	Human prostate cancer cell, LNCaP	Dostalova et al. (2016)
Horse spleen ferritin	C3d peptide	NCAM	Tumor-derived endothelial cells	Crich et al. (2006)
P. furiosus ferritin	Fc-binding peptide + Herceptin	HER2	Human breast cancer cell, SKBR3	Kang et al. (2012)
P. furiosus ferritin	Fc-binding peptide + rabbit antifolate receptor Ab	Folate receptor	Human oral epithelial cancer cell, KB	Kang et al. (2012)
Dps, a bacterial ferritin	RGD (CDCRGDCFC) peptide	$\alpha_v\beta_3$ integrins	Human glioma cell, U87MG	Ahn et al. (2014)

protein cage with specific ligands such as peptides (e.g., RGD) (Uchida et al., 2006), growth factors (e.g., EGF) (Li et al., 2012a), and antibody-binding proteins (e.g., protein G) (Hwang et al., 2013) to improve its targeting capability. For instance, Douglas's group genetically modified human H-ferritin with RGD-4C (CDCRGDCFC) peptide that can specifically target tumor angiogenesis via binding to $\alpha_v\beta_3$ integrin molecules on the vascular endothelium (Kitagawa et al., 2012; Uchida et al., 2006). Furthermore, many groups demonstrated that the RGD-4C modified H-ferritin nanoprobe bound to many types of tumor cells, including human amelanotic melanoma (Uchida et al., 2006), glioblastoma (Lin et al., 2011a; Zhen et al., 2013a, 2013b), lung adenocarcinoma cells (Li et al., 2012b), and mouse breast cancer cells (Huang et al., 2017a, 2014; Zhen et al., 2014). Currently, the RGD modification is the most-used tumor targeting strategy in bioengineered ferritin nanoprobes. Another ligand, a matrix metalloproteinase (MMP) protease substrate, was also introduced onto the exterior surface of the H-ferritin nanocage, and could be cleaved by MMP enriched in the tumor (Lin et al., 2011b). As a matter of fact, any ligand, as long as it is able to target tumor markers, could be used for modification of ferritin nanoprobes, such as melanocyte-stimulating hormone (MSH) and its receptor for mouse melanoma cells (Fantechi et al., 2014; Vannucci et al., 2012), or epidermal growth factor (EGF), EGF-derived peptides and its receptor EGFR for human breast cancer cells (Lee et al., 2015; Li et al., 2012a). In addition, some newly identified peptides, including VNTANST peptide for Vimentin in mouse melanoma cells (Lee et al., 2015), and CGKRK(KLAKLAK)$_2$ peptide for proapoptotic tumor targeting in human breast cancer and promyelocytic leukemia cells (Kim et al., 2016a) have already been genetically modified on the surface of H-ferritin nanoprobes to target tumor cells. An antibody is a powerful tool for tumor targeting. The antibody Fc binding protein-protein G (Hwang et al., 2013), or human IgG-specific binding repebody (Kim et al., 2017) were also genetically introduced to the exterior of H-ferritin nanoprobes, which allowed antibody-functionalized ferritin nanoprobes to target human breast cancer cells or pancreatic cancer cells. These antibody-binding ferritin nanoprobes also possess the potential to bind therapeutic antibodies to treat cancer cells.

Besides genetic modification, chemical conjugation provides useful approaches for functionalization of ferritin nanoprobes. Falvo et al. directly conjugated a monoclonal antibody to the surface of the H-ferritin nanocage to enable the ferritin nanoprobes to specifically target human melanoma cells (Falvo et al., 2013). A fibroblast activation protein-α specific single chain variable fragment (scFv) was conjugated onto the exterior of the ferritin nanocage to target carcinoma-associated fibroblasts of mouse breast cancer (Zhen et al., 2016). In addition to the use of single-ligand modified magnetoferritin, Chen's group recently generated chimeric ferritin nanocages for multiple function loading and multimodal imaging by combination of chemical modification and genetic engineering to visualize tumors with high resolution and sensitivity (Lin et al., 2011a).

Human L-ferritin is another choice for the structural motif of bioengineered ferritin nanoprobes. Kim et al. genetically modified RGD peptide (GRGDSP) onto the exterior of human L-ferritin to target human glioma and fibrosarcoma cells (Kim et al., 2016b). However, compared with human H-ferritin, L-ferritin is less frequently used in tumor targeting ferritin nanoprobes.

Horse spleen ferritin (HosFn) is also employed to develop tumor targeting nanoprobes. Because HosFn is a commercially available natural ferritin, the targeting nanoprobes based on HosFn are typically modified by chemical conjugation. For instance, DRGETGPAC peptide was chemically conjugated onto the surface of HosFn to target the human prostate cancer cell, PC-3, and prostate cancer-associated fibroblasts (Ji et al., 2013). In addition, biotin was chemically introduced to the exterior of HosFn to construct a universal ligand/antibody-binding nanoprobe based on the biotin-streptavidin interaction (Li & Mann, 2004). Biotin-C3d peptide was successfully loaded on the biotin-HosFn nanoprobe through streptavidin, thus making a nanoprobe to specifically target tumor-derived endothelial cells (Crich et al., 2006). With the assistance of gold nanoparticles, HosFn was modified with antibody-binding peptides. Anti-PSMA antibodies were then loaded on the surface of the HosFn-Au-peptide nanoprobe to make this system specifically target human prostate cancer cells (Dostalova et al., 2016).

Recently, a novel kind of ferritin isolated from *Pyrococcus furiosus* (optimal living condition at 100°C) was reported (Tatur, Hagen, & Matias, 2007). With a similar structure to human ferritin, the *P. furiosus* ferritin has been proved to have better thermal stability than human ferritins. Thus, *P. furiosus* ferritin is another candidate for tumor targeting nanoprobes. The Fc-binding peptide was also genetically introduced into the exterior of *P. furiosus* ferritin and allowed the ferritin nanoprobes to target human breast cancer cells (with anti-Her2 antibody) and human oral epithelial cancer cells (with antifolate receptor antibody) (Kang et al., 2012). Because of the high stability of *P. furiosus* ferritin, in vitro applications based on this kind of ferritin may be more suitable.

DNA-binding proteins from starved cells (Dps) are 12-mer bacterial ferritins that protect DNA from oxidative stress, and have been implicated in bacterial survival and virulence. Like other ferritins, Dps proteins have a hollow spherical cage-like structure (Arnold, Zhou, & Barton, 2016). Genetic modification of the Dps with RGD peptide (GRGDSP) is another way to construct tumor targeting nanoprobes. Ahn et al. reported that RGD-modified Dps was able to specifically target human glioma cells (Ahn et al., 2014). Considering its relatively smaller size, the Dps-based tumor targeting nanoprobe may possess advantages (e.g., being easily excreted) in tumor imaging applications (Chakravarty, Goel, & Cai, 2014).

4.2 Direct Targeting of Ferritin Nanoprobes Against Tumors

Because the finding that TfR1 is the membrane receptor of human H-ferritin was reported, the intrinsic tumor targeting property of ferritin nanoprobes has drawn increasing attention from nanomedicine researchers. This finding brings us new insight into the physiological functions of ferritin and allows its application as a powerful nanoplatform for cancer diagnosis and therapy. As shown in Table 4, recent studies tend to use the intrinsic tumor targeting property of ferritin nanoprobes, instead of genetically or chemically modified ferritin nanoprobes.

Seaman's group demonstrated that after binding to TfR1 on the cell surface, the H-ferritin-TfR1 complex is internalized, and transported to endosomes and lysosomes (Li et al., 2010). The TfR1 (also named CD71) is a type II transmembrane glycoprotein, which forms a homodimer on the surface of cells (Daniels et al., 2006a). TfR1 was originally identified as the receptor for transferrin (Tf). The Tf-TfR1 regulated iron uptake pathway is the most important route for cellular iron uptake. In addition, TfR1 is also involved in regulating cell

Table 4: Tumor targeting strategies via ferritin-receptor interactions

Ferritin Types	Ferritin Receptors	Tumor Types	References
Human H-ferritin	Human TfR1	Human clinical tumor tissues: hepatocellular carcinoma, lung squamous cell carcinoma, colonic adenocarcinoma, cervical squamous cell carcinoma, prostate adenocarcinoma, ovarian serous papillary carcinoma, breast ductal carcinoma, thymic carcinoma, esophagus squamous cell carcinoma.	Fan et al. (2012)
Human H-ferritin	Human TfR1	Human hepatocellular carcinoma cell, SMMC-7721; human ovarian cancer cell, SKOV-3; human colon cancer cell, HT-29, HCT-116; human breast cancer cell, MDA-MB-231, MDA-MB-468; human melanoma cell, A375; human pancreatic cancer cell, Panc-1, CFPAC-1; human gastric cancer cell, MGC-803; human squamous cell carcinomas of the head and neck cell, FaDu, A-253, Detroit 562	Cai et al. (2015), Cao et al. (2014), Damiani et al. (2017), Fan et al. (2012), Lei et al. (2016), Liang et al. (2014), Pandolfi et al. (2017), and Tongwei et al. (2017)
Human H-ferritin	Mouse receptor (*need to be identified*)	Mouse breast cancer cell, 4T1; mouse lung cancer cell, 3LL	Huang et al. (2017b) and Mazzucchelli et al. (2016)
Horse spleen ferritin	Mouse Scara5 or human TfR1	Mouse breast cancer cell, TUBO; mouse glioma cell, C6; human hepatocellular carcinoma cell, HepG2; human breast cancer cell, MCF-7, MDA-MB-231; human cervical cancer cell, Hela; human prostate cancer cell, LNCaP; human colon cancer cell, HT-29	Chen et al. (2017b, 2017c), Conti et al. (2016), Dostalova et al. (2017), Geninatti Crich et al. (2015), Ghosh et al. (2016), Li et al. (2016a), Liu et al. (2011, 2012), Turino et al. (2017), Yan et al. (2008), and Yang et al. (2015)
Rat H-ferritin	Human receptor (*need to be identified*)	Human glioblastoma cell, U87MG;	Wang et al. (2016)
Pig pancreas ferritin	Human receptor (*need to be identified*)	Human gastric cancer cells, BGC823; human cervical cancer cell, Hela	Ji, Huang, and Huang (2012)
Pig spleen ferritin	Human receptor (*need to be identified*)	Human hepatocellular carcinoma cell, HepG2; human cervical cancer cell, Hela; human breast cancer cell, MCF-7	Liu et al. (2013)

growth (Neckers & Trepel, 1986; O'Donnell et al., 2006). It has been shown that expression of TfR1 in cancer cells may be up to 100-fold higher than in normal cells (Daniels et al., 2006a, 2006b). This might be due to rapidly proliferating cancer cells requiring more iron. The fact that TfR1 is overexpressed in a variety of malignancies and is efficiently internalized makes it an excellent target for tumor diagnosis and treatment (Daniels et al., 2006b).

The strategy of targeting TfR1 for tumor imaging and therapy has already been summarized by many excellent reviews (Daniels et al., 2006a, 2006b, 2012; Fan et al., 2013). Monoclonal antibodies to TfR1 and its natural ligand transferrin have been successfully used to target malignant cells, either alone or carrying various cytotoxic or imaging agents (Daniels et al., 2006b, 2012). For instance, transferrin (Tf)-conjugated diphtheria toxin, termed Tf-CRM107, has been applied in treatment of brain tumors and has been approved for phase III clinical trials (Laske, Youle, & Oldfield, 1997; Weaver & Laske, 2003). Anti-TfR1 antibody-conjugated ricin A chain, termed 454A12-RTA, is currently in phase II clinical trials (Daniels et al., 2012).

Instead of using transferrin and antibodies to target TfR1 (Daniels et al., 2006b), we have previously demonstrated the use of H-ferritin nanoprobes as a new ligand, and successfully targeted TfR1 on different tumor tissues. We screened 474 clinical samples and found that H-ferritin specifically binds to the nine most common human solid tumors, including liver, lung, colon, cervical, ovarian, prostate, breast, and thymus cancers (Fan et al., 2012). Furthermore, it has also been shown that TfR1 can be used as a prognosis indicator in breast cancer (Yang et al., 2001), leukemia (Das Gupta & Shah, 1990; Habeshaw et al., 1983), lung cancer (Kondo et al., 1990), and bladder cancer (Seymour et al., 1987). Thus, H-ferritin nanoprobes possess intrinsic broad-spectrum tumor targeting properties.

Consistent with our clinical tissue results, H-ferritin has been verified to specifically target human hepatocellular carcinoma cells, ovarian cancer cells, colon cancer cells, breast cancer cells, melanoma cells, pancreatic cancer cells, gastric cancer cells, and squamous cell carcinomas of the head and neck, of which all the tumor cells expressed high levels of TfR1 (Cai et al., 2015; Cao et al., 2014; Damiani et al., 2017; Fan et al., 2012; Lei et al., 2016; Liang et al., 2014; Pandolfi et al., 2017; Tongwei et al., 2017). Intriguingly, Sakamoto et al. recently demonstrated that H-ferritin is preferentially incorporated by TfR1-positive cells in a threshold-dependent manner, whereas there is no threshold for transferrin uptake (Sakamoto et al., 2015). This finding renders H-ferritin a more unique tumor targeting property than other ligands or antibodies of TfR1.

Interestingly, several studies recently reported that human H-ferritin also exhibits specific binding to mouse breast cancer cells (Mazzucchelli et al., 2016) and mouse lung cancer cells (Huang et al., 2017b). Because the receptor of mouse H-ferritin has been identified as TIM-2, according to the homological analysis, the potential receptor for human H-ferritin in mouse cancer cells may be TIM-2, but not mouse TfR1 (Huang et al., 2017b). However, this hypothesis needs to be further confirmed.

As we mentioned in Section 2, the commercially available horse spleen ferritin is extensively used in ferritin nanoprobes for tumor theranostics. After the intrinsic tumor targeting property of human H-ferritin was reported, the targeting property of HosFn was also investigated. To date, HosFn has been reported to specifically target mouse breast cancer (Conti et al., 2016) and glioma cells (Chen et al., 2017b), and human hepatocellular carcinoma cells (Liu et al., 2011, 2012; Yang et al., 2015), breast cancer cells (Conti et al., 2016; Geninatti Crich et al., 2015; Ghosh et al., 2016; Li et al., 2016a; Turino et al., 2017; Yan et al., 2008), cervical cancer cells (Chen et al., 2017c), prostate cancer cells (Dostalova et al., 2017), and colon cancer cells (Yang et al., 2015). To the best of our knowledge, HosFn specifically targets mouse cancer cells via cross-interacting with mouse Scara5 (Li et al., 2009), the receptor of mouse L-ferritin, while HosFn recognizes human cancer cells via binding to TfR1 (Li et al., 2010), the receptor of human H-ferritin.

Unexpectedly, some reports demonstrated that ferritins from other species also specifically bind to human cancer cells. For instance, Wang et al. found that rat H-ferritin targets human glioma cells (Wang et al., 2016). Ji et al. demonstrated that pig pancreatic ferritin specifically recognizes human gastric cancer cells and cervical cancer cells (Ji et al., 2012). Liu et al. reported that pig spleen ferritin specifically binds to human hepatocellular carcinoma cells, cervical cancer cells, and breast cancer cells (Liu et al., 2013). These results indicate that there are potential receptors in human cancer cells for these different ferritins. Considering the highly conserved structure of ferritins from different species and absolutely different receptors of human H-ferritin and mouse H-ferritin, it is necessary to systematically study the cross-interactions of ferritins and their reported receptors.

Taken together, the intrinsic tumor targeting property of natural ferritins will most likely be the key strategy to designing tumor targeting ferritin nanoprobes in the future.

5 Ferritin Nanoprobes for Tumor Imaging

With the emergence of nanotechnology, ferritins can be conveniently mineralized or synthesized to produce many kinds of nanoparticles without disruption of the integrity of the protein shells (Fan et al., 2013; Kashanian et al., 2012; Meldrum et al., 1991; Meldrum, Heywood, & Mann, 1992; Sun et al., 2011b; Suzumoto, Okuda, & Yamashita, 2012). By employing recombinant human H-ferritin as a template, Douglas's group biomimetically synthesized a type of magnetic nanoparticle in 2006 (Uchida et al., 2006). This engineered ferritin molecule contains an iron core in the form of magnetite (Fe_3O_4), which is different from the natural ferritin iron core of ferrihydrite (Fe_2O_3) (Cao et al., 2010; Uchida et al., 2006). This engineered ferritin, which has the same architecture as natural H-ferritin, is termed magnetoferritin. Because of the unique architecture, the ferritin nanoprobe provides an ideal nanoplatform for multifunctional loading to enhance the functionality of its surface (e.g., to target tumors), and metal cations can be encapsulated in the interior (e.g., contrast imaging probes) (Fig. 2).

Because the magnetoferritin nanoprobe is a biomimetic product, this means that natural ferritin nanoparticles exist in the human body. It is possible to directly use endogenous ferritin as the reporter to monitor disease progression in vivo with MRI rather than administering foreign premade ferritin nanoparticles (Cohen et al., 2009; Gilad et al., 2007). Kim et al. and Choia et al. have developed a model which allows tumor cells overexpressing H-ferritin to use ferritin as a reporter for MRI. This model could be used for in vivo tumor imaging and monitoring of tumor metastasis in lymph nodes (Choi et al., 2012; Kim et al., 2010). Lu et al. designed an endogenous ferritin reporter gene for magnetic resonance imaging (MRI) with a tumor-specific promoter (AFP-promoter) to monitor the in vivo tumor targeting of peptide-functionalized superparamagnetic iron oxide (Lu et al., 2017). The advantage of using ferritin as a reporter for MRI is that it does not require an exogenous contrast agent to be delivered to the targeted tumor area and makes it possible to carry out long-term in vivo imaging. However, the level of ferritin expression during tumor development needs to be investigated. Compared with magnetoferritin nanoparticles, the limitation of using endogenous ferritin is its inherent low sensitivity.

Magnetoferritin nanoparticles consist of iron nanocrystals. The superparamagnetism of magnetoferritin makes it an ideal contrast agent for magnetic resonance imaging (MRI) for tumor diagnosis. MRI can provide high spatial resolution and functional anatomic and physiological information with simultaneous noninvasive imaging. Normally, use of magnetoferritin nanoparticles for imaging results in reduced signal intensity in T_2-weighted MRI, which is also called T_2* signal loss (Uchida et al., 2006, 2008). The tumor-bearing area displays weakened intensity in MRI when using ferritin nanoparticles as the contrast agent compared to MRI of normal tissue. This negative change makes it difficult to acquire enough accurate information for tumor diagnosis.

Therefore, ferritin nanoprobe-based multimodality imaging is used, which combines MRI with other imaging modalities, such as fluorescence imaging or positron emission tomography (PET). Based on the special architecture of ferritin nanoprobes, two modification strategies could allow the performance of multimodal imaging in a single examination. One is introduction of functional groups onto the exterior surface of the nanoparticles by chemical or genetic modification of ferritin. For example, a fluorescent dye can be covalently conjugated onto H-ferritin (Uchida et al., 2006), and GFP could be fused into H-ferritin by genetic engineering (Li et al., 2012b). However, fluorescence imaging may not have sufficient signal intensity during noninvasive detection, especially when tumors are located in deep tissue, because the light could not effectively penetrate the skin and deep tissues even using in situ exposure (Terashima et al., 2011). Recently, our group developed a novel ferritin nanoprobe to acquire both MRI and nuclear imaging in a single dose via chemically conjugating magnetoferritin with ^{125}I. This strategically engineered magnetoferritin nanoprobe can image tumors with high sensitivity and specificity using SPECT and MRI in living mice after a single intravenous injection, thus enabling simultaneous functional and morphological tumor imaging without reliance on multiple injections (Zhao et al., 2016). Novel fluorescence imaging technology is also needed to overcome the limitations of MRI based on magnetoferritin. Chen's group has developed near-infrared fluorescence imaging (NIRF) by conjugation of Cy 5.5/ZW800/Alexa 750/IRdye800 dye (Li et al., 2012a; Lin et al., 2011a, 2011b; Zhen et al., 2013a, 2013b, 2016) onto H-ferritin for in vivo imaging of tumor xenografted mice either with NIRF only or combining it with PET for multimodal imaging (Lee et al., 2011; Lin et al., 2011a).

Another route is entrapping functional contrast agents into the core of the nanoparticles by regulating the disassembly/reassembly process under pH control (Huang et al., 2014). Gadolinium (Gd(III)) chelates and Mn^{II} could be loaded into the core of ferritin nanoprobes for in vivo tumor MRI with enhanced T1-weighted signal because of high R1 relaxivity (Aime, Frullano, & Crich, 2002; Crich et al., 2006; Kalman, Geninatti-Crich, & Aime, 2010; Szabo et al., 2012). Radioisotopes are another potential contrast agent that can be loaded into the core of the nanoparticles for multimodal tumor imaging. Chen's group loaded ^{64}Cu into the core of H-ferritin nanocages to carry out in vivo PET for tumor imaging by combining PET with NIRF imaging in a single examination (Lin et al., 2011a). In addition, Huang et al. loaded IR820, a photoacoustic imaging dye, into the cavity of H-ferritin nanocages to perform the in vivo PAI/NIRF multimodal imaging (Huang et al., 2014). The generality of modification and functionalization makes ferritin nanoparticles a promising nanoplatform to achieve multimodal imaging in translational cancer diagnosis.

Recently, our group developed magnetoferritin as a dual-functional reagent, allowing simultaneous targeting and visualization of tumors. Based on this finding, we developed a novel type of immunochemistry (IHC) technique for tumor detection. After verifying the interaction between H-ferritin and its receptor TfR1 in clinical tissue specimens, we found that H-ferritin specifically recognizes nine types of tumor tissues (Fan et al., 2012). In addition, we found that magnetoferritin nanoprobes have intrinsic peroxidase-like activity (Fan et al., 2012; Gao et al., 2007). Both the mineral phase composition and the size of the iron core encapsulated in ferritin protein shells determine the peroxidase activity of the magnetoferritin (Cai et al., 2015; Fan et al., 2012). Biomimetically synthesized magnetoferritin nanoparticles with mineral cores consisting of magnetite or maghemite exhibited a much higher peroxidase activity when compared with natural holoferritin. Thus, magnetoferritin nanoparticles can catalyze peroxidase substrates to produce the same color reaction as peroxidase enzymes. For instance, magnetoferritin nanoparticles react with TMB to produce a blue color and DAB to produce a brown precipitate. Furthermore, Zhang et al. reported that cobalt-doped magnetoferritin nanoprobes can further enhance the peroxidase-like activity of magnetoferritin, thus improving the sensitivity of the magnetoferritin-based IHC method for tumor detection (Tongwei et al., 2017).

With architectures similar to human H-ferritin, other types of ferritins are also extensively employed to develop multimodal imaging nanoprobes for tumor imaging. For horse spleen ferritin, Gd-HPDO3A/Mn^{II}/Melanin-Fe^{3+}/^{64}Cu were loaded into the cavity of HosFn for MRI/PAI/PET multimodal imaging via regulating the disassembly/reassembly process under pH control (Conti et al., 2016; Geninatti Crich et al., 2015; Kalman et al., 2010; Szabo et al., 2012; Yang et al., 2015). FAM was chemically conjugated onto the exterior surface of HosFn for NIRF imaging (Crich et al., 2006; Ji et al., 2013). CuS and ^{64}Cu were loaded into the cavity of a rat H-ferritin nanoprobe for PAI/PET imaging (Wang et al., 2016). For Dps, Cy5.5 dye was conjugated onto the surface of the protein cage for NIRF tumor imaging (Ahn et al., 2014).

As shown in Table 5, the unique architecture and physicochemical properties of ferritin nanoprobes have made it easy to construct multimode imaging nanoprobes for NIRF/MRI/SPECT/PET/PAI or IHC for tumor imaging in the last decades (Fig. 3). Combining optical, photoacoustic and nuclear imaging, and tumor targeting strategies could improve and refine the intraoperative tumor imaging, to accurately excise the tumor lesion with clear margins, and provide targeted theranostics tools for cancer diagnosis and therapy (Fig. 4).

Table 5: Ferritin nanoprobes and their applications in tumor imaging

Ferritin Nanoprobes	Encapsulation/Labeling Imaging Reagents	Imaging Methods	References
Human-H-ferritin	Fe_3O_4 (magnetite/maghemite)/$Co_xFe_{3-x}O_4$	IHC	Cai et al. (2015), Fan et al. (2012), and Tongwei et al. (2017)
Human-H-ferritin	Cy5.5 Fe_3O_4 (magnetite/maghemite) ^{125}I	NIRF MRI SPECT	Cao et al. (2014), Liang et al. (2014), Mazzucchelli et al. (2016), and Uchida et al. (2008)
Surface-modified human H-ferritin	Cy 5.5/ZW800/Alexa 750/IRdye800 ^{64}Cu	NIRF PET	Li et al. (2012a), Lin et al. (2011a, 2011b), and Zhen et al. (2013a, 2013b, 2016)
Surface-modified human H-ferritin	Fe_3O_4 (magnetite/maghemite)	MRI	Li et al. (2012b)
Surface-modified human H-ferritin	IR820	NIRF/PAI	Huang et al. (2014)
Horse spleen ferritin	Gd-HPDO3A/Mn^{II} Melanin/Fe^{3+} ^{64}Cu	MRI PAI PET	Conti et al. (2016), Geninatti Crich et al. (2015), Kalman et al. (2010), Szabo et al. (2012), and Yang et al. (2015)
Surface-modified horse spleen ferritin	FAM Gd-HPDO3A	NIRF MRI	Crich et al. (2006) and Ji et al. (2013)
Rat H-ferritin	CuS ^{64}Cu	PAI PET	Wang et al. (2016)
Surface-modified *Escherichia coli* DNA-binding protein	Cy 5.5	NIRF	Ahn et al. (2014)

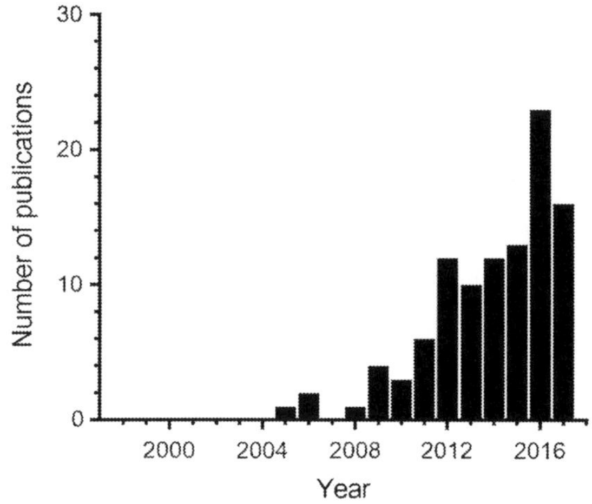

Fig. 3
Number of published papers on ferritin nanoprobes for tumor imaging by the end of September 2017. The data is based on Web of Science.

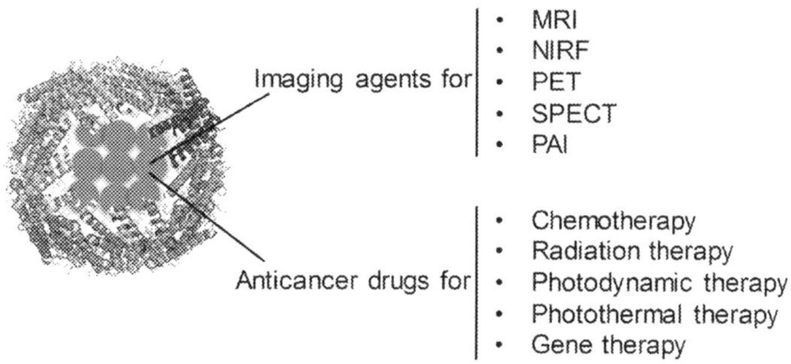

Fig. 4
Encapsulation cargos inside of ferritin nanoprobes for tumor theranostics.

6 Ferritin Nanocarriers for Tumor Therapy

In the last decades, plenty of studies have begun to design ferritin nanoprobes as carriers to deliver drugs for therapeutic purposes (Fig. 5). Instead of a magnetite core, apoferritin nanoparticles can be used to encapsulate chemical drugs in the interior cavity by manipulating disassembly/reassembly of the nanoprobe under pH control or urea concentration control (Kilic, Ozlu, & Calis, 2012; Liang et al., 2014; Simsek & Kilic, 2005). The first anticancer drug encapsulation inside a horse spleen ferritin nanocage was cisplatin and carboplatin, reported by Guo's group, with each ferritin encapsulating around 30 cisplatin molecules and 15 carboplatin molecules (Yang et al., 2007). Functional studies showed that ferritin with entrapped cisplatin could induce apoptosis of cancer cells (Falvo et al., 2013; Ji et al., 2012). However, Doxorubicin

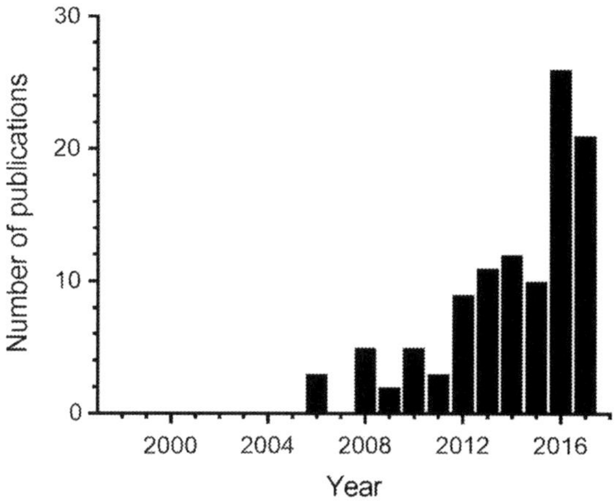

Fig. 5
Number of published papers on ferritin nanoprobes for tumor therapy by the end of September, 2017. The data is based on Web of Science.

(Dox) encapsulation in ferritin nanocages represents the most extensively studied system for tumor-targeted drug delivery (Table 6).

Dox is a widely used clinical anticancer drug for several types of solid tumors, while its use is dose-limited due to its severe side effects. The initial Dox-loading strategy was employing a disassembly/reassembly process under pH control by using HosFn, and the maximum loading was 28 Dox per ferritin molecule, but the yield was very low (Kilic et al., 2012). After analyzing the structure and physicochemical properties of human H-ferritin, we developed, for the first time, a urea-based Dox-loading strategy, improved the loading efficiency of HFn-Dox and evaluated the antitumor activity of HFn-Dox in vivo in three different human tumor xenograft tumor models (Liang et al., 2014). Following our finding, Lei et al. and Mazzucchelli et al. also developed HFn-Dox and tested the antitumor activity in cell lines and tumor models (Lei et al., 2016; Mazzucchelli et al., 2016). Zhen et al. demonstrated that Dox precomplexed with Cu(II) can be effectively loaded into ferritin nanocages, and ferritin-Cu-Dox exhibits significant antitumor activity in vitro and in vivo (Zhen et al., 2013a). However, this heavy-metal dependent drug loading strategy is hard to translate.

Recently, Ceci's group reported that fusion of human H-ferritin with proline, serine, and alanine elements and chemical modification with PEG significantly improve the Dox-loading efficiency (up to 90 Dox per ferritin nanocage) (Table 6) (Falvo et al., 2016). They also demonstrated that this modified ferritin drug delivery system exhibits significant antitumor activity in human fibrosarcoma xenograft model (Fracasso et al., 2016) and head and neck cancer xenograft model (Damiani et al., 2017); however, this modification also brings new challenges to translational research, and the researchers underestimated the potential impact of additional amino acid sequences on biosafety and the in vivo response (Tosi et al., 2016).

Table 6: Ferritin nanocarriers and their antitumor effects

Ferritin Nanoprobes	Encapsulation Drugs	Encapsulating Methods	Loading Efficiency (Drug/Ferritin or wt%)	Treatment Details and Tumor Growth Inhibition (TGI)	References
Horse spleen ferritin	Doxorubicin (anticancer drug)	Disassembly at pH 2.5, then reassembly at pH 7	28 molecules	–	Kilic et al. (2012)
Human H-ferritin	Doxorubicin	Disassembly at 8 M urea, then gradual reassembly from 7 M urea to PBS	33 molecules	Single dose of 20 mg Dox (eqv.)/kg; ~90% TGI for human colon cancer xenograft; ~86.67% TGI for human melanoma xenograft; ~75% TGI for human breast cancer xenograft model	Liang et al. (2014)
Human H-ferritin	Doxorubicin	Disassembly at 8 M urea, then gradual reassembly from 7 M urea to PBS	32.5 molecules	–	Lei et al. (2016)
Human H-ferritin	Doxorubicin	Disassembly at 8 M urea, then gradual reassembly from 7 M urea to PBS	29 molecules	1.24 mg/kg of Dox (eqv.) were i.v. injected every 4 days for 3 weeks; 75% TGI for mouse breast cancer	Mazzucchelli et al. (2016)
Surface-modified human H-ferritin	Doxorubicin	Precomplexation with Cu(II). then incubation with ferritin nanoprobes	73.41 wt%	5 mg Dox (eqv.)/kg) were i.v. injected every 2 days for 2 weeks; 89.6% TGI for human glioma xenograft model	Zhen et al. (2013a)
Surface-modified human H-ferritin	Doxorubicin	Disassembly at pH 2, then reassembly at pH 8.0	90 molecules	5 mg/kg of Dox (eqv.) were i.v. injected twice a week for two weeks; ~90.9% TGI for human fibrosarcoma xenograft model	Falvo et al. (2016) and Fracasso et al. (2016)
Surface-modified human H-ferritin	Doxorubicin	Disassembly at pH 2, then reassembly at pH 8.0	86 molecules	5 mg/kg of Dox (eqv.) were i.v. injected five times twice a week; ~88.88% TGI for human head and neck cancer xenograft model	Damiani et al. (2017)
Surface-modified human H-ferritin	Doxorubicin	Chemical conjugation	88 molecules	Single dose of 0.25 mg/kg Dox (eqv.) was administered intratracheally; 40-fold decrease of tumor BLI signal for mouse lung cancer model	Huang et al. (2017b)
Horse spleen ferritin	Methylene blue (PDT drug)	Disassembly at pH 2, then reassembly at pH 7	1 molecule	–	Yan et al. (2008, 2010)
Horse spleen ferritin	Daunomycin (anticancer drug)	Loading drug at pH 5, assistance with poly-L-aspartic acid	0.2 molecule	–	Ma-Ham et al. (2011)

Continued

Table 6: Ferritin nanocarriers and their antitumor effects—cont'd

Ferritin Nanoprobes	Encapsulation Drugs	Encapsulating Methods	Loading Efficiency (Drug/Ferritin or wt%)	Treatment Details and Tumor Growth Inhibition (TGI)	References
Surface-modified human H-ferritin	ZnF16Pc (PDT drug)	Incubation in PBS	60 wt%	i.v. injection of 1.5 mg ZnF16Pc (eqv.)/kg, with irradiation of 671 nm laser, 0.3 W/cm^2 for 15 min; 83.64% TGI of human glioma xenograft model; 82.65% TGI for mouse breast cancer model with folate receptor target; 85.9% TGI of mouse breast cancer model in combination with Doxil (10 mg/kg); 88.60% TGI for mouse breast cancer model in combination with FAP targeting	Zhen et al. (2013b, 2014, 2015, 2016)
Surface-modified human H-ferritin	IR820 (PTT drug)	Disassembly at pH 2, then reassembly at pH 7.4	17.32 wt%	0.4 mg IR820 (eqv.) with irradiation of 808 nm laser, 0.5–1 W/cm^2 for 10 min; 100% tumor elimination of mouse breast cancer model	Huang et al., 2014
Surface-modified human H-ferritin	Sinoporphyrin sodium (PDT and PTT drug)	Disassembly at pH 2, then reassembly at pH 7.4	66.67 wt%	8 mg/kg of drug (eqv.) was injected i.v., then exposed to a 630-nm laser at a laser dose of 75 J; 100% tumor ablation for mouse breast cancer model	Huang et al. (2017a)
Horse spleen ferritin	Cisplatin (anticancer drug)	Disassembly at pH 2, then reassembly at pH 7	30 molecules	-	Yang et al. (2007)
Horse spleen ferritin	Cisplatin	Disassembly at pH 13, then reassembly at pH 7	20–55 molecules	-	Pontillo et al. (2016)
Pig pancreas ferritin	Cisplatin	Disassembly at pH 13, then reassembly at pH 7	11.26 molecules	-	Ji et al. (2012)
Antibody-conjugated human H-ferritin	Cisplatin	Disassembly at pH 2.2, then reassembly at pH 7.5	50 molecules	3 mg cisplatin (eqv.)/kg, were i.v. injected twice. 64.29% TGI of human melanoma xenograft model	Falvo et al. (2013)
Horse spleen ferritin	Carboplatin	Disassembly at pH 2, then reassembly at pH 7	15 molecules	-	Yang et al. (2007)
Human H-ferritin	Curcumin (anticancer drug)	Disassembly at pH 12.5, then reassembly at pH 7.4	90 molecules	-	Pandolfi et al. (2017)

Human H-ferritin	Curcumin	Disassembly at pH 2, then reassembly at pH 7.5	14.7 molecules	-	Chen et al. (2014b)
Truncated human H-ferritin	Curcumin	Disassembly at pH 4.0, then reassembly at pH 7.5	14 molecules	-	Chen et al. (2016)
Horse spleen ferritin	Curcumin	Disassembly at pH 2, then reassembly at pH 7	9.6 molecules	10 mg Curcumin (eqv.)/kg of were i.v. injected every 3 days for 6 weeks; 60% of mouse breast cancer response to the treatment	Conti et al. (2016), Cutrin et al. (2013), and Geninatti Crich et al. (2015)
Horse spleen ferritin	Gold-based anticancer drugs (Auoxo3/Auoxo4/ Au_2phen)	Disassembly at pH 13, then reassembly at pH 7.4	210 Auoxo3 192 Auoxo4 216 Au_2phen	-	Ferraro et al. (2016) and Monti et al. (2017)
Human H-ferritin	Atropine (anticancer drug)	Disassembly at 8 M urea, then gradual reassembly from 7 M urea to PBS	46.7 molecules	0.30–0.45 mg/kg of atropine (eqv.) were i.v. injected twice a week for 4 weeks; ~86.67% TGI for human pancreatic cancer xenograft model.	Lei et al. (2016)
Human H-ferritin	β-Carotene (anticancer drug)	Disassembly at pH 11, then reassembly at pH 7.5	12.4 molecules	-	Chen et al., 2014a
Horse spleen ferritin	^{235}U (anticancer radiation reagent)	Incubation in PBS	~800 molecules	-	Hainfeld (1992)
Pig spleen ferritin	5-Fluorouracil (anticancer drug)	Incubation with gold nanoparticle loaded ferritin nanoprobes	45.5 molecules	-	Liu et al. (2013)
Human H-ferritin	siRNA	Disassembly at pH 2, then reassembly at pH 8.5	0.2 molecule	-	Li et al. (2016b)
Human H-ferritin	Gefitinib	Incubation in Tris buffer (pH 8.0)	~10 molecules	-	Kuruppu et al. (2015)

In addition to encapsulating chemical drugs in its cavity, the exterior surface is another platform to load protein drugs, such as antibodies, toxins, and peptides, by genetic engineering or chemical conjugation. Interestingly, Huang et al. reported that a human H-ferritin nanoprobe can penetrate airway mucus and tumor tissue to treat lung cancers. Employing chemical conjugation, Huang et al. loaded up to 88 Dox on the surface of H-ferritin, and further modified it with PEG. This airway mucus-penetrating ferritin delivery system exhibits excellent antilung cancer activity (Huang et al., 2017b).

Besides Dox, Daunomycin was successfully loaded inside of horse spleen ferritin with about 0.2 molecules per nanocage by a poly-L-aspartic acid-assisted loading strategy, although its antitumor activity was absent (Ma-Ham et al., 2011). Other types of anticancer drugs, such as curcumin, have been effectively loaded inside of human H-ferritin (Pandolfi et al., 2017), truncated human H-ferritin (Chen et al., 2016), or HosFn (Conti et al., 2016) via disassembly/reassembly of the nanoprobes under pH control. The significant antitumor activity of HosFn-curcumin was demonstrated in a mouse breast cancer model (Conti et al., 2016; Cutrin et al., 2013; Geninatti Crich et al., 2015). Anticancer gold(III) compounds, such as Auoxo3, Auoxo4, and Au_2phen, were effectively encapsulated within the HosFn nanocages and exhibited significant cytotoxic effects on different human cancer cells (Ferraro et al., 2016; Monti et al., 2017). β-carotene was loaded inside of human H-ferritin via a pH control loading strategy, but with no antitumor activity (Chen et al., 2014a). Atropine was abundantly encapsulated inside of the human H-ferritin nanocage by a urea loading strategy, and exhibited excellent antitumor activity in a human pancreatic cancer xenograft model (Lei et al., 2016). 5-Fluorouracil was successfully encapsulated inside of pig spleen ferritin with the assistance of gold nanoparticles, and Gefitinib was loaded inside of human H-ferritin via incubation in Tris buffer. Both of them lack in vivo antitumor studies (Kuruppu et al., 2015; Liu et al., 2013). siRNA was also successfully encapsulated inside of the human H-ferritin nanocage by a pH control loading strategy, while the antitumor activity of HFn-siRNA was only tested in tumor cells in vitro (Li et al., 2016b). This report paves the way for future application of ferritin nanoprobes in gene therapy, but more efforts are needed in this field.

With the unique property of storing metal ions, ferritin-based nanocages could be exploited to deliver radioisotopes for increased loading efficiency and improving pharmacokinetics in vivo. Hainfeld reported that ^{235}U can be effectively loaded (~800 molecules per nanocage) inside of HosFn via incubation in PBS, and the ^{235}U-HosFn nanocages are able to kill tumor cells (Hainfeld, 1992). The clinically used isotopes, such as ^{90}yttrium and ^{177}lutetium, are also able to use this ferritin delivery system to improve the tumor targeting properties and in vivo behaviors.

The drugs used for photodynamic therapy (PDT) or photothermal therapy (PTT), such as Methylene blue, ZnF16Pc, IR820 and Sinoporphyrin sodium, typically have high immunogenicity, nonbiodegradability, long-term toxicity, and poor pharmacokinetics

(Huang et al., 2014; Yan et al., 2008, 2010). Ferritin nanoprobes provide an ideal nanocarrier for PDT and PTT drug delivery. Employing RGD-modified human H-ferritin, Xie's group demonstrated that ZnF16Pc was effectively encapsulated inside of the ferritin nanocage via incubation and exhibits excellent PDT effects in treating a human glioma xenograft model, combining with Doxil to treat a mouse breast cancer model, or treating a mouse breast cancer model in combination with FAP targeting (Zhen et al., 2013b, 2014, 2015, 2016). Huang et al. reported that IR820 could be abundantly encapsulated inside of the surface-modified human H-ferritin via a pH control loading strategy, and the IR820-ferritin nanoprobes exhibit 100% tumor elimination in a mouse breast cancer model (Huang et al., 2014). Recently, Liu's group developed surface-modified human H-ferritin as the nanocarrier of Sinoporphyrin sodium by a pH control loading strategy, and demonstrated that the PDT/PTT treatments exhibit 100% tumor ablation in a mouse breast cancer model (Huang et al., 2017a).

In summary, taking advantage of the disassembly/reassembly of the ferritin nanoprobe under pH control or urea concentration control, anticancer drugs (e.g., Dox, Daunomycin, Cisplatin, Carboplatin, Curcumin, Atropine, β-carotene, 5-fluorouracil and Gefitinib), radioisotopes (^{235}U), siRNA and PDT/PTT drugs (e.g., Methylene blue, ZnF16Pc, IR820 and Sinoporphyrin sodium) have been successfully loaded inside ferritin nanocages, and some of them have been evaluated for antitumor activity in tumor models (Fig. 4). It will be interesting to use this nanoplatform to combine multimodal imaging and targeted drug delivery for cancer diagnosis and therapy.

7 Conclusions and Prospects

Genetically or chemically conjugating targeting motifs onto the surface of ferritin is the major strategy to improve ferritin nanoprobe tumor homing. While the strategy has been demonstrated to be very promising, it also brings drawbacks that make the bioengineered nanoprobes hard to translate from bench to bedside. These modifications may alter the nanoprobes' immunogenicity upon entering the body, significantly reduce the yield of ferritin nanoprobes and increase the complexity of purification of ferritin nanoprobes (Liang et al., 2014; Uchida et al., 2006; Zhao et al., 2016). Thus, the surface-modified ferritin nanoprobes may not be the best choice for the translational studies of ferritin-based tumor theranostics.

The identification of ferritin receptors, especially the reporting of TfR1 as the receptor of human H-ferritin, reveals its intrinsic tumor targeting property (Fan et al., 2012; Li et al., 2010). This intrinsic property, together with its unique architecture, excellent biocompatibility and reversible disassembly/reassembly ability, makes ferritins an ideal nanocarrier for tumor theranostics. Combined with other numerous favorable features of ferritin such as stability, homogeneity, and ease of scale-up, ferritin nanoprobes are promising candidates for translation to clinical use. However, we point out that different ferritins from different species do not share the same receptors. So, when employing ferritin as a nanocarrier to develop an antidisease system, we must clearly recognize which receptor is suitable. Because HosFn is cheap and conveniently supplied, it still has potential for research in the near future. For now, the cross-interaction receptors of HosFn are human TfR1 and mouse Scara5, and other interactions are yet to be confirmed.

Ideally, human ferritins have more advantages in comparison with HosFn, for example, they are of human origin, which means that they have better biocompatibility and no immunogenic reaction. The receptors of human ferritins are more comprehensive, especially H-ferritin, of which the receptor is TfR1, a well-known tumor biomarker in tumor therapy. Moreover, lab production of recombinant human ferritins is straightforward, elementary, robust, and convenient. Thus, we recommend using human ferritin as the nanocarrier for translational studies.

Dealing with the potential toxicity and immune response of the human body to ferritin nanoprobes is important for its translational application in clinical imaging and therapy. Revealing the physiological and pathological functions of ferritin will help us to better understand the potential effects of its use in the human body. Although numerous new findings have provided us with insights to better understand this protein (i.e., human H-ferritin recognizes TfR1; serum ferritin levels correlate with tumor levels), the novel and unexpected functions of ferritin continue to be revealed, and further study of this interesting molecule is warranted. For example, the issue of the source of secretory ferritin and identification of human L-ferritin receptors; the mechanism and function of ferritin in its involvement in many pathological processes; and the exact binding site of the human H-ferritin/TfR1 interaction are aspects that will be important to unravel (Fan et al., 2013). In addition, the influence of human ferritin nanoprobes on iron homeostasis, the biodistribution and clearance of these nanoprobes after entering the human body are the main concern for the current translational studies of ferritin nanoprobes. As ferritin nanoprobes have been recognized as an ideal nanoplatform because of their physiological features and bioengineering versatility, further advances in their clinical applications for tumor diagnosis, imaging, and therapy are expected.

References

Adams, P. C., Powell, L. W., & Halliday, J. W. (1988). Isolation of a human hepatic ferritin receptor. *Hepatology*, *8*(4), 719–721.

Ahn, K. Y., et al. (2014). Engineered protein nanoparticles for in vivo tumor detection. *Biomaterials*, *35*(24), 6422–6429.

Aime, S., Frullano, L., & Crich, S. G. (2002). Compartmentalization of a gadolinium complex in the apoferritin cavity: a route to obtain high relaxivity contrast agents for magnetic resonance imaging. *Angewandte Chemie, International Edition*, *41*(6), 1017.

Alkhateeb, A. A., & Connor, J. R. (2010). Nuclear ferritin: a new role for ferritin in cell biology. *Biochimica et Biophysica Acta-General Subjects*, *1800*(8), 793–797.

Arnold, A. R., Zhou, A., & Barton, J. K. (2016). Characterization of the DNA-mediated oxidation of Dps, a bacterial ferritin. *Journal of the American Chemical Society*, *138*(35), 11290–11298.

Arosio, P., Ingrassia, R., & Cavadini, P. (2009). Ferritins: a family of molecules for iron storage, antioxidation and more. *Biochimica et Biophysica Acta*, *1790*(7), 589–599.

Belletti, D., et al. (2017). Protein cage nanostructure as drug delivery system: magnifying glass on apoferritin. *Expert Opinion on Drug Delivery*, *14*(7), 825–840.

Cai, Y., et al. (2015). Enhanced magnetic resonance imaging and staining of cancer cells using ferrimagnetic H-ferritin nanoparticles with increasing core size. *International Journal of Nanomedicine*, *10*, 2619–2634.

Cao, C., et al. (2014). Targeted in vivo imaging of microscopic tumors with ferritin-based nanoprobes across biological barriers. *Advanced Materials*.

Cao, C. Q., et al. (2010). Magnetic characterization of noninteracting, randomly oriented, nanometer-scale ferrimagnetic particles. *Journal of Geophysical Research. Solid Earth*, *115*.

Carpagnano, G. E., et al. (2012). Could exhaled ferritin and SOD be used as markers for lung cancer and prognosis prediction purposes? *European Journal of Clinical Investigation*, *42*(5), 478–486.

Caulfield, J. B. (1963). Studies on ferritin uptake by isolated tumor cells. *Laboratory Investigation*, *12*(10), 1018–1025.

Cazzola, M., et al. (1985). Immunological reactivity of serum ferritin in patients with malignancy. *Tumori*, *71*(6), 547–554.

Chakravarty, R., Goel, S., & Cai, W. (2014). Nanobody: the "magic bullet" for molecular imaging? *Theranostics*, *4*(4), 386–398.

Chauhan, V. P., & Jain, R. K. (2013). Strategies for advancing cancer nanomedicine. *Nature Materials*, *12*(11), 958–962.

Chen, H., et al. (2016). Engineering protein interfaces yields ferritin disassembly and reassembly under benign experimental conditions. *Chemical Communications (Cambridge)*, *52*(46), 7402–7405.

Chen, H., et al. (2017a). Rethinking cancer nanotheranostics. *Nature Reviews Materials*, *2*, 17024.

Chen, L., et al. (2014a). Encapsulation of β-carotene within ferritin nanocages greatly increases its water-solubility and thermal stability. *Food Chemistry*, *149*, 307–312.

Chen, L. L., et al. (2014b). Encapsulation of curcumin in recombinant human H-chain ferritin increases its water-solubility and stability. *Food Research International*, *62*, 1147–1153.

Chen, T. T., et al. (2005). TIM-2 is expressed on B cells and in liver and kidney and is a receptor for H-ferritin endocytosis. *Journal of Experimental Medicine*, *202*(7), 955–965.

Chen, W., et al. (2017c). H2S activated drug release from protein cages. *ACS Applied Materials & Interfaces*.

Chen, Z., et al. (2017b). Apoferritin nanocage for brain targeted doxorubicin delivery. *Molecular Pharmaceutics*.

Cheng, Z., et al. (2012). Multifunctional nanoparticles: cost versus benefit of adding targeting and imaging capabilities. *Science*, *338*(6109), 903–910.

Choi, S. H., et al. (2012). Imaging and quantification of metastatic melanoma cells in lymph nodes with a ferritin MR reporter in living mice. *NMR in Biomedicine*, *25*(5), 737–745.

Cohen, B., et al. (2009). Ferritin nanoparticles as magnetic resonance reporter gene. *WIREs Nanomedicine and Nanobiotechnology*, *1*(2), 181–188.

Conti, L., et al. (2016). L-Ferritin targets breast cancer stem cells and delivers therapeutic and imaging agents. *Oncotarget*, *7*(41), 66713–66727.

Crich, S. G., et al. (2006). Magnetic resonance visualization of tumor angiogenesis by targeting neural cell adhesion molecules with the highly sensitive gadolinium-loaded apoferritin probe. *Cancer Research*, *66*(18), 9196–9201.

Cutrin, J. C., et al. (2013). Curcumin/Gd loaded apoferritin: a novel "theranostic" agent to prevent hepatocellular damage in toxic induced acute hepatitis. *Molecular Pharmaceutics*, *10*(5), 2079–2085.

Damiani, V., et al. (2017). Therapeutic efficacy of the novel stimuli-sensitive nano-ferritins containing doxorubicin in a head and neck cancer model. *International Journal of Molecular Sciences*, *18*(7).

Daniels, T. R., et al. (2006a). The transferrin receptor. Part I: biology and targeting with cytotoxic antibodies for the treatment of cancer. *Clinical Immunology*, *121*(2), 144–158.

Daniels, T. R., et al. (2006b). The transferrin receptor. Part II: Targeted delivery of therapeutic agents into cancer cells. *Clinical Immunology*, *121*(2), 159–176.

Daniels, T. R., et al. (2012). The transferrin receptor and the targeted delivery of therapeutic agents against cancer. *Biochimica et Biophysica Acta*, *1820*(3), 291–317.

Das Gupta, A., & Shah, V. I. (1990). Correlation of transferrin receptor expression with histologic grade and immunophenotype in chronic lymphocytic leukemia and non-Hodgkin's lymphoma. *Hematologic Pathology*, *4*(1), 37–41.

de Almeida, S. M., et al. (2008). Quantification of cerebrospinal fluid ferritin as a biomarker for CNS malignant infiltration. *Arquivos de Neuro-Psiquiatria*, *66*(3B), 720–724.

Dostalova, S., et al. (2016). Site-directed conjugation of antibodies to apoferritin nanocarrier for targeted drug delivery to prostate cancer cells. *ACS Applied Materials & Interfaces*, *8*(23), 14430–14441.

Dostalova, S., et al. (2017). Apoferritin as an ubiquitous nanocarrier with excellent shelf life. *International Journal of Nanomedicine*, *12*, 2265–2278.

Drysdale, J., et al. (2002). Mitochondrial ferritin: a new player in iron metabolism. *Blood Cells, Molecules & Diseases*, *29*(3), 376–383.

Easty, G. C., Yarnell, M. M., & Andrews, R. D. (1964). The uptake of proteins by normal and tumour cells in vitro. *British Journal of Cancer*, *18*, 354–367.

Elliott, R. L., et al. (1993). Breast-carcinoma and the role of iron-metabolism—a cytochemical, tissue-culture, and ultrastructural-study. *Breast Cancer: From Biology to Therapy*, *698*, 159–166.

Falvo, E., et al. (2013). Antibody-drug conjugates: targeting melanoma with cisplatin encapsulated in protein-cage nanoparticles based on human ferritin. *Nanoscale*, *5*(24), 12278–12285.

Falvo, E., et al. (2016). Improved doxorubicin encapsulation and pharmacokinetics of ferritin-fusion protein nanocarriers bearing proline, serine, and alanine elements. *Biomacromolecules*, *17*(2), 514–522.

Fan, K., Gao, L., & Yan, X. (2013). Human ferritin for tumor detection and therapy. *WIREs Nanomedicine and Nanobiotechnology*, *5*(4), 287–298.

Fan, K., Zhou, M., & Yan, X. (2017). Questions about horse spleen ferritin crossing the blood brain barrier via mouse transferrin receptor 1. *Protein & Cell*, *8*(11), 788–790.

Fan, K., et al. (2012). Magnetoferritin nanoparticles for targeting and visualizing tumour tissues. *Nature Nanotechnology*, *7*(7), 459–464.

Fantechi, E., et al. (2014). A smart platform for hyperthermia application in cancer treatment: cobalt-doped ferrite nanoparticles mineralized in human ferritin cages. *ACS Nano*, *8*(5), 4705–4719.

Fargion, S., et al. (1988). Characteristics and expression of binding sites specific for ferritin H-chain on human cell lines. *Blood*, *71*(3), 753–757.

Ferraro, G., et al. (2016). Gold-based drug encapsulation within a ferritin nanocage: X-ray structure and biological evaluation as a potential anticancer agent of the Auoxo3-loaded protein. *Chemical Communications*, *52*(61), 9518–9521.

Fracasso, G., et al. (2016). Selective delivery of doxorubicin by novel stimuli-sensitive nano-ferritins overcomes tumor refractoriness. *Journal of Controlled Release*, *239*, 10–18.

Gao, L., et al. (2007). Intrinsic peroxidase-like activity of ferromagnetic nanoparticles. *Nature Nanotechnology*, *2*(9), 577–583.

Geninatti Crich, S., et al. (2015). Targeting ferritin receptors for the selective delivery of imaging and therapeutic agents to breast cancer cells. *Nanoscale*, *7*(15), 6527–6533.

Ghosh, S., et al. (2016). Apoferritin nanocage delivers combination of microtubule and nucleus targeting anticancer drugs. *ACS Applied Materials & Interfaces*, *8*(45), 30824–30832.

Gilad, A. A., et al. (2007). Developing MR reporter genes: promises and pitfalls. *NMR in Biomedicine*, *20*(3), 275–290.

Gray, C. P., Arosio, P., & Hersey, P. (2003). Association of increased levels of heavy-chain ferritin with increased CD4+ CD25+ regulatory T-cell levels in patients with melanoma. *Clinical Cancer Research*, *9*(7), 2551–2559.

Greish, K. (2010). Enhanced permeability and retention (EPR) effect for anticancer nanomedicine drug targeting. *Methods in Molecular Biology*, *624*, 25–37.

Habeshaw, J. A., et al. (1983). Correlation of transferrin receptor expression with histological class and outcome in non-Hodgkin lymphoma. *Lancet*, *1*(8323), 498–501.

Hainfeld, J. F. (1992). Uranium-loaded apoferritin with antibodies attached—molecular design for uranium neutron-capture therapy. *Proceedings of the National Academy of Sciences of the United States of America*, *89*(22), 11064–11068.

Harrison, P. M. (1986). The structure and function of ferritin. *Biochemical Education*, *14*(4), 154–162.

Harrison, P. M., & Arosio, P. (1996). The ferritins: molecular properties, iron storage function and cellular regulation. *Biochimica et Biophysica Acta*, *1275*(3), 161–203.

He, D. D., & Marles-Wright, J. (2015). Ferritin family proteins and their use in bionanotechnology. *New Biotechnology*, *32*(6), 651–657.

Heger, Z., et al. (2014). Apoferritin applications in nanomedicine. *Nanomedicine (London, England)*, *9*(14), 2233–2245.

Huang, C., et al. (2017a). Ultra-high loading of sinoporphyrin sodium in ferritin for single-wave motivated photothermal and photodynamic co-therapy. *Biomaterials Science*, *5*(8), 1512–1516.

Huang, P., et al. (2014). Dye-loaded ferritin nanocages for multimodal imaging and photothermal therapy. *Advanced Materials*, *26*(37), 6401–6408.

Huang, X., et al. (2017b). Protein nanocages that penetrate airway mucus and tumor tissue. *Proceedings of the National Academy of Sciences of the United States of America*, *114*(32), E6595–E6602.

Hwang, M. P., et al. (2013). Think modular: a simple apoferritin-based platform for the multifaceted detection of pancreatic cancer. *ACS Nano*, *7*(9), 8167–8174.

Jain, P., et al. (2016). Elevated ferritin predicts for inferior survival in patients with acute leukemia and may be an early marker of a underlying systemic pathologic inflammation. *Blood*, *128*(22), 2791.

Jezequel, P., et al. (2012). Validation of tumor-associated macrophage ferritin light chain as a prognostic biomarker in node-negative breast cancer tumors: a multicentric 2004 national PHRC study. *International Journal of Cancer*, *131*(2), 426–437.

Ji, T., et al. (2013). Tumor fibroblast specific activation of a hybrid ferritin nanocage-based optical probe for tumor microenvironment imaging. *Small*, *9*(14), 2427–2431.

Ji, X. T., Huang, L., & Huang, H. Q. (2012). Construction of nanometer cisplatin core-ferritin (NCC-F) and proteomic analysis of gastric cancer cell apoptosis induced with cisplatin released from the NCC-F. *Journal of Proteomics*, *75*(11), 3145–3157.

Kalman, F. K., Geninatti-Crich, S., & Aime, S. (2010). Reduction/dissolution of a beta-MnOOH nanophase in the ferritin cavity to yield a highly sensitive, biologically compatible magnetic resonance imaging agent. *Angewandte Chemie (International Ed in English)*, *49*(3), 612–615.

Kang, H. J., et al. (2012). Developing an antibody-binding protein cage as a molecular recognition drug modular nanoplatform. *Biomaterials*, *33*(21), 5423–5430.

Kang, S., et al. (2008). Controlled assembly of bifunctional chimeric protein cages and composition analysis using noncovalent mass spectrometry. *Journal of the American Chemical Society*, *130*(49), 16527–16529.

Karimi, M., et al. (2017). Nanocaged platforms: modification, drug delivery and nanotoxicity: opening synthetic cages to release the tiger. *Nanoscale*, *9*(4), 1356–1392.

Kashanian, S., et al. (2012). Biomimetic synthesis and characterization of cobalt nanoparticles using apoferritin, and investigation of direct electron transfer of Co(NPs)-ferritin at modified glassy carbon electrode to design a novel nanobiosensor. *Molecular Biology Reports*, *39*(9), 8793–8802.

Kilic, M. A., Ozlu, E., & Calis, S. (2012). A novel protein-based anticancer drug encapsulating nanosphere: apoferritin-doxorubicin complex. *Journal of Biomedical Nanotechnology*, *8*(3), 508–514.

Kim, H. S., et al. (2010). In vivo imaging of tumor transduced with bimodal lentiviral vector encoding human ferritin and green fluorescent protein on a 1.5T clinical magnetic resonance scanner. *Cancer Research*, *70*(18), 7315–7324.

Kim, J.-w., et al. (2017). Genetically functionalized ferritin nanoparticles with a high-affinity protein binder for immunoassay and imaging. *Analytica Chimica Acta*.

Kim, S., et al. (2016a). Double-chambered ferritin platform: dual-function payloads of cytotoxic peptides and fluorescent protein. *Biomacromolecules*, *17*(1), 12–19.

Kim, S., et al. (2016b). Designing peptide bunches on nanocage for bispecific or superaffinity targeting. *Biomacromolecules*, *17*(3), 1150–1159.

Kitagawa, T., et al. (2012). RGD-conjugated human ferritin nanoparticles for imaging vascular inflammation and angiogenesis in experimental carotid and aortic disease. *Molecular Imaging and Biology*, *14*(3), 315–324.

Knovich, M. A., et al. (2009). Ferritin for the clinician. *Blood Reviews*, *23*(3), 95–104.

Kondo, K., et al. (1990). Transferrin receptor expression in adenocarcinoma of the lung as a histopathologic indicator of prognosis. *Chest*, *97*(6), 1367–1371.

Kuruppu, A. I., et al. (2015). An apoferritin-based drug delivery system for the tyrosine kinase inhibitor gefitinib. *Advanced Healthcare Materials*, *4*(18), 2816–2821.

Laske, D. W., Youle, R. J., & Oldfield, E. H. (1997). Tumor regression with regional distribution of the targeted toxin TF-CRM107 in patients with malignant brain tumors. *Nature Medicine*, *3*(12), 1362–1368.

Lee, E. J., et al. (2015). Engineered proteinticles for targeted delivery of siRNA to cancer cells. *Advanced Functional Materials*, *25*(8), 1279–1286.

Lee, S., et al. (2011). Hybrid ferritin nanoparticles as activatable probes for tumor imaging. *Angewandte Chemie, International Edition*, *50*(7), 1569–1572.

Lei, Y., et al. (2016). Targeted tumor delivery and controlled release of neuronal drugs with ferritin nanoparticles to regulate pancreatic cancer progression. *Journal of Controlled Release*, *232*, 131–142.

Li, J. Y., et al. (2009). Scara5 is a ferritin receptor mediating non-transferrin iron delivery. *Developmental Cell*, *16*(1), 35–46.

Li, K., et al. (2012b). Multifunctional ferritin cage nanostructures for fluorescence and MR imaging of tumor cells. *Nanoscale*, *4*(1), 188–193.

Li, L., et al. (2010). Binding and uptake of H-ferritin are mediated by human transferrin receptor-1. *Proceedings of the National Academy of Sciences of the United States of America*, *107*(8), 3505–3510.

Li, L., et al. (2016b). Ferritin-mediated siRNA delivery and gene silencing in human tumor and primary cells. *Biomaterials*, *98*, 143–151.

Li, M., & Mann, S. (2004). DNA-directed assembly of multifunctional nanoparticle networks using metallic and bioinorganic building blocks. *Journal of Materials Chemistry*, *14*(14), 2260–2263.

Li, X., et al. (2012a). Epidermal growth factor-ferritin H-chain protein nanoparticles for tumor active targeting. *Small*, *8*(16), 2505–2514.

Li, X., et al. (2016a). Protein nanocages for delivery and release of luminescent ruthenium(II) polypyridyl complexes. *ACS Applied Materials & Interfaces*, *8*(35), 22756–22761.

Liang, M., et al. (2014). H-ferritin-nanocaged doxorubicin nanoparticles specifically target and kill tumors with a single-dose injection. *Proceedings of the National Academy of Sciences of the United States of America*, *111*(41), 14900–14905.

Lim, E. K., et al. (2015). Nanomaterials for theranostics: recent advances and future challenges. *Chemical Reviews*, *115*(1), 327–394.

Lin, X., et al. (2011a). Chimeric ferritin nanocages for multiple function loading and multimodal imaging. *Nano Letters*, *11*(2), 814–819.

Lin, X., et al. (2011b). Hybrid ferritin nanoparticles as activatable probes for tumor imaging. *Angewandte Chemie International Edition*, *50*(7), 1569–1572.

Liu, X., et al. (2013). Bio-inspired protein-gold nanoconstruct with core-void-shell structure: beyond a chemo drug carrier. *Journal of Materials Chemistry B*, *1*(25), 3136–3143.

Liu, X. Y., et al. (2011). Apoferritin-camouflaged Pt nanoparticles: surface effects on cellular uptake and cytotoxicity. *Journal of Materials Chemistry*, *21*(20), 7105–7110.

Liu, X. Y., et al. (2012). Apoferritin-CeO2 nano-truffle that has excellent artificial redox enzyme activity. *Chemical Communications*, *48*(26), 3155–3157.

Lobello, N., et al. (2016). Ferritin heavy chain is a negative regulator of ovarian cancer stem cell expansion and epithelial to mesenchymal transition. *Oncotarget*.

Lu, C. Y., et al. (2017). T2-weighted magnetic resonance imaging of hepatic tumor guided by SPIO-loaded nanostructured lipid carriers and ferritin reporter genes. *ACS Applied Materials & Interfaces*.

Ma-Ham, A. H., et al. (2011). Apoferritin-based nanomedicine platform for drug delivery: equilibrium binding study of daunomycin with DNA. *Journal of Materials Chemistry*, *21*(24), 8700–8708.

Matzner, Y., Konijn, A. M., & Hershko, C. (1980). Serum ferritin in hematologic malignancies. *American Journal of Hematology*, *9*(1), 13–22.

Mazzucchelli, S., et al. (2016). Nanometronomic treatment of 4T1 breast cancer with nanocaged doxorubicin prevents drug resistance and circumvents cardiotoxicity. *Oncotarget*.

Meldrum, F. C., Heywood, B. R., & Mann, S. (1992). Magnetoferritin: in vitro synthesis of a novel magnetic protein. *Science*, *257*(5069), 522–523.

Meldrum, F. C., et al. (1991). Synthesis of inorganic nanophase materials in supramolecular protein cages. *Nature*, *349*(6311), 684–687.

Mendes-Jorge, L., et al. (2014). L-ferritin binding to Scara5: a new iron traffic pathway potentially implicated in retinopathy. *PLoS ONE*, *9*(9), e106974.

Meyron-Holtz, E. G., Moshe-Belizowski, S., & Cohen, L. A. (2011). A possible role for secreted ferritin in tissue iron distribution. *Journal of Neural Transmission*, *118*(3), 337–347.

Monti, D. M., et al. (2017). Ferritin nanocages loaded with gold ions induce oxidative stress and apoptosis in MCF-7 human breast cancer cells. *Dalton Transactions*.

Moss, D., et al. (1992). Functional roles of the ferritin receptors of human liver, hepatoma, lymphoid and erythroid cells. *Journal of Inorganic Biochemistry*, *47*(3-4), 219–227.

Neckers, L. M., & Trepel, J. B. (1986). Transferrin receptor expression and the control of cell-growth. *Cancer Investigation*, *4*(5), 461–470.

Ngoune, R., et al. (2016). Accumulating nanoparticles by EPR: a route of no return. *Journal of Controlled Release*, *238*, 58–70.

O'Donnell, K. A., et al. (2006). Activation of transferrin receptor 1 by c-Myc enhances cellular proliferation and tumorigenesis. *Molecular and Cellular Biology*, *26*(6), 2373–2386.

Pandolfi, L., et al. (2017). H-ferritin enriches the curcumin uptake and improves the therapeutic efficacy in triple negative breast cancer cells. *Biomacromolecules*.

Pontillo, N., et al. (2016). Cisplatin encapsulation within a ferritin nanocage: a high-resolution crystallographic study. *Chemical Communications (Cambridge)*, *52*(22), 4136–4139.

Rosager, A. M., et al. (2017). Transferrin receptor-1 and ferritin heavy and light chains in astrocytic brain tumors: expression and prognostic value. *PLoS ONE*, *12*(8), e0182954.

Ryser, H., Caulfield, J. B., & Aub, J. C. (1962). Studies on protein uptake by isolated tumor cells. I. Electron microscopic evidence of ferritin uptake by Ehrlich ascites tumor cells. *Journal of Cell Biology*, *14*, 255–268.

Sakamoto, S., et al. (2015). H-ferritin is preferentially incorporated by human erythroid cells through transferrin receptor 1 in a threshold-dependent manner. *PLoS One*, *10*(10).

Santambrogio, P., et al. (1992). Evidence that a salt bridge in the light chain contributes to the physical stability difference between heavy and light human ferritins. *Journal of Biological Chemistry*, *267*(20), 14077–14083.

Seymour, G. J., et al. (1987). Transferrin receptor expression by human bladder transitional cell carcinomas. *Urological Research*, *15*(6), 341–344.

Shi, J., et al. (2016). Cancer nanomedicine: progress, challenges and opportunities. *Nature Reviews. Cancer*.

Simsek, E., & Kilic, M. A. (2005). Magic ferritin: a novel chemotherapeutic encapsulation bullet. *Journal of Magnetism and Magnetic Materials*, *293*(1), 509–513.

Strebhardt, K., & Ullrich, A. (2008). Paul Ehrlich's magic bullet concept: 100 years of progress. *Nature Reviews. Cancer*, *8*(6), 473–480.

Sun, C., et al. (2011a). Controlling assembly of paired gold clusters within apoferritin nanoreactor for in vivo kidney targeting and biomedical imaging. *Journal of the American Chemical Society*, *133*(22), 8617–8624.

Sun, C. J., et al. (2011b). Controlling assembly of paired gold clusters within apoferritin nanoreactor for in vivo kidney targeting and biomedical imaging. *Journal of the American Chemical Society*, *133*(22), 8617–8624.

Suzumoto, Y., Okuda, M., & Yamashita, I. (2012). Fabrication of zinc oxide semiconductor nanoparticles in the apoferritin cavity. *Crystal Growth & Design*, *12*(8), 4130–4134.

Szabo, I., et al. (2012). Mn loaded apoferritin as an MRI sensor of melanin formation in melanoma cells. *Chemical Communications*, *48*(18), 2436–2438.

Szymendera, J. J., et al. (1985). Clinical usefulness of serum ferritin measurements in patients with testicular germ-cell tumors. *Oncology*, *42*(4), 253–258.

Tatur, J., Hagen, W. R., & Matias, P. M. (2007). Crystal structure of the ferritin from the hyperthermophilic archaeal anaerobe *Pyrococcus furiosus*. *Journal of Biological Inorganic Chemistry*, *12*(5), 615–630.

Terashima, M., et al. (2011). Human ferritin cages for imaging vascular macrophages. *Biomaterials*, *32*(5), 1430–1437.

Theil, E. C. (1987). Ferritin: structure, gene-regulation, and cellular function in animals, plants, and microorganisms. *Annual Review of Biochemistry*, *56*, 289–315.

Thompson, K. J., et al. (2002). Regulation, mechanisms and proposed function of ferritin translocation to cell nuclei. *Journal of Cell Science*, *115*(Pt 10), 2165–2177.

Tongwei, Z., et al. (2017). Enhanced peroxidase activity and tumour tissue visualization by cobalt-doped magnetoferritin nanoparticles. *Nanotechnology*, *28*(4), 045704.

Torti, F. M., & Torti, S. V. (2002). Regulation of ferritin genes and protein. *Blood*, *99*(10), 3505–3516.

Tosi, G., et al. (2016). Apoferritin nanocage as drug reservoir: is it a reliable drug delivery system? *Expert Opinion on Drug Delivery*, *13*(10), 1341–1343.

Truffi, M., et al. (2016). Ferritin nanocages: a biological platform for drug delivery, imaging and theranostics in cancer. *Pharmacological Research.*

Turino, L. N., et al. (2017). Ferritin decorated PLGA/paclitaxel loaded nanoparticles endowed with an enhanced toxicity toward MCF-7 breast tumor cells. *Bioconjugate Chemistry*, *28*(4), 1283–1290.

Uchida, M., et al. (2006). Targeting of cancer cells with ferrimagnetic ferritin cage nanoparticles. *Journal of the American Chemical Society*, *128*(51), 16626–16633.

Uchida, M., et al. (2008). A human ferritin iron oxide nano-composite magnetic resonance contrast agent. *Magnetic Resonance in Medicine*, *60*(5), 1073–1081.

Uchida, M., et al. (2010). The ferritin superfamily: supramolecular templates for materials synthesis. *Biochimica et Biophysica Acta*, *1800*(8), 834–845.

Vannucci, L., et al. (2012). Selective targeting of melanoma by PEG-masked protein-based multifunctional nanoparticles. *International Journal of Nanomedicine*, *7*, 1489–1509.

Venditto, V. J., & Szoka, F. C. (2013). Cancer nanomedicines: so many papers and so few drugs! *Advanced Drug Delivery Reviews*, *65*(1), 80–88.

Volpino, P., et al. (1984). Clinical usefulness of serum ferritin measurements in lung-cancer patients. *The Journal of Nuclear Medicine and Allied Sciences*, *28*(1), 27–30.

Wang, R. B., Billone, P. S., & Mullett, W. M. (2013). Nanomedicine in action: an overview of cancer nanomedicine on the market and in clinical trials. *Journal of Nanomaterials.*

Wang, W., et al. (2010). Serum ferritin: past, present and future. *Biochimica et Biophysica Acta*, *1800*(8), 760–769.

Wang, Z., et al. (2016). Biomineralization-inspired synthesis of copper sulfide-ferritin nanocages as cancer theranostics. *ACS Nano.*

Wang, Z., et al. (2017). Functional ferritin nanoparticles for biomedical applications. *Frontiers of Chemical Science and Engineering*, *11*(4), 633–646.

Weaver, M., & Laske, D. W. (2003). Transferrin receptor ligand-targeted toxin conjugate (Tf-CRM107) for therapy of malignant gliomas. *Journal of Neuro-Oncology*, *65*(1), 3–13.

Weinstein, R. E., et al. (1989). Tissue ferritin concentration and prognosis in carcinoma of the breast. *Breast Cancer Research and Treatment*, *14*(3), 349–353.

Werneke, D., Elliott, R., & Ledford, L. (1990). The significance of increased tissue ferritin concentration in breast-carcinoma. *Clinical Chemistry*, *36*(6), 1062–1063.

Wilhelm, S., et al. (2016). Analysis of nanoparticle delivery to tumours. *Nature Reviews Materials*, *1*, 16014.

Xie, J., Lee, S., & Chen, X. (2010). Nanoparticle-based theranostic agents. *Advanced Drug Delivery Reviews*, *62*(11), 1064–1079.

Yan, F., et al. (2008). Apoferritin protein cages: a novel drug nanocarrier for photodynamic therapy. *Chemical Communications*, (38), 4579–4581.

Yan, F., et al. (2010). Cellular uptake and photodynamic activity of protein nanocages containing methylene blue photosensitizing drug. *Photochemistry and Photobiology*, *86*(3), 662–666.

Yang, D. C., et al. (2001). Expression of transferrin receptor and ferritin H-chain mRNA are associated with clinical and histopathological prognostic indicators in breast cancer. *Anticancer Research*, *21*(1B), 541–549.

Yang, M., et al. (2015). Dragon fruit-like biocage as an iron trapping nanoplatform for high efficiency targeted cancer multimodality imaging. *Biomaterials*, *69*, 30–37.

Yang, Z., et al. (2007). Encapsulation of platinum anticancer drugs by apoferritin. *Chemical Communications*, (33), 3453–3455.

Zhang, L., Chen, Z., & Xu, A. (2017). FTL: a novel predictor in gastric cancer. *International Journal of Clinical and Experimental Pathology*, *10*(7), 7865–7872.

Zhao, Y., et al. (2016). Bioengineered magnetoferritin nanoprobes for single-dose nuclear-magnetic resonance tumor imaging. *ACS Nano*, *10*(4), 4184–4191.

Zhen, Z., et al. (2013a). RGD-modified apoferritin nanoparticles for efficient drug delivery to tumors. *ACS Nano*, *7*(6), 4830–4837.

Zhen, Z., et al. (2013b). Ferritin nanocages to encapsulate and deliver photosensitizers for efficient photodynamic therapy against cancer. *ACS Nano*, *7*(8), 6988–6996.

Zhen, Z., et al. (2014). Tumor vasculature targeted photodynamic therapy for enhanced delivery of nanoparticles. *ACS Nano*, *8*(6), 6004–6013.

Zhen, Z., et al. (2015). Folic acid conjugated ferritins as photosensitizer carriers for photodynamic therapy. *Nanoscale*, *7*(23), 10330–10333.

Zhen, Z., et al. (2016). Protein nanocage mediated FAP-targeted photo-immunotherapy to enhance cytotoxic T cell infiltration and tumor control. *Nano Letters*, *17*(2), 862–869.

CHAPTER 7

The Role of Nanoscience in Cancer Diagnosis

Luca Guerrini[*,#], **Eduardo Garcia-Rico**[†,‡,§,¶], **Ramon Alvarez-Puebla**[*,‖]

[*]Department of Physical and Inorganic Chemistry - EMaS, Universitat Rovira i Virgili, Tarragona, Spain [†]Fundacion de Investigacion HM Hospitales, Madrid, Spain [‡]Centro Integral Oncologico Clara Campal (CIOCC), Madrid, Spain [§]Hospital Universitario HM Torrelodones, Torrelodones, Spain [¶]San Pablo CEU, Madrid, Spain [‖]ICREA, Barcelona, Spain [#]Centre Tecnològic de la Química de Catalunya, Tarragona, Spain

1 Introduction

Cancer is a major global health problem with 8.2 million deaths per year (over 22,000 per day) with a worsening trend, according to the epidemiological evidence (World Health Organization, n.d.). Further, cancer has the lowest success rates of all diseases in conventional clinical trials. This strikingly contrasts the much higher success rates obtained at the preclinical level. In fact, the 48 new regimes approved by the US FDA from 2002 to 2014 conferred a median of 2.1-month overall survival benefit (Wise, 2016). Of the total 5-year survival observed in this period, only 20% were attributable to the new treatments, while the remaining 80%, appeared to be due to developments and improvements in the early diagnosis and/or other interdisciplinary strategies. Thus, despite the huge investment in research and development, new therapies seem to have only marginal positive impact, at least in adults with metastatic cancer. Therefore, it appears to be of utmost importance to rethink the way we tackle cancer, and consequently, to prioritize our investments to maximize return in terms of patient and cost benefits (Cagan & Meyer, 2017).

In recent decades, cancer research has mostly taken an inductive approach by using cellular models and data from lab animals as starting points. This approach, however, tends to exclude two essential phenomena that are related to the complexity of the disease: metastasis and tumor heterogeneity. Notably, 90% of cancer deaths are not directly linked to the primary site or origin of the tumor, but rather, to the spread of these primary tumors and their ability to cause metastasis. The importance of this phenomenon is clear: (a) 30% of tumors have macroscopic metastasis at the time of diagnosis, (b) another 30% have micrometastasis (the limit of the radiological ability

Handbook of Nanomaterials for Cancer Theranostics. https://doi.org/10.1016/B978-0-12-813339-2.00007-4

to distinguish a nodule is one centimeter, corresponding to 10^9 cells in 1 cm^3), and (c) metastasis represents a very early biological event in the disease process and increases the probability of occurrence with its course. Therefore, cancer must be considered a systemic disease (Lambert, Pattabiraman, & Weinberg, 2017). This is further supported by the evidence that a decrease in metastasis because of the treatment of the primary tumor (abscopal effect) is a rare phenomenon (Postow et al., 2012), which has been also related to a systemic, possibly immune, response. On the other hand, tumor heterogeneity implies the presence in tumors of both genotypic and phenotypically different cells which explains the mechanism underlying the onset of the invasive and metastatic phenomenon (Hanahan & Weinberg, 2011).

Clearly, tumors originating in epithelial tissues (carcinomas and adenocarcinomas) progress towards different forms of dedifferentiation and redifferentiation, giving rise to different degrees of malignancy. In these phenotypic transitions, tumor cells come to adopt characteristics like those of the embryonic mesenchymal cells: movement, invasion, migration, deformability, etc. This phenomenon, underlying tumor heterogeneity, has been termed "epithelial mesenchymal transition" (EMT) (Lamouille, Xu, & Derynck, 2014). Tumor heterogeneity is not only spatial but also dynamic, both from an intratumoral (spatial and temporal) point of view and from the point of view of metastasis (primary metastasis or metastasis-metastasis). Thus, this phenomenon constitutes a major limitation to the use of new therapeutic strategies based on molecular targets (personalized medicine). This enormous complexity suggests that a new cancer scope should be grounded in data from human patients or from epidemiological data. There are, therefore, three aspects that are usually not adequately considered in the diagnosis and treatment of cancer. These are the tumor heterogeneity, the metastatic (systemic) dimension of the disease, and the role of the microenvironment/tumor environment (Welch, 2016).

Nanoscience holds enormous potential for overcoming many of the problems associated with conventional diagnostic methods in the detection, diagnosis, and prognosis of cancer (Choi, Kwak, & Park, 2010; Velpurisiva, Gad, Piel, Jadia, & Rai, 2017). Nanodiagnostics, defined as the use of nanoscience for clinical diagnostic purposes, have been developed to meet the demands of clinical diagnostics for earlier detection of the disease, offering an exquisite sensitivity and precision that is difficult to match with any other technology. When appropriately functionalized, nanomaterials can selectively target cancer biomarkers and cancer cells, allowing more sensitive diagnosis, early detection requiring a minimal amount of sample, and monitoring of the progress of therapy and tumor burden over time. However, developing clever strategies based in nanoscience requires the previous knowledge of the target disease to be efficient. Thus, in this chapter, we discuss the new paradigms in the field of oncology and how nanoscience is already helping or can help in diagnosing tumor-based malignancies (Fig. 1).

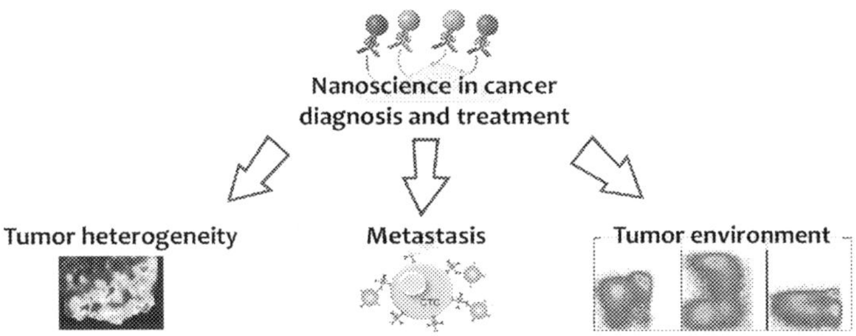

Fig. 1
Schematic overview of the impact of nanotechnology in understanding the role of three major factors in cancer diagnosis and treatment: tumor heterogeneity, the metastatic (systemic) dimension of the disease, and the role of the tumor environment.

2 Modern Paradigms in Cancer

2.1 Cancer Heterogeneity

The current tumor classification systems (UICC, etc.) (Brierley, 2017) are based on classical pathological anatomy sometimes supplemented with molecular techniques. This classification, however, is not able to capture the actual dynamics of the disease in each patient (precision oncology) (Prasad, 2016; Sohal et al., 2016). Therefore, there is a great "intertumoral" heterogeneity between the same tumor in two different patients. Presently, this fact implies a major diagnostic and therapeutic challenge in oncology. On the other hand, a second key challenge consists in spatial and temporal intratumoral heterogeneity. Thus, modern clinical oncology requires technologies capable of accurately revealing these categories on a timely basis.

2.2 Metastasis

The success of cancer treatment (surgical, medical, or radiotherapeutic) depends (exceptions notwithstanding) on an early diagnosis and an adequate extension study. Metastases accounts for 90% of cancer mortality. Currently available technologies for the screening of cancer are based either on imaging (i.e., X-ray, computerized tomography, magnetic resonance imaging, or positron emission imaging) or in (bio)analytical molecular methods on samples obtained from tissue biopsy (histo/immunohistochemistry, HC/IHC). Notably, imaging techniques require a large tumoral mass (1 cm^3), implying a high load of tumor cells (typically 10^9 cells/cm^3) which often are already metastatic. On the other hand, the molecular characterization of tumors, consistently applied in clinical oncology, identifies the disease, adds predictive and prognostic value, and determines the presence of specific therapeutic targets. HC/IHC allows the determination of morphological characteristics and the expression of biomarkers in the tissues, reaching, thus, a diagnosis and prognosis

(Subik et al., 2010). However, not only is this process slow, involving preanalytic (fixation, embedding, processing, and sectioning), analytic (permeation, staining, and visualization) and postanalytic steps (interpretation and diagnostic), a tissue biopsy is often quite challenging as it is costly, painful, or potentially risky for the patient. Further, it cannot track the disease evolution in real time and, in some cases, it is not even feasible. Thus, the development of noninvasive methods to detect, characterize (at the tissue and cellular levels), and monitor tumors is a major challenge in oncology. Several tests, mainly based in molecular immunology, have been well established for monitoring tumor burden in some breast and lung cancers. However, the extension of these tools to other malignancies requires detailed knowledge of the cell molecular biology of the tumor cells to find appropriate selective targets that may be used for classification and, thus, diagnosis, monitoring, and treatment. Most frequently, we find in these cases that the tumor has already metastasized. Therefore, sensitive and specific techniques capable of detecting the disease "before" it is disseminated are necessary. On the other hand, a correct diagnosis of extension is also a determinant for a correct treatment. That is, it is necessary to detect the disease not only in its primary situation, but also that of all secondary metastasis.

2.3 Tumor Environment (Disease as a Complex System)

Tumor environment relates to the understanding of the behavior of a given tumor in the personalized micro- or macroenvironment of the given patient, for example, due to statute and nature of the immune response. Both diagnostic and therapeutic strategies aimed at this aspect are the ones that are giving the best results (Hirata & Sahai, 2017).

3 Cancer Heterogeneity

Nanoparticle-based technology offers diverse valuable tools for characterizing the phenotypic and functional heterogeneity of cancer cells. The basic functional principle relies on the conjugation of a recognition element (antibody, aptamers, peptides, etc.), imparting selectivity toward the cellular target onto the nanoconstruct which, for its part, provides the signal read-out, enabling the unique identification and quantification of the recognition events. For the specific application, multiplexity and sensitivity are explicit requirements to be satisfied by the nanosensing technology. Luminescent semiconductor quantum dots (QDs) are widely exploited as nanomaterials for in vitro and in vivo bioimaging research (Hildebrandt et al., 2017; Wegner & Hildebrandt, 2015). Among different fluorescent probes, QDs offer key advantages, such as tunable photoluminescence over a wide spectral range (from UV to IR), high quantum yields, narrow bandwidths, and limited tendency toward photobleaching (Hildebrandt et al., 2017; Wegner & Hildebrandt, 2015). Thus, QDs are especially suited for multiplexed quantitative molecular profiling. An illustrative example of this application at single-cell level is reported by Zrazhevskiy and Gao (2013) and outlined in Fig. 2. Here,

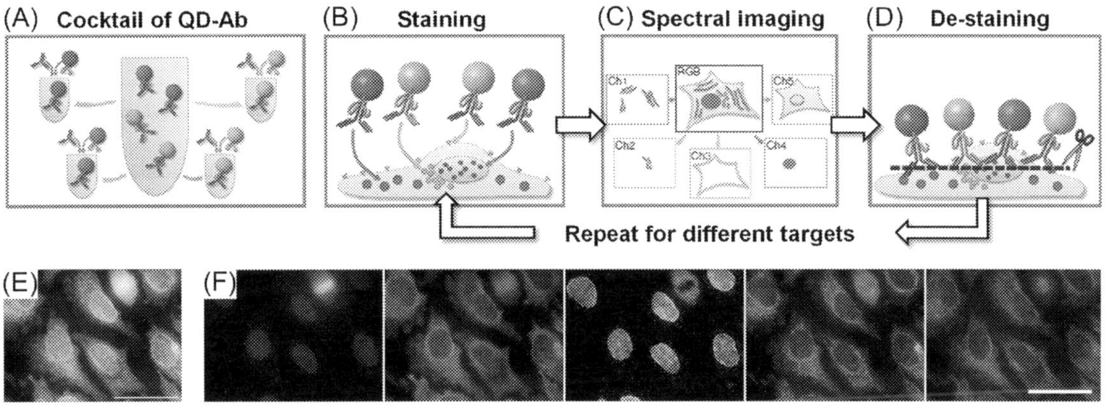

Fig. 2

(A) Functionalization of protein A-modified QDs with Abs in a single purification-free step and subsequent mixing of individual QD-Ab probes into a single cocktail. (B) Single-step, multiplexed staining of cells. (C) Spectral imaging and subsequent unmixing of QD-Ab contributions to generate a quantitative profiling. (D) De-staining of cells to restore the specimen for another staining cycle. (E) Hyperspectral high-resolution fluorescence microscopy of five preassembled QD-Ab probes (emitting at 525, 545, 565, 585, and 605 nm) on HaLa cells, simultaneously targeting Ki-67, HSP90, Lamin A, Cox-4 and β-tubulin, respectively. (F) Spectral unmixing of individual QD-Abs signals. Scale bar, 50 mm. *Adapted with permission from Zrazhevskiy, P., & Gao X. (2013). Quantum dot imaging platform for single-cell molecular profiling. Nature Communications, 4, 1619. Copyright 2013, Nature Publishing Group.*

five different classes or antibody-QD (QD-Ab) nanoparticles were fabricated via direct self-assembly of protein A-modified QDs and the unique Ab. Protein A acts as a universal surface element promoting the noncovalent adhesion of Abs. QD-Abs are mixed together to yield the corresponding cocktail of fluorescent nanoprobes (Fig. 2A) and, then, incubated with cells (Fig. 2B). Fluorescence microscopy is applied to generate the quantitative antigen expression profile resulting from the multiplexed staining (Fig. 2C). Finally, a washing step is performed to remove the fluorescent probes from the sample, which is then available to be submitted to the next full cycle of staining (Fig. 2D). Fig. 2E illustrates the simultaneous staining of HeLa cells with a pool of five diverse QD-Abs (emitting at 525, 545, 565, 585, and 605 nm) targeting nuclear Lamin A, cytoplasmic HSP90, mitochondrial Cox-4, cytoskeletal protein β-tubulin, and Ki-67 (a protein strictly associated with cell proliferation). Once extracted, the individual contributions of each QD-Ab type (Fig. 2F) facilitate the quantitative analysis of staining intensity.

A more recent approach than QD-based technology for engineering powerful nanoprobes for sensitive multiplexed bioimaging that has attracted a great deal of interest is represented by the development of SERS-encoded nanoparticles (SEPs) (Fabris, 2016; Guerrini, Pazos-Perez, Garcia-Rico, & Alvarez-Puebla, 2017; Lane, Qian, & Nie, 2015;

Wang, Yan, & Chen, 2013). Surface-enhanced Raman scattering (SERS) spectroscopy (Schlücker, 2014) is an ultrasensitive vibrational technique that relies on the giant plasmonic nanostructure-mediated amplification of the optical signal (Le Ru & Etchegoin, 2009; Schlücker, 2014). The intrinsic nature of SERS phenomenon can be translated into the fabrication of optical nanoprobes with major advantages over QDs including, among others, (a) higher multiplexing capabilities, facilitated by the acquisition of the unique vibrational fingerprint of Raman spectra comprising well-defined narrow features; (b) higher potential sensitivity, provided by the giant electromagnetic fields concentrated at the nanostructured surfaces; (c) higher photostability; and (d) simultaneous acquisition of the different SERS signatures with a single laser beam (Lane et al., 2015). Typically, SEPs comprise a plasmonic core (mostly, Au or Ag) acting as the optical enhancer, functionalized with molecular labels of high Raman cross-section providing intense and well-defined SERS spectra, and bioconjugated with recognition elements analogous to those for QDs. Finally, an outer protective layer, commonly based on silica or polymeric materials, is also usually integrated into the SEP structure to impart colloidal stability and prevent desorption of the SERS labels from the metallic surfaces. The final size of SEPs mainly falls within the ~20–120 nm range (larger than QDs typical size, ~5–50 nm). A notable example of the potential use of SEPs as contrast agents that is gaining a great deal of interest in the very recent years is represented by topical spray-like applications targeting cancer biomarkers located on exposed surfaces, such as epithelial cancers or surgically exposed/resected tissues (Guerrini et al., 2017; Mallia, McVeigh, Fisher, Veilleux, & Wilson, 2015; Wang et al., 2016; Zavaleta et al., 2013). This "spray & image" approach drastically reduces the issues of toxicity and clearance associated with SEPs administrated via systemic routes (e.g., intravenous injections), while maximizing their potential at revealing the heterogeneous features of tumor tissues with unparalleled level of sensitivity and resolution. For example, Wang et al. (2016) topically applied SEPs onto fresh human breast specimens for the rapid (<15 min) and quantitative molecular phenotyping of the tumor surface, as a potential intraoperative tool to be used in guiding breast-conserving surgeries (Fig. 3A). To remove the problem associated with nonspecific accumulations of SEPs within the tissues, while, at the same time, avoiding the need for time-consuming rinsing cycles (unsuited to live surgical guidance applications), the authors implemented a quantitative radiometric-imaging method. In this approach, the cocktail of contrast agents also included a class of nontargeted nanoparticles in addition of SEPs conjugated with targeting elements (in this case, monoclonal antibodies targeting the epidermal growth factor receptor, EGFR, or the human epidermal growth factor receptor 2, HER2—Fig. 3B). Each class of SEPs has been engineered with different SERS labels providing unique SERS signals (Fig. 3C). The SERS mapping of targeted-SEPs applied on the resected tissues is then presented as a radiometric map against the background signal of nontargeted nanoparticles, thus maximizing the molecular image contrast between tumor and normal tissue (Fig. 3D and E).

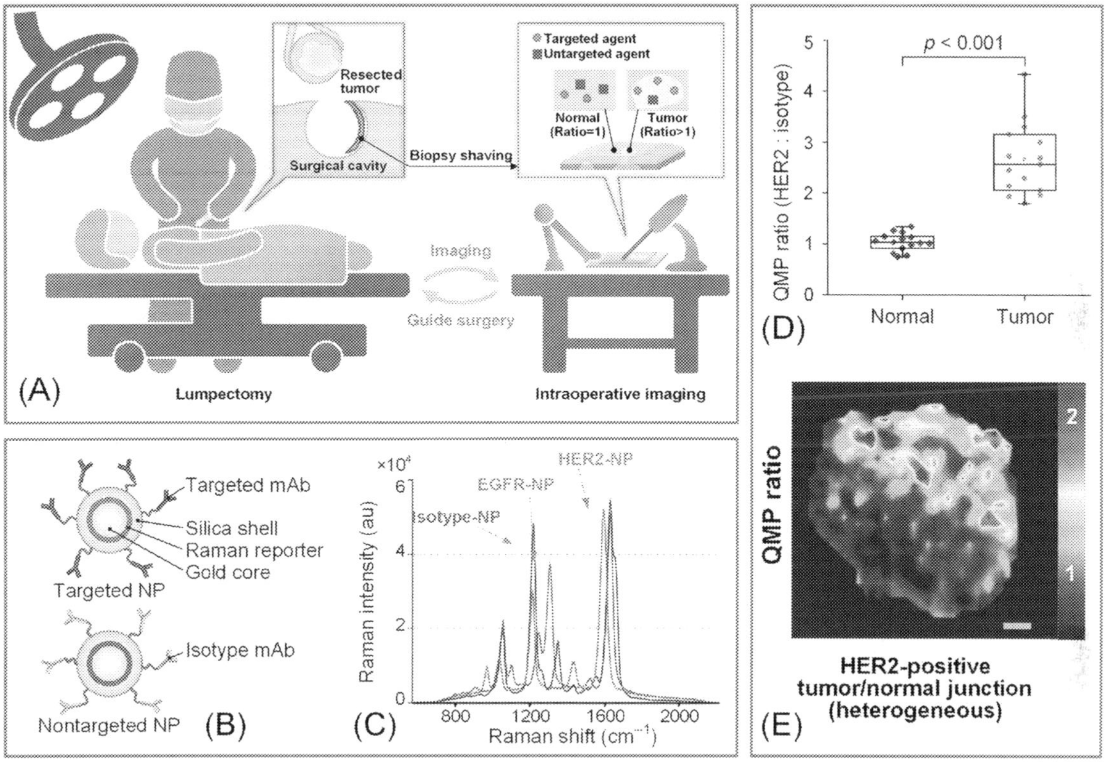

Fig. 3

(A) Conceptual view of the implementation of the SERS-based intraoperative imaging technique to the rapid identification of residual tumors at the margins of freshly resected tissues for guiding breast-conserving surgeries. (B) Structural depiction of the targeted and nontargeted SERS-encoded nanoparticles (SEPs) and (C) their corresponding SERS profiles. (D) Targeted (HER2) vs. nontargeted concentration ratios of SEPs on normal tissues and tumors. (E) Image of the concentration ratio (QMP) of targeted vs. nontargeted SEPs of a tissue specimen containing both tumor and normal tissue regions. *Adapted with permission from Wang, Y., Kang, S., Khan, A., Ruttner, G., Leigh, S. Y., Murray, M., et al. (2016). Quantitative molecular phenotyping with topically applied SERS nanoparticles for intraoperative guidance of breast cancer lumpectomy. Scientific Reports 6, 21242, used under CC BY. Copyright 2016, Nature Publishing Group.*

4 Metastasis

Cancer is a dynamic process that begins at t_0 but which cannot be diagnosed until time t_D (Fig. 4A), when the tumor is composed by approximately 10^9 cells (a stage which is likely metastatic). Thus, the earlier diagnosis, during the preclinical stage of the disease (before t_D), is of paramount importance as it is very likely to increase the rate of survival. Metastatic tumors are responsible for over 90% of cancer deaths (Wittekind & Neid, 2005). The metastatic process is accomplished by a circulating tumor cells (CTCs) successfully

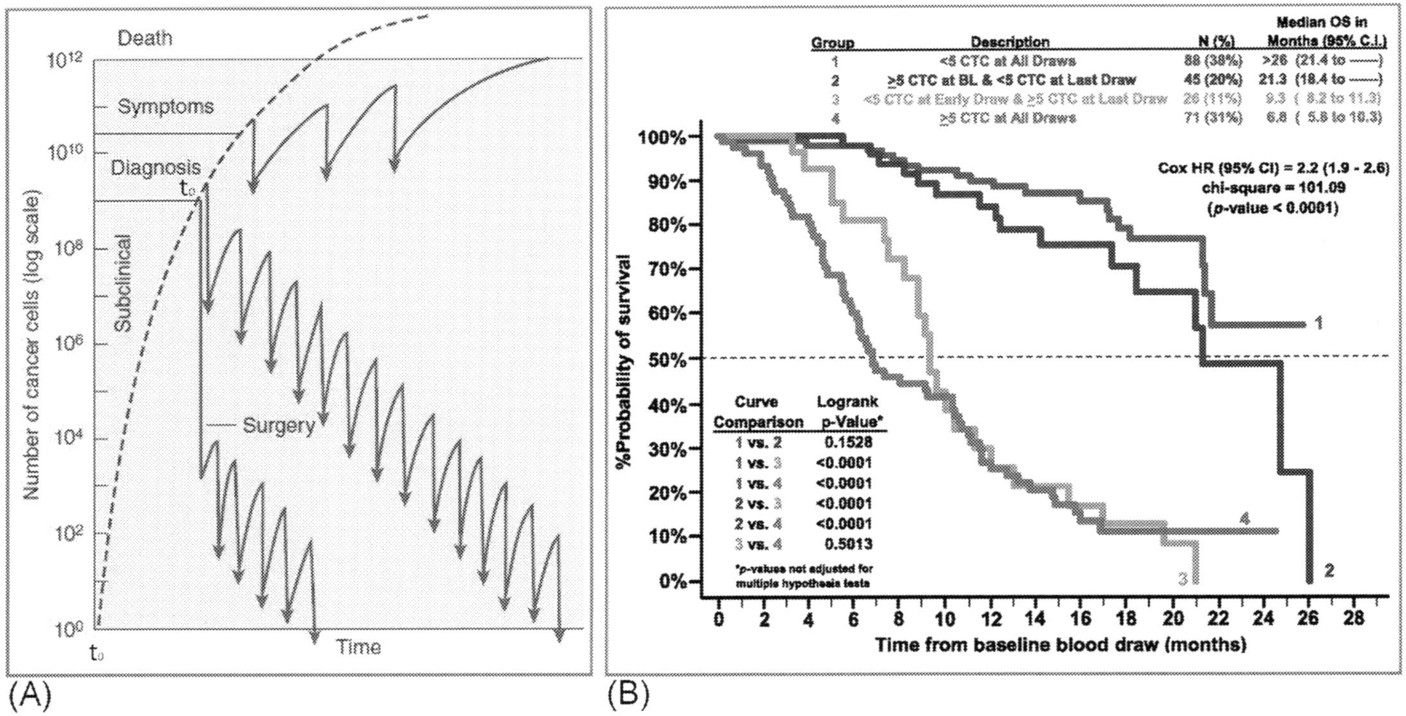

Fig. 4

(A) Log-kill hypothesis-relationship of tumor cell number to time of diagnosis, symptoms, treatment, and survival. (B) CTCs and prognostic prediction. The number of CTCs detected per 7.5 mL blood shows a strong correlation with the survival rate in breast cancer. When more than five CTCs are detected, the survival rate is lowered to <50%. Similar results have been obtained in prostate cancer, colorectal cancer and lung cancer. *Adapted with permission from de Bono, J. S., Scher, H. I., Montgomery, R. B., Parker, C., Miller, M. C., Tissing, H., et al. (2008). Circulating tumor cells predict survival benefit from treatment in metastatic castration-resistant prostate cancer. Clinical Cancer Research, 14(19), 6302–6309. Copyright 2008, AACR Journals.*

carrying out a series of processes. First, the CTCs detach from the primary tumor and intravasate into the blood stream. Once in circulation, the CTCs evade immune detection and extravasate into microvessels of a target tissue (Chambers, Groom, & MacDonald, 2002). The successful formation of a metastatic lesion is dependent upon the CTC ability to adapt, survive, and induce neoangiogenesis in the target tissue. Thus, by monitoring a low-invasive sample such as blood (i.e., liquid biopsy) it is possible to diagnose and monitor the tumor. Although the hypothesis that CTCs are a fundamental prerequisite to metastasis was first proposed in the mid-19th century (Ashworth, 1869), its determination was not possible until very recently. Besides the obvious benefits provided by this information related to the metastatic process and its potential to unmask novel therapeutic targets for the treatment of the tumor, the content of CTCs has been demonstrated to show a strong correlation with the survival rate, both at the diagnosis and in monitoring treatment cycles (Fig. 4B). Progress in CTC detection, however, has been hindered until quite recently by the technical challenges related to their very low concentration in blood.

In this scenario, nanoscience, often coupled to microfluidics, has emerged as a good alternative to the liquid biopsy of CTCs, but also of other mensurable products such as circulating tumor nucleic acids, proteins, exosomes, or platelets, allowing for the design of strategies for early diagnosis or monitoring of cancer in small devices that can be implemented in the point of care (POC). This implementation adds two new dimensions: the patient "as a whole," and the time as POC allows continuous monitoring while facilitating the definition of new surrogate end points.

4.1 Nanoparticles for CTCs Enrichment, Quantification, and Characterization

4.1.1 Nanoparticles for CTC enrichment

Due to the extremely low concentrations of CTCs in blood, many methods use an enrichment process before the quantification step. Current enrichment methods can be divided into two categories: those based on physical properties such as size, density, or deformability, and those resting on biological properties such as protein expressions (Alix-Panabieres & Pantel, 2014). Among approaches based on protein expression of CTCs at the membrane, magnetic separation using magnetic nanoparticles is very common. In these cases, magnetic nanoparticles, usually of iron oxide, are functionalized with antibodies against epithelial cell adhesion molecule (EpCAM). The rationale for this choice lies in the idea that epithelial lineage is not common in blood, and, thus, epithelial cells in blood imply cancer. EpCAM-magnetic particles are exposed to the blood sample, and, after the appropriate time for conjugate binding with targeted cells, extracted by means of a permanent magnetic field. These systems normally use more than one antibody. For example, Cell Search (Jansen Technologies) also uses anti-CD45 attached to a dye as a negative control to detect false positives due to the retention of some anti-EpCAM by the mononuclear cell fraction of blood (Carroll & Al-Rubeai, 2005).

4.1.2 Nanoparticles for detection

The most straightforward alternative for detection relies on the use of QDs and fluorescence. As in the case of concentration, QDs are functionalized with the appropriate antibodies, and detection is carried out either directly in the sample by flowing it through an optofluidic device such as a microfluidic chip or a flow-cytometer, or upon coupling this fluorescence strategy to a previous (or posterior) enrichment stage (Fig. 5B) (Min, Jo, & Kim, 2015).

Notably, the use of fluorescent QDs is restricted by the number of colors that can be used simultaneously with uncertainty. Considering the EMT that tumor cells suffer before intravasate into the blood stream, the use of a single phenomenological positive marker together with a control marker (anti-CD45) is, in many cases, not enough. Thus, in many developed test separated tumor cells require to undergo an extensive HC/IHC characterization which implies time cost, and avoids the concept of POC. In recent years, new approaches have emerged to address these issues, mainly based on plasmonic encoded nanoparticles (Mir-Simon, Reche-Perez, Guerrini, Pazos-Perez, & Alvarez-Puebla, 2015) and SERS, enabling the multiplex quantification of different CTCs (Pallaoro, Hoonejani, Braun, Meinhart, & Moskovits, 2015; Wang et al., 2011; Wu et al., 2015). Advantages of this methodology include ultrahigh analytical resolution combined with a virtually unlimited number of codes, and, thus, markers that can be studied simultaneously. Unfortunately, state-of-the art SERS acquisition times are higher than 10 ms, impeding the use of this technique for real samples where a single CTC is diluted into billions of other cells. In our opinion, a viable solution could be achieved by combining the acquisition speed of fluorescence (down to ns) for a rapid screening and sorting, with the analytical resolution of SERS applied to the smaller cell fraction previously separated by physical or molecular techniques.

4.2 Nanoparticles for Circulating Tumor Nucleic Acid and Oncoprotein Quantification and Characterization

Cancers arise through the sequential alteration of genes that control cell growth. In solid tumors such as those of the colon or breast, it has been shown that, on average, approximately 80 genes harbor subtle mutations that are present in virtually every tumor cell but are not present in normal cells. These somatic mutations thereby have the potential to serve as highly specific biomarkers in blood (Fig. 6). Tumor-derived mutant DNA can also be detected in the cell-free fraction of the blood of patients with cancer and raises the possibility that the circulating mutant DNA fragments can be used to track disease. However, the reliable detection of such mutant DNA fragments is challenging. In fact, circulating mutant DNA represents only a small fraction of the total circulating DNA, sometimes less than 0.01%.

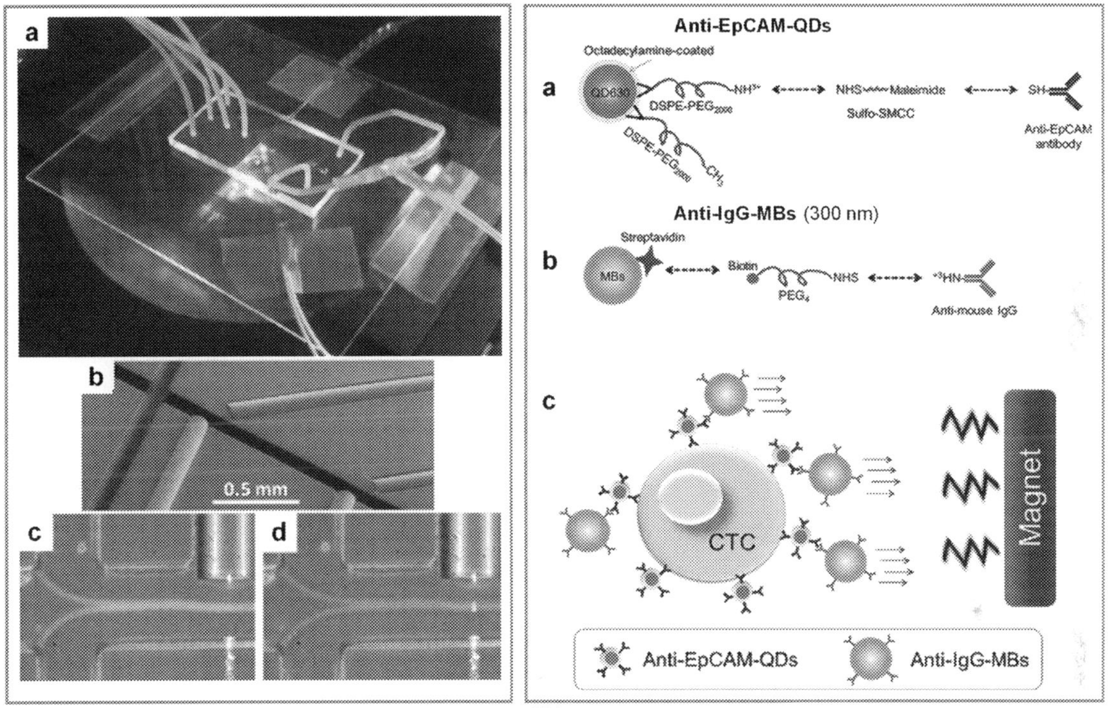

(A) (B)

Fig. 5

(A) Optofluidic microchip. (a) SEM image of the optofluidic chip. (b and c) Confinement of the sample at the central region of the channel. A *blue light* path, impinging from the top (c), is visible only when a cell intersects the pumping beam. Absorption and emission properties of the dyes used to label the antibodies and their performance on the cell membrane. (B) Schematic representation of CTC capturing using quantum dots and magnetic beads. (a) Construction of anti-EpCAM antibody-conjugated quantum dots (anti-EpCAM-QDs) as primary nanoparticles. DSPE-PEG2000-methoxy and DSPE-PEG2000-amine were coated onto the surface of the QDs, and the resulting QDs and antibody were conjugated using a sulfo-SMCC linker. Quantum dot 630 emitting *red* color was used. (b) Construction of antimouse IgG antibody-conjugated magnetic beads (anti-IgG-MBs) as secondary particles. Biotinlyated antimouse IgG was immobilized to the streptavidin-introduced MBs. (c) Schematics for capturing the CTCs using anti-EpCAM-QDs and anti-IgG-MBs. Anti-EpCAM-QDs were first targeted to the cancer cells. The anti-IgG-MBs recognize the anti-EpCAM-QDs and the cancer cells can be isolated through magnetic force. *(A) Adapted with permission from Pedrol, E., Garcia-Algar, M., Massons, J., Nazarenus, M., Guerrini, L., Martínez, J., et al. (2017). Optofluidic device for the quantification of circulating tumor cells in breast cancer. Scientific Reports 7, 3677, used under CC BY. (B) Adapted with permission from Min, H., Jo, S.-M., Kim, H.-S. (2015). Efficient capture and simple quantification of circulating tumor cells using quantum dots and magnetic beads. Small 11(21), 2536–2542. Copyright 2015, Wiley.*

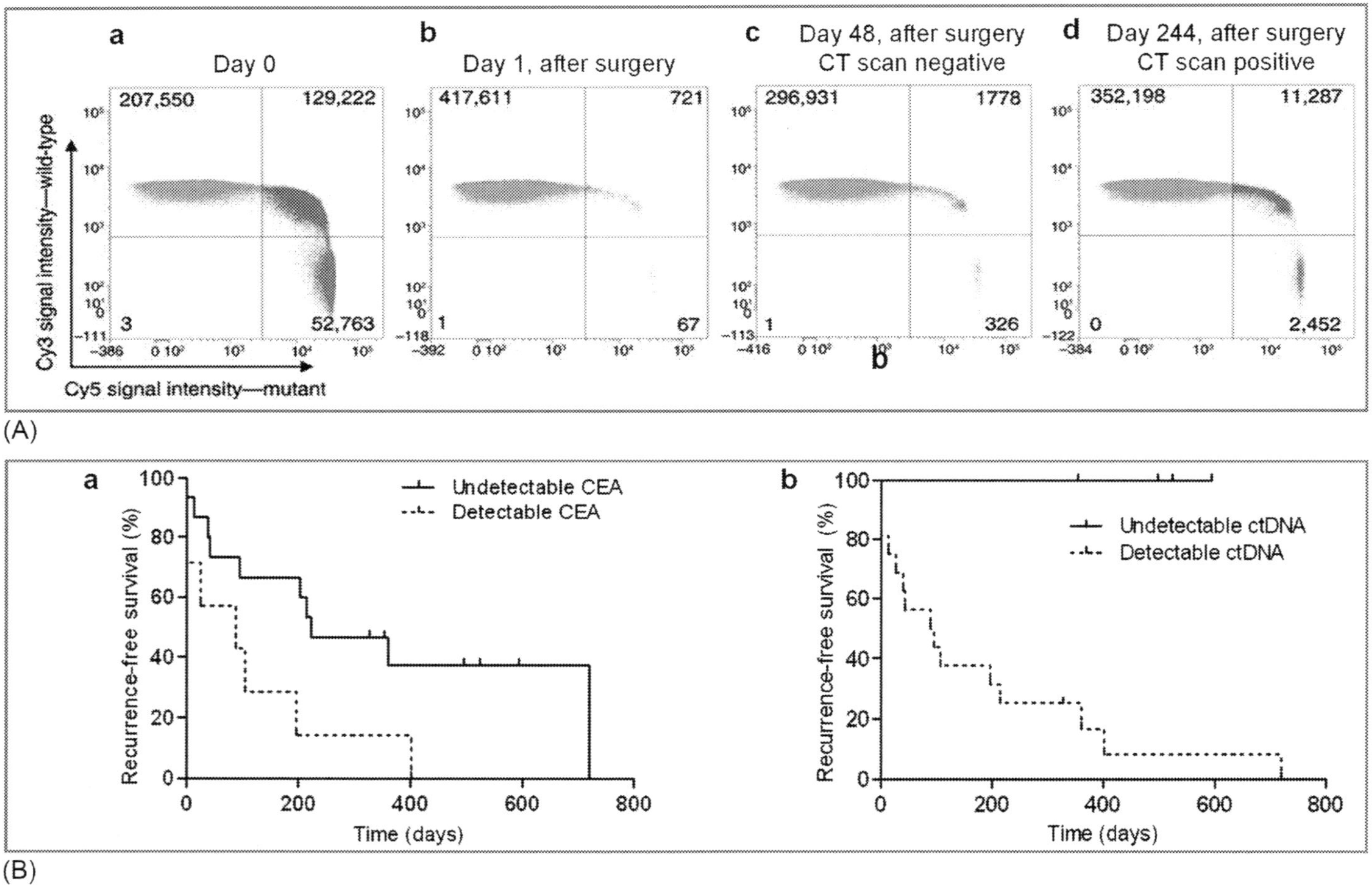

Fig. 6

(A) Representative flow cytometric data obtained from a patient with breast cancer. The four graphs illustrate the data obtained at different time points during treatment. The *green and red dots* (light gray and gray, respectively) represent beads bound to wild-type and mutant (tumor) nucleic acid fragments, respectively. The *blue dots* (dark gray) represent beads bound to both wild-type and mutant fragments. Numbers in each quadrant represent events for each population measured. (a) Before surgery, the fraction of mutant DNA fragments was 13.4%. (b) After surgery (day 3), the fraction of mutant DNA fragments dropped to 0.015%. (c) After surgery (day 48), the fraction of mutant DNA fragments increased to 0.11%, suggesting disease recurrence. (d) On day 244, the subject had progressive disease and the fraction mutant DNA fragments increased further to 0.66%. (B) Recurrence-free survival, as detected by ctDNA and CEA. (a) The difference in recurrence-free survival in subjects with detectable versus undetectable postoperative CEA levels ($P = .03$ by Mantel-Cox log-rank test). (b) The difference in recurrence-free survival in subjects with detectable versus undetectable postoperative ctDNA levels ($P = .006$ by Mantel-Cox log-rank test). *Adapted with permission from Diehl, F., Schmidt, K., Choti, M.A., Romans, K., Goodman, S., Li, M., et al. (2008). Circulating mutant DNA to assess tumor dynamics. Nature Medicine 14(9), 985–990. Copyright 2008, Nature.*

A second, though less investigated, class of blood-based biomarkers relies on alterations of protein expressions in plasma CTCs and peripheral blood mononuclear cells. For instance, the protein c-Jun, a component of the AP-1 transcription complex associated with carcinogenic mechanisms in several human cancers, has been shown to be significantly overexpressed in a large number of malignant transformations (Blau et al., 2012). However, current techniques for the identification and quantification of proteins in biofluids are time-consuming, expensive and, often, lack of the necessary sensitivity. To tackle these issues, our group devised a new SERS-based sensing method for the quantitative proteomic liquid biopsy in real samples (Fig. 7A) (Guerrini et al., 2013; Pazos et al., 2016). Here, we exploited the high affinity and selectivity of the native protein partner c-Fos for binding c-Jun, to synthesize a peptide c-Fos chemoreceptor incorporating a 4-mercaptobenzoyl unit (MB) to the side chain amino group of the N-terminal amino acid residue (Fig. 7A(a)). MB functionality is two-fold: it firmly anchors the peptide receptor to the nanoparticle surfaces and provides a very intense, well-defined, SERS fingerprint. Upon c-Fos/c-Jun heterodimerization, a large structural rearrangement of the chemoreceptor takes place, perturbing, in turn, the MB moiety to produce conformational changes and/or mechanical deformations which can be clearly detected in the SERS spectra (Fig. 7A(b)). The extent of such spectral alterations can be quantitatively correlated to the concentration of c-Jun in the media (Fig. 7A(c)).

In addition to indirect sensing approaches, such as those exemplified by the use of SEPs as contrast agents in biomaging (e.g., Fig. 3) and plasmonic nanostructures functionalized with chemoreceptors for oncoprotein detection (e.g., Fig. 7A), SERS spectroscopy can also be applied by directly acquiring the distinctive SERS signal of the target analytes directly adsorbed onto the nanostructured surface. Notably, direct SERS in combination with chemometrics has been successfully employed for the classification of relevantly clinical mutations in K-Ras oncogene, which are major markers for diagnostics and treatment of tumors, especially in colorectal cancer (Morla-Folch et al., 2017). This methodology exploits the large conformational changes that single- and multiple-base substitutions determine on long single strands when bent and folded onto positively-charged silver nanoparticles (Fig. 7B). This geometrical diversity is reflected in the spectral profile of the DNA SERS spectra, enabling the strand-to-strand discrimination with single-base sensitivity via statistical classification methods.

4.3 Nanoparticles for Circulating Tumor Exosome Quantification and Characterization

Exosomes are ~40 to 100 nm vesicles secreted by a wide range of mammalian cells types. The exosomes consist of a lipid bilayer membrane surrounding a small cytosol and are devoid of cellular organelles. Exosomes contain various molecular constituents of their cell of origin, including proteins and nucleic acid material (all types of RNAs). Although exosomes

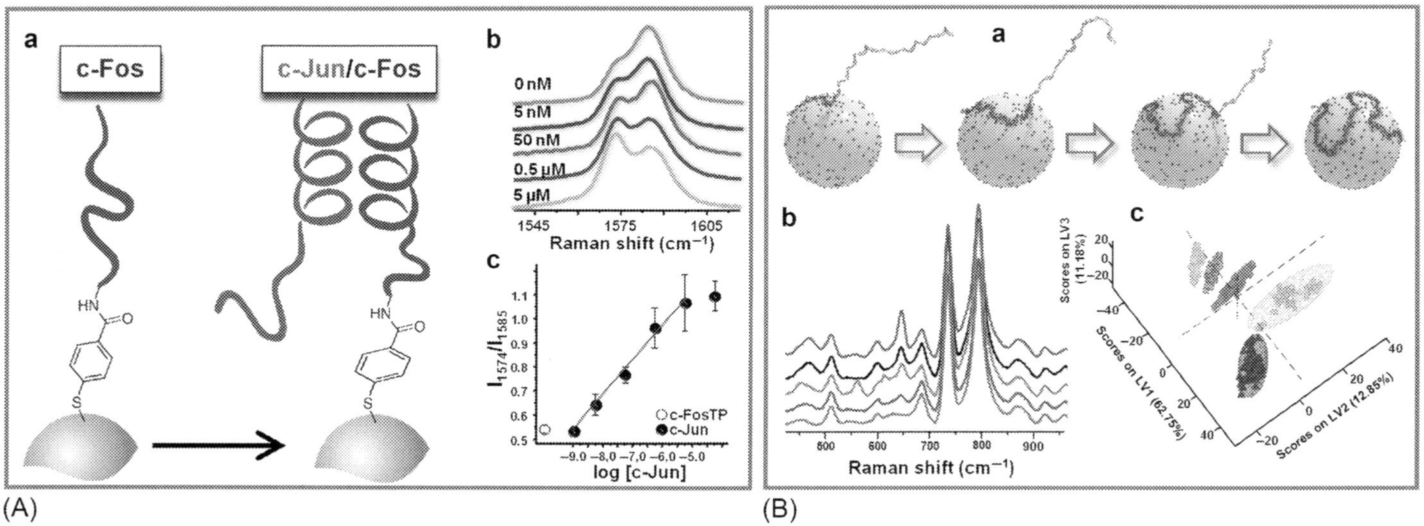

Fig. 7

(A) SERS detection of oncoprotein c-Jun in biofluids. (a) Depiction of the c-Fos/c-Jun dimerization and consequent structural rearrangement of the 4-mercaptobenzoyl moiety bound to the nanoparticle surface. (b) SERS spectra of c-Fos-MB chemoreceptor at increasing c-Jun concentration in the 1540–1615 cm^{-1} spectra range. (c) Intensity ratio of the band at 1574 and 1585 cm^{-1} (I_{1574}/I_{1585}) as a function of c-Jun concentration (logarithmic scale). (B) Direct SERS classification of clinical mutations in K-Ras oncogene. (a) Molecular dynamics simulation of the adsorption and wrapping process of 141-nt single-stranded DNA fragments on positively-charged silver nanoparticles at low salt concentration. (b) SERS spectra of 141-nt ssDNAs with point mutations in the 420–960 cm^{-1} spectra range. (c) Partial least-squares discriminant analysis (PLS-DA) of SERS spectra from 141-nt ssDNAs. *(A) Adapted with permission from Guerrini, L., Pazos, E., Penas, C., Vázquez, M. E., Mascareñas, J. L., Alvarez-Puebla, R. A. (2013). Highly sensitive SERS quantification of the oncogenic protein c-Jun in cellular extracts. Journal of the American Chemical Society 135 (28), 10314–10317. Copyright 2013, American Chemical Society. (B) Adapted with permission from Morla-Folch, J., Gisbert-Quilis, P., Masetti, M., Garcia-Rico, E., Alvarez-Puebla, R. A., & Guerrini, L. (2017). Conformational SERS classification of K-Ras point Mutations for cancer diagnostics. Angewandte Chemie, International Edition 56(9), 2381–2385. Copyright 2017, Wiley VCH.*

contain an evolutionary-conserved common set of protein molecules, their exosomal protein composition varies with the cell and tissue of origin. Thus, their composition varies according to the cell of origin and process of biogenesis, with cancer-cell-derived exosomes generally reflecting the tumor stage of the cell of origin (Raposo & Stoorvogel, 2013). Further, exosomes are stable in circulation, being found in body fluids, including blood, saliva, breast milk, and urine, which indicates that circulating exosomes may be suitable biomarkers for cancer diagnosis and prognosis. Finally, they can alter the phenotype of a recipient cell, being responsible for the tumor microenvironment maturation and cancer progression. Understanding the role of exosomes in tumorigenesis prompted a new era in cancer diagnosis and therapy, taking advantage of the use of circulating exosomes as tumor biomarkers.

Malicious exosomes are enriched in proteins, mRNAs, and miRNAs that are differentially expressed in cancer cells. Therefore, malicious exosomes are potential biomarkers in biological fluids for liquid biopsy. Notably, detection and evaluation of these exosomes can be achieved by using the same techniques, derived from nanoscience, as explained above. These techniques may include the use of antibody functionalized magnetic particles for enrichment (Li, Kaslan, Lee, Yao, & Gao, 2017), optically active encoded particles, either with SERS (Guerrini et al., 2017) or fluorescence (Boriachek et al., 2017), and the myriad of direct and indirect SERS techniques (Guerrini et al., 2017) for the characterization and quantification of nucleic acids and proteins (Feliu et al., 2017).

5 Tumor Environment (Disease as a Complex System)

Cancers can be depicted as complex "rogue" organs interacting with the tumor microenvironment (TME) via a complex and dynamic network of intercellular communication. At the same time, as a result of these interactions, the nonmalignant cells of the TME display, in turn, a dynamic and often tumor-promoting function at all stages of carcinogenesis (Hanahan & Coussens Lisa, 2012). This includes at least three phenomena: proliferation, invasion, and metastasis. Each of these involves an adaptation or response of the environment (not only local but also systemic) to the presence of the tumor. This adaptation, in the case of the immune system, is called immunoediting. The analysis of such adaptations would therefore be a "mirror" of the presence of the tumor. From a clinical point of view, of these adaptations, the most relevant are those of the immune system, fundamentally because, at present, there are therapeutic tools to manipulate this relationship to obtain effective antitumor responses, for example through PD1 or PDL1 antibodies.

Among the most relevant cells in the TME, some of key importance are lymphocytes T and B, NK and NKT cells, tumor-associated macrophages, dendritic cells, cancer-associated fibroblasts or MDSC (myeloid derived suppressor cells) (Balkwill, Capasso, & Hagemann, 2012). Typically, these cells are differentiated based on specific cell-surface molecular markers. These nonmalignant cells, and mediators of their communication, can provide valuable biomarkers for diagnosis and prognosis as well therapeutic targets (Balkwill et al., 2012). Conventional immunohistochemistry is unable to capture this dimension, particularly of MDSCs. Notably, the immune response to cancer involves diverse set of highly specific T lymphocytes within the TME (TIL—tumor infiltrating lymphocytes). In particular, cytotoxic $CD8^+$ memory T cells play a major role at eliminating cells and have been directly linked with a good prognosis (Fridman, Pagès, Sautès-Fridman, & Galon, 2012). The ensemble of these T cell populations can potentially constitute an immune response signature exploitable as a valuable biomarker for cancer diagnosis and monitoring thanks to their stability in blood samples (Anderson & LaBaer, 2005). However, cancer and innate immune system interactions, especially at the early stage of the disease, remain elusive, and their in-depth understanding demands methodologies capable of investigating a large number of parameters simultaneously on single cells.

Integration of QD technologies with flow cytometry has proven to significantly expand the polychromatic capabilities of this technique for simultaneously resolving a large number of diverse fluorescence emissions (Akinfieva, Nabiev, & Sukhanova, 2013; Chattopadhyay et al., 2006). This improvement in multiplexing capability of flow cytometry enables the designing of considerably more complex experiments. Notably, Chattopadhyay et al. (2006) demonstrated this concept by combining eight different classes of QDs with nine traditional fluorochromes into standard multicolor flow cytometry for T cell immunophenotyping. The analysis of multiple cell-surface markers revealed a significant phenotypic heterogeneity among $CD8^+$ T-cell populations of different antigen specificities (no two antigen-specific populations displayed identical phenotype), providing a new insight into the complexity of T-cell immunity. Fig. 8 illustrates the outcome of the corresponding flow cytometry analysis, providing a clear visualization of the phenotypic heterogeneity of different antigen-specific T cell populations contained in the samples, and validating the need for multiparameter technologies.

Cell-tracking using contrast agents also constitutes an important characterization tool for the development of new immunotherapy approaches. Among others, the use of genetically engineered T lymphocytes and natural killer (NK) cells in cancer immunotherapy has already shown promising preclinical and clinical results (Curran, Pegram, & Brentjens, 2012; Jha et al., 2010). In this regard, noninvasive real-time monitoring techniques are necessary to evaluate the tumor-targeting capability of these engineered cells, as well as quantify their cytotoxicity once transferred from in vitro to in vivo TMEs. For instance, Yong Taik, Mi Young, Young-Woock, Jin Woong, and Bong Hyun (2009) exploited QDs as highly sensitive fluorescent labels to track intratumoral injected human natural killer cells (NK92MI) in mice bearing human tumor xenografts and monitor over time their therapeutic effect. However, it is worth stressing that the in vivo application of these nanomaterials at the clinical level is restricted to the generation of new families of biocompatible QDs (Zhou, Yang, & Zhang, 2015), which overcome the intrinsic toxicity of traditional QDs such as those based on cadmium cores.

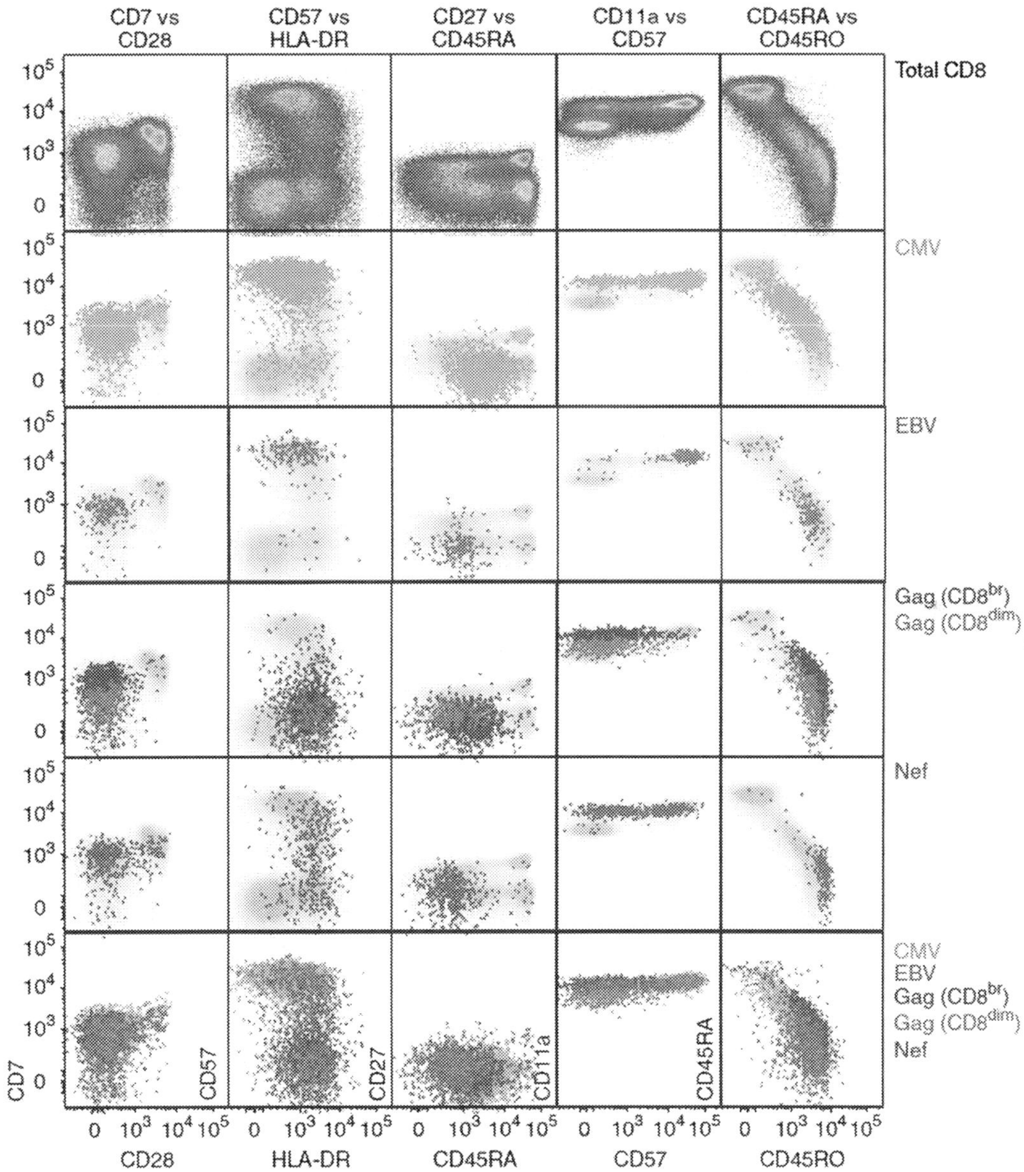

Fig. 8

17-color analysis of antigen-specific T cells using quantum dots. *Top row*: expression of various phenotypic markers within total CD8$^+$ T cells. *Subsequent rows:* comparison of various phenotypic markers expression within each antigen-specific population *(in color)* to the overall CD8$^+$ T-cell population *(light gray)*. *Bottom row:* overlay of the staining patterns of all antigen-specific cell populations studied. Axes indicate the fluorescence intensity of each marker. Within bulk CD8$^+$ T cells (top row), bright, dim and negative populations could be distinguished for nearly every marker. Antigen-specific T cell populations: EBV, CMV, Gag (CD8^{br} and CD8^{dim}) and Nef. Phenotypic markers: CD7, CD28, HLA-DR, CD45RA, CD57, CD45RO. *Adapted with permission from Chattopadhyay, P. K., Price, D. A., Harper, T. F., Betts, M. R., Yu, J., Gostick, E., et al. (2006) Quantum dot semiconductor nanocrystals for immunophenotyping by polychromatic flow cytometry. Nature Medicine 12(8), 972–977. Copyright 2006, Nature Publishing Group.*

6 Conclusions

In the fight against cancer, early detection is likely the key to win the battle. Nanoscience provides new materials to facilitate earlier and more accurate initial diagnosis as well as the continuous monitoring of cancer patient treatment. Also, for cancer, nanodevices are being investigated for the capture of blood borne biomarkers, including cancer-associated proteins CTCs, circulating tumor DNA, and tumor-shed exosomes. Nano-enabled sensors are capable of high sensitivity, specificity, and multiplexed measurements. Recent technological advances in the coupling of complex microfluidics and nanoscale materials have allowed the high-purity capture and downstream functional characterization of CTCs, cell-free tumor DNA, exosomes, proteins, neoantigens, and more.

One of the ultimate goals of nanoscience is POC cancer diagnosis. POC, which is done on site, is a field in which these new family of sensors could have a major impact, allowing the patient and medical staff to get results rapidly and easily. POC will allow for faster diagnosis and has the potential to significantly reduce costs. Although many technologies have been already designed and demonstrated for POC nanosensors, these systems are still at an early stage. However, it is expected that in the next 5–10 years, nanoscience will revolutionize cancer diagnosis and therapy. Integrating nanomaterials and biosensors will allow us to detect disease earlier, improve cancer imaging, and aid in diagnosis/prognosis and advance drug delivery while minimizing adverse reactions. Cancer takes place at the nanoscale level, and it is at the nanoscale that the scientific community should confront the disease. Although the complexity and diversity of cancer has posed many challenges in the medical field, nanobiosensor technology has the potential to provide fast, accurate results, while maintaining cost effectiveness.

References

Akinfieva, O., Nabiev, I., & Sukhanova, A. (2013). New directions in quantum dot-based cytometry detection of cancer serum markers and tumor cells. *Critical Reviews in Oncology/Hematology*, *86*(1), 1–14.

Alix-Panabieres, C., & Pantel, K. (2014). Technologies for detection of circulating tumor cells: facts and vision. *Lab on a Chip*, *14*(1), 57–62.

Anderson, K. S., & LaBaer, J. (2005). The sentinel within: exploiting the immune system for cancer biomarkers. *Journal of Proteome Research*, *4*(4), 1123–1133.

Ashworth, T. R. (1869). A case of cancer in which cells similar to those in the tumors were seen in the blood after death. *The Australasian Medical Journal*, *14*, 146–149.

Balkwill, F. R., Capasso, M., & Hagemann, T. (2012). The tumor microenvironment at a glance. *Journal of Cell Science*, *125*(23), 5591–5596.

Blau, L., Knirsh, R., Ben-Dror, I., Oren, S., Kuphal, S., Hau, P., et al. (2012). Aberrant expression of c-Jun in glioblastoma by internal ribosome entry site (IRES)-mediated translational activation. *Proceedings of the National Academy of Sciences*, *109*(42), E2875–E2884.

Boriachek, K., Islam, M. N., Gopalan, V., Lam, A. K., Nguyen, N.-T., & Shiddiky, M. J. A. (2017). Quantum dot-based sensitive detection of disease specific exosome in serum. *Analyst*, *142*(12), 2211–2219.

Brierley, J. D. (2017). *TNM classification of malignant tumours*. Chichester, UK: John Wiley & Sons Ltd..

Cagan, R., & Meyer, P. (2017). Rethinking cancer: current challenges and opportunities in cancer research. *Disease Models & Mechanisms*, *10*, 349–352. The Company of Biologists Ltd.

Carroll, S., & Al-Rubeai, M. (2005). ACSD labelling and magnetic cell separation: a rapid method of separating antibody secreting cells from non-secreting cells. *Journal of Immunological Methods*, *296*(1), 171–178.

Chambers, A. F., Groom, A. C., & MacDonald, I. C. (2002). Dissemination and growth of cancer cells in metastatic sites. *Nature Reviews Cancer*, *2*, 563.

Chattopadhyay, P. K., Price, D. A., Harper, T. F., Betts, M. R., Yu, J., Gostick, E., et al. (2006). Quantum dot semiconductor nanocrystals for immunophenotyping by polychromatic flow cytometry. *Nature Medicine*, *12*(8), 972–977.

Choi, Y.-E., Kwak, J.-W., & Park, J. W. (2010). Nanotechnology for early cancer detection. *Sensors*, *10*(1), 428–455.

Curran, K. J., Pegram, H. J., & Brentjens, R. J. (2012). Chimeric antigen receptors for T cell immunotherapy: current understanding and future directions. *The Journal of Gene Medicine*, *14*(6), 405–415.

Fabris, L. (2016). SERS tags: the next promising tool for personalized cancer detection? *ChemNanoMat: Chemistry of Nanomaterials for Energy, Biology and More*, *2*(4), 249–258.

Feliu, N., Hassan, M., Garcia Rico, E., Cui, D., Parak, W., & Alvarez-Puebla, R. (2017). SERS quantification and characterization of proteins and other biomolecules. *Langmuir*, *33*(38), 9711–9730.

Fridman, W. H., Pagès, F., Sautès-Fridman, C., & Galon, J. (2012). The immune contexture in human tumours: impact on clinical outcome. *Nature Reviews Cancer*, *12*, 298.

Guerrini, L., Pazos, E., Penas, C., Vázquez, M. E., Mascareñas, J. L., & Alvarez-Puebla, R. A. (2013). Highly sensitive SERS quantification of the oncogenic protein c-Jun in cellular extracts. *Journal of the American Chemical Society*, *135*(28), 10314–10317.

Guerrini, L., Pazos-Perez, N., Garcia-Rico, E., & Alvarez-Puebla, R. (2017). Cancer characterization and diagnosis with SERS-encoded particles. *Cancer Nanotechnology*, *8*(1), 5.

Hanahan, D., & Coussens Lisa, M. (2012). Accessories to the crime: functions of cells recruited to the tumor microenvironment. *Cancer Cell*, *21*(3), 309–322.

Hanahan, D., & Weinberg, R. A. (2011). Hallmarks of cancer: the next generation. *Cell*, *144*(5), 646–674.

Hildebrandt, N., Spillmann, C. M., Algar, W. R., Pons, T., Stewart, M. H., Oh, E., et al. (2017). Energy transfer with semiconductor quantum dot bioconjugates: a versatile platform for biosensing, energy harvesting, and other developing applications. *Chemical Reviews*, *117*(2), 536–711.

Hirata, E., & Sahai, E. (2017). Tumor microenvironment and differential responses to therapy. *Cold Spring Harbor Perspectives in Medicine*, *7*, a026781.

Jha, P., Golovko, D., Bains, S., Hostetter, D., Meier, R., Wendland, M. F., et al. (2010). Monitoring of natural killer cell immunotherapy using noninvasive imaging modalities. *Cancer Research*, *70*(15), 6109–6113.

Lambert, A. W., Pattabiraman, D. R., & Weinberg, R. A. (2017). Emerging biological principles of metastasis. *Cell*, *168*(4), 670–691.

Lamouille, S., Xu, J., & Derynck, R. (2014). Molecular mechanisms of epithelial-mesenchymal transition. *Nature Reviews Molecular Cell Biology*, *15*(3), 178.

Lane, L. A., Qian, X. M., & Nie, S. M. (2015). SERS nanoparticles in medicine: from label-free detection to spectroscopic tagging. *Chemical Reviews*, *115*(19), 10489–10529.

Le Ru, E. C., & Etchegoin, P. G. (2009). *Principles of surface-enhanced Raman spectroscopy*. Amsterdam, The Netherlands: Elsevier.

Li, P., Kaslan, M., Lee, S. H., Yao, J., & Gao, Z. (2017). Progress in exosome isolation techniques. *Theranostics*, *7*(3), 789–804.

Mallia, R. J., McVeigh, P. Z., Fisher, C. J., Veilleux, I., & Wilson, B. C. (2015). Wide-field multiplexed imaging of EGFR-targeted cancers using topical application of NIR SERS nanoprobes. *Nanomedicine*, *10*(1), 89–101.

Min, H., Jo, S.-M., & Kim, H.-S. (2015). Efficient capture and simple quantification of circulating tumor cells using quantum dots and magnetic beads. *Small*, *11*(21), 2536–2542.

Mir-Simon, B., Reche-Perez, I., Guerrini, L., Pazos-Perez, N., & Alvarez-Puebla, R. A. (2015). Universal one-pot and scalable synthesis of SERS encoded nanoparticles. *Chemistry of Materials*, *27*(3), 950–958.

Morla-Folch, J., Gisbert-Quilis, P., Masetti, M., Garcia-Rico, E., Alvarez-Puebla, R. A., & Guerrini, L. (2017). Conformational SERS classification of K-Ras point Mutations for cancer diagnostics. *Angewandte Chemie, International Edition*, *56*(9), 2381–2385.

Pallaoro, A., Hoonejani, M. R., Braun, G. B., Meinhart, C. D., & Moskovits, M. (2015). Rapid identification by surface-enhanced Raman spectroscopy of cancer cells at low concentrations flowing in a microfluidic channel. *ACS Nano*, *9*(4), 4328–4336.

Pazos, E., Garcia-Algar, M., Penas, C., Nazarenus, M., Torruella, A., Pazos-Perez, N., et al. (2016). Surface-enhanced Raman scattering surface selection rules for the proteomic liquid biopsy in real samples: efficient detection of the oncoprotein c-MYC. *Journal of the American Chemical Society*, *138*(43), 14206–14209.

Postow, M. A., Callahan, M. K., Barker, C. A., Yamada, Y., Yuan, J., Kitano, S., et al. (2012). Immunologic correlates of the abscopal effect in a patient with melanoma. *New England Journal of Medicine*, *366*(10), 925–931.

Prasad, V. (2016). Perspective: the precision-oncology illusion. *Nature*, *537*(7619), S63.

Raposo, G., & Stoorvogel, W. (2013). Extracellular vesicles: exosomes, microvesicles, and friends. *The Journal of Cell Biology*, *200*(4), 373–383.

Schlücker, S. (2014). Surface-enhanced Raman spectroscopy: concepts and chemical applications. *Angewandte Chemie, International Edition*, *53*(19), 4756–4795.

Sohal, D. P., Rini, B. I., Khorana, A. A., Dreicer, R., Abraham, J., Procop, G. W., et al. (2016). Prospective clinical study of precision oncology in solid tumors. *Journal of the National Cancer Institute*, *108*(3), pii: djv332.

Subik, K., Lee, J.-F., Baxter, L., Strzepek, T., Costello, D., Crowley, P., et al. (2010). The expression patterns of ER, PR, HER2, CK5/6, EGFR, Ki-67 and AR by immunohistochemical analysis in breast cancer cell lines. *Breast Cancer (Auckl)*, *4*, 35–41.

Velpurisiva, P., Gad, A., Piel, B., Jadia, R., & Rai, P. (2017). Nanoparticle design strategies for effective cancer immunotherapy. *Journal of Biomedicine (Sydney, NSW)*, *2*(2), 64.

Wang, X., Qian, X., Beitler, J. J., Chen, Z. G., Khuri, F. R., Lewis, M. M., et al. (2011). Detection of circulating tumor cells in human peripheral blood using surface-enhanced Raman scattering nanoparticles. *Cancer Research*, *71*(5), 1526–1532.

Wang, Y., Kang, S., Khan, A., Ruttner, G., Leigh, S. Y., Murray, M., et al. (2016). Quantitative molecular phenotyping with topically applied SERS nanoparticles for intraoperative guidance of breast cancer lumpectomy. *Scientific Reports*, *6*, 21242.

Wang, Y. Q., Yan, B., & Chen, L. X. (2013). SERS tags: novel optical nanoprobes for bioanalysis. *Chemical Reviews*, *113*(3), 1391–1428.

Wegner, K. D., & Hildebrandt, N. (2015). Quantum dots: bright and versatile in vitro and in vivo fluorescence imaging biosensors. *Chemical Society Reviews*, *44*(14), 4792–4834.

Welch, D. R. (2016). Tumor heterogeneity—a 'contemporary concept' founded on historical insights and predictions. *Cancer Research*, *76*(1), 4–6.

Wise, P. H. (2016). Cancer drugs, survival, and ethics. *BMJ*, *355*, i5792.

Wittekind, C., & Neid, M. (2005). Cancer invasion and metastasis. *Oncology*, *69*(Suppl. 1), 14–16.

World Health Organization. *Projections of mortality and causes of death, 2015 and 2030*. Available from: http://www.who.int/healthinfo/global_burden_disease/projections/en/.

Wu, X., Luo, L., Yang, S., Ma, X., Li, Y., Dong, C., et al. (2015). Improved SERS nanoparticles for direct detection of circulating tumor cells in the blood. *ACS Applied Materials & Interfaces*, *7*(18), 9965–9971.

Yong Taik, L., Mi Young, C., Young-Woock, N., Jin Woong, C., & Bong Hyun, C. (2009). Near-infrared emitting fluorescent nanocrystals-labeled natural killer cells as a platform technology for the optical imaging of immunotherapeutic cells-based cancer therapy. *Nanotechnology*, *20*(47), 475102.

Zavaleta, C. L., Garai, E., Liu, J. T. C., Sensarn, S., Mandella, M. J., Van de Sompel, D., et al. (2013). A Raman-based endoscopic strategy for multiplexed molecular imaging. *Proceedings of the National Academy of Sciences*, *110*(25), E2288–E2297.

Zhou, J., Yang, Y., & Zhang, C.-y. (2015). Toward biocompatible semiconductor quantum dots: from biosynthesis and bioconjugation to biomedical application. *Chemical Reviews*, *115*(21), 11669–11717.
Zrazhevskiy, P., & Gao, X. (2013). Quantum dot imaging platform for single-cell molecular profiling. *Nature Communications*, *4*, 1619.

Further Reading

de Bono, J. S., Scher, H. I., Montgomery, R. B., Parker, C., Miller, M. C., Tissing, H., et al. (2008). Circulating tumor cells predict survival benefit from treatment in metastatic castration-resistant prostate cancer. *Clinical Cancer Research*, *14*(19), 6302–6309.
Diehl, F., Schmidt, K., Choti, M. A., Romans, K., Goodman, S., Li, M., et al. (2008). Circulating mutant DNA to assess tumor dynamics. *Nature Medicine*, *14*(9), 985–990.
Pedrol, E., Garcia-Algar, M., Massons, J., Nazarenus, M., Guerrini, L., Martínez, J., et al. (2017). Optofluidic device for the quantification of circulating tumor cells in breast cancer. *Scientific Reports*, *7*, 3677.

CHAPTER 8

Nanotheranostics and Their Potential in the Management of Metastatic Cancer

Abi Judit Vázquez-Ríos*, Marta Alonso-Nocelo†,‡,§, Belén López Bouzo*, Juan Ruiz-Bañobre*, María de la Fuente*

**Health Research Institute of Santiago de Compostela (IDIS), SERGAS, CIBERONC, Santiago de Compostela, Spain †Autonomous University of Madrid (UAM), Madrid, Spain ‡"Alberto Sols" Biomedical Research Institute, CSIC-UAM, Madrid, Spain §Ramón y Cajal Health Research Institute (IRYCIS), Madrid, Spain*

1 The Management of Metastatic Cancer, an Unmet Clinical Need

Cancer is a major public health problem worldwide and the second leading cause of death in the United States and Europe. The most common causes of cancer death are lung, colorectal, and prostate cancers in men, and lung, breast, and colorectal cancers in women. In the United States, these four cancers account for 46% of all cancer deaths, with more than one-quarter (26%) attributed to lung cancer (Siegel, Miller, & Jemal, 2017; WHO, 2017).

Metastases are the cause of 90% of human cancer deaths (Taketo, 2011). As an example, the 5-year survival rate for localized-stage colorectal cancer (CRC) is around 90%, declining to 71% and 14% for patients diagnosed with regional and distant stages respectively (American Cancer Society, 2017). In the case of pancreatic cancer, the 5-year survival for localized disease is around 29%, and falls to 3% in the locally advanced or metastatic setting (Ilic & Ilic, 2016). Metastasis is an exceedingly complex process that occurs through a series of sequential steps, as reviewed by Mehlen and Puisieux (2006). Despite the fact that tumors at an early stage can often be surgically removed, there is growing evidence that dissemination could indeed happen at a very early stage in the carcinogenesis process (Martínez-Cardús, Moran, Musulen, et al., 2016) a fact that could explain why, in some tumor types, the 5-year survival even for localized disease is so poor.

With the aim of increasing patients' survival, different drugs and some biologicals are given as single sequential therapies or in combination in different clinical scenarios, palliative, adjuvant, and neoadjuvant settings. The main problems are related to toxicity and secondary side effects and can worsen the quality of life in short and long term (Hall, Cameron, Waters, et al., 2014). Additionally, the clonal evolution of tumors under antineoplastic treatments pressure can lead to therapeutic failure and progress of the disease (Aparicio & Caldas, 2013).

Handbook of Nanomaterials for Cancer Theranostics. https://doi.org/10.1016/B978-0-12-813339-2.00008-6

The development of targeted therapies has resulted in substantial benefits in terms of survival and quality of life for cancer patients. In the last 20 years, different drugs have been approved by the Food and Drug Administration (FDA) and the European Medicines Agency (EMA) for several tumor types, as for example, trastuzumab (breast cancer and gastric/gastroesophageal junction adenocarcinoma) (Amadori, Milandri, Comella, Saracchini, et al., 2011), imatinib (gastrointestinal stromal tumors and dermatofibrosarcoma protuberans) (Sugiura, Fujiwara, Ando, Kawai, et al., 2010), or cetuximab (colorectal cancer and head and neck cancer) (Bertino, McMichael, Mo, Trikha, et al., 2016), among others. In the case of nonsmall cell lung cancer (NSCLC), the treatment breakthrough was the introduction of targeted therapies against EGFR mutations, rearrangements involving the ALK or ROS1 genes, and recently, B-RAF mutations (Novello, Barlesi, Califano, et al., 2016; Planchard, Besse, Groen, et al., 2016). Another milestone is the development of immune-checkpoint inhibitors, a group of antibodies designed to block specific targets present on tumor cells or lymphocyte surface [as for example ipilimumab (Hodi, O'Day, McDermott, et al., 2010), the first approved anti-CTL4 antibody, and nivolumab (Borghaei, Paz-Ares, Horn, et al., 2015; Brahmer, Reckamp, Baas, et al., 2015; Motzer, Escudier, McDermott, et al., 2015; Robert, Long, Brady, et al., 2014), the first approved anti-PD-1 antibody], and consequently boosting the immune system against cancer. Nanomedicines have also being introduced into the clinic, as detailed in Section 3, showing a well-tolerated toxicity profile and good results in terms of efficacy (Thakor & Gambhir, 2013). Due to these novel approaches, the median overall survival (OS) for patients with metastatic cancer has considerably improved. For example, for NSCLC and metastatic colorectal cancer, OS is ~12 and 30 months, respectively, more than double what it was 20 years ago (Reck, Rodríguez-Abreu, Robinson, et al., 2016; Rolfo, Passiglia, Ostrowski, et al., 2015; Van Cutsem, Cervantes, Adam, et al., 2016). Despite this, success is still limited, making it mandatory to focus our efforts on developing novel alternatives specifically tailored to make a breakthrough in the management of disseminated cancer.

2 The Metastatic Cascade

The process of metastasis is defined by a cascade of complex events in which malignant cells detach from the primary tumor, invade the basement membrane, and then migrate into the circulation, either via the blood or lymphatic vessels, to finally spread to distant sites to form metastases (Chambers, Groom, & MacDonald, 2002; Pantel & Brakenhoff, 2004). The metastatic cascade can be divided into three main phases, (i) premetastatic initiation, (ii) dissemination, and (iii) colonization.

Premetastatic initiation starts with local invasiveness, that is, the entry of well-restrained cancer cells into the surrounding tumor-associated stroma. To accomplish this, tumor cells reduce their adhesion to neighbor cells, increasing their motility and degrading the basement membrane (Kedrin, van Rheenen, Hernandez, Condeelis, & Segall, 2007; Weber, 2008). In the next step, tumor cells intravasate into the lymphatics and/or blood vessels, disseminating by hematogenous or lymphatic routes (Pantel & Brakenhoff, 2004;

Sleeman, Cady, & Pantel, 2012). Next, tumor cells extravasate into distant organs and eventually colonize them to generate metastasis (Irmisch & Huelsken, 2013).

The metastatic process is complex and inefficient, with many cancer cells initially achieving access to circulation, but only a few finally succeeding in proliferating as distant metastases (Wong, Lee, Shientag, et al., 2001). Despite of this limited efficiency, stopping the spread of cancer cells from the primary tumor to distant sites continues to be one of the greatest challenges nowadays. Hence, the development of new approaches that target one or more steps in the metastatic cascade is needed, to ultimately improve the outcome of patients with metastatic cancer. Fig. 1 summarizes the metastatic cascade and compiles the key events susceptible of therapeutic intervention.

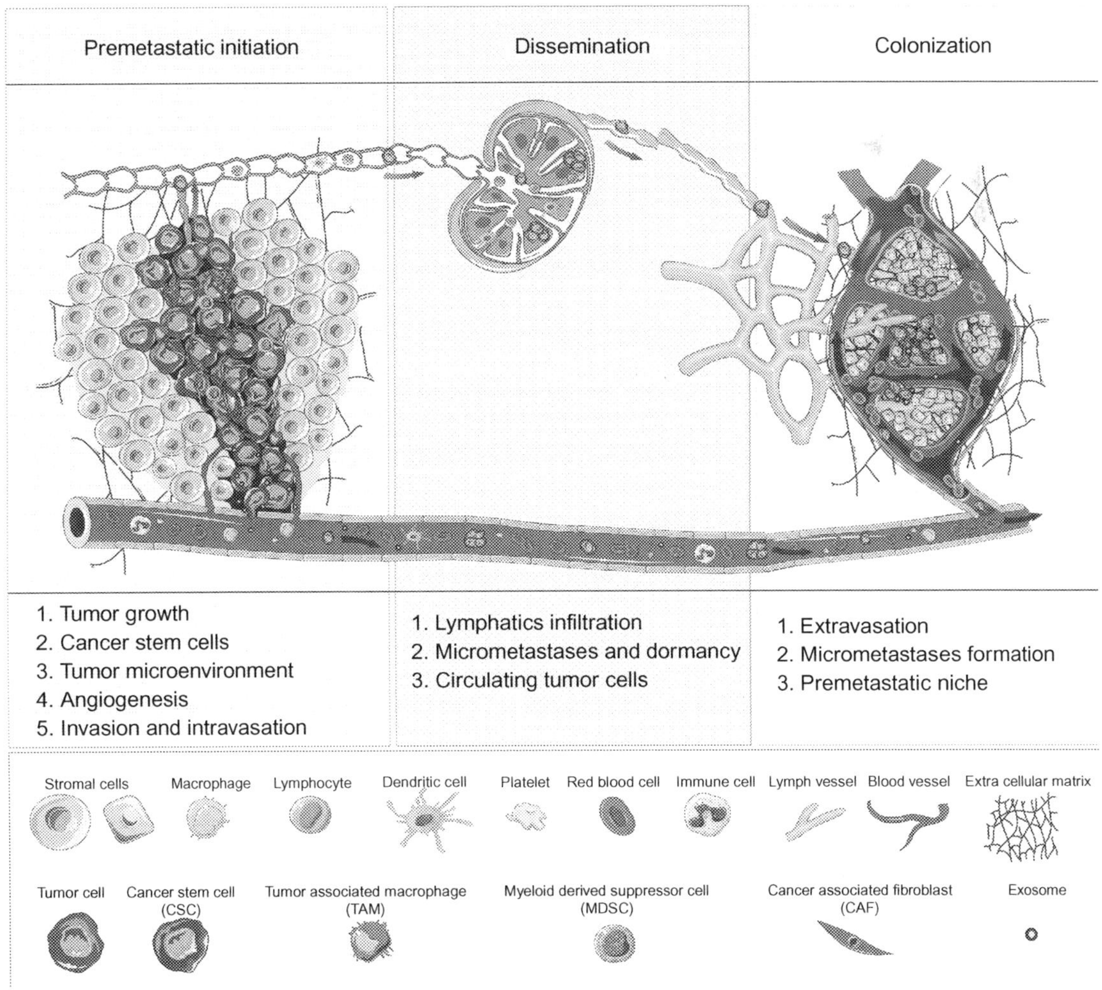

Fig. 1
Schematic representation of the process of metastasis.

3 The Impact of Nanomedicine in Cancer Therapy

Nanotechnology can be defined as the science of manipulating and controlling materials and systems on a molecular scale. The merging of nanotechnology and medicine, nanomedicine, has allowed the development of several nanoparticle delivery vehicles, specifically designed for carrying and delivering their cargo to disease sites for treatment, prevention, and/or diagnosis (Ledet & Mandal, 2012; Wang, Langer, & Farokhzad, 2012).

Nanomedicine has been widely explored in the last decades. Different nanosystems composed of a variety of materials have been proposed for the management of several diseases, as liposomes and other lipid-based nanosystems, polymer-based nanoparticles, micelles, polyplexes, dendrimers, polymersomes, and drug/protein conjugates (Beloqui, Memvanga, Coco, et al., 2016; Cordeiro, Alonso, & de la Fuente, 2015; de la Fuente, Csaba, Garcia-Fuentes, & Alonso, 2008; de la Fuente, Jones, Santander-Ortega, et al., 2015; de la Fuente, Raviña, Sousa-Herves, et al., 2012; de la Fuente, Seijo, & Alonso, 2008; Mazza, Alonso-Sande, Jones, & de la Fuente, 2013; Reimondez-Troitiño, Csaba, Alonso, & de la Fuente, 2015; Swain, Sahu, Beg, & Babu, 2016). Nanotechnology offers many advantages in drug delivery, including (i) protection of drugs from premature degradation, (ii) increased solubility and stability in biological media, (iii) prevention of premature interactions of drugs with the biological environment, (iv) controlled pharmacokinetics and biodistribution, (v) improved delivery of therapeutics across biological barriers, and (vi) targeting of drugs to the diseased area (Peer et al., 2007). Due to these properties and their ability to accommodate various types of drugs and biomolecules, with different physicochemical properties and activities, nanocarriers have emerged as attractive candidates for the development of personalized medicine (Sanna, Pala, & Sechi, 2014).

The application of nanotechnology in cancer therapy is already a reality that enables incremental, localized delivery of chemotherapeutics to the tumor, leading to improved pharmacological activities and significantly minimizing secondary side effects. In 1995, the FDA approved the first nanoparticle for cancer treatment, Doxil, a PEGylated liposome loaded with the chemotherapeutic drug doxorubicin (Barenholz, 2012). From then, some other nanotherapeutics have reached the market for different cancer treatments, being predominantly drug-loaded PEGylated liposomes, and as the main application, the treatment of unresectable locally advanced or metastatic cancer. Current FDA and EMA approved nanotherapeutics are summarized in Table 1. Most cases are examples of passively targeted first-generation PEGylated nanomedicines that mainly rely on modifying the pharmacokinetics and biodistribution of a drug by improving its accumulation in tumors, due to the phenomenon called Enhanced Permeability and Retention (EPR) effect, first described in 1986 by Matsumura and Maeda (1986). The attachment of the polymer polyethylene glycol (PEG) is widely used in nanomedicine to increase circulation lifetimes and reduce doses. The covalent attachment of PEG to proteins (i.e., enzymes, cytokines, antibodies, growth factors) has, for example, brought more than ten polymer-protein drugs into the market for different

Table 1: Clinically approved nanoparticles for cancer therapy

Commercial Name	Type of Nanoparticle	Drug/Therapy	Indication	Approval[a] (Year)
Doxil/Caelyx	Liposomal	Doxorubicin	HIV-related Kaposi sarcoma Breast cancer Ovarian cancer	FDA (1995) FDA (2005) FDA (2008)
DaunoXome	Liposomal	Daunorubicin	Advanced Karposi's sarcoma Acute myeloid leukemia	FDA (1996) EMA (2008)
DepoCyt	Liposomal	Cytarabine	Lymphomatous meningitis	FDA (1996)
Myocet	Liposomal	Doxorubicin	Breast cancer	EMA (2000)
Mepact	Liposomal	Mifamurtide	Osteosarcoma	EMA (2009)
Marqibo	Liposomal	Vincristine	Acute lymphoblastic leukemia	FDA (2012)
Onivyde	Liposomal	Irinotecan	Pancreatic cancer	FDA (2015)
Opaxio	Polimeric nanoparticle	Paclitaxel	Glioblastoma	FDA (2012)
Oncaspar	Polimeric nanoparticle	PEGylated L-Asparaginase	Acute lymphoblastic leukemia	FDA (2006)
Eligard	Polimeric micelles	Leuprolide acetate	Prostate cancer	FDA (2002)
Eloxatin	Micelles	Oxaliplatin	Colon cancer	FDA (2004)
Neulasta	Polymer-protein conjugate	Filgastrim	Chemotherapy induced neutropenia	FDA (2002)
Abraxane	Protein-drug conjugate	Albumin-paclitaxel	Breast cancer Nonsmall-cell lung carcinoma Pancreatic cancer	FDA (2005) FDA (2012) FDA (2013)
Ontak	Protein-drug conjugate	Interleukina-2-diphteria toxin	Cutaneos T-cell lymphoma	FDA (1999)
Adcetris	Antibody-drug conjugate	Brentuximab-monomethyl auristatin E (MMAE)	Hodgkin lymphoma Lymphoma (ALCL)	FDA (2011) FDA (2011)
Kadcyla	Antibody-drug conjugate	Trastuzumab-Emtansine	Breast cancer	FDA (2013)

Sources: https://www.accessdata.fda.gov/scripts/cder/daf/; http://www.ema.europa.eu/ema/ (accessed 10 August 2017).
[a] All the information found in: (1) FDA approved drug products. Drugs@FDA. www.accessdata.fda.gov/scripts/cder/drugsatfda/. Accessed August 10, 2017. (2) European Medicines Agency: http://www.ema.europa.eu/ema/. Accessed August 10, 2017.

diseases (e.g., PEGASYS or PEG-INTRON for chronic hepatitis C, Cimzia for Crohn's disease, or Oncaspar for acute lymphoblastic leukemia) (Alconcel, Baas, & Maynard, 2011). The other mechanism described by which nanoparticles can be accumulated in cancerous tissues is active targeting, which can be achieved through functionalization of nanoparticles' surfaces with bioactive molecules able to recognize cancer-specific targets overexpressed in malignant cells, such as transferrin, engineered antibodies or fragments, peptides, enzymes, and folic acid, among others (Bregoli et al., 2016). Although active targeted nanomedicines are advanced at the preclinical stage, to date, only antibody- and protein-based nanomedicines (Abraxane, Adcetris, Kadcyla, and Ontak) have been FDA approved. Ongoing clinical trials with nanotherapeutics are compiled in Table 2, suggesting that we are prone to see a growth

Table 2: Nanotherapeutics under clinical evaluation (active trials recruiting patients, active not yet recruiting and completed without published results) for the treatment of solid tumors

Investigational Drug Name	Drug/Therapy	Type of Nanoparticle	Disease	Clinical Trial Phase	Trial Status	NCT Number
ALN-VSP02	Anti-KSP and VEGFA siRNAs	Lipid nanoparticle	Solid tumors	1	Completed	NCT01158079
			Solid tumors	1	Completed	NCT00882180
dHER2+AS15	Recombinant HER2 (dHER2) antigen and AS15 adjuvant	Liposome	Metastatic breast cancer	1/2	Completed	NCT00952692
			Breast cancer	1	Completed	NCT00058526
			Metastatic breast cancer	1	Completed	NCT00140738
ABI-009	Rapamycin	Albumin nanoparticle	Refractory central nervous System Cancer	1	Active, not recruiting	NCT02975882
			Advanced genitourinary, gynecological or head and neck cancers	2	Active, not recruiting	NCT02646319
			Solid tumors	1	Completed	NCT00635284
			PEComa	2	Recruiting	NCT02494570
			Non-muscle invasive bladder cancer	1/2	Recruiting	NCT02009332
			Advanced sarcoma	1/2	Recruiting	NCT03190174
anti-EGFR-IL-DOX	Doxorubicin	EGFR-targeting liposome	Solid tumors	1	Completed	NCT01702129
			Breast cancer	2	Recruiting	NCT02833766
Atu027	Antiprotein kinase N3 siRNAs	Liposome	Advanced pancreatic cancer	1/2	Completed	NCT01808638
			Advanced solid tumors	1	Completed	NCT00938574
AuroLase	Photothermal therapy	Silica core with a gold nanoshell	Prostate cancer	Device study	Recruiting	NCT02680535
BIND-014	Docetaxel	PSMA-targeting polymeric nanoparticle	Metastatic castration-resistant prostate cancer	2	Completed	NCT01812746
			Advanced solid tumors	1	Completed	NCT01300533
			Nonsmall cell lung cancer	2	Completed	NCT01792479
			Advanced nonsmall cell lung cancer	2	Completed	NCT02283320
CPX-1	Irinotecan and floxuridine (1:1)	Liposome	Advanced colorectal cancer	2	Completed	NCT00361842

CRLX101	Camptothecin	Cyclodextran conjugated	Ovarian cancer	1/2	Active, not recruiting	NCT02389985
			Solid tumors	1	Active, not recruiting	NCT02648711
			Recurrent ovarian, fallopian tube or primary peritoneal cancers	2	Active, not recruiting	NCT01652079
			Rectal cancer	1/2	Active, not recruiting	NCT02010567
			Advanced solid tumors	1/2	Active, not recruiting	NCT02380677
			Recurrent/persistent ovarian, fallopian tube and primary peritoneal cancers	1/2	Active, not recruiting	NCT02389985
			Solid tumors	1	Active, not recruiting	NCT02648711
			Rectal cancer	1/2	Active, not recruiting	NCT02010567
			Advanced stomach, gastroesophageal or esophageal cancers	Pilot study	Completed	NCT01612546
			Advanced solid tumors	1/2	Completed	NCT00333502
			Metastatic renal cell carcinoma	2	Completed	NCT02187302
			Solid tumors	1/2	Completed	NCT00333502
			Metastatic renal cell carcinoma	2	Completed	NCT02187302
			Refractory/recurrent small cell lung cancer	1/2	Recruiting	NCT02769962
CRLX301	Docetaxel	Polymer conjugated	Advanced solid tumors	1/2	Active, not recruiting	NCT02380677
CYT-6091	rhTNF	Colloidal gold nanoparticle	Advanced solid tumors	1	Completed	NCT00356980
			Solid tumors	1	Completed	NCT00436410
DC-Chol-EGFR	EGFR antisense DNA	Liposome	Head and neck cancer	1	Completed	NCT00009841
DCR-MYC	Anti-MYC Dicer-substrate siRNA	Lipid nanoparticle	Malignant mesothelioma	2	Active, not recruiting	NCT02293005
DOTAP: cholesterol-Fus1	Fus1 gene (insertion)	Liposome	Lung cancer	1	Completed	NCT00059605
			Lung cancer	1/2	Recruiting	NCT01455389
DPX-0907	Multi-tumor associated antigens	Liposome	Adavanced breast, ovarian or prostate cancers	1	Completed	NCT01095848

Continued

Table 2: Nanotherapeutics under clinical evaluation (active trials recruiting patients, active not yet recruiting and completed without published results) for the treatment of solid tumors—cont'd

Investigational Drug Name	Drug/Therapy	Type of Nanoparticle	Disease	Clinical Trial Phase	Trial Status	NCT Number
EndoTAG-1	Paclitaxel	Liposome	Breast cancer	2	Completed	NCT01537536
			Breast cancer	2	Completed	NCT00448305
			Advanced pancreatic cancer	2	Completed	NCT00377936
			Liver metastatic cancer	2	Completed	NCT00542048
			Advanced pancreatic cancer	3	Not yet recruiting	NCT03126435
			Metastatic breast cancer	3	Recruiting	NCT03002103
Genexol-PM	Paclitaxel	Polymeric micelle	Advanced urothelial cancers	2	Completed	NCT01426126
			Advanced nonsmall cell lung cancer	2	Completed	NCT01770795
			Recurrent/metastatic breast cancer	3	Completed	NCT00876486
			Advanced ovarian cancer	1	Completed	NCT00877253
			Advanced pancreatic cancer	2	Recruiting	NCT02739633
			Hepatocellular carcinoma	2	Recruiting	NCT03008512
			Metastatic breast cancer	2	Recruiting	NCT01784120
			Gynecologic cancers	1	Recruiting	NCT02739529
LErafAON	c-Raf antisense oligodeoxynucleotides	Liposome	Advanced solid tumors	1	Completed	NCT00024661
			Advanced solid tumors	1	Completed	NCT00024648
			Advanced solid tumors	1	Completed	NCT00100672
Lipoplatin	Cisplatin	Pegylated liposome	Pleural malignancies	1	Recruiting	NCT02702700
Lipovaxin-MM	Melanoma antigens	Liposome	Melanoma	1	Completed	NCT01052142
MBP-426	Oxaliplatin	Transferrin receptor-targeting liposome	Solid tumors	1	Completed	NCT00355888
MM-302	Doxorubicin	HER2-targeting liposome	Advanced breast cancer	1	Active, not recruiting	NCT01304797
Nal-IRI	Irinotecan	Liposome	Metastatic pancreatic cancer	2	Active, not recruiting	NCT02697058
			Advanced cholangiocarcinoma	2	Not yet recruiting	NCT03043547
			Advanced biliary-tract cancers	2	Not yet recruiting	NCT03044587
			Pancreatic cancer	2	Recruiting	NCT02551991
			Diffuse intrinsic pontine glioma	1	Recruiting	NCT03086616
			Solid tumors	1	Recruiting	NCT01770353

Nanoplatin	Cisplatin	Polymeric micelle	Advanced pancreatic cancer	1/2	Completed	NCT00910741
			Head and neck cancer	1	Recruiting	NCT02817113
NBTXR3	Radiotherapy	Hafnium oxide NP	Soft tissue sarcoma	1	Completed	NCT01433068
			Liver cancers	1/2	Recruiting	NCT02721056
			Prostate cancer	1/2	Recruiting	NCT02805894
			Head and neck cancer	1	Recruiting	NCT01946867
			Soft tissue sarcoma	2/3	Recruiting	NCT02379845
			Head and neck cancer	1/2	Recruiting	NCT02901483
			Rectal cancer	1/2	Recruiting	NCT02465593
NC-4016	Oxaliplatin	Polymeric micelle	Advanced solid tumors	1	Active, not recruiting	NCT03168035
NK012	SN-38 (irinotecan metabolite)	Polymeric micelle	Small cell lung cancer	2	Completed	NCT00951613
			Advanced breast cancer	2	Completed	NCT00951054
			Refractory solid tumors	1	Completed	NCT00542958
			Metastatic breast cancer	1	Completed	NCT01238952
			Solid tumors	1	Completed	NCT01238939
NK105	Paclitaxel	Polymeric micelle	Advanced breast cancer	3	Completed	NCT01644890
CT-2103 (Xyotax)	Paclitaxel	Polymer-drug conjugates (polyglumex	Advanced fallopian tube, ovarian or primary peritoneal cancers	3	Active, not recruiting	NCT00108745
			Metastatic breast cancer	1	Completed	NCT00270907
			Recurrent/persistent ovarian or primary peritoneal cancers	2	Completed	NCT00045682
			Metastatic breast cancer	2	Completed	NCT00265733
			Prostate cancer	2	Completed	NCT00446836
			Ovarian, fallopian tube or primary peritoneal cancers	1	Completed	NCT00060359
			Advanced colorectal cancer	1	Completed	NCT00598247
			Metastatic breast cancer	2	Completed	NCT00148707
			Recurrent ovarian, fallopian tube or primary peritoneal cancers	1/2	Completed	NCT00017017
			Ovarian cancer	2	Completed	NCT00069901
pbi-shRNA STMN1 LP	Antistathmin 1 siRNA	Lipid nanoparticle	Advanced solid tumors	1	Completed	NCT01505153
PNT2258	DNA oligonucleotide against BCL-2	Liposome	Advanced solid tumors	1	Completed	NCT01191775
Promitil	Mitomycin-C	Liposome	Solid tumors	1	Recruiting	NCT01705002

Continued

Table 2: Nanotherapeutics under clinical evaluation (active trials recruiting patients, active not yet recruiting and completed without published results) for the treatment of solid tumors—cont'd

Investigational Drug Name	Drug/Therapy	Type of Nanoparticle	Disease	Clinical Trial Phase	Trial Status	NCT Number
SGT-53	Plasmid encoding normal human WT p53 DNA	Transferrin receptor-targeting liposome	Solid tumors	1	Completed	NCT00470613
			Recurrent glioblastoma	2	Recruiting	NCT02340156
			Refractory/recurrent solid tumors	1	Recruiting	NCT02354547
			Metastatic pancreatic cancer	2	Recruiting	NCT02340117
siRNA-EphA2-DOPC	Anti-EphAS siRNA	Liposome	Advanced cancers	1	Recruiting	NCT01591356
Tecemotide	MUC1 antigen	Liposome	Colorectal carcinoma	2	Active, not recruiting	NCT01462513
			Prostate cancer	2	Completed	NCT01496131
ThermoDox	Doxorubicin	Liposome	Recurrent breast cancer	1/2	Completed	NCT00826085
			Liver metastatic or primary cancers	1	Completed	NCT00441376
			Liver cancers	1	Completed	NCT02181075
			Hepatocellular carcinoma	3	Recruiting	NCT02112656
			Refractory solid tumors	1	Recruiting	NCT02536183
TKM-080301	Anti-PLK1 siRNA	Lipid nanoparticle	Neuroendocrine or adrenocortical tumors	1/2	Completed	NCT01262235
			Advanced hepatocellular carcinoma	1/2	Completed	NCT02191878
			Liver metastatic or primary cancers	1	Completed	NCT01437007

Sources: Thakor, A. S., Gambhir, S. S. (2013). Nanooncology: the future of cancer diagnosis and therapy. *CA: A Cancer Journal for Clinicians, 63*(6), 395–418. doi:10.3322/caac.21199; Shi, J., Kantoff, P. W., Wooster, R., Farokhzad, O. C. (2016). Cancer nanomedicine: progress, challenges and opportunities. *Nature Publishing Group, 17*(1), 20–37. doi:10.1038/nrc.2016.108; Caster, J. M., Patel, A. N., Zhang, T., Wang, A. (2017). Investigational nanomedicines in 2016: a review of nanotherapeutics currently undergoing clinical trials. *Wiley Interdisciplinary Reviews. Nanomedicine and Nanobiotechnology*, *9*(1). doi:10.1002/wnan.1416; https://clinicaltrials.gov (accessed 24 October 2017).

in the number of FDA-approved therapeutic nanoparticles over the foreseeable future, which will also integrate new molecular entities and novel classes of therapeutic agents (e.g., mRNA, siRNA, plasmids, DNA) (Caster, Patel, Zhang, & Wang, 2017; Shi, Kantoff, Wooster, & Farokhzad, 2016; Thakor & Gambhir, 2013).

Recently, the failure in phase II clinical trials of a promising targeted polymeric nanocarrier containing the chemotherapeutic drug docetaxel, BIND-014 (Bobo, Robinson, Islam, Thurecht, & Corrie, 2016; Ledford, 2016) has raised several questions regarding interindividual factors that could determine the efficacy of nanotherapies. According to a recent work, the average tumor uptake of nanoparticles is only about 0.7% of the injected dose (Wilhelm, Tavares, Dai, et al., 2016). This fact supports the idea that models fail to predict accumulation of systemic delivered nanoparticles into the tumor, as they are oversimplified and overlook multiple biological steps in the process. In order to overcome this hurdle, efforts are being made to classify patients by high EPR effect to determine if they are suitable candidates for nanotherapies. Understanding individual predisposition to nanoparticle accumulation in tumor sites is essential for selecting the correct nanotherapy. Some relevant studies related to this are disclosed in Section 5.1.1. Taking this scenario into consideration, we believe that our position should be to exploit the benefits of nanotechnology but recognize its limitations, and even more than before, work together with other modalities such as diagnosis to carefully select the patients most likely to benefit from a given nanotherapy. Nanotheranostics, we believe, would and should play an important role in this new situation.

4 Nanotheranostics

A great deal of effort has been made over the last three decades to develop a plethora of nanocarriers for cancer therapeutic purposes. However, the current low response rates and limited increase of overall patient survival evidenced that the chemotherapeutic revolution nanomedicines promised is still far from reality. It is essential now to move forward in the future toward personalized cancer medicine to produce real clinical impact and translation. Integrating imaging modalities and therapeutic molecules into a single multifunctional nanoparticle, a nanotheranostic, could help in the clinical practice by concurrently diagnosing disease, stratifying patients, guiding focal therapy, tracking drug release and penetration within tumors, monitoring response, and if required, switching treatments (Cui & Wang, 2016). These exceptional applications are particularly important in the era of precision medicine. Different theranostic modalities include photodynamic therapy, photothermal, magnetothermal and electrothermal therapies, photo- and ultrasound-triggered chemotherapeutic release, X-ray and radiofrequency, computed tomography (CT), magnetic resonance imaging (MRI), positron emission tomography (PET), and single-photon emission computed tomography (SPECT) (Park, Aalipour, Vermesh, Yu, & Gambhir, 2017; Sneider, VanDyke, Paliwal, & Rai, 2017; Thakor & Gambhir, 2013).

Considering nanoparticles that have been developed for cancer imaging, there are several examples in clinical practice or under clinical development, as extensively reviewed by Thakor et al. (Thakor & Gambhir, 2013) including superparamagnetic iron oxide nanoparticles and ultrasmall superparamagnetic iron oxide nanoparticles for MRI, heavy metal nanoparticles (i.e., gold, lanthanide and tantalum) for CT, TC^{99m} sulphur colloid nanoparticles for SPECT, I-labelled cRGDY silica nanoparticles for PET, surface-enhanced Raman scattering nanoparticles for optical imaging and single- walled carbon nanotubes for photoacoustic imaging. With respect to nanotheranostics, several nanoscaled materials have been explored during the last years, including liposomes (Dai & Yue, 2017), micelles (Pellosi et al., 2017), dendrimers (Zhu, Fu, Xiong, Shen, & Shi, 2015), solid lipid nanoparticles (Bae, Lee, Lee, Park, & Nam, 2013), polymeric nanoparticles (Alberti, Protti, Franck, et al., 2017), albumin-based nanoparticles (Huang, Yang, Liu, et al., 2017), quantum dots (QDs) (Guo, Qiu, Guo, et al., 2017), metallic nanoparticles (Rajkumar & Prabaharan, 2016; Shi, Ye, He, et al., 2014; Zhu, Zhou, Mao, & Yang, 2017), carbon nanotubes (Wang, Zhang, Yang, et al., 2016), nanobubbles (Duan et al., 2017), and iodinated nanoparticles (Lee, Chung, Cho, & Kim, 2016). However, despite the fact the field of nanotheranostics is drastically emerging, and many experiments show promising results, no formulation has reached the market yet. As far as we know, there is only one formulation undergoing clinical trials (phase I), AGuIX (Activation and Guidance of Irradiation by X-ray), a gadolinium-based nanoparticle of around 5 nm diameter for the treatment of multiple brain metastases (NCT02820454) with a double role, radiosensitizer and imaging agent. At first developed for imaging applications due to its magnetic resonance contrast properties, it was observed that, when combined with X-ray radiation, led to a threefold enhancement in the efficiency of the radiotherapy in mice (Detappe, Kunjachan, Rottmann, et al., 2015; Kotb, Detappe, Lux, et al., 2016).

The big challenge for the translation of all this technology into daily clinical practice is the need for a detailed understanding of the interactions between cancer nanotheranostics and the human body, the tumor, and the microenvironment. Moreover, a long-term assessment of toxicity, and the establishment of regulatory protocols for cancer nanotheranostics, will also be required (Chen, Zhang, Zhu, Xie, & Chen, 2017). As metastatic cancer is a systemic disease, this means that most commonly, patients are diagnosed and monitored by computed tomography (CT), magnetic resonance (MRI), and positron emission tomography (PET) (or PET/CT if available). Therefore, we believe that nanotheranostics designed for these modalities will initially have greater chances at translation.

5 Specific Interventions in the Metastatic Cascade Making Use of Nanotheranostics

As mentioned before, the metastatic process is very complex, with a series of consecutive events susceptible to therapeutic interventions. The biological mechanisms that specifically drive each step of metastasis might be addressable using nanoparticles, and could also benefit from the development of more sophisticated nanotheranostics.

5.1 Premetastatic Initiation

5.1.1 Tumor growth

Targeting the primary tumor to interfere with tumor growth has been the main objective of most of the research work carried out in the nanomedicine field. In most cases, results refer to nanocarriers that encapsulate conventional anticancer drugs (e.g., doxorubicin, paclitaxel, docetaxel) (Chow & Ho, 2013), and are passively accumulated in the tumor, achieving prolonged circulation times, higher areas under the plasma concentration-time curve (AUC), lower clearance rates, decreased toxicities, higher accumulation in tumors (xenograft or orthotopic mice models), and improved therapeutic efficiencies, compared to the free drugs (Gabizon, Shmeeda, & Barenholz, 2003). However, as we mentioned before, there is generally little translation to clinic. One of the main problems is that, in reality, only a tiny dose of the administered drug actually accumulates into the tumor (Wilhelm et al., 2016). Molecular imaging can provide a solution to understand tumor heterogeneity and determine if nanoparticles can indeed accumulate into the tumors, for example by co-administering nanoparticles that can be tracked, in combination with the therapeutic nanoparticles. In this regard, magnetic nanoparticles (ferumoxytol) were administered together with paclitaxel PLGA-PEG nanoparticles to mice bearing subcutaneous human fibrosarcoma, with the purpose of tracking particle distribution within tumors and the microvasculature, and predicting the effectiveness of the therapeutic nanoparticles (Miller, Gadde, Pfirschke, et al., 2015). PET-radiolabeled nanoparticles (Zirconium-89 nanoreporter) were also co-injected with docetaxel-loaded liposomes (Doxil) for quantifying biodistribution, and predicting the therapeutic efficacy of the liposomes in a xenograft breast cancer mouse model (Pérez-Medina, Abdel-Atti, Tang, et al., 2016). Injection of iodine-labelled liposomes in rats bearing breast tumor xenografts has similarly been of utility for assessing tumor uptake by mammography and to classify animals into good and bad prognosis subgroups, previous treatment with liposomal doxorubicin. Interestingly, tumor progression correlated well with the subgroups of good and bad prognosis, confirming the high variability and relevance of the EPR effect status (Karathanasis, Suryanarayanan, Balusu, et al., 2009). This novel approach could, therefore, permit classification of patients that are prone to benefit from nanotherapeutics. Even if the works reported here actually refer to the combination of two individual nanostructures, multimodal all-in-one nanotheranostics could also have a real application for patient stratification, minimizing the delay between diagnosis and treatment, and allowing evaluation in real-time. Along with this, there are several other applications of potential interest for nanotheranostics, such as tracking drug release and penetration within tumors, imaging-guided local therapy, or monitoring therapeutic responses (Chen, Zhang, et al., 2017).

Improving the targeting of nanotheranostics to tumors beyond EPR-mediated effect can be achieved by tumor-specific delivery, upon decoration with molecules such as peptides, aptamers, and antibodies (Diou, Tsapis, & Fattal, 2012). In this regard, Zolata, Abbasi Davani, and Afarideh (2015) proposed the development of a complex multimodal nanoparticle based on minosilane-PEG coated magnetic nanoparticles, loaded with (Quail & Joyce, 2013). In radiolabeled doxorubicin, for bimodal tumor imaging (MRI/SPECT). These nanoparticles were functionalized with trastuzumab for specific targeting to HER2 receptor. Results in mice bearing HER2 positive breast tumors show the potential of this approach for accurate tumor imaging with improved sensitivity, allowing the simultaneous treatment with radio- and chemopharmaceuticals. A different active targeting with theranostic nanoparticles was proposed by Kim, Jeong, and Jon (2010) referring to the functionalization of doxorubicin-loaded gold nanoparticles with a prostate-specific membrane antigen (PSMA) RNA aptamer for CT imaging and targeted chemotherapy, but this formulation has not been tested in vivo yet.

The development of the so-called "smart" nanosystems, that is, nanoparticles that are able to deliver their drug or induce a cytotoxic effect in cancer cells just in response to an externally controlled exogenous stimulus (e.g., light, ultrasound, magnetism, temperature and electric field) is gaining increased attention for the development of nanotheranostics. Remotely triggered nanosystems present promising clinical benefits, as they have the ability to turn "on" or "off" the delivery of therapeutics depending on multiple variables, such as whether the nanoparticle reached the tumor or not, hence reducing the toxicity in "normal" tissue in a noninvasive fashion (Sneider et al., 2017). Exogenous stimuli can be provided by photodynamic therapy, photothermal therapy, and phototriggered chemotherapeutic release. Currently, only photodynamic therapy is clinically approved for cancer treatment, while photothermal and phototriggered modalities are still undergoing clinical trials. An example of a remotely triggered nanotherapy is NanoTherm, a thermoresponsive nanoformulation that effectively converts energy into heat that can also be tracked by CT for imaging purposes. Inorganic materials (i.e., gold, silica, silver) are being widely explored for photothermal therapy (PTT), however, the fact that they are not biodegradable has hampered their clinical translation. To overcome this limitation, a recent study reports the development of biocompatible and bioeliminable magnetoplasmonic nanoassemblies (MPNAs), formed by Fe_3O_4 nanoclusters and gelatin-coated with gold, which are efficient for computed tomography (CT), photoacoustic tomography (PAT), and magnetic resonance (MRI), and also allow photothermal cancer therapy. MPNAs showed extraordinary results, with complete tumor recession in liver-cancer-bearing mice after exposure to near infrared (NIR) light for 3 min (Li, Fu, Chen, et al., 2016; Fig. 2). Stimuli-responsive nanotheranostics are also being developed for gene delivery (Shim & Kwon, 2012). In a recent paper, Gao, Gao, Xu, et al. (2017) reported the development of nanodroplets for targeted ultrasound diagnosis and gene delivery in HER2-overexpressing breast cancer cells.

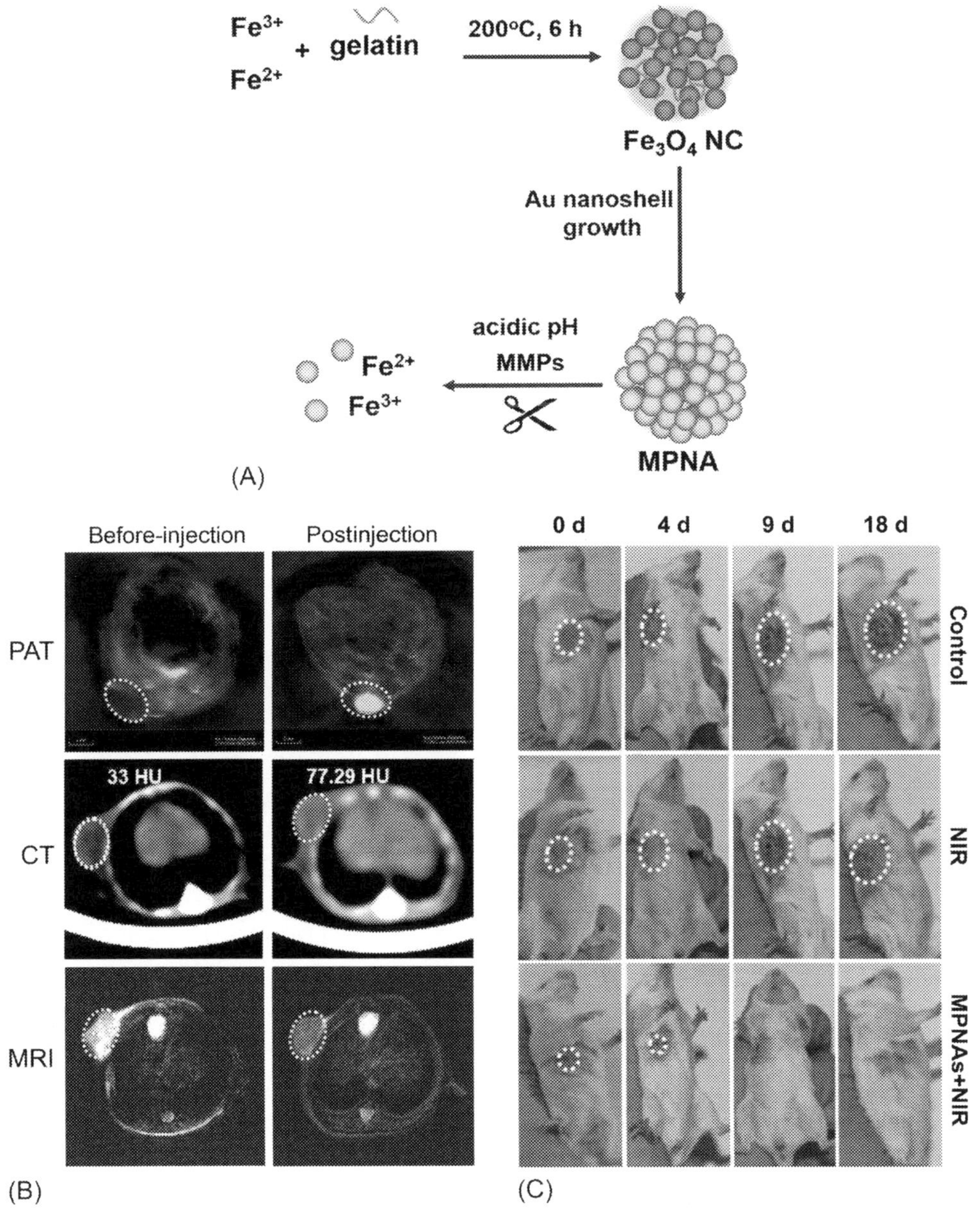

Fig. 2
(A) Schematic illustration of the biodegradable process of magnetoplasmonic nanoassemblies (MPNAs). (B) In vivo trimodal imaging of subcutaneous liver tumor bearing mice before and after 2 h of the intravenous injection of MPNAs as a contrast agent for photoacoustic tomography (PAT), computed tomography (CT) and magnetic resonance imaging (MRI). (C) In vivo photothermal therapy of MPNAs under a near infrared (NIR) light at different time intervals of therapy, compared with the control and NIR light alone. *Data from Li, L., Fu, S., Chen, C., et al. (2016). Microenvironment-driven bioelimination of magnetoplasmonic nanoassemblies and their multimodal imaging-guided tumor photothermal therapy. ACS Nano, 10(7), 7094–7105. doi:10.1021/acsnano.6b03238, with the permission of American Chemical Society, Copyright (2017).*

In general, it can be concluded that the field of nanotheranostics applied to cancer has drastically emerged over the last years. Despite this, further research still needs to be done in order to improve translation and have a clear impact in patient survival. More experiments should also focus on targeting other determinant phases of cancer disease, apart from the primary tumor, providing novel and complementary approaches for the management of metastatic cancer.

5.1.2 Cancer stem cells

Intratumor heterogeneity is a key factor contributing to the devastating outcome of cancer, therapeutic failure, and drug resistance, and should be taken into account for the development of efficient therapeutic strategies (Greaves, 2015; Huang, Shen, Ding, & Geng, 2014; Mcgranahan & Swanton, 2017). Tumors contain genetically and phenotypically distinct subpopulations of cancer cells with unique capacities for growth, differentiation, and invasion (Carey, Starchenko, McGregor, & Reinhart-King, 2013). This variability is present at multiple levels, that is, genetically, histologically, and physiologically (De Palma & Hanahan, 2012). Indeed, only a minority of all cancers are caused by germline mutations, whereas the enormous majority (90%) are associated dynamic processes related to somatic mutations and environmental factors (Grivennikov, Greten, & Karin, 2010). Changes in the subclonal composition can occur, for example, in the context of disease progression, that is, metastasis, or during drug treatment (Campbell, Yachida, Mudie, et al., 2010; Ding, Ley, Larson, et al., 2012). Thus, selective clones with specific phenotypic advantages will survive or predominate in the tumor. Cancer stem cells (CSCs), also known as tumor-initiating cells (TICs), constitute a biologically unique subset of stem-like cells within the bulk tumor cell population with exclusive characteristics, unlimited self-renewal, and pluripotency (Sainz Jr, Carron, Vallespinós, & Machado, 2016) and are related to the generation of resistances, tumor relapse, and formation of micrometastases (Mina & Sledge, 2011).

In the past few decades, different cell signaling pathways (Wnt/β-catenin, Notch, PI3K/AKT, transforming growth factor-β, and hedgehog signaling) and specific markers (aldehyde dehydrogenases, CD44, CD90, and CD133), involved in CSC function and maintenance, have been disclosed (Hong, Jang, Lee, & Nam, 2015; Singh, Minz, & Sahoo, 2017). The fact that CSCs express different biomarkers at the cell surface makes them targetable by surface-decorated nanoparticles for drug delivery purposes (Livney & Assaraf, 2013). This has led to the development of different therapeutic strategies based on organic and inorganic magnetic nanoparticles, directed for example to inhibit ABC transporters overexpressed on the CSCs surface and associated with multidrug resistance, cell signaling pathways, quiescent CSCs, CSCs dedifferentiation, etc. (Gener, Rafael, Fernández, et al., 2016; Hong et al., 2015; Li, Li, Mo, & Dai, 2017; Li, Li, Wu, et al., 2017; Livney & Assaraf, 2013; Shen, Xia, & Wang, 2016; Singh et al., 2017; Steg, Katre, Goodman, et al., 2012; Takeishi et al., 2013). Besides all these encouraging advances, promising future perspectives imply the combination of these innovative therapeutic strategies with different imaging modalities for precise identification of this cell subpopulation. Within this premise, Barth, Altinoglu, Shanmugavelandy, et al. (2011) reported the preparation of calcium phosphosilicate nanoparticles, decorated with a surface biomarker of leukemia stem cells (anti-CD117 antibody) to selectively deliver a photosensitizer (indocyanine green) for diagnostic purposes and photodynamic therapy. This strategy led to a 29% disease-free survival in a myeloid leukemia mouse model, with some mice exhibiting leukemia-free survival, revealing a successful CSCs eradication. In another recent paper, Tang, Zhou, Liu, et al. (2017) reported the design of a multifunctional magnetic nanosystem based on a magnetic Fe_3O_4 core for imaging and hyperthermia treatment, a polypyrrole shell to allow high drug loading capacity (DAPT), and hyaluronic acid for targeting CD44. This multimodal nanosystem allowed monitoring of the drug distribution within the tumor by MRI and photoacoustic imaging, and the combination of chemotherapy and magnetohyperthermia therapy to significantly inhibit tumor growth and blood vessels density, altogether revealing an effective elimination of CSC without causing side effects in 4T1 breast tumor-bearing mice. A novel theranostic strategy, was also developed by Mohanty, Chen, Li, et al. (2017) in this case for glioblastoma multiforme (GBM). Glioblastoma-initiating cells (GIC) are resistant to the limited number of current therapies leading to poor prognosis and 90% of recurrence in GBM patients. Cross-linked iron oxide nanoparticles (CLIO) were loaded with a VDA-prodrug (azademethylcolchicine) and conjugated to a highly potent vascular disrupting agent (ICT) linked to a matrix-metalloproteinase (MMP-14) cleavable peptide. These nanoparticles were followed by MRI and efficiently disrupted the tumor vasculature of MMP-14 expressing GBM, induced GIC apoptosis, and almost led to a complete reduction of orthotopic glioblastoma tumors, with a significant impact on survival.

Despite the encouraging results, much work still needs to be done in order to move nanotheranostics forward to clinical practice for this particular approach. However, we believe that this is a promising strategy to determine the efficiency of the therapy in this particular cell subpopulation, which is more resistant to current treatments and responsible for tumor-relapse and recurrence.

5.1.3 Tumor microenvironment

It is known now that malignant features of tumor cells cannot be revealed without an important interplay between them and their local environment (Alonso-Nocelo, Abuín, López-López, & de la Fuente, 2016; Sounni & Noel, 2013). Communication between cells and their associated stroma is crucial for normal tissue homeostasis and tumor growth, influencing disease initiation, progression, and patient prognosis (Quail & Joyce, 2013). Specifically, tumor cells are surrounded by a complex and diverse stroma composed by different nontransformed cells, such as fibroblasts, endothelial cells and immune cells, constituting the so-called Tumor Microenvironment (TME) (Balkwill & Mantovani, 2001; Hanahan & Coussens, 2012; Junttila & de Sauvage, 2013; Mueller & Fusenig, 2004;

Quail & Joyce, 2013). Infiltrating immune cells produce chemokines, growth factors, and matrix degrading enzymes that can create a permissive environment of chronic inflammation, ideal for tumor growth and progression. Several studies have shown that tumor-associated macrophages (TAMs) facilitate tumor cell dissemination, guiding tumor cells towards the circulation (Bonde, Tischler, Kumar, Soltermann, & Schwendener, 2012; Pollard, 2004; Solinas, Schiarea, Liguori, et al., 2010). Cancer-associated fibroblasts (CAFs) are also present in high numbers and elicit pro-tumorigenic functions, producing pro-inflammatory cytokines and vascular endothelial growth factors (VEGF), which induce vascular permeability and angiogenesis and enhance the metastatic process (Xing, Saidou, & Watabe, 2010). Elucidating the nature of interactions and the mechanisms of co-evolution of tumor cells and their microenvironment is crucial for understanding what drives tumor initiation and progression, and for the development of better cancer preventive and therapeutic interventions.

Stimulated by advances made in the field of material sciences and drug delivery, nanosystem able to deliver their cargo in a spatial-temporal fashion and dosage-controlled manner, taking advantage of the singularities presented in the TME, such as pH variations (Aires, Ocampo, Simões, et al., 2016; Kim, Lee, Oh, Gao, & Bae, 2008; Ling, Xia, Park, et al., 2014; Mo, Jiang, DiSanto, Tai, & Gu, 2014; Mok, Veiseh, Fang, et al., 2010; Mura, Nicolas, & Couvreur, 2013; Sawant, Hurley, Salmaso, et al., 2006; Shen, Huang, Han, et al., 2012; Zou, Yu, Wang, et al., 2010), metabolic state (Tennant, Durán, & Gottlieb, 2010; Vaupel, Kallinowski, & Okunieff, 1989), hypoxic state (Garcia-Mazas, Csaba, & Garcia-Fuentes, 2017; Kanapathipillai, Brock, & Ingber, 2014), and redox state (Cavallaro, Campisi, Licciardi, Ogris, & Giammona, 2006; Kirpotin, Hong, Mullah, Papahadjopoulos, & Zalipsky, 1996; Saito, Swanson, & Lee, 2003; West & Otto, 2005) have been developed. An example with nanotheranostics refers to the development of pH-sensitive modified-dextran/polyvinyl alcohol (PVA) nanoparticles, loaded with near infrared (NIR) emissive probes and doxorubicin. Apart from tracking the nanoparticle distribution, the imaging probe was of use to reveal the release of the drug during the degradation of the nanoparticles in the acidic environment (Yu, Chen, Zhang, et al., 2014). pH-sensitive nanotheranostics, composed of a nanogel functionalized with folic acid for active targeting, labeled with rhodamine B, and loaded with the drug triptolide, have also been proposed for the treatment of hepatocellular carcinoma (HCC). In this case, in vivo results showed a good accumulation in the tumor cells and the release of the drug during endosome maturation (Ling et al., 2014). Triggered delivery can also be achieved by endogenous intracellular concentrations of certain metabolites. In this regards, Qian, Chen, Zhu, et al. (2016) have developed an ATP-responsive and near-infrared (NIR)-emissive conjugated polymer-based nanocarrier for the controlled release of anticancer drugs and real-time monitoring. At the significantly higher intracellular ATP concentrations, the drug doxorubicin was efficiently disassociated, leading to a good pharmacological response in terms of tumor growth. Real-time NIR fluorescent imaging was efficient for monitoring nanoparticle biodistribution. In a different work, Chen, Jia,

Tham, et al. (2017) describe the use of glutathione (GSH)-responsive nanosized vesicles for the synergetic delivery of a drug (camptothecin), siRNA (siP1K1), and a fluorescence turn-on mechanism that allows monitoring of the cellular uptake of the nanovesicles in real-time. Lastly, Feng, Cheng, Dong, et al. (2017) have described the use of a theranostic liposome for encapsulation of a hydrophilic hypoxia-activated prodrug (AQ4N) plus a hydrophobic photosensitizer (hexadecylamine conjugated chlorin e6 (*h*Ce6)), which only demonstrate toxicity when reaching hypoxic conditions in the TME. Moreover, association of $^{64}Cu^{2+}$ allowed tracking of the liposomes by PET, and verification of the successful accumulation of the nanotheranostics into the tumor after intravenous administration to mice bearing breast cancer tumors.

It has been suggested that active targeting of specific tumor microenvironment-infiltrated immune cells and cells from the stroma could be a promising strategy for the development of novel therapeutic approaches that can benefit from nanotechnology (Allavena, Sica, Solinas, Porta, & Mantovani, 2008; Cirri & Chiarugi, 2011; Franco, Shaw, Strand, & Hayward, 2010; Junttila & de Sauvage, 2013; Patel & Janjic, 2015; Serafini, Borrello, & Bronte, 2006). To date, very few reports refer to the specific use of nanotheranostics for targeting immune cells into the tumor microenvironment (de la Fuente, Langer, & Alonso, 2014). An interesting example refers to the use of gold nanoparticles loaded with Alexa Fluor 488-labelled siRNA anti-VEGF and decorated with the M2pep peptide for selective targeting to TAMs. After intratracheal administration to mice bearing orthotopic lung tumors, results showed an efficient accumulation in TAMs, accompanied by a decrease in VEGF TAM-derived protein expression of 80%, thus, reducing their recruitment to the inflammatory tumor environment and leading to a complete tumor regression after 36 days of therapy (Conde, Bao, Tan, et al., 2015). Another recent study refers to the development of PEGylated carboxymethiylcellulose nanoparticles loaded with docetaxel and DiI as fluorescent marker. In vivo studies in human pancreatic xenografts demonstrated a 90% reduction of CAFs after three cycles of treatment, which translates in significant tumor regression (Ernsting, Hoang, Lohse, et al., 2015). As a last example, lipid nanocapsules loaded with both C12 carbon chain modified gemcitabine and DiD were developed for specific target of the monocytic-myeloid derived suppressor cells (M-MDSC) (Sasso, Lollo, Pitorre, et al., 2016). After subcutaneous administration to melanoma-bearing mice, results demonstrated the effective targeting capacity of the formulation to M-MDSCs, and an improved effect in terms of tumor growth.

Considering this information, it is possible to conclude that there is a lot of potential for the application of nanotheranostics to improve the understanding of tumor-stroma interactions in cancer progression and to determine the relevance of eliminating specific subpopulations of immune-infiltrated cells from the tumor equation to block tumor progression and dissemination.

5.1.4 Angiogenesis

Angiogenesis is a crucial process and key step during tumor progression and dissemination, decisive for getting a good supply of nutrients and oxygen (Quail & Joyce, 2013; Weis & Cheresh, 2011). Tumors reaching a size of 2 mm suffer a process of increased interstitial pressure that inhibits the uptake of nutrients and metabolites. At that moment, the establishment of hypoxic areas prompt the sprouting of new blood vessels from the preexisting ones (Danhier, Feron, & Préat, 2010). Activation of oncogenes as *RAS* and *MYC* initiates the induction of a transcriptional program that enhances tumor microenvironment remodeling and an angiogenic switch (Grivennikov et al., 2010). Angiogenesis process also involves cell-to-cell (i.e., vascular endothelial cells, fibroblasts, myeloid cells, monocytes, and pericytes) and cell to ECM interactions (Ruoslahti & Pierschbacher, 1987; Weis & Cheresh, 2011). However, the new tumor vasculature is imperfect and deeply contributes to cells access and intravasation into the circulatory or lymphatic systems (Morikawa et al., 2002).

Selective delivery of drugs to tumor vasculature is a promising strategy that has been widely explored in cancer therapeutics (Byrne, Betancourt, & Brannon-Peppas, 2008; Chen & Cai, 2014; Chung, Lee, & Ferrara, 2010; Desgrosellier & Cheresh, 2010). Moreover, imaging of the tumor vasculature is of high interest for patient stratification and diagnosis purposes (Baetke, Lammers, & Kiessling, 2015; Cai & Chen, 2008; Liu, Miyoshi, & Nakamura, 2007) Over the last years, and paralleled to the relevant advances in the nanotechnology and chemical engineering field, different formulations have emerged as targeted systems to specific biomarkers expressed in the tumor vasculature, including integrins ($\alpha_V\beta_3$, $\alpha_V\beta_5$, $\alpha_5\beta_1$, and $\alpha_3\beta_1$), vascular endothelial growth factor receptors (VGFRs), and endoglin (CD105), among others (Kunjachan, Pola, Gremse, et al., 2014; Seon, Haba, Matsuno, et al., 2011; Weis & Cheresh, 2011). Several types of nanosystems have been developed for antiangiogenesis targeting purposes comprising polymer conjugates (Fonsatti, Nicolay, Altomonte, Covre, & Maio, 2010; Mitra et al., 2005; Oliveira & Faintuch, 2015; Park, Lee, Jung, et al., 2008; Yin, Hui, Yao, et al., 2015), liposomes (Ren, Chen, Li, et al., 2016; Sonali, Singh, Sharma, et al., 2016; Völkel, Hölig, Merdan, Müller, & Kontermann, 2004; Zhang, Feng, Henning, et al., 2009), inorganic nanoparticles (Chen, Ayala-Orozco, Biswal, et al., 2014; Chen, Hong, Zhang, et al., 2013; Gao, Xie, Chen, et al., 2014; Liu, Chen, Li, & Xu, 2016; Schleich, Po, Jacobs, et al., 2014; Yang, Hong, Grailer, et al., 2011; Zhang et al., 2014), and organic nanoparticles (Dahmani et al., 2016; Danhier, Danhier, Schleich, et al., 2016; Du, Zhang, Jing, et al., 2015; Murphy, Majeti, Barnes, et al., 2008; Schleich et al., 2014).

Several peptide-derived sequences have been described to target the integrin family, and could, therefore, be exploited for the functionalization of nanotheranostics. Arginine-glycine-aspartate amino acid sequence, named RGD motif, was the first endothelial binding peptide described in 1987 (Ruoslahti & Pierschbacher, 1987). A recent work published by Dahmani et al. (2016) describes auto-assembled nanoparticles, formed by the conjugation of heparin

(LMWH) with a cyclic RGD-derived peptide c(RGDyK), of use for the encapsulation of the hydrophobic antiangiogenic drug, gambogic acid, and labelled with Cy7 for optical imaging. In vivo studies show the successful delivery of this targeted nanotheranostics to both cancer cells and endothelial cells, and an efficient inhibition of tumor growth in U87MG xenografts (Dahmani et al., 2016). PEGylated liposomes decorated with c(RGDyK) have also been proposed for lung cancer theranostics, upon encapsulation of hydrophobic and hydrophilic anticancer drugs (paclitaxel and carboplatin) and gadodiamide for MRI imaging. In vivo results show a tumor signal 36-fold higher than the commercial contrast agent Omniscan, after 5-h administration, and a synergistic effect due to the combination of both drugs into a single nanostructure (Ren et al., 2016). Another interesting example describes RGD-TPGS decorated liposomes containing docetaxel (DCX) and quantum dots (QDs) for brain cancer, to improve tumor imaging and drug accumulation in the tumor (Sonali et al., 2016). PLA nanoparticles functionalized with a novel peptide, GX1 (CGNSNPKSC) (Hui, Han, Liang, et al., 2008), were developed for encapsulation of a recombinant antiangiogenic endostatin (GPENs) and labeled with IRDye 800CW for optical imaging (Du et al., 2015). Results demonstrate that this nanotheranostic formulation successfully delivered the drug to the tumor vasculature and exhibited an enhanced antitumor efficacy in colorectal tumor-bearing xenografts. Last example refers to the use of mesoporous silica nanoparticles (MSN), decorated with TRC105 (Fab), for targeting endoglin. Dual-labeling with ^{64}Cu and IRDye 800CW allowed successful in vivo PET/NIRF imaging and determining the biodistribution of the nanostructures in 4T1 breast tumor-bearing mice, showing an improved accumulation in the actively-proliferating tumor vasculature (Chen et al., 2013).

We are convinced that many other interesting multicomponent nanotheranostic approaches will be described in the upcoming years, both to interfere with and monitor the angiogenic switch, as well as to stratify patients according to their tumor microvasculature, in an attempt to predict the accumulation and prognosis of therapeutic nanoparticles.

5.1.5 Invasion and intravasation

To initiate invasion, cancer cells must redefine their communication with surrounding cells and the extracellular matrix (ECM). For that purpose, tumor cells overexpress dynamic cell adhesion molecules (CAMs) on their surface, such as integrins (mediating cell-ECM adhesions) and cadherins (involved in cell-cell interactions) (Van Zijl, Krupitza, & Mikulits, 2011). They also secrete cytokines such as vascular endothelial growth factor (VEGF), inducing the formation of leaky blood vessel or causing endothelial cell death by releasing reactive oxygen species (Lin et al., 2011). As a result, tumor cells experiment with changes in their transcriptional program, for example, expression molecules responsible for motility control and migration, to enhance the epithelial-to-mesenchymal transition process (EMT) (Kedrin et al., 2007). The ECM has the ability to influence tumor initiation and progression towards malignancy as well. Different mechanisms have been proposed to mediate cancer cell entry into the circulation such as collagen crosslinking, stiffening of the extracellular matrix (ECM), and increased focal adhesions (Levental, Yu, Kass, et al., 2009). Indeed, primary tumors of different metastatic potential vary in stiffness and composition of both tumor- and stroma-derived ECM components (Branco da Cunha, Klumpers, Li, et al., 2014; Chaudhuri, Koshy, Branco da Cunha, et al., 2014). Immune cells of the TME can also facilitate tumor cell dissemination through a plethora of cytokines and the production of proteases that pave the way degrading the ECM, clearing a path for migration, and guiding tumor cells towards the bloodstream (Bonde et al., 2012; Pollard, 2004; Solinas et al., 2010).

Several therapeutic strategies making use of nanotechnology have been reported during the last few years with the purpose of interfering with these processes, for example, addressing cell adhesion proteins (e.g., cadherins, integrins), proteinases (e.g., metalloproteinases (MMPs), the cofilin family of proteins), ion/water channels (e.g., aquaporins, chloride and calcium channels), transcription factors, and signal transducers (e.g., SNAIL, STAT3, Twist, chemokine receptors), as well as to interfere with the formation of invadopodia (Eckert, Lwin, Chang, et al., 2011; Gallo, Kamaly, Lavdas, et al., 2014; Hatakeyama, Akita, Ishida, et al., 2007; Olmeda, Jordá, Peinado, Fabra, & Cano, 2007; Veiseh, Kievit, Ellenbogen, & Zhang, 2011; Zhu, Kate, & Torchilin, 2012; Zou, He, He, et al., 2015). With respect to the specific use of nanotheranostics in this field, there are only a limited number of works. An example refers to the use of gold nanocages, functionalized with a peptide for selective targeting of prostate cancer cells and subsequent NIR irradiation for both imaging and photothermal therapy (Avvakumova, Galbiati, Sironi, et al., 2016). In vitro results in PC3 cells confirmed a highly selective interaction and, interestingly, a complete loss of surface microvilli and different morphology of treated cells compared to control cells. This loss of microvilli on tumor cells is highly interesting for preventing the disease development and progression, because microvilli play a key role in cell migration, cell attachment, and invasion. In another recent paper, Ali, Hsu, Hsieh, et al. (2016) designed ultra-small superparamagnetic iron oxide nanoparticles conjugated to erlotinib for targeted therapy and noninvasive real-time monitoring of tumors. In vitro results showed a suppression of the expression of MMP-9 and XIAP, and subsequently inhibition of the migration and invasion capabilities of cancer cells. These results were also confirmed in vivo in xenograft bearing mice, additionally revealing the imaging (MRI) capabilities of their nanoparticles. Targeting strategies to MMPs can also include the use of a unique targeting agent, scorpion venom chlorotoxin (El-Ghlban, Kasai, Shigehiro, et al., 2014; Mok et al., 2010; Soroceanu, Gillespie, Khazaeli, & Sontheimer, 1998; Sun, Fang, Stephen, et al., 2008; Veiseh, Gunn, Kievit, et al., 2009). The conjugation of superparamagnetic iron oxide nanoparticles (SPIONs) with the 36-residue peptide of chlorotoxin resulted in the inhibition of cell invasion by deactivation of the membrane-bound matrix metalloproteinase 2 (MMP-2), and highlights the potential of this strategy for noninvasive diagnosis (Veiseh et al., 2009). Another interesting strategy is the incorporation of gene therapies into the formulations, considering that some microRNAs have been reported to mediate the process of tumor cell migration and tissue invasion (e.g., miRNA 218, miRNA

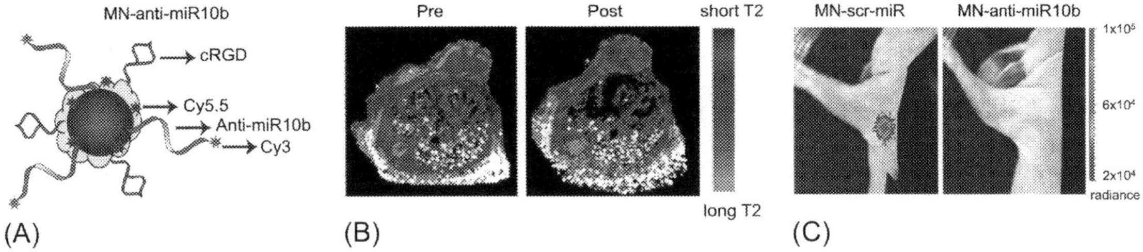

Fig. 3

(A) Schematic representation of magnetic nanoparticles conjugated to Cy5.5, a tumor targeting peptide (cRGD), and the therapeutic knockdown oligonucleotide miRNA-10b with a 3′-Cy3 modification. (B) MRI images of orthotopic breast cancer tumors before (left) and 24 h after the injection of the nanotheranostics (right) showing effective accumulation in the tumor. (C) Bioluminescence images of tumor-bearing mice, showing effective metastasis prevention capacity in lymph nodes for mice treated with the nanodrug MN-anti-miR10b, compared to the inactive control nanoparticles MN-scr-miR. *Data from Yigit, M. V., Ghosh, S. K., Kumar, M., et al. (2013). Context-dependent differences in miR-10b breast oncogenesis can be targeted for the prevention and arrest of lymph node metastasis, Oncogene 32 (12), 1530–1538, doi:10.1038/onc.2012.173, with the permission of Nature Publishing Group, copyright (2017).*

320a, miRNA 520a, miRNA 10b) (Cui et al., 2017; Lv, Wu, Wang, et al., 2016; Tu, Gao, Li, et al., 2013). As an example, Yigit, Ghosh, Kumar, et al. (2013) have developed a theranostic nanosystem for the inhibition of the pro-metastatic microRNA-10b based on ultrasmall magnetic nanoparticles (imaging capability) functionalized with the RGD peptide (targeting to integrins) and decorated with therapeutic nucleic acid oligonucleotides. This nanosystem allowed a noninvasive imaging of the whole body by near infrared fluorescence optical imaging, and a close up of the target tumor by MRI. In vivo results after intravenous injection to breast cancer orthotopic mouse models showed the inhibition of the invasion and migration of cancer cells and, therefore, the complete prevention of metastatic dissemination from the primary tumor to lymph nodes (Fig. 3).

Overall, cancer cell invasion is a highly complex process that involves several molecular pathways leading to the fatal progression of cancer metastasis. Nanotheranostics, multifunctional nanosystems that can simultaneously target several molecular pathways and incorporate imaging agents for noninvasive diagnosis are well suited to improve understanding and tackling of this phenomenon.

5.2 Dissemination

5.2.1 Tumor-infiltrated local and distal lymph nodes

Tumor cells can disseminate by lymphatic or haematogenous routes (Pantel & Brakenhoff, 2004; Sleeman et al., 2012). Because of its particular architecture, characterized by thin-wall large vessels and slow fluid flow, the lymphatic route is mainly accessible for invasive tumor cells in comparison to blood vessels (Abellán-Pose, Csaba, & Alonso, 2016; Sleeman & Thiele, 2009). Additionally, chemokines and other factors (i.e., CCL21, VEGF) produced by tumor or stromal cells play a key role in promoting the entry of tumor cells into the lymphatics (Hirakawa et al., 2006; Tomei, Siegert, Britschgi, Luther, & Swartz, 2009). Tumor-associated lymphatic vessels act as channels by which disseminating tumor cells access regional lymph nodes. Once there, cells can stay in a dormant state before traveling to other organs to form secondary metastases (Li & Li, 2015). Indeed, monitoring and detection of metastases within the first draining lymph node, the sentinel lymph node (SLN), has major prognostic value for patient survival in different types of cancer such as breast and lung, and frequently determines the type of treatment (Nathanson & Mahan, 2011). One of the limitations for detection of infiltrated tumor cells in the lymph nodes is the fact that they can exist in a quiescent state, or as micro-metastases in which proliferation is balanced by cell death through apoptosis. Consequently, detection of micrometastases is a great challenge in oncology, because it can be a promising strategy to stop metastatic growth (Ryan, Kaminskas, & Porter, 2014; Trevaskis, Kaminskas, & Porter, 2015). Mimicking this condition through the development of sophisticated 3D in vitro models has been proposed for the development of tailored nanotherapeutics (Alonso-Nocelo, Abellan-Pose, Vidal, et al., 2016).

Various nanoprobes, mainly based on MRI, PET, NIR, and SPECT/CT imaging techniques, have been developed to date for noninvasive SLN imaging, for example, mesoporous silica-based nanoprobes for trimodal imaging (NIR/MRI/PET), and lipid-calcium phosphate nanoparticles for SPECT/CT imaging (Huang, Zhang, Lee, et al., 2012; Tseng, Xu, Guley, Yuan, & Huang, 2014). On the other side, nanotherapeutic strategies designed to target metastatic cells placed into the lymphatic circulation and/or organs have similarly been proposed (Hirano, Hunt, Strubbe, & MacGregor, 1985; Ryan et al., 2014). An example of eradication of metastatic cells retained into SLN is methotrexate (MTX)-conjugated dendrimers, which efficiently accumulate in draining lymph nodes after subcutaneous injection to rats (Kaminskas, McLeod, Ascher, et al., 2014).

The application of nanotheranostics for targeting disseminated cancer cells and infiltrated lymph nodes seems therefore coherent. Simultaneous detection and eradication could have a big impact on interfering with lymphatic dissemination of the tumor cells and allowing monitoring of the efficacy of the proposed therapies. Some examples of nanotheranostics in this field include nanosystems designed for photothermal therapy. This is the case of phase-transition nanodroplets developed for optical, photoacoustic, and ultrasound imaging modalities, which showed improved efficiencies for detection and elimination of breast cancer cells located in SLN of rabbits (Yang, Cheng, Chen, et al., 2017), nitrogen-doped carbon nanodots, which combine photothermal therapy with photoacoustic imaging and led to complete tumor ablation without recurrence in SLN of rat and mouse models (Lee, Kwon, Beack, et al., 2016), graphene oxides modified with iron oxide nanoparticles for photothermal therapy and MRI and NIR imaging, efficient for imaging and treatment of regional lymph nodes in pancreatic cancer (Wang et al., 2016), and $W_{18}O_{49}$ nanoparticles designed for CT imaging-guided photothermal ablation of HER-2 positive breast malignancy in metastatic lymph nodes (Huo, He, Li, et al., 2014). Some other nanotheranostics suited for different types of therapy are iron oxide nanoparticles, which were co-encapsulated with the chemotherapeutic drug doxorubicin into PLGA microbubbles for dual imaging (US/MRI) and chemotherapy of lymph node metastasis in rabbits (Niu, Wang, Lu, et al., 2013), nanoparticles loaded with indocyanine green for fluorescence imaging and photodynamic therapy of lymph node metastasis of gastric cancer (Tsujimoto, Morimoto, Takahata, et al., 2015), and iron oxide nanoparticles, combined with the immunostimulatory activity of polyinosinic-polycytidylic acid (poly (I:C)) and labeled with the radionucleotide fac-$[^{99m}Tc(OH_2)_3(CO)_3]$, for dual MRI/SPECT imaging and targeted and programmable immunotherapy of metastatic lymph nodes (Cobaleda-Siles, Henriksen-Lacey, De Angulo, et al., 2014).

These examples evidence the potential benefits of using nanotheranostics for disseminated and metastatic cancer cells located in the lymph nodes in future clinical practices, allowing noninvasive imaging techniques, and more specific and local therapies without causing damage to the neighboring tissues.

5.2.2 Circulating tumor cells

During circulation, circulating tumor cells (CTCs) travel the bloodstream and overcome a range of events (e.g., anchorage-free survival, shear stress) in order to reach distant organs, representing the metastatic intermediates between the primary tumor and sites of dissemination. Still, it remains controversial whether CTCs have the ability to actively target certain organs, or if they are passively arrested within capillary beds, such as those in lung and liver, due to the architecture of the vasculature (Miles, Pruitt, van Golen, & Cooper, 2008). CTCs can also form clusters, known as circulating tumor microemboli (CTM), and large aggregates with platelets in order to both protect themselves from shear forces and evade immune detection (Gay & Felding-Habermann, 2011). It has also been suggested that CTCs might fuse with macrophages, providing metastatic cells with myeloid characteristics, advantageous for their survival in circulation (Pawelek & Chakraborty, 2008). Although it is thought that only 0.01% of CTCs are able to survive in circulation and produce metastasis, they are responsible for more than 90% cancer-related deaths (Fidler, 1970). Therefore, effective strategies in order to detect and eradicate CTCs to block metastasis are urgently required. Recent technological advances have facilitated the identification of biomolecules characteristic of CTCs for detection within the bloodstream of carcinoma patients, and some therapeutic drugs (e.g., docetaxel, gemcitabine, doxorubicin) are also being investigated for eradicating CTCs (Aires et al., 2016; Barbazán, Vieito, Abalo, et al., 2012; Chen, Gowda, Newswanger, et al., 2016; Deng, Wu, Zhao, et al., 2015; Muinelo-Romay, Vieito, Abalo, et al., 2014; Yang, Jiang, Zhang, et al., 2015).

In the past decade, most efforts have been focused on ex vivo CTCs detection and isolation from blood samples for diagnosis and monitoring of the disease progression, because the number of CTCs detected in a blood sample has been reported to directly correlate with disease severity (Danila, Pantel, Fleisher, & Scher, 2011). These novel technologies rely on characteristic properties of CTCs, for example, large size, characteristic surface molecules, deformability, and density (Alix-Panabières & Pantel, 2014). For instance, some CTC detection methods in advanced stage of clinical translation are CellSearch, ISET, and CTC-chip. CellSearch, the only currently FDA-approved system for automated CTC detection, relies on inmunomagnetic separation by antibodies against EpCAM; ISET is based on size separation by 8 μm pore filters because of the larger size of CTCs compared to hematological cells, and CTC-chip, is a immunoaffinity-based microfluidic device functionalized with anti-EpCAM antibody-coated microposts. However, due to the rarity of CTCs (around one CTC for every billion or so of normal blood cells) and the small processing capacity of these technologies ex vivo (1–10 mL of blood), the number of detected and isolated CTCs is quite limited and cannot fulfill the requirements for in-depth clinical studies.

Many of the emerging CTC detection techniques use inorganic and organic nanoparticles functionalized with targeting ligands to coat surfaces for ex vivo enrichment of CTCs and further analysis (Huang, O'Connor, & Kwizera, 2017; Liu, Xu, He, Deng, & Zhu, 2013; Myung, Roengvoraphoj, Tam, et al., 2015; Sun, Wang, Ji, et al., 2015; Yoon, Kozminsky, & Nagrath, 2014). These nanoparticles provide devices with an increased sensitivity and imaging properties (photoacoustic, fluorescence, SERS, and high-resolution X-ray imaging) in order to intercept and quantified CTCs. Nanoparticles are also being explored for in vivo enrichment and detection of CTCs. For instance, Galanzha, Shashkov, et al. (2009) have reported the use of surface-decorated magnetic nanoparticles for binding to CTCs after intravenous injection to xenograft mouse models. Once internalized by CTCs, nanoparticles could be concentrated and trapped by an external magnet attached to the skin, above a peripheral blood vessel, for in situ detection by photoacoustics, or for microsurgical extraction of CTCs for ex vivo analysis.

The possibility of in vivo monitoring the number of CTCs without the need for blood sample extraction and preparation would allow the screening of larger volumes of blood (~5 L) compared to the small blood samples taken for ex vivo analysis (1–10 mL) and, therefore, significantly improve sensitivity. In addition, novel in vivo strategies capable of simultaneously detecting, monitoring, and killing CTCs in the entire blood circulation may dramatically enhance the outcome of the disease by blocking the metastatic cascade and inhibit disease progression, making a breakthrough in cancer research. In this sense, functionalized theranostic nanosystems hold great promise for (1) screening the whole blood for intercepting and detecting CTCs in vivo, (2) enrichment of CTCs for biological function assays in vivo or minimally invasive extraction of CTCs for ex vivo analysis (molecular and genetic assessments), (3) monitoring the treatment response in vivo, in situ, and in real-time,

(4) targeting delivery of therapeutics to destroy CTCs and interrupt the metastatic cascade, taking advantage, for example, of the development of remotely triggered nanotheranostic (Kim, Galanzha, Zaharoff, Griffin, & Zharov, 2014; Sneider et al., 2017; Xie, Gao, Zhao, et al., 2015).

The current status of nanotheranostics in CTC detection and therapy is almost unexplored. However, some pioneering experiments show encouraging results. In 2009, Galanzha, Kim, and Zharov (2009) demonstrated, for the first time, in vivo real-time imaging and therapy of CTCs by combining magnetic nanoparticles for capturing circulating tumor cells under a magnet with a second contrast agent, golden carbon nanotubes as super contrast photoacoustic and photothermal agents for improving sensitivity. In vivo results show efficient photoacoustic detection and photothermal ablation of CTCs on a xenograft mouse model of human breast cancer.

Despite all the potential benefits, many challenges still need to be overcome before nanotheranostics can reach their full clinical potential for this application, such as understanding the heterogeneity of the CTC population (there are several CTC phenotypes between epithelial-mesenchymal that can simultaneously coexist in circulation), the formation of clusters of CTCs (different drug resistance among CTCs), and the shielding of CTCs with platelets and other cell types (physical barrier to nanoparticles penetrate in CTCs) (Li, Sharkey, Huang, & King, 2015). If these challenges are overcome, the multimodal nanoplatforms for theranostic purposes to detect and eliminate CTCs may catalyze a paradigm shift in cancer research.

5.3 Colonization

5.3.1 Extravasation and formation of micrometastases

Extravasation of CTCs is one of the critical events in metastasis. This process includes several steps, such as cells getting arrested into the microvasculature, entering into the parenchyma of distant tissues, adapting to survive in foreign microenvironments, and eventually colonizing to generate metastasis (Tsai & Yang, 2013). CTCs can initially show a leukocyte-like rolling behavior on the vascular walls. They create transient interactions with the endothelium, followed by stronger adhesions mediated by selectins and vascular cell adhesion molecules (VCAMs), respectively. Neutrophils and platelets can secrete cytokines to induce the expression of these adhesion proteins on endothelial cells, to allow the establishment of temporary interactions with tumor cells (Kim, Borsig, Han, Varki, & Varki, 1999; Labelle & Hynes, 2012; Laubli, Spanaus, & Borsig, 2009). Their adherence to vascular fibronectin deposits has also been described (Barbazán et al., 2017). CTCs can also be trapped into isolated small vessels into the microcirculation, becoming activated and therefore transmigrating (Chambers et al., 2002; Van Zijl et al., 2011). Moreover, accumulating evidence supports tumor cell plasticity and reprogramming during metastatic colonization, controlling epithelial and mesenchymal traits of tumor cells. So, the EMT genetic program triggered during invasion and tumor dissemination might be suppressed upon arrival at the site of metastasis, and the reciprocal program MET (mesenchymal-to-epithelial transition) would be consequently induced to help cells form extensive macro-metastasis at distant organs (Luo, Brooks, & Wicha, 2015; Tsai & Yang, 2013). After seeding, growth of metastatic tumors requires adaptation to a new microenvironment, but also remodeling it into a supportive niche (Aguirre-Ghiso, 2007). The mechanisms by which tumor cells alter this scenario (ECM remodeling as well as other stromal cellular components) still remains unclear. Recent evidence suggests that the metastatic sites can be transformed prior to tumor cell arrival, creating the so-called "premetastatic niche" (Sceneay, Smyth, & Möller, 2013) that will be discussed in the next section. The local environment might provide suitable factors that support tumor cell growth as well as for the establishment of new vasculature through the production of angiogenic factors and recruitment of vascular cells (Rafii, Lyden, Benezra, Hattori, & Heissig, 2002). This niche would also drive the creation of dormant cells, trapped in a quiescent state, until activated by certain factors or signaling pathways that make the scenario suitable for metastases development.

Once these disseminated tumor cells reach the distant organ and survive in the new environment, they are typically resistant to current therapies, unable to control their speedy proliferation and growth (He, Guoc, Qiana, & Chen, 2015). Therefore, it is of utmost importance to battle metastasis before reaching this point. Early metastasis intervention procedures, such as detection and treatment of micrometastasis or interfering the metastatic niche remodeling, may offer new therapeutic opportunities. Detection of micrometastases, that is, deposits of single tumor cells or very small clusters of cancer cells in a distal organ or lymph nodes, is currently limited by the sensitivity of clinical detection of a tumor mass, and only highly invasive molecular methods that suffer from low efficiency are applicable (McGowan, Kirstein, & Chambers, 2009; Pantel, Alix-Panabières, & Riethdorf, 2009). The development of a rapid, noninvasive detection tool that can give accurate whole-body readouts of occult micrometastases and eliminate them might, therefore, have substantial clinical impact in the management of disseminated disease before the burden of metastases.

Nanotheranostics, we believe, have the potential to increase the sensibility of current diagnosis techniques and deliver new therapeutic entities more selective for treating disseminated cells avoiding unspecific effects in the surrounding tissue (Wan, Pantel, & Kang, 2013). Preclinical studies with a multifunctional nanosystem based on poly(methacrylic acid)-polysorbate 80-grafted-starch for the delivery of anticancer drugs (doxorubicin) and imaging agents (Gadolinium and Hoechst) to the brain highlight the potential of this approach for brain metastasis (Li, Cai, Shalviri, et al., 2014). Upconversion luminescent lanthanide nanoparticles were also proposed for this application due to their unique optical properties and chemo-/photo-stability, which is ideal for photodynamic therapy, and they efficiently inhibited tumor growth and metastasis formation in pulmonary and hepatic mouse models of micrometastases (Li, Tang, Pan, et al., 2016). Zhou, Zhao, Tian, et al. (2015) also designed a theranostic nanoplatform for combinational therapy (radiotherapy and photothermal therapy) based on copper-64-labeled copper sulfide PEG-nanoparticles ([^{64}Cu]CuS NPs) that could efficiently suppress lung metastasis in orthotopic 4T1 breast tumor models through eradication of initiating cells (Fig. 4). This system was tracked by PET/CT imaging showing that more than 90% of the radioactivity remained in the tumor 24 h after intratumor injection, much more higher than the 25% of the aqueous solution of $^{64}CuCl_2$. The combined therapy of

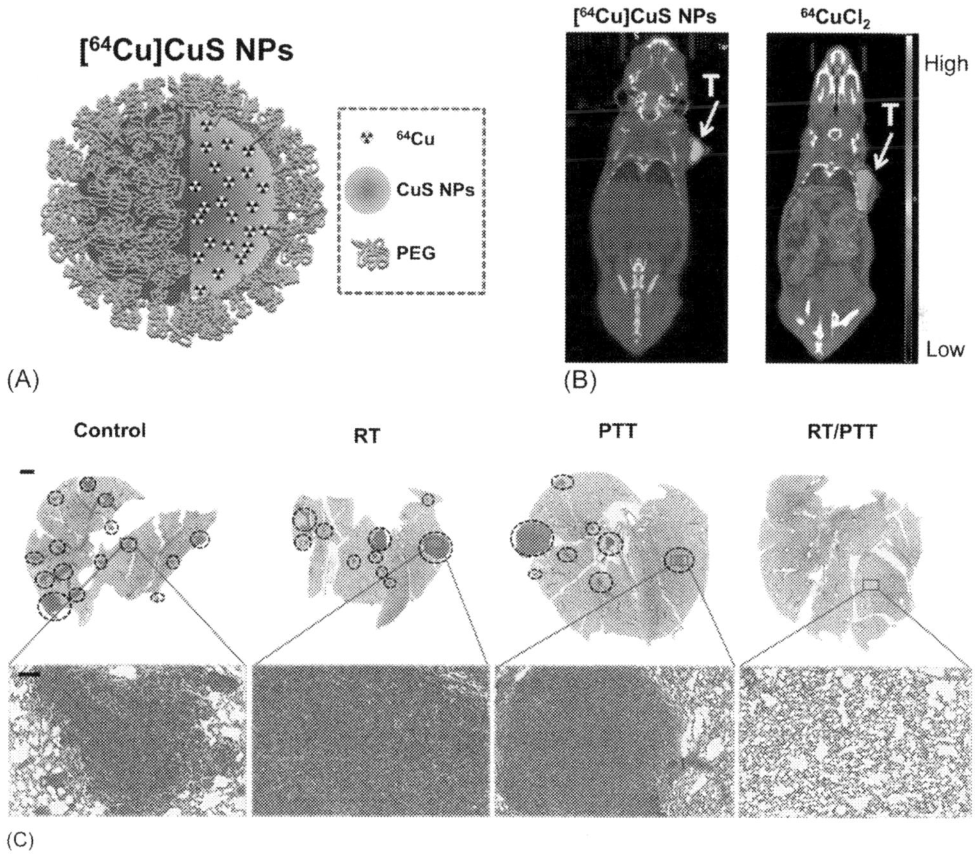

Fig. 4

(A) Schematic representation of the theranostic nanoplatform based on copper-64-labeled copper sulfide PEG-nanoparticles ([^{64}Cu]CuS NPs). (B) PET/CT images of subcutaneous breast tumor-bearing mice 24 h after intratumoral injection of [^{64}Cu]CuS NPs and the aqueous solution of $^{64}CuCl_2$ as control. (C) Histology analysis of 4T1 breast tumor mice showing the number of metastatic nodules after radiotherapy (RT), photothermal therapy (PTT) or RT/PTT therapy compared to the control group. *Data from Zhou, M., Zhao, J., Tian, M., et al. (2015). Radio-photothermal therapy mediated by a single compartment nanoplatform depletes tumor initiating cells and reduces lung metastasis in the orthotopic 4T1 breast tumor model. Nanoscale, 7(46), 19438–19447. doi: 10.1039/C5NR04587H, with the permission of Royal Society of Chemistry, copyright (2017).*

RT/PTT in 4T1 breast tumor mice significantly reduced the number of metastatic nodules in lung compared to the control group and the monotherapies (RT, PTT). In vivo results showed a significantly prolonged survival of breast tumor mice after RT/PTT therapy.

Detection and eradication of micrometastases is still a challenging task, with a high potential for prevention of the onset of metastasis. Nanotheranostics will have a lot to offer, but progress must necessarily be accompanied by the development of more sensitive and sophisticated imaging detection systems as well as improved imaging processing tools.

5.3.2 Premetastatic niche

Evidence suggests that distant organs are not just casual targets of the metastatic process, and that the primary tumor itself is indeed able to modify the environment of secondary organs, facilitating the formation of future supportive microenvironments called "premetastatic niches" (Peinado, Zhang, Matei, et al., 2017; Sceneay et al., 2013). Early on, signals released by the primary tumor such as cytokines, enzymes, and exosomes, are revealed as able to regulate tumor cell homing and assist CTC extravasation during the metastatic cascade (Kaplan, Riba, Zacharoulis, et al., 2005; Massagué & Obenauf, 2016; Peinado, Lavotshkin, & Lyden, 2011). Immune cells, especially myeloid cells, have also been described as key components of the premetastatic niche formation. They can be recruited through chemoattractant inflammatory proteins, such as S100A8 and S100A9, and angiogenic cytokines such as VEGF, forming localized accumulations within the niche (Kaplan et al., 2005). Once there, these cells collaborate with other stromal cells, either recruited or residing in the tissue parenchyma (i.e., fibroblast, immune cells). ECM composition and structure, together with matrix metalloproteinases secretion, are also essential for local tissue remodeling (Høye & Erler, 2016). Several reports have recently highlighted the importance of matrisomal proteins in the establishment of premetastatic and metastatic niches of various cancer types (Oskarsson, Acharyya, Zhang, et al., 2011; Wong, Gilkes, Zhang, et al., 2011).

The role of primary tumor-derived exosomes preparing the premetastatic niche at distant organs for colonization by disseminated cancer cells has been reported in several recent works (Costa-Silva, Aiello, Ocean, et al., 2015; Hoshino, Costa-Silva, Shen, et al., 2015; Steinbichler, Dudás, Riechelmann, & Skvortsova, 2017; Weidle, Birzele, Kollmorgen, & Rüger, 2017). These lipidic extracellular vesicles containing biofunctional molecules (i.e., proteins, lipids, DNA, and RNA), are secreted by most cellular types in healthy and pathological conditions, and can be efficiently transferred to recipient cells at distant sites (Kharaziha, Ceder, Li, & Panaretakis, 2012). Specific integrins expressed on tumor-derived exosomes are related to tissue organotropism of future metastasis, and blocking this exosome communication at some point has been proposed as a promising strategy to stop the premetastatic niche remodeling. One of the proposed strategies includes the use of gold nanoparticles to tackle the biogenesis of exosomes, their circulation pathway, or their uptake by recipient cells (Roma-Rodrigues et al., 2017).

Regarding premetastatic niche formation, much still remains to be understood, but it is undoubtedly a window for new therapeutic strategies against metastasis. For example, Yamamoto, Kikuchi, Ohta, et al. (2008) have reported that blocking the chemoquine CXCR2 and interleukine IL-12 secreted from TK-4 cancer cells with neutralizing antibodies significantly inhibited liver metastasis in orthotopic colon cancer mouse models. In another paper, Belle, Ali, Lonic, et al. (2015) found that PTPN14 (also known as Pez, PTPD2, PTP36) acts as a suppressor of metastasis, particularly to the metastatic niche in the bone, by reducing intracellular protein trafficking through the secretory pathway in xenograft breast cancer models. With respect to the specific use of nanotechnology, Kang, Zhu, Wei, et al. (2017) have recently developed a novel nanoparticle inspired by neutrophils ((poly(latic-co-glycolic acid) nanoparticles coated with neutrophils membranes) due to the ability of neutrophils for targeting both CTCs and the premetastatic niche by intrinsic cell adhesion molecules on their surface membrane. When these nanoparticles were loaded with the drug carfilzomib, CTCs were selectively depleted, and the formation of the metastatic niche inhibited, providing an advanced strategy for metastasis prevention and treatment. Imaging techniques can also help to elucidate the premetastatic niche formation process. Shokeen, Zheleznyak, Wilson, et al. (2012) have recently demonstrated throw a high-affinity PET probe 64Cu-labeled 11-bis(carboxymethyl)-1,4,8,11-tetraazabicyclo[6.6.2] hexadecane (CB-TE2A)-LLP2A that bone marrow-derived (BMD) cells that express vascular endothelial growth factor receptor-1 and very late antigen-4 (VLA-4) arrive first at sites of metastasis to form a receptive environment for tumor cells. Metastatic mouse models were used to perform these experiments by PET/CT.

As we mentioned before, the process of premetastatic niche formation still remains unclear; however, theranostic nanoplatforms combining imaging techniques, targeting capabilities, and therapeutic applications could provide insight into this phenomenon and relevant clues for inhibiting the formation of metastasis.

6 Conclusive Remarks

Cancer nanotechnology is an interdisciplinary research that involves biology, chemistry, engineering, and medicine with the clear objective of leading to major advances in cancer detection, diagnosis, and treatment. In recent decades, an increasing knowledge of functional molecular and biological basis combined to create advances in the materials field, has led to new and modern improvements in the development of innovative nanotechnologies for cancer detection and targeted treatments. Nevertheless, a small number of works have been specifically addressed to improve the management and outcome of metastatic cancer. As highlighted in this chapter, the generation of metastases is a very complex process with a lot of unknowns. A deeper understanding of the tumor biology and the process of metastases will necessarily accompany advances in cancer nanotechnology. Reciprocally, nanotheranostics will certainly play a key role in allowing a more rational development of nanotechnologies that can cover unmet clinical needs and quickly reach translation.

To achieve a breakthrough in the field, it is mandatory to better understand the metastatic process, develop better experimental models for metastasis disease, identify genotypic and phenotypic differences (e.g., targetable receptors) between primary and metastasized tumors, define biomolecules and signatures of CTCs and disseminated cancer cells, and explore novel antimetastatic strategies. In our understanding, strategies for targeting metastatic cancer should tackle one or more processes simultaneously, among the following: (1) targeting the primary tumor to block the metastatic capacity of tumor cells (i.e., antimigration, antiangiogenesis, antiinvasion), (2) targeting the tumor microenvironment, (3) targeting specifically tumor cells subpopulations, related to cancer resistance, metastases, and tumor scape (cancer stem cells), (4) targeting cancer circulating cells before they reach distal tissues, (5) targeting dormant cells in order to move them out of the dormant state so they might be sensitive to conventional chemotherapy or alternatively, maintaining the dormant state to prevent the emergence of metastasis, and (6) targeting established colonization process and the formation of metastatic tumors (oligometastasis or polymetastasis). We believe that within this approach, important advances in the management of metastatic cancer will be achieved in the near future, and this will surely involve the development of tailored nanotheranostics.

Acknowledgments

The authors acknowledge the support of the Carlos III Health Institute/FEDER (PI15/00828), and the Spanish Ministry of Education, Culture, and Sport (FPU15/06595).

References

Abellán-Pose, R., Csaba, N., & Alonso, M. J. (2016). Lymphatic targeting of nanosystems for anticancer drug therapy. *Current Pharmaceutical Design*, *22*(9), 1194–1209. https://doi.org/10.2174/1381612822666151216150809.

Aguirre-Ghiso, J. A. (2007). Models, mechanisms and clinical evidence for cancer dormancy. *Nature Reviews Cancer*, *7*(11), 834–846. https://doi.org/10.1038/nrc2256.

Aires, A., Ocampo, S. M., Simões, B. M., et al. (2016). Multifunctionalized iron oxide nanoparticles for selective drug delivery to CD44-positive cancer cells. *Nanotechnology*, *27*(6), 65103. https://doi.org/10.1088/0957-4484/27/6/065103.

Alberti, D., Protti, N., Franck, M., et al. (2017). Theranostic nanoparticles loaded with imaging probes and rubrocurcumin for combined cancer therapy by folate receptor targeting. *ChemMedChem*, *12*(7), 502–509. https://doi.org/10.1002/cmdc.201700039.

Alconcel, S. N. S., Baas, A. S., & Maynard, H. D. (2011). FDA-approved poly(ethylene glycol)-protein conjugate drugs. *Polymer Chemistry*, *2*(7), 1442. https://doi.org/10.1039/c1py00034a.

Ali, A. A. A., Hsu, F., Hsieh, C., et al. (2016). Erlotinib-conjugated iron oxide nanoparticles as a smart cancer-targeted theranostic probe for MRI *Nature Publishing Group*, *6*(1), 1–16. https://doi.org/10.1038/srep36650.

Alix-Panabières, C., & Pantel, K. (2014). Technologies for detection of circulating tumor cells: facts and vision. *Lab on a Chip*, *14*(1), 57–62. https://doi.org/10.1039/c3lc50644d.

Allavena, P., Sica, A., Solinas, G., Porta, C., & Mantovani, A. (2008). The inflammatory micro-environment in tumor progression: the role of tumor-associated macrophages. *Critical Reviews in Oncology/Hematology, 66*(1), 1–9. https://doi.org/10.1016/j.critrevonc.2007.07.004.

Alonso-Nocelo, M., Abellan-Pose, R., Vidal, A., et al. (2016). Selective interaction of PEGylated polyglutamic acid nanocapsules with cancer cells in a 3D model of a metastatic lymph node. *Journal of Nanobiotechnology, 14*(1), 51. https://doi.org/10.1186/s12951-016-0207-8.

Alonso-Nocelo, M., Abuín, C., López-López, R., & de la Fuente, M. (2016). Development and characterization of a three-dimensional co-culture model of tumor T cell infiltration. *Biofabrication, 8*(2), 25002. https://doi.org/10.1088/1758-5090/8/2/025002.

Amadori, D., Milandri, C., Comella, G., Saracchini, S., et al. (2011). A phase I/II trial of non-pegylated liposomal doxorubicin, docetaxel and trastuzumab as first-line treatment in HER-2-positive locally advanced or metastatic breast cancer. *European Journal of Cancer, 47*(14), 2091–2098. https://doi.org/10.1016/j.ejca.2011.05.005.

American Cancer Society. (2017). *Colorectal cancer facts&figures 2017-2019*. Atlanta: American Cancer Society. https://doi.org/10.1016/S0140-6736(13)61649-9.

Aparicio, S., & Caldas, C. (2013). The implications of clonal genome evolution for cancer medicine. *The New England Journal of Medicine, 368*(9), 842–851. https://doi.org/10.1056/NEJMra1204892.

Avvakumova, S., Galbiati, E., Sironi, L., et al. (2016). Theranostic nanocages for imaging and photothermal therapy of prostate cancer cells by active targeting of neuropeptide-Y receptor. *Bioconjugate Chemistry, 27*(12), 2911–2922. https://doi.org/10.1021/acs.bioconjchem.6b00568.

Bae, K. H., Lee, J. Y., Lee, S. H., Park, T. G., & Nam, Y. S. (2013). Optically traceable solid lipid nanoparticles loaded with siRNA and paclitaxel for synergistic chemotherapy with in situ imaging. *Advanced Healthcare Materials, 2*(4), 576–584. https://doi.org/10.1002/adhm.201200338.

Baetke, S. C., Lammers, T., & Kiessling, F. (2015). Applications of nanoparticles for diagnosis and therapy of cancer. *The British Journal of Radiology, 88*(1054), 1–12. https://doi.org/10.1259/bjr.20150207.

Balkwill, F., & Mantovani, A. (2001). Inflammation and cancer: back to Virchow? *Lancet, 357*(9255), 539–545. https://doi.org/10.1016/S0140-6736(00)04046-0.

Barbazán, J., Alonso-Alconada, L., Elkhatib, N., Geraldo, S., Gurchenkov, V., Glentis, A., et al. (2017). Liver metastasis is facilitated by the adherence of circulating tumor cells to vascular fibronectin deposits. *Cancer Research, 77*(13), 3431–3441. https://doi.org/10.1158/0008-5472.CAN-16-1917.

Barbazán, J., Vieito, M., Abalo, A., et al. (2012). A logistic model for the detection of circulating tumor cells in human metastatic colorectal cancer. *Journal of Cellular and Molecular Medicine, 16*(10), 2342–2349. https://doi.org/10.1111/j.1582-4934.2012.01544.x.

Barenholz, Y. (2012). Doxil®—the first FDA-approved nano-drug: lessons learned. *Journal of Controlled Release, 160*(2), 117–134. https://doi.org/10.1016/j.jconrel.2012.03.020.

Barth, B. M., Altinoglu, E. I., Shanmugavelandy, S. S., et al. (2011). *Targeted indocyanine-green-loaded calcium phosphosilicate nanoparticles for in vivo photodynamic therapy of leukemia*. 5(7), 5325–5337. https://doi.org/10.1021/nn2005766.

Belle, L., Ali, N., Lonic, A., et al. (2015). The tyrosine phosphatase PTPN14 (Pez) inhibits metastasis by altering protein trafficking. *Science Signaling, 8*(364), 1–13. https://doi.org/10.1126/scisignal.2005547.

Beloqui, A., Memvanga, P. B., Coco, R., et al. (2016). A comparative study of curcumin-loaded lipid-based nanocarriers in the treatment of inflammatory bowel disease. *Colloids and Surfaces B, Biointerfaces, 1*(143), 327–335. https://doi.org/10.1016/j.colsurfb.2016.03.038.

Bertino, E., McMichael, E., Mo, X., Trikha, P., et al. (2016). A phase i trial to evaluate antibody-dependent cellular cytotoxicity of cetuximab and lenalidomide in advanced colorectal and head and neck cancer. *Molecular Cancer Therapeutics, 15*(9), 2244–2250. https://doi.org/10.1158/1535-7163.MCT-15-0879.

Bobo, D., Robinson, K. J., Islam, J., Thurecht, K. J., & Corrie, S. R. (2016). Nanoparticle-based medicines: a review of FDA-approved materials and clinical trials to date. *Pharmaceutical Research, 33*(10), 2373–2387. https://doi.org/10.1007/s11095-016-1958-5.

Bonde, A.-K., Tischler, V., Kumar, S., Soltermann, A., & Schwendener, R. A. (2012). Intratumoral macrophages contribute to epithelial-mesenchymal transition in solid tumors. *BMC Cancer, 12*, 35. https://doi.org/10.1186/1471-2407-12-35.

Borghaei, H., Paz-Ares, L., Horn, L., et al. (2015). Nivolumab versus docetaxel in advanced nonsquamous non-small-cell lung cancer. *The New England Journal of Medicine*, *373*(17), 1627–1639. https://doi.org/10.1056/NEJMoa1507643.

Brahmer, J., Reckamp, K. L., Baas, P., et al. (2015). Nivolumab versus docetaxel in advanced squamous-cell non-small-cell lung cancer. *The New England Journal of Medicine*, *2015*, 373123–373135.

Branco da Cunha, C., Klumpers, D. D., Li, W. A., et al. (2014). Influence of the stiffness of three-dimensional alginate/collagen-I interpenetrating networks on fibroblast biology. *Biomaterials*, *35*(32), 8927–8936. https://doi.org/10.1016/j.biomaterials.2014.06.047.

Bregoli, L., Movia, D., Gavigan-Imedio, J. D., Lysaght, J., Reynolds, J., & Prina-Mello, A. (2016). Nanomedicine applied to translational oncology: a future perspective on cancer treatment. *Nanomedicine: Nanotechnology, Biology and Medicine*, *12*(1), 81–103. https://doi.org/10.1016/j.nano.2015.08.006.

Byrne, J. D., Betancourt, T., & Brannon-Peppas, L. (2008). Active targeting schemes for nanoparticle systems in cancer therapeutics. *Advanced Drug Delivery Reviews*, *60*(15), 1615–1626. https://doi.org/10.1016/j.addr.2008.08.005.

Cai, W., & Chen, X. (2008). Multimodality molecular imaging of tumor angiogenesis. *Journal of Nuclear Medicine*, *49*(Suppl_2), 113S–128S. https://doi.org/10.2967/jnumed.107.045922.

Campbell, P. J., Yachida, S., Mudie, L. J., et al. (2010). The patterns and dynamics of genomic instability in metastatic pancreatic cancer. *Nature*, *467*(7319), 1109–1113. https://doi.org/10.1038/nature09460.

Carey, S. P., Starchenko, A., McGregor, A. L., & Reinhart-King, C. A. (2013). Leading malignant cells initiate collective epithelial cell invasion in a three-dimensional heterotypic tumor spheroid model. *Clinical & Experimental Metastasis*, *30*(5), 615–630. https://doi.org/10.1007/s10585-013-9565-x.

Caster, J. M., Patel, A. N., Zhang, T., & Wang, A. (2017). Investigational nanomedicines in 2016: a review of nanotherapeutics currently undergoing clinical trials. *Wiley Interdisciplinary Reviews Nanomedicine and Nanobiotechnology*, *9*(1), e1416. https://doi.org/10.1002/wnan.1416.

Cavallaro, G., Campisi, M., Licciardi, M., Ogris, M., & Giammona, G. (2006). Reversibly stable thiopolyplexes for intracellular delivery of genes. *Journal of Controlled Release*, *115*(3), 322–334. https://doi.org/10.1016/j.jconrel.2006.07.027.

Chambers, A. F., Groom, A. C., & MacDonald, I. C. (2002). Dissemination and growth of cancer cells in metastatic sites. *Nature Reviews Cancer*, *2*(8), 563–572. https://doi.org/10.1038/nrc865.

Chaudhuri, O., Koshy, S. T., Branco da Cunha, C., et al. (2014). Extracellular matrix stiffness and composition jointly regulate the induction of malignant phenotypes in mammary epithelium. *Nature Materials*, *13*(10), 970–978. https://doi.org/10.1038/nmat4009.

Chen, W., Ayala-Orozco, C., Biswal, N. C., et al. (2014). Targeting pancreatic cancer with magneto-fluorescent theranostic gold nanoshells. *Nanomedicine (London, England)*, *9*(8), 1209–1222. https://doi.org/10.2217/nnm.13.84.

Chen, F., & Cai, W. (2014). Tumor vasculature targeting: a generally applicable approach for functionalized nanomaterials. *Small*, *10*(10), 1887–1893. https://doi.org/10.1002/smll.201303627.

Chen, Y.-C., Gowda, R., Newswanger, R. K., et al. (2016). Targeting cholesterol transport in circulating melanoma cells to inhibit metastasis. *International Journal of Laboratory Hematology*, *38*(1), 42–49. https://doi.org/10.1111/ijlh.12426.

Chen, F., Hong, H., Zhang, Y., et al. (2013). In vivo tumor targeting and image-guided drug delivery with antibody-conjugated, radio labeled mesoporous silica nanoparticles. *ACS Nano*, *7*(10), 9027–9039. https://doi.org/10.1021/nn403617J.

Chen, H., Jia, H., Tham, H. P., et al. (2017). Theranostic prodrug vesicles for imaging guided co-delivery of camptothecin and siRNA in synergetic cancer therapy. *ACS Applied Materials & Interfaces*, *9*(28), 23536–23543. https://doi.org/10.1021/acsami.7b06936.

Chen, H., Zhang, W., Zhu, G., Xie, J., & Chen, X. (2017). Rethinking cancer nanotheranostics. *Nature Reviews Materials*, *2*(May), 17024. https://doi.org/10.1038/natrevmats.2017.24.

Chow, E. K., & Ho, D. (2013). Cancer nanomedicine: from drug delivery to imaging. *Science Translational Medicine*, *5*(216), 1–12. https://doi.org/10.1126/scitranslmed.3005872.

Chung, A. S., Lee, J., & Ferrara, N. (2010). Targeting the tumour vasculature: insights from physiological angiogenesis. *Nature Reviews Cancer*, *10*(7), 505–514. https://doi.org/10.1038/nrc2868.

Cirri, P., & Chiarugi, P. (2011). Cancer associated fibroblasts: the dark side of the coin. *American Journal of Cancer Research*, *1*(4), 482–497. http://www.pubmedcentral.nih.gov/articlerender.fcgi?artid=3186047&tool=pmcentrez&rendertype=abstract.

Cobaleda-Siles, M., Henriksen-Lacey, M., De Angulo, A. R., et al. (2014). An iron oxide nanocarrier for dsRNA to target lymph nodes and strongly activate cells of the immune system. *Small*, *10*(24), 5054–5067. https://doi.org/10.1002/smll.201401353.

Conde, J., Bao, C., Tan, Y., et al. (2015). Dual targeted immunotherapy via In vivo delivery of biohybrid RNAi-peptide nanoparticles to tumor-associated macrophages and cancer cells. *Advanced Functional Materials*, *25*, 4183–4194. https://doi.org/10.1002/adfm.201501283.

Cordeiro, A. S., Alonso, M. J., & de la Fuente, M. (2015). Nanoengineering of vaccines using natural polysaccharides. *Biotechnology Advances*, *33*(6 Pt 3), 1279–1293. https://doi.org/10.1016/j.biotechadv.2015.05.010.

Costa-Silva, B., Aiello, N. M., Ocean, A. J., et al. (2015). Pancreatic cancer exosomes initiate pre-metastatic niche formation in the liver. *Nature Cell Biology*, *17*(6), 816–826. https://doi.org/10.1038/ncb3169.

Cui, S., Liu, L., Wan, T., Jiang, L., Shi, Y., & Luo, L. (2017). MiR-520b inhibits the development of glioma by directly targeting MBD2. *American Journal of Cancer Research*, *7*(7), 1528–1539.

Cui, H., & Wang, J. (2016). Progress in the development of nanotheranostic systems. *Theranostics*, *6*(7), 915–917. https://doi.org/10.7150/thno.16153.

Dahmani, F. Z., Xiao, Y., Zhang, J., Yu, Y., Zhou, J., & Yao, J. (2016). Multifunctional polymeric nanosystems for dual-targeted combinatorial chemo/angiogenesis therapy of tumors. *Advanced Healthcare Materials*, *5*(12), 1447–1461. https://doi.org/10.1002/adhm.201600169.

Dai, Z., & Yue, X. (2017). Liposomal nanotechnology for cancer theranostics. *Current Medicinal Chemistry*, *24*(999), 1. https://doi.org/10.2174/0929867324666170306105350.

Danhier, F., Danhier, P., Schleich, N., et al. (2016). Tumor targeting by RGD-grafted PLGA-based nanotheranostics loaded with paclitaxel and superparamagnetic iron oxides. In Y. J. Kang (Ed.), *Methods in pharmacology & toxicology* (pp. 1–29). Switzerland: Springer International Publishing AG. Humana Press. https://doi.org/10.1007/7653_2015_43.

Danhier, F., Feron, O., & Préat, V. (2010). To exploit the tumor microenvironment: passive and active tumor targeting of nanocarriers for anti-cancer drug delivery. *Journal of Controlled Release*, *148*(2), 135–146. https://doi.org/10.1016/j.jconrel.2010.08.027.

Danila, D. C., Pantel, K., Fleisher, M., & Scher, H. I. (2011). Circulating tumors cells as biomarkers: progress toward biomarker qualification. *Cancer Journal*, *17*(6), 438–450. https://doi.org/10.1097/PPO.0b013e31823e69ac.Circulating.

de la Fuente, M., Csaba, N., Garcia-Fuentes, M., & Alonso, M. J. (2008). Nanoparticles as protein and gene carriers to mucosal surfaces. *Nanomedicine (London, England)*, *3*(6), 845–857. https://doi.org/10.2217/17435889.3.6.845.

de la Fuente, M., Jones, M. C., Santander-Ortega, M. J., et al. (2015). A nano-enabled cancer-specific ITCH RNAi chemotherapy booster for pancreatic cancer. *Nanomedicine*, *11*(2), 369–377. https://doi.org/10.1016/j.nano.2014.09.010.

Fuente, de la, Langer, M., & Alonso, M. J. (2014). Nanotechnology approaches for cancer immunotherapy and immunomodulation. In M. José Alonso & M. Garcia-Fuentes (Eds.), *Nano-oncologicals: New targeting and delivery approaches*. (pp. 215–242). Switzerland: Springer. https://doi.org/10.1007/978-3-319-08084-0_8.

de la Fuente, M., Raviña, M., Sousa-Herves, A., et al. (2012). Exploring the efficiency of gallic acid-based dendrimers and their block copolymers with PEG as gene carriers. *Nanomedicine (London, England)*, *7*(11), 1667–1681. https://doi.org/10.2217/nnm.12.51.

de la Fuente, M., Seijo, B., & Alonso, M. J. (2008). Design of novel polysaccharidic nanostructures for gene delivery. *Nanotechnology*, *19*(7), 075105. https://doi.org/10.1088/0957-4484/19/7/075105.

De Palma, M., & Hanahan, D. (2012). The biology of personalized cancer medicine: facing individual complexities underlying hallmark capabilities. *Molecular Oncology*, *6*, 111–127. https://doi.org/10.1016/j.molonc.2012.01.011.

Deng, S., Wu, Q., Zhao, Y., et al. (2015). Biodegradable polymeric micelle-encapsulated doxorubicin suppresses tumor metastasis by killing circulating tumor cells. *Nanoscale*, *7*(12), 5270–5280. https://doi.org/10.1039/C4NR07641A.

Desgrosellier, J. S., & Cheresh, D. A. (2010). Integrins in cancer: biological implications and therapeutic opportunities. *Nature Reviews Cancer*, *10*(1), 9–22. https://doi.org/10.1038/nrc2748.

Detappe, A., Kunjachan, S., Rottmann, J., et al. (2015). AGuIX nanoparticles as a promising platform for image-guided radiation therapy. *Cancer Nanotechnology*, *6*(1), 4. https://doi.org/10.1186/s12645-015-0012-3.

Ding, L., Ley, T. J., Larson, D. E., et al. (2012). Clonal evolution in relapsed acute myeloid leukaemia revealed by whole-genome sequencing. *Nature*, *481*(7382), 506–510. https://doi.org/10.1038/nature10738.

Diou, O., Tsapis, N., & Fattal, E. (2012). Targeted nanotheranostics for personalized cancer therapy. *Expert Opinion on Drug Delivery*, *9*(12), 1475–1487. https://doi.org/10.1517/17425247.2012.736486.

Du, Y., Zhang, Q., Jing, L., et al. (2015). GX1-conjugated poly(lactic acid) nanoparticles encapsulating endostar for improved in vivo anticolorectal cancer treatment. *International Journal of Nanomedicine*, *10*, 3791–3802. https://doi.org/10.2147/IJN.S82029.

Duan, S., Guo, L., Shi, D., Shang, M., Meng, D., & Li, J. (2017). Development of a novel folate-modified nanobubbles with improved targeting ability to tumor cells. *Ultrasonics Sonochemistry*, *37*, 235–243. https://doi.org/10.1016/j.ultsonch.2017.01.013.

Eckert, M. A., Lwin, T. M., Chang, A. T., et al. (2011). Twist1-induced invadopodia formation promotes tumor metastasis. *Cancer Cell*, *19*(3), 372–386. https://doi.org/10.1016/j.ccr.2011.01.036.

El-Ghlban, S., Kasai, T., Shigehiro, T., et al. (2014). Chlorotoxin-Fc fusion inhibits release of MMP-2 from pancreatic cancer cells. *BioMed Research International*, *2014*, 152659. https://doi.org/10.1155/2014/152659.

Ernsting, M. J., Hoang, B., Lohse, I., et al. (2015). Targeting of metastasis-promoting tumor-associated fibroblasts and modulation of pancreatic tumor-associated stroma with a carboxymethylcellulose-docetaxel nanoparticle. *Journal of Controlled Release*, *206*, 122–130. https://doi.org/10.1016/j.jconrel.2015.03.023.

Feng, L., Cheng, L., Dong, Z., et al. (2017). Theranostic liposomes with hypoxia-activated prodrug to effectively destruct hypoxic tumors post-photodynamic therapy. *ACS Nano*, *11*(1), 927–937. https://doi.org/10.1021/acsnano.6b07525.

Fidler, I. J. (1970). Metastasis: quantitative analysis of distribution and fate of tumour emboli labeled with 125I-5-iodo-2[prime]-deoxyuridine. *Journal of the National Cancer Institute*, *45*, 773–782.

Fonsatti, E., Nicolay, H. J. M., Altomonte, M., Covre, A., & Maio, M. (2010). Targeting cancer vasculature via endoglin/CD105: a novel antibody-based diagnostic and therapeutic strategy in solid tumours. *Cardiovascular Research*, *86*(1), 12–19. https://doi.org/10.1093/cvr/cvp332.

Franco, O. E., Shaw, A. K., Strand, D. W., & Hayward, S. W. (2010). Cancer associated fibroblasts in cancer pathogenesis. *Seminars in Cell & Developmental Biology*, *21*(1), 33–39. https://doi.org/10.1016/j.semcdb.2009.10.010.

Gabizon, A., Shmeeda, H., & Barenholz, Y. (2003). Pharmacokinetics of pegylated liposomal doxorubicin: review of animal and human studies. *Clinical Pharmacokinetics*, *42*(5), 419–436. https://doi.org/10.2165/00003088-200342050-00002.

Galanzha, E. I., Kim, J. W., & Zharov, V. P. (2009). Nanotechnology-based molecular photoacoustic and photothermal flow cytometry platform for in-vivo detection and killing of circulating cancer stem cells. *Journal of Biophotonics*, *2*(12), 725–735. https://doi.org/10.1002/jbio.200910078.

Galanzha, E. I., Shashkov, E. V., Kelly, T., Kim, J.-W., Yang, L., & Zharov, V. P. (2009). In vivo magnetic enrichment and multiplex photoacoustic detection of circulating tumour cells. *Nature Nanotechnology*, *4*(12), 855–860. https://doi.org/10.1038/nnano.2009.333.

Gallo, J., Kamaly, N., Lavdas, I., et al. (2014). CXCR4-targeted and MMP-responsive iron oxide nanoparticles for enhanced magnetic resonance imaging. *Angewandte Chemie, International Edition*, *53*(36), 9550–9554. https://doi.org/10.1002/anie.201405442.

Gao, D., Gao, J., Xu, M., et al. (2017). Targeted ultrasound-triggered phase transition nanodroplets for Her2-overexpressing breast cancer diagnosis and gene transfection. *Molecular Pharmaceutics*, *14*(4), 984–998. https://doi.org/10.1021/acs.molpharmaceut.6b00761.

Gao, Y., Xie, J., Chen, H., et al. (2014). Nanotechnology-based intelligent drug design for cancer metastasis treatment. *Biotechnology Advances*, *32*(4), 761–777. https://doi.org/10.1016/j.biotechadv.2013.10.013.

Garcia-Mazas, C., Csaba, N., & Garcia-Fuentes, M. (2017). Biomaterials to suppress cancer stem cells and disrupt their tumoral niche. *International Journal of Pharmaceutics*, *523*(2), 490–505. https://doi.org/10.1016/j.ijpharm.2016.12.013.

Gay, L. J., & Felding-Habermann, B. (2011). Contribution of platelets to tumour metastasis. *Nature Reviews Cancer*, *11*(2), 123–134. https://doi.org/10.1038/nrc3004.

Gener, P., Rafael, D. F., Fernández, Y., et al. (2016). Cancer stem cells and personalized cancer nanomedicine. *Nanomedicine*, *11*(3), 307–320. https://doi.org/10.2217/nnm.15.200.

Greaves, M. (2015). Evolutionary determinants of cancer. *Cancer Discovery*, *5*(8), 806–820. https://doi.org/10.1158/2159-8290.CD-15-0439.

Grivennikov, S. I., Greten, F. R., & Karin, M. (2010). Immunity, inflammation, and cancer. *Cell*, *140*(6), 883–899. https://doi.org/10.1016/j.cell.2010.01.025.

Guo, W., Qiu, Z., Guo, C., et al. (2017). Multifunctional theranostic agent of $Cu_2(OH)PO_4$ quantum dots for photoacoustic image-guided photothermal/photodynamic combination cancer therapy. *ACS Applied Materials & Interfaces*, *9*(11), 9348–9358. https://doi.org/10.1021/acsami.6b15703.

Hall, E., Cameron, D., Waters, R., et al. (2014). Comparison of patient reported quality of life and impact of treatment side effects experienced with a taxane-containing regimen and standard anthracycline based chemotherapy for early breast cancer: 6 year results from the UK TACT trial (CRUK/01/001). *European Journal of Cancer*, *50*, 2375–2389. https://doi.org/10.1016/j.ejca.2014.06.007.

Hanahan, D., & Coussens, L. M. (2012). Accessories to the crime: functions of cells recruited to the tumor microenvironment. *Cancer Cell*, *21*(3), 309–322. https://doi.org/10.1016/j.ccr.2012.02.022.

Hatakeyama, H., Akita, H., Ishida, E., et al. (2007). Tumor targeting of doxorubicin by anti-MT1-MMP antibody-modified PEG liposomes. *International Journal of Pharmaceutics*, *342*(1–2), 194–200. https://doi.org/10.1016/j.ijpharm.2007.04.037.

He, Q., Guoc, S., Qiana, Z., & Chen, X. (2015). Development of individualized anti-metastasis strategies by engineering nanomedicines. *Chemical Society Reviews*, *44*(17), 6258–6286. https://doi.org/10.1039/c4cs00511b.

Hirakawa, S., Brown, L. F., Kodama, S., Paavonen, K., Alitalo, K., & Detmar, M. (2006). VEGF-C-induced lymphangiogenesis in sentinel lymph nodes promotes tumor metastasis to distant sites. *Blood*, *109*(3), 1010–1017. https://doi.org/10.1182/blood-2006-05-021758.

Hirano, K., Hunt, C. A., Strubbe, A., & MacGregor, R. D. (1985). Lymphatic transport of liposome-encapsulated drugs following intraperitoneal administration-effect of lipid composition. *The American Association of Pharmaceutical Scientists Journal*, *2*(6), 271–278. https://doi.org/10.1023/A:1016337500364.

Hodi, F. S., O'Day, S. J., McDermott, D. F., et al. (2010). Improved survival with ipilimumab in patients with metastatic melanoma. *The New England Journal of Medicine*, *363*(8), 711–723. https://doi.org/10.1056/NEJMoa1003466.

Hong, I.-S., Jang, G.-B., Lee, H.-Y., & Nam, J.-S. (2015). Targeting cancer stem cells by using the nanoparticles. *International Journal of Nanomedicine*, *10*(Special Issue), 251–260. https://doi.org/10.2147/IJN.S88310.

Hoshino, A., Costa-Silva, B., Shen, T.-L., et al. (2015). Tumour exosome integrins determine organotropic metastasis. *Nature*, *527*(7578), 329–335. https://doi.org/10.1038/nature15756.

Høye, A. M., & Erler, J. T. (2016). Structural ECM components in the premetastatic and metastatic niche. *American Journal of Physiology Cell Physiology*, *310*(11), C955–C967. https://doi.org/10.1152/ajpcell.00326.2015.

Huang, X., O'Connor, R., & Kwizera, E. A. (2017). Gold nanoparticle based platforms for circulating cancer marker detection. *Nanotheranostics*, *1*(1), 80–102. https://doi.org/10.7150/ntno.18216.

Huang, M., Shen, A., Ding, J., & Geng, M. (2014). Molecularly targeted cancer therapy: some lessons from the past decade. *Trends in Pharmacological Sciences*, *35*(1), 41–50. https://doi.org/10.1016/j.tips.2013.11.004.

Huang, H., Yang, D.-P., Liu, M., et al. (2017). pH-sensitive Au–BSA–DOX–FA nanocomposites for combined CT imaging and targeted drug delivery. *International Journal of Nanomedicine*, *12*, 2829–2843. https://doi.org/10.2147/IJN.S128270.

Huang, X., Zhang, F., Lee, S., et al. (2012). Long-term multimodal imaging of tumor draining sentinel lymph nodes using mesoporous silica-based nanoprobes. *Biomaterials*, *33*(17), 4370–4378. https://doi.org/10.1016/j.biomaterials.2012.02.060.

Hui, X., Han, Y., Liang, S., et al. (2008). Specific targeting of the vasculature of gastric cancer by a new tumor-homing peptide CGNSNPKSC. *Journal of Controlled Release*, *131*(2), 86–93. https://doi.org/10.1016/j.jconrel.2008.07.024.

Huo, D., He, J., Li, H., et al. (2014). X-ray CT guided fault-free photothermal ablation of metastatic lymph nodes with ultrafine HER-2 targeting W18O49 nanoparticles. *Biomaterials*, *35*(33), 9155–9166. https://doi.org/10.1016/j.biomaterials.2014.07.034.

Ilic, M., & Ilic, I. (2016). Epidemiology of pancreatic cancer. *World Journal of Gastroenterology*, *22*(44), 9694–9705. https://doi.org/10.3748/wjg.v22.i44.9694.

Irmisch, A., & Huelsken, J. (2013). Metastasis: new insights into organ-specific extravasation and metastatic niches. *Experimental Cell Research*, *319*, 1604–1610. https://doi.org/10.1016/j.yexcr.2013.02.012.

Junttila, M. R., & de Sauvage, F. J. (2013). Influence of tumour micro-environment heterogeneity on therapeutic response. *Nature*, *501*(7467), 346–354. https://doi.org/10.1038/nature12626.

Kaminskas, L. M., McLeod, V. M., Ascher, D. B., et al. (2014). Methotrexate-conjugated PEGylated dendrimers show differential patterns of deposition and activity in tumor-burdened lymph nodes after intravenous and subcutaneous administration in rats. *Molecular Pharmaceutics*, *12*(2), 432–443. https://doi.org/10.1021/mp500531e.

Kanapathipillai, M., Brock, A., & Ingber, D. E. (2014). Nanoparticle targeting of anti-cancer drugs that alter intracellular signaling or influence the tumor microenvironment. *Advanced Drug Delivery Reviews*, *79*, 107–118. https://doi.org/10.1016/j.addr.2014.05.005.

Kang, T., Zhu, Q., Wei, D., et al. (2017). Nanoparticles coated with neutrophil membranes can effectively treat cancer metastasis. *ACS Nano*, *11*(2), 1397–1411. https://doi.org/10.1021/acsnano.6b06477.

Kaplan, R. N., Riba, R. D., Zacharoulis, S., et al. (2005). VEGFR1-positive haematopoietic bone marrow progenitors initiate the pre-metastatic niche. *Nature*, *438*(7069), 820–827. https://doi.org/10.1038/nature04186.

Karathanasis, E., Suryanarayanan, S., Balusu, S. R., et al. (2009). Imaging nanoprobe for prediction of outcome of nanoparticle chemotherapy by using mammography. *Radiology*, *250*(2), 398–406. https://doi.org/10.1148/radiol.2502080801.

Kedrin, D., van Rheenen, J., Hernandez, L., Condeelis, J., & Segall, J. E. (2007). Cell motility and cytoskeletal regulation in invasion and metastasis. *Journal of Mammary Gland Biology and Neoplasia*, *12*(2–3), 143–152. https://doi.org/10.1007/s10911-007-9046-4.

Kharaziha, P., Ceder, S., Li, Q., & Panaretakis, T. (2012). Tumor cell-derived exosomes: a message in a bottle. *Biochimica et Biophysica Acta, Reviews on Cancer*, *1826*(1), 103–111. https://doi.org/10.1016/j.bbcan.2012.03.006.

Kim, Y. J., Borsig, L., Han, H.-L., Varki, N. M., & Varki, A. (1999). Distinct selectin ligands on colon carcinoma mucins can mediate pathological interactions among platelets, leukocytes, and endothelium. *The American Journal of Pathology*, *155*(2), 461–472. https://doi.org/10.1016/S0002-9440(10)65142-5.

Kim, J.-W., Galanzha, E. I., Zaharoff, D. A., Griffin, R. J., & Zharov, V. P. (2014). Nanotheranostics of circulating tumor cells, infections and other pathological features. *In Vivo*, *10*(3), 813–830. https://doi.org/10.1021/mp300577s.Nanotheranostics.

Kim, D., Jeong, Y. Y., & Jon, S. (2010). A drug loaded aptamer-gold nanoparticle bioconjugate for combined CT imaging and therapy of prostate cancer. *ACS Nano*, *4*(7), 3689–3696. https://doi.org/10.1021/nn901877h.

Kim, D., Lee, E. S., Oh, K. T., Gao, Z., & Bae, Y. H. (2008). Doxorubicin-loaded polymeric micelle overcomes multidrug resistance of cancer by double-targeting folate receptor and early endosomal pH. *Small*, *4*(11), 2043–2050. https://doi.org/10.1002/smll.200701275.

Kirpotin, D., Hong, K., Mullah, N., Papahadjopoulos, D., & Zalipsky, S. (1996). Liposomes with detachable polymer coating: destabilization and fusion of dioleoylphosphatidylethanolamine vesicles triggered by cleavage of surface-grafted poly(ethylene glycol). *FEBS Letters*, *388*(2–3), 115–118. https://doi.org/10.1016/0014-5793(96)00521-2.

Kotb, S., Detappe, A., Lux, F., et al. (2016). Gadolinium-based nanoparticles and radiation therapy for multiple brain melanoma metastases: proof of concept before phase I trial. *Theranostics*, *6*(3), 418–427. https://doi.org/10.7150/thno.14018.

Kunjachan, S., Pola, R., Gremse, F., et al. (2014). Passive versus active tumor targeting using RGD- and NGR-modified polymeric nanomedicines. *Nano Letters*, *14*(2), 972–981. https://doi.org/10.1021/nl404391r.

Labelle, M., & Hynes, R. O. (2012). The initial hours of metastasis: the importance of cooperative host-tumor cell interactions during hematogenous dissemination. *Cancer Discovery*, *2*(12), 1091–1099. https://doi.org/10.1158/2159-8290.CD-12-0329.

Laubli, H., Spanaus, K.-S., & Borsig, L. (2009). Selectin-mediated activation of endothelial cells induces expression of CCL5 and promotes metastasis through recruitment of monocytes. *Blood*, *114*(20), 4583–4591. https://doi.org/10.1182/blood-2008-10-186585.

Ledet, G., & Mandal, T. K. (2012). Nanomedicine: emerging therapeutics for the 21st century. *US Pharmacist*, *37*(3), 7–11.

Ledford, H. (2016). Bankruptcy filing worries developers of nanoparticle cancer drugs. *Nature*, *533*(7603), 304–305. https://doi.org/10.1038/533304a.

Lee, J.-Y., Chung, S.-J., Cho, H.-J., & Kim, D.-D. (2016). Iodinated hyaluronic acid oligomer-based nanoassemblies for tumor-targeted drug delivery and cancer imaging. *Biomaterials*, *85*, 218–231. https://doi.org/10.1016/j.biomaterials.2016.01.060.

Lee, C., Kwon, W., Beack, S., et al. (2016). Biodegradable nitrogen-doped carbon nanodots for non-invasive photoacoustic imaging and photothermal therapy. *Theranostics*, *6*(12), 2196–2208. https://doi.org/10.7150/thno.16923.

Levental, K. R., Yu, H., Kass, L., et al. (2009). Matrix crosslinking forces tumor progression by enhancing integrin signaling. *Cell*, *139*(5), 891–906. https://doi.org/10.1016/j.cell.2009.10.027.

Li, J., Cai, P., Shalviri, A., et al. (2014). A multifunctional polymeric nanotheranostic system delivers doxorubicin and imaging agents across the blood brain barrier targeting brain metastases of breast cancer. *ACS Nano*, *8*(10), 9925–9940. https://doi.org/10.1021/nn501069c.

Li, L., Fu, S., Chen, C., et al. (2016). Microenvironment-driven bioelimination of magnetoplasmonic nanoassemblies and their multimodal imaging-guided tumor photothermal therapy. *ACS Nano*, *10*(7), 7094–7105. https://doi.org/10.1021/acsnano.6b03238.

Li, S., & Li, Q. (2015). Cancer stem cells, lymphangiogenesis, and lymphatic metastasis. *Cancer Letters*, *357*(2), 438–447. https://doi.org/10.1016/j.canlet.2014.12.013.

Li, B., Li, Q., Mo, J., & Dai, H. (2017). Drug-loaded polymeric nanoparticles for cancer stem cell targeting. *Frontiers in Pharmacology*, *8*(February), 1–12. https://doi.org/10.3389/fphar.2017.00051.

Li, L., Li, X., Wu, Y., et al. (2017). Multifunctional nucleus-targeting nanoparticles with ultra-high gene transfection efficiency for *in vivo* gene therapy. *Theranostics*, *7*(6), 1633–1649. https://doi.org/10.7150/thno.17588.

Li, J., Sharkey, C. C., Huang, D., & King, M. R. (2015). Nanobiotechnology for the therapeutic targeting of cancer cells in blood. *Cellular and Molecular Bioengineering*, *8*(1), 137–150. https://doi.org/10.1007/s12195-015-0381-z.

Li, Y., Tang, J., Pan, D. X., et al. (2016). A versatile imaging and therapeutic platform based on dual-band luminescent lanthanide nanoparticles toward tumor metastasis inhibition. *ACS Nano*, *10*(2), 2766–2773. https://doi.org/10.1021/acsnano.5b07873.

Lin, R.-Z., Wang, T.-P., Hung, R.-J., Chuang, Y.-J., Chien, C.-C.M., & Chang, H.-Y. (2011). Tumor-induced endothelial cell apoptosis: roles of NAD(P)H oxidase-derived reactive oxygen species. *Journal of Cellular Physiology*, *226*(7), 1750–1762. https://doi.org/10.1002/jcp.22504.

Ling, D., Xia, H., Park, W., et al. (2014). pH-sensitive nanoformulated triptolide as a targeted therapeutic strategy for hepatocellular carcinoma. *ACS Nano*, *8*(8), 8027–8039.

Liu, Z., Chen, W., Li, Y., & Xu, Q. (2016). Integrin $\alpha_v \beta_3$–targeted C-dot nanocomposites as multifunctional agents for cell targeting and photoacoustic imaging of superficial malignant tumors. *Analytical Chemistry*, *88*(23), 11955–11962. https://doi.org/10.1021/acs.analchem.6b03927.

Liu, Y., Miyoshi, H., & Nakamura, M. (2007). Nanomedicine for drug delivery and imaging: a promising avenue for cancer therapy and diagnosis using targeted functional nanoparticles. *International Journal of Cancer*, *120*(12), 2527–2537. https://doi.org/10.1002/ijc.22709.

Liu, H., Xu, S., He, Z., Deng, A., & Zhu, J. J. (2013). Supersandwich cytosensor for selective and ultrasensitive detection of cancer cells using aptamer-DNA concatamer-quantum dots probes. *Analytical Chemistry*, *85*(6), 3385–3392. https://doi.org/10.1021/ac303789x.

Livney, Y. D., & Assaraf, Y. G. (2013). Rationally designed nanovehicles to overcome cancer chemoresistance. *Advanced Drug Delivery Reviews*, *65*(13–14), 1716–1730. https://doi.org/10.1016/j.addr.2013.08.006.

Luo, M., Brooks, M., & Wicha, M. S. (2015). Epithelial-mesenchymal plasticity of breast cancer stem cells: implications for metastasis and therapeutic resistance. *Current Pharmaceutical Design*, *21*(10), 1301–1310.

Lv, G., Wu, M., Wang, M., et al. (2016). miR-320a regulates high mobility group box 1 expression and inhibits invasion and metastasis in hepatocellular carcinoma. *International Journal of Laboratory Hematology*, *38*(1), 42–49. https://doi.org/10.1111/ijlh.12426.

Martínez-Cardús, A., Moran, S., Musulen, E., et al. (2016). Epigenetic homogeneity within colorectal tumors predicts shorter relapse-free and overall survival times for patients with locoregional cancer. *Gastroenterology*, *151*(5), 961–972. https://doi.org/10.1053/j.gastro.2016.08.001.

Massagué, J., & Obenauf, A. C. (2016). Metastatic colonization by circulating tumour cells. *Nature*, *529*(7586), 298–306. https://doi.org/10.1038/nature17038.

Matsumura, Y., & Maeda, H. (1986). A new concept for macromolecular therapeutics in cancer chemotherapy: mechanism of tumoritropic accumulation of proteins and the antitumor agent smancs. *Cancer Research*, *46*(12 Pt 1), 6387–6392.

Mazza, M., Alonso-Sande, M., Jones, M. C., & de la Fuente, M. (2013). The potential of nanoemulsions in biomedicine. In I. F. Uchegbu, A. G. Schätzlein, W. P. Cheng, & A. Lalatsa (Eds.), *Fundamentals of harmaceutical nanoscience*. (pp. 117–155). Switzerland: Springer. https://doi.org/10.1007/978-1-4614-9164-4.

McGowan, P. M., Kirstein, J. M., & Chambers, A. F. (2009). Micrometastatic disease and metastatic outgrowth: clinical issues and experimental approaches. *Future Oncology*, *5*(7), 1083–1098. https://doi.org/10.2217/fon.09.73.

Mcgranahan, N., & Swanton, C. (2017). Clonal heterogeneity and tumor evolution: past, present, and the future. *Cell*, *168*(4), 613–628. https://doi.org/10.1016/j.cell.2017.01.018.

Mehlen, P., & Puisieux, A. (2006). Metastasis: a question of life or death. *Nature Reviews Cancer*, *6*(6), 449–458. https://doi.org/10.1038/nrc1886.

Miles, F. L., Pruitt, F. L., van Golen, K. L., & Cooper, C. R. (2008). Stepping out of the flow: capillary extravasation in cancer metastasis. *Clinical & Experimental Metastasis*, *25*(4), 305–324. https://doi.org/10.1007/s10585-007-9098-2.

Miller, M. A., Gadde, S., Pfirschke, C., et al. (2015). Predicting therapeutic nanomedicine efficacy using a companion magnetic resonance imaging nanoparticle. *Science Translational Medicine*, *7*(314), 314ra183. https://doi.org/10.1126/scitranslmed.aac6522.

Mina, L. A., & Sledge, G. W. (2011). Rethinking the metastatic cascade as a therapeutic target. *Nature Reviews Clinical Oncology*, *8*(6), 325–332. https://doi.org/10.1038/nrclinonc.2011.59.

Mitra, A., Mulholland, J., Nan, A., McNeill, E., Ghandehari, H., & Line, B. R. (2005). Targeting tumor angiogenic vasculature using polymer-RGD conjugates. *Journal of Controlled Release*, *102*(1), 191–201. https://doi.org/10.1016/j.jconrel.2004.09.023.

Mo, R., Jiang, T., DiSanto, R., Tai, W., & Gu, Z. (2014). ATP-triggered anticancer drug delivery. *Nature Communications*, *5*, 3364. https://doi.org/10.1038/ncomms4364.

Mohanty, S., Chen, Z., Li, K., et al. (2017). A novel theranostic strategy for MMP-14 expressing glioblastomas impacts survival. *Molecular Cancer Therapeutics*, *16*(9), 1909–1921. https://doi.org/10.1158/1535-7163.MCT-17-0022.

Mok, H., Veiseh, O., Fang, C., et al. (2010). pH-sensitive siRNA nanovector for targeted gene silencing and cytotoxic effect in cancer cells. *Molecular Pharmaceutics*, *7*(6), 1930–1939.

Morikawa, S., Baluk, P., Kaidoh, T., Haskell, A., Jain, R. K., & McDonald, D. M. (2002). Abnormalities in pericytes on blood vessels and endothelial sprouts in tumors. *The American Journal of Pathology*, *160*(3), 985–1000. https://doi.org/10.1016/S0002-9440(10)64920-6.

Motzer, R. J., Escudier, B., McDermott, D. F., et al. (2015). Nivolumab versus everolimus in advanced renal-cell carcinoma. *The New England Journal of Medicine*, *373*(19), 1803–1813. https://doi.org/10.1056/NEJMoa1510665.

Mueller, M. M., & Fusenig, N. E. (2004). Friends or foes–bipolar effects of the tumour stroma in cancer. *Nature Reviews Cancer*, *4*(11), 839–849. https://doi.org/10.1038/nrc1477.

Muinelo-Romay, L., Vieito, M., Abalo, A., et al. (2014). Evaluation of circulating tumor cells and related events as prognostic factors and surrogate biomarkers in advanced NSCLC patients receiving first-line systemic treatment. *Cancers*, *6*(1), 153–165. https://doi.org/10.3390/cancers6010153.

Mura, S., Nicolas, J., & Couvreur, P. (2013). Stimuli-responsive nanocarriers for drug delivery. *Nature Materials*, *12*(11), 991–1003. https://doi.org/10.1038/nmat3776.

Murphy, E. A., Majeti, B. K., Barnes, L. A., et al. (2008). Nanoparticle-mediated drug delivery to tumor vasculature suppresses metastasis. *Proceedings of the National Academy of Sciences of the United States of America*, *105*(27), 9343–9348. https://doi.org/10.1073/pnas.0803728105.

Myung, J. H., Roengvoraphoj, M., Tam, K. A., et al. (2015). Effective capture of circulating tumor cells from a transgenic mouse lung cancer model using Dendrimer surfaces immobilized with anti-EGFR. *Analytical Chemistry*, *87*(19), 10096–10102. https://doi.org/10.1021/acs.analchem.5b02766.

Nathanson, S. D., & Mahan, M. (2011). Sentinel lymph node pressure in breast cancer. *Annals of Surgical Oncology*, *18*(13), 3791–3796. https://doi.org/10.1245/s10434-011-1796-y.

Niu, C., Wang, Z., Lu, G., et al. (2013). Doxorubicin loaded superparamagnetic PLGA-iron oxide multifunctional microbubbles for dual-mode US/MR imaging and therapy of metastasis in lymph nodes. *Biomaterials*, *34*(9), 2307–2317. https://doi.org/10.1016/j.biomaterials.2012.12.003.

Novello, S., Barlesi, F., Califano, R., et al. (2016). Metastatic non-small-cell lung cancer: ESMO clinical practice guidelines for diagnosis, treatment and follow-up. *Annals of Oncology*, *27*(Supplement 5), V1–V27. https://doi.org/10.1093/annonc/mdw326.

Oliveira, E. A., & Faintuch, B. L. (2015). Radiolabeling and biological evaluation of the GX1 and RGD-GX1 peptide sequence for angiogenesis targeting. *Nuclear Medicine and Biology*, *42*(2), 123–130. https://doi.org/10.1016/j.nucmedbio.2014.09.004.

Olmeda, D., Jordá, M., Peinado, H., Fabra, A., & Cano, A. (2007). Snail silencing effectively suppresses tumour growth and invasiveness. *Oncogene*, *26*(13), 1862–1874. https://doi.org/10.1038/sj.onc.1209997.

Oskarsson, T., Acharyya, S., Zhang, X. H., et al. (2011). Breast cancer cells produce tenascin C as a metastatic niche component to colonize the lungs. *Nature Medicine*, *17*(7), 867–874. https://doi.org/10.1038/nm.2379.

Pantel, K., Alix-Panabières, C., & Riethdorf, S. (2009). Cancer micrometastases. *Nature Reviews Clinical Oncology*, *6*(6), 339–351. https://doi.org/10.1038/nrclinonc.2009.44.

Pantel, K., & Brakenhoff, R. H. (2004). Dissecting the metastatic cascade. *Nature Reviews Cancer*, *4*(6), 448–456. https://doi.org/10.1038/nrc1370. pii: nrc1370.

Park, S., Aalipour, A., Vermesh, O., Yu, J. H., & Gambhir, S. S. (2017). Towards clinically translatable in vivo nanodiagnostics. *Nature Reviews Materials*, *2*, 1–20. 17014. https://doi.org/10.1038/natrevmats.2017.14.

Park, J. A., Lee, J. J., Jung, J. C., et al. (2008). Gd-DOTA conjugate of RGD as a potential tumor-targeting MRI contrast agent. *ChemBioChem*, *9*(17), 2811–2813. https://doi.org/10.1002/cbic.200800529.

Patel, S. K., & Janjic, J. M. (2015). Macrophage targeted theranostics as personalized nanomedicine strategies for inflammatory diseases. *Theranostics*, *5*(2), 150–172. https://doi.org/10.7150/thno.9476.

Pawelek, J. M., & Chakraborty, A. K. (2008). Fusion of tumour cells with bone marrow-derived cells: a unifying explanation for metastasis. *Nature Reviews Cancer*, *8*(5), 377–386. https://doi.org/10.1038/nrc2371.

Peer, D., Karp, J. M., Hong, S., Farokhzad, O. C., Margalit, R., & Langer, R. (2007). Nanocarriers as an emerging platform for cancer therapy. *Nature Nanotechnology*, *2*(12), 751–760.

Peinado, H., Lavotshkin, S., & Lyden, D. (2011). The secreted factors responsible for pre-metastatic niche formation: old sayings and new thoughts. *Seminars in Cancer Biology*, *21*(2), 139–146. https://doi.org/10.1016/j.semcancer.2011.01.002.

Peinado, H., Zhang, H., Matei, I. R., et al. (2017). Pre-metastatic niches: organ-specific homes for metastases. *Nature Reviews Cancer*, *17*(5), 302–317. https://doi.org/10.1038/nrc.2017.6.

Pellosi, D. S., Calori, I. R., de Paula, L. B., Hioka, N., Quaglia, F., & Tedesco, A. C. (2017). Multifunctional theranostic Pluronic mixed micelles improve targeted photoactivity of Verteporfin in cancer cells. *Materials Science and Engineering: C*, *71*, 1–9. https://doi.org/10.1016/j.msec.2016.09.064.

Pérez-Medina, C., Abdel-Atti, D., Tang, J., et al. (2016). Nanoreporter PET predicts the efficacy of anti-cancer nanotherapy. *Nature Communications*, *7*(May), 11838. https://doi.org/10.1038/ncomms11838.

Planchard, D., Besse, B., Groen, H. J. M., et al. (2016). Dabrafenib plus trametinib in patients with previously treated BRAFV600E-mutant metastatic non-small cell lung cancer: an open-label, multicentre phase 2 trial. *The Lancet Oncology*, *17*(7), 984–993. https://doi.org/10.1016/S1470-2045(16)30146-2.

Pollard, J. W. (2004). Tumour-educated macrophages promote tumour progression and metastasis. *Nature Reviews Cancer*, *4*(1), 71–78. https://doi.org/10.1038/nrc1256.

Qian, C., Chen, Y., Zhu, S., et al. (2016). ATP-responsive and near-infrared-emissive nanocarriers for anticancer drug delivery and real-time imaging. *Theranostics*, *6*(7), 1053–1064. https://doi.org/10.7150/thno.14843.

Quail, D. F., & Joyce, J. A. (2013). Microenvironmental regulation of tumor progression and metastasis. *Nature Medicine*, *19*(11), 1423–1437. https://doi.org/10.1038/nm.3394.

Rafii, S., Lyden, D., Benezra, R., Hattori, K., & Heissig, B. (2002). Vascular and haematopoietic stem cells: novel targets for anti-angiogenesis therapy? *Nature Reviews Cancer*, *2*(11), 826–835. https://doi.org/10.1038/nrc925.

Rajkumar, S., & Prabaharan, M. (2016). Theranostics based on iron oxide and gold nanoparticles for imaging-guided photothermal and photodynamic therapy of cancer. *Current Topics in Medicinal Chemistry*, (November). http://www.ncbi.nlm.nih.gov/pubmed/27875977. Accessed May 31, 2017.

Reck, M., Rodríguez-Abreu, D., Robinson, A. G., et al. (2016). Pembrolizumab versus chemotherapy for PD-L1-positive non-small-cell lung cancer. *The New England Journal of Medicine*, *375*(19), 1823–1833. https://doi.org/10.1056/NEJMoa1606774.

Reimondez-Troitiño, S., Csaba, N., Alonso, M. J., & de la Fuente, M. (2015). Nanotherapies for the treatment of ocular diseases. *European Journal of Pharmaceutics and Biopharmaceutics*, *95*(Pt B), 279–293. https://doi.org/10.1016/j.ejpb.2015.02.019.

Ren, L., Chen, S., Li, H., et al. (2016). MRI-guided liposomes for targeted tandem chemotherapy and therapeutic response prediction. *Acta Biomaterialia*, *35*, 260–268. https://doi.org/10.1016/j.actbio.2016.02.011.

Robert, C., Long, G. V., Brady, B., et al. (2014). Nivolumab in previously untreated melanoma without BRAF mutation. *The New England Journal of Medicine*, *372*(4), 320–330. https://doi.org/10.1056/NEJMoa1412082.

Rolfo, C., Passiglia, F., Ostrowski, M., et al. (2015). Improvement in lung cancer outcomes with targeted therapies: an update for family physicians. *Journal of American Board of Family Medicine*, *28*(1). 124–133. https://doi.org/10.3122/jabfm.2015.01.140072.

Roma-Rodrigues, C., Raposo, L. R., Cabral, R., Paradinha, F., Baptista, P. V., & Fernandes, A. R. (2017). Tumor microenvironment modulation via gold nanoparticles targeting malicious exosomes: implications for cancer diagnostics and therapy. *International Journal of Molecular Sciences*, *18*(1), 162–188. https://doi.org/10.3390/ijms18010162.

Ruoslahti, E., & Pierschbacher, M. (1987). New perspectives in cell adhesion: RGD and integrins. *Science*, *238*(4826), 491–497. https://doi.org/10.1126/science.2821619.

Ryan, G. M., Kaminskas, L. M., & Porter, C. J. H. (2014). Nano-chemotherapeutics: maximising lymphatic drug exposure to improve the treatment of lymph-metastatic cancers. *Journal of Controlled Release*, *193*, 241–256. https://doi.org/10.1016/j.jconrel.2014.04.051.

Sainz, B., Jr., Carron, E., Vallespinós, M., & Machado, H. L. (2016). Cancer stem cells and macrophages: implications in tumor biology and therapeutic strategies. *Mediators of Inflammation*, *2016*, 1–15. https://doi.org/10.1155/2016/9012369.

Saito, G., Swanson, J. A., & Lee, K. D. (2003). Drug delivery strategy utilizing conjugation via reversible disulfide linkages: role and site of cellular reducing activities. *Advanced Drug Delivery Reviews*, *55*(2), 199–215. https://doi.org/10.1016/S0169-409X(02)00179-5.

Sanna, V., Pala, N., & Sechi, M. (2014). Targeted therapy using nanotechnology: focus on cancer. *International Journal of Nanomedicine*, *9*, 467–483. https://doi.org/10.2147/IJN.S36654.

Sasso, M. S., Lollo, G., Pitorre, M., et al. (2016). Low dose gemcitabine-loaded lipid nanocapsules target monocytic myeloid-derived suppressor cells and potentiate cancer immunotherapy. *Biomaterials*, *96*, 47–62. https://doi.org/10.1016/j.biomaterials.2016.04.010.

Sawant, R. M., Hurley, J. P., Salmaso, S., et al. (2006). "SMART" drug delivery systems: double-targeted pH-responsive pharmaceutical nanocarriers. *Bioconjugate Chemistry*, *17*(4), 943–949. https://doi.org/10.1021/bc060080h.

Sceneay, J., Smyth, M. J., & Möller, A. (2013). The pre-metastatic niche: finding common ground. *Cancer Metastasis Reviews*, *32*(3–4), 449–464. https://doi.org/10.1007/s10555-013-9420-1.

Schleich, N., Po, C., Jacobs, D., et al. (2014). Comparison of active, passive and magnetic targeting to tumors of multifunctional paclitaxel/SPIO-loaded nanoparticles for tumor imaging and therapy. *Journal of Controlled Release*, *194*, 82–91. https://doi.org/10.1016/j.jconrel.2014.07.059.

Seon, B. K., Haba, A., Matsuno, F., et al. (2011). Endoglin-targeted cancer therapy. *Current Drug Delivery*, *8*(1), 135–143. http://www.ingentaconnect.com/content/ben/cdd/2011/00000008/00000001/art00009.

Serafini, P., Borrello, I., & Bronte, V. (2006). Myeloid suppressor cells in cancer: recruitment, phenotype, properties, and mechanisms of immune suppression. *Seminars in Cancer Biology*, *16*(1), 53–65. https://doi.org/10.1016/j.semcancer.2005.07.005.

Shen, M., Huang, Y., Han, L., et al. (2012). Multifunctional drug delivery system for targeting tumor and its acidic microenvironment. *Journal of Controlled Release*, *161*(3), 884–892. https://doi.org/10.1016/j.jconrel.2012.05.013.

Shen, S., Xia, J.-X., & Wang, J. (2016). Nanomedicine-mediated cancer stem cell therapy. *Biomaterials*, *74*, 1–18. https://doi.org/10.1016/j.biomaterials.2015.09.037.

Shi, J., Kantoff, P. W., Wooster, R., & Farokhzad, O. C. (2016). Cancer nanomedicine: progress, challenges and opportunities. *Nature Publishing Group*, *17*(1), 20–37. https://doi.org/10.1038/nrc.2016.108.

Shi, H., Ye, X., He, X., et al. (2014). Au@Ag/Au nanoparticles assembled with activatable aptamer probes as smart "nano-doctors" for image-guided cancer thermotherapy. *Nanoscale*, *6*(15), 8754–8761. https://doi.org/10.1039/c4nr01927j.

Shim, M. S., & Kwon, Y. J. (2012). Stimuli-responsive polymers and nanomaterials for gene delivery and imaging applications. *Advanced Drug Delivery Reviews*, *64*(11), 1046–1059. https://doi.org/10.1016/j.addr.2012.01.018.

Shokeen, M., Zheleznyak, A., Wilson, J. M., et al. (2012). Molecular imaging of very late antigen-4 ($\alpha 4\beta 1$ Integrin) in the premetastatic niche. *Journal of Nuclear Medicine*, *53*, 779–786. https://doi.org/10.2967/jnumed.111.100073.

Siegel, R., Miller, K., & Jemal, A. (2017). Cancer statistics, 2017. *CA: a Cancer Journal for Clinicians*, *67*(1), 7–30. https://doi.org/10.3322/caac.21254.

Singh, D. Q., Minz, A. P., & Sahoo, S. K. (2017). Nanomedicine-mediated drug targeting of cancer stem cells. *Drug Discovery Today*, *22*(6), 952–959. https://doi.org/10.1016/j.drudis.2017.04.005.

Sleeman, J. P., Cady, B., & Pantel, K. (2012). The connectivity of lymphogenous and hematogenous tumor cell dissemination: biological insights and clinical implications. *Clinical & Experimental Metastasis*, *29*(7), 737–746. https://doi.org/10.1007/s10585-012-9489-x.

Sleeman, J. P., & Thiele, W. (2009). Tumor metastasis and the lymphatic vasculature. *International Journal of Cancer*, *125*(12), 2747–2756. https://doi.org/10.1002/ijc.24702.

Sneider, A., VanDyke, D., Paliwal, S., & Rai, P. (2017). Remotely triggered nano-theranostics for cancer applications. *Nanotheranostics*, *1*(1), 1–22. https://doi.org/10.1038/nbt.3121.ChIP-nexus.

Solinas, G., Schiarea, S., Liguori, M., et al. (2010). Tumor-conditioned macrophages secrete migration-stimulating factor: a new marker for M2-polarization, influencing tumor cell motility. *Journal of Immunology*, *185*(1), 642–652. https://doi.org/10.4049/jimmunol.1000413.

Sonali, Singh, R. P., Sharma, G., et al. (2016). RGD-TPGS decorated theranostic liposomes for brain targeted delivery. *Colloids and Surfaces B, Biointerfaces*, *147*, 129–141. https://doi.org/10.1016/j.colsurfb.2016.07.058.

Soroceanu, L., Gillespie, Y., Khazaeli, M. B., & Sontheimer, H. (1998). Use of chlorotoxin for targeting of primary brain tumors. *Cancer Research*, *58*(21), 4871–4879. http://cancerres.aacrjournals.org/content/58/21/4871.short.

Sounni, N. E., & Noel, A. (2013). Targeting the tumor microenvironment for cancer therapy. *Clinical Chemistry*, *59*(1), 85–93. https://doi.org/10.1373/clinchem.2012.185363.

Steg, A. D., Katre, A. A., Goodman, B., et al. (2012). Targeting the notch ligand Jagged1 in both tumor cells and stroma in ovarian cancer. *Clinical Cancer Research*, *17*(17), 5674–5685. https://doi.org/10.1158/1078-0432.CCR-11-0432.Targeting.

Steinbichler, T. B., Dudás, J., Riechelmann, H., & Skvortsova, I. I. (2017). The role of exosomes in cancer metastasis. *Seminars in Cancer Biology*, *44*, 170–181. https://doi.org/10.1016/j.semcancer.2017.02.006.

Sugiura, H., Fujiwara, Y., Ando, M., Kawai, A., et al. (2010). Multicenter phase II trial assessing effectiveness of imatinib mesylate on relapsed or refractory KIT-positive or PDGFR-positive sarcoma. *Journal of Orthopaedic Science*, *15*(5), 654–660. https://doi.org/10.1007/s00776-010-1506-9.

Sun, C., Fang, C., Stephen, Z., et al. (2008). Tumor-targeted drug delivery and MRI contrast enhancement by chlorotoxin-conjugated iron oxide nanoparticles. *Nanomedicine*, *3*(4), 495–505. https://doi.org/10.2217/17435889.3.4.495.

Sun, N., Wang, J., Ji, L., et al. (2015). A cellular compatible chitosan nanoparticle surface for isolation and in situ culture of rare number CTCs. *Small*, *11*(40), 5444–5451. https://doi.org/10.1002/smll.201501718.

Swain, S., Sahu, P. K., Beg, S., & Babu, S. M. (2016). Nanoparticles for cancer targeting: current and future directions. *Current Drug Delivery*, *13*(8), 1290–1302.

Takeishi, S., Matsumoto, A., Onoyama, I., Naka, K., Hirao, A., & Nakayama, K. I. (2013). Ablation of Fbxw7 eliminates leukemia-initiating cells by preventing quiescence. *Cancer Cell*, *23*(3), 347–361. https://doi.org/10.1016/j.ccr.2013.01.026.

Taketo, M. M. (2011). Reflections on the spread of metastasis to cancer prevention. *Cancer Prevention Research (Philadelphia, PA)*, *4*(3), 324–328. https://doi.org/10.1158/1940-6207.CAPR-11-0046.

Tang, J., Zhou, H., Liu, J., et al. (2017). Dual-mode imaging-guided synergistic chemo- and magnetohyperthermia therapy in a versatile nanoplatform to eliminate cancer stem cells. *ACS Applied Materials & Interfaces*, *9*(28), 23497–23507. https://doi.org/10.1021/acsami.7b06393.

Tennant, D. A., Durán, R. V., & Gottlieb, E. (2010). Targeting metabolic transformation for cancer therapy. *Nature Reviews Cancer*, *10*(4), 267–277. https://doi.org/10.1038/nrc2817.

Thakor, A. S., & Gambhir, S. S. (2013). Nanooncology: the future of cancer diagnosis and therapy. *CA: a Cancer Journal for Clinicians*, *63*(6), 395–418. https://doi.org/10.3322/caac.21199.

Tomei, A. A., Siegert, S., Britschgi, M. R., Luther, S. A., & Swartz, M. A. (2009). Fluid flow regulates stromal cell organization and CCL21 expression in a tissue-engineered lymph node microenvironment. *Journal of Immunology*, *183*(7), 4273–4283. https://doi.org/10.4049/jimmunol.0900835.

Trevaskis, N. L., Kaminskas, L. M., & Porter, C. J. H. (2015). From sewer to saviour-targeting the lymphatic system to promote drug exposure and activity. *Nature Reviews Drug Discovery*, *14*(11), 781–803. https://doi.org/10.1038/nrd4608.

Tsai, J. H., & Yang, J. (2013). Epithelial-mesenchymal plasticity in carcinoma metastasis. *Genes & Development*, *27*(20), 2192–2206. https://doi.org/10.1101/gad.225334.113.

Tseng, Y. C., Xu, Z., Guley, K., Yuan, H., & Huang, L. (2014). Lipid-calcium phosphate nanoparticles for delivery to the lymphatic system and SPECT/CT imaging of lymph node metastases. *Biomaterials*, *35*(16), 4688–4698. https://doi.org/10.1016/j.biomaterials.2014.02.030.

Tsujimoto, H., Morimoto, Y., Takahata, R., et al. (2015). Theranostic photosensitive nanoparticles for lymph node metastasis of gastric cancer. *Annals of Surgical Oncology*, *22*(December), 923–928. https://doi.org/10.1245/s10434-015-4594-0.

Tu, Y., Gao, X., Li, G., et al. (2013). MicroRNA-218 inhibits glioma invasion, migration, proliferation, and cancer stem-like cell self-renewal by targeting the polycomb group gene Bmi1. *Cancer Research*, *73*(19), 6046–6055. https://doi.org/10.1158/0008-5472.CAN-13-0358.

Van Cutsem, E., Cervantes, A., Adam, R., et al. (2016). ESMO consensus guidelines for the management of patients with metastatic colorectal cancer. *Annals of Oncology*, *27*(8), 1386–1422.

Van Zijl, F., Krupitza, G., & Mikulits, W. (2011). Initial steps of metastasis: cell invasion and endothelial transmigration. *Mutation Research, Reviews in Mutation Research*, *728*(1–2), 23–34. https://doi.org/10.1016/j.mrrev.2011.05.002.

Vaupel, P., Kallinowski, F., & Okunieff, P. (1989). Blood flow, oxygen and nutrient supply, and metabolic microenvironment human tumors: a review. *Cancer Research*, *49*(23), 6449–6465.

Veiseh, O., Gunn, J. W., Kievit, F. M., et al. (2009). Inhibition of tumor-cell invasion with chlorotoxin-bound superparamagnetic nanoparticles. *Small*, *5*(2), 256–264. https://doi.org/10.1002/smll.200800646.

Veiseh, O., Kievit, F. M., Ellenbogen, R. G., & Zhang, M. (2011). Cancer cell invasion: treatment and monitoring opportunities in nanomedicine. *Advanced Drug Delivery Reviews*, *63*(8), 582–596. https://doi.org/10.1016/j.addr.2011.01.010.

Völkel, T., Hölig, P., Merdan, T., Müller, R., & Kontermann, R. E. (2004). Targeting of immunoliposomes to endothelial cells using a single-chain Fv fragment directed against human endoglin (CD105). *Biochimica et Biophysica Acta, Biomembranes*, *1663*(1–2), 158–166. https://doi.org/10.1016/j.bbamem.2004.03.007.

Wan, L., Pantel, K., & Kang, Y. (2013). Tumor metastasis: moving new biological insights into the clinic. *Nature Medicine*, *19*(11), 1450–1464. https://doi.org/10.1038/nm.3391.

Wang, A. Z., Langer, R., & Farokhzad, O. C. (2012). Nanoparticle delivery of cancer drugs. *Annual Review of Medicine*, *63*(1), 185–198. https://doi.org/10.1146/annurev-med-040210-162544.

Wang, S., Zhang, Q., Yang, P., et al. (2016). Manganese oxide-coated carbon nanotubes as dual-modality lymph mapping agents for photothermal therapy of tumor metastasis. *ACS Applied Materials & Interfaces*, *8*(6), 3736–3743. https://doi.org/10.1021/acsami.5b08087.

Weber, G. F. (2008). Molecular mechanisms of metastasis. *Cancer Letters*, *270*(2), 181–190. https://doi.org/10.1016/j.canlet.2008.04.030.

Weidle, U. H., Birzele, F., Kollmorgen, G., & Rüger, R. (2017). The multiple roles of exosomes in metastasis. *Cancer Genomics & Proteomics*, *14*(1), 1–16. https://doi.org/10.21873/cgp.20015.

Weis, S. M., & Cheresh, D. A. (2011). Tumor angiogenesis: molecular pathways and therapeutic targets. *Nature Medicine*, *17*(11), 1359–1370. https://doi.org/10.1038/nm.2537.

West, K. R., & Otto, S. (2005). Reversible covalent chemistry in drug delivery. *Current Drug Discovery Technologies*, *2*(3), 123–160. https://doi.org/10.2174/1570163054866882.

WHO. (2017). *Cancer*. Geneva, Switzerland: WHO. http://www.who.int/topics/cancer/en/.

Wilhelm, S., Tavares, A. J., Dai, Q., et al. (2016). Analysis of nanoparticle delivery to tumours. *Nature Reviews Materials*, *1*, 1–12. https://doi.org/10.1038/natrevmats.2016.14.

Wong, C.C.-L., Gilkes, D. M., Zhang, H., et al. (2011). Hypoxia-inducible factor 1 is a master regulator of breast cancer metastatic niche formation. *Proceedings of the National Academy of Sciences of the United States of America*, *108*(39), 16369. https://doi.org/10.1073/pnas.1113483108.

Wong, C. W., Lee, A., Shientag, L., et al. (2001). Apoptosis: an early event in metastatic inefficiency. *Cancer Research*, *61*(1), 333–338.

Xie, J., Gao, Y., Zhao, R., et al. (2015). Ex vivo and in vivo capture and deactivation of circulating tumor cells by dual-antibody-coated nanomaterials. *Journal of Controlled Release*, *209*, 159–169. https://doi.org/10.1016/j.jconrel.2015.04.036.

Xing, F., Saidou, J., & Watabe, K. (2010). Cancer associated fibroblasts (CAFs) in tumor microenvironment. *Frontiers in Bioscience*, *15*, 166–179.

Yamamoto, M., Kikuchi, H., Ohta, M., et al. (2008). TSU68 prevents liver metastasis of colon cancer xenografts by modulating the premetastatic niche. *Cancer Research*, *68*(23), 9754–9762. https://doi.org/10.1158/0008-5472.CAN-08-1748.

Yang, L., Cheng, J., Chen, Y., et al. (2017). Phase-transition nanodroplets for real-time photoacoustic/ultrasound dual-modality imaging and photothermal therapy of sentinel lymph node in breast cancer. *Scientific Reports*, *7*(March), 45213. https://doi.org/10.1038/srep45213.

Yang, X., Hong, H., Grailer, J. J., et al. (2011). CRGD-functionalized, DOX-conjugated, and 64Cu-labeled superparamagnetic iron oxide nanoparticles for targeted anticancer drug delivery and PET/MR imaging. *Biomaterials*, *32*(17), 4151–4160. https://doi.org/10.1016/j.biomaterials.2011.02.006.

Yang, N., Jiang, Y., Zhang, H., et al. (2015). Active targeting docetaxel-PLA nanoparticles eradicate circulating lung cancer stem-like cells and inhibit liver metastasis. *Molecular Pharmaceutics*, *12*(1), 232–239. https://doi.org/10.1021/mp500568z.

Yigit, M. V., Ghosh, S. K., Kumar, M., et al. (2013). Context-dependent differences in miR-10b breast oncogenesis can be targeted for the prevention and arrest of lymph node metastasis. *Oncogene*, *32*, 1530–1538.

Yin, J., Hui, X., Yao, L., et al. (2015). Evaluation of Tc-99 m labeled dimeric GX1 peptides for imaging of colorectal cancer vasculature. *Molecular Imaging and Biology*, *17*(5), 661–670. https://doi.org/10.1007/s11307-015-0838-4.

Yoon, H. J., Kozminsky, M., & Nagrath, S. (2014). Emerging role of nanomaterials in circulating tumor cell isolation and analysis. *American Chemical Society*, *8*(3), 1995–2017. https://doi.org/10.1021/nn5004277.

Yu, J., Chen, Y., Zhang, Y., et al. (2014). pH-responsive and near-infrared-emissive polymer nanoparticles for simultaneous delivery, release, and fluorescence tracking of doxorubicin in vivo. *Chemical Communications (Cambridge, England)*, *50*(36), 4699–4702. https://doi.org/10.1039/c3cc49870k.

Zhang, D., Feng, X. Y., Henning, T. D., et al. (2009). MR imaging of tumor angiogenesis using sterically stabilized Gd-DTPA liposomes targeted to CD105. *European Journal of Radiology*, *70*(1), 180–189. https://doi.org/10.1016/j.ejrad.2008.04.022.

Zhang, S., Gong, M., Zhang, D., Yang, H., Gao, F., & Zou, L. (2014). Thiol-PEG-carboxyl-stabilized Fe2O3/Au nanoparticles targeted to CD105: synthesis, characterization and application in MR imaging of tumor angiogenesis. *European Journal of Radiology*, *83*(7), 1190–1198). https://doi.org/10.1016/j.ejrad.2014.03.034.

Zhou, M., Zhao, J., Tian, M., et al. (2015). Radio-photothermal therapy mediated by a single compartment nanoplatform depletes tumor initiating cells and reduces lung metastasis in the orthotopic 4T1 breast tumor model. *Nanoscale*, *7*(46), 19438–19447. https://doi.org/10.1039/C5NR04587H.

Zhu, J., Fu, F., Xiong, Z., Shen, M., & Shi, X. (2015). Dendrimer-entrapped gold nanoparticles modified with RGD peptide and alpha-tocopheryl succinate enable targeted theranostics of cancer cells. *Colloids and Surfaces B, Biointerfaces*, *133*, 36–42. https://doi.org/10.1016/j.colsurfb.2015.05.040.

Zhu, L., Kate, P., & Torchilin, V. P. (2012). Matrix metalloprotease 2-responsive multifunctional liposomal nanocarrier for enhanced tumor targeting. *ACS Nano*, *6*(4), 3491–3498. https://doi.org/10.1021/nn300524f.

Zhu, L., Zhou, Z., Mao, H., & Yang, L. (2017). Magnetic nanoparticles for precision oncology: theranostic magnetic iron oxide nanoparticles for image-guided and targeted cancer therapy. *Nanomedicine*, *12*(1), 73–87. https://doi.org/10.2217/nnm-2016-0316.

Zolata, H., Abbasi Davani, F., & Afarideh, H. (2015). Synthesis, characterization and theranostic evaluation of indium-111 labeled multifunctional superparamagnetic iron oxide nanoparticles. *Nuclear Medicine and Biology*, *42*(2), 164–170. https://doi.org/10.1016/j.nucmedbio.2014.09.007.

Zou, Z., He, X., He, D., et al. (2015). Programmed packaging of mesoporous silica nanocarriers for matrix metalloprotease 2-triggered tumor targeting and release. *Biomaterials*, *58*, 35–45. https://doi.org/10.1016/j.biomaterials.2015.04.034.

Zou, P., Yu, Y., Wang, Y. A., et al. (2010). Superparamagnetic iron oxide nanotheranostics for targeted cancer cell imaging and pH-dependent intracellular drug release. *Molecular Pharmaceutics*, *7*(6), 1974–1984.

CHAPTER 9

Targeted Nanotheranostics for Selective Drug Delivery in Cancer

Livia P. Mendes*,†, Eliana M. Lima‡, Vladimir P. Torchilin*

*Northeastern University, Boston, MA, United States †CAPES Foundation, Ministry of Education of Brazil, Brasilia, Brazil ‡Federal University of Goiás, Goiânia, Brazil

1 Introduction

The term "theranostics" was originally employed in the early 2000s as a reference to the combination of therapeutic and diagnostic into the same treatment system (Kelkar & Reineke, 2011). Nanotheranostics has been used to refer to the association of therapeutic and diagnostic agents in a nanostructured drug delivery system, which is frequently targeted by means of specific ligands (Fig. 1). The therapeutic agent can be a drug, siRNA, or other oligonucleotides, whereas the diagnostic agent is a molecule, such as an imaging contrast agent, that allows the visualization of diseased tissue with better resolution and specificity (Muthu, Leong, Mei, & Feng, 2014).

The concept of nanotheranostics, however, is much broader than that. The combination drug-imaging agent allows not only for simultaneous treatment and diagnosis of diseases at an early stage, but also leads to the development of therapeutic strategies using a rational approach in a controlled manner (Chen, Zhang, Zhu, Xie, & Chen, 2017; Jo, Ku, Won, Kim, & Kwon, 2016). These strategies have been widely applied to the field of oncology, adding to the benefits of the use of nanomedicines to overcome the drawbacks of conventional chemotherapy (Belyanina et al., 2017; Ma, Huang, Song, Chen, & Zhang, 2016).

Cancer is a complex disease with distinct manifestations in each patient. Despite some common genetic traces and symptoms, solid tumors are generally heterogeneous. This fact has been closely related to the failure of chemotherapy and the failed translation of the newly developed platforms to the clinic (Janku, 2014; Maeda, 2015). The medication will not work precisely in the same manner in all individuals. The use of nanotheranostics for personalized medicine also aims to prevent these failures (Chen, Zhang, et al., 2017).

Nanoparticles (NP) have been approved for clinical use in cancer treatment for over 20 years (Dawidczyk et al., 2014). Doxil was the first nanostructured formulation approved by the

Handbook of Nanomaterials for Cancer Theranostics. https://doi.org/10.1016/B978-0-12-813339-2.00009-8

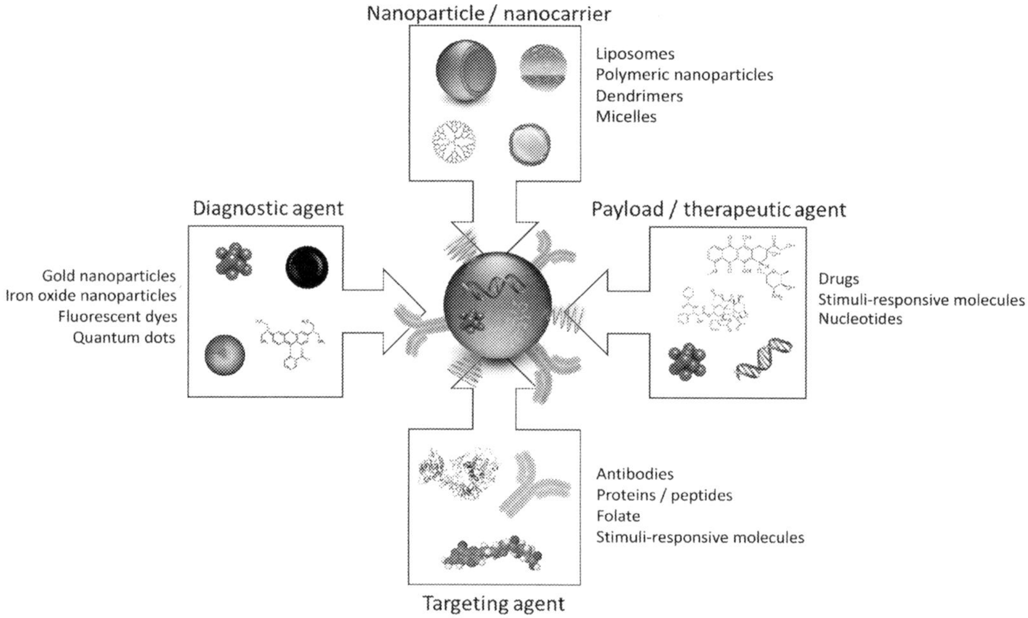

Fig. 1
Targeted nanotheranostics: the combination of various molecules in multifunctional nanocarriers enables their use in imaging diagnosis and therapeutic approaches.

FDA for the treatment of Kaposi's sarcoma in 1995, and later for metastatic ovarian cancer and multiple myeloma. They are comprised of polyethylene glycol (PEG) coated liposomes loaded with doxorubicin (DOX) (Barenholz, 2012). PEG coating imparts long-circulation properties (Suk, Xu, Kim, Hanes, & Ensign, 2016). This characteristic results in enhanced tumor accumulation and lower off-site effects, which decreases the adverse effects of doxorubicin such as cardiomyopathy (Barenholz, 2012; Yildirim et al., 2008). Doxil approval was followed by other nanomedicines for cancer therapy, including Abraxane (albumin-bound paclitaxel nanoparticles), Marqibo (vincristine sulfate liposomes), DaunoXome (daunorubicin citrate liposomes), and other DOX-containing liposomes (Calyx and Myocet) (Dawidczyk et al., 2014). The main effect introduced by the use of NP was the greater accumulation of drugs in the target tissue, which consequently decreased the side effects and enhanced the therapeutic response (Shi, Kantoff, Wooster, & Farokhzad, 2017). Other advantages of using nanoparticulate delivery systems include the possibility of the attachment of molecules to their surface to target specific receptors or trigger a biorecognition event, which would improve their specificity and their versatility and allow for the encapsulation of more than one payload to enhance the response through a synergistic effect (Mendes et al., 2014; Sriraman, Pan, Sarisozen, Luther, & Torchilin, 2016).

In addition to their use as drug delivery carriers, NP have also been explored as a tool for diagnosis due to their molecular imaging potential (Chen, Zhen, Todd, Chu, & Xie, 2013; Kang et al., 2017). Materials at the nanometer scale present exclusive features that improve signal amplification and quantification, promoting a better visualization of the target tissue (Toy, Bauer, Hoimes, Ghaghada, & Karathanasis, 2014). Furthermore, NP may contribute to overcoming common drawbacks related to conventional diagnostic methods, resulting in decreased toxic side effects and increase of the concentration of the imaging agents when associated to NP. Also, surface functionalization and prolonged circulation time contribute to increased tissue-targeted delivery (Estelrich, Sánchez-Martín, & Busquets, 2015). These characteristics, allied to controlled biodistribution that can be achieved with NP, highlight the advantages of this approach.

Advances in the oncology field and the discovery of biomarkers exclusively or preferentially expressed by tumor cells have helped improving tumor specificity and therapy outcome. Modifying the surface of NP with targeting moieties helps to overcome therapy drawbacks related to the side effects that arise from the similarities between healthy and tumor cells (Bazak, Houri, Achy, Kamel, & Refaat, 2015; Chitgupi, Qin, & Lovell, 2017).

2 Nanoparticles in Theranostics

Regardless of the composition of the various NP explored in cancer theranostics, the ultimate objectives are reaching the target tissue and achieving better efficacy of the therapy and diagnostic to improve pharmacokinetics and biodistribution of the delivered molecules to increase stability and circulation time in the blood (Ma, Huang, et al., 2016). With these purposes in mind, NP can be designed to target both therapeutic and imaging moieties, in an "all-in-one" platform. As demonstrated in Fig. 2, the number of studies with nanoparticles have significantly increased in the past decades, including those using theranostic.

Although there are no NP in the clinic allying both the diagnostic and the therapeutic functions, there are hundreds of clinical trials of nanomedicines in cancer, and we will present in Table 1 only those which relate to theranostics and are in advanced stages of developments.

Depending on the materials used to obtain the NP, they can be classified as organic or inorganic NP. In the past, mainly organic platforms have reached the market, although not within the scope of theranostic applications (Anselmo & Mitragotri, 2016).

2.1 Organic Platforms

Various drug-containing organic NP formulations, mainly liposomes, have been approved and made available in the clinic. A significant percentage is utilized in cancer therapy. In addition, their use as nanotheranostic devices has been widely explored because of their capacity to improve stability, bioavailability, and efficacy of contrast agents (Ma, Huang, et al., 2016). The encapsulation of contrast agents within NP allows their application in several diagnostic techniques, including magnetic resonance imaging (MRI) (Belyanina et al., 2017), ultrasound (US) (Martin & Dayton, 2013), nuclear imaging and fluorescence (Martelli, Dico, Diceglie, Lucignani, & Ottobrini, 2016).

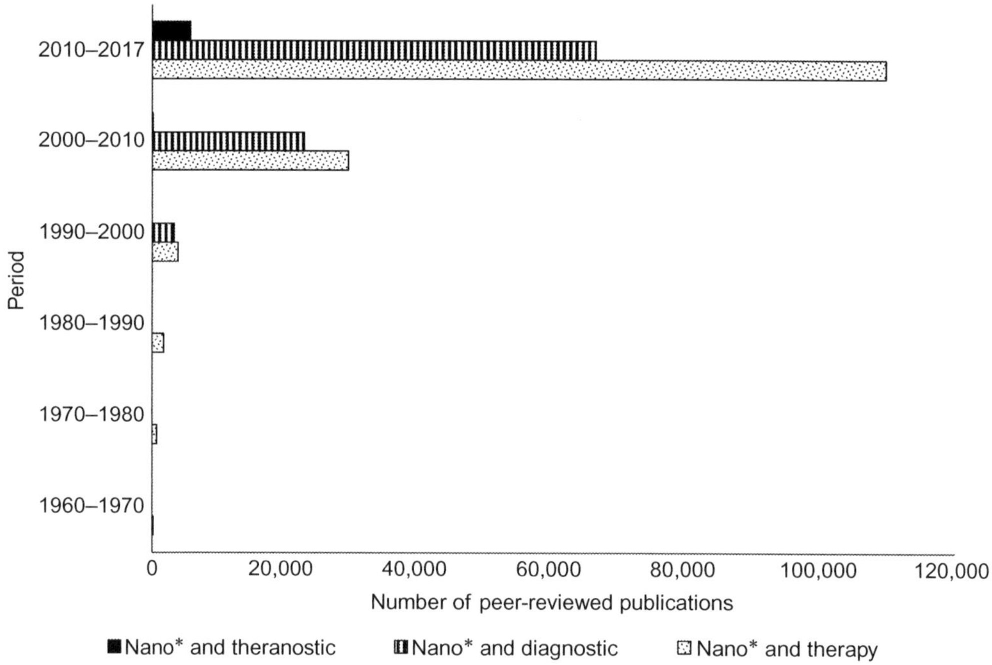

Fig. 2
Number of peer-reviewed publications crossing any form of nanoparticle or nanostructured device and therapy, diagnostics, or theranostics.

Table 1: Recent examples of advanced clinical studies related to nanotheranostics in cancer

Nanomedicine	Indication	Phase	Status of the Trial	ClinicalTrials. gov Identifier[a]
Vincristine liposome	CD20+ aggressive B-cell lymphoma	3	Recruiting	NCT01478542
ThermoDox (doxorubicin liposome)	Nonresectable hepatocellular carcinoma	3	Recruiting	NCT02112656
NBTXR3 (hafnium oxide nanoparticles)	Soft tissue sarcoma	2 and 3	Recruiting	NCT02379845
Feraheme (SPION)	Preoperative staging of pancreatic cancer	4	Completed	NCT00920023
Brentuximab vedotin (antibody-drug conjugate)	Refractory Hodgkin lymphoma	4	Ongoing	NCT01990534
Trastuzumab emtansine (antibody-drug conjugate)	HER2+ breast cancer	3	Ongoing	NCT01702571
Paclitaxel-incorporating micellar nanoparticle	Metastatic or recurrent breast cancer	3	Completed	NCT01644890
Nab-paclitaxel	Stage IIIC and IV breast cancer	3	Ongoing	NCT00785291
Doxorubicin immunoliposome	Triple negative EGFR+ breast cancer	2	Recruiting	NCT02833766
AuroShell (gold/silica NP)	Prostate cancer ablation	–	Recruiting	NCT02680535

Table 1: Recent examples of advanced clinical studies related to nanotheranostics in cancer—cont'd

Nanomedicine	Indication	Phase	Status of the Trial	ClinicalTrials.gov Identifier[a]
Silica nanoparticles	Real-time image-guided intraoperative mapping of nodal metastases	1 and 2	Recruiting	NCT02106598
^{89}Zr-nanocolloidal albumin	Preoperative imaging of sentinel lymph node in colon cancer	2 and 3	Completed	NCT02850783
^{89}Zr-cetuximab	Image-guided treatment optimization of metastatic colorectal cancer	1 and 2	Recruiting	NCT02117466
AGuIX-NP (polysiloxane Gd-chelates based NP)	Brain metastases	1	Recruiting	NCT02820454
Ferumoxtran-10 (ultra-small superparamagnetic iron oxide)	MRI to detect lymph node metastases in rectal and breast cancer	3	Recruiting	NCT02751606

[a] Information retrieved from clinicaltrials.gov accessed on November 1st, 2017.

2.1.1 Lipid-based nanoparticles

Liposomes are NP comprised of phospholipid bilayers that self-assemble in an aqueous medium to form vesicles with an aqueous interior (Aryasomayajula, Salzano, & Torchilin, 2017; Sriraman & Torchilin, 2014). Due to their great versatility, they are the most widely investigated nanosystem for drug delivery. Lipids in liposomes are generally biocompatible and biodegradable, with very limited or no toxicity. They can entrap hydrophilic molecules into their aqueous interior, while certain quantity of hydrophobic molecules can be entrapped into their bilayer lipid membrane. The versatility of liposomes include their relatively easy surface modification that improves their stability, and increases their blood circulation time and targetability to specific receptors (Deshpande, Biswas, & Torchilin, 2013; Eloy, Petrilli, Trevizan, & Chorilli, 2017; Pattni, Chupin, & Torchilin, 2015). Liposomes can enhance apparent drug solubility, allowing for drug delivery in safer vehicles once encapsulated in liposomes (Hoogevest, Leigh, & Fahr, 2013).

Liposomes have been successfully used as carriers for drugs and imaging agents (Ren et al., 2015; Xing, Hwang, & Lu, 2016). Encapsulating imaging probes into targeted liposomes leads to advantages due to favorable liposome features, notably lower toxicity compared to the administration of other commonly used contrast agents, such as gadolinium (Martínez-González, Estelrich, & Busquets, 2016), and even greater accumulation in the tissue of interest (Erdogan & Torchilin, 2017). Additionally, some imaging probes, for instance, superparamagnetic iron oxide nanoparticles (SPIONs), may be encapsulated in liposomes to improve their stability for biological application (Cintra et al., 2009) and can be used as a trigger mechanism for drug release (Shah et al., 2016). These nanoparticles, known as magnetoliposomes, are potential carriers in nanotheranostics because they ally the advantages provided by the drug delivery ability of liposomes and the diagnostic capacity of SPIONs (Shirmardi Shaghasemi, Virk, & Reimhult, 2017).

Solid lipid nanoparticles (SLN) are a different type of lipid-based nanosystem. They are colloidal systems made of a solid lipid matrix and can be used as carriers for hydrophobic and hydrophilic molecules either by internalization or adsorption on the surface of the nanoparticle (Geszke-Moritz & Moritz, 2016; Muller, Mader, & Gohla, 2000). Although less explored than liposomes for diagnostic imaging strategies, they have been studied in association with magnetic nanoparticles to promote localized accumulation by application of a magnetic field (Grillone et al., 2015; Zhao, Zhang, Han, Wang, & Yang, 2015) and for magnetic hyperthermia (Muñoz de Escalona, Sáez-Fernández, Prados, Melguizo, & Arias, 2016).

Even though no lipid-based formulation has been approved for theranostics applications, they show good potential, especially when the possibility of utilizing them together with antitumor drugs to obtain a controlled release pattern is considered.

2.1.2 Polymeric nanoparticles

Polymeric nanoparticles are also very popular delivery systems and some of those have been approved as drug carriers in cancer therapy (Bobo, Robinson, Islam, Thurecht, & Corrie, 2016). The advantages of their use are similar to other organic nanoparticles and are related to improved characteristics in the drug delivery process, including biocompatibility, loading rate, controlled release, and modified biodistribution and pharmacokinetics (Li, Li, Mo, & Dai, 2017). The term polymeric nanoparticles represent several types of devices. NP obtained with synthetic polymers include nanocapsules and nanospheres, polymeric micelles, drug-polymer conjugates, dendrimers, polymersomes, and polyplexes (Prabhu, Patravale, & Joshi, 2015). They can also be obtained from natural polysaccharides, including dextran, chitosan, and alginate (Wang, Niu, & Chen, 2014). The most common FDA-approved polymers utilized to prepare NP are poly-(D,L-lactic-co-glycolic acid)-(PLGA), polylactic acid (PLA), and PEG.

Nanocapsules (NC) and nanospheres (NS) are similar systems. The former consists of a polymeric shell with an oil or aqueous core. The latter is a polymeric matrix platform. These are the most explored polymeric NP for drug delivery due to their versatility and biocompatibility and the fact they can be obtained by simple methods, especially when using presynthesized polymers using self-assembly methods, including nanoprecipitation, emulsification/solvent evaporation, and salting-out methods (Mora-Huertas, Fessi, & Elaissari, 2010). The method used to obtain the NPs and their polymer composition allows control of their properties such as size and release kinetics of the encapsulated agent (Kamaly, Yameen, Wu, & Farokhzad, 2016), which can be advantageous when applied for controlled release from theranostics. Polymeric nanocapsules have been designed to encapsulate cyanine green dye (IR820), paclitaxel (PTX) and SPIONs for multifunctional theranostics (Liao et al.,

2017). A magnetic field applied externally can be used to target the tumor due to the presence of SPIONs. IR820, useful as an imaging contrast agent for NIR imaging and MRI, also served as a photosensitizer. Balb/C mice bearing 4T1 breast cancer xenograft tumors treated with NC-SPIONs-IR820-PTX and an external stimulus of a magnetic field showed higher inhibition of tumor growth even when compared to the group without the stimulus, highlighting the synergism of the photochemotherapy and targeting ability via magnetic field.

Dendrimers are branched polymeric NP with small uniform sizes, biocompatibility, and nonimmunogenicity. They present mostly strong positive charges, which confers on them great usefulness as transfection agents, and the surface-containing terminal groups that allow functionalization with various groups, including antibodies, peptides, folate, and other targeting molecules (Palmerston Mendes, Pan, & Torchilin, 2017). They have been widely explored as drug and gene carriers, but also as theranostic NPs because the presence of the surface functional groups favors the encapsulation or conjugation of a wide variety of molecules (Ma, Mou, Wang, Zhu, & Yan, 2016; Wei et al., 2016). The administration of dendrimers entrapping gold NP and complexing siRNA, for instance, allows computed tomography (CT) imaging of tumors. Administration of photothermal therapy allied to gene therapy, potentiates the outcome through the synergism of both approaches (Wei et al., 2016; Zhu, Fu, Xiong, Shen, & Shi, 2015).

Unlike matrix systems such as NC, NS, and dendrimers, polymer-drug conjugates are structures where the polymer is covalently conjugated to a drug or imaging agent to work as the delivery agent (Wang et al., 2014). The polymer backbone provides the platform for drug conjugation and also for functionalization of targeting moieties. The high molecular weight of these structures favors their long circulation time and an EPR effect, and thus an increased accumulation in the tumor (Larson & Ghandehari, 2012; Xu, Ma, et al., 2015). Proteins like albumin, as naturally occurring polymers, are used to prepare this type of nanosystems, because they promote increased circulation half-life and improved intracellular delivery of the drugs (Larsen, Kuhlmann, Hvam, & Howard, 2016). Abraxane is a well-known example of a paclitaxel-albumin conjugate that is now employed in the clinic (Hirsh, 2014; Micha, Goldstein, Birk, Rettenmaier, & Brown, 2006). In a preclinical study, Wang et al. (2017) showed the potential of albumin-based NP loaded with gadolinium as a contrast agent to enhance MRI imaging, combined with their good biocompatibility in vivo.

2.2 Inorganic Nanoparticles

Inorganic NP have shown promising applications for both cancer diagnosis and therapy (Anselmo & Mitragotri, 2015), but due to toxicity concerns, organic nanoplatforms have usually been preferred (Rajagopalan & Yakhmi, 2017). Nevertheless, their relatively simple synthesis and versatility give these nanodevices great potential in theranostics. Examples of inorganic NPs include iron oxide NPs, mainly superparamagnetic NPs (SPIONs), gold NPs

(AuNP), quantum dots (QD), and silica-based NPs (Anselmo & Mitragotri, 2015; Ehlerding, Chen, & Cai, 2016; Wang, Li, Cheng, & Yuan, 2016). Inorganic NPs have been explored widely as imaging contrast agents, mainly for MRI and positron emission tomography (PET) scans, and also in association with other conventional drugs for theranostic purposes. In addition, some of these NPs can also be utilized as therapeutic agents themselves. In specific therapies, for instance, hyperthermia and photothermal ablation, AuNPs and SPIONs are used to simultaneously enhance the imaging contrast and promote cell death upon application of an extrinsic stimulus, such as radiation or a magnetic field (Abadeer & Murphy, 2016; Ahmad, Fu, He, & Li, 2016; Fernandes Rodrigues et al., 2017). Recent studies have associated inorganic NPs with organic ones, such as liposomes (magnetoliposomes) (Cintra et al., 2009; Ferreira et al., 2016; Guo et al., 2017) and polymeric NPs (Costa Lima, Gaspar, Reis, & Duraes, 2017; Mosafer, Abnous, Tafaghodi, Mokhtarzadeh, & Ramezani, 2017; Oliveira et al., 2012) to combine them with other therapeutic agents to achieve a theranostic outcome.

SPIONs are one of the most explored inorganic NPs in the field of theranostics (Li, Nejadnik, & Daldrup-Link, 2017; Yen, Padmanabhan, & Selvan, 2013). They are biocompatible and biodegradable, and their superparamagnetic characteristics enable tunable control over their behavior (Teleki, Haufe, Hirt, Pratsinis, & Sotiriou, 2016). The magnetic moment of SPIONS becomes aligned under the influence of a magnetic field, but no residual magnetism remains when the magnetic field ceases, unlike other ferromagnetic materials. Magnetite (Fe_3O_4) and maghemite (Fe_2O_3) are the most common forms of iron oxide nanoparticles (Revia & Zhang, 2016). They are currently being used as imaging agents, mainly for MRI (Jin, Lin, Li, & Ai, 2014), and as therapeutic molecules in hyperthermia procedures (Quinto, Mohindra, Tong, & Bao, 2015). When an alternating magnetic field is applied to these NPs, the temperature in the area rises, and the heat generated in the tumor is able to kill the cancer cells (Bañobre-López, Teijeiro, & Rivas, 2013; Dennis & Ivkov, 2013).

QDs are semiconductor nanocrystals with fluorescent properties that have unique optical and electronic properties. They are small NPs, ranging between 3 and 10 nm, and can be used in imaging diagnosis due to their strong fluorescence, with a wide emission spectrum (the fluorescence spectra depend on their size), especially in the near-infrared (NIR) region (wavelength >700 nm). These features promote more stability to the NPs than other organic fluorophores (Selvan & Narayanan, 2016) and allow imaging utilizing optical approaches (Ho & Leong, 2010). The surfaces of QDs can be modified to enhance their stability and biocompatibility, modify their solubility properties and functionalization to provide better specificity for targets (Michalet et al., 2005).

AuNPs have been used in cancer therapy and diagnosis for many years. Their biocompatibility, optical properties, small size, easy synthesis, and convenience for functionalization make them promising NPs for therapeutic and theranostic applications. A wide variety of NPs have been obtained, including nanospheres, nanorods, nanoshells, and nanocages (Abadeer & Murphy, 2016).

For therapy, they have been used as photothermal agents. Once excited by specific light, they generate heat that kills the cancer cells (Yeo et al., 2017), a similar process of ablation imparted by SPIONs with hyperthermia. Gold NPs can be functionalized to achieve passive and active targeting, to promote enhanced tumor accumulation. AuNPs are employed as X-ray contrast agents, mainly in computerized tomography for a higher X-ray attenuation (Cole, Ross, Tilley, Vargo-Gogola, & Roeder, 2015).

Silica-based NPs, as well as most inorganic NPs, are mostly still in the preclinical stages of development, in spite of showing promising results as drug delivery systems, with low toxicity and favorable biocompatibility (Mai & Meng, 2013). Mesoporous silica NPs (MSNP) are the most explored silica-based nanodevices due to their high loading capacity. The porous structure of MSNPs allows a high-yield encapsulation of hydrophobic and hydrophilic molecules, including drugs and imaging agents (Bharti, Nagaich, Pal, & Gulati, 2015). Nonporous NPs are also used in drug delivery and allow conjugation of drugs and functionalization due to their high surface area (Xu, Ma, et al., 2017). Cornell dots (C-dots) are PEGylated silica-based NPs currently in clinical trials for use in cancer diagnosis (Table 1). They encapsulate a NIR fluorophore and a radioisotope to enhance the signal for PET scans. In addition, these NPs were functionalized with a target ligand (cRGDY), which promoted higher accumulation in $\alpha\nu\beta3$-integrin expressing tumors and consequently improved the target-to-background ratio compared to free dyes (Anselmo & Mitragotri, 2015; Ehlerding et al., 2016).

3 Diagnostic Approach

Diagnosis of cancer takes into consideration several factors, including the patient's symptoms, medical history, and physical exam. Nevertheless, the confirmation can be obtained by performing laboratory tests, biopsy, and imaging procedures (National Cancer Institute, 2015). Biomedical imaging allows not only detection but also estimation of the stage of the disease and the prognosis for the patient. Furthermore, during treatment, imaging procedures facilitate monitoring the response to therapy, indicating the effectiveness of the chosen strategy (Fass, 2008). The techniques currently available for molecular imaging include MRI, CT, US, single-photon emission computed tomography (SPECT), PET, and optical imaging (Frangioni, 2008; Janib, Moses, & Mackay, 2010). Early diagnosis is pivotal to a positive outcome, and for this reason, these imaging techniques contribute significantly to successful therapy. The main advantages and drawbacks of these modalities will be presented, and their applications in nanotheranostics are discussed further.

3.1 Magnetic Resonance Imaging

MRI has been widely explored as an imaging tool due to its many advantages. It is noninvasive, has high spatial resolution, no penetration depth limits and real-time responsiveness (Baetke, Lammers, & Kiessling, 2015). It is also believed to be safer than other imaging modalities that expose the patient to X-rays because it operates with nonionizing radiation. The technique is based on a process in which a sample is submitted to a magnetic field. The hydrogen nuclei from water molecules emit a signal upon application of radio-frequency pulses. The signal is then captured and converted into an image (Bushong & Clarke, 2014). MRI can be performed without the use of contrasts, but these compounds have the ability to increase the differentiation between tissues and thus the resolution of the images. The most common agents used as contrast enhancers used in the clinic are gadolinium and manganese (Janib et al., 2010). The main problem related to the utilization of these contrast agents is that they must be utilized at high concentrations to achieve enhanced signal, and that can lead to considerable toxicity related to their accumulation in the organs (Chen et al., 2015; Janib et al., 2010). SPIONs, although not yet approved as contrast agents, have also been employed as such in clinical trials because of their promising safety profile, as compared to gadolinium and other chelating imaging agents (Jin et al., 2014; Kim et al., 2016). Other disadvantages of MRI include high cost and excessive time consumption (Wang, 2017).

3.2 Computed Tomography

Compared to MRI, CT is a quick and low-cost procedure. Features similar to MRI include high spatial resolution and unlimited penetration depth, with good signal quantification when contrast agents are employed (Ma et al., 2017). The technique consists of the measurement of the absorption of X-rays that pass through the tissues, allowing cross-sectional tomographic images. The difference in the tissue density will influence the X-ray absorption, where the more dense tissue will incur in higher absorption (Kalender, 2011; Lusic & Grinstaff, 2013). Low molecular weight agents are typically used as contrasts, such as iodine-, gold-, bismuth sulfide-, and lanthanide-based materials. All of these are currently being clinically employed (Lusic & Grinstaff, 2013). The major drawbacks of X-ray CT scans are the exposure to radiation and the low soft tissue contrast without the use of imaging agents. Additionally, the low molecular weight contrasts commonly used are rapidly cleared from the body, which can impair the imaging evaluation (Baetke et al., 2015). The encapsulation of these contrast agents in lipid or polymer NP can overcome this obstacle due to their longer blood circulation characteristic (Liu, Ai, & Lu, 2012; Naha et al., 2014). PEGylated liposomes prepared with iodinated CT contrasts and a NIR fluorescent dye enabled CT imaging the characterization of the heterogeneity of the distribution of this nanopreparation within the tumor (Ekdawi et al., 2015). This strategy can be used in theranostic approaches to predict the outcome of therapies using similar liposomal formulations.

3.3 Ultrasound

The use of US as an imaging method for diagnosis is widespread because of several advantages, including its low cost, noninvasiveness, capacity to penetrate deep into the tissues, real-time imaging, fast examination, high temporal and spatial resolution, and nonexposure to ionizing radiation. Additionally, the fact that it uses a portable device facilitates its application (Martin & Dayton, 2013). Although it does not provide a "big picture," as does CT or MRI modalities, because it is only able to evaluate small areas at a time, its good resolution encourages its use when contrast agents are not indicated due to patient's condition, such as hypersensitivity reactions or renal failure.

3.4 Single-Photon Emission Computed Tomography and Positron Emission Tomography

SPECT and PET are both imaging techniques that utilize the gamma rays originated from the decay of radioactive sources in the body. Thus, these modalities require the administration of radionuclides to generate the signal for the images (Velikyan, 2012). Among the advantages of these imaging approaches are sensitivity, noninvasiveness, little background noise, and the generation of three-dimensional images in a real-time manner. PET is more sensitive than SPECT and is many times associated with other techniques like CT to obtain anatomical contrast (Satterlee & Huang, 2016). Both techniques are extremely useful not only for the early detection of the disease, but also to monitor the progress of therapy. One of the main disadvantages of these techniques is the half-lives of the radionuclides, which are usually short, limiting the time of analysis, and the exposure to radiation. Common radioisotopes for PET are ^{18}F, ^{15}O, ^{13}N, and ^{11}C, while for SPECT, they are ^{99m}Tc, ^{131}I, ^{123}I, and ^{111}In (Key & Leary, 2014). NPs provide a great platform for the attachment of radioisotopes to be used for early diagnosis of diseases (Abadjian, Choi, & Anderson, 2017) and allow the combination of in vivo imaging and drug delivery (Chakravarty, Chen, Dash, & Cai, 2017). Furthermore, the use of radiolabeled NPs enhances accumulation of the radioisotopes in the tumor by EPR effect, reducing the amount of these agents in nontarget tissues, generating an even higher sensitivity to the techniques (Kunjachan, Ehling, Storm, Kiessling, & Lammers, 2015).

3.5 Optical Imaging

Optical imaging is an approach that allows noninvasive imaging of the body based on photons emitted from bioluminescent or fluorescent probes. The fact that there is no need for the use of radiation is one of its main advantages, along with being an affordable technique and having a good spatial resolution in the visible and near-infrared (NIR) spectra. However, its biggest limitation lies in poor penetration into the tissues and high noise/background due to the light absorption and autofluorescence of the tissues. Probes that allow optical imaging include AuNP, QD, fluorescent molecules such as Cy5.5 and Cy 7, and Luciferase, a widely employed reporter gene with bioluminescent detection (Martelli et al., 2016). AuNPs, for instance, are inorganic NPs that can be used in their different sizes and structures as direct probes for optical imaging due to their surface plasmon absorption and scattering properties (Huang & El-Sayed, 2010). Additionally, they can be assembled in other nanocarriers to improve their stability, functionalization, and targeting ability, and they hold the possibility of association with therapeutic agents for imaging-guided therapy (Yang, Song, et al., 2017; Zhong, Yang, Wen, & Xing, 2016). The same approach is possible with QDs (Alibolandi et al., 2016; Liu et al., 2016) and fluorescent molecules (Choi et al., 2012).

4 Targeting Tumors

At an initial stage, basically all NPs relied on the enhanced permeability and retention (EPR) effect to reach the tumor microenvironment. This effect, first reported by Maeda (Maeda, 2001; Matsumura & Maeda, 1986), is due to the characteristics presented by the newly developed vasculature around the tumor. As a tumor grows, new vessels are formed to supply the nutrients required for further growth. Their rapid development comes with gaps between the endothelial cells, resulting in a leaky vasculature through which macromolecules and NPs may escape. Healthy tissues, in general, do not present these gaps, which makes the penetration of NPs more difficult. Thus, NPs preferentially accumulate in tumors, rather than in healthy tissues (Maeda, 2015). To achieve the EPR effect, however, NPs need to circulate longer and avoid being captured and cleared from the circulation by the mononuclear phagocytic system (MPS). PEGylation has been the gold-standard procedure imparting stealth properties to the NPs, which allows them to avoid the MPS, increases their blood circulation time, and increases their chances of reaching the tumor (Jokerst, Lobovkina, Zare, & Gambhir, 2011; Suk et al., 2016). This strategy is referred to as passive targeting, and although extensively utilized, it can enhance specificity when combined with an active targeting approach (Xu, Ho, Zhang, Bertrand, & Farokhzad, 2015).

Actively targeted NPs exhibit higher specificity to the targeted tissue. For cancer treatment, the modification of NP usually involves the use of antibodies or antibody fragments, peptides, aptamers, small molecules, and proteins, whose targets are overexpressed in tumor cells (Bazak et al., 2015). Within this scope, promising targets include tumor vasculature (integrins) (Marelli, Rechenmacher, Sobahi, Mas-Moruno, & Kessler, 2013), folate (Peres-Filho et al., 2017; Zwicke, Mansoori, & Jeffery, 2012) and transferrin receptors (Daniels et al., 2012), CD44 (Arabi, Badiee, Mosaffa, & Jaafari, 2015), vascular endothelial growth factor (VEGF) (Ambasta, Sharma, & Kumar, 2011), and human epidermal growth factor receptor 2 (HER2) (Tai, Mahato, & Cheng, 2010). Another strategy to improve the specificity of the treatment is the development of NPs that can be triggered to release their payload as a response to different stimuli, which may be intrinsic to the tumor microenvironment (presence of enzymes overexpressed by the tumor, pH changes, and redox conditions) or external (use of hyperthermia, ultra-sound, and magnetic field) (Torchilin, 2014).

Both VEGF and $\alpha v\beta 3$ integrin are usually overexpressed in cancer cells and play a crucial role in tumor angiogenesis (Somanath, Malinin, & Byzova, 2009), turning them into useful targets for nanotheranostics. The peptide sequence Arg-Gly-Asp (RGD) was found to bind to $\alpha v\beta 3$ and other integrins and has been used in active targeting strategies, mainly in the cyclic form due to its higher stability (Danhier, LE Breton, & Préat, 2012). Radiolabeled RGD peptides can be used as PET tracers and may be utilized in the clinic for tumor staging, selection for antiangiogenic therapy and monitoring the treatment response (Chen, Niu, Wu, & Chen, 2016). When used as NP surface modification, RGD imparts targetability toward integrin-overexpressing cancer cells and improves NP internalization of them, thus optimizing the therapy or the specificity and signaling intensity for imaging strategies (Amin et al., 2015; Wu, Onodera, et al., 2017; Yoshimoto, Hayakawa, Yamaguchi, Kimura, & Fujii, 2016). Similar therapeutic and imaging outcomes can be achieved with the use of anti-VEGF antibodies, such as bevacizumab, to decorate and actively target NP (Abakumov et al., 2015; Nevala et al., 2016; Zhu et al., 2017). Another monoclonal antibody (mAb) extensively used in active targeting approaches is trastuzumab (TZ), which binds to HER2 receptors, increasing the specificity of the nanocarrier for cancer cells overexpressing this receptor (Cai et al., 2016, 2017; Espelin, Leonard, Geretti, Wickham, & Hendriks, 2016; Steinhauser, Spänkuch, Strebhardt, & Langer, 2006).

The use of transferrin (Tf) receptors as a target for cancer therapy is of great interest because its expression by normal cells is low compared to the levels in the tumor (Daniels et al., 2012). In a distinctive approach, Tf was used both as a targeting ligand and drug carrier when conjugated to IR780, a hydrophobic NIR dye (Wang, Zhang, et al., 2016). In fact, other protein-based platforms, mainly albumin ones, have been previously used as drug carriers (Kratz, 2014; Schettini, Giuliano, DE Placido, & Arpino, 2016). Wang, Zhang, et al. (2016) conjugated IR780 iodide with Tf for simultaneous photodynamic (PDT) and photothermal (PTT) therapies, and results showed great efficacy in Ct26 colon cancer cells. Because IR780 can also be used as a NIR probe for imaging, it allowed evaluation of in vivo biodistribution of NPs, showing that higher tumor accumulation was achieved 48 h after intravenous administration. Normal cells (fibroblasts), however, presented lower uptake of the NP, indicating the potential of such NP in an imaging-guided theranostic use.

The cellular uptake of nanoparticles via specific biorecognition mechanisms might be strongly affected by the adsorbed protein corona (Dutta et al., 2007). The loss of selectivity in receptor-mediated endocytosis has been demonstrated following protein adsorption on targeted silica nanoparticles (Salvati et al., 2013). The protein corona on the surface of nanoparticles might also affect immunogenicity and drug release properties (Obst et al., 2017). Tissue uptake and disposition of nanoparticles are key parameters for their effectiveness as nanotheranostics, including circulation time and clearance, because reaching the target is a pivotal step. Surface hydrophilicity appears to have a key impact on the early clearance of nanoparticles, but other circulating proteins, such as apolipoproteins, also seem to have an effect on nanoparticle circulation time and clearance mechanisms (Bertrand et al., 2017), impacting their performance as theranostics nanodevices.

5 Stimuli-Responsive Nanoparticles in Theranostics

Besides the ability to combine drugs and imaging agents, other benefits can be explored with nanosystems, including the preparation of stimuli-sensitive devices, which are activated only in the presence of a specific trigger. This approach has been successfully used in the drug delivery field to improve and facilitate the interaction with cells to enhance the intracellular delivery of the drugs (Torchilin, 2014). The tumor microenvironment is one of the most explored regions, the intrinsic characteristics of which serve as stimuli to cause the release of a payload from NPs (Du, Lane, & Nie, 2015). The characteristics used as stimuli include higher acidity, hypoxic conditions, and the presence of enzymes overexpressed in most solid tumors, such as matrix metalloproteinases (MMP) (Chen, Dai, et al., 2017; Zhu & Torchilin, 2013). For instance, temperature and pH-responsive polymeric NPs modified with folic acid (FA) for tumor targeting were prepared to deliver DOX and mixed ferrite NP (MFNP). The presence of FA increased the receptor-mediated endocytosis of NPs in folate-receptor overexpressing cancer cells, and the acidic pH in the endosomes triggered the release of DOX and MFNP. This theranostic approach favored the promotion of a cytotoxic effect by the drug, whereas the MFNPs increased the signal for MRI analysis (Bhattacharya et al., 2016).

External stimuli are also very efficient and include the application of light, temperature, irradiation, magnetic field, or ultrasound (Manouras & Vamvakaki, 2017; Zardad et al., 2016). The incorporation of MNPs in organic nanoparticles, for example, can be utilized as a strategy to trigger the release of the drug upon the exposure to a magnetic field, allowing the precise control of the release kinetics of the drug (Shirmardi Shaghasemi et al., 2017).

With theranostics, the advantages of these tunable nanosystems can be expanded. Some NPs generally used as imaging agents, and sometimes even as the therapeutic molecule, can also serve as the trigger mechanism for the release of the drug. Baek et al. (2016) developed a theranostic nanocarrier with both pH and temperature responsiveness containing a core of Au nanorods, coated by MSNP loaded with DOX and a final protection with poly(*N*-isopropylacrylamide)-based *N*-butyl imidazolium copolymer, resulting in the hybrid system Au@mSi-DOX@P. By utilizing a NIR light source, the thermo-responsive polymer allows the controlled release of DOX and the activation of photothermal therapy due to the presence of Au nanorods. The AuNPs also worked as contrast agents to allow imaging of the tumor with X-ray CT. Zhao, Yang, Chen, and Yan (2017) also developed a AuNP-based targeted theranostic nanosystem with stimuli responsiveness, but without the association of a chemotherapeutic agent. In this study, Au nanorods (AuNRs) were functionalized with asymmetric cyanine (Acy) for fluorescence imaging via an MMP-sensitive peptide and with glycosyl for tumor targeting. When released in the tumor microenvironment due to the slightly acidic pH of 6.0, upon application of NIR light, Acy emitted a fluorescent signal, whereas in neutral pH the fluorescence was quenched, avoiding a false positive signal from nontumor tissues. The laser irradiation was also used for activating the AuNRs to induce the photothermal therapy. In a different approach, Zhang et al. (2014) utilized NIR laser irradiation as a targeting strategy to enhance tumor accumulation of thermoresponsive polymer-encapsulated AuNRs. The NP was also loaded with DOX, which was released with the increase in temperature enabled by the laser irradiation of AuNRs.

MNP are also used in stimuli-sensitive nanosystems to trigger drug release from temperature-responsive materials upon application of an alternating magnetic field (AMF). In addition to triggering the drug release, the increase in temperature can cause tumor cell death by thermal ablation (Ling & Hyeon, 2017). Yang, Wang, et al. (2017) have developed PLGA implants loaded with cisplatin and magnetite (Fe_3O_4) for synergistic magnetic thermal ablation and cytotoxic treatment and evaluated their efficacy in xenograft hepatic carcinoma in rabbits. The Fe_3O_4 NP allowed US and CT visualization of the tumors pre- and postinjection and after the thermal ablation, highlighting the advantage of using a nanotheranostic system for real-time guidance and monitoring of the treatment. The efficacy of the chemohyperthermia ablation strategy was superior to that of strategies utilizing cisplatin or hyperthermia alone.

Although chemotherapy is one of the most widely used nonsurgical anticancer treatments, recent approaches involving phototherapy have shown appealing results (Rajkumar & Prabaharan, 2017; Wu, Mao, et al., 2017). PTT and PDT are treatments in which a photosensitizer (PS) (a nontoxic molecule) is administered to the patient, who is then exposed to irradiation with light at a specific wavelength. The light source excites the PS, which will then cause damage to the target tissue (Fig. 3). In the case of PTT, this damage is caused due to the heat promoted by the excitation (de Melo-Diogo, PAIS-Silva, Dias, Moreira, & Correia, 2017), whereas in PDT it is derived from the generation of reactive oxygen species (Spring, Rizvi, Xu, & Hasan, 2015).

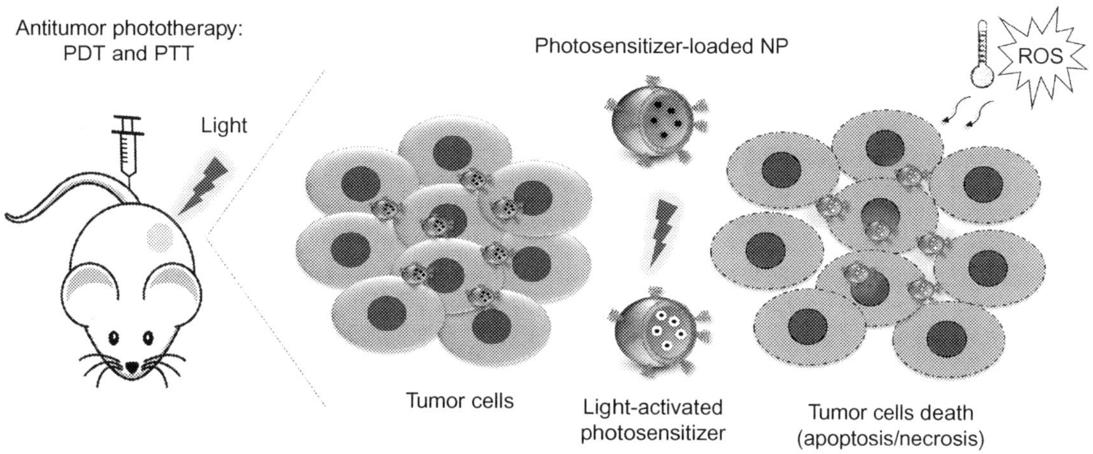

Fig. 3
Schematic representation of the mechanism of action of PDT and PTT on tumors.

Besides their therapeutic outcome, a variety of molecules used for this therapy allow imaging using different techniques, including CT scans (Curry, Kopelman, Shilo, & Popovtzer, 2014) and optical imaging (Mallidi, Spring, & Hasan, 2015) and are, therefore, ideal for theranostic use. iRGD-modified indocyanine green (ICG) liposomes were prepared and utilized for imaging-guided tumor phototherapy against a breast cancer mouse model in vivo (Yan et al., 2016). Greater accumulation of these targeted liposomes was visualized in the tumors compared to nontargeted ones using NIR molecular imaging (Fig. 4). Upon excitation with single wavelength NIR laser, ICG was able to achieve PDT and PTT effects. Furthermore, the

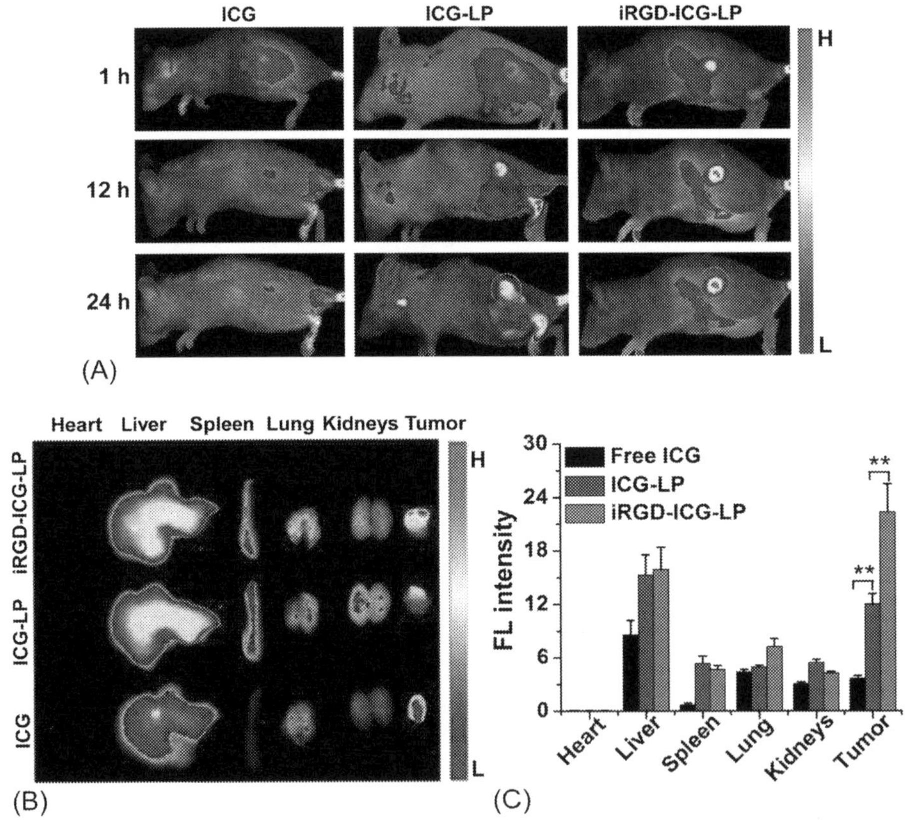

Fig. 4

4T1 tumor-bearing mice were injected with free ICG, ICG-LPs or iRGD-ICG-LPs. (A) Fluorescence signal was obtained at different time points following administration of each formulation. (B) Organs were extracted 24 h postinjection and ex vivo fluorescence images were obtained at 24 h postinjection for each group. (C) Fluorescence intensity for the organs and tumors was semiquantitatively analyzed, showing higher signal intensity in the tumor of animals administered with iRGD-ICG-LPs than the other groups. $**P<0.01$. *Reprinted from Yan, F., Wu, H., Liu, H., Deng, Z., Liu, H., Duan, W., et al. (2016). Molecular imaging-guided photothermal/photodynamic therapy against tumor by iRGD-modified indocyanine green nanoparticles. Journal of Controlled Release, 224, 217–228. Copyright (2015), with permission from Elsevier.*

higher accumulation of the targeted liposomes in the tumor translated into significant tumor growth inhibition and longer overall survival rate.

While PDT and PTT show great potential as anticancer therapies, the penetration depth of the light source is limited in deep tissues, and they are ineffective against metastatic tumors due to their localized characteristic (Agostinis et al., 2011; Zou et al., 2016). Given this scenario, better treatment outcomes can be achieved when PDT and PTT are associated with chemotherapy (Huang et al., 2016).

Most of the recent successful theranostic approaches ally both external and intrinsic stimuli to obtain more precise and efficient results. This demonstrates how the many aspects and specific features of each tumor may influence the choice of the treatment and consequently, the outcome of the therapy. It is also evident how versatile theranostic NPs can be when designed to be remotely triggered and to work as imaging and therapeutic agents.

6 Therapeutic Outcomes

The association of drugs and imaging probes in the same nanotheranostics devices is very useful for simultaneously monitoring the biodistribution of the NPs and the localization of the drug, among other advantages (Fig. 5) (Jo et al., 2016).

In the field of oncology, the main drugs that have been investigated as the pharmacological agent payload entrapped into nanoparticles include doxorubicin, vincristine, cisplatin, and paclitaxel. Liposomal doxorubicin (Doxil) has been in clinical use since 1995, with significantly improved therapeutic index (Gabizon, Shmeeda, & Barenholz, 2003). Theranostic approaches have been investigated with both nontargeted and targeted nanocarriers, including triggered responsive nanocarriers (Xue et al., 2017). Table 2 summarizes the main characteristics of some examples of nanotheranostic platforms addressed in this review.

Because molecular biorecognition and specific cellular or tissue distribution represent important advantages to cancer nanomedicines, targeting has been widely investigated as a strategy to optimize therapeutic response. Antibodies have been employed in cancer therapy mainly in association with cytotoxic drugs either conjugated or unconjugated since the late 1990s (Adams & Weiner, 2005; Wang & Jia, 2016). In addition to their role as targeting moieties in the drug delivery field (Bazak et al., 2015), they also influence the decrease of the downstream signaling cascade and restore the regular metabolism of the cells, making them susceptible to cytotoxic drugs (Weiner, Surana, & Wang, 2010). They can also be employed in inhibition processes related to tumor progression, such as angiogenesis and other features of the tumor microenvironment (Wang & Jia, 2016). Moreover, they can be utilized to aid the immunomodulatory response of cancer vaccines (Melero et al., 2014). In nanotheranostics, antibodies can be employed as active targeting molecules to increase the specificity and

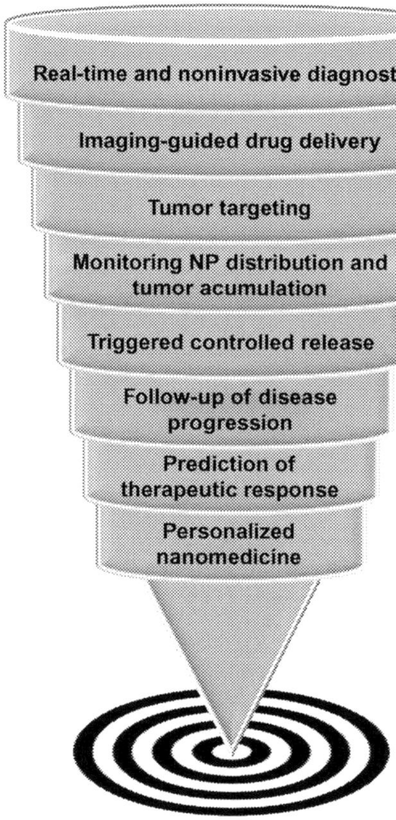

Fig. 5
Advantages promoted by the use of targeted nanotheranostics in cancer therapy.

accumulation of nanoparticles in the tumor. As final goals, when associated with imaging agents, these targeted nanoparticles allow the evaluation of the biodistribution and dosimetry prior to the therapy and allow prediction of the response of the disease to the proposed therapy (Fleuren et al., 2014).

Tumor-target antibody therapeutics are designed for receptors overexpressed by certain tumors, which include the epidermal growth factor receptor (EGFR) (Nakai, Hung, & Yamaguchi, 2016), human epidermal growth factor receptor 2 (HER2) (Swain et al., 2015), vascular endothelial growth factor (VEGF) (Vasudev & Reynolds, 2014), and antigens (such as CD20, CD52, CD19) expressed on hematopoietic cells for treatment of hematologic malignancies (Palanca-Wessels & Press, 2014).

Examples of the application of antibodies as theranostic agents include Bexxar and Zevalin, two monoclonal antibodies directed against CD20 conjugated to radionuclides. These conjugates were approved by the FDA in the early 2000s for the treatment of non-Hodgkin's

Table 2: Examples of different types of NP and their theranostic applications

Type of NP	Targeting Molecule	Theranostic Application	Therapeutic Agent	Diagnostic Agent	Reference
Liposomes	RGD	MRI + chemotherapy	Paclitaxel	Gd-DOTA	Ren et al. (2015)
Liposomes	MNPs and methotrexate	MRI/fluorescence imaging + light/magnetic hyperthermia + chemotherapy	Methotrexate	Cy5.5 + MNPs	Guo et al. (2017)
Liposomes	RGD	NIR imaging + PDT and PTT	Indocyanine green	Indocyanine green	Yan et al. (2016)
Polymeric NC	-	NIR imaging and MRI + photo-chemotherapy	Paclitaxel	IR820 + SPIONs	Liao et al. (2017)
Polymeric NP	Folate	MRI + chemotherapy	Doxorubicin	MNPs	Bhattacharya et al. (2016)
Protein-based NP	Transferrin	NIR imaging + PDT and PTT	IR780	IR780	Wang, Li, et al. (2016) and Wang, Zhang, et al. (2016)
Antibody-drug conjugate	Trastuzumab	NIR imaging + photoimmunotherapy	Emtansine	IR700	Ito et al. (2017)
Dendrimer-stabilized Au nanostars	RGD	CT imaging + PTT	Au nanostars + siRNA anti-VEGF	Au nanostars	Wei et al. (2016)

lymphoma, and are used in targeted radionuclide therapy (TRT) and radioimmunotherapy (Zukotynski, Jadvar, Capala, & Fahey, 2016). However, Bexxar was discontinued in 2014 by the manufacturer. Other studies are being conducted on antibody radionuclide conjugates for TRT owning to the promising potential of the combination of specificity of molecular targeting allied with the cytotoxic role from ionizing radiation and the imaging capacity of the radionuclides (Blakkisrud et al., 2017; Gill, Falzone, Du, & Vallis, 2017; Jauw et al., 2016).

HER2 is overexpressed in 20%–30% of patients with breast cancer, who would thus benefit from immunotherapy. Trastuzumab was first approved in 1998 as a targeted therapy for breast cancer in combination with paclitaxel or as a single agent (Vu & Claret, 2012). Also in 1998, HercepTest was approved as a semiquantitative immunohistochemistry assay as an antibody theranostic for solid tumors (Taylor, 2014). It allows identifying whether a patient will respond or not to the therapy (one of the applications of theranostics). In 2013, the antibody-drug conjugate T-DM1 (trastuzumab-emtansine) was approved for patients with HER2-positive metastatic breast cancer (Amiri-Kordestani et al., 2014). Clinical trials have shown that T-DM1 had good activity and better safety compared with standard chemotherapy allied to trastuzumab and increased overall survival compared to other established therapies (Krop et al., 2015; Peddi & Hurvitz, 2013). In a preclinical study, this antibody-drug conjugate was modified with IR700, a photoabsorbing phthalocyanine dye, for NIR-photoimmunotherapy (NIR-PT) and increased cytotoxicity based on the association of the therapies (Ito et al., 2017). The NIR-PT alone showed significant tumor growth inhibition of 3T3/HER2 cells, but the combination of trastuzumab, DM1 and IR700 had an even better outcome, with prolonged survival. Fluorescence imaging was enabled by the presence of IR700 and showed the high selectivity of the photoactivatable antibody-drug conjugates to the HER2-overexpressing tumor. Rainone et al. (2017) developed silica NP radiolabeled with ^{99m}Tc and functionalized with trastuzumab and demonstrated higher biodistribution of the nanosystems in HER2-positive SK-BR-3 tumor xenografts compared to nontargeted NP.

The process of neoangiogenesis involved in tumor progression is another important target that can be explored by the use of mAb, including the anti-VEGF bevacizumab (Sullivan & Brekken, 2010). ^{89}Zr-labeled bevacizumab was used for PET imaging to assess the in vivo heterogeneity of biodistribution of the mAb in children with a brain tumor. The results showed noticeable heterogeneity in drug delivery besides the enhanced uptake of ^{89}Zr-bevacizumab compared to that of a healthy brain. This result highlights the importance of the use of nanotheranostics to screen and determine which patients have better chances to benefit from a specific treatment (Jansen et al., 2017).

The use of nanotheranostic systems incorporating nucleic acids and other oligonucleotides, although only modestly investigated compared to those with small molecules and antibodies, is also an alternative in medical oncology (Shahbazi, Ozpolat, & Ulubayram, 2016). Several studies have demonstrated the potential of gene therapy in cancer treatment and the increase in stability of these labile molecules when encapsulated in NPs (Leite Nascimento et al., 2016; Salzano et al., 2016; Sarisozen, Salzano, & Torchilin, 2016), reaffirming their role in theranostics. Bruniaux et al. (2017) have prepared PEGylated magnetic siRNA nanovectors (S-MSN) for siRNA transfection of MDA-MB-231 breast cancer cells expressing green fluorescence protein (GFP). The NPs were able to downregulate GFP in vitro, and the presence of SPIONs in their composition could allow their use for MRI diagnostic in a possible future theranostic approach. In a more complex study, Wei et al. (2016) designed a theranostic nanoplatform of gold nanostars (AuNS) stabilized with dendrimers and targeted with cyclic RGD (RGD-Au DSNSs) for siRNA delivery of anti-VEGF and PTT in U87MG tumor-bearing mice. The presence of AuNS enables CT imaging, as well as thermal imaging upon laser irradiation. RGD-Au DSNSs/siRNA showed effective tumor growth inhibition when associated with PTT, suggesting that this platform holds great theranostic potential for combinatorial PTT and gene therapy.

7 Personalized Medicine

Personalized medicine is believed to be one of the most promising strategies for the future of cancer therapy. The most recent example of this is the approval by the FDA of Kymriah, the first gene therapy in the United States for use in children and young adults with B-cell acute lymphoblastic leukemia (US Food and Drug Administration, 2017). The treatment consists of the collection of T-cells from the patient, which are then genetically modified to express a particular protein that will target and kill leukemia cells expressing CD19. After the gene modification, these cells are reinserted in the patient as treatment. The therapy was given to patients who had failed all previous conventional therapies and showed 82.5% overall remission, an extraordinary outcome (Ledford, 2017). This approval brings even more light to the importance of personalized medicine, especially in the oncology field.

The implementation of NP-based formulations in oncology treatment is a useful tool for personalized medicine for treatment and follow-up of the disease. As one example, this application of nanotheranostics aims to monitor and optimize the therapy according to each patient's response to the treatment. If patients do not respond or show toxic side effects, the treatment is terminated (Kim, Lee, & Chen, 2013; Mura & Couvreur, 2012). When using stimuli-triggered NPs, it also enables, at an initial stage, tracking the fate of NPs in a real-time and noninvasive manner. The advantage of this type of nanotheranostic is the control of the release of the drug only if sufficient NPs reach the tumor while avoiding healthy tissues, enabled by the presence of an imaging agent and the drug in a single NP. Thus, it leads to a precise and more efficient approach (Zhao et al., 2017).

As mentioned, one of the greatest promises of nanotheranostics is the real-time and noninvasive monitoring of the disease. This monitoring can involve determining the stage of the disease and the therapeutic response, besides allowing the assessment of the risk of side effects if the NP ultimately do not accumulate preferentially in the tumor.

Also included in the personalized medicine strategy is the use of radiolabeled NPs to screen patients that may be more susceptible to certain therapies. The use of mAbs for the treatment of cancer has proven to be a great success. They are important molecules for targeted therapy. Radiolabeled antibodies are suitable vehicles for imaging-guided drug delivery. The prediction of the tumor delivery and accumulation using radiolabeled antibodies associated with imaging techniques, such as SPECT and PET, could avoid the administration of targeted therapies that ultimately fail due to the variations and heterogeneity of the tumors among different individuals.

8 Conclusions and Perspectives

Various studies have shown the potential of nanotheranostics in the field of oncology. The effectiveness of the proof-of-concept and the increase in the number of research groups working on theranostics are of great importance to the translation of this strategy into clinical trials. An ideal design, taking into consideration circulation time, pharmacokinetics, and biodistribution, for instance, must be achieved to synchronize the desired features of both the therapeutic and diagnostic agents. Moreover, safety concerns related to the theranostic NPs, (mainly inorganic ones) must be fully addressed before taking these platforms to the clinic. Nevertheless, targeted nanotheranostics for drug delivery hold great promise to overcome the present hurdles encountered in cancer therapy by accomplishing a more precise assessment and monitoring of the disease.

As we gain new understanding of tumor microenvironment physiopathology and biomolecular interactions, new possibilities for developing multicomponent nanoparticles with theranostic capabilities become feasible. Polymer chemistry allows the preparation of a variety of macromolecules that can be used as nanotheranostics scaffolds with a number of different functionalities and stimuli-responsive behaviors. The constant search for the "magic bullet" finds new roadmaps with molecular biology advancements, oncoproteomics, and the identification of biomarkers, peptide moieties, or target receptors which are either specific to or overexpressed by the tumor.

After more than 30 years since the EPR effect was first described, the mechanism of nanoparticle accumulation in the tumor tissue is still mostly described as a result of the enhanced permeability and retention effect. EPR has been widely demonstrated in animal models for several tumors. However, clinical translation of nanoparticle tumor accumulation does not follow the same behavior and is rather unpredictable.

To date, most nanoparticulate drug delivery systems are liposome-based and with a relatively simple composition. Most of the therapeutic advantages from Doxil, for instance, rely on the marked altered pharmacokinetics of doxorubicin, which also leads to a lower nonspecific organ toxicity, rather than on tumor accumulation. Its innovative formulation came mainly from the remote drug-loading via transmembrane gradient approach. However, marketed liposomal doxorubicin formulations are composed of either hydrogenated soy phosphatidylcholine, cholesterol, and PEG-modified phosphatidylethanolamine or egg phosphatidylcholine and cholesterol. Even with this straightforward formulation, it took nearly two decades for a generic version of Doxil to be approved in the United States by the FDA. This example allows the extrapolation of certain properties of simple liposomal formulations to the highly complex composition of theranostic nanoparticles; in both cases, the pharmacological activity is dependent on the drug release following administration. Also, the drug's biodistribution will be governed by the biodistribution characteristics of the nanoparticles. In light of that, nanoparticle components, which influence circulation time and biodistribution, will strongly affect toxicity and efficacy of the nanomedicine.

Nanotheranostics whose targeting moiety is based on antibodies, aptamers, or other peptides have yielded promising results following in vitro studies or from in vivo models often based on knockout nude mice. These data might be casting a shadow on possible unforeseen immune responses in the clinic, which can hinder the theoretical advantages from the engineered nanoparticle.

In addition to the aforementioned absence of clinical observation of the EPR effect, stimuli-responsive and biorecognition-based nanotheranostics depend on the tumor microenvironment to fully exert their function. This highlights the strong susceptibility of these systems to interindividual variations or even intraindividual variations resulting from distinct tumor and metastasis sites, hence, distinct microenvironments.

Regulatory issues regarding the complexity of the formulations and the high number of parameters that must be controlled and standardized for each product cannot be overlooked. A combination of these factors is likely the reason (even after extensive knowledge and technological development in this field) that not one single clinical trial based on theranostic nanoparticles has been approved or initiated, unlike the hundreds of ongoing trials with less complex nanoparticle drug-delivery systems.

Growing investments in and worldwide initiatives toward translational nanomedicines are certainly going to facilitate the transition from preclinical success and promise to the clinical application of these highly elaborated nanomedicines. The new era of personalized medicine will possibly encompass some of the future advances of nanotheranostics.

Acknowledgment

The authors thank Dr. William Hartner for helpful comments.

References

Abadeer, N. S., & Murphy, C. J. (2016). Recent progress in cancer thermal therapy using gold nanoparticles. *The Journal of Physical Chemistry C*, *120*, 4691–4716.

Abadjian, M.-C.Z., Choi, J., & Anderson, C. J. (2017). Nanoparticles for PET imaging of tumors and cancer metastasis. In J. W. M. Bulte & M. M. J. Modo (Eds.), *Design and applications of nanoparticles in biomedical imaging*. Cham: Springer International Publishing.

Abakumov, M. A., Nukolova, N. V., Sokolsky-Papkov, M., Shein, S. A., Sandalova, T. O., Vishwasrao, H. M., et al. (2015). VEGF-targeted magnetic nanoparticles for MRI visualization of brain tumor. *Nanomedicine*, *11*, 825–833.

Adams, G. P., & Weiner, L. M. (2005). Monoclonal antibody therapy of cancer. *Nature Biotechnology*, *23*, 1147–1157.

Agostinis, P., Berg, K., Cengel, K. A., Foster, T. H., Girotti, A. W., Gollnick, S. O., et al. (2011). Photodynamic therapy of cancer: An update. *CA: A Cancer Journal for Clinicians*, *61*, 250–281.

Ahmad, R., Fu, J., He, N., & Li, S. (2016). Advanced gold nanomaterials for photothermal therapy of cancer. *Journal of Nanoscience and Nanotechnology*, *16*, 67–80.

Alibolandi, M., Abnous, K., Sadeghi, F., Hosseinkhani, H., Ramezani, M., & Hadizadeh, F. (2016). Folate receptor-targeted multimodal polymersomes for delivery of quantum dots and doxorubicin to breast adenocarcinoma: In vitro and in vivo evaluation. *International Journal of Pharmaceutics*, *500*, 162–178.

Ambasta, R. K., Sharma, A., & Kumar, P. (2011). Nanoparticle mediated targeting of VEGFR and cancer stem cells for cancer therapy. *Vascular Cell*, *3*, 26.

Amin, M., Mansourian, M., Koning, G. A., Badiee, A., Jaafari, M. R., & TEN Hagen, T. L. M. (2015). Development of a novel cyclic RGD peptide for multiple targeting approaches of liposomes to tumor region. *Journal of Controlled Release*, *220*, 308–315.

Amiri-Kordestani, L., Blumenthal, G. M., Xu, Q. C., Zhang, L., Tang, S. W., Ha, L., et al. (2014). FDA approval: Ado-trastuzumab emtansine for the treatment of patients with HER2-positive metastatic breast cancer. *Clinical Cancer Research*, *20*, 4436.

Anselmo, A. C., & Mitragotri, S. (2015). A review of clinical translation of inorganic nanoparticles. *The AAPS Journal*, *17*, 1041–1054.

Anselmo, A. C., & Mitragotri, S. (2016). Nanoparticles in the clinic. *Bioengineering & Translational Medicine*, *1*, 10–29.

Arabi, L., Badiee, A., Mosaffa, F., & Jaafari, M. R. (2015). Targeting CD44 expressing cancer cells with anti-CD44 monoclonal antibody improves cellular uptake and antitumor efficacy of liposomal doxorubicin. *Journal of Controlled Release*, *220*, 275–286.

Aryasomayajula, B., Salzano, G., & Torchilin, V. P. (2017). Multifunctional liposomes. *Methods in Molecular Biology*, *1530*, 41–61.

Baek, S., Singh, R. K., Kim, T.-H., Seo, J.-W., Shin, U. S., Chrzanowski, W., et al. (2016). Triple hit with drug carriers: pH- and temperature-responsive theranostics for multimodal chemo- and photothermal therapy and diagnostic applications. *ACS Applied Materials & Interfaces*, *8*, 8967–8979.

Baetke, S. C., Lammers, T., & Kiessling, F. (2015). Applications of nanoparticles for diagnosis and therapy of cancer. *The British Journal of Radiology*, *88*, 20150207.

Bañobre-López, M., Teijeiro, A., & Rivas, J. (2013). Magnetic nanoparticle-based hyperthermia for cancer treatment. *Reports of Practical Oncology & Radiotherapy*, *18*, 397–400.

Barenholz, Y. (2012). Doxil(R)—The first FDA-approved nano-drug: Lessons learned. *Journal of Controlled Release*, *160*, 117–134.

Bazak, R., Houri, M., Achy, S. E., Kamel, S., & Refaat, T. (2015). Cancer active targeting by nanoparticles: A comprehensive review of literature. *Journal of Cancer Research and Clinical Oncology*, *141*, 769–784.

Belyanina, I., Kolovskaya, O., Zamay, S., Gargaun, A., Zamay, T., & Kichkailo, A. (2017). Targeted magnetic nanotheranostics of cancer. *Molecules*, *22*, 975.

Bertrand, N., Grenier, P., Mahmoudi, M., Lima, E. M., Appel, E. A., Dormont, F., et al. (2017). Mechanistic understanding of in vivo protein corona formation on polymeric nanoparticles and impact on pharmacokinetics. *Nature Communications*, *8*, 777.

Bharti, C., Nagaich, U., Pal, A. K., & Gulati, N. (2015). Mesoporous silica nanoparticles in target drug delivery system: A review. *International Journal of Pharmaceutical Investigation*, *5*, 124–133.

Bhattacharya, D., Behera, B., Sahu, S. K., Ananthakrishnan, R., Maiti, T. K., & Pramanik, P. (2016). Design of dual stimuli responsive polymer modified magnetic nanoparticles for targeted anti-cancer drug delivery and enhanced MR imaging. *New Journal of Chemistry*, *40*, 545–557.

Blakkisrud, J., Holtedahl, J. E., Londalen, A., Dahle, J., Bach-Gansmo, T., Holte, H., et al. (2017). Biodistribution and dosimetry results from a phase 1 trial of 177Lu-lilotomab satetraxetan antibody-radionuclide conjugate therapy. *Journal of Nuclear Medicine*, https://doi.org/10.2967/jnumed.117.195347.

Bobo, D., Robinson, K. J., Islam, J., Thurecht, K. J., & Corrie, S. R. (2016). Nanoparticle-based medicines: A review of FDA-approved materials and clinical trials to date. *Pharmaceutical Research*, *33*, 2373–2387.

Bruniaux, J., Djemaa, S. B., Hervé-Aubert, K., Marchais, H., Chourpa, I., & David, S. (2017). Stealth magnetic nanocarriers of siRNA as platform for breast cancer theranostics. *International Journal of Pharmaceutics*, *532*, 660–668.

Bushong, S., & Clarke, G. (2014). *Magnetic resonance imaging: Physical and biological principles*. St. Louis, MO: Elsevier.

Cai, Z., Chattopadhyay, N., Yang, K., Kwon, Y. L., Yook, S., Pignol, J.-P., et al. (2016). ^{111}In-labeled trastuzumab-modified gold nanoparticles are cytotoxic in vitro to HER2-positive breast cancer cells and arrest tumor growth in vivo in athymic mice after intratumoral injection. *Nuclear Medicine and Biology*, *43*, 818–826.

Cai, Z., Yook, S., Lu, Y., Bergstrom, D., Winnik, M. A., Pignol, J.-P., et al. (2017). Local radiation treatment of HER2-positive breast cancer using trastuzumab-modified gold nanoparticles labeled with 177Lu. *Pharmaceutical Research*, *34*, 579–590.

Chakravarty, R., Chen, F., Dash, A., & Cai, W. (2017). Theranostic nanoplatforms for PET image-guided drug delivery. In J. W. M. Bulte & M. M. J. Modo (Eds.), *Design and applications of nanoparticles in biomedical imaging*. Cham: Springer International Publishing.

Chen, B., Dai, W., He, B., Zhang, H., Wang, X., Wang, Y., et al. (2017). Current multistage drug delivery systems based on the tumor microenvironment. *Theranostics*, *7*, 538–558.

Chen, R., Ling, D., Zhao, L., Wang, S., Liu, Y., Bai, R., et al. (2015). Parallel comparative studies on mouse toxicity of oxide nanoparticle- and gadolinium-based T1 MRI contrast agents. *ACS Nano*, *9*, 12425–12435.

Chen, H., Niu, G., Wu, H., & Chen, X. (2016). Clinical application of radiolabeled RGD peptides for PET imaging of integrin $\alpha(v)\beta(3)$. *Theranostics*, *6*, 78–92.

Chen, H., Zhang, W., Zhu, G., Xie, J., & Chen, X. (2017). *Rethinking cancer nanotheranostics*. *2*, 17024.

Chen, H., Zhen, Z., Todd, T., Chu, P. K., & Xie, J. (2013). Nanoparticles for improving cancer diagnosis. *Materials Science & Engineering R, Reports: A Review Journal*, *74*, 35–69.

Chitgupi, U., Qin, Y., & Lovell, J. F. (2017). Targeted nanomaterials for phototherapy. *Nanotheranostics*, *1*, 38–58.

Choi, K. Y., Jeon, E. J., Yoon, H. Y., Lee, B. S., Na, J. H., Min, K. H., et al. (2012). Theranostic nanoparticles based on PEGylated hyaluronic acid for the diagnosis, therapy and monitoring of colon cancer. *Biomaterials*, *33*, 6186–6193.

Cintra, E. R., Ferreira, F. S., Santos Junior, J. L., Campello, J. C., Socolovsky, L. M., Lima, E. M., et al. (2009). Nanoparticle agglomerates in magnetoliposomes. *Nanotechnology*, *20*, 045103.

Cole, L. E., Ross, R. D., Tilley, J. M., Vargo-Gogola, T., & Roeder, R. K. (2015). Gold nanoparticles as contrast agents in X-ray imaging and computed tomography. *Nanomedicine (London, England)*, *10*, 321–341.

Costa Lima, S. A., Gaspar, A., Reis, S., & Duraes, L. (2017). Multifunctional nanospheres for co-delivery of methotrexate and mild hyperthermia to colon cancer cells. *Materials Science & Engineering C: Materials for Biological Applications*, *75*, 1420–1426.

Curry, T., Kopelman, R., Shilo, M., & Popovtzer, R. (2014). Multifunctional theranostic gold nanoparticles for targeted CT imaging and photothermal therapy. *Contrast Media & Molecular Imaging*, *9*, 53–61.

Danhier, F., LE Breton, A., & Préat, V. (2012). RGD-based strategies to target alpha(v) beta(3) integrin in cancer therapy and diagnosis. *Molecular Pharmaceutics*, *9*, 2961–2973.

Daniels, T. R., Bernabeu, E., Rodríguez, J. A., Patel, S., Kozman, M., Chiappetta, D. A., et al. (2012). Transferrin receptors and the targeted delivery of therapeutic agents against cancer. *Biochimica et Biophysica Acta*, *1820*, 291–317.

Dawidczyk, C. M., Kim, C., Park, J. H., Russell, L. M., Lee, K. H., Pomper, M. G., et al. (2014). State-of-the-art in design rules for drug delivery platforms: Lessons learned from FDA-approved nanomedicines. *Journal of Controlled Release*, *187*, 133–144.

de Melo-Diogo, D., PAIS-Silva, C., Dias, D. R., Moreira, A. F., & Correia, I. J. (2017). Strategies to improve cancer photothermal therapy mediated by nanomaterials. *Advanced Healthcare Materials*, *6*, 1700073.

Dennis, C. L., & Ivkov, R. (2013). Physics of heat generation using magnetic nanoparticles for hyperthermia. *International Journal of Hyperthermia*, *29*, 715–729.

Deshpande, P. P., Biswas, S., & Torchilin, V. P. (2013). Current trends in the use of liposomes for tumor targeting. *Nanomedicine (London, England)*, *8*, 1509–1528.

Du, J., Lane, L. A., & Nie, S. (2015). Stimuli-responsive nanoparticles for targeting the tumor microenvironment. *Journal of Controlled Release*, *219*, 205–214.

Dutta, D., Sundaram, S. K., Teeguarden, J. G., Riley, B. J., Fifield, L. S., Jacobs, J. M., et al. (2007). Adsorbed proteins influence the biological activity and molecular targeting of nanomaterials. *Toxicological Sciences*, *100*, 303–315.

Ehlerding, E. B., Chen, F., & Cai, W. (2016). Biodegradable and renal clearable inorganic nanoparticles. *Advanced Science*, *3*, 1500223.

Ekdawi, S. N., Stewart, J. M. P., Dunne, M., Stapleton, S., Mitsakakis, N., Dou, Y. N., et al. (2015). Spatial and temporal mapping of heterogeneity in liposome uptake and microvascular distribution in an orthotopic tumor xenograft model. *Journal of Controlled Release*, *207*, 101–111.

Eloy, J. O., Petrilli, R., Trevizan, L. N. F., & Chorilli, M. (2017). Immunoliposomes: A review on functionalization strategies and targets for drug delivery. *Colloids and Surfaces B: Biointerfaces*, *159*, 454–467.

Erdogan, S., & Torchilin, V. P. (2017). Gadolinium-loaded polychelating polymer-containing tumor-targeted liposomes. *Methods in Molecular Biology*, *1522*, 179–192.

Espelin, C. W., Leonard, S. C., Geretti, E., Wickham, T. J., & Hendriks, B. S. (2016). Dual HER2 targeting with trastuzumab and liposomal-encapsulated doxorubicin (MM-302) demonstrates synergistic antitumor activity in breast and gastric cancer. *Cancer Research*, *76*, 1517–1527.

Estelrich, J., Sánchez-Martín, M. J., & Busquets, M. A. (2015). Nanoparticles in magnetic resonance imaging: From simple to dual contrast agents. *International Journal of Nanomedicine*, *10*, 1727–1741.

Fass, L. (2008). Imaging and cancer: A review. *Molecular Oncology*, *2*, 115–152.

Fernandes Rodrigues, H., Capistrano, G., Mello, F., Zufelato, N., Silveira-Lacerda, E., & Bakuzis, A. (2017). Precise determination of the heat delivery during in vivo magnetic nanoparticle hyperthermia with infrared thermography. *Physics in Medicine and Biology*, *62*(10), 4062–4082.

Ferreira, R. V., Martins, T. M., Goes, A. M., Fabris, J. D., Cavalcante, L. C., Outon, L. E., et al. (2016). Thermosensitive gemcitabine-magnetoliposomes for combined hyperthermia and chemotherapy. *Nanotechnology*, *27*, 085105.

Fleuren, E. D. G., Versleijen-Jonkers, Y. M. H., Heskamp, S., Van Herpen, C. M. L., Oyen, W. J. G., Van der Graaf, W. T. A., et al. (2014). Theranostic applications of antibodies in oncology. *Molecular Oncology*, *8*, 799–812.

Frangioni, J. V. (2008). New technologies for human cancer imaging. *Journal of Clinical Oncology*, *26*, 4012–4021.

Gabizon, A., Shmeeda, H., & Barenholz, Y. (2003). Pharmacokinetics of pegylated liposomal doxorubicin. *Clinical Pharmacokinetics*, *42*, 419–436.

Geszke-Moritz, M., & Moritz, M. (2016). Solid lipid nanoparticles as attractive drug vehicles: Composition, properties and therapeutic strategies. *Materials Science & Engineering C, Materials for Biological Applications*, *68*, 982–994.

Gill, M. R., Falzone, N., Du, Y., & Vallis, K. A. (2017). Targeted radionuclide therapy in combined-modality regimens. *The Lancet Oncology*, *18*, e414–e423.

Grillone, A., Riva, E. R., Mondini, A., Forte, C., Calucci, L., Innocenti, C., et al. (2015). Active targeting of sorafenib: Preparation, characterization, and in vitro testing of drug-loaded magnetic solid lipid nanoparticles. *Advanced Healthcare Materials*, *4*, 1681–1690.

Guo, Y., Zhang, Y., Ma, J., Li, Q., Li, Y., Zhou, X., et al. (2017). Light/magnetic hyperthermia triggered drug released from multi-functional thermo-sensitive magnetoliposomes for precise cancer synergetic theranostics. *Journal of Controlled Release*, *272*, 145–158.

Hirsh, V. (2014). Nab-paclitaxel for the management of patients with advanced non-small-cell lung cancer. *Expert Review of Anticancer Therapy*, *14*, 129–141.

Ho, Y.-P., & Leong, K. W. (2010). Quantum dot-based theranostics. *Nanoscale*, *2*, 60–68.

Hoogevest, P. V., Leigh, M., & Fahr, A. (2013). Liposomes as intravenous solubilizers for poorly water-soluble drugs. In *Drug delivery strategies for poorly water-soluble drugs*. Somerset, NJ: John Wiley & Sons Ltd.

Huang, X., & El-Sayed, M. A. (2010). Gold nanoparticles: Optical properties and implementations in cancer diagnosis and photothermal therapy. *Journal of Advanced Research*, *1*, 13–28.

Huang, H.-C., Mallidi, S., Liu, J., Chiang, C.-T., Mai, Z., Goldschmidt, R., et al. (2016). Photodynamic therapy synergizes with irinotecan to overcome compensatory mechanisms and improve treatment outcomes in pancreatic cancer. *Cancer Research*, *76*, 1066.

Ito, K., Mitsunaga, M., Nishimura, T., Saruta, M., Iwamoto, T., Kobayashi, H., et al. (2017). Near-infrared photochemoimmunotherapy by photoactivatable bifunctional antibody–drug conjugates targeting human epidermal growth factor receptor 2 positive cancer. *Bioconjugate Chemistry*, *28*, 1458–1469.

Janib, S. M., Moses, A. S., & Mackay, J. A. (2010). Imaging and drug delivery using theranostic nanoparticles. *Advanced Drug Delivery Reviews*, *62*, 1052–1063.

Janku, F. (2014). Tumor heterogeneity in the clinic: Is it a real problem? *Therapeutic Advances in Medical Oncology*, *6*, 43–51.

Jansen, M. H., Veldhuijzen van Zanten, S. E. M., Van Vuurden, D. G., Huisman, M. C., Vugts, D. J., Hoekstra, O. S., et al. (2017). Molecular drug imaging: 89Zr-bevacizumab PET in children with diffuse intrinsic pontine glioma. *Journal of Nuclear Medicine*, *58*, 711–716.

Jauw, Y., Zijlstra, J., Hoekstra, O., Zweegman, S., Van Dongen, G., & Huisman, M. (2016). Impact of injected dose on accuracy and precision of tumor and organ uptake quantification of 89Zr-labeled anti-CD20 antibodies with PET. *Journal of Nuclear Medicine*, *57*, 1411.

Jin, R., Lin, B., Li, D., & Ai, H. (2014). Superparamagnetic iron oxide nanoparticles for MR imaging and therapy: Design considerations and clinical applications. *Current Opinion in Pharmacology*, *18*, 18–27.

Jo, S. D., Ku, S. H., Won, Y. Y., Kim, S. H., & Kwon, I. C. (2016). Targeted nanotheranostics for future personalized medicine: Recent progress in cancer therapy. *Theranostics*, *6*, 1362–1377.

Jokerst, J. V., Lobovkina, T., Zare, R. N., & Gambhir, S. S. (2011). Nanoparticle PEGylation for imaging and therapy. *Nanomedicine (London, England)*, *6*, 715–728.

Kalender, W. A. (2011). *Computed tomography: Fundamentals, system technology, image quality, applications*. Erlangen, Germany: Publicis Publishing.

Kamaly, N., Yameen, B., Wu, J., & Farokhzad, O. C. (2016). Degradable controlled-release polymers and polymeric nanoparticles: Mechanisms of controlling drug release. *Chemical Reviews*, *116*, 2602–2663.

Kang, T., Li, F., Baik, S., Shao, W., Ling, D., & Hyeon, T. (2017). Surface design of magnetic nanoparticles for stimuli-responsive cancer imaging and therapy. *Biomaterials*, *136*, 98–114.

Kelkar, S. S., & Reineke, T. M. (2011). Theranostics: Combining imaging and therapy. *Bioconjugate Chemistry*, *22*, 1879–1903.

Key, J., & Leary, J. F. (2014). Nanoparticles for multimodal in vivo imaging in nanomedicine. *International Journal of Nanomedicine*, *9*, 711–726.

Kim, T. H., Lee, S., & Chen, X. (2013). Nanotheranostics for personalized medicine. *Expert Review of Molecular Diagnostics*, *13*, 257–269.

Kim, S. J., Lewis, B., Steiner, M. S., Bissa, U. V., Dose, C., & Frank, J. A. (2016). Superparamagnetic iron oxide nanoparticles for direct labeling of stem cells and in vivo mri tracking. *Contrast Media & Molecular Imaging*, *11*, 55–64.

Kratz, F. (2014). A clinical update of using albumin as a drug vehicle: A commentary. *Journal of Controlled Release*, *190*, 331–336.

Krop, I. E., Lin, N. U., Blackwell, K., Guardino, E., Huober, J., Lu, M., et al. (2015). Trastuzumab emtansine (T-DM1) versus lapatinib plus capecitabine in patients with HER2-positive metastatic breast cancer and central nervous system metastases: A retrospective, exploratory analysis in EMILIA. *Annals of Oncology*, *26*, 113–119.

Kunjachan, S., Ehling, J., Storm, G., Kiessling, F., & Lammers, T. (2015). Non-invasive imaging of nanomedicines and nanotheranostics: Principles, progress and prospects. *Chemical Reviews*, *115*, 10907–10937.

Larsen, M. T., Kuhlmann, M., Hvam, M. L., & Howard, K. A. (2016). Albumin-based drug delivery: Harnessing nature to cure disease. *Molecular and Cellular Therapies*, *4*, 3.

Larson, N., & Ghandehari, H. (2012). Polymeric conjugates for drug delivery. *Chemistry of Materials : A Publication of the American Chemical Society*, *24*, 840–853.

Ledford, H. (2017). *Engineered cell therapy for cancer gets thumbs up from FDA advisers* (Online). https://www.nature.com/articles/d41586-017-01536-3: (https://www.nature.com/articles/d41586-017-01536-3, 02 September 2017) Nature. Available from: https://www.nature.com/articles/d41586-017-01536-3 (Accessed 02 September 2017).

Leite Nascimento, T., Hillaireau, H., Vergnaud, J., Rivano, M., Deloménie, C., Courilleau, D., et al. (2016). Hyaluronic acid-conjugated lipoplexes for targeted delivery of siRNA in a murine metastatic lung cancer model. *International Journal of Pharmaceutics*, *514*, 103–111.

Li, B., Li, Q., Mo, J., & Dai, H. (2017). Drug-loaded polymeric nanoparticles for cancer stem cell targeting. *Frontiers in Pharmacology*, *8*, 51.

Li, K., Nejadnik, H., & Daldrup-Link, H. E. (2017). Next-generation superparamagnetic iron oxide nanoparticles for cancer theranostics. *Drug Discovery Today*, *22*, 1421–1429.

Liao, J., Wei, X., Ran, B., Peng, J., Qu, Y., & Qian, Z. (2017). Polymer hybrid magnetic nanocapsules encapsulating IR820 and PTX for external magnetic field-guided tumor targeting and multifunctional theranostics. *Nanoscale*, *9*, 2479–2491.

Ling, D., & Hyeon, T. (2017). Magnetic nanomaterials for therapy. In *Magnetic nanomaterials: Fundamentals, synthesis and applications*. Weinheim, Germany: Wiley-VCH Verlag GmbH & Co. KGaA.

Liu, Y., Ai, K., & Lu, L. (2012). Nanoparticulate X-ray computed tomography contrast agents: From design validation to in vivo applications. *Accounts of Chemical Research*, *45*, 1817–1827.

Liu, X., Braun, G. B., Zhong, H., Hall, D. J., Han, W., Qin, M., et al. (2016). Tumor-targeted multimodal optical imaging with versatile cadmium-free quantum dots. *Advanced Functional Materials*, *26*, 267–276.

Lusic, H., & Grinstaff, M. W. (2013). X-Ray computed tomography contrast agents. *Chemical Reviews*, *113*. https://doi.org/10.1021/cr200358s.

Ma, Y., Huang, J., Song, S., Chen, H., & Zhang, Z. (2016). Cancer-targeted nanotheranostics: Recent advances and perspectives. *Small*, *12*, 4936–4954.

Ma, Y.-Y., Jin, K.-T., Wang, S.-B., Wang, H.-J., Tong, X.-M., Huang, D.-S., et al. (2017). Molecular imaging of cancer with nanoparticle-based theranostic probes. *Contrast Media & Molecular Imaging*, *2017*, 11.

Ma, Y., Mou, Q., Wang, D., Zhu, X., & Yan, D. (2016). Dendritic polymers for theranostics. *Theranostics*, *6*, 930–947.

Maeda, H. (2001). The enhanced permeability and retention (EPR) effect in tumor vasculature: The key role of tumor-selective macromolecular drug targeting. *Advances in Enzyme Regulation*, *41*, 189–207.

Maeda, H. (2015). Toward a full understanding of the EPR effect in primary and metastatic tumors as well as issues related to its heterogeneity. *Advanced Drug Delivery Reviews*, *91*, 3–6.

Mai, W. X., & Meng, H. (2013). Mesoporous silica nanoparticles: A multifunctional nano therapeutic system. *Integrative Biology (Cambridge)*, *5*, 19–28.

Mallidi, S., Spring, B. Q., & Hasan, T. (2015). Optical imaging, photodynamic therapy and optically-triggered combination treatments. *Cancer Journal (Sudbury, Mass)*, *21*, 194–205.

Manouras, T., & Vamvakaki, M. (2017). Field responsive materials: Photo-, electro-, magnetic- and ultrasound-sensitive polymers. *Polymer Chemistry*, *8*, 74–96.

Marelli, U. K., Rechenmacher, F., Sobahi, T. R. A., Mas-Moruno, C., & Kessler, H. (2013). Tumor targeting via integrin ligands. *Frontiers in Oncology*, *3*, 222.

Martelli, C., Dico, A. L., Diceglie, C., Lucignani, G., & Ottobrini, L. (2016). Optical imaging probes in oncology. *Oncotarget*, *7*, 48753–48787.

Martin, K. H., & Dayton, P. A. (2013). Current status and prospects for microbubbles in ultrasound theranostics. *Wiley Interdisciplinary Reviews Nanomedicine and Nanobiotechnology*, *5*, https://doi.org/10.1002/wnan.1219.

Martínez-González, R., Estelrich, J., & Busquets, M. A. (2016). Liposomes loaded with hydrophobic iron oxide nanoparticles: Suitable T(2) contrast agents for MRI. *International Journal of Molecular Sciences*, *17*, 1209.

Matsumura, Y., & Maeda, H. (1986). A new concept for macromolecular therapeutics in cancer chemotherapy: Mechanism of tumoritropic accumulation of proteins and the antitumor agent smancs. *Cancer Research*, *46*, 6387.

Melero, I., Gaudernack, G., Gerritsen, W., Huber, C., Parmiani, G., Scholl, S., et al. (2014). Therapeutic vaccines for cancer: An overview of clinical trials. *Nature Reviews Clinical Oncology*, *11*, 509–524.

Mendes, L. P., Gaeti, M. P., de Avila, P. H., De Sousa Vieira, M., Dos Santos Rodrigues, B., DE Avila Marcelino, R. I., et al. (2014). Multicompartimental nanoparticles for co-encapsulation and multimodal drug delivery to tumor cells and neovasculature. *Pharmaceutical Research*, *31*, 1106–1119.

Micha, J. P., Goldstein, B. H., Birk, C. L., Rettenmaier, M. A., Brown, J. V., 3rd. (2006). Abraxane in the treatment of ovarian cancer: The absence of hypersensitivity reactions. *Gynecologic Oncology*, *100*, 437–438.

Michalet, X., Pinaud, F. F., Bentolila, L. A., Tsay, J. M., Doose, S., Li, J. J., et al. (2005). Quantum dots for live cells, in vivo imaging, and diagnostics. *Science*, *307*, 538–544.

Mora-Huertas, C. E., Fessi, H., & Elaissari, A. (2010). Polymer-based nanocapsules for drug delivery. *International Journal of Pharmaceutics*, *385*, 113–142.

Mosafer, J., Abnous, K., Tafaghodi, M., Mokhtarzadeh, A., & Ramezani, M. (2017). In vitro and in vivo evaluation of anti-nucleolin-targeted magnetic PLGA nanoparticles loaded with doxorubicin as a theranostic agent for enhanced targeted cancer imaging and therapy. *European Journal of Pharmaceutics and Biopharmaceutics*, *113*, 60–74.

Muller, R. H., Mader, K., & Gohla, S. (2000). Solid lipid nanoparticles (SLN) for controlled drug delivery: A review of the state of the art. *European Journal of Pharmaceutics and Biopharmaceutics*, *50*, 161–177.

Muñoz de Escalona, M., Sáez-Fernández, E., Prados, J. C., Melguizo, C., & Arias, J. L. (2016). Magnetic solid lipid nanoparticles in hyperthermia against colon cancer. *International Journal of Pharmaceutics*, *504*, 11–19.

Mura, S., & Couvreur, P. (2012). Nanotheranostics for personalized medicine. *Advanced Drug Delivery Reviews*, *64*, 1394–1416.

Muthu, M. S., Leong, D. T., Mei, L., & Feng, S.-S. (2014). Nanotheranostics: Application and further development of nanomedicine strategies for advanced theranostics. *Theranostics*, *4*, 660–677.

Naha, P. C., AL Zaki, A., Hecht, E., Chorny, M., Chhour, P., Blankemeyer, E., et al. (2014). Dextran coated bismuth-iron oxide nanohybrid contrast agents for computed tomography and magnetic resonance imaging. *Journal of Materials Chemistry B*, *2*, 8239–8248.

Nakai, K., Hung, M.-C., & Yamaguchi, H. (2016). A perspective on anti-EGFR therapies targeting triple-negative breast cancer. *American Journal of Cancer Research*, *6*, 1609–1623.

National Cancer Institute. (2015). *Cancer Diagnosis (Online)*. National Cancer Institute. Available from: https://www.cancer.gov/about-cancer/diagnosis-staging/diagnosis.

Nevala, W. K., Buhrow, S. A., Knauer, D. J., Reid, J. M., Atanasova, E. A., & Markovic, S. N. (2016). Antibody-targeted chemotherapy for the treatment of melanoma. *Cancer Research*, *76*, 3954–3964.

Obst, K., Yealland, G., Balzus, B., Miceli, E., Dimde, M., Weise, C., et al. (2017). Protein corona formation on colloidal polymeric nanoparticles and polymeric nanogels: Impact on cellular uptake, toxicity, immunogenicity, and drug release properties. *Biomacromolecules*, *18*, 1762–1771.

Oliveira, R. R., Ferreira, F. S., Cintra, E. R., Branquinho, L. C., Bakuzis, A. F., & Lima, E. M. (2012). Magnetic nanoparticles and rapamycin encapsulated into polymeric nanocarriers. *Journal of Biomedical Nanotechnology*, *8*, 193–201.

Palanca-Wessels, M. C., & Press, O. W. (2014). Advances in the treatment of hematologic malignancies using immunoconjugates. *Blood*, *123*, 2293–2301.

Palmerston Mendes, L., Pan, J., & Torchilin, V. (2017). Dendrimers as nanocarriers for nucleic acid and drug delivery in cancer therapy. *Molecules*, *22*, 1401.

Pattni, B. S., Chupin, V. V., & Torchilin, V. P. (2015). New developments in liposomal drug delivery. *Chemical Reviews*, *115*, 10938–10966.

Peddi, P. F., & Hurvitz, S. A. (2013). Trastuzumab emtansine: The first targeted chemotherapy for treatment of breast cancer. *Future Oncology (London, England)*, *9*. https://doi.org/10.2217/fon.13.7.

Peres-Filho, M. J., Dos Santos, A. P., Nascimento, T. L., DE Avila, R. I., Ferreira, F. S., Valadares, M. C., et al. (2017). Antiproliferative activity and VEGF expression reduction in MCF7 and PC-3 cancer cells by paclitaxel and imatinib co-encapsulation in folate-targeted liposomes. *AAPS PharmSciTech*.

Prabhu, R. H., Patravale, V. B., & Joshi, M. D. (2015). Polymeric nanoparticles for targeted treatment in oncology: Current insights. *International Journal of Nanomedicine*, *10*, 1001–1018.

Quinto, C. A., Mohindra, P., Tong, S., & Bao, G. (2015). Multifunctional superparamagnetic iron oxide nanoparticles for combined chemotherapy and hyperthermia cancer treatment. *Nanoscale*, *7*, 12728–12736.

Rainone, P., Riva, B., Belloli, S., Sudati, F., Ripamonti, M., Verderio, P., et al. (2017). Development of (99m) Tc-radiolabeled nanosilica for targeted detection of HER2-positive breast cancer. *International Journal of Nanomedicine*, *12*, 3447–3461.

Rajagopalan, R., & Yakhmi, J. V. (2017). Nanotechnological approaches toward cancer chemotherapy A2—Ficai, Anton. In A. M. Grumezescu (Ed.), *Nanostructures for cancer therapy*. Amsterdam, Netherlands: Elsevier. [Chapter 8].

Rajkumar, S., & Prabaharan, M. (2017). Theranostics based on iron oxide and gold nanoparticles for imaging- guided photothermal and photodynamic therapy of cancer. *Current Topics in Medicinal Chemistry*, *17*, 1858–1871.

Ren, L., Chen, S., Li, H., Zhang, Z., Ye, C., Liu, M., et al. (2015). MRI-visible liposome nanovehicles for potential tumor-targeted delivery of multimodal therapies. *Nanoscale*, *7*, 12843–12850.

Revia, R. A., & Zhang, M. (2016). Magnetite nanoparticles for cancer diagnosis, treatment, and treatment monitoring: Recent advances. *Materials Today*, *19*, 157–168.

Salvati, A., Pitek, A. S., Monopoli, M. P., Prapainop, K., Bombelli, F. B., Hristov, D. R., et al. (2013). Transferrin-functionalized nanoparticles lose their targeting capabilities when a biomolecule corona adsorbs on the surface. *Nature Nanotechnology*, *8*, 137–143.

Salzano, G., Costa, D. F., Sarisozen, C., Luther, E., Mattheolabakis, G., Dhargalkar, P. P., et al. (2016). Mixed nanosized polymeric micelles as promoter of doxorubicin and miRNA-34a co-delivery triggered by dual stimuli in tumor tissue. *Small*, *12*, 4837–4848.

Sarisozen, C., Salzano, G., & Torchilin, V. P. (2016). Lipid-based siRNA delivery systems: Challenges, promises and solutions along the long journey. *Current Pharmaceutical Biotechnology*, *17*, 728–740.

Satterlee, A. B., & Huang, L. (2016). Current and future theranostic applications of the lipid-calcium-phosphate nanoparticle platform. *Theranostics*, *6*, 918–929.

Schettini, F., Giuliano, M., DE Placido, S., & Arpino, G. (2016). Nab-paclitaxel for the treatment of triple-negative breast cancer: Rationale, clinical data and future perspectives. *Cancer Treatment Reviews*, *50*, 129–141.

Selvan, S. T., & Narayanan, K. (2016). Semiconducting nanoparticles or quantum dots for theranostics. In *Introduction to nanotheranostics*. Singapore: Springer Singapore.

Shah, S. A., Aslam Khan, M. U., Arshad, M., Awan, S. U., Hashmi, M. U., & Ahmad, N. (2016). Doxorubicin-loaded photosensitive magnetic liposomes for multi-modal cancer therapy. *Colloids and Surfaces B: Biointerfaces*, *148*, 157–164.

Shahbazi, R., Ozpolat, B., & Ulubayram, K. (2016). Oligonucleotide-based theranostic nanoparticles in cancer therapy. *Nanomedicine*, *11*, 1287–1308.

Shi, J., Kantoff, P. W., Wooster, R., & Farokhzad, O. C. (2017). Cancer nanomedicine: Progress, challenges and opportunities. *Nature Reviews Cancer*, *17*, 20–37.

Shirmardi Shaghasemi, B., Virk, M. M., & Reimhult, E. (2017). Optimization of magneto-thermally controlled release kinetics by tuning of magnetoliposome composition and structure. *Scientific Reports*, *7*, 7474.

Somanath, P. R., Malinin, N. L., & Byzova, T. V. (2009). Cooperation between integrin α(v)β(3) and VEGFR2 in angiogenesis. *Angiogenesis*, *12*, 177–185.

Spring, B. Q., Rizvi, I., Xu, N., & Hasan, T. (2015). The role of photodynamic therapy in overcoming cancer drug resistance. *Photochemical & Photobiological Sciences: Official Journal of the European Photochemistry Association and the European Society for Photobiology*, *14*, 1476–1491.

Sriraman, S. K., Pan, J., Sarisozen, C., Luther, E., & Torchilin, V. (2016). Enhanced cytotoxicity of folic acid-targeted liposomes co-loaded with C6 ceramide and doxorubicin: In vitro evaluation on HeLa, A2780-Adr, and H69-AR cells. *Molecular Pharmaceutics*, *13*, 428–437.

Sriraman, S. K., & Torchilin, V. P. (2014). Recent advances with liposomes as drug carriers. In *Advanced biomaterials and biodevices*. Hoboken, NJ: John Wiley & Sons, Inc.

Steinhauser, I., Spänkuch, B., Strebhardt, K., & Langer, K. (2006). Trastuzumab-modified nanoparticles: Optimisation of preparation and uptake in cancer cells. *Biomaterials*, *27*, 4975–4983.

Suk, J. S., Xu, Q., Kim, N., Hanes, J., & Ensign, L. M. (2016). PEGylation as a strategy for improving nanoparticle-based drug and gene delivery. *Advanced Drug Delivery Reviews*, *99*, 28–51.

Sullivan, L. A., & Brekken, R. A. (2010). The VEGF family in cancer and antibody-based strategies for their inhibition. *MAbs*, *2*, 165–175.

Swain, S. M., Baselga, J., Kim, S.-B., Ro, J., Semiglazov, V., Campone, M., et al. (2015). Pertuzumab, trastuzumab, and docetaxel in HER2-positive metastatic breast cancer. *New England Journal of Medicine*, *372*, 724–734.

Tai, W., Mahato, R., & Cheng, K. (2010). The role of HER2 in cancer therapy and targeted drug delivery. *Journal of Controlled Release: Official Journal of the Controlled Release Society*, *146*, 264–275.

Taylor, C. R. (2014). Predictive biomarkers and companion diagnostics. The future of immunohistochemistry—'In situ proteomics', or just a 'stain'? *Applied Immunohistochemistry & Molecular Morphology: AIMM/Official Publication of the Society for Applied Immunohistochemistry*, *22*, 555–561.

Teleki, A., Haufe, F. L., Hirt, A. M., Pratsinis, S. E., & Sotiriou, G. A. (2016). Highly scalable production of uniformly-coated superparamagnetic nanoparticles for triggered drug release from alginate hydrogels. *RSC Advances*, *6*, 21503–21510.

Torchilin, V. P. (2014). Multifunctional, stimuli-sensitive nanoparticulate systems for drug delivery. *Nature Reviews Drug Discovery*, *13*, 813–827.

Toy, R., Bauer, L., Hoimes, C., Ghaghada, K. B., & Karathanasis, E. (2014). Targeted nanotechnology for cancer imaging. *Advanced Drug Delivery Reviews*, *76*, 79–97.

U.S. Food and Drug Administration. (2017). *FDA approval brings first gene therapy to the United States (Online)*. Washington, DC: U.S. Food and Drug Administration. Available from: https://www.fda.gov/NewsEvents/Newsroom/PressAnnouncements/ucm574058.htm Accessed September 2, 2017.

Vasudev, N. S., & Reynolds, A. R. (2014). Anti-angiogenic therapy for cancer: Current progress, unresolved questions and future directions. *Angiogenesis*, *17*, 471–494.

Velikyan, I. (2012). Molecular imaging and radiotherapy: Theranostics for personalized patient management. *Theranostics*, *2*, 424–426.

Vu, T., & Claret, F. X. (2012). Trastuzumab: Updated mechanisms of action and resistance in breast cancer. *Frontiers in Oncology*, *2*, 62.

Wang, L. (2017). Early diagnosis of breast cancer. *Sensors (Basel, Switzerland)*, *17*, 1572.

Wang, S., & Jia, M. (2016). Antibody therapies in cancer. In S. Zhang (Ed.), *Progress in cancer immunotherapy*. Dordrecht: Springer Netherlands.

Wang, F., Li, C., Cheng, J., & Yuan, Z. (2016). Recent advances on inorganic nanoparticle-based cancer therapeutic agents. *International Journal of Environmental Research and Public Health*, *13*, 1182.

Wang, L., Lin, H., Ma, L., Jin, J., Shen, T., Wei, R., et al. (2017). Albumin-based nanoparticles loaded with hydrophobic gadolinium chelates as T1-T2 dual-mode contrast agents for accurate liver tumor imaging. *Nanoscale*, *9*, 4516–4523.

Wang, Z., Niu, G., & Chen, X. (2014). Polymeric materials for theranostic applications. *Pharmaceutical Research*, *31*, 1358–1376.

Wang, K., Zhang, Y., Wang, J., Yuan, A., Sun, M., Wu, J., et al. (2016). Self-assembled IR780-loaded transferrin nanoparticles as an imaging, targeting and PDT/PTT agent for cancer therapy. *Scientific Reports*, *6*, 27421.

Wei, P., Chen, J., Hu, Y., Li, X., Wang, H., Shen, M., et al. (2016). Dendrimer-stabilized gold nanostars as a multifunctional theranostic nanoplatform for CT imaging, photothermal therapy, and gene silencing of tumors. *Advanced Healthcare Materials*, *5*, 3203–3213.

Weiner, L. M., Surana, R., & Wang, S. (2010). Monoclonal antibodies: Versatile platforms for cancer immunotherapy. *Nature Reviews Immunology*, *10*, 317–327.

Wu, W., Mao, D., Hu, F., Xu, S., Chen, C., Zhang, C.-J., et al. (2017). A highly efficient and photostable photosensitizer with near-infrared aggregation-induced emission for image-guided photodynamic anticancer therapy. *Advanced Materials*, *29*, 1700548.

Wu, P. H., Onodera, Y., Ichikawa, Y., Rankin, E. B., Giaccia, A. J., Watanabe, Y., et al. (2017). Targeting integrins with RGD-conjugated gold nanoparticles in radiotherapy decreases the invasive activity of breast cancer cells. *International Journal of Nanomedicine*, *12*, 5069–5085.

Xing, H., Hwang, K., & Lu, Y. (2016). Recent developments of liposomes as nanocarriers for theranostic applications. *Theranostics*, *6*, 1336–1352.

Xu, X., Ho, W., Zhang, X., Bertrand, N., & Farokhzad, O. (2015). Cancer nanomedicine: From targeted delivery to combination therapy. *Trends in Molecular Medicine*, *21*, 223–232.

Xu, Z., Ma, X., Gao, Y.-E., Hou, M., Xue, P., Li, C. M., et al. (2017). Multifunctional silica nanoparticles as a promising theranostic platform for biomedical applications. *Materials Chemistry Frontiers*, *1*, 1257–1272.

Xu, H., Ma, H., Yang, P., Zhang, X., Wu, X., Yin, W., et al. (2015). Targeted polymer-drug conjugates: Current progress and future perspective. *Colloids and Surfaces B: Biointerfaces*, *136*, 729–734.

Xue, P., Sun, L., Li, Q., Zhang, L., Xu, Z., Li, C. M., et al. (2017). PEGylated magnetic Prussian blue nanoparticles asa multifunctional therapeutic agent for combined targeted photothermal ablation and pH-triggered chemotherapy of tumour cells. *Journal of Colloid and Interface Science*, *509*, 384–394.

Yan, F., Wu, H., Liu, H., Deng, Z., Liu, H., Duan, W., et al. (2016). Molecular imaging-guided photothermal/ photodynamic therapy against tumor by iRGD-modified indocyanine green nanoparticles. *Journal of Controlled Release*, *224*, 217–228.

Yang, Z., Song, J., Dai, Y., Chen, J., Wang, F., Lin, L., et al. (2017). Self-assembly of semiconducting-plasmonic gold nanoparticles with enhanced optical property for photoacoustic imaging and photothermal therapy. *Theranostics*, *7*, 2177–2185.

Yang, Y., Wang, F., Zheng, K., Deng, L., Yang, L., Zhang, N., et al. (2017). Injectable PLGA/Fe(3)O(4) implants carrying cisplatin for synergistic magnetic hyperthermal ablation of rabbit VX2 tumor. *PLoS One*, *12*, e0177049.

Yen, S. K., Padmanabhan, P., & Selvan, S. T. (2013). Multifunctional iron oxide nanoparticles for diagnostics, therapy and macromolecule delivery. *Theranostics*, *3*, 986–1003.

Yeo, E. L. L., Cheah, J. U. J., Lim, B. Y., Thong, P. S. P., Soo, K. C., & Kah, J. C. Y. (2017). Protein corona around gold nanorods as a drug carrier for multimodal cancer therapy. *ACS Biomaterials Science & Engineering*, *3*, 1039–1050.

Yildirim, Y., Gultekin, E., Avci, M. E., Inal, M. M., Yunus, S., & Tinar, S. (2008). Cardiac safety profile of pegylated liposomal doxorubicin reaching or exceeding lifetime cumulative doses of 550 mg/m^2 in patients with recurrent ovarian and peritoneal cancer. *International Journal of Gynecological Cancer*, *18*, 223–227.

Yoshimoto, M., Hayakawa, T., Yamaguchi, M., Kimura, S., & Fujii, H. (2016). Characterization of RGD-modified liposomes for multimodal molecular imaging of$\alpha v\beta 3$ integrin-expressing pancreatic cancer. *Journal of Nuclear Medicine*, *57*, 1206.

Zardad, A.-Z., Choonara, Y., DU Toit, L., Kumar, P., Mabrouk, M., Kondiah, P., et al. (2016). A review of thermo- and ultrasound-responsive polymeric systems for delivery of chemotherapeutic agents. *Polymer*, *8*, 359.

Zhang, Z., Wang, J., Nie, X., Wen, T., Ji, Y., Wu, X., et al. (2014). Near infrared laser-induced targeted cancer therapy using thermoresponsive polymer encapsulated gold nanorods. *Journal of the American Chemical Society*, *136*, 7317–7326.

Zhao, X., Yang, C.-X., Chen, L.-G., & Yan, X.-P. (2017). Dual-stimuli responsive and reversibly activatable theranostic nanoprobe for precision tumor-targeting and fluorescence-guided photothermal therapy. *Nature Communications*, *8*, 14998.

Zhao, S., Zhang, Y., Han, Y., Wang, J., & Yang, J. (2015). Preparation and characterization of cisplatin magnetic solid lipid nanoparticles (MSLNs): Effects of loading procedures of Fe_3O_4 nanoparticles. *Pharmaceutical Research*, *32*, 482–491.

Zhong, J., Yang, S., Wen, L., & Xing, D. (2016). Imaging-guided photoacoustic drug release and synergistic chemo-photoacoustic therapy with paclitaxel-containing nanoparticles. *Journal of Controlled Release*, *226*, 77–87.

Zhu, J., Fu, F., Xiong, Z., Shen, M., & Shi, X. (2015). Dendrimer-entrapped gold nanoparticles modified with RGD peptide and alpha-tocopheryl succinate enable targeted theranostics of cancer cells. *Colloids and Surfaces B: Biointerfaces*, *133*, 36–42.

Zhu, L., & Torchilin, V. P. (2013). Stimulus-responsive nanopreparations for tumor targeting. *Integrative Biology: Quantitative Biosciences From Nano to Macro*, *5*, https://doi.org/10.1039/c2ib20135f.

Zhu, R., Wang, Z., Liang, P., He, X., Zhuang, X., Huang, R., et al. (2017). Efficient VEGF targeting delivery of DOX using bevacizumab conjugated SiO2@LDH for anti-neuroblastoma therapy. *Acta Biomaterialia*, *63*, 163–180.

Zou, L., Wang, H., He, B., Zeng, L., Tan, T., Cao, H., et al. (2016). Current approaches of photothermal therapy in treating cancer metastasis with nanotherapeutics. *Theranostics*, *6*, 762–772.

Zukotynski, K., Jadvar, H., Capala, J., & Fahey, F. (2016). Targeted radionuclide therapy: Practical applications and future prospects. *Biomarkers in Cancer*, *8*, 35–38.

Zwicke, G. L., Mansoori, G. A., & Jeffery, C. J. (2012). Utilizing the folate receptor for active targeting of cancer nanotherapeutics. *Nanotechnology Reviews*, *3*, https://doi.org/10.3402/nano.v3i0.18496.

Further Reading

Dos Santos Ferreira, D., Jesus de Oliveira Pinto, B. L., Kumar, V., Cardoso, V. N., Fernandes, S. O., Souza, C. M., et al. (2017). Evaluation of antitumor activity and cardiac toxicity of a bone-targeted ph-sensitive liposomal formulation in a bone metastasis tumor model in mice. *Nanomedicine: Nanotechnology, Biology and Medicine*, *13*, 1693–1701.

Palanca-Wessels, M. C., Booth, G. C., Convertine, A. J., Lundy, B. B., Berguig, G. Y., Press, M. F., et al. (2016). Antibody targeting facilitates effective intratumoral siRNA nanoparticle delivery to HER2-overexpressing cancer cells. *Oncotarget*, *7*, 9561–9575.

Ravar, F., Saadat, E., Gholami, M., Dehghankelishadi, P., Mahdavi, M., Azami, S., et al. (2016). Hyaluronic acid-coated liposomes for targeted delivery of paclitaxel, in-vitro characterization and in-vivo evaluation. *Journal of Controlled Release*, *229*, 10–22.

Sriraman, S. K., Salzano, G., Sarisozen, C., & Torchilin, V. (2016). Anti-cancer activity of doxorubicin-loaded liposomes co-modified with transferrin and folic acid. *European Journal of Pharmaceutics and Biopharmaceutics*, *105*, 40–49.

Tang, Y., Soroush, F., Tong, Z., Kiani, M. F., & Wang, B. (2017). Targeted multidrug delivery system to overcome chemoresistance in breast cancer. *International Journal of Nanomedicine*, *12*, 671–681.

Xu, X., Saw, P. E., Tao, W., Li, Y., Ji, X., Bhasin, S., et al. (2017). ROS-responsive polyprodrug nanoparticles for triggered drug delivery and effective cancer therapy. *Advanced Materials*, *29*, 1700141.

Zuo, T., Guan, Y., Chang, M., Zhang, F., Lu, S., Wei, T., et al. (2016). RGD(Arg-Gly-Asp) internalized docetaxel-loaded pH sensitive liposomes: Preparation, characterization and antitumor efficacy in vivo and in vitro. *Colloids and Surfaces B: Biointerfaces*, *147*, 90–99.

CHAPTER 10

Considerations for the Human Health Implications of Nanotheranostics

Michael J. Burgum[a], Stephen J. Evans[a], Gareth J. Jenkins, Shareen H. Doak, Martin J.D. Clift
Swansea University Medical School, Institute of Life Sciences, Swansea, Wales, United Kingdom

1 Introduction

Nanotechnology is an exciting discipline in science and engineering defined as the design, production, characterization and application of materials or devices which possess at least one physical dimension within the nanometer range (1–100 nm) (Silva, 2004). At this submicroscopic scale, the fundamental properties (optical, structural, reactivity) of bulk materials have been shown to change dramatically resulting in novel uses for drug delivery and enhanced biological imaging technologies which, when applied to medical science, result in applications for disease diagnosis, therapeutic delivery systems, and enhanced imaging capabilities (Saini, Saini, & Sharma, 2010). The size reduction of commonly sourced materials, for example, iron oxide particles, to the nano-size range dramatically changes their surface area, increasing reactivity and how these particles absorb light, thus giving rise to applications the bulk material does not possess. With an estimated $18.1 billion invested in nanotechnology in 2014 alone (LUX research), the United States plunges large donations towards the medical, technological, and consumer aspects of nanotechnology. There are numerous examples in which the field of nanoscience has garnered much attention for consumer products containing elements of nanotechnology including skin-care products (Gupta, Bansal, Jindal, & Jindal, 2013), sporting equipment (Institute of Medicine Roundtable on Environmental Health Sciences, Research, and Medicine, 2005), clothing (Kulthong, Srisung, Boonpavanitchakul, Kangwansupamonkon, & Maniratanachote, 2010), and food items (Srinivas et al., 2010). It is the medical field, however, which warrants the greatest degree of scientific interest, continued research, and, most importantly, concern with regard to nanotheranostics. The great benefit of utilizing nanoparticles, with their high surface-area-to-volume ratio as opposed to larger particles, is that they possess unique bioavailability (Jia, 2005), that is, they can be targeted

[a] Authors contributed equally as first author.

Handbook of Nanomaterials for Cancer Theranostics. https://doi.org/10.1016/B978-0-12-813339-2.00010-4

to specific sites for specific applications. The advantage of this scenario is that specific pharmacokinetics can be engineered and tailored to each molecule to reduce the interaction between therapeutic nanoparticles and undesired cellular targets: healthy cells and tissue. A vast array of nanomaterials (e.g., quantum dots (QDs), carbon nanotubes, nanoparticles) have medical applications, and thus far, have been incorporated into cell labeling, contrast agents, drug delivery vehicles, and antimicrobial agents. Therein lies the attraction to the hybrid discipline of nanotheranostics, the monitoring and treatment of human health through a combination of therapeutics and diagnostics, with the goal of increasing drug efficacy, safety, and biocompatibility (Wang, Chuang, & Ho, 2012).

Yet, despite an elevated interest in nanotheranostics and their proposed advantageous applications, there remains a growing public concern as to the potentially detrimental health hazards associated with engineered nanomaterial (ENM) exposure (Digesu, Hofferberth, Grinstaff, & Colson, 2016). Nanoparticles may access the body as a nanotheranostic via four main portals; inhalation, intravenous injection, translocation across the epidermal layer, or ingestion (Medina, Santos-Martinez, Radomski, Corrigan, & Radomski, 2007). In the case of each exposure route, there remains a risk of nanoparticles entering the bloodstream and subsequently translocating and accumulating in other tissues and organs of the body (e.g., liver and kidneys) (Reddy, Krishna, Reddy, & Himabindu, 2010). Nanoparticle portal routes are of particular concern when examining the toxicological aspect of their nature (Yah, Simate, & Iyuke, 2012), as the specificity associated with entry into the body may lead to diverse biological effects from intravenous injection to direct skin-contact approaches. The rather limited data available on unintentional nanomaterial ingestion, due to the numerous host defense mechanisms in place to prevent such localization, still advises that strong caution should be exercised as particle uptake is still a strong possibility in regions of the gut, Peyer's patches for instance (Bergin & Witzmann, 2013). The ideal approach for nanotheranostics would emphasize the need for intravenous injection and skin portals for maximum efficiency, ease of access, and localized treatment. The emphasis, therefore, is currently placed on the urgent need to assess the biocompatibility of medically targeted materials with nanotheranostic potential, utilizing nano(geno)toxicology as a screening tool. Summarizing, the purpose of this chapter is to provide an overview of the current knowledge encompassing nanotheranostics, and incorporating nanotoxicology, outline the biocompatibility and subsequent hazard associated with current and projected nanotheranostic options available.

2 Nanotheranostics

In conjunction with nanomedicine (application of nanotechnology in a clinical setting) (Farrell, Ptak, Panaro, & Grodzinski, 2011; Webster, 2006), we are striving for the next big development in managing human health. Within the nanomedicine sector, drug delivery dominates the marketplace, with in vitro diagnostics the second leading field of nanomedicine (Morigi et al., 2012), which, in 2012, was worth an estimated $4.8 billion

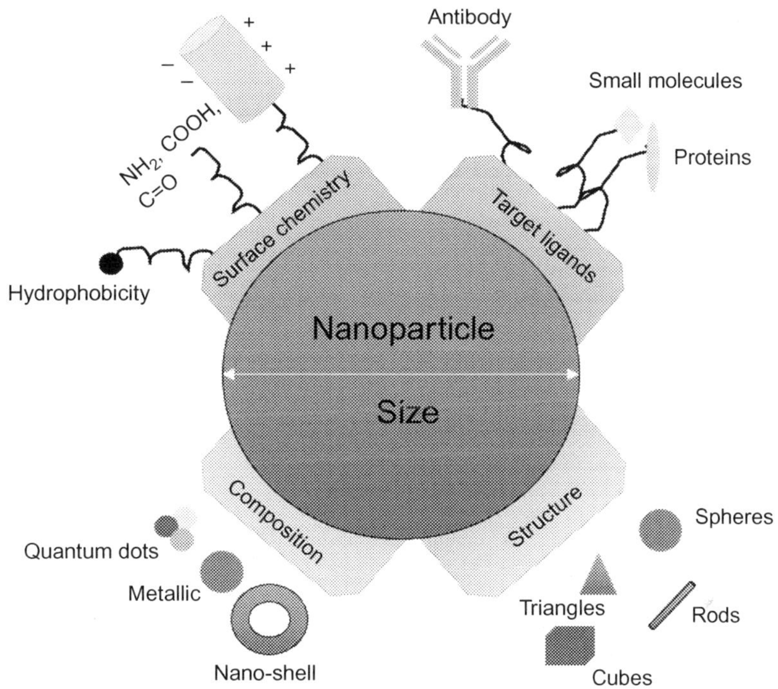

Fig. 1

ENP & ENM design can be highly dynamic in shape, surface coating, structure, and function. Resulting nanotheranostics can, therefore, adopt unique and highly specific forms to suit the medical application, whether that be drug delivery or translocation to desired cells for optical purposes.*Adapted from Chou, L. Y., Ming, K., & Chan, W. C. (2011). Strategies for the intracellular delivery of nanoparticles. Chemical Society Reviews, 40, 233–245.*

(Zhao & Castranova, 2011). Given the extensive funds available to the nanotheranostic sector, product manufacture and distribution have increased exponentially; metal oxide, carbon-based and inorganic nanoparticles represent a select few examples. Nanomaterials' specifically engineered and medically relevant physicochemical characteristics can be further functionalized with surface coatings bearing specific pharmacokinetics, enhancing therapeutic potency, or enabling the potential molecule-targeting benefits of such modifications (Fig. 1).

Current applications in nanotheranostics rely heavily on the physicochemical characteristics (PCCs) of the nanomaterials involved being known beforehand. Furthermore, the successful application of these nanomaterials as a theranostic tool is inherently linked to the ability to manipulate the PCCs of the nanomaterial (de Jong & Borm, 2008), particularly where passage across the blood-brain barrier or enhanced distribution is required. The successful application of nanoparticles as a nanotheranostic could be entirely based upon their biodistribution capabilities; this quality of nanoparticles is dependent, however, on features such as morphology, shape, and charge. Fortunately, modern advances in nanoparticle functionalization mean that additional targeting biodistribution can be engineered or "programmed" into the material. Gold nanoparticles, for instance, have been explicitly studied due to their low toxicity, bio-imaging, delivery, and conjugation characteristics (Tiwari et al., 2014). With simple modifications to the surface of these particles using cell penetrating peptides (CPP) derived from glycoproteins, the intracellular uptake and biodistribution were vastly enhanced in mammalian cell lines (HEp-2 &HeLa) quantified with flow cytometry and transmission electron microscopy (TEM). In the same study, the biodistribution in mice was then monitored using both functionalized and nonfunctionalized gold nanoparticles administered via intravenous injection at low doses of 63.72 μg (200 μL) and high doses of 79.65 μg (250 μL); the results showed a higher accumulation of functionalized particles in the liver and spleen of the mice, indicative of higher biodistribution capabilities, allowing for their translocation throughout the body and resulting in their eventual deposition in the liver, primarily where nanoparticle clearance can be initiated (Tiwari et al., 2014).

Nanotheranostic materials can differ greatly in reactivity as opposed to their macroscopic counterparts; for example, superparamagnetic iron oxide nanoparticles magnetic potential increases as size decreases (Lu, Salabas, & Schuth, 2007), and graphene edge reactivity increases as atomic layer number approaches one (Sharma, Baik, Perera, & Strano, 2010). It is these novel, tunable properties on which theranostics is based, but these features could be inadvertently altered (reprogrammed) in a biological environment, potentially corrupting their intended therapeutic function. Size, morphology, surface area, and surface charge may become significantly changed once the theranostic enters a biological environment (Gatoo et al., 2014); the now-altered nanomaterial may exhibit unintentional toxicology through increased cellular affinity or an innate ability to trigger an immune response (Zolnik, Gonzalez-Fernandez, Sadrieh, & Dobrovolskaia, 2010). Considering the degree of variability between nanomaterials (e.g., size, shape, aspect ratio functional groups), the safety of these materials is paramount. It follows, therefore, that all nanomedicines should be screened for their ability to induce biological damage at a cellular, immune-compromising, and genetic level before they can be disclosed for managing, treating, and curing human health conditions. At present, the risk assessment of ENMs and exposure hazards may still present significant risks after clearing Phase I, Phase II, and Phase III clinical trials (Resnik & Tinkle, 2007). Associated with oral, dermal, inhalation, and injection routes of exposure, there are currently three crucial elements to toxicity screen testing nanomaterials: physicochemical characterization, in vitro assays (both cellular and acellular), and in vivo assays (Oberdorster et al., 2005). Each of these present unique advantages in robustly screening nanomaterial hazards; physicochemical characteristics may inherently be associated with toxicity dependent upon shape, charge, distribution, and porosity. However, many of these features are routinely absent in some nanoparticle orientated publications. In vitro techniques provide opportunity for specific mechanisms or pathways to be elucidated in a controlled environment, aspects which are not feasible when introducing in vivo testing. Finally, Tier 1 in vivo testing (oral, dermal, and injection) and Tier 2 (inhalation) exposures provide a more true-to-life scenario accounting for elements not present under in vitro conditions. Tier 1 studies tend to focus on inflammation, proliferation, and oxidative stress, which can be specifically monitored and quantified with in vitro experiments, while Tier 2 may then elaborate on deposition, translocation, bio-persistence and toxicokinetics (Oberdorster et al., 2005).

Currently, there exists a battery of assays which can successfully screen nanotheranostics toxicology in cellular systems both in vitro and in vivo. The number of genetic toxicology assays has grown in recent years to the present day, where there exists a tiered screening system for new chemicals and nanomaterials. The HPRT forward mutation assay can be performed in mammalian cells; this requires more time and a more complex operating procedure but provides reliable results (Johnson, 2012). Especially useful assays are able to quantify genetic damage but also perform analyses in a high throughput manner, the cytokinesis block micronucleus (CBMN) assay being a key example. The CBMN assay relies on cell division to exclude micronuclei from the parent nucleus; the frequency with which micronuclei are captured during cell scoring provides an indicator of genetic damage, and can be applied to nanomaterial exposures (Doak, Manshian, Jenkins, & Singh, 2012). It follows that there is a clear necessity for nanotoxicology research to inform the scientific development of nanotheranostic therapies as the ever-growing demand for more efficient and potent theranostic treatments with higher efficacy increases in the near future.

3 Nanotheranostic Biocompatibility

The vital feature of any nanotheranostic material is its biocompatibility that is, its ability to be introduced into a biological environment without the induction of adverse toxic endpoints. Ensuring a biomedical NM is biocompatible can however be an issue, as the physicochemical properties that evolve due the material being on the nanoscale can make it inherently toxic. Due to this issue, the first step of nanotoxicology study is full physicochemical characterization of the test material. These characteristics are defined as being either primary or secondary, referring to properties of the pristine material and the properties when the material is under experimental/biological conditions, respectively. Primary NM characteristics include size, morphology, surface area, composition/purity, surface charge, and surface chemistry. Surface charge and chemistry, however, will become altered upon introduction into a biological environment, and so infer new secondary characteristics alongside agglomeration state and protein corona formation (Johnston, Brown, Kermanizadeh, Gubbins, & Stone, 2012). It is essential that both primary and secondary physicochemical characteristics of a nanotheranostic material are evaluated. This ensures it possesses the properties required for its intended function but also allows a correlation to be drawn with any adverse toxicological endpoints.

Upon NM administration to the human body, secondary characteristics will evolve upon interaction with biological fluid and biomolecules prior to initial cellular contact (Monopoli et al., 2011). Biomolecules including proteins, lipids, and sugars are adsorbed on to a NM surface resulting in the formation of bimolecular interface roughly divided into two components known as the hard and soft coronas (Monopoli, Aberg, Salvati, & Dawson, 2012). The soft corona is highly dynamic, interchangeable, and attached to the tightly bound hard corona, which is comprised of those biomolecules with the greatest affinity for the material. Although the hard corona is tightly bound to the material, dissociation does occur, but its association does ensue for many hours, and so, is regarded as inferring the identity of a NM within a biological environment (Carrillo-Carrion, Carril, & Parak, 2017). The cascade of interactions that occurs during the process of corona formation at the NM surface can cause alteration of the secondary structure of attached proteins resulting exposure of new epitopes, potentially giving rise to unexpected biological response (Fleischer & Payne, 2014). Moreover, corona formation will be a determining factor in NM hydrodynamic diameter, agglomeration state, and surface charge. Consequently, NM-corona formation is a vital consideration when evaluating nanotheranostic biocompatibility, as it will be a determining factor with regard to cell surface interactions, cellular uptake, and translocation though the body.

A prime example of how corona formation how can influence cellular interaction/uptake of a NM is the absorption of opsonins such as immunoglobulins and complement factors, which facilitates phagocytic uptake by leukocytes (Chen et al., 2017; Schöttler, Landfester, & Mailänder, 2016). If a nanotheranostic is designed to specifically target leukocytes (for example, vaccine delivery), then this opsonization may enhance the material's function (Lameijer, Tang, Nahrendorf, Beelen, & Mulder, 2013). Conversely, enhanced leukocyte uptake may cause the initiation of a chronic immune response, potentially leading to downstream toxicity (Evans et al., 2016). Various studies have concluded that NM-corona formation, in fact, inhibits cellular uptake rather than enhances it (Cheng et al., 2015; Doak et al., 2009; Guo et al., 2016). For example, Cheng et al. (2015) investigated how protein corona formation influenced uptake of Au NPs in RAW 264.7 (murine) macrophages and HepG2 cells (human hepatocytes). Corona formation by incubation in serum resulted not only in reduced NP uptake, but also alteration of the primary uptake mechanism by obstructing NP-scavenger receptor recognition and promoting recognition by clathrin.

Alteration of NP uptake mechanism due to corona formation introduces a major challenge in the design of targeted therapeutic NPs. In order to facilitate cell specific targeting by an NP, its surface is typically functionalized with antibodies or other biomolecules (Mout, Moyano, Rana, & Rotello, 2012). NP-corona formation on a functionalized NP is likely to block or impair targeting moieties by causing structural or conformational disruption and obscuring surface recognition (Saha, Moyano, & Rotello, 2014; Zanganeh, Spitler, Erfanzadeh, Alkilany, & Mahmoudi, 2016). We must also consider the effect of corona formation on the drug release profile from nanocarriers with surface-loaded drugs. For example, it has been demonstrated that corona formation on gold nanorods capped with cetyltrimethylammonium bromide causes a greater hold capacity on the drug payload compared to nanorods without the presence of a corona (Kah, Chen, Zubieta, & Hamad-Schifferli, 2012).

4 Exposure Routes and Translocation

The typical routes of nanotheranostic exposure are via inhalation, intravenous injection, via dermal exposure and the gastrointestinal (GI) tract (El-Sherbiny, El-Baz, & Yacoub, 2015; Jatana & DeLouise, 2014; Shi, Kantoff, Wooster, & Farokhzad, 2017; Sim et al., 2016). For medical applications, the degree of this exposure will vary depending on the treatment or diagnostic required, varying from a one-off dose (e.g., for an imaging technique) to extensive long-term treatment (e.g., for treatment of a chronic condition) (Nalwa, 2014; Singh et al., 2012).

Prior to the advent of nanomedicine, inhalable drugs have been widely available not only to specifically target the lung, but also as an alternative to intravenous injection for systematic delivery (Corkery, 2000). Drug delivery via the lung offers the advantages of avoiding the pharmacokinetic issue of presystemic metabolism, being easy to self-administer, and being noninvasive (Rau, 2005). When NPs are administered via the respiratory tract, they offer the advantage over their microcounterparts of being able to penetrate deeper into the alveolar region and potentially avoid clearance by alveolar macrophages (Thorley & Tetley, 2013). This ability to penetrate deeper into the respiratory tree than microparticles means that there are numerous potential applications of nanotheranostics via the inhalation route. An exciting example of this is the development of vaccine nanoformulations such as alternatives to the standard BCG vaccine (Ballester et al., 2011). The study entailed conjugating the TB antigen Ag85B to 30 nm polypropylene NPs and demonstrated increased vaccine efficacy when administered by inhalation compared to intradermal delivery in murine models. Nanomedicines can also be constructed to facilitate drug delivery in the bronchial regions of the lung, for example, polyethylene glycol (PEG) NP coatings have been shown to permit NP mucus penetration if it is constructed with a high density and low molecular weight (Liu, Zhang, Shan, & Huang, 2015). There is, however, a significant toxicological risk when administering nanomedicine via the respiratory tract therefore full toxicity assessment and physicochemical characterization is essential. Once within the lung a NM has the potential to promote a chronic inflammatory response and subsequently lung injury. And at the cellular level promote oxidative stress, inflammatory signaling, mitochondrial damage, protein denaturation, impairment of phagocytosis, endothelial dysfunction, cell cycle alteration, and DNA damage (Bakand, Hayes, & Dechsakulthorn, 2012).

Self-medication is most commonly undertaken by oral administration due to its convenience and noninvasiveness. However, taking drugs via the gastrointestinal tract is not particularly efficient due to low bioavailability, because of the dissolution rate-limited absorption typical of many drugs, first pass metabolism effects, food effects, and short half-lives (Gershanik & Benita, 2000). These are particular challenges when treating chronic conditions such HIV, diabetes, or psychiatric illnesses (e.g., schizophrenia), challenges that may be resolved by the use of nanodrug delivery systems (Dening, Rao, Thomas, & Prestidge, 2016; Giardiello et al., 2016). A suitable example is the use of solid lipid nanoparticles (SLNs) which offer the ability to encapsulate lipophilic and hydrophilic drug molecules for controlled release and also offer protection from the environment of the gastrointestinal tract (Das & Chaudhury, 2011). Indeed, SNLs have been developed that provide a notable increase in drug bioavailability; by encapsulating the antipsychotic clozapine in SNLs, the drug was absorbed by lymphatic transport pathways, thereby avoiding the hepatic portal vein system, preventing the first-pass effect (Dening et al., 2016). A further example of a NM increasing oral drug bioavailability is the development of insulin conjugation chondroitin sulfate capped AuNPs, which like SNLs provide protection against the harsh environment of the GI tract (Cho et al., 2014). From a toxicological standpoint, however, there is significant potential for an adverse response, as with the airway tract with the additional consideration of the gut's microbiome and the effect of its perturbation on the host (Bergin and Witzmann, 2013).

Medical application of NMs to the skin maybe either for a topical treatment or by transdermal drug delivery (TDD) for systematic treatment. AgNPs, for instance, may be applied as topical antimicrobial agents to control microbial growth, and, consequently, would be required to stay on the dermal surface (Jacobs, van de Poel, & Osseweijer, 2010; Pal, Yoon, Tak, Choi, & Song, 2009). Conversely, TDD is required to cross the dermal barrier and reach the circulatory system or be retained in the skin. TDD offers the benefit of being less painful than standard needle injection and the ability to bypass first pass metabolism (Palmer & DeLouise, 2016). The toxicological risk an NM possesses is dependent on its potential to penetrate the dermal layer. Typically, NMs may potentially penetrate the skin via three different pathways: intracellularly, or via dermal structures such as hair follicles (Baroli et al., 2007).

Depending on its physicochemical characteristics, NM may not be confined to its target region and maybe capable of distributing from the site of exposure to a number of secondary organs via the cardiovascular, central nervous, renal, or hepatic systems (Kermanizadeh, Balharry, Wallin, Loft, & Møller, 2015). Depending on the NMs' intended function, this may or may not be a desirable outcome. For example, if the intended function of an NM was the treatment of a systemic decease by inhalation, then translocation across the air/blood barrier would be essential (Thorley & Tetley, 2013). It is of note, however, that the exact mechanism of NM translocation though the air/blood barrier is unclear. It is thought that alveolar macrophages play a role in NM translocation by transporting NMs to the thoracic lymph nodes (Shaw et al., 2016; Zhao et al., 2011). It is also speculated that formation of an NM-protein corona via interaction with lung lining proteins may allow shuttling of the NM across the air/blood barrier

(Kermanizadeh et al., 2015). As with airway administration, nanomedicines administered via the GI tracts also have to utilize transepithelial absorbance mechanisms to facilitate systemic distribution. The major pathways that have been implicated in the uptake of NMs across the GI wall are through M-cells lining intestinal Peyer's patches and through intestinal enterocytes (Bellmann et al., 2015). Following translocation through the GI epithelial layer, an NM will enter into the lymph fluid and potentially drain into the systemic cardiovascular system or be absorbed into the capillaries of the cardiovascular circulatory system (Cuong & Hsieh, 2011). In comparison, the rate of NM translocation following intravenous injection will differ significantly from inhalation and oral ingestion. Intravenous administration permits direct access to the circulatory system and, consequently, rapid distribution throughout the body. A recent three-part in vivo study by Kreyling et al. has evaluated the quantitative biokinetics of radioactive titanium dioxide TiO_2 NPs administered in rats by intravenous injection, oral application, and inhalation (Kreyling, Holzwarth, Haberl, et al., 2017a, 2017b; Kreyling, Holzwarth, Schleh, et al., 2017). This large study noted that that 24 h following intravenous administration, 95.5% of the NPs administered located to the liver, 2.3% to the spleen, 0.7% the skeletal system, 0.5% remained in blood circulation, and there were detectable NPs in all other organs; interestingly, NPs were still undergoing hepatic clearance 28 days after treatment (Kreyling, Holzwarth, Haberl, et al., 2017a). Comparatively, part two of the study demonstrated a very different biokinetic pattern when the same NPs were administered orally, with <0.6% of the applied dose absorbed across the GI barrier, an NP faction >0.001% was present still in most organs 7 days after treatment (Kreyling, Holzwarth, Schleh, et al., 2017). Finally, part three of the study again demonstrated different biokinetic patters compared to the intravenous administration and one more similar to the GI tract (Kreyling, Holzwarth, Haberl, et al., 2017b). After 1 h, 4% of the applied NP dose was able to pass through the air/blood barrier, and 0.3% was retained in the animal after 28 days with the highest fractions present in the liver and kidneys (0.03%). This study demonstrated the variation in NP translocation depending on exposure route; however, biopersistence of the test material, regardless of how it was administered, is important, also. This consequently shows how toxicological consideration must also be given to regions of the body other than the specific target of a nanotheranostic material due to its potential biokinetic distribution. In particular, its effect on the hepatic and renal systems must be considered (He, Aker, Huang, Watts, & Hwang, 2015).

5 Metal-Based Nanomaterials

In recent years, significant emphasis has been placed upon the development of metal- and metal oxide-based nanotheranostic materials. There is a great variety of different metal-based NMs being developed for usage in the medical field for purposes such as photodynamic therapy, targeted drug delivery, and the enhancement of imaging techniques (Estelrich, Sánchez-Martín, & Busquets, 2015; He et al., 2015; Jain, Jain, & Mahajan, 2015). However, this discussion will focus on the use and potential toxicity of iron oxide

NPs; there has been extensive focus on their development for nanotheranostic approaches in addition to assessment of their toxicology profile.

5.1 SPION Uses and Characteristics

The unique physicochemical characteristics of superparamagnetic iron oxide nanoparticles (SPIONs) have made them an ideal candidate for several biomedical applications. These include magnetic resonance imaging (MRI) enhancement, targeted drug/gene delivery systems, and hyperthermia therapy (Bulte & Kraitchman, 2004; Dizaj, Jafari, & Khosroushahi, 2014; Gupta & Gupta, 2005; Laurent, Dutz, Häfeli, & Mahmoudi, 2011; Mahmoudi, Sant, Wang, Laurent, & Sen, 2011). SPION for MRI enhancement can be used as both a negative and positive contrast agent by allowing for reducing signal deterioration that results in darker or brighter images respectively compared to more traditional contrast chemicals (Dolci et al., 2013). This is achievable due to the superparamagnetic behavior SPIONs exert. The induced magnetic movement of the particles creates a local magnetic field that affects nearby water molecules. The latitudinal and longitudinal relaxation times of those water molecules are detected by the MRI equipment such that the resulting relaxation time map produces an image of the tissue under investigation (Neuwelt et al., 2015). In terms of targeted drug delivery, SPIONs' superparamagnetic properties can allow localization of a drug to a specific target site within the body using a magnetic field. This enhances the specificity of drug release at a defined pathological site, allowing the reduction of dosage and potential adverse drug toxicity (Mahmoudi et al., 2011). Similarly, the ability of the material to be localized in this manner potentially makes it suitable as a gene delivery vector; SPIONs coated with PEG-grafted polyethylenimine have been shown an efficient MRI-visible vector for gene delivery into human adipose derived mesenchymal stem cells (Pang et al., 2014). Fe_3O_4 has a cubic crystalline lattice inverse spinel structure comprised of closely packed O-2 ions; Fe^{2+} ions are in half of the octahedral sites, and Fe^{3+} ions are in the remaining octahedral and tetrahedral sites. Pure γ-Fe_2O_3 has a similar cubic structure, but with the absence of Fe^{2+} ions. Due to the presence of Fe^{2+} in Fe_3O_4 it is considered to be thermodynamically unstable, as the Fe^{2+} can readily undergo oxidation. It is, therefore, not uncommon for SPIONs to exist in a nonstoichiometric state in between γ-Fe_2O_3 and Fe_3O_4 when in suspension. SPIONs' chemical state, among their other physicochemical properties, is a vital consideration when evaluating their toxicological profile.

5.2 SPION Safety Assessment

With potential widespread usage of SPION, assessment of the NM potential toxicity is vital. Indeed, various studies have demonstrated SPION to promote the formation of apoptotic bodies, activate an immune response, generate reactive oxygen species, and cause DNA damage. Early studies of SPION toxicity focused primarily on evaluation of cytotoxicity; a study by Gupta and Gupta (2005), for instance, evaluated the toxicity of pullulan-coated SPION (40–45 nm). The investigation determined the test SPION to be cytotoxic and a promoter of cytoskeleton disruption in human dermal fibroblasts (Gupta & Gupta, 2005). Similarly, dimercaptosuccinic acid (DMSA) coated SPIONs promoted acute cytotoxicity in growing rat neurons (PC12). Various studies have, however, also shown SPIONs to be noncytotoxic (Yu et al., 2008).

A multitude of investigations have shown SPIONs to be capable of inducing an immunoloigcal response. For example, elevated IL-8 and IL-6 expression occurs in normal epidermal keratinocytes (HEK) following exposure to SPION. Furthermore, whole blood treated with both polyacrylic acid (PAA)-coated and noncoated SPION induced an increase of 6 cytokines; interleukins 1β, 6, 8, 10 and the tumor necrosis factor (IL-1β, IL-10, IL-6, IL-8, and TNF) via activation of the TAK1, p38 MAPK and JNK pathways. In vivo, a dose-dependent increase in IL-1β and TNF expression was quantified in the lungs of Sprague-Dawley rats following inhalation of bare γ-Fe2O3 NPs (~72 nm) (Zhang, Xia, Zhao, Liu, & Zhang, 2010). The same study noted increased lung oxidative stress identified by elevations in a glutathione disulfate (GSSG) to glutathione (GSH) ratio and a reduction in ferric/reducing antioxidant power. Similarly, aerosol exposure of uncoated Fe_3O_4 NPs of Wistar rats over 4 weeks resulted in substantial neutrophil influx into the alveolar region of the lung alongside oxidative stress quantified by increased 8-hydroxy-2′-deoxyguanosine (8-OHdG) levels. The ability of SPION to activate immunological pathways demonstrates the importance of considering the materials' interaction with the cellular components of the innate immune system and the consequential influence of this on other cell types within tissue, particularly in relation to the onset of genotoxicity.

Iron has long been associated with the induction of carcinogenesis. Iron ions are able to catalyze the formation of •OH radicals via the Fenton reaction, potentially causing a redox imbalance and, consequently, oxidative stress. Certainly, studies have proved SPIONs to be capable of inducing genotoxicity in this manner; Singh et al. (2012) demonstrated SPION genotoxicity to be redox state-dependent. Investigating genotoxic potential of dSPION in both γ-Fe_2O_3 and Fe_3O_4 forms via the in vitro micronucleus assay showed only γ-Fe_2O_3 capable of inducing double-stranded DNA breaks as a direct consequence of oxidative stress in the MCL-5 B-lymphoblastoid cell line (Singh et al., 2012). This was correlated with the ability of γ-Fe_2O_3 dSPION to undergo cellular uptake by endocytosis and ion dissociation to occur within lysosomes. Similarly, noncoated γ-Fe_2O_3 SPION have been shown to cause oxidative DNA damage in liver hepatocellular cells (HEPG2). Moreover, a recent study comparing metal oxide NP genotoxicity in primary human lymphocytes, uncoated γ-Fe_2O_3 promoted ROS generation and subsequent genotoxicity quantified by the single-cell gel electrophoresis assay (comet assay). Fe_3O_4 SPIONs have, however, also been shown to be genotoxic, differing from the study conducted by Singh et al. (2012). Oleate-coated Fe_3O_4, for instance, promotes DNA damage in the TK6 lymphoblastoid cell line (measured by the comet assay); the same study did, however, state that identical SPION without the oleate coating did not induce genotoxicity.

Not all investigations have proven SPIONs to be genotoxic. For example, L-glutamic acid coated γ-Fe_2O_3 that underwent genotoxic assessment by the comet and micronucleus assay demonstrated no significant genotoxic response. Similarly, an investigation comparing iron oxide NPs with bulk iron oxide claimed that neither nanosized nor bulk γ-Fe_2O_3 and Fe_3O_4 caused genotoxicity in Syrian hamster embryo cells. Differences between SPION genotoxicity studies are the result of numerous factors, the most prominent being variation between physicochemical characteristics. It can, however, be summarized that SPIONs (dependent upon form) pose a substantial risk of causing genotoxicity and, ultimately, potential downstream carcinogenesis; hence, assessment of their risk is essential.

A vast number of nanotoxicology studies in the literature determine metal and metal-oxide NPs such as SPIONs are inherently toxic, and, as such, imply they are fit for their intended biomedical purpose. It should, however, be of note that rather than using these studies to dismiss the use of an NM, they should, in fact, be used to identify the specific properties that cause toxicity so they can be rectified in future designs.

6 Carbon-Based Nanomaterials

Carbon-derived nanostructures, arguably, are the most diverse in application and design with respect to drug delivery and cancer treatment within the field of nanomedicine (Elhissi, Ahmed, Hassan, Dhanak, & D'emanuele, 2012; Madani, Naderi, Dissanayake, Tan, & Seifalian, 2011). Carbon nanostructures are typically highly ordered and possess characteristics engineered for a specific purpose, for example, carbon black pigments in printer toner (Pirela et al., 2015), carbon nanotube rigidity as cell scaffolds in tissue engineering (Edwards, Werkmeister, & Ramshaw, 2009), and lastly, graphene and few-layer graphene with hydrogen storage capabilities (Kostoglou et al., 2015). Graphene also holds tremendous potential as a smart therapeutic delivery system of photodynamic treatments (Wei, Zhou, Zhang, Chen, & Xing, 2016) which can be targeted to tumor growths with monoclonal antibodies to deliver doses of phototoxicity. Carbon black (and, particularly, carbon nanotubes) has garnered significant attention for its ability to elicit genotoxic responses (Patlolla, Knighten, & Tchounwou, 2010; Toyokuni, 2013). Recently, however, graphene-based genotoxicity is being reported in literature (Ma-Hock et al., 2013), largely due to the physicochemical characteristics and aspect ratio similarities shared with carbon nanotubes. Carbon-based nanomaterials exhibit sp^2 and sp^3 electronic valences commonly which provides opportunities for fluorescence and absorbance exploration in biological imaging. Nanodiamond and carbon dots, by virtue of their optoelectronic properties, may display natural fluorescence (Wang et al., 2012), beneficial to theranostic tracking in vivo.

(i) *Graphene*, an inorganic nanomaterial, is becoming more commonly associated with drug delivery and has been extensively explored as a chemotherapeutic agent and theranostic (Rahman et al., 2015) due to its high tunability, high surface area, and adsorption properties (Li et al., 2016). A large number of molecules, polymers, and antibodies may be functionalized with graphene in order to enhance their targeting or potency as a theranostic treatment. With every atom exposed, graphene exposes an extremely large surface area (>1000 m^2/g) efficient for therapeutic loading and bioconjugation (Yang, Feng, Shi, & Liu, 2013). As a nanotheranostic, the PCCs of graphene have been exploited, in particular, the near-infra-red (NIR) optical absorbance which (when utilized during in vivo treatments) can deliver photothermal therapy, and which has already achieved excellent antitumor effects (Yang et al., 2013). As with all nanotheranostics, however, significant attention is being placed on the toxicity of such treatments, where it is abundantly clear that size and surface chemistry directly influence toxicity (Chang et al., 2011; Nezakati, Cousins, & Seifalian, 2014; Yang et al., 2013). Graphene also shows ultrahigh in vivo tumor uptake, only being surpassed in uptake potential by PEGylated CNTs using mouse models, utilizing the NIR absorbance properties in phototherapy (Fig. 2).

The functionalization of graphene, graphene-oxide (GO) and reduced GO (rGO) have been widely explored as potential anticancer therapeutics with coatings including polyethylene glycol (PEG) to maintain biocompatibility as well as acutely potent drugs such as doxorubicin. The various combinations of graphene and targeted applications can be summarized below as each pertain to the treatment of cancer (Table 1) (Orecchioni, Cabizza, Bianco, & Delogu, 2015).

Using current literature as a guide (2014 saw three times more publications released than in 2012, Orecchioni et al., 2015), there appears to be an increasing interest in

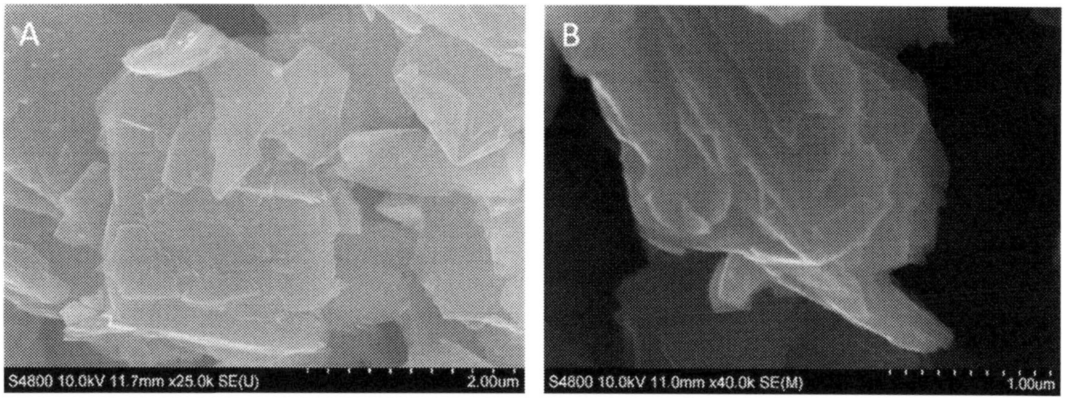

Fig. 2

Scanning electron micrographs of dry few-layer graphene as presented immediately postmanufacture within an occupational environment. (A) Overview. (B) Magnified view. Analysis was performed using the Hitachi Ultra High-Resolution field emission (FE)-SEM model number: S-4800 at Swansea University, images captured and provided by Michael J. Burgum.

Table 1: The various applications of functionalized graphene as nanotheranostics

Application	Cancer Type	Model	Graphene Therapeutic
Drug delivery & imaging	Burkitt's lymphoma	Human in vitro	GO-Doxorubicin + Rituxan (Yang et al., 2010)
Imaging & photothermal therapy	Breast cancer	Mouse in vivo	PEGylated nano-graphene sheets (nGS-PEG) (Yang et al., 2012)
Drug delivery	Breast cancer	Human in vitro	GO-Adryamicin (Wu et al., 2012)

The nature of the material and the therapeutic agent it carries can specifically alter the targeting of the material as well as the function and target cell type.
Adapted from Orecchioni, M., Cabizza, R., Bianco, A., & Delogu, L. G. (2015). Graphene as cancer theranostic tool: progress and future challenges. *Theranostics*, 5, 710–723.

the use of graphene as an anticancer theranostic. The double-exposed surface sides of graphene present unique opportunities for modifications via π-π stacking; however, if nanomaterial clearance cannot be achieved by host immune cells, potent toxicity issues are likely. Moreover, graphene surfaces present a platform for multiple anticancer drug functionalization for targeted therapeutics. Initially the graphene is decorated with sulfonic acid groups rendering the molecule stable in physiological solution, followed by folic acid (FA) conjugation allowing targeted delivery to human breast cancer MCF-7 cells and subsequent release of two commonly used therapeutic treatments, doxorubicin (DOX) and camptothecin (CPT) (Zhang et al., 2010). The folic acid-conjugated GO bearing DOX and CPT demonstrated high affinity for MCF-7 cells and high cytotoxicity as opposed to FA-GO bearing only one of the anticancer drugs, leading to the conclusion that, in the future, the combined use of DOX and CPT aboard FA-GO delivers a far more potent cytotoxic dose to breast cancer cells (Zhang et al., 2010).

(ii) *Carbon nanotubes (CNTs)*, similarly to graphene, possess numerous PCCs which make them choice selections as theranostics, due to their high surface area for functionalization and NIR optical properties (Wang et al., 2012), in delivering phototoxic doses to tumor cells (Robinson et al., 2010). In the study carried out by (Robinson et al., 2010) the first dual function for PEGylated CNTs was exploited whereby the biologically nontoxic nanotubes administered intravenously delivered photothermal treatment for tumor elimination at 808 nm and acted as photoluminescent agents for in vivo imaging. In the experiment, 70 μg CNTs/mouse (equivalent to 3.6 mg/kg) was administered and revealed significant tumor elimination with no toxic side effects; notably, this result closely resembled a comparable experiment repeated with gold nanorods. However, the nanorod dose required 10 times the concentration to achieve similar results, and showed more irradiation, suggesting significant benefits of utilizing CNTs for combined cancer therapy and treatment (Fig. 3).

It has been established that introducing nanomaterials to the human body, or in vivo animal testing poses toxicity issues, which can stem from nanoparticle size, surface area, and

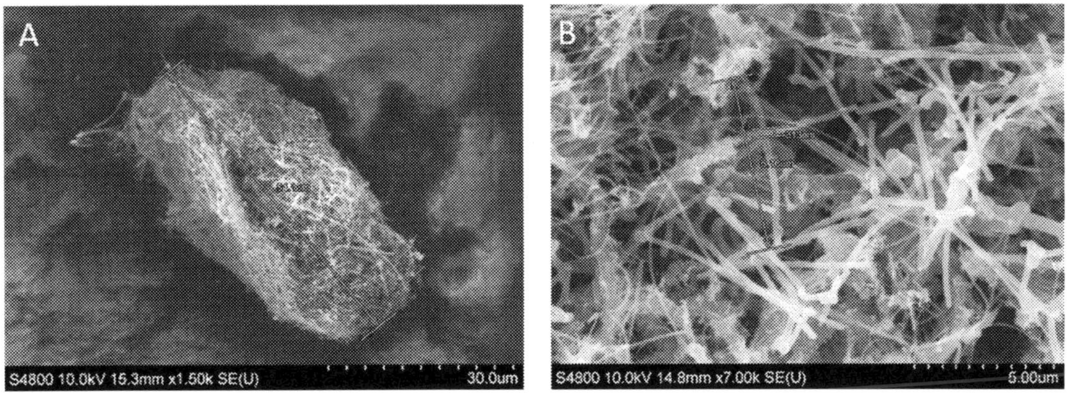

Fig. 3
Scanning electron microscope images of (A) water-dispersed CNTs 30 μm and (B) 5 μm resolution. CNTs dispersed in this manner commonly associate into "bird nest" agglomerates and pose a significant biological hazard in this biopersistent arrangement, making enzymatic clearance difficult. Analysis was performed using the Hitachi Ultra High-Resolution field emission (FE)-SEM model number: S-4800 at Swansea University; images captured and provided by Michael J. Burgum.

functionalization. Carbon nanomaterials, particularly graphene and CNTs, possess these toxicity capabilities; however, they pose greater risk owing to their morphology and aspect ratio which can induce frustrated phagocytosis (Boyles et al., 2015), hinder nanomaterial clearance (Roberts et al., 2016), and potentially exacerbate immune responses. The question remains: should these nanostructures be introduced as nanotheranostics when the majority of literature hints at their toxicity capabilities and biopersistent tendencies? In vitro exposure studies with graphene and CNTs have revealed cytotoxicity, genotoxicity, immune-cell activation, and oxidative stress to be common themes throughout, with biopersistence likely a contributing factor (Sanchez, Pietruska, Miselis, Hurt, & Kane, 2009). The manipulation of carbon-based nanotheranostics in vivo, therefore, becomes pivotal at two control points: bio-persistence and clearance (both of which tightly associate with their pathogenicity, similar to asbestos). The biopersistence of these materials is fundamentally correlated with their tendency to agglomerate, thus making nanomaterial clearance by macrophages and neutrophils much more difficult. This degradation, largely in the form of peroxidase enzymes released by immune neutrophils (Kotchey, Zhao, Kagan, & Star, 2013), is generally accepted in literature as the primary means of carbon-based material degradation in vivo; however, other mechanisms likely play a role in assisted clearance. It has also been reported that controlled respiratory burst by macrophages containing lignin peroxidase can induce biodegradation in single-walled carbon nanotubes (SWCNTs), and it was later hypothesized that the biodegraded products would induce a weaker immune response (Hou et al., 2016). Following the use of respiratory burst, the authors noted that in vitro, mimicking the enzymatic reactions of the biodegradation the SWCNTs order of accelerated degradation was heavily reliant on functionalization of the order: OH-SWCNTs, > ox-SWCNTs, >> pristine-SWCNTs. It should also be noted that defect sites on the CNTs were likely to accelerate the biodegradation process by allowing the enzymes access into the carbon-based structures. These findings are likely to be beneficial during the design and safe manufacture of clinically-relevant CNTs in the future whereby their increased biodegradability, efficient clearance, and lower toxicity will allow them to be potential theranostic agents.

It is evident that size, aspect ratio, morphology, composition, and functionalization play a crucial role in altering carbon-based theranostics toxicity both in positive and (in some instances) detrimental ways. As a result, carbon-based nanomaterials should always be treated with extreme caution, and extensive theoretical investigation coupled to robust toxicological testing prior to their use in clinical applications.

7 Quantum Dots

Where modern biological imaging is concerned, there are few choices available which surpass quantum dots, brightly fluorescent nano crystals spanning just a few nanometers in diameter and can be engineered across a broad spectrum of light frequencies (Smith & Nie, 2009). These semiconductor nano-scale quantum dots encompass an exciting part of inorganic fluorophores with use in optoelectronics as well as biological imaging (Chen et al., 2008) with unique advantages over molecular dyes with highly resistant photo-bleaching capabilities (Medintz, Mattoussi, & Clapp, 2008). In addition to the increased stability of quantum dots over conventional fluorescent dyes used in traditional biological assays, quantum dots are engineered to exhibit a wider range of excitation and emission wavelengths (Lovric et al., 2005). Microscopic observations of these particles have revealed an on-off blinking omission characteristic; however, current advances in this technology specifically engineering alloyed composition from the core to the quantum dot surface can greatly enhance the omission to now remain consistently "on" (Smith & Nie, 2009). Quantum dot composition can fluctuate widely in accordance with intended use and manufacture preference, fundamentally a single quantum dot consists of a fluorescent core and an outer crystal shell which can vary in thickness, the function of which is to insulate the core from ionization processes within a biological environment (Chen et al., 2008). As Chen et al concluded, however, there are limitations as to the photoluminescence of single and bulk nanocrystal quantum dots (NQDs) under experimental conditions. These problems stem from the surface chemistry of the crystalline shell which, after interacting with the environmental physiology, can greatly affect the emission spectra of these molecules. This study overcame the issue by manufacturing "giant" cadmium selenide (CdSe) quantum dots possessing an overly large shell consequently placing a sufficiently large gap between photoluminescent function and surface chemistry. Herein, however, lies a major issue with quantum dots; they quite often contain core components, chemicals, and nanostructures which can offer a degree of toxicity. Cadmium in this case has been extensively studied for its genotoxicity (Celik et al., 2009). The paradox remains, therefore, that to utilize these nanotheranostics overcoming the innate toxicity associated with nanostructures and chemicals is crucial.

Biological applications of quantum dots require their stable dispersion in aqueous media, tolerance of neutral, acidic, and basic pH ranges, and (ideally) water solubility. With advances in nanotheranostics in mind, there now exist numerous effective strategies for producing hydrophilic quantum dots postsynthesis. These two categories are complete cap exchange and native surface modification. Complete cap exchange encompasses hydrophobic ligand displacement in favor of hydrophilic groups which coordinate the outermost chains in the surrounding shell of the quantum dot. Typically, this can be as primitive as the removal of amines and addition of thiols conjugated to hydrophilic moieties which act both to improve water solubility and enhance stability (Chan & Nie, 1998). Designing a multifunctional nanotheranostic utilizing quantum dots can theoretically accomplish precision diagnosis with effective treatment typically as antitumor agents; this is the case with tungsten sulfide quantum dots, which remain under optimization in delivering radiotherapy/photothermal synergistic treatment (Yong et al., 2015). Quantum dots in clinical applications would be administered intravenously for maximum efficiency, with tumor targeting in mind to deliver acute doses of radiotherapy and photothermal exposures. Moreover, in the study by Yong et al, the administration of 3 nm tungsten sulfide quantum dots revealed no toxicity, confirmed through eosin staining, blood hematology, and biological assays either in vitro or in vivo, which demonstrates high levels of biocompatibility and therapeutic efficacy. Quantum dot toxicity appears to be inherently linked to size, as is the case with cadmium telluride (CdTe), which at sizes below 4 nm showed a greater ability to localize and induce toxic effects on PC12 and N9 cells at concentrations as relatively low as 10 μg/mL, quantified with chromatin condensation and membrane blebbing. Moreover, the surface charge of quantum dots bearing positively charged groupings induced higher toxicity opposed to neutral-charge quantum dots (Lovric et al., 2005). In the 2005 study by Lovric et al, the conclusions drawn were that size and surface chemistry greatly influence quantum dot localization and subcellular distribution; however, with the addition of antioxidant reagents (*N*-acetylcysteine) and bovine serum albumin, the toxicity of 3–4 nm quantum dots can be mitigated significantly.

Because it has been well established that the relative toxicity of nanotheranostics can stem from their biopersistence, the degradation of quantum dots presents a unique challenge, especially if this process takes place in vivo as the potential byproducts would be heavy metals, along with potentially modified nanostructures of the shell. The following study by Mancini et al offers novel methods of quantum dot degradation, in which they have been encapsulated in a polymer coating to better target the therapeutic, in which reactive oxygen species (ROS) a collection of short-lived, high-energy radicals produced naturally through cellular metabolism actively break down the quantum dots (Mancini, Kairdolf, Smith, & Nie, 2008). This particular finding also facilitates fluorescence quenching, with hypochlorous acid (HOCl) and hydrogen peroxide proving the most efficient at crossing the polymer coating and instigating degradation and simultaneous quenching, leading to the oxidation of either selenium ions in the core or other forms of heavy metal constituent. This may prove to be a hindrance as a nanotheranostic tool; however, as quantum dots, either at the time of delivery or over a chronic exposure period may generate ROS or activate an immune response capable of introducing more oxidative stress to a localized site promoting the premature removal of the quantum dots.

8 Summary

Presently, despite numerous studies displaying medical promise with a vast array of nanotheranostics ranging from metal oxide, quantum dots, and carbon-based materials, there remains a pressing issue. While the mechanistic science of toxicology and (pro)-inflammation (pathway elucidation) is well understood from in vitro exposures in monoculture and co-cultures as well as basic in vivo exposures (mice, rats), the pressing issue of nanomaterial fate still remains. As this is often driven by physicochemical characteristics during in vitro exposures, extrapolating this to in vivo conditions rarely yields similar trends. Currently, however, there appears to be much conflict of interest within publications as to the potential toxicity of the nanomaterials discussed and their intended use in nanotheranostics. Ideally, a nanomaterial which boasts low toxicity provides a good platform when considering nanotheranostic applications which incorporates high biocompatibility, no biopersistence, and can be specifically targeted. A summary of the advantages and disadvantages posed by the nanotheranostics discussed in this chapter can be seen in Table 2, which also details the use of various cell models used in nano(geno)toxicology to provide a robust risk assessment posed by novel ENMs.

Unfortunately, it seems at the nano-scale, the majority of nanomaterials pose some biological hazard, either through toxicity or biopersistence in nontarget cells inducing downstream, damaging effects. With the heightened interest and influx of funding, nanotheranostics will likely receive far more exposure and attention moving forward; however, it is essential that the correlation between successful application and minimal detrimental health effects remains paramount.

Conflict of Interest Statement

The authors are responsible for all content and the views portrayed within this chapter.

Table 2: Suitability of ENMs applied as nanotheranostics

Criteria for Assessing Nanotheranostic Potential	Formulation for Potential Nanotheranostic		
	Metallic NPs	**Carbon-Based ENMs**	**QDs**
Advantages for use in human clinical trials	Superparamagnetic behavior would allow for development of targeted drug delivery systems and enhanced medical imaging techniques with high resolution capabilities.	Highly diverse in structure, chemistry and potential applications. High surface area (graphene & CNTs) provides ample functionalization space, targeted therapeutics. Resilient to damage, long-lasting therapeutic potential of slow release drug	Exceptional fluorescent capabilities, vast array of excitation and emission wavelengths possible permitting tailored applications. Surface modifications can enhance uptake and targeted theranostics.
Disadvantages for use in clinical trials	Evidence within the literature that transition metal NPs can possess adverse toxicological profiles. The primary risk highlighted is the induction of oxidative stress, which can cause an adverse immunogenic response and genotoxicity.	Largely untested in complex cell models, with conflicting data in the literature regarding genotoxicity. Surface chemistry greatly governs the overall toxicity of carbon-based ENMs, problematic in targeted therapeutic design. Imaging, tracking of carbon-based ENMs difficult, require additional surface modifications which alter toxicity.	Outer shell encapsulating the nanocrystal core can be greatly affected by physiological conditions. Nanocrystal composition typically contains structures, chemicals which offer a degree of toxicity (cadmium).
Cell model advised for nanotheranostic development	Co-culture model representative of the region(s) of exposure. It is vital that these models incorporate immune cell types to elucidate complex cell-to-cell interactions.	Extensive airway co-culture modeling required to elucidate complex cell-cell interactions (immunotoxicity).	Complex 3D structures, highly representative, premade specific tissue/organ. Allows screening potential for drug delivery within a specific tissue while maintaining optical properties for diagnostics.

References

Bakand, S., Hayes, A., & Dechsakulthorn, F. (2012). Nanoparticles: a review of particle toxicology following inhalation exposure. *Inhalation Toxicology*, *24*.

Ballester, M., Nembrini, C., Dhar, N., de Titta, A., de Piano, C., Pasquier, M., et al. (2011). Nanoparticle conjugation and pulmonary delivery enhance the protective efficacy of Ag85B and CpG against tuberculosis. *Vaccine*, *29*, 6959–6966.

Baroli, B., Ennas, M. G., Loffredo, F., Isola, M., Pinna, R., & Lopez-Quintela, M. A. (2007). Penetration of metallic nanoparticles in human full-thickness skin. *The Journal of Investigative Dermatology*, *127*, 1701–1712.

Bellmann, S., Carlander, D., Fasano, A., Momcilovic, D., Scimeca, J. A., Waldman, W. J., et al. (2015). Mammalian gastrointestinal tract parameters modulating the integrity, surface properties, and absorption of food-relevant nanomaterials. *Wiley Interdisciplinary Reviews: Nanomedicine and Nanobiotechnology*, *7*, 609–622.

Bergin, I. L., & Witzmann, F. A. (2013). Nanoparticle toxicity by the gastrointestinal route: evidence and knowledge gaps. *International Journal of Biomedical Nanoscience and Nanotechnology*, *3*, https://doi.org/10.1504/IJBNN.2013.054515.

Boyles, M. S., Young, L., Brown, D. M., Maccalman, L., Cowie, H., Moisala, A., et al. (2015). Multi-walled carbon nanotube induced frustrated phagocytosis, cytotoxicity and pro-inflammatory conditions in macrophages are length dependent and greater than that of asbestos. *Toxicology In Vitro*, *29*, 1513–1528.

Bulte, J. W., & Kraitchman, D. L. (2004). Iron oxide MR contrast agents for molecular and cellular imaging. *NMR in Biomedicine*, *17*, 484–499.

Carrillo-Carrion, C., Carril, M., & Parak, W. J. (2017). Techniques for the experimental investigation of the protein corona. *Current Opinion in Biotechnology*, *46*, 106–113.

Celik, A., Buyukakilli, B., Cimen, B., Tasdelen, B., Ozturk, M. I., & Eke, D. (2009). Assessment of cadmium genotoxicity in peripheral blood and bone marrow tissues of male Wistar rats. *Toxicology Mechanisms and Methods*, *19*, 135–140.

Chan, W. C., & Nie, S. (1998). Quantum dot bioconjugates for ultrasensitive nonisotopic detection. *Science*, *281*, 2016–2018.

Chang, Y., Yang, S. T., Liu, J. H., Dong, E., Wang, Y., Cao, A., et al. (2011). In vitro toxicity evaluation of graphene oxide on A549 cells. *Toxicology Letters*, *200*, 201–210.

Chen, F., Wang, G., Griffin, J. I., Brenneman, B., Banda, N. K., Holers, V. M., et al. (2017). Complement proteins bind to nanoparticle protein corona and undergo dynamic exchange in vivo. *Nature Nanotechnology*, *12*(4), 387–393. https://doi.org/10.1038/nnano.2016.269.

Chen, Y., Vela, J., Htoon, H., Casson, J. L., Werder, D. J., Bussian, D. A., et al. (2008). "Giant" multishell CdSe nanocrystal quantum dots with suppressed blinking. *Journal of the American Chemical Society*, *130*, 5026–5027.

Cheng, X., Tian, X., Wu, A., Li, J., Tian, J., Chong, Y., et al. (2015). Protein corona influences cellular uptake of gold nanoparticles by phagocytic and nonphagocytic cells in a size-dependent manner. *ACS Applied Materials & Interfaces*, *7*, 20568–20575.

Cho, H.-J., Oh, J., Choo, M.-K., Ha, J.-I., Park, Y., & Maeng, H.-J. (2014). Chondroitin sulfate-capped gold nanoparticles for the oral delivery of insulin. *International Journal of Biological Macromolecules*, *63*, 15–20.

Corkery, K. (2000). Inhalable drugs for systemic therapy. *Respiratory Care*, *45*, 831–835.

Cuong, N. V., & Hsieh, M. F. (2011). Molecular targeting of liposomal nano-particles to lymphatic system. *Current Cancer Drug Targets*, *11*, 147–155.

Das, S., & Chaudhury, A. (2011). Recent advances in lipid nanoparticle formulations with solid matrix for oral drug delivery. *AAPS PharmSciTech*, *12*, 62–76.

de Jong, W. H., & Borm, P. J. (2008). Drug delivery and nanoparticles: applications and hazards. *International Journal of Nanomedicine*, *3*, 133–149.

Dening, T. J., Rao, S., Thomas, N., & Prestidge, C. A. (2016). Oral nanomedicine approaches for the treatment of psychiatric illnesses. *Journal of Controlled Release*, *223*, 137–156.

Digesu, C. S., Hofferberth, S. C., Grinstaff, M. W., & Colson, Y. L. (2016). From diagnosis to treatment: clinical applications of nanotechnology in thoracic surgery. *Thoracic Surgery Clinics*, *26*, 215–228.

Dizaj, S., Jafari, S., & Khosroushahi, A. (2014). A sight on the current nanoparticle-based gene delivery vectors. *Nanoscale Research Letters*, *9*, 1–9.

Doak, S. H., Griffiths, S. M., Manshian, B., Singh, N., Williams, P. M., Brown, A. P., et al. (2009). Confounding experimental considerations in nanogenotoxicology. *Mutagenesis*, *24*, 285–293.

Doak, S. H., Manshian, B., Jenkins, G. J., & Singh, N. (2012). In vitro genotoxicity testing strategy for nanomaterials and the adaptation of current OECD guidelines. *Mutation Research*, *745*, 104–111.

Dolci, S., Ierardi, V., Gradisek, A., Jaglicic, Z., Remskar, M., Apih, T., et al. (2013). Precursors of magnetic resonance imaging contrast agents based on cystine-coated iron-oxide nanoparticles. *Current Physical Chemistry*, *3*, 493–500.

Edwards, S. L., Werkmeister, J. A., & Ramshaw, J. A. (2009). Carbon nanotubes in scaffolds for tissue engineering. *Expert Review of Medical Devices*, *6*, 499–505.

Elhissi, A. M., Ahmed, W., Hassan, I. U., Dhanak, V. R., & D'emanuele, A. (2012). Carbon nanotubes in cancer therapy and drug delivery. *Journal of Drug Delivery*, *2012*, 837327.

El-Sherbiny, I. M., El-Baz, N. M., & Yacoub, M. H. (2015). Inhaled nano- and microparticles for drug delivery. *Global Cardiology Science & Practice*, *2015*, 2.

Estelrich, J., Sánchez-Martín, M. J., & Busquets, M. A. (2015). Nanoparticles in magnetic resonance imaging: from simple to dual contrast agents. *International Journal of Nanomedicine*, *10*, 1727–1741.

Evans, S. J., Clift, M. J., Singh, N., de Oliveira Mallia, J., Burgum, M., Wills, J. W., et al. (2016). Critical review of the current and future challenges associated with advanced in vitro systems towards the study of nanoparticle (secondary) genotoxicity. *Mutagenesis*. gew054.

Farrell, D., Ptak, K., Panaro, N. J., & Grodzinski, P. (2011). Nanotechnology-based cancer therapeutics—promise and challenge—lessons learned through the NCI Alliance for Nanotechnology in Cancer. *Pharmaceutical Research*, *28*, 273–278.

Fleischer, C. C., & Payne, C. K. (2014). Secondary structure of corona proteins determines the cell surface receptors used by nanoparticles. *The Journal of Physical Chemistry B*, *118*, 14017–14026.

Gatoo, M. A., Naseem, S., Arfat, M. Y., Dar, A. M., Qasim, K., & Zubair, S. (2014). Physicochemical properties of nanomaterials: implication in associated toxic manifestations. *BioMed Research International*, *2014*, 498420.

Gershanik, T., & Benita, S. (2000). Self-dispersing lipid formulations for improving oral absorption of lipophilic drugs. *European Journal of Pharmaceutics and Biopharmaceutics*, *50*, 179–188.

Giardiello, M., Liptrott, N. J., Mcdonald, T. O., Moss, D., Siccardi, M., Martin, P., et al. (2016). Accelerated oral nanomedicine discovery from miniaturized screening to clinical production exemplified by paediatric HIV nanotherapies. *Nature Communications*, 7.

Guo, L., Feng, Z., Cai, L., Xi, Z., Mou, X., Wang, T., et al. (2016). Effects of a protein-corona on the cellular uptake of ferroferric oxide nanoparticles. *Journal of Nanoscience and Nanotechnology*, *16*, 7125–7128.

Gupta, A. K., & Gupta, M. (2005). Synthesis and surface engineering of iron oxide nanoparticles for biomedical applications. *Biomaterials*, *26*, 3995–4021.

Gupta, S., Bansal, R., Jindal, N., & Jindal, A. (2013). Nanocarriers and nanoparticles for skin care and dermatological treatments. *Indian Dermatology Online Journal*, *4*, 267–272.

He, X., Aker, W. G., Huang, M. J., Watts, J. D., & Hwang, H. M. (2015). Metal oxide nanomaterials in nanomedicine: applications in photodynamic therapy and potential toxicity. *Current Topics in Medicinal Chemistry*, *15*, 1887–1900.

Hou, J., Wan, B., Yang, Y., Ren, X. M., Guo, L. H., & Liu, J. F. (2016). Biodegradation of single-walled carbon nanotubes in macrophages through respiratory burst modulation. *International Journal of Molecular Sciences*, *17*, 409.

Institute of Medicine Roundtable on Environmental Health Sciences, Research, and Medicine. (2005). The national academies collection: reports funded by National Institutes of Health. In L. Goldman & C. Coussens (Eds.), *Implications of Nanotechnology for Environmental Health Research*. Washington, DC: National Academies Press, National Academy of Sciences.

Jacobs, J. F., van de Poel, I., & Osseweijer, P. (2010). Sunscreens with titanium dioxide (TiO(2)) nano-particles: a societal experiment. *NanoEthics*, *4*, 103–113.

Jain, V., Jain, S., & Mahajan, S. C. (2015). Nanomedicines based drug delivery systems for anti-cancer targeting and treatment. *Current Drug Delivery*, *12*, 177–191.

Jatana, S., & DeLouise, L. A. (2014). Understanding engineered nanomaterial skin interactions and the modulatory effects of ultraviolet radiation skin exposure. *Wiley Interdisciplinary Reviews Nanomedicine and Nanobiotechnology*, *6*, 61–79.

Jia, L. (2005). Nanoparticle formulation increases oral bioavailability of poorly soluble drugs: approaches experimental evidences and theory. *Current Nanoscience*, *1*, 237–243.

Johnson, G. E. (2012). Mammalian cell HPRT gene mutation assay: test methods. *Methods in Molecular Biology*, *817*, 55–67.

Johnston, H., Brown, D., Kermanizadeh, A., Gubbins, E., & Stone, V. (2012). Investigating the relationship between nanomaterial hazard and physicochemical properties: informing the exploitation of nanomaterials within therapeutic and diagnostic applications. *Journal of Controlled Release*, *164*, 307–313.

Kah, J. C., Chen, J., Zubieta, A., & Hamad-Schifferli, K. (2012). Exploiting the protein corona around gold nanorods for loading and triggered release. *ACS Nano*, *6*, 6730–6740.

Kermanizadeh, A., Balharry, D., Wallin, H., Loft, S., & Møller, P. (2015). Nanomaterial translocation—the biokinetics, tissue accumulation, toxicity and fate of materials in secondary organs—a review. *Critical Reviews in Toxicology*, *45*, 837–872.

Kostoglou, N., Tzitzios, V., Kontos, A. G., Giannakopoulos, K., Tampaxis, C., Papavasiliou, A., et al. (2015). Synthesis of nanoporous graphene oxide adsorbents by freeze drying or microwave radiation: characterization and hydrogen storage properties. *International Journal of Hydrogen Energy*, *40*, 6844–6852.

Kotchey, G. P., Zhao, Y., Kagan, V. E., & Star, A. (2013). Peroxidase-mediated biodegradation of carbon nanotubes in vitro and in vivo. *Advanced Drug Delivery Reviews*, *65*, 1921–1932.

Kreyling, W. G., Holzwarth, U., Haberl, N., Kozempel, J., Hirn, S., Wenk, A., et al. (2017a). Quantitative biokinetics of titanium dioxide nanoparticles after intravenous injection in rats: Part 1. *Nanotoxicology*, 1–9.

Kreyling, W. G., Holzwarth, U., Haberl, N., Kozempel, J., Wenk, A., Hirn, S., et al. (2017b). Quantitative biokinetics of titanium dioxide nanoparticles after intratracheal instillation in rats: Part 3. *Nanotoxicology*, 1–11.

Kreyling, W. G., Holzwarth, U., Schleh, C., Kozempel, J., Wenk, A., Haberl, N., et al. (2017). Quantitative biokinetics of titanium dioxide nanoparticles after oral application in rats: Part 2. *Nanotoxicology*, 1–11.

Kulthong, K., Srisung, S., Boonpavanitchakul, K., Kangwansupamonkon, W., & Maniratanachote, R. (2010). Determination of silver nanoparticle release from antibacterial fabrics into artificial sweat. *Particle and Fibre Toxicology*, *7*, 8.

Lameijer, M. A., Tang, J., Nahrendorf, M., Beelen, R. H. J., & Mulder, W. J. M. (2013). Monocytes and macrophages as nanomedicinal targets for improved diagnosis and treatment of disease. *Expert Review of Molecular Diagnostics*, *13*, 567–580.

Laurent, S., Dutz, S., Häfeli, U. O., & Mahmoudi, M. (2011). Magnetic fluid hyperthermia: focus on superparamagnetic iron oxide nanoparticles. *Advances in Colloid and Interface Science*, *166*, 8–23.

Li, B., Fan, K., Ma, X., Liu, Y., Chen, T., Cheng, Z., et al. (2016). Graphene-based porous materials with tunable surface area and CO2 adsorption properties synthesized by fluorine displacement reaction with various diamines. *Journal of Colloid and Interface Science*, *478*, 36–45.

Liu, M., Zhang, J., Shan, W., & Huang, Y. (2015). Developments of mucus penetrating nanoparticles. *Asian Journal of Pharmaceutical Sciences*, *10*, 275–282.

Lovric, J., Bazzi, H. S., Cuie, Y., Fortin, G. R., Winnik, F. M., & Maysinger, D. (2005). Differences in subcellular distribution and toxicity of green and red emitting CdTe quantum dots. *Journal of Molecular Medicine*, *83*, 377–385.

Lu, A. H., Salabas, E. L., & Schuth, F. (2007). Magnetic nanoparticles: synthesis, protection, functionalization, and application. *Angewandte Chemie (International Ed in English)*, *46*, 1222–1244.

Madani, S. Y., Naderi, N., Dissanayake, O., Tan, A., & Seifalian, A. M. (2011). A new era of cancer treatment: carbon nanotubes as drug delivery tools. *International Journal of Nanomedicine*, *6*, 2963–2979.

Mahmoudi, M., Sant, S., Wang, B., Laurent, S., & Sen, T. (2011). Superparamagnetic iron oxide nanoparticles (SPIONs): development, surface modification and applications in chemotherapy. *Advanced Drug Delivery Reviews*, *63*, 24–46.

Ma-Hock, L., Strauss, V., Treumann, S., Kuttler, K., Wohlleben, W., Hofmann, T., et al. (2013). Comparative inhalation toxicity of multi-wall carbon nanotubes, graphene, graphite nanoplatelets and low surface carbon black. *Particle and Fibre Toxicology*, *10*, 23.

Mancini, M. C., Kairdolf, B. A., Smith, A. M., & Nie, S. (2008). Oxidative quenching and degradation of polymer-encapsulated quantum dots: new insights into the long-term fate and toxicity of nanocrystals in vivo. *Journal of the American Chemical Society*, *130*, 10836–10837.

Medina, C., Santos-Martinez, M. J., Radomski, A., Corrigan, O. I., & Radomski, M. W. (2007). Nanoparticles: pharmacological and toxicological significance. *British Journal of Pharmacology*, *150*, 552–558.

Medintz, I. L., Mattoussi, H., & Clapp, A. R. (2008). Potential clinical applications of quantum dots. *International Journal of Nanomedicine*, *3*, 151–167.

Monopoli, M. P., Aberg, C., Salvati, A., & Dawson, K. A. (2012). Biomolecular coronas provide the biological identity of nanosized materials. *Nature Nanotechnology*, *7*, 779–786.

Monopoli, M. P., Walczyk, D., Campbell, A., Elia, G., Lynch, I., Baldelli Bombelli, F., et al. (2011). Physical-chemical aspects of protein corona: relevance to in vitro and in vivo biological impacts of nanoparticles. *Journal of the American Chemical Society*, *133*, 2525–2534.

Morigi, V., Tocchio, A., Bellavite Pellegrini, C., Sakamoto, J. H., Arnone, M., & Tasciotti, E. (2012). Nanotechnology in medicine: from inception to market domination. *Journal of Drug Delivery*, *2012*(389485).

Mout, R., Moyano, D. F., Rana, S., & Rotello, V. M. (2012). Surface functionalization of nanoparticles for nanomedicine. *Chemical Society Reviews*, *41*, 2539–2544.

Nalwa, H. S. (2014). A special issue on reviews in biomedical applications of nanomaterials, tissue engineering, stem cells, bioimaging, and toxicity. *Journal of Biomedical Nanotechnology*, *10*, 2421–2423.

Neuwelt, A., Sidhu, N., Hu, C. A., Mlady, G., Eberhardt, S. C., & Sillerud, L. O. (2015). Iron-based superparamagnetic nanoparticle contrast agents for MRI of infection and inflammation. *AJR American Journal of Roentgenology*, *204*, 12733.

Nezakati, T., Cousins, B. G., & Seifalian, A. M. (2014). Toxicology of chemically modified graphene-based materials for medical application. *Archives of Toxicology*, *88*, 1987–2012.

Oberdorster, G., Maynard, A., Donaldson, K., Castranova, V., Fitzpatrick, J., Ausman, K., et al. (2005). Principles for characterizing the potential human health effects from exposure to nanomaterials: elements of a screening strategy. *Particle and Fibre Toxicology*, *2*, 8.

Orecchioni, M., Cabizza, R., Bianco, A., & Delogu, L. G. (2015). Graphene as cancer theranostic tool: progress and future challenges. *Theranostics*, *5*, 710–723.

Pal, S., Yoon, E. J., Tak, Y. K., Choi, E. C., & Song, J. M. (2009). Synthesis of highly antibacterial nanocrystalline trivalent silver polydiguanide. *Journal of the American Chemical Society*, *131*, 16147–16155.

Palmer, B. C., & DeLouise, L. A. (2016). Nanoparticle-enabled transdermal drug delivery systems for enhanced dose control and tissue targeting. *Molecules*, 21.

Pang, P., Li, B., Hu, X., Kang, Z., Guan, S., Gong, F., et al. (2014). Multifunctional nano-vector for gene delivery into human adipose derived mesenchymal stem cells and in vitro cellular magnetic resonance imaging. *Zhonghua Yi Xue Za Zhi*, *94*, 1021–1024.

Patlolla, A., Knighten, B., & Tchounwou, P. (2010). Multi-walled carbon nanotubes induce cytotoxicity, genotoxicity and apoptosis in normal human dermal fibroblast cells. *Ethnicity & Disease*, *20*, . S1-65-72.

Pirela, S. V., Sotiriou, G. A., Bello, D., Shafer, M., Bunker, K. L., Castranova, V., et al. (2015). Consumer exposures to laser printer-emitted engineered nanoparticles: a case study of life-cycle implications from nano-enabled products. *Nanotoxicology*, *9*, 760–768.

Rahman, M., Akhter, S., Ahmad, M. Z., Ahmad, J., Addo, R. T., Ahmad, F. J., et al. (2015). Emerging advances in cancer nanotheranostics with graphene nanocomposites: opportunities and challenges. *Nanomedicine (London, England)*, *10*, 2405–2422.

Rau, J. L. (2005). The inhalation of drugs: advantages and problems. *Respiratory Care*, *50*(3), 367–382.

Reddy, A. R., Krishna, D. R., Reddy, Y. N., & Himabindu, V. (2010). Translocation and extra pulmonary toxicities of multi wall carbon nanotubes in rats. *Toxicology Mechanisms and Methods*, *20*, 267–272.

Resnik, D. B., & Tinkle, S. S. (2007). Ethics in nanomedicine. *Nanomedicine (London, England)*, *2*, 345–350.

Roberts, J. R., Mercer, R. R., Stefaniak, A. B., Seehra, M. S., Geddam, U. K., Chaudhuri, I. S., et al. (2016). Evaluation of pulmonary and systemic toxicity following lung exposure to graphite nanoplates: a member of the graphene-based nanomaterial family. *Particle and Fibre Toxicology*, *13*, 34.

Robinson, J. T., Welsher, K., Tabakman, S. M., Sherlock, S. P., Wang, H., Luong, R., et al. (2010). High performance in vivo near-IR (>1 mum) imaging and photothermal cancer therapy with carbon nanotubes. *Nano Research*, *3*, 779–793.

Saha, K., Moyano, D. F., & Rotello, V. M. (2014). Protein coronas suppress the hemolytic activity of hydrophilic and hydrophobic nanoparticles. *Materials Horizons*, *2014*, 102–105.

Saini, R., Saini, S., & Sharma, S. (2010). Nanotechnology: the future medicine. *Journal of Cutaneous and Aesthetic Surgery*, *3*, 32–33.

Sanchez, V. C., Pietruska, J. R., Miselis, N. R., Hurt, R. H., & Kane, A. B. (2009). Biopersistence and potential adverse health impacts of fibrous nanomaterials: what have we learned from asbestos? *Wiley Interdisciplinary Reviews Nanomedicine and Nanobiotechnology*, *1*, 511–529.

Schöttler, S., Landfester, K., & Mailänder, V. (2016). Controlling the stealth effect of nanocarriers through understanding the protein corona. *Angewandte Chemie International Edition*, *55*, 8806–8815.

Sharma, R., Baik, J. H., Perera, C. J., & Strano, M. S. (2010). Anomalously large reactivity of single graphene layers and edges toward electron transfer chemistries. *Nano Letters*, *10*, 398–405.

Shaw, C. A., Mortimer, G. M., Deng, Z. J., Carter, E. S., Connell, S. P., Miller, M. R., et al. (2016). Protein corona formation in bronchoalveolar fluid enhances diesel exhaust nanoparticle uptake and pro-inflammatory responses in macrophages. *Nanotoxicology*, *10*, 981–991.

Shi, J., Kantoff, P. W., Wooster, R., & Farokhzad, O. C. (2017). Cancer nanomedicine: progress, challenges and opportunities. *Nature Reviews Cancer*, *17*, 20–37.

Silva, G. A. (2004). Introduction to nanotechnology and its applications to medicine. *Surgical Neurology*, *61*, 216–220.

Sim, T., Lim, C., Hoang, N. H., Joo, H., Lee, J. W., Kim, D.-W., et al. (2016). Nanomedicines for oral administration based on diverse nanoplatform. *Journal of Pharmaceutical Investigation*, *46*, 351–362.

Singh, N., Jenkins, G. J. S., Nelson, B. C., Marquis, B. J., Maffeis, T. G. G., Brown, A. P., et al. (2012). The role of iron redox state in the genotoxicity of ultrafine superparamagnetic iron oxide nanoparticles. *Biomaterials*, *33*, 163–170.

Smith, A. M., & Nie, S. (2009). Next-generation quantum dots. *Nature Biotechnology*, *27*, 732–733.

Srinivas, P. R., Philbert, M., Vu, T. Q., Huang, Q., Kokini, J. L., Saltos, E., et al. (2010). Nanotechnology research: applications in nutritional sciences. *The Journal of Nutrition*, *140*, 119–124.

Thorley, A. J., & Tetley, T. D. (2013). New perspectives in nanomedicine. *Pharmacology & Therapeutics*, *140*, 176–185.

Tiwari, P. M., Eroglu, E., Bawage, S. S., Vig, K., Miller, M. E., Pillai, S., et al. (2014). Enhanced intracellular translocation and biodistribution of gold nanoparticles functionalized with a cell-penetrating peptide (VG-21) from vesicular stomatitis virus. *Biomaterials*, *35*, 9484–9494.

Toyokuni, S. (2013). Genotoxicity and carcinogenicity risk of carbon nanotubes. *Advanced Drug Delivery Reviews*, *65*, 2098–2110.

Wang, L. S., Chuang, M. C., & Ho, J. A. (2012). Nanotheranostics—a review of recent publications. *International Journal of Nanomedicine*, *7*, 4679–4695.

Webster, T. J. (2006). Nanomedicine: what's in a definition? *International Journal of Nanomedicine*, *1*, 115–116.

Wei, Y., Zhou, F., Zhang, D., Chen, Q., & Xing, D. (2016). A graphene oxide based smart drug delivery system for tumor mitochondria-targeting photodynamic therapy. *Nanoscale*, *8*, 3530–3538.

Wu, J., Wang, Y. S., Yang, X. Y., Liu, Y. Y., Yang, J. R., Yang, R., et al. (2012). Graphene oxide used as a carrier for adriamycin can reverse drug resistance in breast cancer cells. *Nanotechnology*, *23*(355101).

Yah, C. S., Simate, G. S., & Iyuke, S. E. (2012). Nanoparticles toxicity and their routes of exposures. *Pakistan Journal of Pharmaceutical Sciences*, *25*, 477–491.

Yang, K., Feng, L., Shi, X., & Liu, Z. (2013). Nano-graphene in biomedicine: theranostic applications. *Chemical Society Reviews*, *42*, 530–547.

Yang, K., Wan, J., Zhang, S., Tian, B., Zhang, Y., & Liu, Z. (2012). The influence of surface chemistry and size of nanoscale graphene oxide on photothermal therapy of cancer using ultra-low laser power. *Biomaterials*, *33*, 2206–2214.

Yang, K., Zhang, S., Zhang, G., Sun, X., Lee, S. T., & Liu, Z. (2010). Graphene in mice: ultrahigh in vivo tumor uptake and efficient photothermal therapy. *Nano Letters*, *10*, 3318–3323.

Yong, Y., Cheng, X., Bao, T., Zu, M., Yan, L., Yin, W., et al. (2015). Tungsten Sulfide Quantum Dots as Multifunctional Nanotheranostics for In Vivo Dual-Modal Image-Guided Photothermal/Radiotherapy Synergistic Therapy. *ACS Nano*, *9*, 12451–12463.

Yu, M. K., Jeong, Y. Y., Park, J., Park, S., Kim, J. W., Min, J. J., et al. (2008). Drug-loaded superparamagnetic iron oxide nanoparticles for combined cancer imaging and therapy in vivo. *Angewandte Chemie International Edition*, *47*, 5362–5365.

Zanganeh, S., Spitler, R., Erfanzadeh, M., Alkilany, A. M., & Mahmoudi, M. (2016). Protein corona: opportunities and challenges. *The International Journal of Biochemistry & Cell Biology*, *75*, 143–147.

Zhang, L., Xia, J., Zhao, Q., Liu, L., & Zhang, Z. (2010). Functional graphene oxide as a nanocarrier for controlled loading and targeted delivery of mixed anticancer drugs. *Small*, *6*, 537–544.

Zhao, F., Zhao, Y., Liu, Y., Chang, X., Chen, C., & Zhao, Y. (2011). Cellular uptake, intracellular trafficking, and cytotoxicity of nanomaterials. *Small*, *7*, 1322–1337.

Zhao, J., & Castranova, V. (2011). Toxicology of nanomaterials used in nanomedicine. *Journal of Toxicology and Environmental Health Part B, Critical Reviews*, *14*, 593–632.

Zolnik, B. S., Gonzalez-Fernandez, A., Sadrieh, N., & Dobrovolskaia, M. A. (2010). Nanoparticles and the immune system. *Endocrinology*, *151*, 458–465.

Further Reading

Chou, L. Y., Ming, K., & Chan, W. C. (2011). Strategies for the intracellular delivery of nanoparticles. *Chemical Society Reviews*, *40*, 233–245.

CHAPTER 11

Nanomaterials for Theranostics of Gastric Cancer

Daxiang Cui
Shanghai Jiao Tong University, Shanghai, PR China

1 Introduction

Gastric cancer (GC) is the fourth most common cancer and the second leading cause of cancer-related mortality worldwide (Ferlay et al., 2010). GC is the second most common cancer, and the third leading cause of cancer-related death in China (Jemal et al., 2011). Although gastroscopy has been broadly used for screening of early GC patients, GC remains very difficult to cure effectively, primarily because most patients present advanced stages of the disease. Therefore, detecting early GC has become a great challenge.

In 1999, we first proposed to develop a GC prewarning and early diagnosis system (Takahashi, Saikawa, & Kitagawa, 2013). Then, we tried to realize GC prewarning and early diagnosis by screening early GC biomarkers, developing a new detection method for GC biomarkers, establishing a GC prewarning database including molecular imaging database, biomarker database, and information treatment platform (Dicken et al., 2005; Uemura et al., 2001).

Since Weissleder and colleagues presented the concept of molecular imaging, the method has made great advances, and researchers have been actively investigating potential applications for it such as tumor early diagnosis, in vivo genotyping, targeted imaging and simultaneous therapy, and surgery navigation (Comis & Carter, 1974). These advances highly depend on the advances of nanomaterials and nanotechnology.

Nanotechnology has made an important contribution towards cancer prevention, diagnosis, imaging, and treatment (Kuo, Chao, & Li, 2014). It not only provides unprecedented capability for carrying multiple diagnostic and therapeutic payloads in the same package, but also facilitates targeting delivery into specific sites across complex biological barriers (Proserpio et al., 2014). The multifunctional integrated system combines different properties such as tumor targeting, imaging, and selective therapy in an all-in-one system which will provide more useful multimodal approaches in the battle against cancer.

Handbook of Nanomaterials for Cancer Theranostics. https://doi.org/10.1016/B978-0-12-813339-2.00011-6

Current GC therapeutic methods mainly include surgery, radiation, and chemotherapies, which are generally very effective for early and in situ GCs, but advanced and metastatic cases do not respond to chemo- or radiation therapies (Zhang & Fan, 2010a). Resistance to chemotherapy-induced apoptosis is a major cause for the failure of conventional therapies. The current prognosis of GC is very poor, with 5-year survivals of less than 24% (Cui et al., 2005). Therefore, recognizing, tracking, and killing early GC cells is a great challenge in the treatment of early GC patients.

A lot of studies show that GC is not sensitive to traditional therapies, especially to chemotherapy agents, which seems to be closely related to numerous intrinsic or acquired properties of gastric cancer stem cells (GCSCs) (Zhang et al., 2014). GCSCs can cause GC and are responsible for invasion, metastasis, and resistance to conventional therapies, which has profound implications for anti-cancer targeted therapies (Wang et al., 2011). Therefore, GCSC-based targeted therapy should be one of the most effective anticancer strategies.

Immunological studies illustrate and summarize the reasons for metastasis and recurrence of GC (Ruan et al., 2012). The intrinsic antigenicity weakness of tumor cells means that immunological surveillance of the host cannot identify and eliminate the malignant cells that are distributed out of the resection field and the peripheral lymphoid organs. Because of immunological surveillance defects or dysfunction of the host, after traditional therapy was done, residual tumor cells which had escaped from immune surveillance revived again. Due to the substantial toxicities of most traditional curative intents, coupled with the compromised, extremely feeble immunity of the patient, makes it necessary to develop new therapeutic strategies that could enhance the host's immune surveillance and/or improve immunogenicity of the tumor cell.

Herein, we review the main advances in nanomaterials for theranostics of GC, explore their clinical translational prospects, and discuss the concepts, issues, approaches, and challenges involved.

2 Nanosensor for Breath Biomarker Detection to screen Early Gastric Cancer Patients

Gas chromatography-mass spectrometry (GC-MS) coupled with solid-phase microextraction (SPME) was used to screen out 14 volatile organic compound (VOC) biomarkers in breath to distinguish early gastric cancer (EGC) and advanced gastric cancer (AGC) patients from healthy persons. Then, a breath analysis approach based on a surface enhanced Raman scattering (SERS) sensor was developed to detect those biomarkers. As shown in Fig. 1, using hydrazine vapor adsorbed in graphene oxide (GO) film, the clean SERS sensor was prepared by in situ forming Au nanoparticles (AuNPs) on reduced graphene oxide (RGO). In the SERS sensor, RGO can selectively adsorb and enrich the identified biomarkers from breath as

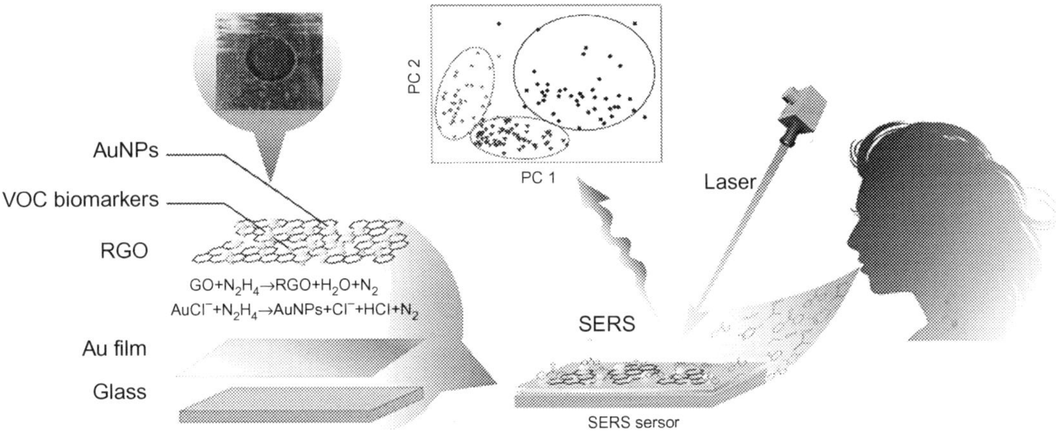

Fig. 1
Schematic diagrams of SERS sensor and overview of the processes involved in breath.
Permission is from ACS.

SPME fiber, and AuNPs dispersing well on RGO endow this SERS sensor with an effective detection of adsorbed biomarkers. Fourteen Raman bands associated with the biomarkers are selected as the fingerprints of biomarker patterns to distinguish the persons in different states. The approach has successfully analyzed and distinguished different simulated breath samples and 200 breath samples of clinical patients with a sensitivity of higher than 83% and a specificity of more than 92%. The VOC biomarkers and established breath analysis approach can not only diagnose GC, but can also distinguish EGC and AGC. This sensor has great potential for clinical translation in primary screening diagnosis and stage-determination of GC in the near future.

3 A Giant Magnetoimpedance-Based Microfluidic System for Serum Biomarker Detection

A detection method in which microfluidic chips are used in combination with giant magnetoimpedance (GMI)-related technology has developed quickly over the past few decades. It is reported that a GMI-based microfluidic system for screening multiplex GC biomarkers was designed and fabricated by micro-electromechanical systems (MEMS) technology. As shown in Fig. 2, this system can analyze eight GC protein biomarkers simultaneously in less than 25 min and offers a more stable detection signal than conventional enzymological or fluorescent methods. The microfluidic chip was then tested in 150 clinical specimens and compared with enzyme-linked immunosorbent assay (ELISA) method. The results indicated excellent agreement. The GMI-based microfluidic system for detection of eight biomarkers shows promising potentials for diagnosis of GC.

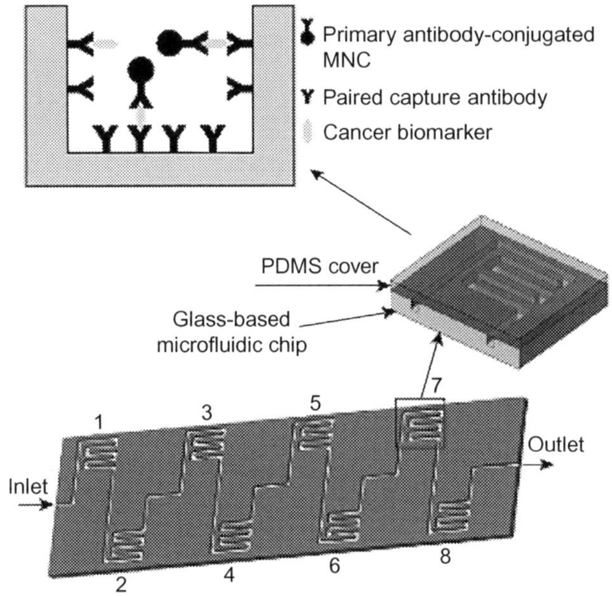

Fig. 2
The schematic illustration of GMI-based microfluidic chip (detection area 1: CA199; 2: CEA; 3: CA125; 4: VEGF; 5: Gastrin 17; 6: CA724; 7: PGI; 8: PGII).

4 Fluorescent Magnetic Nanoprobes for Targeted Imaging and Therapy of Gastric Cancer

In our previous work, we controllably synthesized silica-coated CdTe quantum dots and superparamagnetic nanoparticle composites (FMNPs) with strong fluorescent signals and excellent magnetic properties (He et al., 2011). We also confirmed that as-prepared nanoparticles own good biocompatibility and stability (Li et al., 2010).

4.1 BRCAA1 Antibody-Conjugated FMNPs for Targeted Imaging of Gastric Cancer

In 1999, we screened out and cloned BRCAA1 gene (breast cancer associated antigen 1 gene) from breast cancer cell line MCF-7 cells [AF208045, also called ARID4B (AT-rich interactive domain-containing protein 4B)], and identified its antigen epitope peptide SSKKQKRSHK (Huang et al., 2012; Zhou et al., 2013). We also prepared BRCAA1 polyclonal antibody, and observed that the BRCAA1 protein exhibited over-expression in almost 65% of clinical specimens of GC tissues (Zhang et al., 2013). We also observed that BRCAA1 antigen is over-expressed in GC cell lines such as MKN-1, MKN-74, SGC-7901, KATO-III, and MGC803 cells (Yang, Raufi, & Klempner, 2014). Therefore, we predict that BRCAA1 protein may be one potential targeting molecule for in vivo GC cells.

Based on our previous work, we fully exploited the advantages of FMNPs and BRCAA1 antigen, prepared monoclonal antibody against BRCAA1 protein, and prepared BRCAA1 monoclonal antibody-conjugated fluorescent magnetic nanoprobes (BRCAA1-FMNPs), and, employing a nude mice model loaded with GC of 5 mm in diameter and IVIS imaging system and Magnetic Resonance Imaging System, investigated the feasibility of as-prepared nanoprobes for in vivo targeted dual modal imaging of GC.

As shown in Table 1, we successfully obtained two BRCAA1 monoclonal antibody cell lines with high specificity. As shown in Table 2, BRCAA1 monoclonal antibody conjugated FMNPs nanoprobes were successfully prepared with coupling ratio of more than 80%.

By monitoring real-time fluorescence intensity in the whole body, the tumor-targeting character of the anti-BRCAA1-FMNPs probe was easily determined in the nude mice loaded with GC MGC803 cells. As shown in Fig. 3A, the whole animals produced fluorescent signals within 30 min postinjection with the nanoprobes, and the subcutaneous tumor tissues could be clearly delineated from the surrounding background tissue between 1 and 12 h postinjection, with maximum contrast occurring at 6 h postinjection. Strong fluorescence signals were still being detected in the tumor site at 6 h postinjection, which indicated that the anti-BRCAA1-FMNPs nanoprobes were preferentially accumulated in the tumor tissues. Indeed, based on the results in Fig. 3B, the higher tumor-to-background ratio (TBR) value highly suggested that as-prepared nanoprobes preferentially accumulated in tumor tissues compared to normal control tissues. This was confirmed in fluorescence images, which

Table 1: Titers of BRCAA1 monoclonal antibodies in ascites fluid induced by hybridoma clone cells by ELISA

Clone	Antibody Titer[a]			
	BRCAA1 (C)-OVA[b]	BRCAA1 (C)-BSA[b]	BSA[b]	OVA[b]
S-200-5	1,024,000	1,024,000	<1000	<1000
S-335-5	128,000	512,000	<1000	<1000

[a]The reciprocal of ascites fluid dilution, the first dilution of ascites fluid was 1:1000.
[b]The antigens were coated on ELISA plate.

Table 2: Coupling rate measurement of FMNPs-BRCAA1 antibody

	Total Concentration of the Anti-BRCAA1 Antibody (ng/μL)	The Concentration of BRCAA1 Antibody in Residual Reaction Mixture (ng/μL)	Coupling Rate (%)
1	1000.0	197.3	80.27
2	1000.0	191.2	80.88
3	1000.0	203.0	79.70

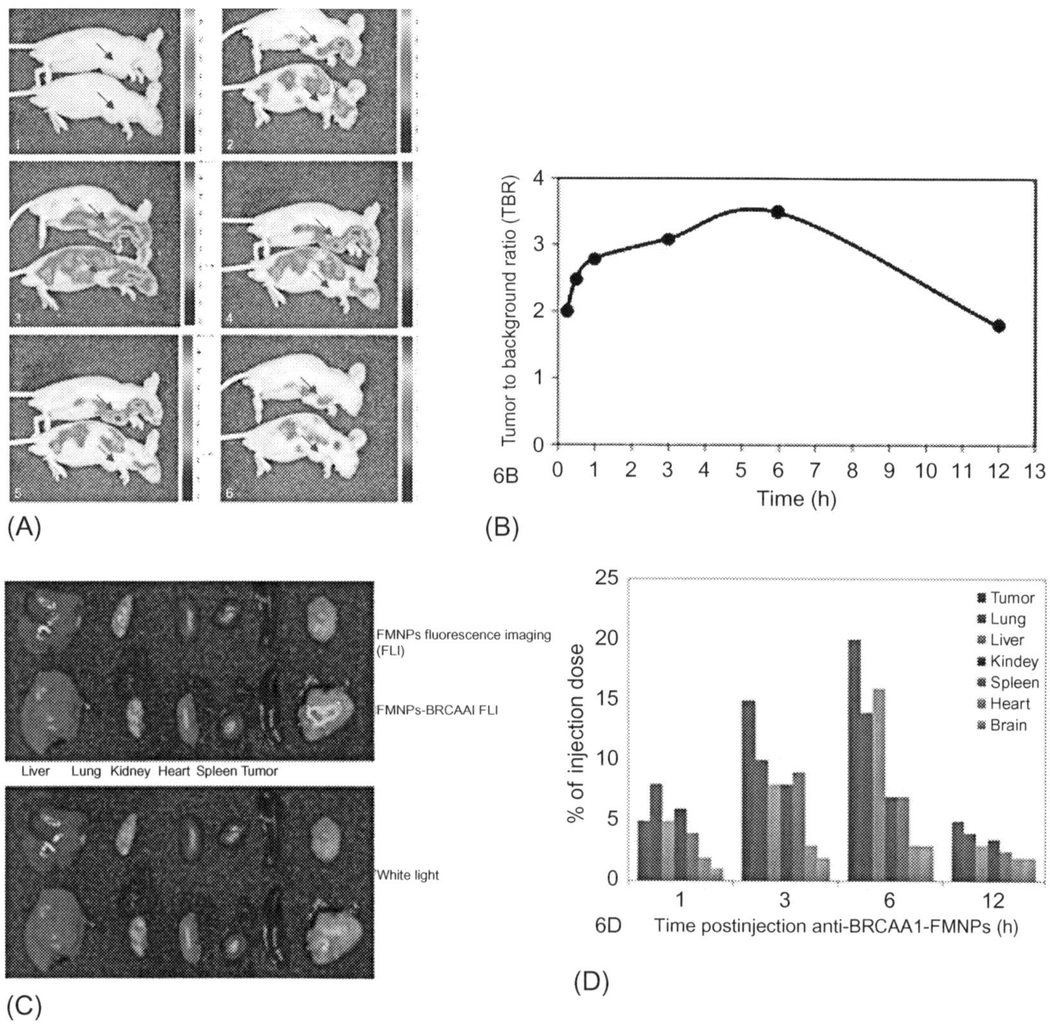

Fig. 3

In vivo fluorescence images of tumor accumulation and tissue distribution for anti-BRCAA1-FMNPs nanoprobes in MGC803 human gastric tumor-bearing athymic nude mice. (A) In vivo fluorescence images of athymic nude mice bearing MGC803 human gastric tumor were obtained after injection of anti-BRCAA1-FMNPs nanoprobes at different time point. The tumor location is specified with an arrow. A-1: 0 h, A-2: 0.5 h, A-3: 1 h, A-4: 3 h, A-5: 6 h, A-6: 12 h. (B) TBR [tissue to background (muscle) ratio] value. The TBR value was determined as follows: TBR = (tumor signal-background signal)/(background signal). (C) ex vivo fluorescence images of dissected organs and tumors of mice bearing MGC803 human gastric tumor sacrificed at 12 h after injection of anti-BRCAA1-FMNPs nanoprobes. The fluorescence images of dissected organs and tumor were obtained using a fluorescence imaging system with a 630 nm emission filter. (D) Biodistribution of anti-BRCAA1-FMNPs in mice after intravenous injection. Several time points after injection, iron amounts in tissue samples were evaluated by ICP mass spectrometry ($n = 3$). *Permission is from Springer Press.*

showed that the fluorescence signal of as-prepared nanoprobes in the tumor site was strongest among all mice organs, as shown in Fig. 3C. In addition, after 12 h postinjection of anti-BRCAA1-FMNPs nanoprobes, fluorescence intensity in the tumor was still observed clearly, while the uptake of prepared nanoprobes in normal organs was not obvious. These data highly suggest that prepared nanoprobes can highly efficiently target tumor tissues inside nude mice loaded with GC. We also observed that those nanoprobes in the whole mouse body almost completely disappeared at 12 h postinjection, we also detected the partial nanoprobes exited out from the cholecyst system, the time-dependent cholecyst clearance of nanoprobes highly suggest that as-prepared nanoprobes cannot stay inside nude mice for a longer time, thus, as-prepared nanoprobes have good biosafety.

In vivo MR imaging was performed on nude mice loaded with subcutaneous GC at 12 h postinjection. Representative images of T2 maps were shown in Fig. 4, after injecting the nanoprobes, a significant change in signal intensity was observed in some regions of the tumors, indicating that there was accumulation of the nanoprobes in the tumor site as shown in Fig. 4B, as the arrow shows. As a control, after the mice models with GC were injected FMNPs for 12 h, the mice were performed MR imaging, which did not show intensive signal in tumor area (Fig. 4A).

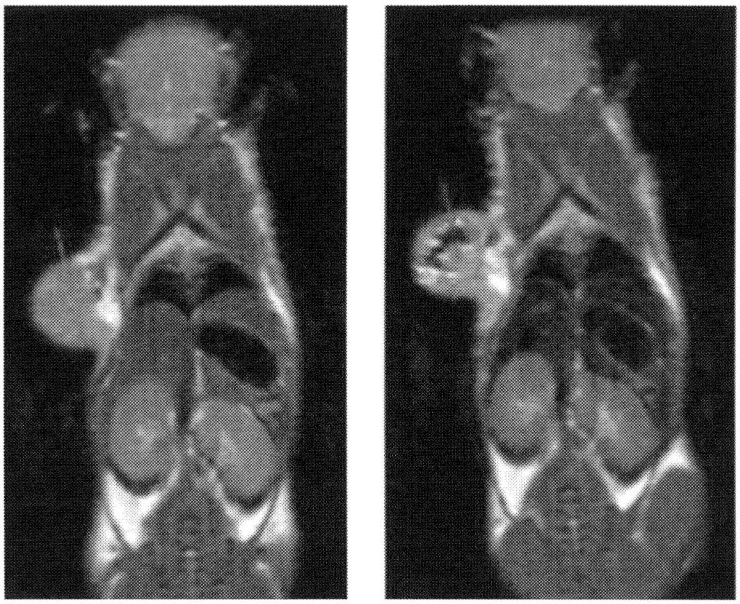

(A) (B)

Fig. 4

In vivo MR imaging of anti-BRCAA1-FMNPs nanoprobes. (A) MR imaging of the mice model with gastric cancer injected with FMNPs for 12 h. (B) MR imaging of the mice model with gastric cancer injected with anti-BRCAA1-FMNPs for 12 h. *Permission is from Springer Press.*

In vitro evaluation of excised major tissues (including liver, lung, spleen, kidney, and heart) and the tumor itself indicated that the anti-BRCAA1-FMNPs probes were mainly up-taken by the tumor tissues, which exhibited strong fluorescence signals, as shown in Fig. 3D, whereas other tissues including liver, lung, spleen, and heart up-took far fewer anti-BRCAA1-FMNPs nanoprobes, which further indicates that as-prepared anti-BRCAA1-FMNPs nanoprobes can target GC tissues. We also used HE staining to check all organs, and no obvious damage was observed in important organs.

Therefore, we successfully prepared a novel anti-BRCAA1-FMNPs nanoprobe which can be used for in vivo bimodal imaging such as fluorescent imaging and magnetic resonance imaging (MRI), and which has an obviously specific targeting ability toward GC tissues with 5 mm in diameter during 0.5 and 12 h of postinjection, and has good biocompatibility. The as-prepared multifunctional nanoprobes also can be used for hyperthermia therapy of GC under in vitro alternating magnetic field irradiation, and have great potential in applications such as simultaneous imaging and targeting therapy of clinical GC in near future (Shen et al., 2012).

4.2 HAI-178 Antibody-Conjugated FMNPs for Targeted Imaging and Therapy

We also prepared another important nanoprobe, that is, HAI-178 antibody-conjugated fluorescent magnetic nanoprobes, which was successfully used for targeted imaging and hyperthermia therapy of GC. Dr. Jian Ni's group prepared a specific monoclonal antibody against α-subunit of ATP synthase, named HAI-178 antibody, and provided HAI-178 antibody to my group. Our primary studies showed that α-subunit of ATP synthase also exhibited over-expression in GC cells and 94.7% clinical GC tissues, and had no or very low expression in normal gastric mucous tissues, and should, therefore, be a potential biomarker with diagnostic value.

We prepared HAI-178 antibody-conjugated fluorescent magnetic nanoparticles (HAI-178-FMNPs), and co-incubated with GC MGC803 cells and gastric mucous GES-1 cells. GC-bearing nude mice models were established and were injected with prepared HAI-178-FMNPs via tail vein, and were then imaged by MRI and small animal fluorescent imaging system. Results showed that α-subunit of ATP synthase exhibited high expression in 94.7% GC tissues. Prepared HAI-178-FMNPs could actively target MGC803 cells, realized fluorescent imaging and MRI of in vivo GC, and actively inhibited growth of GC cells. HAI-178-antibody-conjugated fluorescent magnetic nanoparticles show great potential in applications such as targeted imaging and simultaneous therapy of in vivo early GC cells in the near future (Pan et al., 2009b).

4.3 MNPs-PEG$_{2K}$-FA@Ce6 for Targeted Imaging and Photodynamic Therapy

As shown in Fig. 5, a novel multifunctional photo/magnet-diagnostic nanoprobe (MNPs-PEG$_{2K}$-FA@Ce6) was developed. This nanoprobe is built using folic acid (FA), bifunctional polyethylene glycol (PEG$_{2K}$), and photosensitizer chlorin e6 (Ce6). The MNPs-PEG$_{2K}$-FA@Ce6

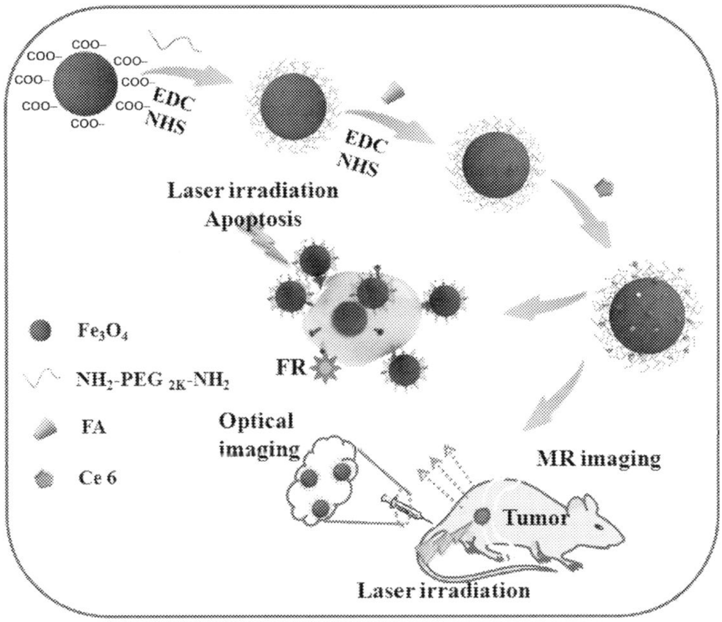

Fig. 5
Schematic of MNPs-PEG$_{2K}$-FA@Ce6 nanoprobes for in vitro/in vivo imaging. *Permission is from Nature Press Group.*

nanoprobes are superparamagnetic, can be synthesized on a large scale by a one-pot hydrothermal process without further surface modification, and are stable in an aqueous environment for eight months. Compared with free Ce6 nanoprobes in vitro studies, the MNPs-PEG$_{2K}$-FA@Ce6 nanoprobes significantly enhance cellular uptake efficiency and promote the effectiveness of photodynamic therapy (PDT) with the help of 633 nm laser irradiation. The unique nanoprobes show superior penetration and a retention time of more than 6 days with less accumulation in the liver, allowing highly effective tumor recognition and monitoring. Additionally, there was little damage to healthy organs or tissues. These exciting new nanoprobes could be potential building blocks to excite new clinical therapies and translational medicine.

4.4 FMNPs-Labeled MSCs for Targeted Imaging and Hyperthermia Therapy

We also prepared FMNPs-labeled human MSCs, and realized targeted imaging and hyperthermia therapy of GC. We first proposed the concept of stem cell nanotechnology (an emerging interdisciplinary field in which nanotechnology is applied to stem cell research and development). Although great advances have been and continue to be made in the stem cell field, several obstacles must be overcome before their therapeutic application can be realized. These include the development of advanced techniques to understand and control the functions of microenvironmental signals and novel methods to track and guide transplanted stem cells. The application of nanomaterials and nanotechnology to stem cell research and development exhibits attractive technological prospects which provide a new chance to solve current problems facing stem cell research and development (Murphy et al., 2011; Qi et al., 2012).

Mesenchymal stem cells (MSCs) are multipotent stem cells that can differentiate into a variety of cell types, including osteoblasts (bone cells), chondrocytes (cartilage cells), and adipocytes (fat cells). MSCs possess immunosuppressive or immunomodulatory properties and have the characteristics of home to the sites of active tumorigenesis. Thus, MSCs can be considered a candidate for cell-based tissue engineering, cancer therapeutics, and regenerative medicine applications. However, the distribution and final fate of MSCs cells inside human body are still not well clarified, which urgently requires novel labeling and in vivo tracking technology.

As shown in Fig. 6, MSCs were characterized. The silica-coated fluorescent supermagnetic nanoparticles (FMNPs) were prepared, were evaluated their effects on MSCs, and were used to label MSCs, and then observed the distribution and final sites of the labeled MSCs in vivo in nude models loaded with GC. Some chemoattractant cytokines were analyzed by ELISA and Western blotting. As shown in Fig. 7, MSCs were labeled with FMNPs efficiently and kept stable fluorescent signal and magnetic properties within 14 days, FMNP-labeled MSCs could target and image in vivo GC cells after being intravenously injected for 14 days, FMNP-labeled MSCs could significantly inhibit the growth of in vivo GC because of hyperthermia effects, and CCL19/CCR7 and CXCL12/CXCR4 axis loops may play key roles in the targeting of MSCs to in vivo GC. FMNPs-based labeling and tracking technology have great potential in applications such as labeling and tracking implanted cells, evaluating cell therapeutic effects, and recognizing and mapping early GC cells (Yu & Pishko, 2011).

5 Multifunctional QDs Probes for Fluorescent Imaging, Genotyping and Therapy of Gastric Cancer

5.1 RGD-Conjugated RNase A Associated dRQDs for Targeted Imaging and Therapy

Colloidal semiconductor nanocrystals (NCs), also known as quantum dots (QDs), have gained immense attention due to their unique optical properties, including a broad absorption with narrow photoluminescence spectra, high quantum yields, low photobleaching, and size-dependent emission-wavelength tenability (Zhou, Shum, Burnett, & Rossi, 2013). These fascinating properties make QDs a promising candidate for biological imaging and labeling probes, with significant advantages over conventional fluorescent dyes. However, the toxicity associated with cadmium-containing QDs turned out to be a chief hurdle for in vitro and in vivo biomedical diagnostics (Guo, 2010). For this reason, a variety of attempts

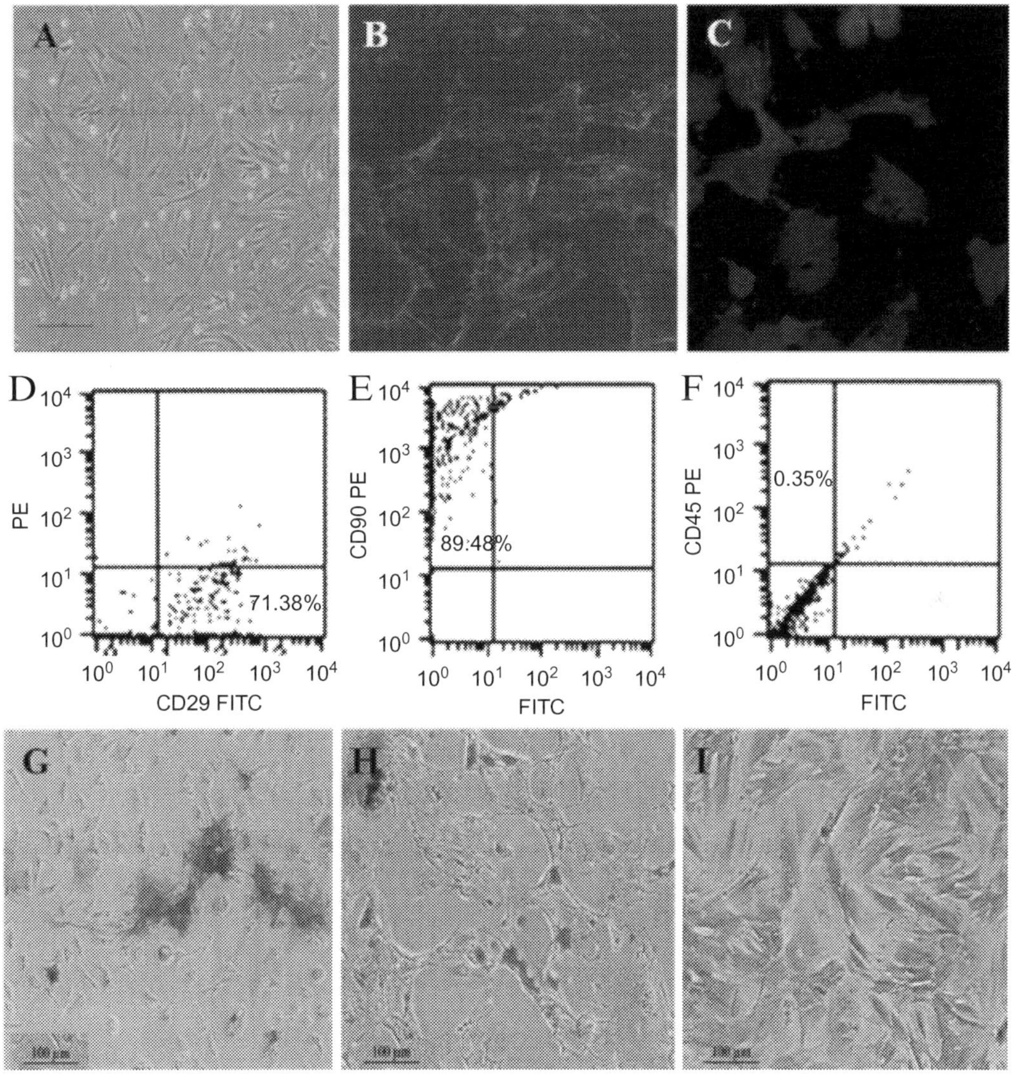

Fig. 6

Morphology and immunophenotypic characterization of MSCs. (A) The fibroblastic morphology of passage 3 MSCs (magnification = ×100). (B) MSCs stained with FITC-conjugated CD29 antibody (×200). (C) MSCs stained with PE-conjugated CD90 antibody (×200). (D)–(F) MSCs analyzed by FACS for the positive expression of CD29 (D) and CD90 (E) and negative expression of CD45 (F). (G) Differentiated osteoblasts tested with alkaline phosphatase staining (×100). (H) Differentiated adipocytes characterized by oil red O staining(×100); (I) Differentiated chondrocytes verified by toluidine blue staining (×100). *Permission is from Springer Press.*

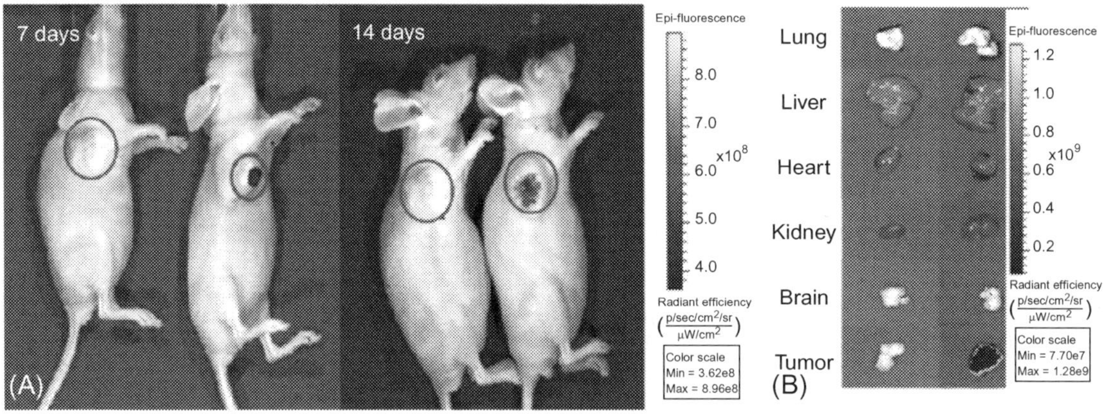

Fig. 7

Fluorescent imaging of FMNP-labeled MSCs targeting gastric cancer cells in vivo. (A) The in vivo fluorescent images show that the tumor sites of the mice in the test group had fluorescent signals after postinjection of FMNP-labeled MSCs at 7 and 14 days (right), and tumor sites of the mice in the control group had no fluorescent signal after postinjection of FMNPs at 7 and 14 days (left). (B) The fluorescent imaging of major organs show that no signal was detected in the tumor and organs of the control group (left), and obviously fluorescent signals were detected in the tumor tissues of the test group (right). *Permission is from Springer Press.*

have been made to reduce the toxicity of QDs and improve their biocompatibility, which has become a rapidly growing area of nanoscience research. Although the surface modification strategy proves to be prosperous, selecting an appropriate surface coating material is not easy. More recently, due to their excellent biocompatibility and bioabsorbility, some proteins have been immobilized on the surface of QDs to produce low-toxicity protein-QDs hybrid nanomaterials. Additionally, protein molecules have also been adopted to serve as the biomolecular templating agent to regulate the formation of inorganic nanostructures (Guo, Haque, Hallahan, Reif, & Li, 2012). We reported the first application of bovine pancreatic ribonuclease A (RNase A) as a biomolecular templating agent for preparing CdTe QD clusters in aqueous phase via a chemical reduction approach. We have chosen RNase A due to its unique features. RNase A is a low-molecular-weight protein (124 residues, 13.7 kDa, p*I* 9.4) with a globular configuration (2.2, 2.8, and 3.2 nm). The outstanding thermal stability of the protein is well known, and they are able to endure the high temperature in our synthesis process. Recently, other groups have reported that RNase A was able to inhibit the development of cancerous cells depending on the enzymatic activity of degrading cellular RNA, and has been used as a chemotherapeutic agent in clinical trials. It was found that RNase A inevitably lost its native conformation to form denatured RNase A (dRNase A)-capping CdTe QD clusters (dRQDs). Moreover, dRNase A not only effectively reduced the cytotoxicity of CdTe QDs, but also unexpectedly induced a temporal evolution of the luminescence during a 50-day storage period under ambient conditions, as shown in

Fig. 5. As shown in Fig. 6, prepared RNase A-QDs composed of QD clusters, the potential synthesis mechanism is shown in Fig. 7. Inspired by these fascinating properties, dRQDs was associated with native RNase A by the physical absorption, which was then coupled with cyclic arginine-glycine-aspartic acid (RGD) peptide c(RGDfK) that served as the targeting biomolecule. A novel multifunctional nanosystem RGD-conjugated RNase A associated dRQDs (RGD-R-dRQDs) was built up for synchronous targeted cellular imaging and therapeutic applications (Guo, Zhang, Chen, Garver, & Trottier, 1998).

5.2 Her2 Antibody Conjugated RQDs for Targeted Imaging and Therapy of In Situ Gastric Cancer

One kind of multifunctional HER2 monoclonal antibody conjugated RNase A-associated CdTe quantum dot cluster (HER2-RQDs) nanoprobes was prepared, and its cytotoxicity was evaluated. Subcutaneous GC nude mouse models and in situ GC SCID mouse models were established, and were intravenously injected HER2-RQDs nanoprobes, the biodistribution and therapeutic effects of HER2-RQDs in vivo were evaluated. Results showed that HER2-RQDs nanoprobes could selectively kill GC MGC803 cells, as shown in Fig. 8, could target imaging subcutaneous GC cells at 3 h postinjection, and in situ GC cells at 6 h postinjection, and could inhibit the growth of GC tissues and extended survival time of GC bearing mouse models, which is closely associated with destroying functional RNAs in cytoplasm by RNase A released from HER2-RQDs nanoprobes, preventing protein synthesis and inducing cell apoptosis, as shown in Fig. 9. High performance HER2-RQDs nanoprobes exhibit great potential in applications such as in situ GC targeted imaging, and selective therapy in the near future (Shu, Moll, Deng, Mao, & Guo, 2004) (Figs. 10–12).

6 Multifunctional Upper Conversion Nanoprobes for Targeted Imaging and Therapy of Gastric Cancer

Upconversion nanoparticles (UCNPs), especially lanthanide-doped nanocrystals, have stolen the limelight due to their seductive optical and chemical features such as low toxicity, large Stokes shifts, and high resistance to photobleaching and photochemical degradation (Abdelmawla et al., 2011; Shu, Shu, Haque, Abdelmawla, & Guo, 2011). Moreover, near infrared (NIR) excitation contributes to auto-fluorescence minimization, deeper penetration, and less harmfulness to cells compared with traditional ultraviolet (UV) excitation. On the other hand, the composites of UCNPs with other materials exert superior performance in a sense, broadening the range of biological applications from multimodality imaging and targeted therapy to biodetection and bioassays. My group first reported the OA/ionic liquid two-phase system combining the merits of thermal decomposition method, the ionic liquid (IL)-based strategy, and the two-phase approach is introduced to synthesize high-quality lanthanide-doped NaGdF 4 upconversion nanocrystals with different crystal-phases in OA-phase and IL-phase

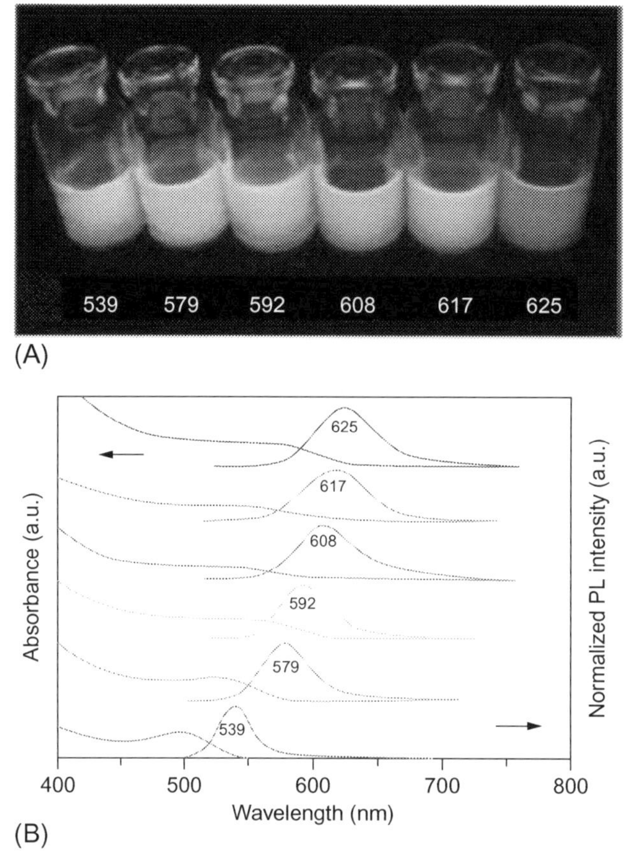

Fig. 8
(A) Photograph of freshly prepared CdTe QDs suspension with different emitting colors in the presence of RNase A as a biomolecular templating agent under a hand-held UV lamp. (B) Corresponding UV-vis absorption and normalized photoluminescence (PL) spectra. *Permission is from Wiley Press.*

through a one-step controllable reaction. Oil-dispersible cubic-phase NaGdF 4:Yb, Er (Ho, Tm) nanocrystals with ultra-small size (~5 nm) and monodispersity are obtained in the OA phase of the two-phase system via an IL-based reaction. More importantly, water-soluble hexagonal-phase NaGdF 4:Yb, Er nanocrystals are obtained in the same system simply by adopting an extremely facile method to complete the dual phase-transition (crystal-phase transition and OA-phase to IL-phase transition) simultaneously. The synthesized lanthanide-doped NaGdF4 upconversion nanocrystals are effective for dual-mode UCL imaging and CT imaging in vivo (Zhang et al., 2013). On this basis, we also improved the synthesis condition, and developed phase- and size-controllable synthesis of hexagonal upconversion rare-earth fluoride nanocrystals through an oleic acid/ionic liquid two-phase system (Shu, Shu, Haque, & Guo, 2013), prepared nanoparticles were successfully used as magnetic nanorattle materials for magnetism-directed targeting imaging and chemical therapy (Haque et al., 2012) (Fig. 13).

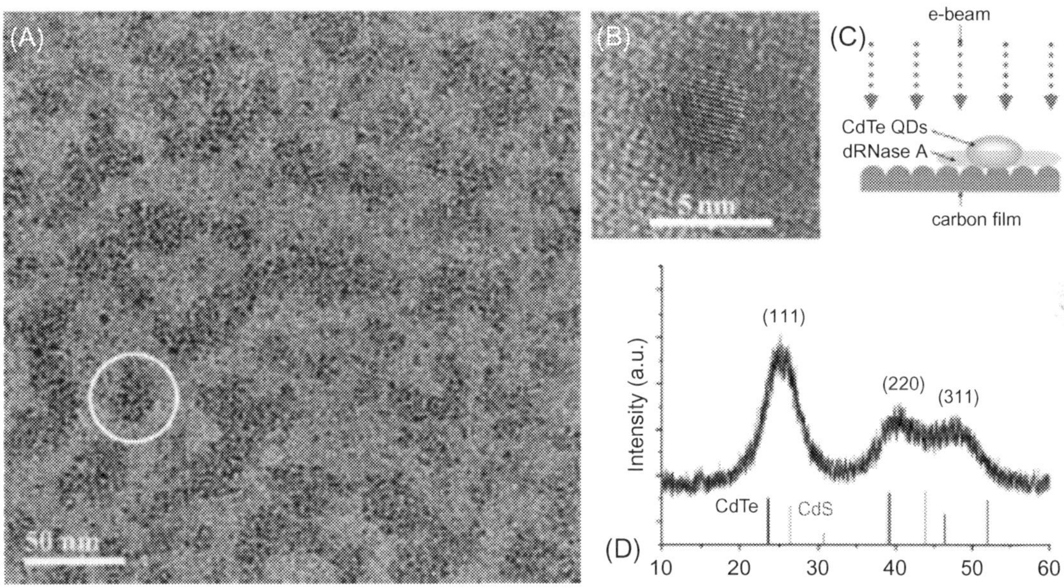

Fig. 9
(A) TEM image of as-prepared dRQD539. A typical cluster is labeled with a white circle. (B) High-resolution TEM image (HR-TEM) of an individual CdTe QD. (C) Schematic diagram to explain the difference in image contrast in the HR-TEM image during imaging in the vertical section. (D) Powder X-ray diffraction (XRD) pattern of dRQD539. Standard diffraction lines of cubic CdTe and cubic CdS are shown for comparison. *Permission is from Wiley Press.*

6.1 Folic Acid-Conjugated Silica-Modified Upconversion Nanoprobes for Targeted UCL and CT Imaging

Folic acid-conjugated silica-modified LaF_3:Yb,Tm upconversion nanoparticles (UCNPs@SiO_2-FA) with high La content in single particles were strategically designed and prepared for simultaneous targeting dual-modality imaging of upconversion luminescence (UCL) and X-ray computed tomography (CT) (Ma et al., 2012). LaF_3 UCNPs were synthesized by a novel oleic acid (OA)/ionic liquid (IL) two-phase system. Afterwards, the folic acid molecule was covalently anchored on the surface of UCNPs with silane coupling agent. The UCNPs@SiO_2-FA exhibits good stability, water dispersibility and solubility, noncytotoxicity, good biocompatibility, highly selective targeting, excellent X-ray attenuation, and UCL emission under excitation at 980 nm. In vivo UCL and CT images of mice show the UCNPs@SiO_2-FA can be used in targeting dual-modality imaging. These results suggest that the as-prepared nanoprobe is a good candidate with excellent imaging and targeting ability for targeting dual-modality imaging of UCL and CT.

Multimodal contrast agents based folic acid-conjugated silica-modified LaF_3:Yb,Tm upconversion nanoparticles (UCNPs@SiO_2-FA) with high La content in single particle were

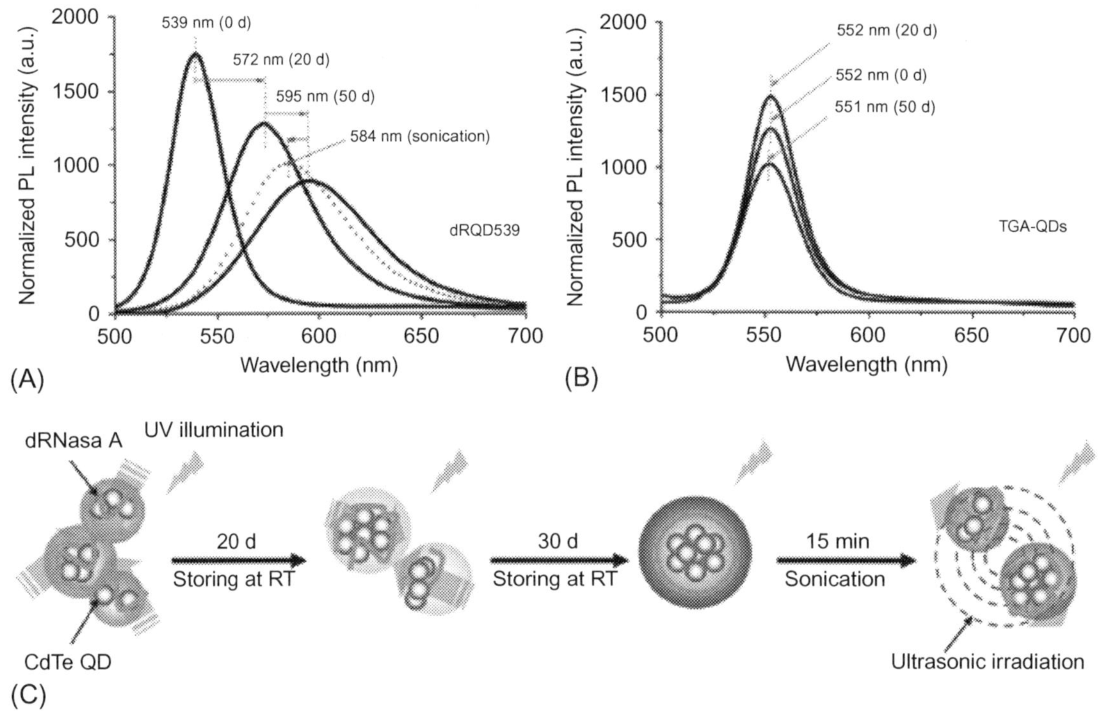

Fig. 10

(A) Normalized photoluminescence (PL) spectra of dRQD539 (excitation wavelength λ_{ex} = 400 nm). The three solid lines are the corresponding PL spectra of the same sample stored for 0 day (0 d), 20 days (20 d), and 50 days (50 d), respectively. The blue dotted line is the PL spectrum of the fifty day-stored dRQD539 after sonicating for ~15 min in an ultrasonic cell crusher. (B) Normalized PL spectra of TGA-capping CdTe QDs stored for 0, 20, and 50 d, respectively. (C) Schematic illustration of the potential mechanism for the red shift of emission peaks of dRQD539 during the storage period under ambient conditions. *Permission is from Wiley Press.*

strategically designed and developed. We have demonstrated that UCNPs@SiO_2-FA is a good candidate with excellent imaging and targeting ability for UCL imaging and CT imaging in vivo. The prepared UCNPs@SiO_2-FA exhibits good stability, water dispersibility and solubility, noncytotoxicity, good biocompatibility, highly selective targeting, excellent X-ray attenuation, and UCL emission under excitation at 980 nm. Our results indicated that the synthesized UCNPs@SiO_2-FA is effective for simultaneous targeting dual-modality imaging of UCL and CT. It could bring novel opportunities to the next generation of theranostic nanoprobes for simultaneous diagnosis and treatment in vivo.

We also developed a novel synthesis method of UCPs, hydrothermal system in which an anion induces the phase transition process to give simultaneous control over the size, morphology, phase, and emission properties. We first confirm that the crystal cell oriented-rotation driven by an anion in a hydrothermal system promoted the phase transition, and the

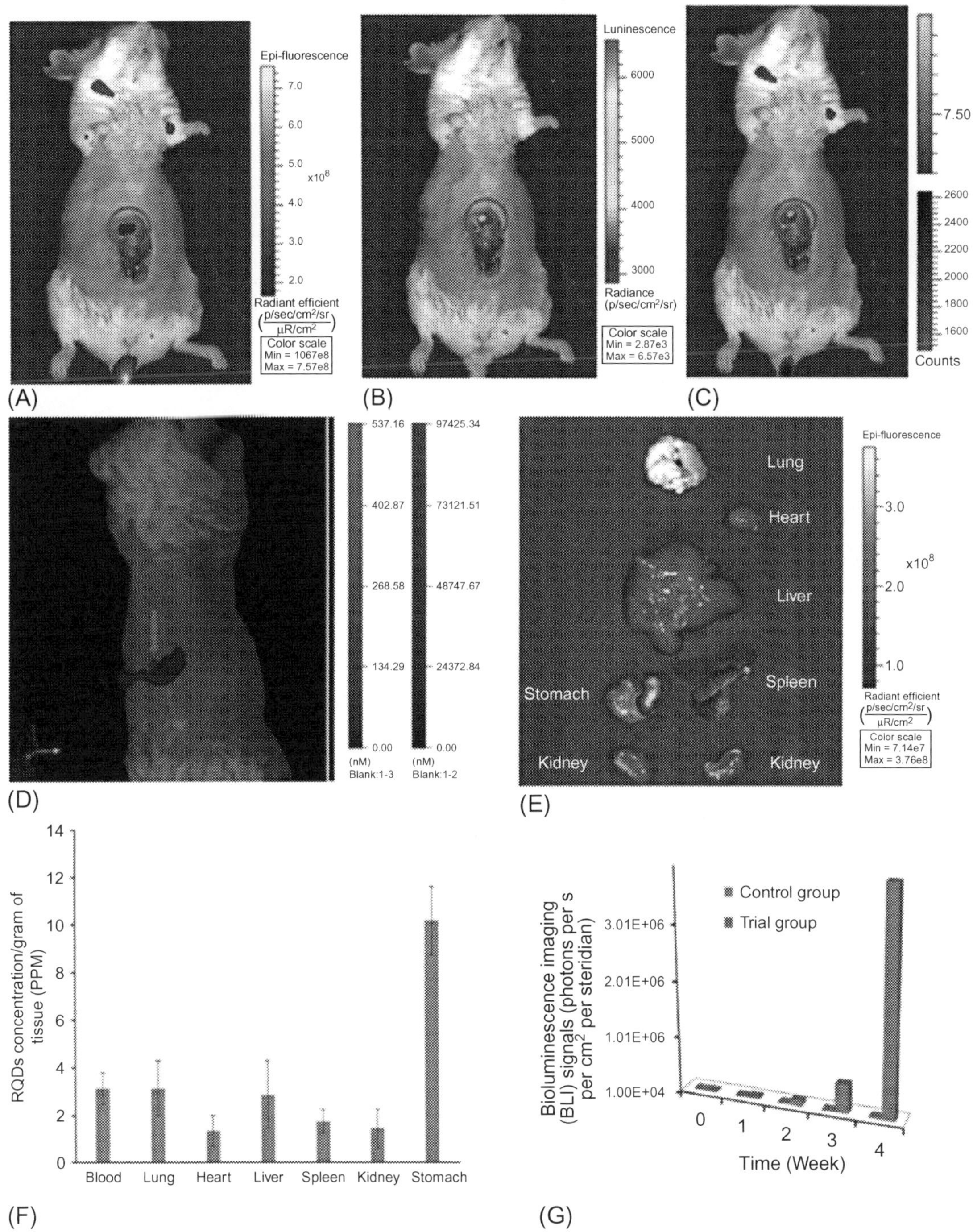

See the legend on next page.

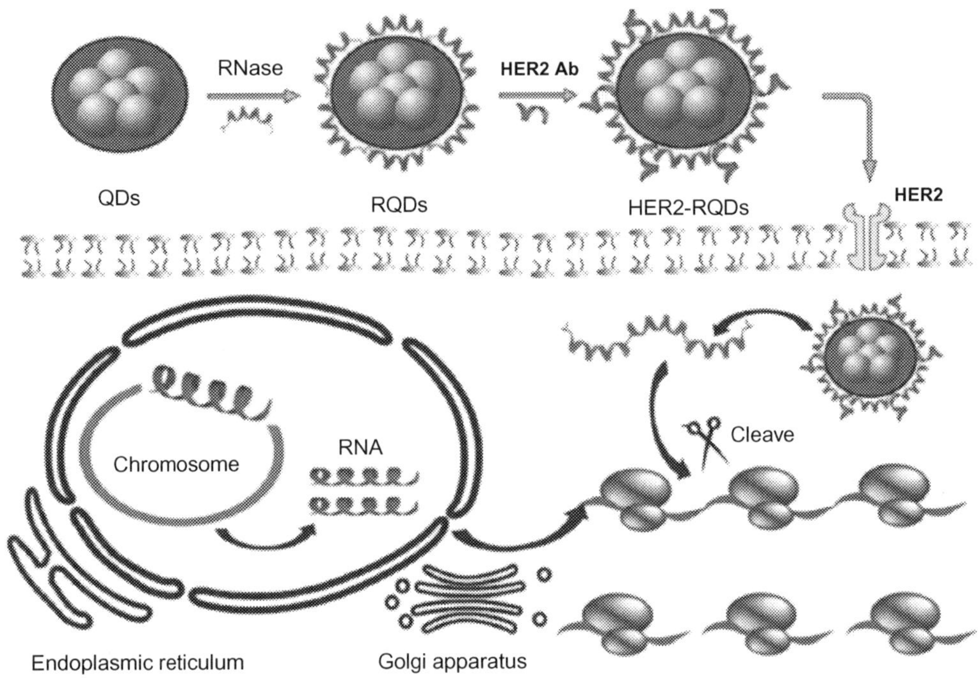

Fig. 12

Therapeutic mechanism of as-prepared nanoprobes: HER2-RQDs nanoprobes could specifically enter into gastric cancer cells, RNase A was released from the HER2-RQDs nanoprobes under the acid environment, which degraded total RNAs inside gastric cancer cells, inhibiting RNA translation and protein synthesis, and finally induced tumor cells' apoptosis. *Permission is from Elsevier Press.*

energy zones figure of the phase transition from cubic to hexagonal structure has been figured out. We have successfully applied the structural mechanics finite element calculations to validate the reaction process. We have also demonstrated that porous UNs can be rationally tuned in size (down to fifteen nanometers), phase (cubic or hexagonal) and emission properties at precisely defined conditions, and were effective for in vitro and in vivo CT imaging (Kalli et al., 2008) (Figs. 14 and 15).

Fig. 11

Optical images of HER2-RQDs nanoprobe in in-situ gastric cancer mice. (A) Fluorescence image of HER2-RQDs nanoprobe targeted in situ gastric cancer tissue. (B) Bioluminescence image of in-situ gastric cancer tissue. (C) Overlay image of fluorescence and bioluminescence imaging. (D) 3D image of HER2-RQDs nanoprobes targeted gastric cancer tissue. (E) Ex vivo fluorescence image of the dissected organs of the mice sacrificed at 24 h postinjection of HER2-RQDs nanoprobes. (F) Biodistribution of HER2-RQDs in mice after intravenous injection. The amounts of RQDs in tissue samples were evaluated by ICP mass spectrometry (*n* ¼ 3). (G) Bioluminescence imaging signals of gastric tumor tissues in in situ gastric cancer mice after treated with HER2-RQDs for 4 weeks. *Permission is from Elsevier Press.*

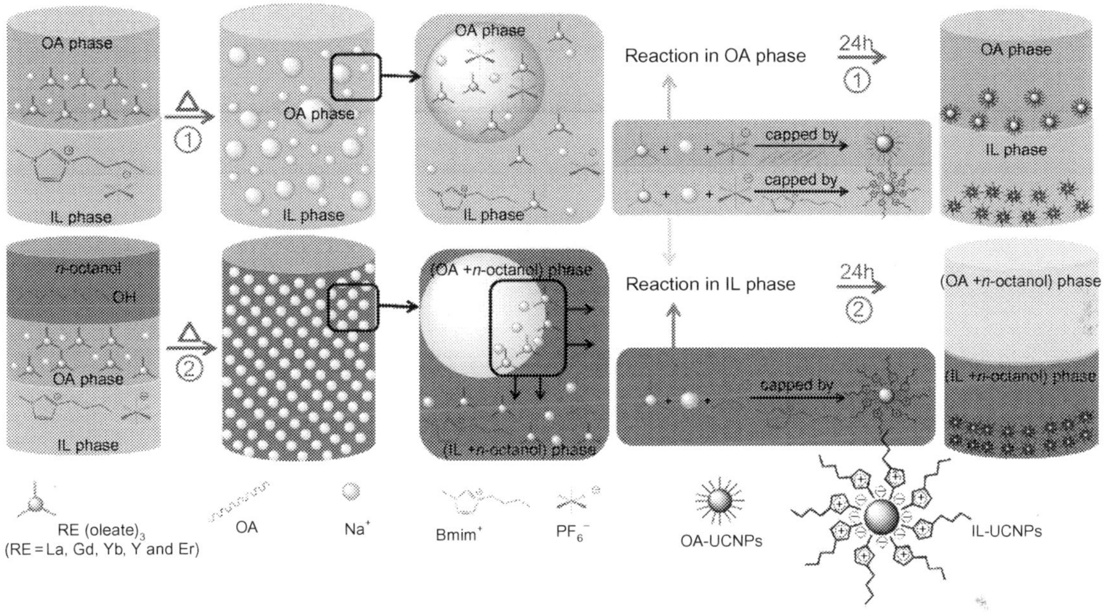

Fig. 13
Schematic diagram showing the formation of RE fluoride nanocrystals in the OA/IL and n-octanol-induced two-phase systems. *Permission is from Wiley Press.*

7 Multifunctional Gold Nanoprobes for Targeted Imaging and Therapy

GNPs are within the same size domain as many biomaterials, including enzymes, antibodies, and protein receptors. Combined with the unique properties of materials in the nanoscale range, gold nanoparticles provide scope for making measurements more efficiently than with existing molecular materials. Since light absorption from biologic tissue components is minimized at NIR wavelengths, most NPs for in vivo imaging have been designed in order to strongly absorb in the NIR region so as to be used as effective contrast agents (Teng, Xie, Teng, & Lee, 2012).

Optical coherence tomography (OCT) is an imaging modality that provides cross-sectional subsurface imaging of biological tissues at micrometer-scale resolution. Briefly, the OCT system can generate a signal based on refractive index mismatches and scattering events. The extra scattering is achieved by using AuNPs, which possess both absorption and scattering properties in the NIR region and provide optical contrast for improved diagnostic imaging at higher light intensities. This property has been used in order to increase the contrast of tumors in mice. Au nanoshells accumulate preferentially in the tumor and so increase contrast more than in other tissues (Ly, Hoyt, Crowell, & Kim, 2012). In situ OCT based on spherical AuNPs in ophthalmology and dental imaging has also been reported (Gao, Xiang, Meng,

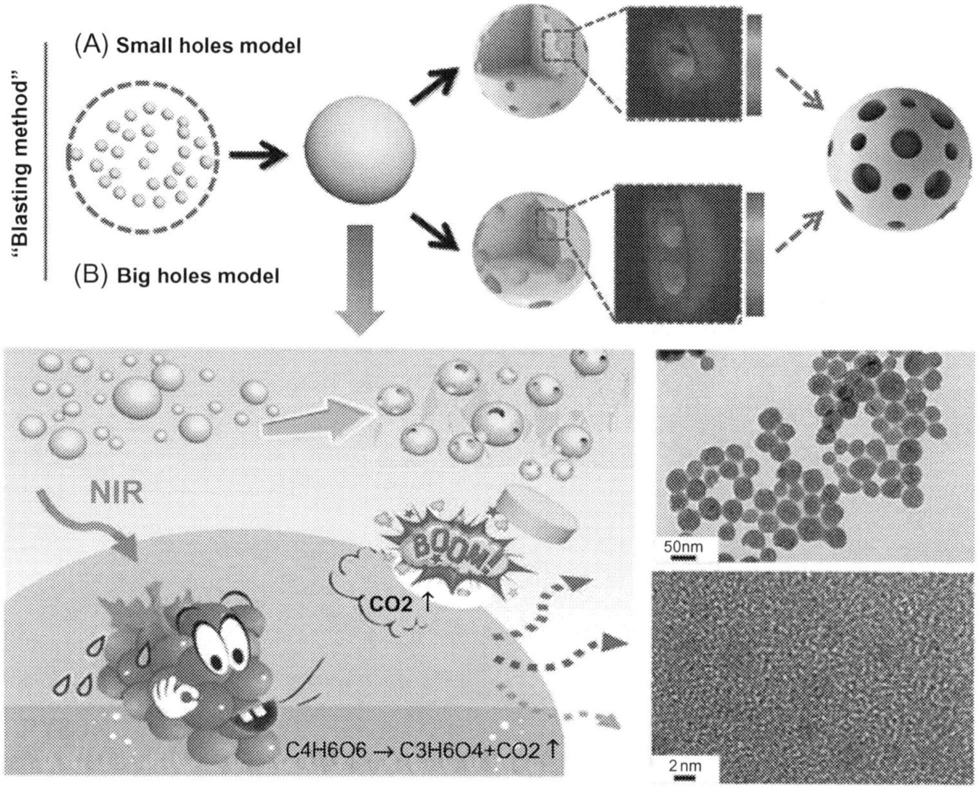

Fig. 14
Illustration of the "blasting" method for fabricating porous UNs. *Permission is from RSC Press.*

Liu, & Qi, 2013). Moreover, an extension of OCT—photothermal OCT—can fill the spatial niche for in vivo imaging of NP delivery (Shi et al., 2013).

X-ray, computed tomography (CT), and micro-CT techniques can offer excellent improvements in terms of medical diagnosis. The basic process of these techniques is to detect the X-rays that pass through a sample by exposing a charge-coupled device detector. Due to their high atomic number and electrodensity (e.g., 79 and 19.32 g/cm^3, respectively, compared to the typical X-ray contrast agent, iodine, with values of 53 and 4.9 g/cm^3, respectively), AuNPs have higher attenuation coefficients and are proposed to be used as better contrast agents for X-ray imaging, CT and micro-CT (Huang et al., 2011). AuNPs have demonstrated greater contrast than the iodine-based agents that are currently used in the clinic, as well as reduced toxicity and prolonged circulation times. Moreover, imaging Au at 80–100 keV reduces the signal interference derived from bone absorption, and therefore, would allow for dose reduction. In addition, fluorescent bioimaging is also achieved with luminescent AuNCs, which can be used for multimodal fluorescence/CT imaging without the need for having two different elements. Via glutathione capping and folic acid conjugation,

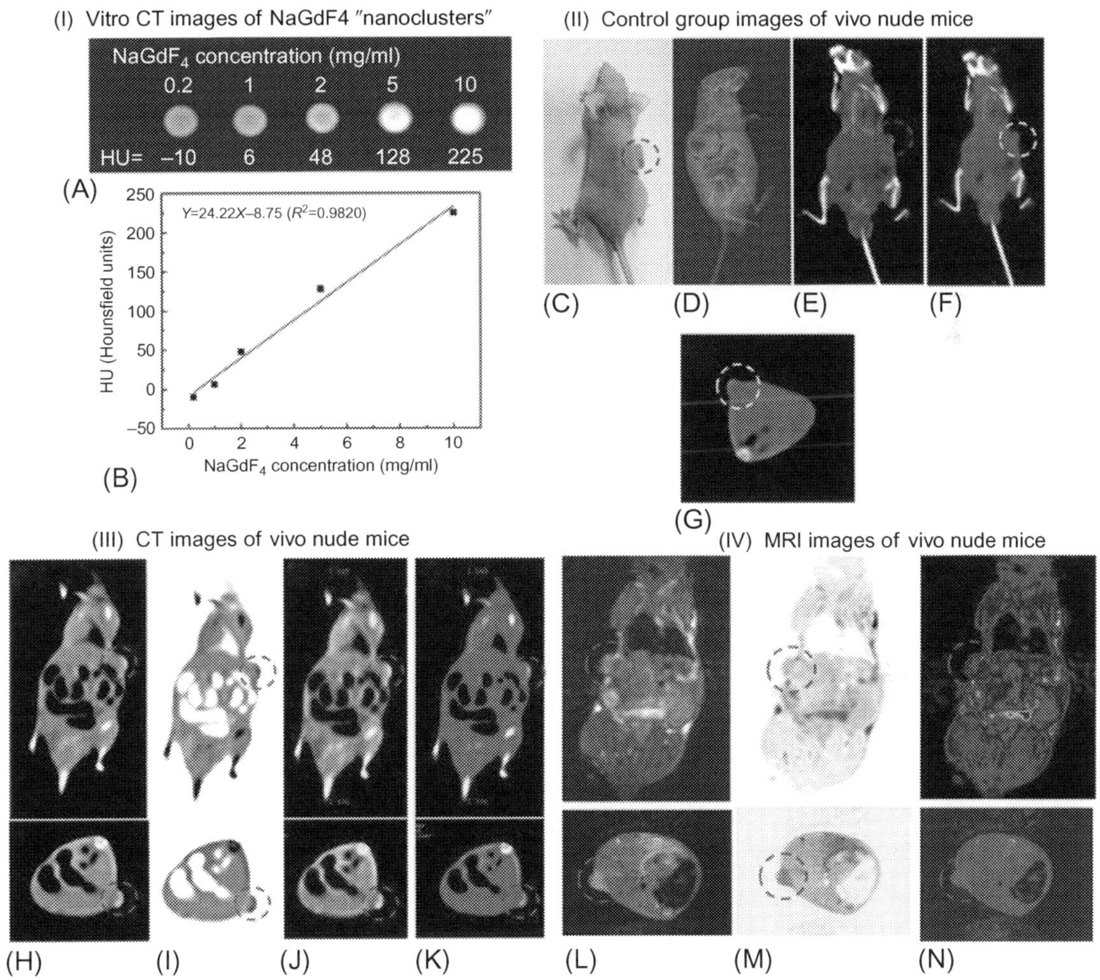

Fig. 15

(I) In vitro CT images (A) of lanthanide-doped NaGdF4 upconversion "nanoclusters" (sub-5 nm) suspended in aqueous solution. CT attenuation (HU) plot (B) of NaGdF4 changes along with the concentration of each sample from 0.2 to 10 mg mL^{-1} to further investigate the CT contrast effect. (II) Images of control group before injection: (C) Photograph of the nude mouse model loaded with gastric cancer MGC-803 cells; (D) X-ray image and (E)–(G) CT images of nude mouse as the control group. (III) CT images (H)–(K) and (IV) MRI images (L)–(N) with NaGdF4 upconversion "nanoclusters" through intravenous injection under the function of passive targeting. *Permission is from RSC Press.*

the as-prepared nanoprobe could target a xenotransplanted tumor model in nude mice, as evaluated by fluorescence/CT dual-mode imaging. Although it is not very common, x-ray fluorescent imaging of AuNPs represents a bioimaging technique with great potential. This method could be of great interest once there are imaging systems in place that have sufficient sensitivity to NP concentrations and, at the same time, are capable of distribution measurements at appropriate tissue depths for in vivo and in vitro studies. Consequently, Ricketts et al. reported that a high detecting sensitivity of AuNPs can be achieved using X-ray fluorescence, enabling a greater depth imaging in comparison with optical modalities.

Photoacoustic imaging (PAI) is a hybrid imaging technique; however, it has only become an efficient method of biomedical imaging in recent decades. The development for PAI accounts for the improvement in the machining process and innovative technology relating to the contrast agent. Briefly, after excitation by a laser, a photoabsorber (for example, an internal photoabsorber, such as hemoglobin, or an external photoabsorber, such as photoacoustic contrast agents) is heated, leading to thermoelastic expansion. Such expansion can cause acoustic pressure elevations and therefore produce ultrasound waves (usually a rise of 1 mK can cause an 800 Pa pressure rise), which are then detected by a sound transducer before finally being reconstructed as photoacoustic images. NPs, especially AuNPs with NIR absorbance due to their intrinsic LSPR, are very promising for use as photoacoustic signal amplifiers. Two important commercialized, small animal PAI systems are MSOT (iThera Medical GmbH, Munich, Germany) and Nexus 128 (Endra Life Sciences, MI, United States). The iThera system is based on multispectral optoacoustic tomography and Nexus 128 is based on hemispherical PAI. Despite the adoption of different strategies for 3D imaging, they are both excellent examples of focusing on a crucial problem in bioimaging and thus obtaining images with high resolution and even in-depth functional information. Unlike OCT, PAI is speckle free. For example, PEG-coated Au nanoshells have been used as contrast agents in order to image their circulating distribution in the vasculature in the rat brain using PAI. The images of the distribution of nanoshells circulating in the vasculature of a rat brain achieved by deep-penetrating NIR light presented a gradual enhancement of the optical absorption in the brain vessels. Au nanorods and Au nanoprisms have also been extensively used in bioimaging due to their ability to have their maximum plasmon resonance tuned further into the NIR region. Although PAI is promising for applications in biomedical imaging, drug delivery, and PTT, there are some challenges for PAI based on the AuNP platform. First, although its application in cancer imaging has been abundant over the last 5 years, most of the results are obtained from subcutaneous cancer models and are confined within angiography. However, the advantage of PAI is its deep penetration at a high resolution and the potential for in situ cancer diagnostics. The reason for this is the limitation of the laser energy. According to the American National Standards Institute (ANSI), the energy for skin exposure is advised to be lower than 20 mJ/cm^2, but deep penetration is difficult to achieve in this range, so the detection limit needs to be improved by enhancing the NIR absorption of AuNPs. Au nanorods, nanocages, nanoprisms, or nanostars are particularly useful as PAI contrast agents. Second, the scan speed needs to be improved for PAI. Although this mainly needs the improvement of the mechanical drive, AuNPs at the range of 100 nm, which can prolong the circulation time, may solve this problem via another method. Diagnostics usually require the combination of anatomic imaging as well as functional imaging at the same time. The former can provide the precise location of organs, and the latter is for imaging the anatomical structures of the area of interest. This goal is achieved by either multimodal bioimaging or bioimaging techniques that can achieve both.

Theranostics refers to the combination of diagnostics and therapeutics. Due to their LSPR, the strong NIR light absorption of AuNPs gives them great potential application in PTT, which possesses efficacy and noninvasiveness. There are numerous PTTs based on AuNPs (e.g., antibody- or small-molecule-conjugated AuNPs in tumor therapy). Combined with the aforementioned PAI, AuNPs have great potential to be used as theranostic platforms.

7.1 Folic Acid-Conjugated Gold Nanorods for Targeted Imaging and Therapy

Multifunctional nanoprobes are designed to have various functions such as tumor targeting, imaging, and selective therapy, which offer great promise for the future of cancer prevention, diagnosis, imaging, and treatment. Herein, silica was applied to replace cetyltrimethylammonium bromide (CTAB) molecules on the surface of gold nanorods (GNRs) by the classic Stöber method, thus eliminating their cytotoxicity and improving their biocompatibility. Folic acid molecule was covalently anchored on the surface of GNRs with silane coupling agent. The resultant folic acid-conjugated silica-modified GNRs show highly selective targeting, enhanced radiation therapy (RT) and photothermal therapy (PTT) effects on MGC803 GC cells, and also exhibited strong X-ray attenuation for in vivo X-ray and computed tomography (CT) imaging. In conclusion, the as-prepared nanoprobe is a good candidate with excellent imaging and targeting ability for X-ray/CT imaging-guided targeting dual-mode enhanced RT and PTT (Li et al., 2010) (Figs. 16–20).

7.2 Human-Induced Pluripotent Stem Cells for Tumor Targeted Delivery of Gold Nanorods and Enhanced Photothermal Therapy

Improving effective accumulation and intratumoral distribution of plasmonic gold nanoparticles has become a great challenge for photothermal therapy of tumors. Herein, we reported a nanoplatform with photothermal therapeutic effects by fabricating Au nanorods@ SiO_2@CXCR4 nanoparticles and loading the prepared nanoparticles into the human induced

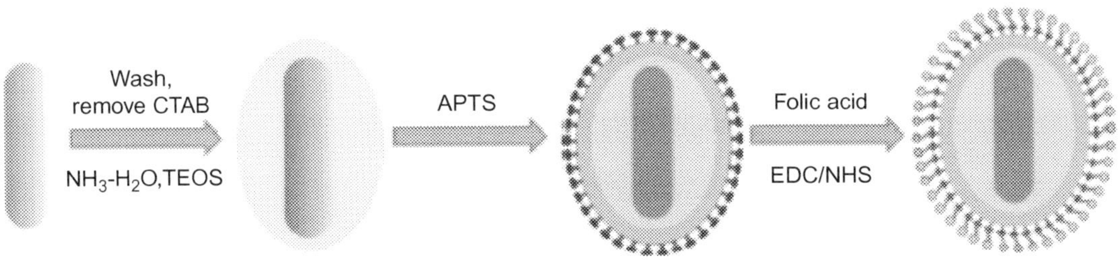

Fig. 16
Synthetic procedure of GNR-SiO2-FA. *Permission is from Elsevier Press.*

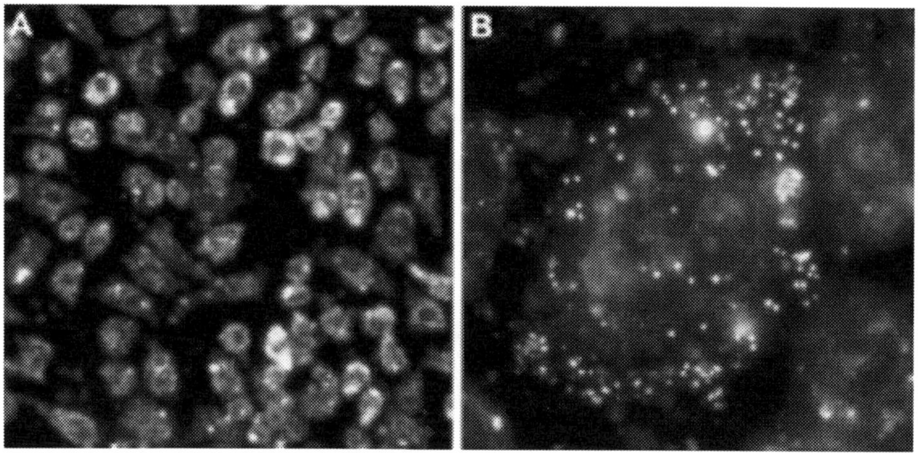

Fig. 17

Dark-field scattering images. (A) Low magnification image of targeted MGC803 cells incubated with 50 mM GNR-SiO_2-FA for 2 h. (B) High magnification image of targeted MGC803 cells incubated with 50 mM GNR-SiO_2-FA for 30 min, monitored by dark-field microscopy. *Permission is from Elsvier Press.*

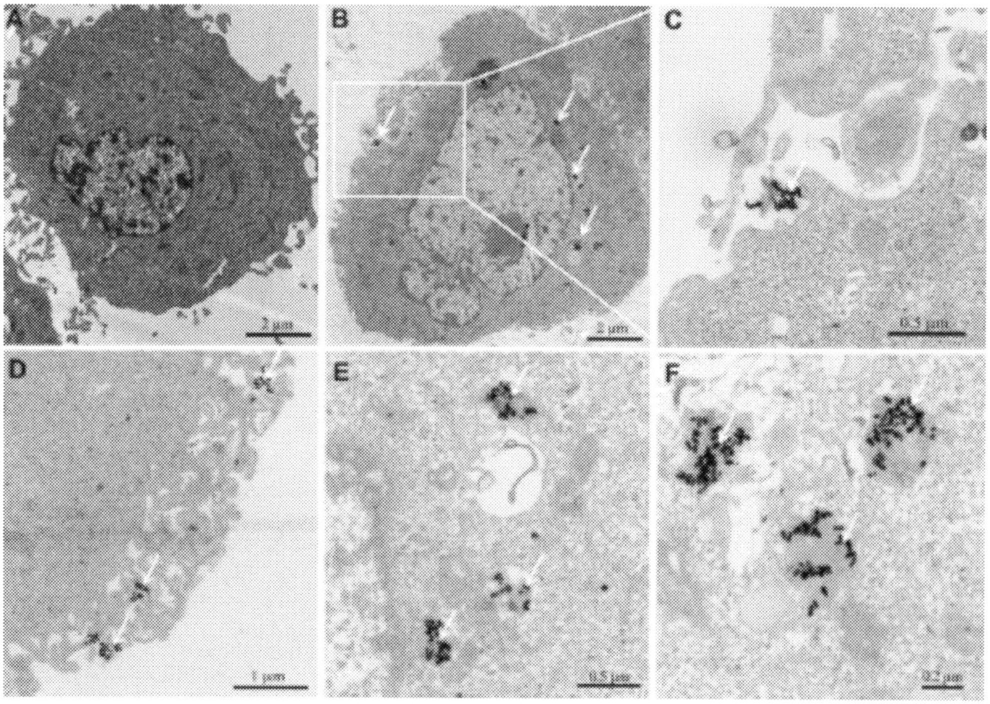

Fig. 18

The thin-section TEM images of MGC803 cells incubated with 50 mM of GNR-SiO2-FA for 2 h. (A) Control MGC803 cells, (B–F) representative ultrastructures of MGC803 cells incubated with GNR-SiO_2-FA. Arrows denote the GNR-SiO_2-FA or their aggregates. *Permission is from Elsevier Press.*

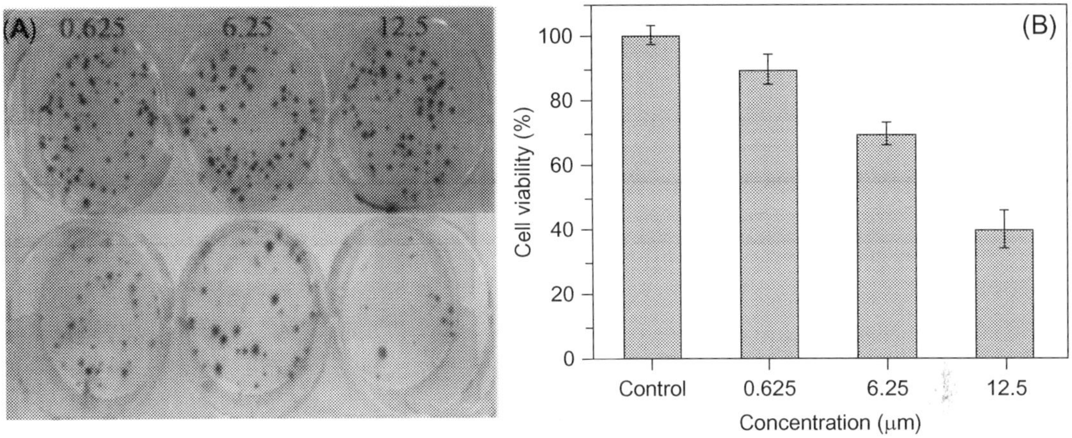

Fig. 19
The proliferation photographs (A) and cell viability (B) of MGC803 cells incubated with 100 mL of varying concentration of GNR-SiO2-FA for 24 h upon 6 Gy of X-ray irradiation. In the photographs (A), the top column of the cells without irradiation, the bottom column of the cells with irradiation. *Permission is from Elsevier Press.*

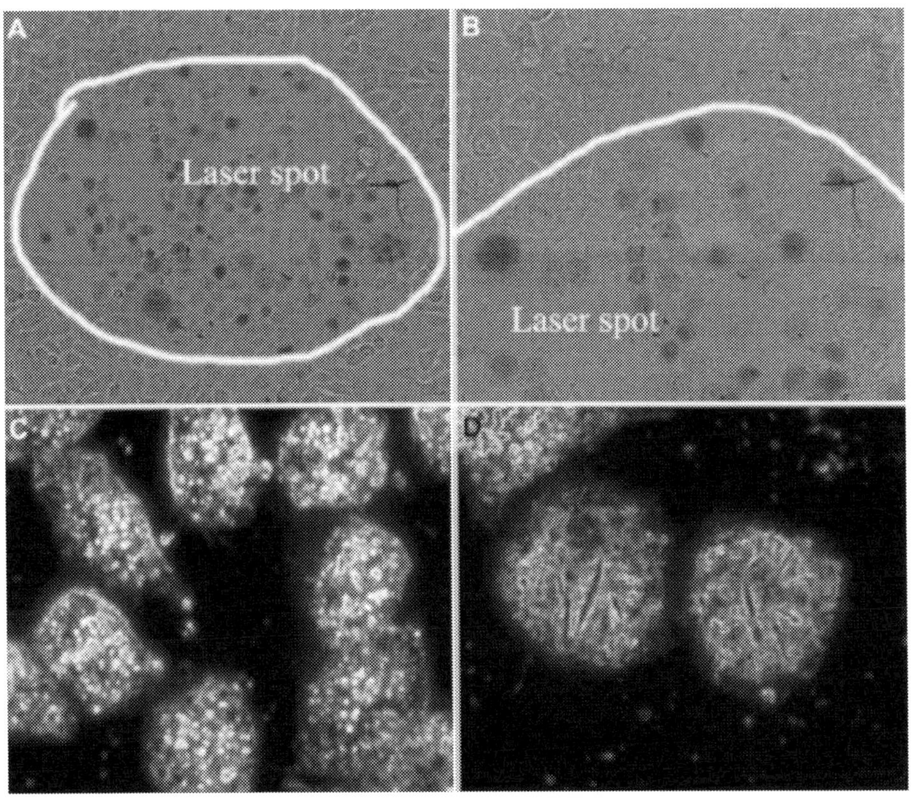

Fig. 20
Photo-thermal therapy effects on MGC803 cells incubated with 12.5 mM of GNR-SiO2-FA for 24 h at 37°C in the dark prior to irradiation for 3 min with 808 nm laser. (A) MGC803 cells on the laser spot, (B) MGC803 cells on the boundary of laser spot, and (C, D) dark-field images of MGC803 cells after irradiation. *Permission is from Elsevier Press.*

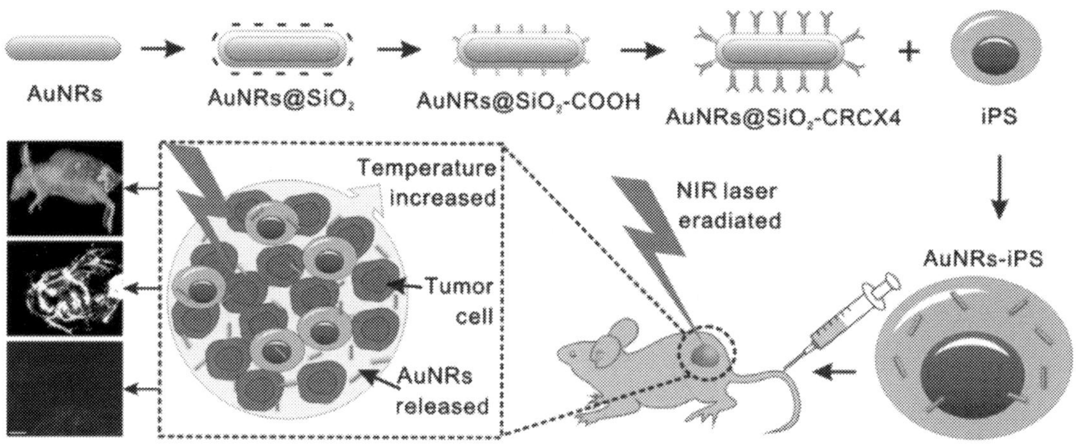

Fig. 21
AuNRs@SiO$_2$@CXCR4-loaded human iPS cells for intratumoral homogeneous distribution of AuNRs and targeted fluorescence/photoacoustic imaging guided photothermal therapy.

pluripotent stem cells (AuNRs-iPS). In virtue of the prominent optical properties of Au nanorods@SiO$_2$@CXCR4 and remarkable tumor target migration ability of iPS cells, the Au nanorods (AuNRs) delivery mediated by iPS cells via the nanoplatform AuNRs-iPS was found to have a prolonged retention time and spatially even distribution in MGC803 tumor-bearing nude mice observed by photoacoustic tomography and two-photon luminescence. Based on these improvements, the nanoplatform displayed a robust migration capacity to target the tumor site and to improve photothermal therapeutic efficacy on inhibiting the growth of tumors in xenograft mice under a low laser power density. The combination of gold nanorods with human iPS cells as a theranostic platform paves an alternative road for cancer theranostics and holds great promise for clinical translation in the near future (Fig. 21).

7.3 BRCAA1 Antibody-Conjugated Gold Nanoprisms for Targeted Photoacoustic Imaging and Photothermal Therapy

The development of high-resolution, nanosized photoacoustic contrast agents is an exciting, as well as challenging, technological advance. Herein, two functionalized gold nanoprisms (AuNprs), PEG/RGD-AuNprs and PEG/Brcaa1-AuNprs were synthesized and used as a combinatorial methodology for in situ photoacoustic imaging, angiography, and local hyperthermia using orthotopic and subcutaneous murine gastric carcinoma models. PEG/RGD-AuNprs are available for tumor angiography, and PEG/Brcaa1-AuNprs are used to target and for in situ imaging of gastric carcinoma orthotopic tumor models. In situ photoacoustic imaging allowed for anatomical and functional imaging at the tumor site. In vivo tumor angiography imaging showed enhancement of the photoacoustic signal in

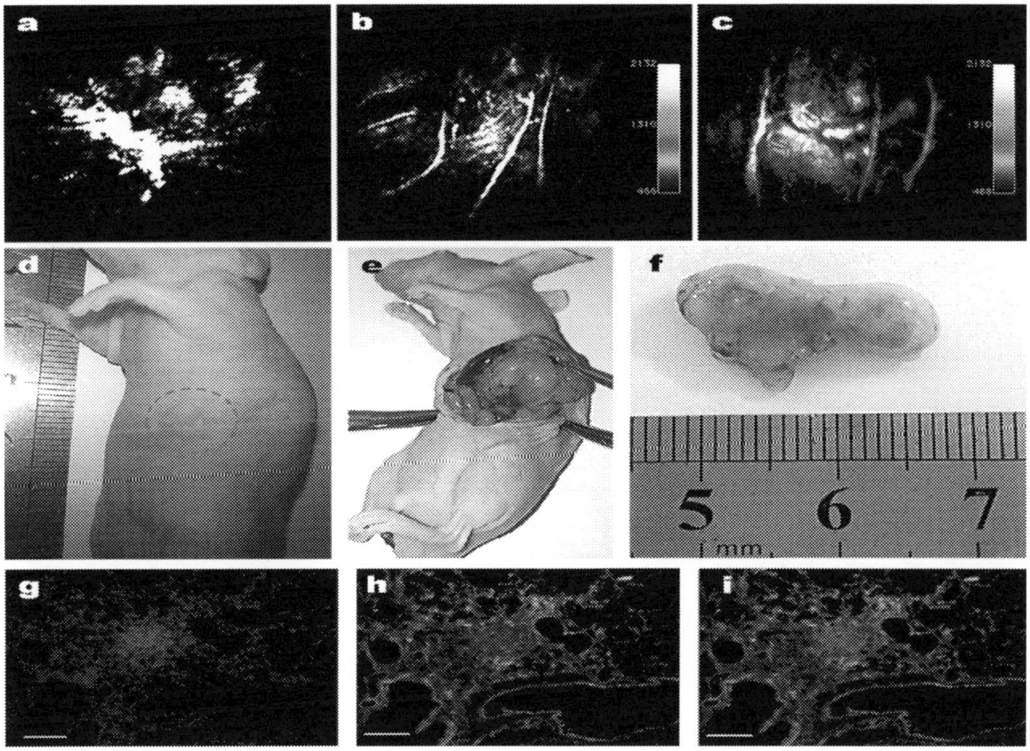

Fig. 22
In vivo photoacoustic imaging study of subcutaneous transplanted tumors. (A–D) Sequential PA maximum intensity projection (MIP) frames after tail injected with RGD-AuNprs. (A) Before injection with RGD-AuNprs; (B) injection with RGD-AuNprs after 1 h; (C) injection with RGD-AuNprs after 3 h; and (D) injection with RGD-AuNprs after 6 h (unpublished data).

a time-dependent way. Furthermore, photoacoustic imaging showed that tumor vessels were clearly damaged after local hyperthermia. This is the first proof-of-concept using two gold nanoprisms probes as high sensitive contrasts and therapeutic agents for in situ tumor detection and inhibition. These smart antibody/peptide AuNprs can be used as an efficient nanotheranostic platform for in vivo tumor detection with high sensitivity as well as for tumor targeting therapy, which results in tumor size reduction and increase mice survival after local hyperthermia with a single-dose injection (Song et al., 2010) (Fig. 22).

7.4 Human CIK Cells for Tumor Targeted Delivery of Gold Nanoprisms and Enhanced Photoacoustic Imaging, Immunotherapy and Photothermal Therapy

We fully exploited the advantages of human CIK cells and gold nanoprisms, with the aim of enhancing synergistic CIK cell therapeutic effects combined with PTT of gold nanoprisms; we prepared human CIK cells, synthesized PEG modified gold nanoprisms, and then

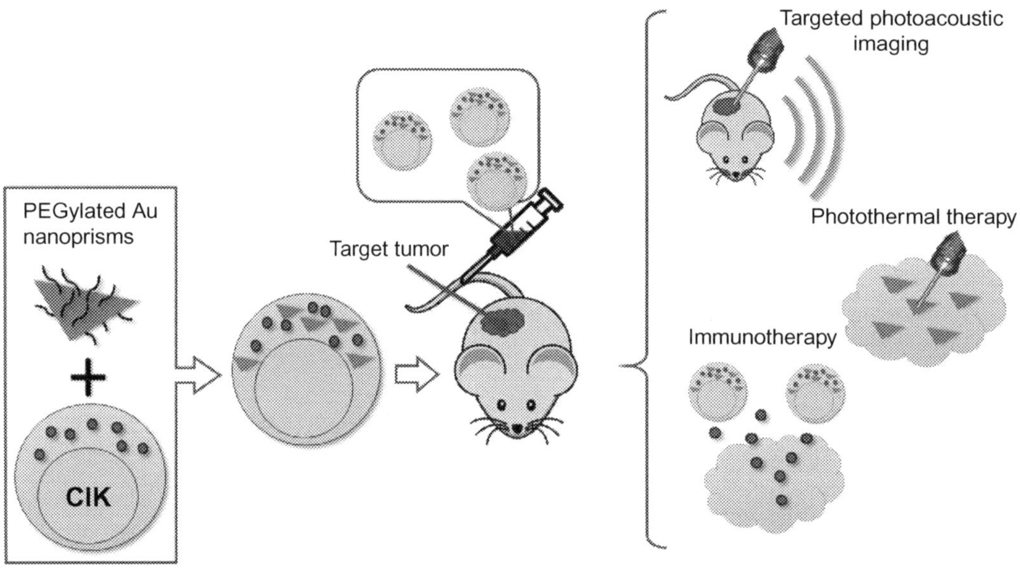

Fig. 23
Schematic of the proposed mechanism of CIK labeled with PEGylated gold nanoprisms for targeted photoacoustic imaging, immunotherapy, and photothermal therapy.

prepared human CIK cells labeled with PEGylated gold nanoprisms, and investigated in vivo distribution and tumor targeting ability and PTT therapeutic efficacy. The results showed that human CIK cells labeled with PEGylated gold nanoprisms could target GC in vivo, and realized immunotherapy and photoacoustic imaging, as well as photothermal therapy of GC, markedly enhancing CIK therapeutic efficacy. Our results provide a new strategy to treat GC, and offer proof-of principle of practicality of CIK combined with PTT based on human CIK cells labeled with PEGylated gold nanoprisms, exhibit potential clinical applications such as theranostics of GCs in middle and late stages and postoperation tumor patients, and a highly efficient delivery system for genes or drugs in the near future (Fig. 23).

7.5 CD44v6 Antibody-Conjugated Gold Nanostars for Photoacoustic Imaging and Photothermal Therapy of Gastric Cancer Stem Cells

Developing safe and effective nanoprobes for targeted imaging and selective therapy of GCSCs has become one of the most promising anticancer strategies (Clarke et al., 2006; Gilbertson & Cancer, 2012; Liang et al., 2015; Vinogradov & Wei, 2012; Zhang & Fan, 2010b). Herein, gold nanostars-based PEGylated multifunctional nanoprobes were prepared with conjugated CD44v6 monoclonal antibodies (CD44v6-GNS) as the targeting ligands. It was observed that the prepared nanoprobes had high affinity towards GCSC spheroid colonies and destroyed them completely with a low power density upon near-infrared (NIR) laser treatment (790 nm, 1.5 W/cm^2, 5 min) in vitro experiment. Orthotopic and subcutaneously xenografted nude mice models of human GC were established. Subsequently, biodistribution and photothermal therapeutic effects after being intravenously injected with the prepared nanoprobes were assessed. Photoacoustic imaging revealed that CD44v6-GNS nanoprobes could target the GC vascular system actively at 4 h postinjection, while the probes inhibited tumor growth remarkably upon NIR laser irradiation, and even extended survivability of the GC-bearing mice. The CD44v6-GNS nanoprobes exhibited great potential for applications of GC targeted imaging and photothermal therapy in the near future (Brabletz, Jung, Spaderna, Hlubek, & Kirchner, 2005; Chen et al., 2010, 2012, 2013; Duan et al., 2013; Heider, Kuthan, Stehle, & Munzert, 2004; Jang, Li, Graham, & Cen, 2011; Li, Wu, Wei, & Tian, 2013; Liu et al., 2013; Marhaba et al., 2008; Misra et al., 2011; Naor, Sionov, & Ish-Shalom, 1997; Prud'Homme, 2012; Takaishi et al., 2009; Xue et al., 2012; Yoshida, Yasuda, Hiramitsu, Ito, & Nakamura, 2008; Zhang, Li, He, Cai, & Yang, 2011) (Figs. 24 and 25).

7.6 Folic Acid/Ce6-Conjugated Gold Nanoclusters for NIR Fluorescent Imaging and Photodynamic Therapy

Gold nanoclusters (GNCs) attract increasing attention due to their potential applications in sensing, catalysis, optoelectronics, and biomedicine. Herein, the formation of highly fluorescent glutathione (GSH)-capped GNCs is achieved through the delicate control of the reduction kinetics and thermodynamic selection of the Au(I)-SG complexes. Furthermore,

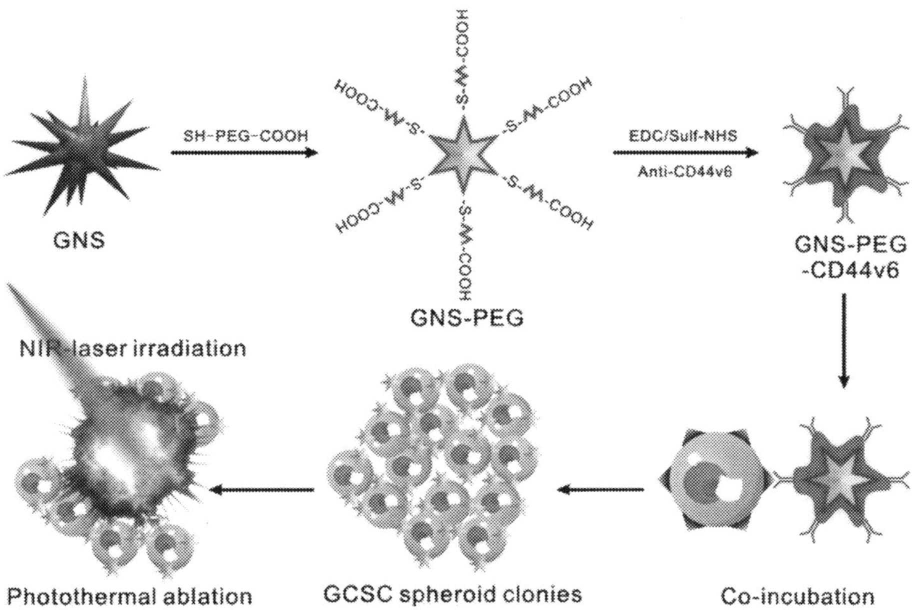

Fig. 24
Schematic illustration of the GNS-PEG-CD44v6 nanocomplex synthesis process and the mechanism of GCSC targeted PTT. *Permission is from IVY press.*

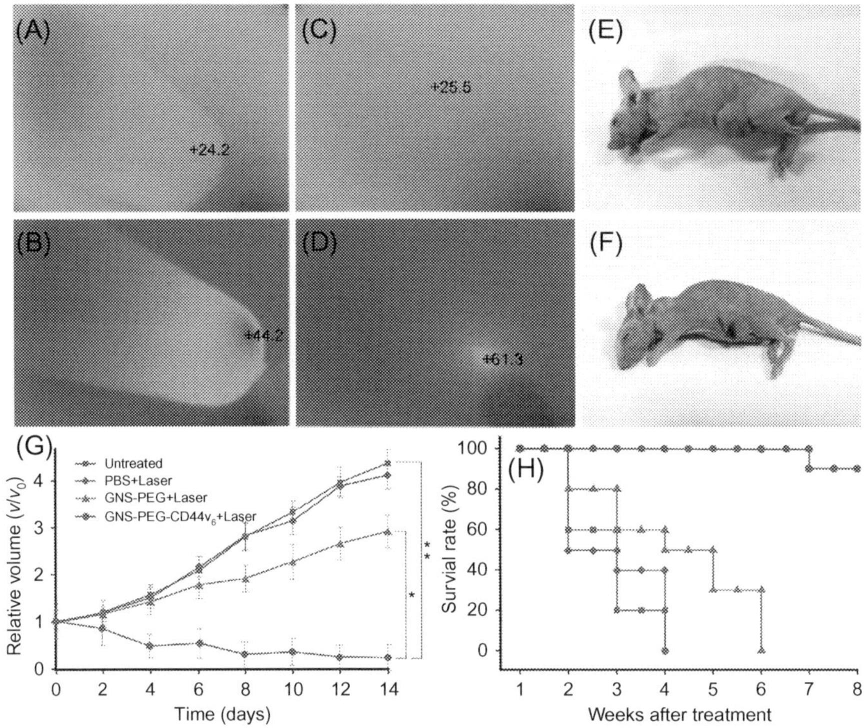

Fig. 25

Infrared microscopic imaging: deionized water (A) and GNS (B) in a tube upon NIR laser irradiation (790 nm, 0.3 W/cm^2, 3 min); subcutaneous tumor of GC without (C) and with (D) injecftion of GNS-PEG-CD44v6 upon NIR laser irradiation (790 nm, 1.5 W/cm^2, 3 min). The nude mouse of GC subcutaneous xenograft, injected with nanoparticles before (E) and after (F) laser irradiation treatment (790 nm, 1.5 W/cm^2, 3 min); (G) tumor growth curves of four groups after treatment with GNS-PEG-CD44v6, GNS-PEG and PBS respectively upon NIR laser irradiation (790 nm, 0.8 W/cm^2, 5 min), and the untreated control group. Error bars represent standard deviation (n = 10/group). $^*P < 0.05$,$^{**}P < 0.01$, (Student's t test); (H) survival rate of GC tumor-bearing mice within eight weeks after treatment. *Permission is from IVY Press.*

the GNCs-based nanoprobes are developed by the covalent coupling folic acid (FA) and PEG (polyethylene glycol) on the surface of GNCs directly, followed by trapping photosensitizer (chlorin e6, Ce6) within PEG networks and attaching to the GNCs surface. The fabricated nanoprobes (Ce6@GNCs-PEG 2K-FA) possess a uniform particle size (hydrodynamic diameter $\approx$6.1 $\pm$ 1.2 nm), without affecting the yield of singlet oxygen of the trapped Ce6. In vitro studies show enhanced cellular uptake and satisfactory PDT effectiveness toward MGC-803 cells when compared with free Ce6. The biodistribution and excretion pathway studies of the nanoprobes in MGC-803 tumor-bearing nude mice reveal their superior penetration and retention behavior in tumors, while the preserved features of renal clearance and stealthy to reticuloendothelial system are mainly attributed to their small hydrodynamic diameters and the FA-capped PEGylated ligands. The enhanced PDT efficacy and nontoxicity to mice provide an exciting new nanoplatform with promising clinical translational potential (Chen et al., 2011; Huang et al., 2011a; Zhang et al., 2015) (Figs. 26 and 27).

7.7 RNA Nanoparticles for Targeted Imaging and siRNA Therapy of Gastric Cancer

In recent years, several new nanodelivery systems with different materials and physicochemical properties have been developed. However, effective strategies to block tumor progression and prevent metastasis are still very few, and there are several challenges including specific cancer targeting, tissue penetration, intracellular delivery, toxicities and side effects due to organ accumulation, nonspecific cell entry, particle heterogeneity, aggregation, dissociation due to dilution after systemic injection, and unfavorable pharmacological profiles. In recent years, RNA nanotechnology (an emerging nanoplatform) has shown great prospect in clinical theranostic applications. RNA nanoparticles can be fabricated with precise control of shape, size, and stoichiometry, as demonstrated by the packaging RNA (pRNA) of the bacteriophage phi29 DNA packaging motor, which forms dimmers, trimers, and hexamers, via hand-in-hand interactions of the interlocking loops

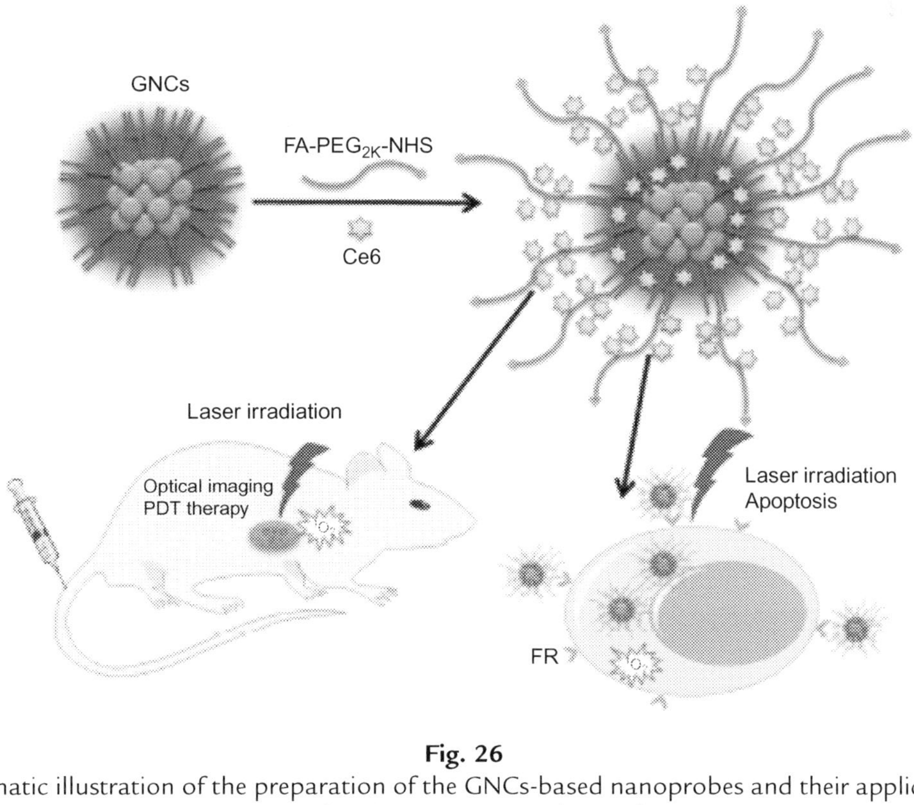

Fig. 26
Schematic illustration of the preparation of the GNCs-based nanoprobes and their applications in vitro and in vivo. *Permission is from Wiley Press.*

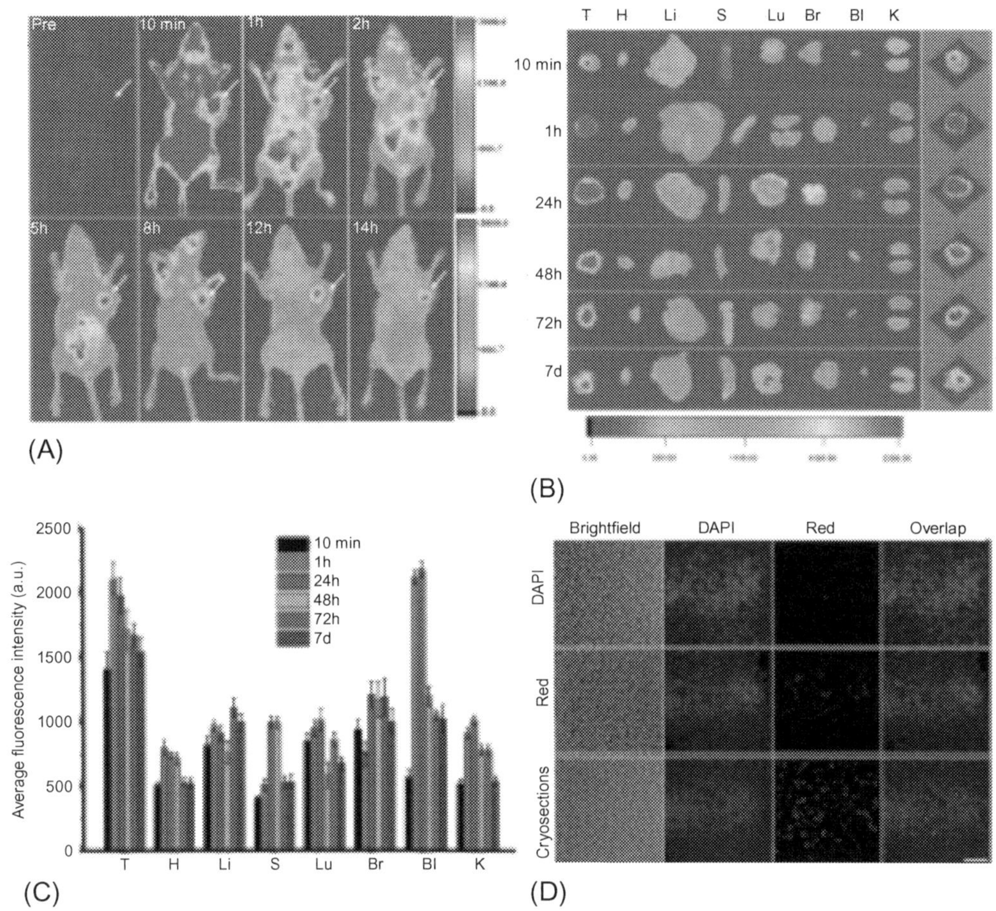

Fig. 27

(A) In vivo fluorescence imaging of MGC-803 tumor-bearing mouse after intravenous injection of Ce6@GNCs-PEG 2K -FA over a period of 24 h. (B) Representative ex vivo fluorescence images of major organs and tumors of mice after different postinjection times. T (tumor), H (heart), Li (liver), S (spleen), Lu (lung), Br (brain), Bl (bladder), K (kidney). The right columns were the corresponding 3D surface plot images of pixel intensities obtained from the tumor regions using NIH Image J software. (C) Quantitative biodistribution of Ce6@GNCs-PEG 2K-FA in different organs and tumors based on ROIs analysis of the average fluorescence intensity from the tumors and normal tissues (3 mice per group). (D) Representative fluorescence images of tumor cryosections (10 μm). Images presented are bright-field, blue fluorescence (DAPI), red fluorescence (>590 nm), and merged (overlap) images. Scale bar, 100 μm. *Permission is from Wiley Press.*

(Clarke et al., 2006; Zhang & Fan, 2010b). The pRNA contains an ultra-stable three-way junction (3WJ) motif which can be assembled from three short fragments with extremely high affinity. Recently, Prof. Peixuan Guo's group has obtained the crystal structure of the pRNA-3WJ motif, and a variety of therapeutic RNA nanoparticles using the pRNA-3WJ and pRNA-X motifs as scaffolds have been constructed. The pRNA-3WJ nanoparticles display thermodynamically stable properties, including high melting temperature with low free energy, resistance to denaturation in 8M urea, and resistance to dissociation at very low concentrations in the blood. Boiling-resistant RNA nanoparticles with controllable shapes and defined stoichiometry have recently been reported. Various functional groups such as imaging agents, targeting ligands such as receptor binding aptamers, and therapeutic modules such as siRNA, miRNA, or ribozymes can be integrated into the 3WJ scaffold without affecting the folding and functionality of the core motif and incorporated functional moieties. Upon 2'-Fluoro (2'-F) modifications of uracil (U) and cytosine (C) nucleotides, the RNA nanoparticles become resistant to RNase degradation with enhanced in vivo half-life while retaining authentic functions of the incorporated modules. Furthermore, the pRNA nanoparticles are nontoxic and nonimmunogenic, and display favorable biodistribution and pharmacokinetic profiles in mice. These favorable findings prompted the use of this novel platform for the treatment of stomach cancer, which is one of the most challenging tasks in clinical oncology.

Such targeted delivery systems need a specific ligand-receptor pair that exists in cancer cells. Many, but not all, cancer cells (including stomach, ovary, lung, breast, kidney, endometrium, colon, and hematopoietic cells) over-expressed folate receptors (FRs) compared to normal cells for high uptake of folate, since folate is an essential component during DNA replication and methylation in highly proliferating cells. Folic acid (FA), a synthetic oxidized form of folate, has been widely used as a ligand conjugate in various cancer-targeting materials. BRCAA1 (breast cancer-associated antigen 1, AF208045) has been confirmed to exhibit over-expression in breast cancer and GC, and no or lower expression in normal gastric mucosa and normal breast tissues. Our previous studies have demonstrated that GC MGC803 cells were transfected with constructed plasmids of shRNA-BRCAA1, the cell growth was greatly inhibited, and the rate of cell apoptosis was significantly higher than those of the untransfected group and a mock plasmid transfected group. We also screened out a new antigen epitope, SSKKQKRSHK, and also screened out and matched two monoclonal antibody cell lines, and successfully prepared monoclonal antibody-conjugated fluorescent magnetic nanoparticles, and realized the targeted imaging and hyperthermal therapy of in vivo GC. Therefore, the *BRCAA1* gene is a potential therapeutic target for GC.

My group cooperated with Prof. Peixuan Guo's group and adopted an innovative RNA nanotechnology approach to overcome some of the aforementioned challenges, and report for the first time a new strategy to target and deliver therapeutic BRCAA1 siRNA to in vivo stomach cancer tissues using FA-conjugated pRNA-3WJ nanoparticles. Our objective is to construct multifunctional, thermodynamically, and chemically stable RNA nanoparticles harboring sequences that allow specific binding to stomach-cancer-specific cell surface antigens or receptors, resulting in the internalization of RNA nanoparticles into target cells and delivery of the siRNA, miRNA, and drugs for attaining synergistic effects for the treatment of stomach cancer; we also investigated the effects of prepared RNA nanoparticles on the regression of GC tissues in vivo, and potential molecular mechanism, with the aim of laying a foundation for further clinical application in the near future (Cui et al., 2015; Khisamutdinov, Jasinski, & Guo, 2014) (Figs. 28–30).

RNA interference via the use of small-interfering RNA (siRNA) is a powerful and useful tool to block gene function through sequence-specific posttranscriptional gene silencing, playing an important role in the down-regulation of gene expression. siRNAs can be transfected into mammalian cells by a variety of methods that influence the strength and duration of the

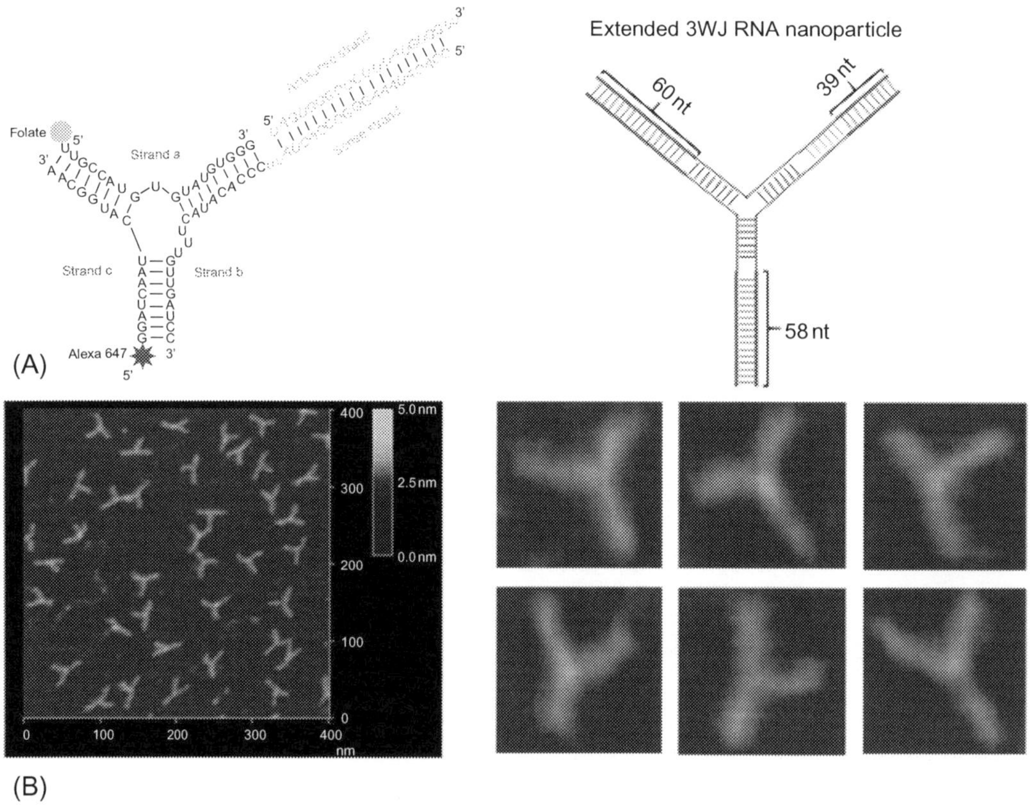

Fig. 28

Global structure of the therapeutic RNA nanoparticles with BRCAA1 siRNA. (A) Design of the RNA nanoparticles. On the left is the one used in the animal trial. To the right is the extended one used to prepare the AFM images. (B) AFM image of extended 3WJ RNA nanoparticles. The RNA complex to the left of a is estimated to be around 10 nm. Due to convolution of the tip size (~10 nm in diameter) in AFM images, features smaller than the size of the tip cannot be resolved. To characterize the structure of the RNA constructs, the 3WJ nanoparticles were extended by 39–60 base-pairs (in red), which is within the persistence length of dsRNA and will not affect the 3WJ folding as described before 31, to generate the AFM image as shown. *Permission is from Nature Press.*

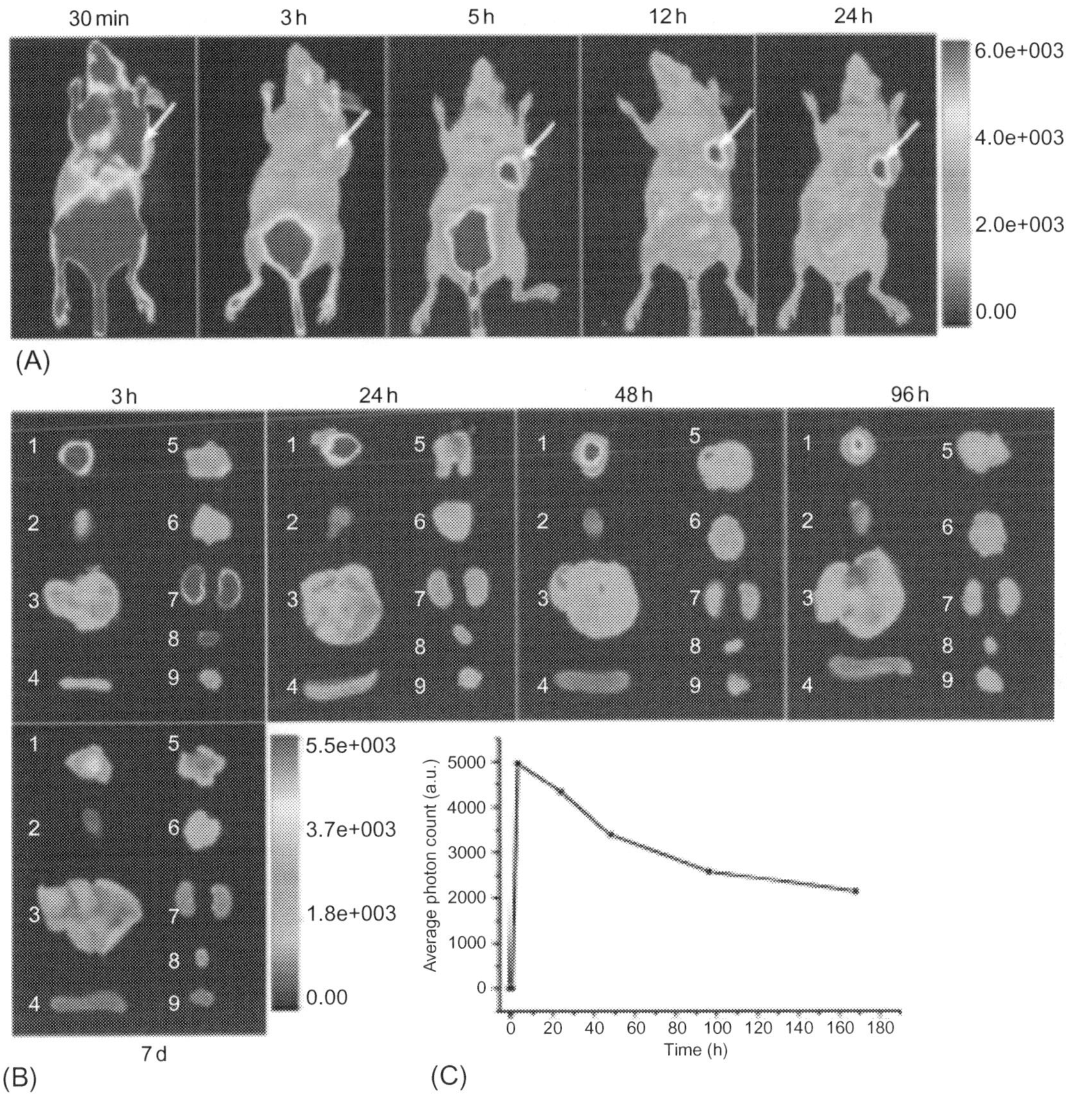

Fig. 29

(A) Representative in vivo fluorescence images of MGC803-tumor-bearing mouse after IV-injected with FA-AlexaFluor647-labeled pRNA nanoparticle. The tumor areas are indicated with arrows. (B) Representative ex vivo images of tumors and organs. Labels: 1, tumor; 2, heart; 3, liver; 4, spleen; 5, lung; 6, stomach; 7, kidneys; 8, bladder; 9, muscle. (C) The average fluorescence intensities from the tumor areas of postinjection. *Permission is from Nature Press.*

silencing response, which in turn, is affected by the amount of siRNA effectively delivered and by the potential of each siRNA to suppress its target. Nevertheless, naked siRNAs show extremely short half-lives due to RNases activity, poor chemical stability, and dissociation from the vector. In fact, the major obstacle to clinical application is the uncertainty about how to deliver siRNA with maximal therapeutic impact (Wang, Yang, Chen, & Shin, 2008).

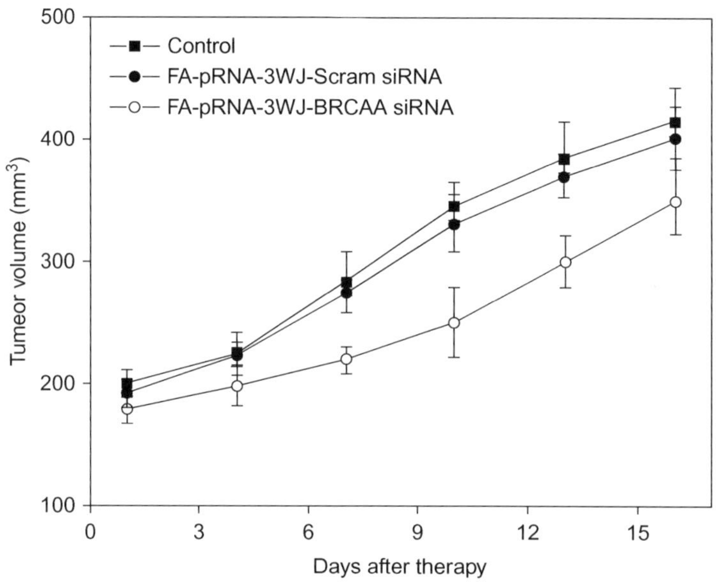

Fig. 30
Tumor size curve as the posttreatment time increases. *Permission is from Nature Press.*

8 Carbon Dots-Based Nanoprobes for Targeted Imaging and Photodynamic Therapy

Carbon dots (C-dots), a very recent rising star within the class of zero-dimensional carbon nanomaterials, possess some of the same major advantageous characteristics of quantum dots (QDs), such as high photostability, tunable emission, and large two-photon excitation cross-sections. Moreover, C-dots exhibit nonblinking fluorescence, excellent water solubility, and are cheaply produced. In particular, C-dots without heavy metal content are more environmentally friendly and can be much safer for biological applications. We strategically designed and prepared novel multifunctional chlorin e6-conjugated C-dots (C-dots-Ce6) as the light-triggered theranostics for simultaneous enhanced near-infrared fluorescent imaging and PDT by FRET mechanism (Huang et al., 2012) (Fig. 31).

Carbon dots exhibit great potential in applications such as molecular imaging and in vivo molecular tracking. However, enhancing the fluorescence intensity of carbon dots has become a great challenge. For the first time, we report a new strategy for synthesizing fluorescent carbon dots (C-dots) with high quantum yields by using ribonuclease A (RNase A) as a biomolecular templating agent under microwave irradiation. The synthesized RNase A-conjugated carbon dots (RNase A@C-dots) exhibited quantum yields of 24.20%. The fluorescent color of the RNase A@C-dots can easily be adjusted by varying the microwave reaction time and microwave power. Moreover, the emission wavelength and intensity

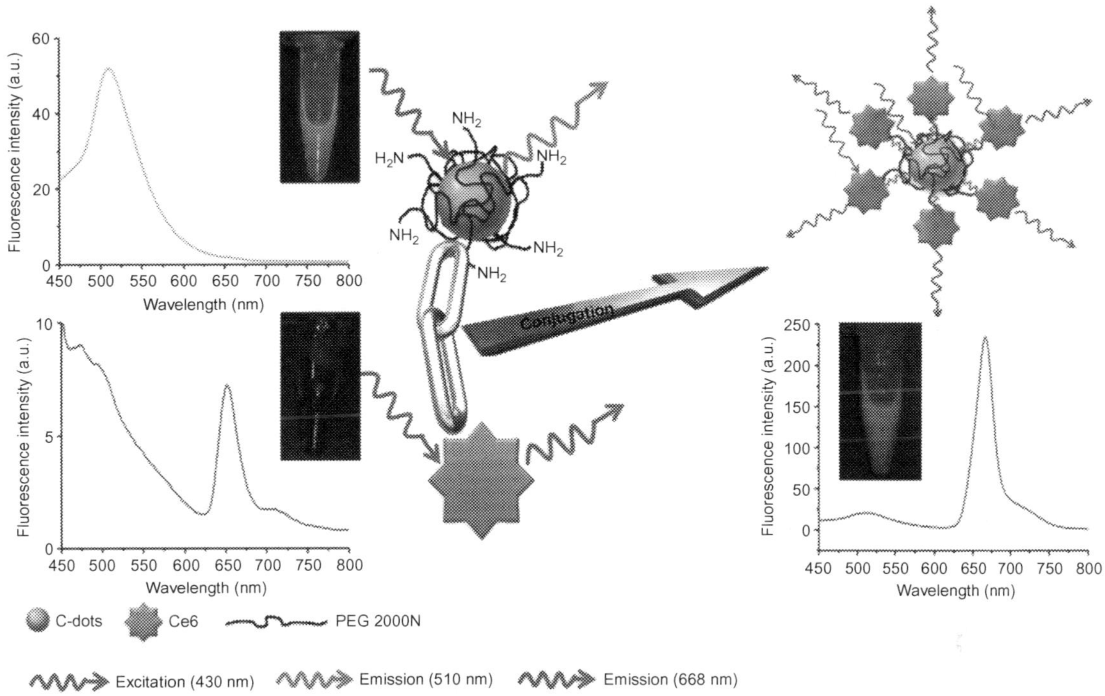

Fig. 31

Förster (fluorescence) resonance energy transfer (FRET) process between C-dots and Ce6. *Permission is from Wiley Press.*

of RNase A@C-dots displayed a marked excitation wavelength-dependent character. As the excitation wavelength alters from 300 to 500 nm, the photoluminescence (PL) peak exhibits gradually red-shifts from 450 to 550 nm, and the intensity reaches its maximum at an excitation wavelength of 380 nm. Its Stokes shift is about 80 nm. Notably, the PL intensity gradually decreases as the pH increases, in an almost linearly dependent manner, and it reaches the maximum at a pH = 2 condition; the emission peaks also clearly show a red-shift, which may be caused by the high activity and perfective dispersion of RNase A in a lower pH solution. In high pH solution, RNase A tends to form RNase A warped carbon dot nanoclusters. Cell imaging confirmed that the RNase A@C-dots could enter into the cytoplasm through cell endocytosis. 3D confocal imaging and transmission electron microscopy observation confirmed partial RNase A@C-dots located inside the nucleus. MTT and real-time cell electronic sensing (RT-CES) analysis showed that the RNase A@C-dots could effectively inhibit the growth of MGC-803 cells. Intratumor injection tests of RNase A@C-dots showed that RNase A@C-dots could be used for imaging in vivo GC cells. The as-prepared RNase A@C-dots are suitable for simultaneous therapy and in vivo fluorescence imaging of nude mice loaded with GC or other tumors. We have developed a microwave-assisted one-step synthesis of C-dots with citric acid as carbon source and tryptophan (Trp) as both nitrogen source and passivation agent. The C-dots with uniform size show superior water solubility, excellent biocompatibility, and high quantum yield. Afterwards, the PEI (polyethylenimine)-adsorbed C-dots nanoparticles (Cdots@PEI) were applied to deliver Survivin siRNA into human GC cell line MGC-803. The results have confirmed that the nanocarrier exhibited excellent biocompatibility and a significant increase in cellular delivery of siRNA, inducing efficient knockdown for Survivin protein to 6.1%. In addition, PEI@ Cdots complexes mediated Survivin silencing, the arrested cell cycle progression in G1 phase as well as cell apoptosis was observed. (Chen et al., 2013; Kim et al., 2011; Park et al., 2013a; Wang et al., 2013; Yuan et al., 2012) Carbon dots may be a good contrast agent and drug delivery system.

9 Clinical Translational Prospects of Multifunctional Nanoprobes

There have been thousands of publications regarding nanoprobes in the last decade, and very few of the assessed nanoprobes progressed from bench to bedside. According to Drolet and Lorenzi's "continuum" theory, when a new basic science discovery transfers into proposed human application, proven clinical application, clinical practice, and finally, impacts public health, there are three "translation chasm" gaps that must be bridged. These gaps are human (T1), clinical trials (T2), and translation to practice (T3). Although there have been very promising applications of nanoprobes in many areas, they are still at an early stage of development on the way to reaching the bedside; in other words, they are mostly in the T1 stage, although they may come into clinical trials.

Up to date, there are several nanoprobes to enter into clinical trials or obtain FDA permission. For example, in 2011, a clinical trial of insulin-coated AuNPs was approved by Midatech Ltd (UK). In addition, AuroShell (Au@silica nanoshells) from Nanospectra (Nanospectra Biosciences, TX, United States), which is used for thermotherapy, was approved by the FDA in 2012. This has been proven to be effective at an earlier stage in a murine glioma model and in the ablation of prostate cancer in a canine model. Another example is from McNeil's laboratory; after their reports of the biodistribution, immune response, and interactions of AuNPs with human blood, they used AuNPs as a drug-delivery system in order to impair the toxicity of TNF-α. It is confirmed to be safe to inject up to three times the amount that had been lethal with previous versions. The modified drug Aurimune (CYT-6091) from CytImmune Sciences (MD, United States) is a TNFα-bound, PEGylated colloidal AuNP. Aurimmune has passed through a Phase I clinical trial and is entering a Phase II trial. It is reported that the systemic TNF toxicity of Aurimune was decreased dramatically compared with TNF alone (e.g., the fever side effect was managed), and the target effect was realized. Another product, AuriTol (CYT-21001), also from CytImmune Sciences, has reached the preclinical stage (Huang et al., 2011b; Li et al., 2010; Lu, Yan, Zhao, & Xu, 2015; Pan et al., 2009a; Weissleder, 2001).

An example of a fast, in vitro, Au-based biomedical application is the FDA-approved Verigene detector device (Nanosphere, IL, United States), which utilizes oligonucleotide-coated AuNPs in order to capture DNA. The Verigene system enables clinicians to rapidly identify and treat the bacteria and viruses that are responsible for some of the most complicated, costly, and fatal infectious diseases (Nie & Chen, 2014).

My group established the new method of RNase A-assisted carbon dots synthesis, and finished biosafety evaluation of carbon dots in national drug safety evaluation center, confirming them to be very safe under the dosage of 100 mg/kg body weight; we also prepared hyaluron acid/RGD-conjugated carbon dots and realized targeted imaging and operation boundary identification of in situ GC, and finished 40 specimens of clinical trials, showing the prepared nanoprobes can image the metastasis of lymph nodes and display the surgery boundary of GC.

Although some nanoprobes got through the clinical trials and obtained FDA permission to enter into clinical application, clinical application for inorganic nanoparticles-based multifunctional nanoprobes remains greatly disputed, the reasons being their inability to degrad in vivo, their long-term residual presence in the body, and the fact that their further toxicological mechanism is not yet well-clarified. Development of safe, highly efficient, targeting nanoprobes has become a great challenge. Only if we are able to resolve these issues concerning the biosafety of nanoprobes will it be possible for multifunctional nanoprobes to enter clinical application in the near future.

10 Conclusion Remark

In summary, GC is a common type of tumor worldwide, and realization of the theranostics for GC is a great challenge. Current advances show that GC therapeutic strategies should focus on early diagnosis and operation therapy, enhanced immunotherapy, and killing GC stem cells to overcome multidrug resistance (MDR). Mindful of these strategies, we designed and prepared multiple series of multifunctional nanoprobes for targeted imaging and therapy of GC, including series of fluorescent magnetic nanoprobes, quantum dots nanoprobes and carbon dots, gold nanoprobes, upconversion nanoprobes, RNA nanoprobes, and nanoprobes for killing GCSCs and enhanced immunotherapeutic efficacy. To date, carbon dots-based nanoprobes have been evaluated to confirm their safety, and have been used for identifying the boundaries of GC and tracking metastasis in lymph nodes, exhibiting clinical application prospects. In the future, the main challenges in translational research of these nanoprobes will focus on gaining a better understanding of the interactions between nanoparticles and biomacromolecules and the immune system, the homogeneity of the material preparations, the paucity of knowledge regarding pertinent biomarkers, and concerns regarding biocompatibility. Green synthesis of nanoparticles, multimodality bioimaging (especially

PAI), SERS analysis, and theranostics are particular fields of interest. However, designers must possess a systematic view from the beginning as they approach the goal of nanoprobes for translational medical research (Chen, Frey, Aglyamov, & Emelianov, 2012; Li et al., 2014; Nie et al., 2014; Sykes, Chen, Zheng, & Chan, 2014).

Acknowledgments

This work was supported by Chinese Key Basic Research Program (973 Project) (No. 2010CB933901 and 2017FYA0205300), the National Natural Scientific Foundation of China (Grant Nos. 81225010 and 81327002), and 863 Project of China (No. 2014AA020700), Shanghai Science and Technology Fund (15DZ2252000).

References

Abdelmawla, S., Guo, S., Zhang, L., Pulukuri, S. M., Patankar, P., Conley, P., et al. (2011). Pharmacological characterization of chemically synthesized monomeric phi29 pRNA nanoparticles for systemic delivery. *Molecular Therapy: The Journal of the American Society of Gene Therapy*, *19*, 1312–1322.

Brabletz, T., Jung, A., Spaderna, S., Hlubek, F., & Kirchner, T. (2005). Opinion: Migrating cancer stem cells—An integrated concept of malignant tumour progression. *Nature Reviews Cancer*, *5*(9), 744.

Chen, F., Huang, P., Zhu, Y., Wu, J., Zhang, C., & Cui, D. X. (2011). The photoluminescence, drug delivery and imaging properties of multifunctional Eu3t/Gd3t dual-doped hydroxyapatite nanorods. *Biomaterials*, *32*, 9031–9039.

Chen, R., Wang, X., Yao, X., Zheng, X., Wang, J., & Jiang, X. (2013). Near-IR-triggered photothermal/photodynamic dual-modality therapy system via chitosan hybrid nanospheres. *Biomaterials*, *34*, 8314–8322.

Chen, T., Yang, K., Yu, J., Meng, W., Yuan, D., Bi, F., et al. (2012). Identification and expansion of cancer stem cells in tumor tissues and peripheral blood derived from gastric adenocarcinoma patients. *Cell Research*, *22*, 248–258.

Chen, W., Zhang, X., Chu, C., Cheung, W. L., Ng, L., Lam, S., et al. (2013). Identification of CD44+ cancer stem cells in human gastric cancer. *Hepato-Gastroenterology*, *60*, 949–954.

Chen, Y., Huang, K., Li, X., Lin, X., Zhu, Z., & Wu, Y. (2010). Generation of a stable anti-human CD44v6 scfv analysis of its cancer-targeting ability in vitro. *Cancer Immunology, Immunotherapy*, *59*, 933–942.

Chen, Y. S., Frey, W., Aglyamov, S., & Emelianov, S. (2012). Environment-dependent generation of photoacoustic waves from plasmonic nanoparticles. *Small*, *8*, 47–52.

Clarke, M. F., Dick, J. E., Dirks, P. B., Eaves, C. J., Jamieson, C. H., Jones, D. L., et al. (2006). Cancer stem cells-perspectives on current status and future directions: Aacr workshop on cancer stem cells. *Cancer Research*, *66*, 9339–9344.

Comis, R. L., & Carter, S. K. (1974). A review of chemotherapy in gastric cancer. *Cancer*, *34*, 1576–1586.

Cui, D., Zhang, C. L., Liu, B., Shu, Y., Du, T., Li, C., et al. (2015). Regression of gastric cancer by Systemic Injection of RNA Nanoparticles Carrying both Ligand and siRNA. *Scientific Reports*, *5*, 10732.

Cui, D. X., Zhang, L., Yan, X. J., Zhang, L. X., Xu, J. R., Guo, Y. H., et al. (2005). A microarray-based gastric carcinoma prewarning system. *World Journal of Gastroenterology*, *11*, 1273–1282.

Dicken, B. J., Bigam, D. L., Cass, C., Mackey, J. R., Joy, A. A., & Hamilton, S. M. (2005). Gastric adenocarcinoma: Review and considerations for future directions. *Annals of Surgery*, *241*, 27–39.

Duan, J. J., Qiu, W., Xu, S. L., Wang, B., Ye, X. Z., Ping, Y. F., et al. (2013). Strategies for isolating and enriching cancer stem cells: Well begun is half done. *Stem Cells and Development*, *22*, 2221–2239.

Ferlay, J., Shin, H. R., Bray, F., Forman, D., Mathers, C., & Parkin, D. M. (2010). Estimates of worldwide burden of cancer in 2008: GLOBOCAN 2008. *International Journal of Cancer*, *127*, 2893–2917.

Gao, W., Xiang, B., Meng, T. T., Liu, F., & Qi, X. R. (2013). Chemotherapeutic drug delivery to cancer cells using a combination of folate targeting and tumor microenvironment-sensitive polypeptides. *Biomaterials*, *34*, 4137–4149.

Gilbertson, R. J., & Cancer, G. T. A. (2012). Resolving the stem-cell debate. *Nature*, *488*, 462–463.
Guo, P. (2010). The emerging field of RNA nanotechnology. *Nature Nanotechnology*, *5*, 833–842.
Guo, P., Haque, F., Hallahan, B., Reif, R., & Li, H. (2012). Uniqueness, advantages, challenges, solutions, and perspectives in therapeutics applying RNA nanotechnology. *Nucleic Acid Therapeutics*, *22*, 226–245.
Guo, P., Zhang, C., Chen, C., Garver, K., & Trottier, M. (1998). Inter-RNA interaction of phage phi29 pRNA to form a hexameric complex for viral DNA transportation. *Molecular Cell*, *2*, 149–155.
Haque, F., Shu, D., Shu, Y., Shlyakhtenko, L. S., Rychahou, P. G., Evers, B. M., et al. (2012). Ultrastable synergistic tetravalent RNA nanoparticles for targeting to cancers. *Nano Today*, *7*, 245–257.
He, M., Huang, P., Zhang, C. L., Hu, H. Y., Bao, C. C., Gao, G., et al. (2011). Dual phase-controlled synthesis of uniform lanthanide-doped NaGdF4 upconversion nanocrystals via an OA/ionic liquid two-phase system for in vivo dual-modality imaging. *Advanced Functional Materials*, *21*, 4470–4477.
Heider, K. H., Kuthan, H., Stehle, G., & Munzert, G. (2004). Cd44v6: A target for antibody-based cancer therapy. *Cancer Immunology, Immunotherapy*, *53*, 567–579.
Huang, P., Bao, L., Zhang, C., Lin, J., Luo, T., Yang, D., et al. (2011). Folic acid-conjugated silica-modified gold nanorods for X-ray/CT imaging-guided dual-mode radiation and photo-thermal therapy. *Biomaterials*, *32*, 9796–9809.
Huang, P., Lin, J., Wang, X. S., Wang, Z., Zhang, C. L., He, M., et al. (2012). Light-triggered theranostics based on photosensitizer-conjugated carbon dots for simultaneous enhanced-fluorescence imaging and photodynamic therapy. *Advanced Materials*, *24*, 5104–5110.
Huang, P., Xu, C., Lin, J., Wang, C., Wang, X., Zhang, C., et al. (2011a). Folic acid-conjugated graphene oxide loaded with photosensitizers for targeting photodynamic therapy. *Theranostics*, *1*, 240–250.
Huang, P., Xu, C., Lin, J., Wang, C., Wang, X., Zhang, C., et al. (2011b). Folic acid-conjugated graphene oxide loaded with photosensitizers for targeting photodynamic therapy. *Theranostics*, *1*, 240–250.
Huang. (2012). Light-triggered theranostic based on photosensitizer-conjugated carbon dots for simultaneous enhanced-fluorescence imaging and photodynamic therapy. *Advanced Materials*, *24*, 5104–5110.
Jang, B. I., Li, Y., Graham, D. Y., & Cen, P. (2011). The role of CD44 in the pathogenesis, diagnosis, and therapy of gastric cancer. *Gut Liver*, *5*, 397–405.
Jemal, A., Bray, F., Center, M. M., Ferlay, J., Ward, E., & Forman, D. (2011). Global cancer statistics. *CA: A Cancer Journal for Clinicians*, *61*, 69–90.
Kalli, K. R., Oberg, A. L., Keeney, G. L., Christianson, T. J., Low, P. S., Knutson, K. L., et al. (2008). Folate receptor alpha as a tumor target in epithelial ovarian cancer. *Gynecologic Oncology*, *108*, 619–626.
Khisamutdinov, E. F., Jasinski, D. L., & Guo, P. (2014). RNA as a boiling-resistant anionic polymer material to build robust structures with defined shape and stoichiometry. *ACS Nano*, *8*, 4771–4781.
Kim, C., Song, H. M., Cai, X., Yao, J., Wei, A., & Wang, L. V. (2011). In vivo photoacoustic mapping of lymphatic systems with plasmon-resonant nanostars. *Journal of Materials Chemistry*, *21*, 2841–2844.
Kuo, C. Y., Chao, Y., & Li, C. P. (2014). Update on treatment of gastric cancer. *Journal of the Chinese Medical Association: JCMA*, *77*(7), 345–353.
Li, R., Wu, X., Wei, H., & Tian, S. (2013). Characterization of side population cells isolated from the gastric cancer cell line SGC-7901. *Oncology Letters*, *5*, 877–883.
Li, W., Sun, X., Wang, Y., Niu, G., Chen, X., Qian, Z., et al. (2014). In vivo quantitative photoacoustic microscopy of gold nanostar kinetics in mouse organs. *Biomedical Optics Express*, *5*, 2679–2685.
Li, Z., Huang, P., Zhang, X., Lin, J., Yang, S., Liu, B., et al. (2010). RGD-conjugated dendrimer-modified gold nanorods for in vivo tumor targeting and photothermal therapy. *Molecular Pharmaceutics*, *7*, 94–104.
Li, Z. M., Huang, P., He, R., Lin, J., Yang, S., Zhang, X. J., et al. (2010). Aptamer-conjugated dendrimer-modified quantum dots for cancer cell targeting and imaging. *Materials Letters*, *64*, 375–378.
Li, Z. M., Huang, P., Zhang, X. J., Lin, J., Yang, S., Liu, B., et al. (2010). RGD-conjugated dendrimer-modified gold nanorods for in vivo tumor targeting and photothermal therapy. *Molecular Pharmaceutics*, *7*, 94–104.
Liang, S. J., Li, C., Zhao, C. L., Chen, Y. S., Xu, L., Bao, C. C., et al. (2015). CD44v6 monoclonal antibody-conjugated gold nanostars for targeted photoacoustic imaging and plasmonic photothermal therapy of gastric cancer stem-like cells. *Theranostics*, *5*, 879–881.

Liu, J., Ma, L., Xu, J., Liu, C., Zhang, J., Liu, J., et al. (2013). Spheroid body-forming cells in the human gastric cancer cell line MKN-45 possess cancer stem cell properties. *International Journal of Oncology*, *42*, 453–459.

Lu, L., Yan, G. Z., Zhao, K., & Xu, F. (2015). An implantable telemetry platform system with ASIC for in vivo monitoring of gastrointestinal physiological information. *IEEE Sensors Journal*, *12*, 3524–3534.

Ly, A., Hoyt, L., Crowell, J., & Kim, Y. I. (2012). Folate and DNA methylation. *Antioxidants and Redox Signaling*, *17*, 302–326.

Ma, J. B., Zhou, Z. J., Zhang, C. L., Gao, G., Li, C., & Cui, D. (2012). Folic acid-conjugated LaF_3:Yb,Tm@ SiO_2 nanoprobes for targeting dual-modality imaging of upconversion luminescence and X-ray computed tomography. *Journal of Physical Chemistry C*, *116*, 14062–14070.

Marhaba, R., Klingbeil, P., Nuebel, T., Nazarenko, I., Buechler, M. W., & Zoeller, M. (2008). CD44 and EPCAM: Cancer-initiating cell markers. *Current Molecular Medicine*, *8*, 784–804.

Misra, S., Heldin, P., Hascall, V. C., Karamanos, N. K., Skandalis, S. S., Markwald, R. R., et al. (2011). Hyaluronan-CD44 interactions as potential targets for cancer therapy. *The FEBS Journal*, *278*, 1429–1443.

Murphy, E. A., Majeti, B. K., Mukthavaram, R., Acevedo, L. M., Barnes, L. A., & Cheresh, D. A. (2011). Targeted nanogels: A versatile platform for drug delivery to tumors. *Molecular Cancer Therapeutics*, *10*, 972–982.

Naor, D., Sionov, R. V., & Ish-Shalom, D. (1997). Cd44: Structure, function, and association with the malignant process. *Advances in Cancer Research*, *71*, 241–319.

Nie, L., & Chen, X. (2014). Structural and functional photoacoustic molecular tomography aided by emerging contrast agents. *Chemical Society Reviews*, *43*, 7132–7170.

Nie, L., Huang, P., Li, W., Yan, X., Jin, A., Wang, Z., et al. (2014). Early-stage imaging of nanocarrier-enhanced chemotherapy response in living subjects by scalable photoacoustic microscopy. *ACS Nano*, *8*, 12141–12150.

Pan, B., Cui, D., Xu, P., Ozkan, C., Feng, G., Ozkan, M., et al. (2009a). Synthesis and characterization of polyamidoamine dendrimer-coated multi-walled carbon nanotubes and their application in gene delivery systems. *Nanotechnology*, *20*, 125101.

Pan, B. F., Cui, D. X., Xu, P., Ozkan, C., Feng, G., Ozkan, M., et al. (2009b). Synthesis and characterization of polyamidoamine dendrimer-coated multi-walled carbon nanotubes and their application in gene delivery systems. *Nanotechnology*, *20*, 125101.

Park, J., Ku, M., Kim, E., Park, Y., Hong, Y., Haam, S., et al. (2013a). CD44-specific supramolecular hydrogels for fluorescence molecular imaging of stem-like gastric cancer cells. *Integrative Biology (Cambridge)*, *5*, 669–672.

Proserpio, I., Rausei, S., Barzaghi, S., Frattini, F., Galli, F., Iovino, D., et al. (2014). Multimodal treatment of gastric cancer. *World Journal of Gastrointestinal Surgery*, *6*, 55–58.

Prud'Homme, G. J. (2012). Cancer stem cells and novel targets for antitumor strategies. *Current Pharmaceutical Design*, *18*, 2838–2849.

Qi, L., Wu, L., Zheng, S., Wang, Y., Fu, H., & Cui, D. X. (2012). Cell-penetrating magnetic nanoparticles for highly efficient delivery and intracellular imaging of siRNA. *Biomacromolecules*, *13*, 2723–2730.

Ruan, J., Song, H., Qian, Q. R., Li, C., Wang, K., Bao, C. C., et al. (2012). HER2 monoclonal antibody conjugated RNase-A-associated CdTe quantum dots for targeted imaging and therapy of gastric cancer. *Biomaterials*, *33*, 7093–7102.

Shen, M., Huang, Y., Han, L., Qin, J., Fang, X., Wang, J., et al. (2012). Multifunctional drug delivery system for targeting tumor and its acidic microenvironment. *Journal of Controlled Release: Official Journal of the Controlled Release Society*, *161*, 884–892.

Shi, J., Zhang, H., Wang, L., Li, L., Wang, H., Wang, Z., et al. (2013). PEI-derivatized fullerene drug delivery using folate as a homing device targeting to tumor. *Biomaterials*, *34*, 251–261.

Shu, D., Moll, W. D., Deng, Z., Mao, C., & Guo, P. (2004). Bottom-up assembly of RNA arrays and superstructures as potential parts in nanotechnology. *Nano Letters*, *4*, 1717–1723.

Shu, D., Shu, Y., Haque, F., Abdelmawla, S., & Guo, P. (2011). Thermodynamically stable RNA three-way junction for constructing multifunctional nanoparticles for delivery of therapeutics. *Nature Nanotechnology*, *6*, 658–667.

Shu, Y., Shu, D., Haque, F., & Guo, P. (2013). Fabrication of pRNA nanoparticles to deliver therapeutic RNAs and bioactive compounds into tumor cells. *Nature Protocols*, *8*, 1635–1659.

Song, H., He, R., Wang, K., Ruan, J., Bao, C. C., Li, N., et al. (2010). Anti-HIF-1 alpha antibody-conjugated pluronic triblock copolymers encapsulated with Paclitaxel for tumor targeting therapy. *Biomaterials*, *31*, 2302–2312.

Sykes, E. A., Chen, J., Zheng, G., & Chan, W. C. (2014). Investigating the impact of nanoparticle size on active and passive tumor targeting efficiency. *ACS Nano*, *8*, 5696–5706.

Takahashi, T., Saikawa, Y., & Kitagawa, Y. (2013). Gastric cancer: Current status of diagnosis and treatment. *Cancers (Basel)*, *5*, 48–63.

Takaishi, S., Okumura, T., Tu, S., Wang, S. S., Shibata, W., Vigneshwaran, R., et al. (2009). Identification of gastric cancer stem cells using the cell surface marker CD44. *Stem Cells*, *27*, 1006–1020.

Teng, L., Xie, J., Teng, L., & Lee, R. J. (2012). Clinical translation of folate receptor-targeted therapeutics. *Expert Opinion on Drug Delivery*, *9*, 901–908.

Uemura, N., Okamoto, S., Yamamoto, S., Matsumura, N., Yamaguchi, S., Yamakido, M., et al. (2001). Helicobacter pylori infection and the development of gastric cancer. *New England Journal of Medicine*, *345*, 784–789.

Vinogradov, S., & Wei, X. (2012). Cancer stem cells and drug resistance: The potential of nanomedicine. *Nanomedicine (London, England)*, *7*, 597–615.

Wang, K., Ruan, J., Qian, Q., Song, H., Bao, C. C., Kong, Y. F., et al. (2011). BRCAA1 monoclonal antibody conjugated fluorescent magnetic nanoparticles for in vivo targeted magnetofluorescent imaging of gastric cancer. *Journal of Nanobiotechnology*, *9*, 23.

Wang, S., Huang, P., Nie, L., Xing, R., Liu, D., Wang, Z., et al. (2013). Single continuous wave laser induced photodynamic/plasmonic photothermal therapy using photosensitizer-functionalized gold nanostars. *Advanced Materials*, *25*, 3055–3061.

Wang, X., Yang, L., Chen, Z. G., & Shin, D. M. (2008). Application of nanotechnology in cancer therapy and imaging. *CA: A Cancer Journal for Clinicians*, *58*, 97–110.

Weissleder, R. (2001). A clearer vision for in vivo imaging. *Nature Biotechnology*, *19*, 316–317.

Xue, Z., Yan, H., Li, J., Liang, S., Cai, X., Chen, X., et al. (2012). Identification of cancer stem cells in vincristine preconditioned SGC-7901 gastric cancer cell line. *Journal of Cellular Biochemistry*, *113*, 302–312.

Yang, W., Raufi, A., & Klempner, S. J. (2014). Targeted therapy for gastric cancer: Molecular pathways and ongoing investigations. *Biochimica et Biophysica Acta*.

Yoshida, M., Yasuda, T., Hiramitsu, T., Ito, H., & Nakamura, T. (2008). Induction of apoptosis by anti-CD44 antibody in human chondrosarcoma cell line SW1353. *Biomedical Research*, *29*, 47–52.

Yu, X., & Pishko, M. V. (2011). Nanoparticle-based biocompatible and targeted drug delivery: Characterization and in vitro studies. *Biomacromolecules*, *12*, 3205–3212.

Yuan, H., Khoury, C. G., Hwang, H., Wilson, C. M., Grant, G. A., & Vo-Dinh, T. (2012). Gold nanostars: Surfactant-free synthesis, 3D modelling, and two-photon photoluminescence imaging. *Nanotechnology*, *23*, 075102.

Zhang, C., Li, C., He, F., Cai, Y., & Yang, H. (2011). Identification of CD44+CD24+ gastric cancer stem cells. *Journal of Cancer Research and Clinical Oncology*, *137*, 1679–1686.

Zhang, C. L., Zhou, Z. J., Qian, Q. R., Gao, G., Li, C., Feng, L. L., et al. (2013). Glutathione-capped fluorescent gold nanoclusters for dual-modal fluorescence/X-ray computed tomography imaging. *Journal of Materials Chemistry B*, *1*, 5045–5053.

Zhang, C. L., et al. (2015). Folic acid/CE6 conjugated gold nanoclusters for NIR fluorescent imaging and photodynamic therapy with enhanced permission and retention. *Advanced Functional Materials*, *28*, 1314–1325.

Zhang, D., & Fan, D. (2010a). New insights into the mechanisms of gastric cancer multidrug resistance and future perspectives. *Future Oncology*, *6*, 527–537.

Zhang, D., & Fan, D. (2010b). New insights into the mechanisms of gastric cancer multidrug resistance and future perspectives. *Future Oncology*, *6*, 527–537.

Zhang, H., Endrizzi, J. A., Shu, Y., Haque, F., Sauter, C., Shlyakhtenko, L. S., et al. (2013). Crystal structure of 3WJ core revealing divalent ion-promoted thermostability and assembly of the Phi29 hexameric motor pRNA. *RNA (New York, New York)*, *19*, 1226–1237.

Zhang, Y. X., Gao, G., Liu, H. J., Fu, H. L., Fan, J., Wang, K., et al. (2014). Identification of volatile biomarkers of gastric cancer cells and ultrasensitive electrochemical detection based on sensing interface of Au-Ag alloy coated MWCNTs. *Theranostics*, *4*, 154–162.

Zhou, J., Shum, K. T., Burnett, J. C., & Rossi, J. J. (2013). Nanoparticle-based delivery of RNAi therapeutics: Progress and challenges. *Pharmaceuticals (Basel, Switzerland)*, *6*, 85–107.

Zhou, Z. J., Zhang, C. L., Qian, Q. R., Ma, J. B., He, M., Pan, L. Y., et al. (2013). Folic acid-conjugated silica capped gold nanoclusters for targeted fluorescence/X-ray computed tomography imaging. *Journal of Nanobiotechnology*, *11*, 17.

Further Reading

Burke, A. R., Singh, R. N., Carroll, D. L., Wood, J. C., RB, D.'.A., Jr., Ajayan, P. M., et al. (2012). The resistance of breast cancer stem cells to conventional hyperthermia and their sensitivity to nanoparticle-mediated photothermal therapy. *Biomaterials*, *33*, 2961–2970.

Chen, L., Zheng, J., Zhang, Y., Yang, L., Wang, J., Ni, J., et al. (2011). Tumor-specific expression of MicroRNA-26a suppresses human hepatocellular carcinoma growth via cyclin-dependent and -independent pathways. *Molecular Therapy*, *19*, 1521–1528.

Choi, J., Yang, J., Bang, D., Park, J., Suh, J. S., Huh, Y. M., et al. (2012). Targetable gold nanorods for epithelial cancer therapy guided by near-IR absorption imaging. *Small*, *8*, 746–753.

Code, J., Tian, F. R., Hemandez, Y., Bao, C. C., Baptisa, P., Cui, D., et al. (2015). RNAi-based glyconanoparticles trigger apoptotic pathways for in vitro and in vivo enhanced cancer-cell killing. *Nanoscale*, *7*, 9083–9091.

Cui, D., Jin, G., Gao, T., Sun, T., Tian, F., Estrada, G. G., et al. (2004). Characterization of BRCAA1 and its novel antigen epitope identification. *Cancer Epidemiology, Biomarkers & Prevention: A Publication of the American Association for Cancer Research, Cosponsored by the American Society of Preventive Oncology*, *13*, 1136–1145.

Fu, H. L., Ma, Y., Lu, L. G., Hou, P., Li, B. J., Jin, W. L., et al. (2014). TET1 exerts its tumor suppressor function by interacting with p53-EZH2 pathway in gastric cancer. *Journal of Biomedical Nanotechnology*, *10*, 1217–1230.

Jin, L., Hope, K. J., Zhai, Q., Smadja-Joffe, F., & Dick, J. E. (2006). Targeting of CD44 eradicates human acute myeloid leukemic stem cells. *Nature Medicine*, *12*, 1167–1174.

Li, C., Yang, J., Wang, C., Liang, S., Zhang, C., Chen, F., et al. (2014). BRCAA1 antibody- and Her2 antibody-conjugated amphiphilic polymerengineered CdSe/ZnS quantum dots for targeted imaging of gastric cancer. *Nanoscale Research Letters*, *9*, 244.

Li, C., et al. (2015). DC integrated inactive gastric cancer cell fused vaccine for targeted imaging and enhanced immunotherapeutic efficacy of gastric cancer. *Biomaterials*, *35*, 177–187.

Liu, Y., Yang, M., Zhang, J. P., Zhi, X., Li, C., Zhang, C. L., et al. (2016). Human induced pluripotent stem cells for tumor targeted delivery of gold nanorods and enhanced photothermal therapy. *ACS Nano*, *10*, 2375–2385.

Park, J., Ku, M., Kim, E., Park, Y., Hong, Y., Haam, S., et al. (2013b). CD44-specific supramolecular hydrogels for fluorescence molecular imaging of stem-like gastric cancer cells. *Integrative Biology*, *5*, 669–672.

Pelaz, B., Alexiou, C., Alvarez-Puebla, R. A., Alves, F., Andrews, A. M., et al. (2017). Diverse applications of nanomedicine. *ACS Nano*, *11*, 2313–2381.

Shu, Y., Haque, F., Shu, D., Li, W., Zhu, Z., Kotb, M., et al. (2013). Fabrication of 14 different RNA nanoparticles for specific tumor targeting without accumulation in normal organs. *RNA (New York, New York)*, *19*, 767–777.

Van de Broek, B., Devoogdt, N., D'Hollander, A., Gijs, H. L., Jans, K., Lagae, L., et al. (2011). Specific cell targeting with nanobody conjugated branched gold nanoparticles for photothermal therapy. *ACS Nano*, *5*, 4319–4328.

Wang, Z., Ruan, J., & Cui, D. X. (2009). Advances and prospect of nanotechnology in stem cells. *Nanoscale Resarch Letters*, *4*, 593–605.

Yin, T., Zhang, Q., Wu, H., Gao, G., Shapter, J. G., et al. (2016). Superparamagnetic Fe_3O_4-PEG(2K)-FA@Ce6 nanoprobes for in vivo dual-mode imaging and targeted photodynamic therapy. *Scientific Reports*, *6*, 36187.

Yin, T., Zhang, Q., Wu, H., Gao, G., Shapter, J. G., et al. (2017a). In vivo high-efficiency targeted photodynamic therapy of ultra-small Fe_3O_4@polymer-NPO/PEG-Glc@Ce6 nanoprobes based on small size effect. *NPG Asia Materials*, *9*, e383.

Yin, T., Zhang, Q., Wu, H., Gao, G., Shapter, J. G., et al. (2017b). In vivo targeted therapy of gastric tumors via the mechanical rotation of a flower-like Fe_3O_4@Au nanoprobe under an alternating magnetic field. *NPG Asia Materials*, *9*, e408.

Yuan, H., Khoury, C. G., Wilson, C. M., Grant, G. A., Bennett, A. J., & Vo-Dinh, T. (2012). In vivo particle tracking and photothermal ablation using plasmon-resonant gold nanostars. *Nanomedicine*, *8*, 1355–1363.

CHAPTER 12

Nanomaterial Based Photo-Triggered Drug Delivery Strategies for Cancer Theranostics

Roshni Iyer*, Joseph Wolf*, Daria Zhukova*, Dona Padanilam*, Kytai T. Nguyen*,†

*University of Texas at Arlington, Arlington, TX, United States †University of Texas Southwestern Medical Center, Dallas, TX, United States

1 Introduction

Cancer is one of the leading causes of deaths worldwide. Conventional anti-cancer therapies such as surgery, chemotherapy, and radiation therapy, although successful, suffer from several limitations; most importantly, their inability to spare healthy tissues from damage during the treatment (Yao, Zhang, et al., 2016). Additionally, tumor and patient heterogeneity further complicates cancer therapy (Chen, Zhang, Zhu, Xie, & Chen, 2017). Nanomedicine is rapidly revolutionizing cancer therapeutics due to its unique properties, particularly the ability to selectively and specifically release its payload at the tumor regions (Sortino, 2012). Researchers worldwide are now focusing on developing integrated theranostic devices that are a single nano-device that encompasses early detection of cancer, in addition to controlled delivery of therapeutic agents. The long-term aim of such an integrated device is to accelerate therapeutic efficacy while improving patient outcomes, minimize side-effects, and reduce costs (Ahmed, Fessi, & Elaissari, 2012). The diagnostic aspects usually involve contrast agents such as iron oxide, quantum dots (QDs), carbon nanotubes, gold nanoparticles (NPs), and fluorescent dyes (Ahmed et al., 2012). Imaging modalities such as magnetic resonance imaging (MRI), optical imaging, and fluorescence imaging have been used to locate and diagnose diseased tissues, particularly cancer (Hussain & Nguyen, 2014).

One such class of theranostic nano-devices is those made from stimuli-responsive nanomaterials that allow controlled release of their payloads in response to stimuli changes, with the goal of maximizing therapeutic efficacy and avoiding off-target side effects (Mura, Nicolas, & Couvreur, 2013). Various types of internal and external stimuli have been used to trigger drug release in NPs, such as pH, temperature, and enzymes, but light has its appeal as it allows for a more precise control when it comes to time and space of release (Sortino, 2012). Although employed since ancient times, the possibility of using light

Handbook of Nanomaterials for Cancer Theranostics. https://doi.org/10.1016/B978-0-12-813339-2.00012-8

for therapeutic benefits gained popularity when pioneers such as Niels Finsen received a Nobel prize in 1903 for his research in the application of light for the treatment of lupus vulgaris (Roelandts, 2002). Today, photo-triggered drug delivery (PDD) and photodynamic therapy (PDT) have been accepted worldwide for the treatment of various cancers and inflammatory diseases, among others. Nanomedicine has further revolutionized this treatment modality. Nanomaterial-based PDD to stimulate drug release is an attractive approach, as it allows precise and minimally invasive treatment via on-demand and controlled drug release (Sortino, 2012). The spatiotemporal aspect of nanomaterial-based PDD and its ability to easily remote-control drug delivery has greatly advanced research and clinical applications. Moreover, parameters such as light intensity, beam diameter, wavelength and exposure time can be adjusted in order to achieve the desired release profile (Fomina, Sankaranarayanan, & Almutairi, 2012). Additionally, drug release can be induced in vivo through radiation in the near-infrared (NIR) range, visible spectrum, and/or ultraviolet (UV) range (Fomina et al., 2012).

PDD uses light to generate a cytotoxic effect on cells; however, there are several ways to induce this cytotoxic effect. Some of those include PDT, where the encapsulated drug is a photo-sensitizer, which produces reactive oxygen species (ROS) that could kill cancer cells when irradiated with a specific wavelength of light (Lucky, Soo, & Zhang, 2015; Mallidi et al., 2016). Another is photo-thermal therapy (PTT), which is a version of PDT that uses a specific band of light to cause heat-induced ablation in tumor cells (Zou et al., 2016). The photochemical mechanism behind PDT is called photosensitization-induced oxidation (Foote & Dobrowolski, 1984). Here, through the illumination of the encapsulated photosensitizer (PS), singlet oxygen as a strong oxidizing agent is formed, and it can induce cancer cell death by disrupting its cell membrane (Foote & Dobrowolski, 1984). Photo-isomerization is another technique that can be used where an organic molecule, such as azobenzene, is incorporated into liposomes, which upon contact with UV light, isomerizes from a *trans* orientation to a *cis*, thereby disrupting and causing leaky outlets for the drug to release (Bisby, Mead, & Morgan, 2000).

Several nanomaterials with photo-sensitive properties, including those made from noble metals such as gold and QDs, and those made of biodegradable and biocompatible polymers, have been widely studied (Menon et al., 2013). Additionally, these materials can be loaded with fluorescent dyes such as indocyanine green (ICG) or NIR dyes such as cyanine-7 for imaging applications. The versatility of nanomaterials allows multi-functionality, where they can simultaneously perform therapeutic and diagnostic functions, with the aim to improve therapeutic efficacy and therefore improve patient outcomes. The therapeutic aspects to PDD for cancer therapy include photo-triggered release of chemotherapeutic agents or PS, and hyperthermia via PTT (Rai et al., 2010). The integration of these modalities into a single system has rendered PDD to be one powerful tool for cancer diagnosis and treatment.

PDD has several advantages over conventional cancer therapy, namely low systemic toxicity, because photo-sensitive materials are only activated upon light exposure (Juarranz, Jaen, Sanz-Rodriguez, Cuevas, & Gonzalez, 2008). Other advantages include the ability to administer the therapeutic agents for PDD systemically and topically, tumor-specific drug delivery and treatment, and the versatility of PDD when combined with other cancer therapies such as chemotherapy, radiation therapy, and immunotherapy. However, to be able to target the irradiated light and ablate tumors located in deep regions of the body is one of the main challenges to PDD. The attenuation of light due to the various layers of tissues ultimately debilitates the cytotoxicity in the tissue of interest and reduces the overall efficacy of PDD (Mallidi et al., 2016). Besides this issue, PDD also faces other challenges such as toxicity of its materials, achieving an effective biodistribution and bioactivity of drugs and PS, and light and/or material tissue interactions. In this chapter, we will review some nanomaterial-based PDD and their applications for cancer diagnosis and therapy (Fig. 1). Additionally, we will discuss the mechanisms involved in PDD, the factors to be considered while developing a PDD system, and future perspectives in PDD.

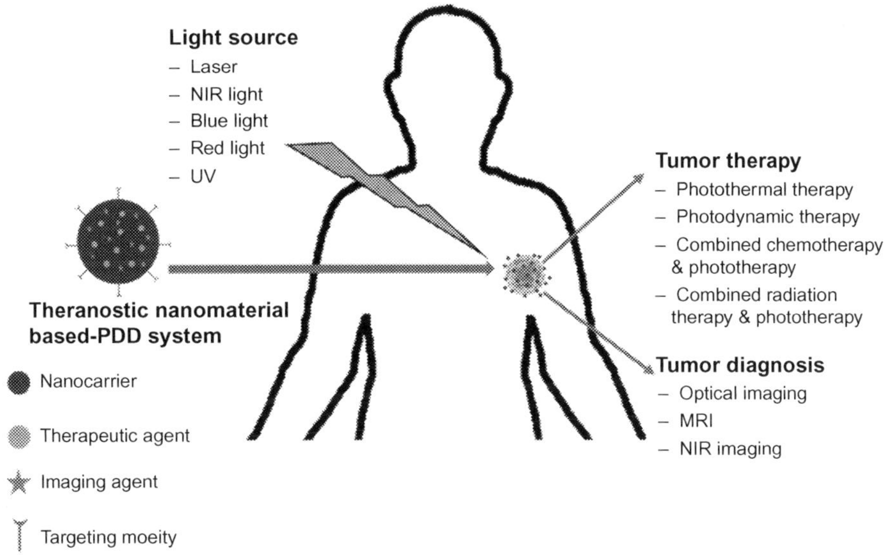

Fig. 1
Schematic of nanomaterial-based photo-triggered drug delivery systems for cancer theranostics.

2 Types of Photo-Sensitive Materials

Photo-sensitive materials encompass components that serve as light responsive therapeutic and/or imaging agents. Major photo-sensitive materials are photosensitizers for therapeutic applications such as porphyrins (e.g., 5-aminolevulinic acid (ALA), verteporfin, and photofrin) with photodynamic properties (Berg et al., 2005), and those for imaging applications such as pyropheophorbide (Rai et al., 2010). Photo-thermal agents that generate heat in response to light irradiation include NIR dyes (e.g., ICG (Lajunen et al., 2016)), metal-based photo-thermal agents (e.g., gold NPs (Park et al., 2009)) and carbon nanotubes (Rai et al., 2010). Additionally, materials such as gold have successfully displayed their imaging and therapeutic capabilities (Huang et al., 2011). Carbon nanotubes can absorb NIR radiation to produce molecular oscillations capable of generating heat in the surrounding tissues (Rai et al., 2010). Another class of light-sensitive materials includes up-conversion NPs (UCNPs) made of lanthanide- or actinide-doped transition metals that absorb two photons of infrared light and release energy in the visible range, subsequently activating associated PS (Chatterjee, Fong, & Zhang, 2008). Photo-sensitive materials are also those that undergo structural modifications to facilitate drug release, such as azobenzenes that transition from their *trans* form to an unstable *cis* form when activated with light to release their cargo (Baroncini, Ragazzon, Silvi, Venturi, & Credi, 2015). Materials such as *O*-nitrobenzyl, on the other hand, get cleaved when irradiated with light and release drugs incorporated within them (Fan, Cheng, & Ho, 2012). The applications of these materials will be discussed in detail in the following sections.

3 Mechanisms of PDD

Over the past few decades, a search for an efficient, on-demand drug delivery platform has been ongoing. Photosensitive nanocarriers offer such advantages, alongside minimal invasiveness and possible spatiotemporal control. Nanocarriers for phototherapy can either be used to carry a photosensitive material such as PS, or be photo-responsive themselves. Nanocarriers themselves possess highly desirable advantages, namely high drug loading efficiency, controllable drug release, and highly customizable surface and chemical modifications (Chatterjee et al., 2008). These advantages have helped nanocarriers outshine other conventional drug delivery systems. Herein, three major mechanisms of photo-triggered release are described: photo-isomerization, photo-cleavage, and PDT.

3.1 Photo-Isomerization

Photo-isomerization can be seen in examples with structures containing chemical groups that are able to change their conformation in response to light irradiation (Fig. 2A). Such chemical groups include azobenzenes and pyrenes (Mura et al., 2013; Sun et al., 2014). These molecules are predominantly composed of double bonds that undergo *trans-cis* transitions in

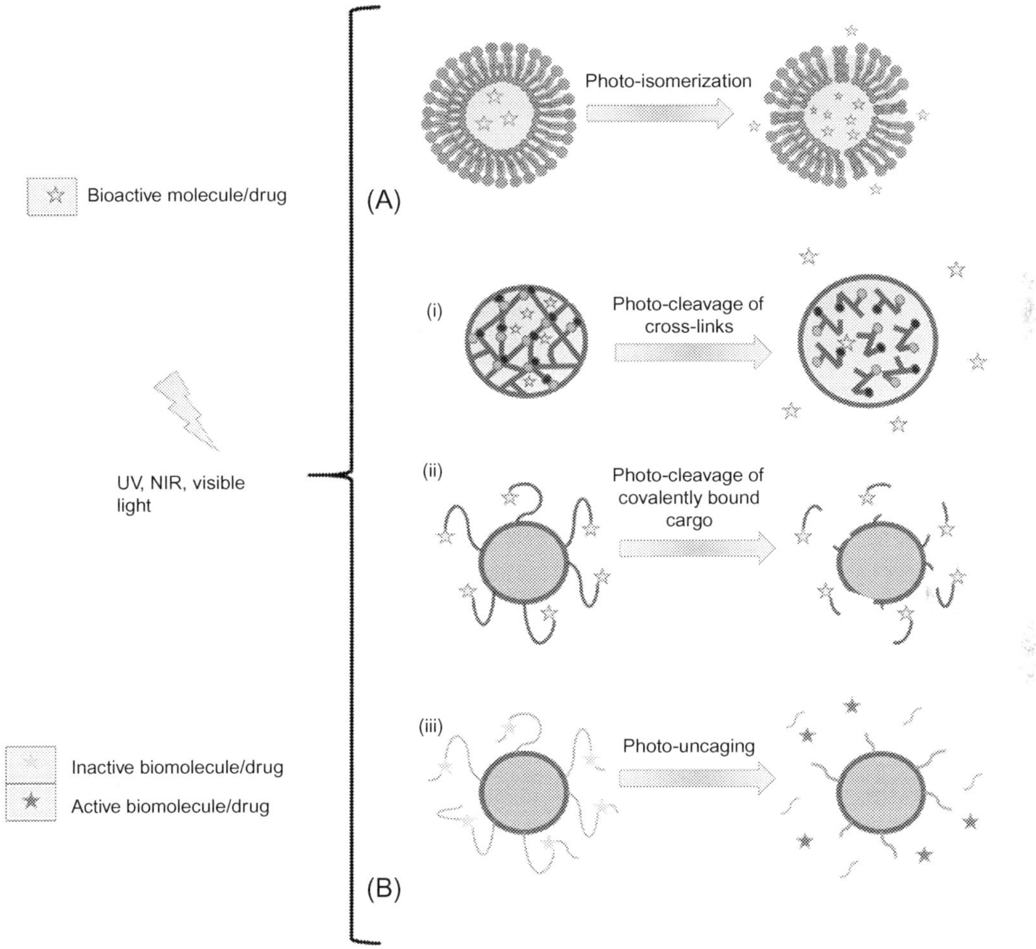

Fig. 2
Schematic of photo-triggered drug release mechanisms in nanocarriers: (A) photo-isomerization and (B) photo-cleavage: of (i) crosslinkers, (ii) cargo, and (iii) photo-uncaging.

response to light. Moreover, due to the reversibility of the *trans-cis* conformational changes, such compounds offer "on-demand" light-triggered capabilities (Bansal & Zhang, 2014; Mura et al., 2013).

Azobenzenes are the most common photosensitive *cis-trans* molecules (Baroncini et al., 2015). Incorporation of azobenzene molecules into nanocarriers has been shown to display photo-responsive and controlled drug release profiles (Bansal & Zhang, 2014; Pearson et al., 2015; Yao, Wang, et al., 2016). Planar and hydrophobic *trans*-azobenzenes usually undergo a conformational change to hydrophilic and non-planar *cis* form upon irradiation by UV light at wavelengths ranging from 300 to 380 nm (Bansal & Zhang, 2014; Sun et al., 2014). Azobenzenes at *cis* conformation then revert back to *trans* conformation in the absence of light due to the instability of *cis* to form at body temperature (Alvarez-Lorenzo, Bromberg, & Concheiro, 2009). This ability of hydrophobic-hydrophilic transitions of azobenzenes has allowed their use in amphiphilic block copolymer micelle disruptions (Alvarez-Lorenzo et al., 2009; Sun et al., 2014). Azobenzenes have been widely researched for PDT-based cancer therapy applications. Pearson et al. (2015) incorporated azobenzenes into self-assembled glycol-polymer micelles for light-responsive delivery of the hydrophobic compound Nile red to melanoma cells in response to UV irradiation. To test the photo-responsiveness of the azobenzenes, the samples were first exposed to UV-vis excitation of 355 nm corresponding to the Π-Π transition of the *trans* isomer, followed by a second irradiation in the perpendicular direction, resulting in their shift to the unstable *cis* form. The conversion of azobenzene to *cis* conformation was achieved post 2 min of treatment, and reversion was observed following several hours in the dark. With a specific galectin-3 receptor targeting, it was shown to potentially provide an effective targeted photo-responsive drug delivery platform via the glycol-polymer/azobenzene nanocarrier for skin cancer. Likewise, azobenzene constituents have been utilized for formulation of light-responsive liposomal carriers (Yao, Wang, et al., 2016). Azobenzene-incorporated liposomal nanostructures encapsulating QD "up-conversion nanoparticles" were able to respond to NIR radiation at 980 nm and emit UV and blue light at 360 and 450 nm, respectively (Yao, Wang, et al., 2016). The emission of the UV light triggered isomerization of the azobenzene groups in the liposomal structures and induced a higher drug release rate (~57 wt%) when compared with the non-irradiated liposomes (5 wt%). This design was particularly interesting due to its use of NIR for irradiation, because UV- and light-triggered delivery systems are often restricted by depth of tissue penetration (~10 mm at wavelength <700 nm) due to tissue scattering and inherent light absorbance by lipid, water, and oxy/deoxy hemoglobin molecules, while NIR (700–1000 nm) offers minimal and non-invasive tissue toxicity, lower scattering capacity, and deeper penetration (Mura et al., 2013).

3.2 Photo-Cleavage and Photo-Uncaging

Photo-cleavage of photo-liable groups within the NP can allow for the release of encapsulated molecules upon activation by light (Fig. 2B(i)). For example, coumarin-containing materials undergo photo-cleavage when irradiated with UV light (200–280 nm) (Fu, Cheng, Zhang, & Shi, 2008). Additionally, small molecules can be covalently conjugated onto the vehicle through photo-sensitive linkers for a more controlled delivery to the specific therapeutic site, as represented in Fig. 2B(ii) (Bansal & Zhang, 2014; Sortino, 2012). Moreover, bioactive molecules can be selectively inactivated, or "caged" into the material, through covalent blockage of their functional groups, including carboxyl and amino groups, by conjugation with photo-sensitive moieties (Fig. 2B(iii)) (Bansal & Zhang, 2014). These molecules can only be activated by light, and hence act like a pro-drug or inactive molecule in the absence of light (Mayer & Heckel, 2006). In principle, activation by light causes an irreversible phenomenon where the photo-labile group is cleaved to release the encapsulated drug (Mayer & Heckel, 2006). The advantage of using photo-caging is that the hydrophilic nature of photo-labile groups and their end products are mostly non-toxic (Mayer & Heckel, 2006). Exceptions include *ortho*-nitrobenzyl, which produces toxic nitroso aldehyde as a by-product of photolysis (Mayer & Heckel, 2006). Upon irradiation with a light source, the photo-linkers are cleaved, resulting in the exposure of the bio-molecule's active sites (Bansal & Zhang, 2014). Fan et al. (2012) devised a folate photo-caged nanocarrier platform, combined with UV irradiation, in the hopes of reducing non-specific toxicity. Photo-caged folate groups were synthesized by caging cancer targeting moieties such as folate within photo-liable *o*-nitrobenzyl groups to reduce the interaction of folate with healthy cells. These caged folate groups were then incorporated with poly (D, L-lactide-*co*-glycolide) (PLGA)-lipid hybrid NPs encapsulating paclitaxel, a chemotherapeutic drug. Upon encountering a tumor, the folate residues were unmasked/uncaged after irradiation with UV light. The exposed folate residues then allowed attachment to the folate receptor expressed on the tumor cells, leading to more specific delivery of NPs to cancer cells. In this study, the tumor cell viability after UV irradiation, followed by treatment with caged-folate conjugated NPs, dramatically decreased from 83% before treatment to 51% after treatment, indicating the superior therapeutic efficacy of the caged NPs. Other examples of photo-caging molecules that can be incorporated are coumarinyl methyl esters (Shibu et al., 2013), benzylic coumaryl (Neveu et al., 2008), *cis*-cinnamoyl imidazole (Mayer & Heckel, 2006), and caged ATP (Mayer & Heckel, 2006).

3.3 Photosensitizers and Photodynamic Therapy

PS are agents that are involved in the in situ creation of cytotoxic compounds, or as a trigger in drug delivery, especially for PDT (Chan et al., 2017; Sortino, 2012). PDT is one of the most widely studied and used forms of PDD. The PS Photofrin® was the first FDA approved PS agent (Huang, 2005). Other approved PS agents include Foscan®, Visudyne®, Levulan®, and Metvix®, and several others in clinical trials (Huang, 2005). Extensive research utilizing Photofrin set the foundation for modern PDT. In total, there are three types of PS agents: porphyrin-based PS (e.g., photofrin and ALA), chlorophyll-based PS (e.g., chlorins and purpurins), and dyes (e.g., phthalocyanine and naphthalocyanine) (Huang, 2005). Among the porphyrins, Photofrin is a first-generation porphyrin, whereas Photofrin derivatives such as ALA constitute second-generation porphyrins, and when conjugated to either biological motifs such as antibodies or synthetic materials such as liposomes, they make up the third generation of porphyrins (Huang, 2005).

Modern PS agents generally consist of a heterocyclic ring structure such as hemoglobin or chlorophyll (Huang, 2005). Light excitation of these PS materials leads to the conversion of light energy into chemical energy, with molecular oxygen serving as a catalyst (Huang, 2005). This leads to the generation of singlet oxygen, nitric oxide, and heat as by-products, all of

Table 1: Types of photosensitizers, their initiator light source, and their downstream cyto-toxic products

Photo-Sensitizer	Reactive Light Source	Cytotoxic Product
Carbon nitride quantum dots (Chan et al., 2017)	NIR (808 nm)	Reactive oxygen species (ROS)
Purpurin-18 (Yoon et al., 2013)	NIR (700 nm)	Singlet oxygen (ROS)
Purpurinimide (Yoon et al., 2013)	NIR (706–718 nm)	Singlet oxygen (ROS)
Ketobacteriochlorins (Yoon et al., 2013)	NIR (750–800 nm)	Singlet oxygen (ROS)
Gold nanoparticles (Yoon et al., 2013)	NIR (700–850 nm)	Localized hyperthermia
Sulfonated porphyrins (Chilakamarthi & Giribabu, 2017)	NIR (695 nm)	Singlet oxygen (ROS)
Phthalocyanines (Chilakamarthi & Giribabu, 2017)	NIR (600–850 nm)	Singlet oxygen (ROS)
Hypericin (Chilakamarthi & Giribabu, 2017)	NIR (600 nm)	Singlet oxygen (ROS)
Hypocrellin (Chilakamarthi & Giribabu, 2017)	Visible light (460 nm)	Singlet oxygen (ROS)

which are major cytotoxic agents (Chilakamarthi & Giribabu, 2017; Yoon, Li, & Shim, 2013). Of these three agents, O_2 (singlet oxygen) is of the greatest interest, as ROS play a large role in the cytotoxic reactions, and are able to cause irreversible damage within the cellular environment (Yoon et al., 2013). In particular, ROS are able to damage the cell membrane, destroy DNA and nucleic acids, and cause destruction of essential proteins (Hemnani & Parihar, 1998; Yoon et al., 2013). PS materials are currently used to produce ROS, but some of them are utilized to create hyperthermic conditions (Kelleher, Thews, Scherz, Salomon, & Vaupel, 2003; Table 1).

Several types of PS agents have been widely studied for cancer therapy in a clinical setting. Photofrin and Foscan-PDT have been clinically used, either individually or in combination with other anti-cancer therapies, to treat head and neck cancer (Feyh, 1996). A Photosan-3 PDT study on head and neck cancer observed a significantly improved patient outcome with more than 90% success 13 months after PDT treatment (Feyh, 1996). The advantage of using PDT for head and neck cancer lies in its ability to cause minimal functional or cosmetic damage to the patient (Huang, 2005). A comprehensive clinical review published in 2010 reported superior treatment efficacy of PDT compared with conventional treatment strategies such as surgery or chemotherapy, for treatment of cancer actinic keratosis, Bowen's disease, and basal cell carcinoma (Fayter, Corbett, Heirs, Fox, & Eastwood, 2010). PDT is also in use for treatment of non-small cell lung cancer (NSCLC). Clinical studies have observed enhanced effects of PDT in combination with radiation for the treatment of NSCLC compared with radiation only in terms of extending the time required to bronchus re-occlusion (Agostinis et al., 2011). Jin et al. (2000) combined PDT (light exposure of 650 nm) with ultrasound therapy, known as sonodynamic therapy (SDT), to improve therapeutic efficacy in a skin squamous cell carcinoma murine model. SDT via ultrasound in combination with PS or sonosensitizers utilizes the superior tissue penetration ability of ultrasound (Jin et al., 2000). The combination of PDT with SDT is expected to provide higher cytotoxic effects in the deeper tissues and/or tumor regions, thus improving cancer therapy. Here a combination of two different PS agents, PH-1126 (a pheophorbide-a derivative) and ATX-70 (gallium porphyrin analogue), were used to study the combined effects of PDT and SDT on tumor therapy. To test the in vivo cytotoxic effect, four groups of mice were designated as control, PDT only, SDT only, and PDT + SDT. The mice treated with the PS were irradiated with laser light at 650 nm (44 J/cm^2), ultrasound, or laser + ultrasound, respectively. Twenty days after treatment, the tumor volumes on mice treated with PDT + SDT were significantly smaller than those treated with PDT alone, SDT alone, and the control group (0.041 vs 0.454, 1.068 and 1.859 cm^3 respectively), thus indicating the superior cancer killing efficacy of the combined treatment.

Nanomaterial-based PDT overcomes the disadvantages associated with PS, such as their limited solubility and aggregation in aqueous solutions (Samia, Chen, & Burda, 2003). Nanomaterial-based PS delivery is classified as either active or passive, depending on the inclusion of targeting moiety. Active nanomaterial based-PDT involves conjugation of antibodies or targeting ligands specific to desired tissues and/or interested cells. The importance of active targeting will be discussed in subsequent sections in this chapter. Nanocarrier encapsulation has successfully improved the delivery and therapeutic outcomes. Meso-tetra (hydroxyphenyl) porphyrin (p-THPP) was encapsulated into biodegradable NPs made of PLGA (either 50:50 or 75:25 ratio) or poly (D, L-lactide), to test their therapeutic capabilities when exposed to red light (655 nm) (Konan, Berton, Gurny, & Allemann, 2003). These NPs, particularly 50:50 PLGA, were observed to produce maximal phototoxicity, allowing lower drug doses and shorter intervals between drug administration and light irradiation. Additionally, the p-THPP loaded NPs showed a PS dose and incubation time dependent phototoxicity when combined with a 6 J/cm^2 light irradiation. The 50:50 PLGA NPs displayed significantly higher phototoxicity compared with free p-THPP (approximately 20% vs 90% cell viability after light treatment at a p-THPP dose of 3 μg/mL).

4 Applications of PDD

Photo-triggered theranostic nanoconstructs are extremely versatile tools that permit the early recognition of diseases via imaging modalities such as MRI or fluorescence optical imaging using contrast agents such as NIR and fluorescent dyes, followed by controlled release of its therapeutic agents in response to light irradiation. The therapeutic agents could be either PS or chemotherapeutic agents. These theranostic agents provide a non-invasive treatment option that can control drug release and prevent over- or under-dosing (Rai et al., 2010). Additionally, PDD can also be coupled with radiation for enhanced anti-tumor therapy (Agostinis et al., 2011). In this section, we will discuss the use of phototherapy for cancer diagnosis, cancer therapy, and cancer theranostics. Additionally, we will review the use of phototherapy in combination with other treatment modalities and its applications in other medical fields.

4.1 Cancer Diagnostics

The use of NPs for cancer detection allows the capacity to image at the molecular level before phenotypic changes occur (Thakor & Gambhir, 2013). Imaging modalities using NPs currently under research include MRI, computed tomography (CT), optical imaging, and photoacoustic imaging (PAI) (Table 2). Many of these particles can be formulated from various methods to be used for a certain modality, or even multimodal, where they can be imaged by two or more modalities (Rai et al., 2010). Multiplexing, the ability of a NP to detect multiple targets, is an area of interest in order to create better disease profiles of cancers to better design effective therapies (Rai et al., 2010; Thakor & Gambhir, 2013).

Table 2: Diagnostic applications of light-based imaging modalities

Nanocarrier	Imaging Agent	Imaging Modality	Study Model	Results
Chitosan NPs (Nam et al., 2010)	Cy5.5, gadolinium (Gd)	Optical imaging, MRI	SCC7 tumor-bearing nude mice	High NP accumulation in tumor tissue characterized by strong signal intensity by MRI (Gd) and optical imaging (Cy5.5)
Dextran NPs (Ding et al., 2017)	Chlorin e6, Fe_3O_4	Optical imaging, MRI	MCF-7 tumor-bearing nude mice	Enhanced redox responsive fluorescence signaling (Ce6) and improved local MRI signal intensity (Fe_3O_4) in tumor tissues
$NaLuF_4$ nanocrystals co-doped with Yb^{3+} and Tm^{3+} (Sun et al., 2013)	Lutetium (Lu^{3+})	Optical imaging, CT	Nude mice	Accurate and highly enhanced images of the lymphatic network
Gold coated CNTs (GNT) (Kim et al., 2009)	Gold, CNT	Photoacoustic imaging	Nude mice	Enhanced in vivo photoacoustic signal intensity in the lymphatic wall from 0.89 to 64.73 a.u. 60 min after GNT administration

MRI, magnetic resonance imaging; *CT*, computed tomography; *CNT*, carbon nanotubes.

4.1.1 Optical imaging

Cancer diagnosis using optical imaging techniques is a non-invasive and safe technique for in vivo imaging. For example, ICG, an FDA-approved NIR fluorophore conjugated with tumor cell-specific antibodies such as daclizumab, trastuzumab, and panitumumab, consists of a selective fluorescence activation ability upon NIR irradiation (Ogawa, Kosaka, Choyke, & Kobayashi, 2009). NIR optical imaging using these conjugates could selectively image the endolysosomes of cells or locate specific cells via their antibodies. Besides fluorophores, PS are also being extensively researched for their applications in optical imaging. For example, ALA and its derivative 5-ALA Hexylester (Hexvix®) is a PS approved material for bladder cancer imaging (Ray et al., 2009). Derivatives of porphyrin such as Photofrin have also been used to detect various malignant and premalignant tumors, when irradiated with a 405 nm full-color endoscopic fluorescence imaging system (Ohsaki et al., 2001). Additionally, the red fluorescence from Photofrin overcame the green autofluorescence from tumors. However, PS are not ideal for imaging deep tumor tissues due to poor absorption at longer wavelengths (Rai et al., 2010). Lanthanide-doped, UCNPs have been used to improve optical imaging due to their up-conversion properties, including better absorption and higher energy photons emitted in the UV to NIR range (Shen, Zhao, & Han, 2013). This property has opened an exciting and promising field in contrast agents for optical imaging. UCNPs have several advantages, such as low background fluorescence, narrow emission bandwidth, low photo-bleaching, less light scattering, and deep tissue penetration (Shen et al., 2013).

4.1.2 Multimodal imaging

Multimodal imaging, a combination of optical imaging with either MRI or CT, is advantageous as it overcomes the limitations associated with photo-imaging modalities, including poor depth of penetration and low sensitivity for detection in deeper tissues (Rai et al., 2010). The combination of multiple modalities allows researchers to use the advantage of one modality to overcome the limitation of another (Key & Leary, 2014). Advantages of multimodal imaging over using a single entity is the ability to incorporate imaging at greater depths, higher anatomical resolution, and greater precision in tracking NPs once injected in the body (Key & Leary, 2014). MRI can also provide noninvasive imaging while providing high anatomical resolution and information regarding physiological and molecular events (Key & Leary, 2014). Nam et al. (2010) loaded chitosan-based NPs with two different imaging agents: Cy5.5, an NIR dye for the optical imaging aspect, and gadolinium (Gd (III)) ions, a contrast agent for MRI. MRI and optical phantom studies were used to determine the efficacy of the NP to act as a dual-modal imaging NP. A 1 mg/mL solution of NPs in phosphate buffer saline (PBS) created a strong near infrared fluorescence (NIRF) signal, demonstrating an ability to be utilized in vivo. T_1-weighted MRI images displayed a progressive increase in signals with an increase with NP concentrations. These phantom tests confirmed the ability of chitosan-based NPs loaded with both Cy5.5 and Gd (III) ions to act as a contrast agent for both optical and MR imaging modalities, respectively. Continuing their study in vivo, the NPs were injected into SCC7 tumor-bearing mice. High spatial resolution tumor tissue was observed in T1 weight MRI images in which the tumor was easily outlined against healthy tissues. A strong fluorescence image was also seen using NIR imaging, which indicates the ability to locate tumors for biopsy and the ability of chitosan-based NPs loaded with separate contrast agents, Cy5.5 and Gd (III) ions, to detect and monitor cancerous tissues. Superparamagnetic iron oxide nanoparticles (SPIONs) are also currently in use clinically as imaging agents for MRI. SPIONs have been shown to be effective in imaging several types of cancers of small masses. SPIONs also allow for multiple targeting molecules to be conjugated to them for enhancing the signal-to-noise ratio (Thakor & Gambhir, 2013). An example of multimodal imaging applications of PDD was a nanoconstruct comprised of chlorin e6 (Ce6)-conjugated dextran NPs encapsulating Fe_3O_4 NPs (Ding et al., 2017). These NPs were capable of both MRI and optical imaging via Fe_3O_4 NPs and Ce6, respectively. Specifically, the redox responsiveness of the NPs triggered the switching "ON" of the fluorescence signal from Ce6 in the intracellular-reducing environment within the cells, thus permitting NIR fluorescence imaging. Additionally, MR imaging followed by an additional magnetic field permitted the enhanced visualization of the NPs localized within the tumor, and improved the uptake of the NPs into the cells, thus potentially improving PDT.

Besides MRI, the use of a dual-modality application combining fluorescent imaging and CT could prove beneficial. CT allows for better resolution (50 μm) and supplies more imaging details at an atomic level in vivo (Key & Leary, 2014). Sun, Peng, Feng, and Li (2013) modified $NaLuF_4$-based nanocrystals by co-doping it with Yb^{3+} and Tm^{3+} ions and incorporating citrate acid for water solubility (cit-$NaLuF_4$: Yb,Tm) to study their CT and optical imaging capabilities when exposed to laser light (980 nm). Yb^{3+} ions act as a PS, while Tm^{3+} ions act as an emitter to enhance the luminescence from lanthanide-NPs. Lu^{3+} ions have both a large atomic number (71) and electron density (9.85 g/cm^3), both of which create a higher X-ray attenuation coefficient, thus rendering them suitable contrast agents for CT imaging. $NaLuF_4$ NPs have been shown to have sterling up-conversion luminescence (UCL). Real-time lymphatic vessel imaging is incredibly important in understanding cancer metastasis, so both UCL and CT imaging of lymphatic vessels were conducted in nude mice (Zhang, Niu, Lu, & Chen, 2011). To perform the imaging, 0.05 mL of the modified nanocrystals were injected intra-dermally in mice and imaged 30 min later using a 980 nm laser at 50 mW/cm^2 with the goal of visualizing the lymphatic vessel. UCL imaging detected strong signals in the lymphatic vessels due to lymphatic drainage of the NPs. The same mice were then scanned using a micro-CT scanner for 20 min, in order to study the anatomical features of the lymphatic network. CT lymphography revealed the same lymphatic network as observed by UCL imaging.

Several other agents have also been investigated for use as CT agents, but the two most researched are gold NPs and iodinated organic compounds loaded into a wide variety of NP types (Hahn, Singh, Sharma, Brown, & Moudgil, 2011). Enhancing iodine concentrations in a localized area results in a higher contrast compared with conventional water-soluble compounds (Jackson, Rahman, Wong, Ackerly, & Geso, 2010). NPs loaded with iodinated contrast agents were found to produce equivalent enhancement in X-ray contrast for many types of disease and to have imaging equivalent to that of gold NPs (Jackson et al., 2010). Additionally, doping of a wide range of NP materials with iodinated compounds has been shown to increase X-ray contrast (Kong et al., 2007). However, iodinated compounds have several risk factors, namely nausea, skin irritation, and kidney disorders such as neuropathy (Jackson et al., 2010). Gold NPs have become of great interest and usage as both imaging and PDT agents for various diseases, including cancer. Gold has many advantages over iodinated compounds as a CT imaging agent, especially stronger X-ray attenuation (about 2.7-fold greater compared with that of iodine), and a higher atomic number (79 vs 53) and electron density (19.32 g/cm^3 vs 4.9 g/cm^3) compared with that of iodine (Lusic & Grinstaff, 2013). Gold can also be functionalized with numerous molecules for coating and targeting purposes, while also being highly biocompatible and having a strong X-ray attenuation, as seen by Popovtzer et al. (2008). However, CT scans utilize radiation to acquire images, thus the risk of potential harm due to this exposure needs to be considered (Key & Leary, 2014).

4.1.3 Photoacoustic imaging (PAI)

PAI is a non-invasive and non-ionizing imaging modality that has high sensitivity and excellent resolution (Hahn et al., 2011; Wang, 2009). A sample tissue is radiated by a short-pulsed laser causing absorption and ensuing thermoelastic expansion of the absorbent (Wang, 2009). This expansion creates an ultrasonic acoustic signal that can be detected by wideband transducers. The two methods typically employed are photoacoustic microscopy (PAM) and photoacoustic computed tomography (PAT), where PAM uses an ultrasonic detector capable of generating three-dimensional (3-D) micrographs without the use of any algorithms (Hahn et al., 2011). PAT, on the other hand, scans the irradiated object in a circular pattern and uses algorithms to further translate these scans into 3-D reconstructed images (Hahn et al., 2011). PAI overcomes the problem of optical imaging techniques because it is able to penetrate tissue due to tissue and hemoglobin absorption because of PAI's lower ultrasonic scattering coefficients of absorbents (Hahn et al., 2011). This allows for penetration of about 50 mm with a resolution >1 mm (Hahn et al., 2011).

In PAI, carbon-based materials are beginning to be used extensively, especially single-walled carbon nanotubes (SWNTs) with optical properties in the NIR regions, which are becoming important agents in PAI (Hahn et al., 2011). However, these materials have very low absorption in the NIR regions, thus studies investigating the gold plating of SWNTs in order to enhance their relatively low absorption coefficients have shown promise (Kim, Galanzha, Shashkov, Moon, & Zharov, 2009). In one such study, Kim et al. (2009) deposited a thin layer of gold over SWNTs, creating golden carbon nanotubes (GNTs) to perform combined photoacoustic and photo-thermal contrast imaging. Further, they added anti-LYVE-1 antibodies specific to the lymphatic endothelial hyaluronan receptor, which comes into contact with tumor cells within the lymph nodes, thus imparting a targeting property to the GNTs. The PAI signal amplitude (a.u.) of the GNTs (100 nm in length and 11 nm in diameter) was compared with several other NPs, including gold nanospheres, uncoated SWNTs, gold nanorods (GNRs), and gold nanoshells. The samples were excited using a 532 nm laser wavelength for the gold nanospheres and 850 nm laser for the others. Upon irradiation, the GNTs had a higher PAI signal (~1100 a.u.) than gold NPs (~10 a.u.), GNRs (~1000 a.u.), and SWNTs (~20 a.u.), but a lower signal than gold nanoshells (~1300 a.u.). In vivo mapping of nude mouse mesentery was done using a laser beam with a diameter of 2–10 μm. Signal amplitude was measured before GNT application (background) and at the 60th minute after GNT administration. GNT administration significantly enhanced the signal amplitude from about 0.89 a.u. at the background (before GNT administration) to about 65 a.u. at the 60th minute after GNT application. The unconjugated GNTs, on the other hand, displayed diminishing signals and random signal fluctuations starting at 15 min after administration, possibly due to GNT wash-off by lymphatic flow. Thus, gold-coated SWNTs can potentially be used for targeted PAI applications with enhanced imaging capabilities.

4.2 Cancer Therapy

Nanomaterial-based photo-therapy involves either light to trigger the release of therapeutic agents from a light-sensitive NP, light to stimulate the generation of ROS from PS, or light to trigger the production of heat in the tumor microenvironment. Importantly, nanocarriers improve the solubility of therapeutic agents, especially those that are hydrophobic, thus improving their biodistribution and preventing off-target drug accumulation and side effects associated with these agents (Lajunen et al., 2016). Some examples of light-based therapeutic modalities are presented in Table 3.

4.2.1 Light-triggered drug release

Light-triggered drug release overcomes the limitations of conventional nanocarriers, namely poor and uncontrolled drug release, by providing an on-demand drug release (Lajunen et al., 2016). One form of light-triggered drug release involves the use of light to produce heat from gold NPs that permeabilizes the liposome membrane to release its encapsulated cargo (Paasonen et al., 2010). For example, liposomes encapsulating GNRs or gold nanostars and calcein as a model therapeutic agent were synthesized to provide NIR light-triggered calcein release from the liposomes (Lajunen et al., 2015). The liposomes released very low concentrations of calcein in the absence of light, but the liposomes carrying GNRs released more than 75% of calcein within 20 min, and 88% within 30 min of light exposure. Gold nanostar-loaded liposomes had an even higher calcein release of more than 86% and 99% within 20 and 30 min of light irradiation, respectively. In vitro release of calcein in

Table 3: Therapeutic applications of photo-triggered drug delivery

Nanocarrier	PS/Thermal Agent	Drug	Treatment Modality	Light Source (Wavelength)	Study Model	Results
Liposomes (Lajunen et al., 2015)	Gold nanorod (GNR)/gold nanostars (GNS)	Calcein	Light-triggered drug release	NIR (N/A)	In vitro: ARPE-19 and HUVEC cells	- An 88% and 99% calcein release within 30 min in liposomes with GNR and GNS, respectively - Light activation induced higher cytosolic calcein release in ARPE-19 and HUVECs compared to control NPs without gold NPs
Liposomes (Lajunen et al., 2016)	Indocyanine green	Calcein	Light-triggered drug release	NIR (808 nm)	None	- About 46% and 94% drug release within 1 and 4 min of 1 W light exposure - 3 W light exposure caused 91% drug release within 15 s - Absence of light exposure caused 0%–2% drug release
SNSC micelles (Cao et al., 2013)	2-Nirobenzyl alcohol	Cypate	Light-triggered drug release	NIR (N/A)	In vivo: kunming mice	- 71% and 23% drug release occurred within 2 h with and without NIR irradiation - Micelles displayed low in vivo toxicity in kunming mice
Gold NPs (Huang et al., 2007)	Gold NPs	-	Photo-thermal therapy	Femtosecond Ti:Sapphire laser (800 nm)	In vitro: HSC 3 cells	- Comparable HSC 3 tumor cells death at an NIR power of 1.1 mW combined with gold NPs and about 22.2 mW for laser treatment alone

Continued

Table 3: Therapeutic applications of photo-triggered drug delivery—cont'd

Nanocarrier	PS/Thermal Agent	Drug	Treatment Modality	Light Source (Wavelength)	Study Model	Results
PLGA NPs with gold half-shell (Park et al., 2009)	Gold NPs	Dox	Photo-thermal therapy	NIR (N/A)	In vitro: HeLa cells	- HeLa cell viability reduced to <20% compared to that of control samples
Polyacrylamide NPs (Gao et al., 2006)	TMPyP	-	Two-photon photodynamic therapy	NIR (780 nm)	In vitro: C6 glioma cells	- 1 s light exposure showed comparable rat C6 glioma cell death at 90 and 120 min for NP and control groups - 1 min light exposure showed cell death at 30 min, with maximum death at 120 min
NaYF4 NPs co-doped with $Y6^{3+}$ and Er^{3+} (Chatterjee & Yong, 2008)	Zinc phthalocyanin	-	Photodynamic therapy	NIR (980 nm)	In vitro: HT29 cells	- HT29 cell viability after treatment with NIR and NaYF4 NPs decreased fivefold compared to untreated control

Dox, doxorubicin; *SNSC*, *N*-succinyl-*N'*-4-(2-nitrobenzyloxy)-succinyl-chitosan; *TMPyP*, tetra (*p*-toluene-sulfate).

ARPE-19 (retinal epithelial cells) and human umbilical vein endothelial cells (HUVECs) were significantly higher after light irradiation, compared with that of control NPs without gold NPs. Light activation resulted in up to 1.75-times more fluorescence intensities from calcein, compared with cells that did not receive light exposure.

Light-sensitive dyes such as ICG can also convert incident NIR light energy into thermal energy (Fernandez-Fernandez et al., 2012; Lajunen et al., 2016). Recently, Lajunen et al. (2016) reported that ICG-loaded liposomes could provide laser light-triggered release of calcein. Samples heated to 37°C, followed by light irradiation at 808 nm, resulted in a higher calcein release from the liposomes. Additionally, increasing light exposure time increased calcein release from the liposomes, where 46% of the drug was released within 1 min of light exposure, further increasing to 94% within 4 min of light irradiation. Absence of light exposure resulted in about 0%–2% drug release, emphasizing the NIR light sensitivity of the liposomes. Thus, incorporation of light-sensitive materials into nanocarriers can potentially induce light-triggered release of their cargo to provide on-trigger drug release at the target site. Besides these materials, polymers containing light-sensitive moieties can also be employed for light-triggered drug release, as discussed earlier. One such NIR light-sensitive group that can be incorporated into the backbone of polymers to render them light-sensitive for drug release is 2-nitrobenzyl alcohol (Cao et al., 2013). For example, micelles generated from *N*-succinyl-*N′*-4-(2-nitrobenzyloxy)-succinyl-chitosan containing 2-nitrobenzyl alcohol in its backbone were able to dissociate and release cypate, an NIR dye with about 52% released within 10 min of NIR exposure, in addition to significantly increasing the uptake of the micelles into MDA-MB-231 breast cancer cells (Cao et al., 2013). The micelles also exerted low toxicity in normal nude mice with similar changes in body weight, and biochemical parameters as the control group; however, their therapeutic efficacy remains to be investigated.

4.2.2 Photodynamic therapy (PDT)

PDT uses light to trigger ROS generation in a PS, which subsequently induces apoptosis in cells (Agostinis et al., 2011). Biodegradable NPs hold several advantages for application in PDT as photosensitizing carriers, including improvement in the dispersibility and biodistribution of the hydrophobic photosensitizing agents, high PS carrying capacity, controlled degradation for PS release, readily available materials, and ease in manufacturing techniques (Konan-Kouakou, Boch, Gurny, & Allemann, 2005). Additionally, NP size has also been shown to be a great determinant in photo-toxicity in vitro, wherein the rate of drug release has been shown to be less with larger particles, thus lowering the effectiveness of PDT (Konan-Kouakou et al., 2005; Nagayasu, Uchiyama, & Kiwada, 1999). To test this effect, two different-sized PLGA (75:25) NPs (167 and 370 nm) were loaded with the hydrophobic PS verteporfin to test whether PLGA enhanced the phototherapeutic index and curtailed side effects (Konan-Kouakou et al., 2005). Photo-toxicity was measured using

EMT-6 mammary mouse tumor cells against a solution containing either free verteporfin or one of the NP formulations. After 1 h of incubation with the samples, the cells were irradiated at a wavelength of 690 nm. Considerable cell death was shown by the 167 nm particles at a concentration below 60 ng/mL, but no effect was observed with the 370 nm particle and free drug solution. At 70 ng/mL, the 167 nm particles showed a 69% cell death, while the free drug and 370 nm particles showed an 11% and 29% reduction in cell viability, respectively. The photo-cytotoxic efficiency of the 167 nm particles was about 1.3- and 1.8-fold higher than those of the free drug and 370 nm NPs. This can be explained by reports that show small NPs have higher cellular uptakes than larger NPs and microparticles (Nagayasu et al., 1999). The release profile of verteporfin is also greatly influenced by the mean particle size. Studies have shown that smaller particles exhibit greater drug release rates due to having a greater drug fraction exposed to the leaching medium as a result of an increase in a total particle surface area (Konan-Kouakou et al., 2005).

4.2.3 Photo-thermal therapy

While tumors are known to have a lower heat tolerance than healthy tissues, the use of hyperthermic treatment has been limited due to collateral damage that can occur to surrounding tissues (Thakor & Gambhir, 2013). Temperatures ranging from 41°C to 45°C have been shown to cause thermal sensitivity and make cancer cells more responsive to other treatments, while temperatures >45°C cause thermal ablation (Thakor & Gambhir, 2013). The use of NPs to selectively target cancerous tissues would provide the local heat needed to utilize nano-photo-thermal therapy (NPTT) or PTT more effectively (Beik et al., 2016). Further, the efficiency of NPTT can be enhanced by conjugation of targeting ligands onto the nanocarriers, such as antibodies against epidermal growth factor receptors (EGFR) or folic acid (Beik et al., 2016; Thakor & Gambhir, 2013). Certain cancer cells highly express receptors for folate and EGFR, so NPs conjugated with these targeting agents would provide the selective binding to cancerous tissues for effective NPTT. Monoclonal antibodies (MoAb) such as Mucin7 have also been used for the targeting of urothelial cancers (Beik et al., 2016). Materials that have shown promise for NPTT are gold (Park et al., 2009), copper (Li et al., 2010), carbon-based nanomaterials (Beik et al., 2016), and palladium NPs (Mcgrath et al., 2015). These materials are advantageous because they have strong absorbance in the NIR ranges from 650 to 900 nm, due to surface plasmon resonance (SPR) (Thakor & Gambhir, 2013). This range allows for an increase in the depth of penetration (Thakor & Gambhir, 2013). These materials also demonstrate very good light absorption and efficient conversion of light to heat (Beik et al., 2016; Thakor & Gambhir, 2013). In order to gain adequate light to heat conversion, particles need to be in the size range from tens to hundreds of nm, which can result in clearance by the reticuloendothelial system (RES) (Thakor & Gambhir, 2013). Researchers are attempting to overcome this limitation by using smaller noble metal NPs that can evade the RES and self-assemble at the tumor site (Thakor & Gambhir, 2013; Zharov, Galitovskaya, Johnson, & Kelly, 2005).

Gold represents a category of NPs known as plasmonic NPs. These types of particles work very well for NPTT due to their augmented optical properties, which come about from the SPR fields (Thakor & Gambhir, 2013). Several shapes of gold nanocarriers have been observed to act as photo-absorbing agents in NPTT. Gold nanospheres allow for an ease of synthesis and bioconjugation of various ligands and polymers for targeting of cancer cells and longer retention, in addition to permitting tunable resonance in the red and NIR wavelengths, making them ideal nanocomponents for therapeutic and imaging purposes (Huang et al., 2011). A popular polymer to conjugate to gold NPs is PEG, which has been shown to increase retention of particles in the blood stream, which allows for higher accumulations of particles in cancer cells due to the enhanced permeation and retention effect (Seol, Kim, Jung, Chang, & Kim, 2013).

Several research groups have extensively studied and applied gold NPs or other variations of gold-based nanocomposites for cancer PTT. Huang, Qian, El-Sayed, and El-Sayed (2007) conjugated anti-epidermal growth factor receptor antibodies (EGFR) to spherical gold NPs (30 nm in diameter) in order to selectively target cancer cells while using NIR laser pulses, in order to alleviate healthy tissue destruction while increasing cancerous tissue death with smaller dosages. HSC 3 oral cancer cells were either left naked or treated with EGFR-gold NPs in vitro and flashed with a femtosecond Ti:Sapphire laser at 800 nm with a pulse duration of 100 fs and a repetition rate of 1 kHz, for 2 min at various powers ranging from 0.7 to 140.0 mW. While the control group (lacking NPs) showed cell death at 22.2 mW and above, cells exposed to the EGFR-gold NPs showed cell death at 1.1 mW, a 20-times lower energy level. Cell death was indicated by trypan blue dye accumulated within the cells, indicating that the power level was low enough to not cause mechanical stress, so death must have occurred due to thermal effects. A fivefold increase in NP concentration led to a 14-times higher cell death. Park et al. (2009) developed a theranostic agent made of DOX loaded PLGA NPs, coated with a gold half-shell. The gold shell on the NP generates hyperthermia by absorbing the incident NIR light (0.7 W/cm^2) and converting it to cytotoxic heat. This composite combines the ability of gold to produce hyperthermia and the chemotherapeutic drug DOX to enhance anti-cancer therapy. When comparing to free DOX or NPs without NIR irradiation, the NIR irradiated PLGA NP group was able to reduce HeLa cell viability to $<20\%$ as a result of the increased temperature to 44°C and increased DOX release. In another setup, Hirsch et al. (2003) generated NIR responsive silica-gold nanoshells. When exposed to NIR light (820 nm, 4 W/cm^2) for 4–6 min, these nanoshells generated photo-thermal heat, increasing the tissue temperature up to 37 ± 6.6°C, well above the therapeutic index, which resulted in severe tumor tissue damage. The nanoshell-free controls, on the other hand, increased the temperature to 9.1 ± 4.7°C after light exposure, rendering the tumor tissue undamaged. Generation of heat using NIR can also be used as a photo-trigger for de-hybridization of bioactive molecules (Mura et al., 2013). Xiao et al. (2012) designed a gold NP system incorporating DOX-bound targeting DNA helices and functionalized with PEG-folic acid. Upon exposure to NIR, the gold NPs generated photo-thermal heat that caused the

DNA helices to denature and release DOX into the tumor site. NPs were incubated with KB cells and irradiated with NIR light for 0, 5, and 10 min; they displayed a fluorescence signal at 5 and 10 min within NIR exposure, with a notable shift in the fluorescence signal due to complete release of DOX within 10 min. As expected, BALB/c nude mice injected with KB tumor cells observed a significantly higher tumor volume reduction when treated with gold NP-DNA-DOX and laser (808 nm, 600 mW for 10 min) compared with those of mice that received the NPs without DOX or PBS as a control (relative tumor volume of about 1.8, 2.5, and 5.5 mm^3 respectively after treatment on day 14).

The questionable biocompatibility of gold particles due to their non-degradable properties in vivo has resulted in the development of systems with heat transducing chromophores in more biocompatible biomaterials (Mura et al., 2013; Sun et al., 2014). Zheng et al. (2013) developed a multi-functional chemotherapeutic photo-thermal PLGA NP drug delivery platform for treatment of breast cancer, in conjunction with an 808 nm laser light. These NPs encapsulated ICG, a photo-thermal agent and optical imaging probe. In addition to ICG, the NPs also encapsulated DOX to provide combined chemotherapeutic and photo-thermal effects. The in vivo therapeutic efficacy study revealed that the combined therapy of the dual drug-loaded NPs and laser exposure led to complete tumor remission with 100% survival at the 30 days' time point in mice bearing MCF-7/ADR tumors. The tumor-bearing mice that received laser treatment combined with either ICG, ICG-loaded NPs, or free ICG/DOX successfully suppressed tumor growth, but could improve survival by only 33%, 50%, and 50%, respectively.

4.2.4 Self-lighting photodynamic therapy

Self-lighting PDT (SLPDT) is a novel method to further reach deep areas to attack cancer cells, as well as supplement radiation therapy, thereby reducing radiation doses for therapy and the side-effects associated with radiation therapy (Chatterjee et al., 2008). Using scintillating luminescent NPs, which are stimulated by ionizing radiation such as X-rays, researchers are able to emit scintillation or persistent luminescence in order to activate PS and create singlet oxygen (Chatterjee et al., 2008). SLPDT uses lower radiation doses and is simpler and more inexpensive than PDT, or a combination of PDT and radiation therapy (Bechet et al., 2008; Chatterjee et al., 2008). Persistent luminescence NPs are a category of scintillating luminescent NPs. These NPs are comprised of porphyrins attached to doped NPs such as $LaF_3:Ce^{3+}$, $LuF_3:Ce^{3+}$, $CaF_2:Mn^{2+}$, $CaF_2:Eu^{2+}$, $BaFBr:Eu^{2+}$, $BaFBr:Mn^{2+}$, and $CaPO_4:Mn^{2+}$ (Bechet et al., 2008). This sub-category continues to emit photons to activate PS after the source of excitation has ended. Lower radiation dosages can then be used because the application of radiation can be dramatically shortened, and the NPs still have the ability to release enough photons for PDT (Chen & Zhang, 2006). The mechanism of energy transfer used is typically forster resonance energy transfer (FRET) (Chen & Zhang, 2006). The theory behind this system is sound, and further research is being done to design precise protocols for utilizing this system; however, in vitro and in vivo studies of these materials for cancer therapy is still forthcoming (Chatterjee et al., 2008).

4.2.5 Two-photon absorption induced excitation and UCNPs

One of the weaknesses of PDT is the inability to produce cytotoxic species more than a few millimeters below the skin's surface (Bechet et al., 2008; Thakor & Gambhir, 2013). Two mechanisms being researched to overcome this limitation are two-photon absorption (TPA)-induced excitation and UCNPs (Bechet et al., 2008). Both mechanisms employ NIR light for tissue excitation, which allows deep tissue penetration in addition to being relatively less harmful to cells. TPA is based on the premise of using certain PS that can simultaneously absorb two photons at lower energies (higher wavelengths). These energies recreate an excitation state that a single photon at higher energies (lower wavelengths) would typically use in order to create singlet oxygen species (Bechet et al., 2008). Tissue absorption at these lower energies is negligible, so a deeper seepage of light is possible (Bechet et al., 2008). One drawback with two-photon dyes is their potential toxicity upon contact with cells (Gao, Agayan, Xu, Philbert, & Kopelman, 2006). Gao et al. (2006) attempted to overcome this drawback by loading tetra (*p*-toluene-sulfate) (TMPyP) into polyacrylamide-based NPs (20–30 nm) and exposing them to laser light (780 nm). Cell death visualized by fluorescence microscopy revealed that cells treated with TMPyP NPs and exposed to light for 1 s showed cell death within 90 min, but those exposed to light only were mostly viable after 120 min. When the light exposure was further extended to about 1 min, the time for cell death was shortened to 30 min, followed by maximal cell death at 120 min, thus confirming the activity of the NPs.

UCNPs use the same principle as TPA, with the exception that NPs are used to absorb two photons of lower energy in the range of infrared light, and emit higher energy in the range of visible light, to activate the loaded or conjugated PS (Chatterjee et al., 2008). The absorption of two low-energy photons causes a transition of the NP from its ground state to an excited state. This, in turn, emits a single photon of higher energy to activate PS. The NPs, coined "photon up-converting nanoparticles", utilize lanthanide- or actinide-doped transition metals (Bechet et al., 2008; Chatterjee et al., 2008). Additionally, performance of the NP is chiefly defined by the host lattice, activator ions, and sensitizer ions. One of the first reports on the use of up-converting NPs involved the use of sodium yttrium fluoride (1:1:4) (NaYF4) NPs co-doped with Yb^{3+} and Er^{3+} ions (ytterbium and erbium) (Chatterjee et al., 2008). This combination of materials has since been shown to have potentially the highest up-conversion efficiency, and is being used more often as the core material for up-converting NPs. For example, Chatterjee and Yong (2008) attached zinc phthalocyanine (ZnPC) to NPs contrived of NaYF4 co-doped with Yb^{3+} and Er^{3+} and conjugated with folic acid to study the enhanced targeting and therapeutic effects of these NPs when irradiated with a laser light of 980 nm. Human colon adenocarcinoma (HT29) cells incubated with these NPs at a concentration of

440 μg/mL and irradiated for 30 min showed a more than fivefold decrease in cell viability compared with untreated controls. Another study utilized photo-uncaging coupled with UCNP to generate ROS in response to NIR irradiation (Liu et al., 2017). This study was comprised of an NIR-responsive photochromic spiropyran anchored onto mesoporous, silica-coated UCNPs (UCNP-SP). These NPs were loaded with curcumin as a model hydrophobic chemotherapeutic drug. The photo-uncaging property of the NPs is rendered by spiropyran, which forms a hydrophobic "close" state on the NPs in the absence of NIR, but when exposed to NIR irradiation, the spiropyran acquires a hydrophilic "open" state, allowing release of the encapsulated drug, thus avoiding any modification of the drug, such as attachment with photolabile groups (as discussed in Section 3.2). Furthermore, the UV/visible light generated by the UCNP can induce ROS production by curcumin to stimulate therapeutic effects. The study observed that a 20-min NIR exposure triggered a 17% release of curcumin within 1 h, and an exposure up to 3 h resulted in a 78% drug release, while those without NIR irradiation released only about 7% of the drug. The 4T1 cells treated with the NPs or free curcumin showed a dose-dependent cell death, with only about 15% and 50% cells remaining viable, respectively. A similar trend in therapeutic efficacy was observed in vivo, wherein 4T1 tumor-bearing mice showed complete tumor inhibition when exposed to UCNP-curcumin + NIR, while those that received UCNP-curcumin NPs + UV irradiation also displayed slower tumor growth profiles (relative tumor volume up to 3 V/V_0 within 15 days). In contrast, the mice that received the NPs, but no NIR irradiation, displayed limited tumor inhibition, while those that received free curcumin (no NIR) displayed a slower, but more significant tumor growth (relative tumor volume of about 5 V/V_0). Thus, photo-uncaging coupled with UCNP can significantly enhance tumor therapy, while enabling on-demand drug release at the interested location.

4.3 Cancer Theranostics

Phototherapy, in combination with imaging modalities, is a powerful tool for enhanced cancer detection and targeted therapy. Incorporating a light-triggered property to the NP and drug reduces the invasiveness of the modality and minimizes patient discomfort (Menon et al., 2013). In this section, we will discuss several light-based theranostic strategies that have been investigated for cancer treatment (Table 4). Gobin et al. (2007) developed metallic core-shell NPs made of a silica core coated with gold that had both light absorbing and scattering properties, designed to enhance optical imaging in the NIR range. Tumors injected with these nanoshells showed a 56% increase in signal intensity, compared with that of PBS controls. The presence of gold coating on the nanoshell also imparted photo-thermal properties. When tumor-bearing mice were irradiated with an 808 nm laser, the mice that received the nanoshells showed a significant regression in tumor size, with an almost 50% reduction in tumor volume and prolonged survival compared with those that received PBS or sham treatment. Lin et al. (2016) developed a single system incorporating several imaging and phototherapy modalities, namely positron emission tomography (PET), MR and PAI

Table 4: Applications of phototherapy in cancer theranostics

Nanocarrier	Imaging Agent/ Modality	Therapeutic Agent/ Modality	Light Source (Wavelength nm)	Study Model	Results
Gold coated silica nanoshells (Gobin et al., 2007)	Gold nanoshell/ optical imaging	Gold/photo-thermal therapy	Laser (808 nm)	In vivo: CT-26 colon cancer	- Enhanced absorption and scattering by nanoshells allow imaging in NIR - Gold coating generated hyperthermia in response to 808 nm laser and resulted in 50% reduction in tumor volume and prolonged survival in mice bearing CT-26 colon carcinoma
Fe_3O_4 NPs coated with melanin and ^{64}Cu (MMN) (Lin et al., 2016)	Fe_3O_4 and ^{64}Cu/MR and PET imaging	Fe_3O_4/photo-thermal therapy	Laser (808 nm)	In vivo: U87 MG glioma cancer	- Fe_3O_4 and ^{64}Cu enhanced contrast in MRI and PET images in a U87 MG xenograft model - Significant temperature increases in the tumor due to MMN combined with laser resulted in complete tumor elimination
Silica coated gold (Au)-silver (Ag) nanocage (Khlebtsov et al., 2011)	Yb-HP/IR luminescence imaging	Au-Ag nanocage, Yb-HP/photo-thermal therapy	Laser (630 nm)	In vitro: HeLa cells In vivo: Ehrlich tumor	- Nanocage + laser treatment significantly decreased HeLa cell viability to 30% compared to nanocage only (80%) and light only (75%) - Enhanced tumor signal intensity of about 0.98 a.u. for the nanocomposite compared to free Yb-HP (0.69 a.u.) observed in a subcutaneous Ehrlich carcinoma model

Continued

Table 4: Applications of phototherapy in cancer theranostics—cont'd

Nanocarrier	Imaging Agent/ Modality	Therapeutic Agent/ Modality	Light Source (Wavelength nm)	Study Model	Results
Silica modified-gold nanorods (GNR) (Huang et al., 2011)	GNR/X-ray, CT imaging	GNR/photo-thermal therapy	Laser (808 nm)	In vitro and in vivo: MGC803 gastric cancer	- Enhanced radiosensitivity and hyperthermia resulting in decreased MGC803 gastric cancer cell proliferation - Generated tumor images with sharp contrast and higher Hounsfield numbers, in mice bearing MGC803 gastric tumors
Graphene oxide nanoparticles (Gdots) (Nurunnabi et al., 2014)	Quantum dots/ optical imaging	Gdots/photodynamic and photo-thermal therapy	Laser (670 nm)	In vivo: MDA-MB 231 breast cancer	- Temperature increase from 25°C to 51°C within 30 min after treatment with Gdots and laser confirms photo-thermal activity - Significant reduction in MDA-MB231 tumor volume (160 and 80 mm^3 for Gdots without and with laser irradiation respectively) - Optical imaging allowed biodistribution analysis of Gdots

Yb-HP, Yb-2,4 dimethoxyhematoporphyrin; *PET*, positron emission tomography.

for cancer detection, along with PTT and shielding against UV and gamma (γ) irradiation. The NPs were composed of iron oxide (Fe_3O_4) nanoparticles (MNP) coated with melanin and functionalized with radioisotope ^{64}Cu (MMN) to impart enhanced contrast for imaging modalities such as MRI and PET and generate hyperthermia when irradiated with a laser light at 808 nm. MR imaging revealed darker (enhanced contrast) T_2-weighted images in the case of the MMN group, compared with bare MNP and commercially available MNPs, whereas PET was able to generate images with clear tumor detection. Greater U87MG cell death in the MMNs + laser group was also observed, compared with other groups: MMNs, laser, and untreated control. The relative tumor volume (V/V_0) in a murine U87MG glioblastoma model was completely terminated in the MMNs + laser group (0 V/V_0), compared with the other groups (~30 V/V_0), in addition to prolonged survival after 30 days without tumor reoccurrence. Another positive finding was the absence of inflammatory response in this group. Additionally, human embryonic kidney-293 cells incubated with MMNs and exposed to UV irradiation showed time-dependent UV protection. Healthy BALB/c mice exposed to ^{60}Co generated γ-irradiation showed longer survival and lower body weight loss when administered with MMNs, compared with PBS control. These results further fortify the therapeutic and radio-protective features of these MMNs.

Materials that have been extensively studied and hold promise for cancer theranostics include SPIONs, gold, silica-based NPs, and carbon nanotubes. These materials have already shown promise in either one or both areas of therapy and/or imaging. Research is ongoing to utilize their unique properties to conjugate and load agents for simultaneous therapy and imaging. Khlebtsov et al. (2011) created a novel NP composite made of a gold-silver nanocage core with a silica shell that was functionalized with the PS Yb-2,4 dimethoxyhematoporphyrin (Yb-HP). This multipurpose NP showed the ability to generate heat for PTT when irradiated with a laser at a 750–800 nm wavelength, IR-luminescence imaging, and the generation of singlet oxygen when excited at 630 nm. The tumor site (Ehrlich Carcinoma tumors) showed an enhanced IR signal intensity, indicating the enhanced imaging capabilities of these nanocomposites. The nanocomposite with light was significantly more cytotoxic to HeLa cells, with a cell viability of about 30%. Samples without light had a cell viability >80%, while cells exposed to light only and light with free Yb-HP were the same at around 75% cell viability. In another setup, silica modified-GNRs conjugated with folic acid were used for enhanced X-ray/CT guided PTT of MGC803 gastric cancer cells in vitro and in MGC803 implanted murine models (Huang et al., 2011). Folic acid provided specific targeting to the cancer cells via folic acid receptors on the cancer cell surface. The imaging capability of the GNRs was tested by injecting them into mice, followed by imaging via X-ray and CT. The area where these GNRs were injected displayed a sharp positive contrast and increased Hounsfield numbers compared with other tissues, indicating that these nanoconstructs can be used for both X-ray and CT imaging. Silica coating has several advantages to the design, such as being a simple coating technique, tunable thickness and porosity, biocompatibility, stability against changes in pH, and finally, the presence of abundant functional groups for conjugation of antibodies or targeting moieties. Most importantly, the silica coating on the GNRs enhances the photo-thermal properties of GNRs while retaining its optical properties. When combined with 6Gy X-ray irradiation, the GNRs enhanced the radio-sensitivity of the irradiated cells by decreasing cell proliferation and cancer cell colony formation in a concentration-dependent manner. In fact, cells treated with 12.5 μM of the nanorods and exposed to 6Gy radiation observed about a twofold reduction in cell viability compared with that of the untreated control. Additionally, the irradiated GNRs produced photo-thermal cell death, as indicated by PI staining, and severe damage to cell morphology.

Recently, graphene oxide nanoconstructs have gained popularity as theranostic systems due to their high surface-to-volume ratio, and a planar surface that allows drug loading on the top and bottom of graphene, allowing enhanced drug loading capacity (Nurunnabi, Khatun, Reeck, Lee, & Lee, 2014). Additionally, toxic fluorescent agents such as QDs, can be loaded onto these nanoconstructs (Gdots) to minimize the toxicity associated with QDs (Nurunnabi et al., 2014). The principle of QDs is similar to that of a PS; that is, they absorb irradiated light and act as energy donors (Bakalova, Ohba, Zhelev, Ishikawa, & Baba, 2004). The energy transfer between QDs results in the production of ROS species such as singlet oxygen, which stimulates cell apoptosis (Bakalova et al., 2004). Additionally, due to their high photostability, broad excitation, and narrow emission spectra, they can produce a broad spectrum of fluorescent colors that can be detected via non-invasive optical imaging modalities (Derfus, Chan, & Bhatia, 2004; Gao, Cui, Levenson, Chung, & Nie, 2004). Reduced graphene oxide also produces photo-thermal effects when irradiated by laser or NIR light (Cheon, Bae, & Chung, 2016; Nurunnabi et al., 2014). Thus, Gdots are non-toxic theranostic agents that can produce photodynamic and photo-thermal therapeutic effects, in addition to intrinsic imaging capabilities. Nurunnabi et al. (2014) demonstrated the use of Gdots in combination with laser irradiation (670 nm, 0.3 W/cm^2) for simultaneous in vivo optical imaging, photodynamic, and PTT. Biodistribution of Gdots post-intravenous injection was studied using optical imaging, and there was substantial accumulation of Gdots at the tumor site, even though a large amount of Gdots accumulated in the lung, kidneys, and liver. However, images taken at several time points observed an increase in photo-bleaching, with an increase in time, possibly because Gdots absorb energy from the laser source. Laser irradiation successfully increased the temperature of a Gdot sample from 25°C to 51°C within 30 min, confirming the photo-thermal effects of Gdots. In vivo phototherapy in an MDA-MB-231 breast cancer cell-implanted mouse model showed significant reduction in tumor volume when Gdots were used in conjunction with laser irradiation (tumor volume was estimated to be about 160 and 80 mm^3 when exposed to Gdots without and with laser irradiation, respectively).

Metallic NPs such as iron oxide NPs (Fe_3O_4) offer various advantages, particularly photothermal properties, when excited by NIR light, in addition to being biocompatible and acting as a contrast agent for MRI. Shen et al. (2017) developed a yolk-shell NP comprised of a silica functionalized-Fe_3O_4 core enveloped in a thermo-responsive poly(*N*-isopropylacrylamide)-*co*-1-vinyl-2-pyrrolidone (p(NIPAM-*co*-NVP)) shell. These Fe_3O_4-PNIPAM NPs were also loaded with the chemotherapeutic drug DOX in the interlayers between the yolk and shell. The presence of Fe_3O_4 allowed NIR-mediated heat production, leading to the shrinking of the p(NIPAM-*co*-NVP) shell and subsequent release of the encapsulated drug. This hypothesis was confirmed by the significant increase in DOX release when irradiated with an NIR laser for 6 h (~35% and 51% release at pH 7.4 and 5.4, respectively), compared with the release in the absence of NIR (<20% release). The imaging capability of these NPs was evaluated using T_2-weighted MR imaging, wherein S180-tumor-bearing mice injected with the NPs showed a darker, higher-contrast tumor as a result of enhanced local NP accumulation in the tumor regions. Multispectral optoacoustic tomography revealed strong photoacoustic signals in the tumors up to 24 h post-injection of the NPs. S180 tumor-bearing mice administered with free DOX, DOX-loaded NPs, bare NPs + NIR, and DOX-loaded NPs + NIR exhibited a tumor inhibition rate of 36%, 40%, 65%, and 92%, respectively. Thus, the DOX-Fe_3O_4-PNIPAM NPs were capable of integrated diagnostic and therapeutic capabilities for enhanced tumor therapy.

PTT can be improved by incorporating targeting ligands on the surface of the NPs to increase targeting and accumulation of NPs in the tumor environment. Gao et al. (2017) recently developed RGD conjugated hollow copper sulfide (CuS) NPs encapsulating thermoresponsive vinyl azide (VA). RGD peptide allows specific targeting of the NPs to vascular endothelial cells via its affinity to $\alpha_v\beta_3$, which is highly expressed on the surface of endothelial cells. When exposed to NIR irradiation, the copper sulfide NPs generate heat, which triggers VA to generate N_2 bubbles that burst and disrupt the tumor vasculature, ultimately inducing tumor necrosis. Additionally, the copper sulfide NPs generate a strong photoacoustic signal (PA) that allows photoacoustic image-guided cancer therapy. Photoacoustic imaging observed a significant increase in PA signal with time up to 4 h post-NP injection (7.99 ± 0.34, 9.44 ± 0.29, and 11.98 ± 0.55 at 1, 2 and 4 h time points, respectively), with a subsequent drop in signal intensity at the 6 h time point, indicating maximum NP accumulation at 4 h post-injection. Furthermore, HeLa tumor-bearing mice injected with RGD@HCuS(VA) NPs observed an increase in tumor temperature up to 48°C, 2 min post-laser irradiation. On the other hand, mice injected with free VA, blank HCuS, and HCuS(VA) NPs observed slight increases in temperatures, not exceeding 39°C. PA imaging revealed vascular deformation in mice injected with RGD@HCuS(VA) NPs and exposed to 10 laser cycles, while those injected with the controls observed minimal damage to the vasculature. RGD@HCuS(VA) + laser groups showed complete tumor regression with no regrowth for 44 days, while those treated with RGD@HCuS(VA) only, and the untargeted HCuS(VA) NPs + laser group, observed an increase in tumor volumes (relative tumor volume of about 25 and 12 V/V_0, respectively) within 44 days. Thus, incorporation of targeting ligands onto NPs can enhance the therapeutic efficacy of PDD, while preventing off-target side-effects.

4.4 Other Applications

The principle of PDT involves rapid uptake of the PS into highly proliferative cells, thus PDT is not limited to cancer treatment only. PDT is a popular technique for treating other diseases related to the eyes, skin, rheumatology, and vasculature (Dolmans, Fukumura, & Jain, 2003; Ortu, Lamuraglia, Roberts, Flotte, & Hasan, 1992). Chloroaluminum-sulfonated phthalocyanine (CASPc)-based PDT was used for the treatment of intimal hyperplasia in a rat balloon injury model (Ortu et al., 1992). CASPc, in combination with a 675-nm laser, significantly decreased intimal hyperplasia in the arteries, compared with those irradiated with a laser only (0.04 mm^2 vs 0.16 mm^2). Interestingly, CASPc administration before or after the induction of intimal hyperplasia did not demonstrate a significant difference in the inhibition profile, indicating that time for drug administration relative to development of the disease does not play a major role in the efficacy of PDT. PDT has also been widely used in the field of cosmetology and dermatology to treat skin ailments such as acne vulgaris and rosacea or photo-rejuvenation for anti-aging applications (Wan & Lin, 2014). Researchers have also explored the use of PDT for treatment of fungal diseases such as *Trichophyton rubrum* (Liang, Lu, Chen, & Lin, 2016). PDT for treatment of rheumatoid arthritis employs the phototoxicity of drugs such as Epirubicin, Chloroquine, and Methotrexate, which have been observed to show enhanced toxicity when irradiated at 351 nm (Hendrich & Siebert, 1997).

5 Limitations of PDD and Strategies to Overcome These Limitations

Several factors, as shown in Fig. 3, have been found to affect the efficacy of PDD, such as toxicity, biodistribution, and availability of PS and tissue properties. Various strategies to overcome these limitations have been widely studied to improve the efficacy of PDD (Mallidi et al., 2016).

5.1 Toxicity to Healthy Cells

Besides killing cancer cells, the drugs and PS associated with PDD might exert some toxicity on the neighboring healthy cells. Although PDD is generally considered to be non-toxic, side-effects of PDD might occur due to the low tumor-to-tissue ratio, and local hyperthermia leading to occluded blood vessels (Dolmans et al., 2003). These side-effects typically manifest at the dermal layers, often accompanied by severe erythema, edema, and development of lesions (Wan & Lin, 2014). PDD has also been reported to severely damage the respiratory tract and airways due to tissue necrosis. Besides the damage caused by light itself, the materials used for PDD might also induce toxicity. QDs, for example, are comprised of toxic core components such as cadmium or selenium (Derfus et al., 2004). Some QDs acquire cytotoxicity only after oxidative or photolytic degradation of their core,

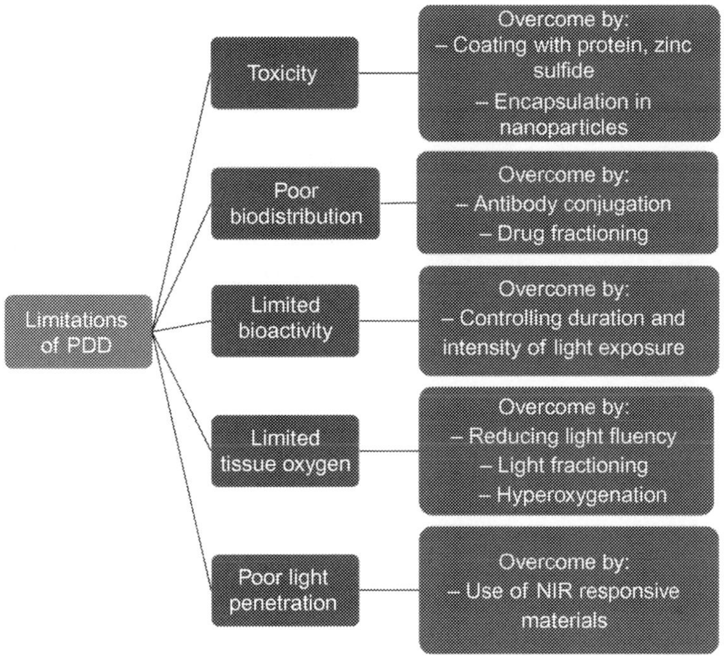

Fig. 3
Limitations of photo-triggered drug delivery and strategies to overcome them.

and thus, are fairly non-toxic in the absence of light (Bakalova et al., 2004; Hardman, 2006). For example, EtNBS is a PS that causes dark toxicity, that is, toxicity in the absence of light (Hung et al., 2016). Photofrin® is another PS that induces tumor cell death by exponentially increasing the temperature. Additionally, Photofrin consists of several complex constituents, many of which are unidentified, and lack selectivity and targeting abilities to the tissue of interest (Konan, Gurny, & Allémann, 2002). Furthermore, due to its poor half-life, the administered dosage required to obtain the desired phototoxicity is significantly higher. Also, because skin is the primary site for accumulation of PS, patients often suffer from acute light sensitivity up to 6–8 weeks after PDD, requiring protection from sunlight. Second-generation porphyrins, such as phthalocyanines and naphthalocyanines, although well characterized and less toxic, are hydrophobic and thus cannot be administered via intravenous routes. In fact, porphyrins treated with 5-ALA or its ester derivatives did not produce any skin photosensitization when applied topically (Berg et al., 2005).

Several strategies have been successfully implemented to reduce the cytotoxicity associated with PS. Strategies to reduce QD toxicity include coating the QD with a protein such as Metallothionein, which detoxifies cadmium to an inert complex (Derfus et al., 2004). Derfus et al. (2004) observed reduced toxicity of cadmium selenide QDs after capping with zinc sulfide (ZnS) as a result of reduced oxidation of the core elements. Additionally, the coated QDs displayed significantly higher albumin production, as a measure of cell viability compared with uncoated QDs, where albumin production reached 70 μg/mL/day for the coated QDs and about 60 μg/mL/day for the uncoated QDs within 13 days in an in vitro hepatocyte cell culture. Nanocarriers have also been used to further reduce the side-effects associated with PS. Recently, PLGA NP encapsulation of EtNBS was implemented to reduce its dark toxicity (Hung et al., 2016). While free EtNBS showed toxicity at a dose of 10 μM, PLGA NPs at the same equivalent dose produced significantly lower toxicity (approximately 10% for free EtNBS vs 90% cell viability for EtNBS-loaded PLGA NPs, respectively). Similarly, the toxicity of QDs can be avoided by grafting onto graphene oxide nanoconstructs, as discussed earlier (Nurunnabi et al., 2014).

5.2 Biodistribution of Therapeutic Agents

Biodistribution of drugs and PS and their availability at the tumor site can profoundly affect the therapeutic efficacy of PDD. Among the factors affecting biodistribution, the property of the tissue to be treated plays a major role in determining the biodistribution of therapeutic agents and at modulating the efficacy of PDD. Because the area of interest for treatment is superficial, such as the skin, free therapeutic agents and NPs encapsulating drugs for PDD are often administered intravenously, subcutaneously, or topically. However, once absorbed, the biodistribution of the PS changes dramatically over time. Thus, it is difficult to predict the timing and duration of light exposure (Dolmans et al., 2003). For example, a short interval only permits drug accumulation in the vasculature; whereas, a longer time interval permits drug accumulation in the extravascular compartment of the tumor, and thus leads to localized tumor killing. Additionally, because the half-life of the singlet oxygen and ROS generated during PDD is very small, the location and dosage of PS and duration of light exposure majorly affect the functioning of PDD (Dolmans et al., 2003). It is also important to note that due to their hydrophobic nature, PS tend to aggregate in aqueous solutions. This further limits their bioavailability, and also leads to poor light absorption (Konan et al., 2002).

Tumor localization can be achieved by conjugation of specific targeting ligands onto the nanomaterials (Konan et al., 2002). This strategy allows improved localization and lowers the dosage required for treatment. Antibody targeting involves conjugation of tumor-specific monoclonal antibody (MoAb) on the surface of the NP or conjugation to the PS itself. The anti-myosarcoma (M1) antibody conjugated to PS hematoporphyrin (Hp) combined with incandescent light exposure was observed to kill 95% of M-1 tumor cells, compared with unconjugated Hp (Mew, Wat, Towers, & Levy, 1983). Additionally, the M1-Hp conjugates were able to curb tumor growth at a dose almost 10-fold lower than free Hp. Similarly, silver NPs grafted onto graphene oxide (GO@Ag), and carrying a chemotherapeutic drug DOX, were functionalized with DSPE-PEG2000-NGR peptide to selectively target the tumor vasculature. This design took advantage of an 808 nm NIR laser (2 W/cm^2) triggered DOX release from the GO@Ag NPs, in addition to enhanced tumor targeting, evidenced by about an eightfold higher DOX uptake and twofold higher DOX release at the tumor site, compared with free DOX (Shi et al., 2014). Roadblocks to antibody-targeted drug delivery include poor conjugation of the MoAb to the therapeutic payload or NP, inactivity of the payload during or after conjugation, insufficient antigen population for binding, and most importantly, immune rejection (Konan et al., 2002). Strategies employing low-density lipoproteins (LDL) have also gained popularity, due to the high number of lipoprotein receptors on the surface of tumor cells and rapid uptake of PS by highly replicating tumor cells (Konan et al., 2002). For example, Hp when conjugated to LDL, displayed higher uptake into tumor tissues, compared with that of a free drug (Barel et al., 1986). Liposomes have been observed to improve the accumulation of PS in cancer cells (Chen, Pogue, & Hasan, 2005; Juarranz et al., 2008; Liang et al., 2016). Additionally, when conjugated with LDL, they exhibit specific tumor cell targeting and uptake (Juarranz et al., 2008). Drug fractioning also permits long-term drug availability and activity compared with a single dose of PDT. A recent study observed that while a single dose of PS MV6401 was capable of delaying tumor growth, its short half-life and accumulation in the interstitial compartment within 2–6 h of administration impaired its long-term activity (Dolmans et al., 2002). Two doses of MV6401 injected 15 min and 4 h before light administration, on the other hand, permitted localization in the vascular and tumor cells, and thus, delayed the progression of tumor growth more effectively than a single dose.

5.3 Bioactivity of Therapeutic Agents

Due to poor half-life, poor insolubility, and structural instability of the therapeutic agents, uncontrolled drug administration, or their release from nanomaterials, affects their bioactivity and leads to sub-therapeutic results (Gunasekaran, Haile, Nigusse, & Dhanaraju, 2014). Temporal control permits the user to control the drug dosage as required, by controlling the intensity, time, and duration of the trigger, while delaying their release from the nanomaterial and reducing the unwanted toxicity, in addition to preserving their bioactivity (Liu, Zhan, & Kohane, 2017; Uhrich, Cannizzaro, Langer, & Shakesheff, 1999; Yoo, Doshi, & Mitragotri, 2011). Carling, Viger, Nguyen Huu, Garcia, and Almutairi (2015) recently developed photocleavable polymeric NPs synthesized from (2-(40-(*N*,*N*-dimethylamino)-4-nitro-[1,10-biphenyl]-3-yl) propyl carbonyl) (ABNP) loaded with Nile red as the model drug. These NPs followed the principle of photocaging, wherein the polymer degraded in response to blue light, and subsequently released its payload. The NPs released about 57% of their payload when irradiated for 90 min, compared with 2% under no light exposure. Additionally, dexamethasone (dex)-loaded ABNP NPs were evaluated for their therapeutic efficacy in a carrageenan-induced hind paw inflammatory model. Therapeutic efficacy was analyzed as a measure of swelling in terms of increase in thickness of the hind paw. The irradiated NPs produced a significant decrease in paw thickness (paw thickness increased from 0.28 to 0.71 mm for PBS, 0.63 mm for free dex, 0.72 mm for NP only and 0.44 mm for NP + light irradiated groups).

5.4 Tissue Oxygenation

Tissue properties such as oxygenation and the availability of oxygen at the tissue site during light exposure can greatly affect the property of PS, and subsequently, the efficacy of PDT itself. Because the principle of PS involves conversion of oxygen to produce free radicals and ROS, an anoxic or hypoxic tissue environment can mitigate the effects of the PS and eventually negate its anti-tumor effects (Dolmans et al., 2003; Huang et al., 2008). Conversely, the process of PDT can also severely consume local tissue oxygen and greatly limit oxygenation in the tissue, and further reduce the potency of PDT (Huang et al., 2003, 2008). Thus, pre-existing tumor tissue hypoxia and hypoxia produced during PDT have a profound effect on the generation of ROS during PDT and can result in literally no cancer cell death (Huang et al., 2003). The concentration of oxygen (pO_2) in tumor tissue serves as a diagnostic biomarker to detect the efficacy of PDT (Walsh et al., 2014). Tumor tissue pO_2 can be measured by several techniques. Oxygen probes and luminescence-based optodes can directly measure the tissue pO_2 level (Woodhams, Macrobert, & Bown, 2007). Recently, Napp et al. (2011) developed fluorescent polymeric nanoprobes doped with an oxygen sensitive dye, palladium meso-tetraphenylporphyrin. These nanoprobes released an NIR signal in response to a decrease in oxygen concentration in vivo. Besides these imaging techniques, other techniques include PET and photosensitized blood oxygen level-dependent MRI (Hallac et al., 2014; Woodhams et al., 2007). The challenges faced due to poor tissue oxygenation led to the development of PDTs that use a smaller light fluence rate and fractionate the PDT and oxygen supplementation to allow re-oxygenation, or oxygen re-perfusion of local tissues, which is discussed as follows (Dolmans et al., 2003; Huang et al., 2008).

5.4.1 Light fluency

Light fluency is the rate at which light is delivered to the tissue. A high fluence rate can drastically increase oxygen usage during PDT, often exceeding the rate at which oxygen is supplied, and leading to oxygen deprivation in the tissues (Henderson, Busch, & Snyder, 2006). Henderson et al. (2004) observed that a low fluence rate PDT ($14\,mW/cm^2$) was able to cure about 70%–80% of tumors, and induced higher levels of cancer cell apoptosis. A higher fluence rate ($112\,mW/cm^2$), on the other hand, was successful at curing only 10%–15% of tumors. It is well known that inflammatory response mediators, namely neutrophils, and pro-inflammatory cytokines, such as tumor necrosis factor α and interleukins (IL)-1β and 6, are up-regulated as a result of PDT-induced inflammatory activation (Korbelik, 1996). These neutrophils greatly exert tumor suppression after PDT, thus serving as a useful marker to study tumor progression after PDT (De Vree et al., 1996). Additionally, fluence rate and subsequent tumor oxygenation has been found to immensely affect the expression of inflammatory markers, such as IL-6 activated during PDT (Henderson et al., 2004). In fact, a low fluence rate of $48\,J/cm^2$ significantly increased IL-6 expression levels, compared with a higher fluence rate of $128\,J/cm^2$ at the same light intensity.

5.4.2 Light fractioning

Light fractioning uses alternate light/dark intervals in order to allow time for the tissue to reperfuse with oxygen (Huang et al., 2008). Light fractioning using a single long dark interval of about 2 h has been implicated to improve the efficacy of PDT in several pre-clinical and clinical studies, improving the PDT therapeutic response rate to 88% compared with 75% by single-dose PDT (De Bruijn, Brooks, Ten Hagen, De Haas, & Robinson, 2016; De Vijlder, Sterenborg, Neumann, Robinson, & De Haas, 2012). Interestingly, it has been observed that the enhanced efficacy of light fractionation is not due to accumulation of PS, but in fact, is due to the increased sensitivity of exposed cells during the first light fraction, rendering them susceptible to the toxicity of the PS during the second light fraction (De Bruijn et al., 2016). However, reducing light fluency or fractioning light and drug dose only improves oxygen depletion as a result of PDT, and not by tumor cell hypoxia (Huang et al., 2008). Additionally, reduced light fluency and fractioning significantly increases the treatment time (Huang et al., 2003).

5.4.3 Hyper-oxygenation

Pre-existing tumor hypoxia can be overcome by hyper-oxygenation during light exposure. Huang et al. (2003) observed the effects of different concentrations of oxygen perfusion on the efficacy of photofrin during light exposure. Hyper-oxygenation significantly decreased the tumor cell survival fraction (about 20% tumor cell survival), compared with those exposed to room air (about 50% tumor cell survival). Incidentally, the mice breathing room air and exposed to light at a standard light fluence rate (150 mW/cm^2) or reduced light fluence rate (75 mW/cm^2) did not show any significant difference in tumor cell survival. This further confirms that light fluence does not affect the efficacy of PDT in pre-existing tumor hypoxia. Chen et al. (2002) also employed hyperbaric oxygen (3 atp) to hyper-oxygenate tumor tissue during Photofrin-based PDT treatment in conjunction with a 630 nm laser light (Chen et al., 2002). They observed that hyperbaric oxygen successfully stopped tumor regrowth in nearly 80% of mice, while normal oxygen concentration could stop tumor regrowth in only 20% of the cases. Additionally, hyper-oxygen was able to restore tumor pO_2 levels and initiate a compensatory oxygen supply against its depletion by PDT. Tomaselli et al. (2001) also observed similar outcomes when they combined PDT using Hp with hyperbaric oxygenation for patients suffering from inoperable non-small cell bronchogenic carcinoma.

5.5 Depth of Light Penetration

Light-tissue interactions determine the ideal optical parameters required to achieve full therapeutic benefits from the exposed light (Yun & Kwok, 2017). Light penetration determines the distance light can travel through tissues (Yun & Kwok, 2017). Different light sources display profound variation in the depth of penetration, primarily due to attenuation by tissues (Yun & Kwok, 2017). Therefore, location of the tissue plays a major role in determining the ideal light source for therapy (Mallidi et al., 2016). Additionally, the depth of penetration increases with an increase in the wavelength of light (Wan & Lin, 2014). For example, melanin in skin attenuates over 70% of the incident light in the range of 500–700 nm (Mallidi et al., 2016). Thus, green light (~500–550 nm) is mostly absorbed by melanin, and can travel only about a few hundred micrometers (Yun & Kwok, 2017). Blue light, on the other hand, can travel about 1.5 mm, while red light and NIR light have the highest depth of penetration, about 1–3 mm (Liang et al., 2016; Yun & Kwok, 2017). Thus, red light can be used for deeper targets, such as thick lesions (Wan & Lin, 2014). The Blu-U® device that emits blue light has been approved by the FDA for use in conjunction with ALA, while red light has been used in combination with MAL (Wan & Lin, 2014). However, only a few organic chromophores actually absorb in the NIR region, and even fewer are able to trigger a drug release by converting the absorbed energy into a thermal or chemical response (Fomina et al., 2012). Recent research proves that the incorporation of gold nanocarriers result in stronger absorption in the NIR region, thereby enabling an effective and minimally invasive laser therapy (Fomina et al., 2012). The suitable light source for the application of interest also depends on the PS being used. The spectral output of a light source should be equivalent to the excitation wavelength of the chosen PS; for example, protoporphyrin IX has excitation peaks in the range of 404–420 nm (Wan & Lin, 2014).

6 Conclusion and Future Perspectives

Nanomaterial-based, PDD has displayed multiplicity in diagnosing and treating various diseases, especially cancer. The spatial and temporal control aspect of PDT has helped it surpass conventional cancer treatment modalities, such as surgery and chemotherapy. Additionally, theranostic PDD has revolutionized the "one drug fits all" approach, via personalized nanocarriers that can be combined with various diagnostic tools to diagnose, image, and treat cancer in a synchronized fashion. Dynamic research in this field has led to more targeted and localized PDD modalities that have reduced systemic and/or off-target side effects, in addition to being able to access solid tumors, while not being limited by poor light penetration. Use of nanomaterials for PDD has made remarkable improvements in the efficacy of PDD, by providing controlled release of therapeutic agents in response to light, improved drug biodistribution, bioavailability, and cancer cell and deep tissue targeting. Despite these positive outcomes, PDD suffers from limitations, such as toxicity of PS, that limit the dosage of PS used for treatment, and poor depth of light penetration that only permits treatment of superficial diseases, namely those that can be directly illuminated (Mura et al., 2013). Although PDD is being used clinically, many of the concepts discussed here are still in the nascent stage, and require considerable improvements both in vivo and clinically. Profound research in techniques to develop biocompatible PS and photosensitive materials for PDD, in addition to sensitive and specialized equipment capable of generating focused light irradiation, can potentially overcome these limitations. Future research sectors in PDD should also include development of better light sources with deeper tissue penetration and combination therapies to target aspects such as tumor heterogeneity, cancer immunotherapy, and overcoming multi-drug resistance. The new directions could potentially open new avenues for PDD research, with the goal of developing better diagnostic and treatment techniques for cancer.

Acknowledgments

This work is partly supported by NIH HL118498 (KTN). The content is solely the responsibility of the authors and does not necessarily represent the official views of the National Institutes of Health.

References

Agostinis, P., Berg, K., Cengel, K. A., Foster, T. H., Girotti, A. W., Gollnick, S. O., et al. (2011). Photodynamic therapy of cancer: an update. *CA: A Cancer Journal for Clinicians*, *61*(4), 250–281.

Ahmed, N., Fessi, H., & Elaissari, A. (2012). Theranostic applications of nanoparticles in cancer. *Drug Discovery Today*, *17*(17–18), 928–934.

Alvarez-Lorenzo, C., Bromberg, L., & Concheiro, A. (2009). Light-sensitive intelligent drug delivery systems. *Photochemistry and Photobiology*, *85*(4), 848–860.

Bakalova, R., Ohba, H., Zhelev, Z., Ishikawa, M., & Baba, Y. (2004). Quantum dots as photosensitizers? *Nature Biotechnology*, *22*(11), 1360–1361.

Bansal, A., & Zhang, Y. (2014). Photocontrolled nanoparticle delivery systems for biomedical applications. *Accounts of Chemical Research*, *47*(10), 3052–3060.

Barel, A., Jori, G., Perin, A., Romandini, P., Pagnan, A., & Biffanti, S. (1986). Role of high-, low- and very low-density lipoproteins in the transport and tumor-delivery of hematoporphyrin in vivo. *Cancer Letters*, *32*(2), 145–150.

Baroncini, M., Ragazzon, G., Silvi, S., Venturi, M., & Credi, A. (2015). The eternal youth of azobenzene: new photoactive molecular and supramolecular devices. *Pure and Applied Chemistry*, *87*(6), 537–545.

Bechet, D., Couleaud, P., Frochot, C., Viriot, M. L., Guillemin, F., & Barberi-Heyob, M. (2008). Nanoparticles as vehicles for delivery of photodynamic therapy agents. *Trends in Biotechnology*, *26*(11), 612–621.

Beik, J., Abed, Z., Ghoreishi, F. S., Hosseini-Nami, S., Mehrzadi, S., Shakeri-Zadeh, A., et al. (2016). Nanotechnology in hyperthermia cancer therapy: from fundamental principles to advanced applications. *Journal of Controlled Release*, *235*, 205–221.

Berg, K., Selbo, P. K., Weyergang, A., Dietze, A., Prasmickaite, L., Bonsted, A., et al. (2005). Porphyrin-related photosensitizers for cancer imaging and therapeutic applications. *Journal of Microscopy*, *218*(Pt 2), 133–147.

Bisby, R. H., Mead, C., & Morgan, C. G. (2000). Active uptake of drugs into photosensitive liposomes and rapid release on UV photolysis. *Photochemistry and Photobiology*, *72*(1), 57–61.

Cao, J., Huang, S., Chen, Y., Li, S., Li, X., Deng, D., et al. (2013). Near-infrared light-triggered micelles for fast controlled drug release in deep tissue. *Biomaterials*, *34*(26), 6272–6283.

Carling, C.-J., Viger, M. L., Nguyen Huu, V. A., Garcia, A. V., & Almutairi, A. (2015). In vivo visible light-triggered drug release from an implanted depot. *Chemical Science*, *6*(1), 335–341.

Chan, M.-H., Pan, Y.-T., Lee, I. J., Chen, C.-W., Chan, Y.-C., Hsiao, M., et al. (2017). Minimizing the heat effect of photodynamic therapy based on inorganic nanocomposites mediated by 808 nm near-infrared light. *Small*, 1700038.

Chatterjee, D. K., Fong, L. S., & Zhang, Y. (2008). Nanoparticles in photodynamic therapy: an emerging paradigm. *Advanced Drug Delivery Reviews*, *60*(15), 1627–1637.

Chatterjee, D. K., & Yong, Z. (2008). Upconverting nanoparticles as nanotransducers for photodynamic therapy in cancer cells. *Nanomedicine (London, England)*, *3*(1), 73–82.

Chen, Q., Huang, Z., Chen, H., Shapiro, H., Beckers, J., & Hetzel, F. W. (2002). Improvement of tumor response by manipulation of tumor oxygenation during photodynamic therapy. *Photochemistry and Photobiology*, *76*(2), 197–203.

Chen, B., Pogue, B. W., & Hasan, T. (2005). Liposomal delivery of photosensitising agents. *Expert Opinion on Drug Delivery*, *2*(3), 477–487.

Chen, W., & Zhang, J. (2006). Using nanoparticles to enable simultaneous radiation and photodynamic therapies for cancer treatment. *Journal of Nanoscience and Nanotechnology*, *6*(4), 1159–1166.

Chen, H., Zhang, W., Zhu, G., Xie, J., & Chen, X. (2017). Rethinking cancer nanotheranostics. *Nature Reviews Materials*, *2*, 17024.

Cheon, Y. A., Bae, J. H., & Chung, B. G. (2016). Reduced graphene oxide nanosheet for chemo-photothermal therapy. *Langmuir*, *32*(11), 2731–2736.

Chilakamarthi, U., & Giribabu, L. (2017). Photodynamic therapy: past, present and future. *Chemical Record*, *17*(8), 775–802.

De Bruijn, H. S., Brooks, S., Ten Hagen, T. L., De Haas, E. R., & Robinson, D. J. (2016). Light fractionation significantly increases the efficacy of photodynamic therapy using BF-200 ALA in normal mouse skin. *PLoS One*, *11*(2), e0148850.

De Vijlder, H. C., Sterenborg, H. J., Neumann, H. A., Robinson, D. J., & De Haas, E. R. (2012). Light fractionation significantly improves the response of superficial basal cell carcinoma to aminolaevulinic acid photodynamic therapy: five-year follow-up of a randomized, prospective trial. *Acta Dermato-Venereologica*, *92*(6), 641–647.

De Vree, W. J. A., Essers, M. C., De Bruijn, H. S., Star, W. M., Koster, J. F., & Sluiter, W. (1996). Evidence for an important role of neutrophils in the efficacy of photodynamic therapy. *Cancer Research*, *56*(13), 2908–2911.

Derfus, A. M., Chan, W. C. W., & Bhatia, S. N. (2004). Probing the cytotoxicity of semiconductor quantum dots. *Nano Letters*, *4*(1), 11–18.

Ding, Z., Liu, P., Hu, D., Sheng, Z., Yi, H., Gao, G., et al. (2017). Redox-responsive dextran based theranostic nanoparticles for near-infrared/magnetic resonance imaging and magnetically targeted photodynamic therapy. *Biomaterials Science*, *5*(4), 762–771.

Dolmans, D.E.J.G.J., Fukumura, D., & Jain, R. K. (2003). Photodynamic therapy for cancer. *Nature Reviews Cancer*, *3*(5), 380–387.

Dolmans, D.E.J.G.J., Kadambi, A., Hill, J. S., Flores, K. R., Gerber, J. N., Walker, J. P., et al. (2002). Targeting tumor vasculature and cancer cells in orthotopic breast tumor by fractionated photosensitizer dosing photodynamic therapy. *Cancer Research*, *62*(15), 4289–4294.

Fan, N.-C., Cheng, F.-Y., Ho, J.-A., & Yeh, C.-S. (2012). Photocontrolled targeted drug delivery: photocaged biologically active folic acid as a light-responsive tumor-targeting molecule. *Angewandte Chemie, International Edition*, *51*(35), 8806–8810.

Fayter, D., Corbett, M., Heirs, M., Fox, D., & Eastwood, A. A. (2010). Systematic review of photodynamic therapy in the treatment of pre-cancerous skin conditions, Barrett's oesophagus and cancers of the biliary tract, brain, head and neck, lung, oesophagus and skin. *Health Technology Assessment*, *14*(37), 1–288.

Fernandez-Fernandez, A., Manchanda, R., Lei, T., Carvajal, D. A., Tang, Y., Kazmi, S. Z., et al. (2012). Comparative study of the optical and heat generation properties of IR820 and indocyanine green. *Molecular Imaging*, *11*(2), 99–113.

Feyh, J. (1996). Photodynamic treatment for cancers of the head and neck. *Journal of Photochemistry and Photobiology B: Biology*, *36*(2), 175–177.

Fomina, N., Sankaranarayanan, J., & Almutairi, A. (2012). Photochemical mechanisms of light-triggered release from nanocarriers. *Advanced Drug Delivery Reviews*, *64*(11), 1005–1020.

Foote, C. S., & Dobrowolski, D. C. (1984). Singlet oxygen production from photodynamic sensitizers. *Journal of Photochemistry*, *25*(2), 549.

Fu, Q., Cheng, L., Zhang, Y., & Shi, W. (2008). Preparation and reversible photo-crosslinking/photo-cleavage behavior of 4-methylcoumarin functionalized hyperbranched polyester. *Polymer*, *49*(23), 4981–4988.

Gao, D., Agayan, R. R., Xu, H., Philbert, M. A., & Kopelman, R. (2006). Nanoparticles for two-photon photodynamic therapy in living cells. *Nano Letters*, *6*(11), 2383–2386.

Gao, X., Cui, Y., Levenson, R. M., Chung, L. W. K., & Nie, S. (2004). In vivo cancer targeting and imaging with semiconductor quantum dots. *Nature Biotechnology*, *22*(8), 969–976.

Gao, W., Li, S., Liu, Z., Sun, Y., Cao, W., Tong, L., et al. (2017). Targeting and destroying tumor vasculature with a near-infrared laser-activated "nanobomb" for efficient tumor ablation. *Biomaterials*, *139*, 1–11.

Gobin, A. M., Lee, M. H., Halas, N. J., James, W. D., Drezek, R. A., & West, J. L. (2007). Near-infrared resonant nanoshells for combined optical imaging and photothermal cancer therapy. *Nano Letters*, *7*(7), 1929–1934.

Gunasekaran, T., Haile, T., Nigusse, T., & Dhanaraju, M. D. (2014). Nanotechnology: an effective tool for enhancing bioavailability and bioactivity of phytomedicine. *Asian Pacific Journal of Tropical Biomedicine*, *4*(Suppl. 1), S1–S7.

Hahn, M. A., Singh, A. K., Sharma, P., Brown, S. C., & Moudgil, B. M. (2011). Nanoparticles as contrast agents for in-vivo bioimaging: current status and future perspectives. *Analytical and Bioanalytical Chemistry*, *399*(1), 3–27.

Hallac, R. R., Zhou, H., Pidikiti, R., Song, K., Stojadinovic, S., Zhao, D., et al. (2014). Correlations of noninvasive BOLD and TOLD MRI with PO_2 and relevance to tumor radiation response. *Magnetic Resonance in Medicine*, *71*(5), 1863–1873.

Hardman, R. (2006). A toxicologic review of quantum dots: toxicity depends on physicochemical and environmental factors. *Environmental Health Perspectives*, *114*(2), 165–172.

Hemnani, T., & Parihar, M. S. (1998). Reactive oxygen species and oxidative DNA damage. *Indian Journal of Physiology and Pharmacology*, *42*(4), 440–452.

Henderson, B. W., Busch, T. M., & Snyder, J. W. (2006). Fluence rate as a modulator of PDT mechanisms. *Lasers in Surgery and Medicine*, *38*(5), 489–493.

Henderson, B. W., Gollnick, S. O., Snyder, J. W., Busch, T. M., Kousis, P. C., Cheney, R. T., et al. (2004). Choice of oxygen-conserving treatment regimen determines the inflammatory response and outcome of photodynamic therapy of tumors. *Cancer Research*, *64*(6), 2120–2126.

Hendrich, C., & Siebert, W. E. (1997). Photodynamic therapy for rheumatoid arthritis? *Lasers in Surgery and Medicine*, *21*(4), 359–364.

Hirsch, L. R., Stafford, R. J., Bankson, J. A., Sershen, S. R., Rivera, B., Price, R. E., et al. (2003). Nanoshell-mediated near-infrared thermal therapy of tumors under magnetic resonance guidance. *Proceedings of the National Academy of Sciences of the United States of America*, *100*(23), 13549–13554.

Huang, Z. A. (2005). Review of progress in clinical photodynamic therapy. *Technology in Cancer Research & Treatment*, *4*(3), 283–293.

Huang, P., Bao, L., Zhang, C., Lin, J., Luo, T., Yang, D., et al. (2011). Folic acid-conjugated silica-modified gold nanorods for X-ray/CT imaging-guided dual-mode radiation and photo-thermal therapy. *Biomaterials*, *32*(36), 9796–9809.

Huang, Z., Chen, Q., Shakil, A., Chen, H., Beckers, J., Shapiro, H., et al. (2003). Hyperoxygenation enhances the tumor cell killing of photofrin-mediated photodynamic therapy. *Photochemistry and Photobiology*, *78*(5), 496–502.

Huang, X., Qian, W., El-Sayed, I. H., & El-Sayed, M. A. (2007). The potential use of the enhanced nonlinear properties of gold nanospheres in photothermal cancer therapy. *Lasers in Surgery and Medicine*, *39*(9), 747–753.

Huang, Z., Xu, H., Meyers, A. D., Musani, A. I., Wang, L., Tagg, R., et al. (2008). Photodynamic therapy for treatment of solid tumors – potential and technical challenges. *Technology in Cancer Research & Treatment*, *7*(4), 309–320.

Hung, H. I., Klein, O. J., Peterson, S. W., Rokosh, S. R., Osseiran, S., Nowell, N. H., et al. (2016). PLGA nanoparticle encapsulation reduces toxicity while retaining the therapeutic efficacy of EtNBS-PDT in vitro. *Scientific Reports*, *6*, 33234.

Hussain, T., & Nguyen, Q. T. (2014). Molecular imaging for cancer diagnosis and surgery. *Advanced Drug Delivery Reviews*, *66*, 90–100.

Jackson, P. A., Rahman, W. N., Wong, C. J., Ackerly, T., & Geso, M. (2010). Potential dependent superiority of gold nanoparticles in comparison to iodinated contrast agents. *European Journal of Radiology*, *75*(1), 104–109.

Jin, Z. H., Miyoshi, N., Ishiguro, K., Umemura, S., Kawabata, K., Yumita, N., et al. (2000). Combination effect of photodynamic and sonodynamic therapy on experimental skin squamous cell carcinoma in C3h/HeN mice. *Journal of Dermatology*, *27*(5), 294–306.

Juarranz, A., Jaen, P., Sanz-Rodriguez, F., Cuevas, J., & Gonzalez, S. (2008). Photodynamic therapy of cancer. Basic principles and applications. *Clinical & Translational Oncology*, *10*(3), 148–154.

Kelleher, D. K., Thews, O., Scherz, A., Salomon, Y., & Vaupel, P. (2003). Combined hyperthermia and chlorophyll-based photodynamic therapy: tumour growth and metabolic microenvironment. *British Journal of Cancer*, *89*(12), 2333–2339.

Key, J., & Leary, J. F. (2014). Nanoparticles for multimodal in vivo imaging in nanomedicine. *International Journal of Nanomedicine*, *9*, 711–726.

Khlebtsov, B., Panfilova, E., Khanadeev, V., Bibikova, O., Terentyuk, G., Ivanov, A., et al. (2011). Nanocomposites containing silica-coated gold–silver nanocages and Yb–2,4-dimethoxyhematoporphyrin: multifunctional capability of IR-luminescence detection, photosensitization, and photothermolysis. *ACS Nano*, *5*(9), 7077–7089.

Kim, J.-W., Galanzha, E. I., Shashkov, E. V., Moon, H.-M., & Zharov, V. P. (2009). Golden carbon nanotubes as multimodal photoacoustic and photothermal high-contrast molecular agents. *Nature Nanotechnology*, *4*(10), 688–694.

Konan, Y. N., Berton, M., Gurny, R., & Allemann, E. (2003). Enhanced photodynamic activity of meso-tetra(4-hydroxyphenyl)porphyrin by incorporation into sub-200 nm nanoparticles. *European Journal of Pharmaceutical Sciences*, *18*(3–4), 241–249.

Konan, Y. N., Gurny, R., & Allémann, E. (2002). State of the art in the delivery of photosensitizers for photodynamic therapy. *Journal of Photochemistry and Photobiology B: Biology*, *66*(2), 89–106.

Konan-Kouakou, Y. N., Boch, R., Gurny, R., & Allemann, E. (2005). In vitro and in vivo activities of verteporfin-loaded nanoparticles. *Journal of Controlled Release*, *103*(1), 83–91.

Kong, W. H., Lee, W. J., Cui, Z. Y., Bae, K. H., Park, T. G., Kim, J. H., et al. (2007). Nanoparticulate carrier containing water-insoluble iodinated oil as a multifunctional contrast agent for computed tomography imaging. *Biomaterials*, *28*(36), 5555–5561.

Korbelik, M. (1996). Induction of tumor immunity by photodynamic therapy. *Journal of Clinical Laser Medicine & Surgery*, *14*(5), 329–334.

Lajunen, T., Kontturi, L.-S., Viitala, L., Manna, M., Cramariuc, O., Róg, T., et al. (2016). Indocyanine green-loaded liposomes for light-triggered drug release. *Molecular Pharmaceutics*, *13*(6), 2095–2107.

Lajunen, T., Viitala, L., Kontturi, L.-S., Laaksonen, T., Liang, H., Vuorimaa-Laukkanen, E., et al. (2015). Light induced cytosolic drug delivery from liposomes with gold nanoparticles. *Journal of Controlled Release*, *203*, 85–98.

Li, Y., Lu, W., Huang, Q., Huang, M., Li, C., & Chen, W. (2010). Copper sulfide nanoparticles for photothermal ablation of tumor cells. *Nanomedicine (London, England)*, *5*(8), 1161–1171.

Liang, Y. I., Lu, L.-M., Chen, Y., & Lin, Y.-K. (2016). Photodynamic therapy as an antifungal treatment. *Experimental and Therapeutic Medicine*, *12*(1), 23–27.

Lin, J., Wang, M., Hu, H., Yang, X., Wen, B., Wang, Z., et al. (2016). Multimodal-imaging-guided cancer phototherapy by versatile biomimetic theranostics with UV and gamma-irradiation protection. *Advanced Materials*, *28*(17), 3273–3279.

Liu, Q., Zhan, C., & Kohane, D. S. (2017). Phototriggered drug delivery using inorganic nanomaterials. *Bioconjugate Chemistry*, *28*(1), 98–104.

Liu, C., Zhang, Y., Liu, M., Chen, Z., Lin, Y., Li, W., et al. (2017). A NIR-controlled cage mimicking system for hydrophobic drug mediated cancer therapy. *Biomaterials*, *139*, 151–162.

Lucky, S. S., Soo, K. C., & Zhang, Y. (2015). Nanoparticles in photodynamic therapy. *Chemical Reviews*, *115*(4), 1990–2042.

Lusic, H., & Grinstaff, M. W. (2013). X-ray computed tomography contrast agents. *Chemical Reviews*, *113*(3), https://doi.org/10.1021/cr200358s.

Mallidi, S., Anbil, S., Bulin, A.-L., Obaid, G., Ichikawa, M., & Hasan, T. (2016). Beyond the barriers of light penetration: strategies, perspectives and possibilities for photodynamic therapy. *Theranostics*, *6*(13), 2458–2487.

Mayer, G., & Heckel, A. (2006). Biologically active molecules with a "light switch". *Angewandte Chemie, International Edition*, *45*(30), 4900–4921.

Mcgrath, A. J., Chien, Y.-H., Cheong, S., Herman, D. J., Watt, J., Henning, A. M., et al. (2015). Gold over branched palladium nanostructures for photothermal cancer therapy. *ACS Nano*, *9*(12), 12283–12291.

Menon, J. U., Jadeja, P., Tambe, P., Vu, K., Yuan, B., & Nguyen, K. T. (2013). Nanomaterials for photo-based diagnostic and therapeutic applications. *Theranostics*, *3*(3), 152–166.

Mew, D., Wat, C. K., Towers, G. H., & Levy, J. G. (1983). Photoimmunotherapy: treatment of animal tumors with tumor-specific monoclonal antibody-hematoporphyrin conjugates. *Journal of Immunology*, *130*(3), 1473–1477.

Mura, S., Nicolas, J., & Couvreur, P. (2013). Stimuli-responsive nanocarriers for drug delivery. *Nature Materials*, *12*(11), 991–1003.

Nagayasu, A., Uchiyama, K., & Kiwada, H. (1999). The size of liposomes: a factor which affects their targeting efficiency to tumors and therapeutic activity of liposomal antitumor drugs. *Advanced Drug Delivery Reviews*, *40*(1–2), 75–87.

Nam, T., Park, S., Lee, S.-Y., Park, K., Choi, K., Song, I. C., et al. (2010). Tumor targeting chitosan nanoparticles for dual-modality optical/MR cancer imaging. *Bioconjugate Chemistry*, *21*(4), 578–582.

Napp, J., Behnke, T., Fischer, L., Würth, C., Wottawa, M., Katschinski, D. M., et al. (2011). Targeted luminescent near-infrared polymer-nanoprobes for in vivo imaging of tumor hypoxia. *Analytical Chemistry*, *83*(23), 9039–9046.

Neveu, P., Aujard, I., Benbrahim, C., Le Saux, T., Allemand, J. F., Vriz, S., et al. (2008). A caged retinoic acid for one- and two-photon excitation in zebrafish embryos. *Angewandte Chemie International Edition in English*, *47*(20), 3744–3746.

Nurunnabi, M., Khatun, Z., Reeck, G. R., Lee, D. Y., & Lee, Y.-K. (2014). Photoluminescent graphene nanoparticles for cancer phototherapy and imaging. *ACS Applied Materials & Interfaces*, *6*(15), 12413–12421.

Ogawa, M., Kosaka, N., Choyke, P. L., & Kobayashi, H. (2009). In vivo molecular imaging of cancer with a quenching near infrared fluorescent probe using conjugates of monoclonal antibodies and indocyanine green. *Cancer Research*, *69*(4), 1268–1272.

Ohsaki, Y., Takeyama, K., Nakao, S., Tanno, S., Toyoshima, E., Nakanishi, K., et al. (2001). Detection of photofrin fluorescence from malignant and premalignant lesions in the bronchus using a full-color endoscopic fluorescence imaging system. *Diagnostic and Therapeutic Endoscopy*, *7*(3–4), 187–195.

Ortu, P., Lamuraglia, G. M., Roberts, W. G., Flotte, T. J., & Hasan, T. (1992). Photodynamic therapy of arteries. A novel approach for treatment of experimental intimal hyperplasia. *Circulation*, *85*(3), 1189–1196.

Paasonen, L., Sipilä, T., Subrizi, A., Laurinmäki, P., Butcher, S. J., Rappolt, M., et al. (2010). Gold-embedded photosensitive liposomes for drug delivery: triggering mechanism and intracellular release. *Journal of Controlled Release*, *147*(1), 136–143.

Park, H., Yang, J., Lee, J., Haam, S., Choi, I.-H., & Yoo, K.-H. (2009). Multifunctional nanoparticles for combined doxorubicin and photothermal treatments. *ACS Nano*, *3*(10), 2919–2926.

Pearson, S., Vitucci, D., Khine, Y. Y., Dag, A., Lu, H., Save, M., et al. (2015). Light-responsive azobenzene-based glycopolymer micelles for targeted drug delivery to melanoma cells. *European Polymer Journal*, *69*, 616–627.

Popovtzer, R., Agrawal, A., Kotov, N. A., Popovtzer, A., Balter, J., Carey, T. E., et al. (2008). Targeted gold nanoparticles enable molecular CT imaging of cancer. *Nano Letters*, *8*(12), 4593–4596.

Rai, P., Mallidi, S., Zheng, X., Rahmanzadeh, R., Mir, Y., Elrington, S., et al. (2010). Development and applications of photo-triggered theranostic agents. *Advanced Drug Delivery Reviews*, *62*(11), 1094–1124.

Ray, E. R., Chatterton, K., Khan, M. S., Thomas, K., Chandra, A., & O'brien, T. S. (2009). Hexylaminolaevulinate 'blue light' fluorescence cystoscopy in the investigation of clinically unconfirmed positive urine cytology. *BJU International*, *103*(10), 1363–1367.

Roelandts, R. (2002). The history of phototherapy: something new under the sun? *Journal of the American Academy of Dermatology*, *46*(6), 926–930.

Samia, A. C. S., Chen, X., & Burda, C. (2003). Semiconductor quantum dots for photodynamic therapy. *Journal of the American Chemical Society*, *125*(51), 15736–15737.

Seol, S. K., Kim, D., Jung, S., Chang, W. S., & Kim, J. T. (2013). One-step synthesis of PEG-coated gold nanoparticles by rapid microwave heating. *Journal of Nanomaterials*, *2013*, 6.

Shen, S., Ding, B., Zhang, S., Qi, X., Wang, K., Tian, J., et al. (2017). Near-infrared light-responsive nanoparticles with thermosensitive yolk-shell structure for multimodal imaging and chemo-photothermal therapy of tumor. *Nanomedicine*, *13*(5), 1607–1616.

Shen, J., Zhao, L., & Han, G. (2013). Lanthanide-doped upconverting luminescent nanoparticle platforms for optical imaging-guided drug delivery and therapy. *Advanced Drug Delivery Reviews*, *65*(5), 744–755.

Shi, J., Wang, L., Zhang, J., Ma, R., Gao, J., Liu, Y., et al. (2014). A tumor-targeting near-infrared laser-triggered drug delivery system based on GO@ag nanoparticles for chemo-photothermal therapy and X-ray imaging. *Biomaterials*, *35*(22), 5847–5861.

Shibu, E. S., Ono, K., Sugino, S., Nishioka, A., Yasuda, A., Shigeri, Y., et al. (2013). Photouncaging nanoparticles for MRI and fluorescence imaging in vitro and in vivo. *ACS Nano*, *7*(11), 9851–9859.

Sortino, S. (2012). Photoactivated nanomaterials for biomedical release applications. *Journal of Materials Chemistry*, *22*(2), 301–318.

Sun, Y., Peng, J., Feng, W., & Li, F. (2013). Upconversion nanophosphors Naluf(4):Yb,Tm for lymphatic imaging in vivo by real-time upconversion luminescence imaging under ambient light and high-resolution X-ray CT. *Theranostics*, *3*(5), 346–353.

Sun, T., Zhang, Y. S., Pang, B., Hyun, D. C., Yang, M., & Xia, Y. (2014). Engineered nanoparticles for drug delivery in cancer therapy. *Angewandte Chemie, International Edition*, *53*(46), 12320–12364.

Thakor, A. S., & Gambhir, S. S. (2013). Nanooncology: the future of cancer diagnosis and therapy. *CA: A Cancer Journal for Clinicians*, *63*(6), 395–418.

Tomaselli, F., Maier, A., Sankin, O., Anegg, U., Stranzl, U., Pinter, H., et al. (2001). Acute effects of combined photodynamic therapy and hyperbaric oxygenation in lung cancer—a clinical pilot study. *Lasers in Surgery and Medicine*, *28*(5), 399–403.

Uhrich, K. E., Cannizzaro, S. M., Langer, R. S., & Shakesheff, K. M. (1999). Polymeric systems for controlled drug release. *Chemical Reviews*, *99*(11), 3181–3198.

Walsh, J. C., Lebedev, A., Aten, E., Madsen, K., Marciano, L., & Kolb, H. C. (2014). The clinical importance of assessing tumor hypoxia: relationship of tumor hypoxia to prognosis and therapeutic opportunities. *Antioxidants & Redox Signaling*, *21*(10), 1516–1554.

Wan, M. T., & Lin, J. Y. (2014). Current evidence and applications of photodynamic therapy in dermatology. *Clinical, Cosmetic and Investigational Dermatology*, *7*, 145–163.

Wang, L. V. (2009). Multiscale photoacoustic microscopy and computed tomography. *Nature Photonics*, *3*(9), 503–509.

Woodhams, J. H., Macrobert, A. J., & Bown, S. G. (2007). The role of oxygen monitoring during photodynamic therapy and its potential for treatment dosimetry. *Photochemical & Photobiological Sciences*, *6*(12), 1246–1256.

Xiao, Z., Ji, C., Shi, J., Pridgen, E. M., Frieder, J., Wu, J., et al. (2012). DNA self-assembly of targeted near-infrared-responsive gold nanoparticles for cancer thermo-chemotherapy. *Angewandte Chemie International Edition*, *51*(47), 11853–11857.

Yao, C., Wang, P., Li, X., Hu, X., Hou, J., Wang, L., et al. (2016). Near-infrared-triggered azobenzene-liposome/upconversion nanoparticle hybrid vesicles for remotely controlled drug delivery to overcome cancer multidrug resistance. *Advanced Materials*, *28*(42), 9341–9348.

Yao, C., Zhang, L., Wang, J., He, Y., Xin, J., Wang, S., et al. (2016). Gold nanoparticle mediated phototherapy for cancer. *Journal of Nanomaterials*, *2016*, 29.

Yoo, J.-W., Doshi, N., & Mitragotri, S. (2011). Adaptive micro and nanoparticles: temporal control over carrier properties to facilitate drug delivery. *Advanced Drug Delivery Reviews*, *63*(14–15), 1247–1256.

Yoon, I., Li, J. Z., & Shim, Y. K. (2013). Advance in photosensitizers and light delivery for photodynamic therapy. *Clinical Endoscopy*, *46*(1), 7–23.

Yun, S. H., & Kwok, S. J. (2017). Light in diagnosis, therapy and surgery. *Nature Biomedical Engineering*, *1*, 0008.

Zhang, F., Niu, G., Lu, G., & Chen, X. (2011). Preclinical lymphatic imaging. *Molecular Imaging and Biology*, *13*(4), 599–612.

Zharov, V. P., Galitovskaya, E. N., Johnson, C., & Kelly, T. (2005). Synergistic enhancement of selective nanophotothermolysis with gold nanoclusters: potential for cancer therapy. *Lasers in Surgery and Medicine*, *37*(3), 219–226.

Zheng, M., Yue, C., Ma, Y., Gong, P., Zhao, P., Zheng, C., et al. (2013). Single-step assembly of DOX/ICG loaded lipid–polymer nanoparticles for highly effective chemo-photothermal combination therapy. *ACS Nano*, *7*(3), 2056–2067.

Zou, L., Wang, H., He, B., Zeng, L., Tan, T., Cao, H., et al. (2016). Current approaches of photothermal therapy in treating cancer metastasis with nanotherapeutics. *Theranostics*, *6*(6), 762–772.

CHAPTER 13

Nanoengineering of Soft Polymer Particles for Exploring Bio-Nano Interactions

Ling Wang[*], **Mattias Björnmalm**[†], **Jingcheng Hao**[*], **Jiwei Cui**[*]
[*]*Shandong University, Jinan, China* [†]*Imperial College London, London, United Kingdom*

1 Introduction

The application of nanotechnology for the design of particle-based cancer therapeutics has achieved great progress in the last few decades (Chen, Zhang, Zhu, Xie, & Chen, 2017; Farokhzad & Langer, 2009; Petros & DeSimone, 2010; Shi, Kantoff, Wooster, & Farokhzad, 2017). The increasing ability to control physicochemical properties of therapeutic particles has revolutionized our understanding of interactions between nanostructured particles and biological systems (bio-nano interactions), and helped improve their therapeutic efficacy (Blanco, Shen, & Ferrari, 2015; Mitragotri, Burke, & Langer, 2014). For example, the influence of size, shape, surface chemistry, and biodegradability of particles on their biological behavior has been widely studied (Albanese, Tang, & Chan, 2012; Cui, Richardson, Björnmalm, Faria, & Caruso, 2016; Fox, Szoka, & Fréchet, 2009; Petros & DeSimone, 2010). Recent reports have indicated that stiffness is a key parameter for governing bio-nano interactions, including cell interactions, biodistribution, circulation time, and cell targeting (Best, Yan, & Caruso, 2012; Cui et al., 2015; Zhang et al., 2012). However, it is an easily overlooked parameter as it is typically difficult to tune for many types of particles (e.g., inorganic particles), and it can be challenging to characterize stiffness, especially for nanomaterials. Recently, several strategies for the nanoengineering of polymer particles with tunable stiffness have emerged, for example, by changing composition, cross-linking density, and particle structure. Due to the flexibility of available fabrication techniques and the precise control of the resulting particle properties, soft polymer particles are receiving increasing interest in the field of cancer theranostics (Anselmo & Mitragotri, 2017; Cui, 2017; Xu et al., 2013).

In this chapter, we first introduce the characterization methods that are used to (semi) quantify the stiffness of polymer particles, which is essential for the development of the field, but remains largely unexplored (Fig. 1). Then, we focus on strategies for the

Handbook of Nanomaterials for Cancer Theranostics. https://doi.org/10.1016/B978-0-12-813339-2.00013-X

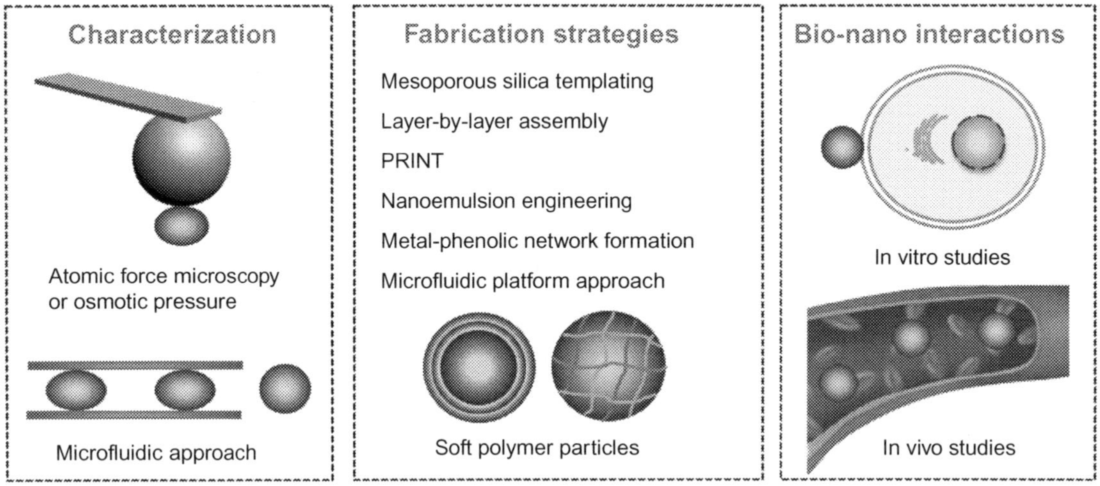

Fig. 1
Overview of the main focus of this chapter on the characterization, fabrication strategies, and bio-nano interactions of soft polymer particles, as well as the related importance and application for nanomedicine.

fabrication of soft polymer particles via mesoporous silica (MS) templating, layer-by-layer (LbL) assembly, particle replication in nonwetting templates (PRINT), microemulsion engineering, metal-phenolic network (MPN) formation, and microfluidic approaches. Finally, the influence of stiffness on bio-nano interactions and its importance for applications within biomedicine is discussed. We believe this chapter will provide insightp useful for researchers in cancer theranostics, and help guide future developments and innovation.

2 Characterization Methods for Investigating the Stiffness of Polymer Particles

Characterization methods for determining the stiffness of particles vary depending on the type of polymer particle. For example, the membrane elasticity of polymersomes can be probed by micropipette aspiration techniques (Bermudez, Brannan, Hammer, Bates, & Discher, 2002). This technique is mainly for assemblies of amphiphilic lipids or copolymers. A "universal testing system" for both tensile and compression applications (Merkel, Jones, Herlihy, Kersey, & DeSimone, 2011) and rheometers (Anselmo et al., 2015) have been used to measure the stiffness of bulk hydrogels. These are indirect methods for investigating the stiffness of polymer particles. Herein, we will introduce the three most commonly used methods for directly determining the stiffness of polymer particles.

2.1 Colloidal Probe-Atomic Force Microscopy (CP-AFM)

AFM is a powerful and precise tool for imaging and mechanical analysis of soft materials, including planar films (Schmidt et al., 2010), hollow capsules (Markutsya, Jiang, Pikus, & Tsukruk, 2005), hydrogel particles (Best, Javed, et al., 2013), and cells (Bolshakova, Kiselyova, & Yaminsky, 2004). When the probe is moved close to the sample surface, interactions between the probe and sample induce a deflection of the cantilever (Fig. 2A). The deflection signal is transferred into an electric signal to generate force curves, which can be used to determine local Young's modulus (E_Y) and stiffness. Compared with sharp-tip probes, colloidal probes with a higher interaction area, for example as developed by Ducker, Senden, and Pashley (1991) can significantly improve the sensitivity of force measurements. The diameter of the colloidal probe is usually in the range of micrometers to tens of micrometers, depending on the size of the particle to be measured. To measure soft polymer particles, it is often better to use soft cantilevers with spring constants of less than $0.1\,\mathrm{N\,m^{-1}}$, as this can improve the measurement accuracy. Because the measurements are based on an in situ technique, mechanical tests of polymer particles can be performed at different solution conditions (e.g., pH, temperature, ionic strength) (Lulevich & Vinogradova, 2004).

The combination of AFM with confocal laser scanning microscopy (CLSM) can help increase our understanding of sample deformation, and inform the development of soft polymer particles toward potential biomedical applications (e.g., to increase the efficacy of various medical treatments) (Fig. 2A). Fig. 2B shows typical force curves with force versus relative deformation of LbL capsules, where the loaded force increases along with increasing deformation, which can also be observed from the CLSM images. The nanoscale stiffness of polymer particles is typically evaluated using the linear gradient of the force/deformation curve in small deformation profiles. Several models have been developed to calculate E_Y or stiffness of the measured samples. For example, the Reissner model (Reissner, 1946) has been used to evaluate the E_Y of LbL assembled capsules (Best, Neubauer, et al., 2013), while the Hertz model (Hertz, 1882) is typically used for polymer hydrogel particles (Best, Javed, et al., 2013; Cui et al., 2013). The application of Leissner's formula is limited for capsules with a ratio of shell thickness and a radius below 1/20 (Gregory, Milac, & Wan, 1998). When using the Hertz model to evaluate polymer particles, the E_Y of the colloidal probe (e.g., silica particles) should be much greater than that of the samples. Three-dimensional E_Y maps can be generated through force map scanning of particles, which can help provide information about spatial heterogeneity of mechanical properties, as well as illustrate mechanical properties of the samples more visually (Cui et al., 2014). For many previously reported studies, the indentation of capsules or particles during measurements has usually not been considered. Although this effect may not influence relative comparisons within a study, absolute measurements and determinations of properties such as the E_Y may be underpredicted if these effects are not considered (Berry, Mettu, & Dagastine, 2017).

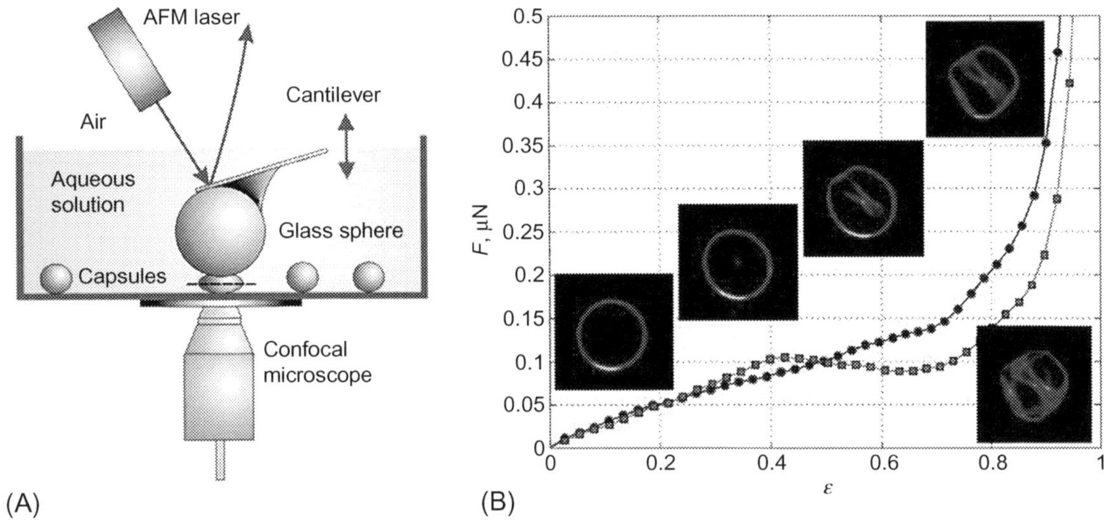

Fig. 2
(A) Schematic illustration of a stiffness measurement performed by combining an AFM with a CLSM (for fluorescence imaging). (B) Force (F) versus relative deformation (ε) of LbL capsules. Adsorption time of polyelectrolytes for capsule assembly is 20 *(circles)* and 40 *(squares)* min, respectively. Insets are CLSM images in XY confocal mode focusing on the "equator" of the capsules. *Reproduced with permission from Vinogradova, O. I., Lebedeva, O. V., Vasilev, K., Gong, H., Garcia-Turiel, J., & Kim, B.-S. (2005). Multilayer DNA/poly(allylamine hydrochloride) microcapsules: assembly and mechanical properties. Biomacromolecules, 6, 1495.*

2.2 Microfluidic Approach

Microfluidic devices have been used to prepare particles with different morphologies and to mimic blood capillaries, for example to study particle deformation for qualitative investigation of particle stiffness over microscale dimensions. One simple approach for investigating flow-based particle deformation is using filters, for example, to check whether particles with different stiffness can pass through filter pores (Zhang et al., 2012). The DeSimone group has synthesized red blood cell (RBC)-like polymer particles (about 5.5 μm in diameter and 1.4 μm thick) with controllable stiffness by changing the cross-linking density, and then investigated their deformation in a 3 × 3 μm microchannel (Merkel et al., 2011). The 1% cross-linked polymer particles could stretch their shape and pass through the channels at a constant flow rate of 0.06 mL min^{-1}, while 2% cross-linked particles were easily stuck at the entrance of the microfluidic channels. In vivo results were consistent with the microfluidic results, that is, softer particles did not accumulate much in the lung, which contains many small capillaries and possesses longer circulation time compared with stiffer particles, which were trapped, to a larger extent, in the lung.

Conventional microfluidic devices are commonly designed with one inlet and one outlet where particles dispersions are pushed from the inlet to the outlet while conducting flow-based deformation studies (Haghgooie, Toner, & Doyle, 2010; She, Li, Shan, Tong, & Gao, 2013; She, Xu, Yin, Tong, & Gao, 2012). This can make the differential pressure and pressure drops difficult to tune. In addition, the microchannels may be easily blocked when using stiffer particles. We recently designed a four-point microfluidic device with two inlets and two outlets, which facilitates the establishment of well-controlled differential pressures and pressure drops across the microchannels, as the pressure on each side can be independently tuned (Fig. 3) (Cui et al., 2014). This design mimics the in vivo environment

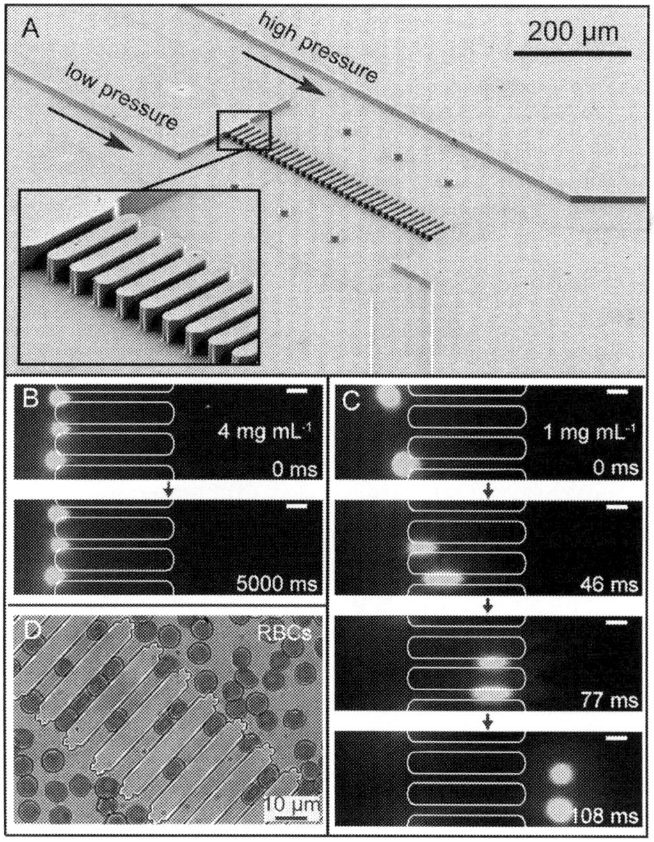

Fig. 3
(A) Scanning electron microscopy (SEM) image of a four-point microfluidic device with two inlets and two outlets. (B, C) Fluorescence microscopy images of PEG particles labeled with Alexa Fluor 488 prepared with different cross-linker concentration stopping or passing through the microchannels. (D) Bright-field microscopy image of RBCs passing through microchannels. *Reproduced with permission from Cui, J., Björnmalm, M., Liang, K., Xu, C., Best, J. P., Zhang, X., et al. (2014). Super-soft hydrogel particles with tunable elasticity in a microfluidic blood capillary model. Advanced Materials, 26, 7295.*

where a high-pressure arterial side is connected to a low-pressure venous side through capillary networks. This type of microfluidic device was then used to investigate the flow-based deformability behavior of soft poly(ethylene glycol) (PEG) particles. The stiffness of these types of PEG particles could be tuned by adjusting the cross-linking density. PEG particles prepared using less cross-linker could more easily pass through the microchannels, a behavior that was found to be similar to that of human RBCs (Fig. 3). Interestingly, polymer particles and capsules with similar stiffness at the nanoscale dimension, determined using CP-AFM, were found to exhibit significantly different deformation behaviors in the microfluidic device, where deformation occurs over the microscale dimension (Sun, Björnmalm, et al., 2015). It was observed that polymer capsules with a hollow interior behaved like RBCs, while polymer particles with a matrix-structured interior had difficulty passing through the microchannels. This difference was attributed to the fact that the hollow capsule shares some similarities with the RBC structure, such as a vesicle of fluid wrapped by a flexible membrane, which facilitates better shape recovery by having a more fluid-like deformation (Palange, Palomba, Rizzuti, Ferreira, & Decuzzi, 2017). The observations indicate that the deformation ability of polymer particles is related to both stiffness and morphology, and flow-based microscale deformation studies can therefore be useful to complement nanoscale stiffness measurements to get an increased understanding of the deformability of soft polymer particles.

2.3 Osmotic Pressure Method

The shells of polymer capsules are typically impermeable to polymers with high molecular weight, which is dependent on the composition and structure (e.g., number of layers for LbL assembled capsules) of the polymer capsules. When polymer capsules are dispersed in a polymer solution, the difference of polymer concentration in the hollow structure and exterior of capsules generates an osmotic pressure that can cause soft capsules to deform. This osmotic pressure method can be used to investigate mechanical properties of polymer capsules (Chen et al., 2016; Gao, Donath, Moya, Dudnik, & Möhwald, 2001). In this approach, isotropic force is applied, in contrast with CP-AFM, where anisotropic force is applied.

The modulus is usually calculated using the following equation:

$$P_c = 4\mu\left(\frac{\delta}{R}\right)^2$$

where P_c is the critical osmotic pressure, which indicates that half of the capsules are deformed at a certain concentration of the polymer solution. δ is the capsule shell thickness, which can be measured by AFM (e.g., the height of shell thickness for collapsed capsules) (Cui, Liu, & Hao, 2009; Donath, Sukhorukov, Caruso, Davis, & Möhwald, 1998). R is the radius of the capsules and μ is the modulus that can then be calculated from the measured parameters. For capsules of the same size, the critical osmotic pressure increases linearly with the square of capsule shell thickness. The advantage of the osmotic pressure method is that it facilitates studies on a large number of capsules. However, it is challenging for capsules smaller than 200 nm, due to the resolution and accuracy of fluorescence microscopy (although this can partially be addressed using superresolution microscopy).

3 Strategies for the Fabrication of Soft Polymer Particles

3.1 MS Templating Method

The MS templating method has proven to be a versatile way to prepare polymer replica particles with tunable size, shape, structure, stiffness, and degradability for drug and vaccine delivery (Cui, Richardson, et al., 2016; Gause et al., 2016; Suma et al., 2017). The construction of soft polymer particles is typically achieved by infiltrating polymers into the pores of the MS particles, followed by cross-linking of the polymer chains, and subsequent removal of the hard templates (Fig. 4) (Björnmalm et al., 2017; Cui et al., 2013; Cui, Yan, Wang, & Caruso, 2012; Wang, Yu, & Caruso, 2005). The stiffness of the resulting polymer replica particles is

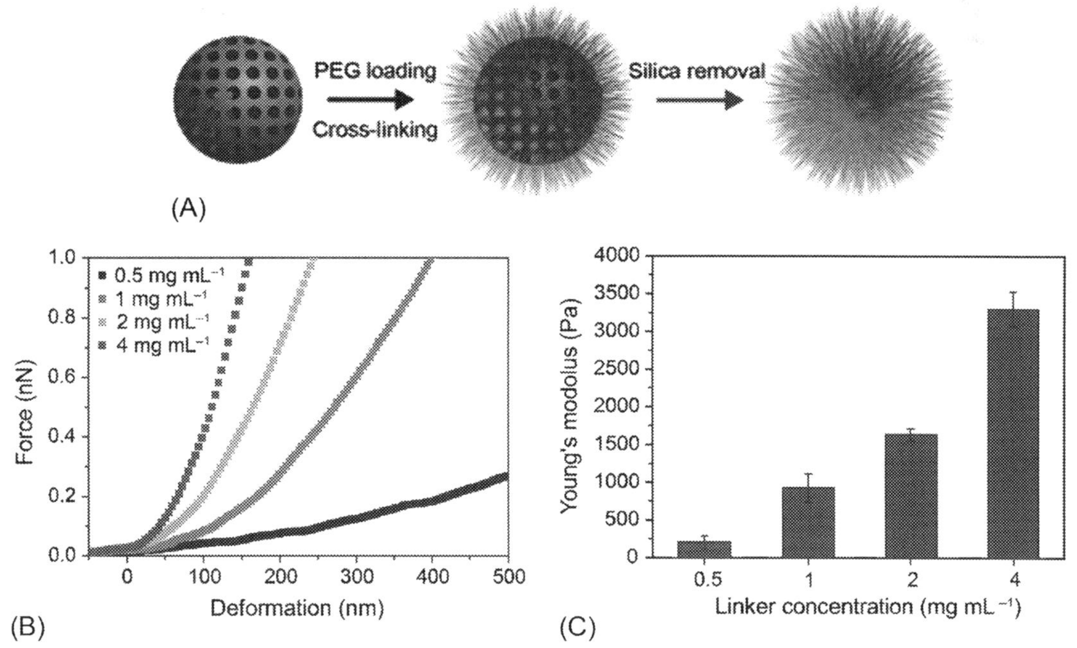

Fig. 4
(A) Schematic illustration of the preparation of polymer particles via MS templating method. Representative F-δ curves (B) and (C) E_Y for the PEG particles with increased cross-linker concentrations. *Reproduced with permission from Cui, J., Björnmalm, M., Liang, K., Xu, C., Best, J. P., Zhang, X., et al. (2014). Super-soft hydrogel particles with tunable elasticity in a microfluidic blood capillary model. Advanced Materials, 26, 7295.*

typically tuned by varying the concentration of cross-linker used at the cross-linking step. For example, poly(L-glutamic acid) (PGA) particles conjugated with the vaccine adjuvant CpG were prepared using the MS templating method combined with thiol-disulfide exchange chemistry (Cui et al., 2013). The E_Y of the resulting PGA particles varied from 2 to 25 kPa, as the cross-linker concentration was increased from 0.2 to 2 mg mL^{-1}. The increased cross-linker concentration increased the degree of cross-linking density, and therefore the particle stiffness, which also resulted in an increase in the CpG loading amount. Similarly, poly(methacrylic acid) (PMA) particles with tunable stiffness were obtained based on similar chemistry as for the PGA particle preparation (Best, Cui, Müllner, & Caruso, 2013). The amount of cross-linker (thiolated PMA, PMA_{SH}) was quantified by determining the PMA_{SH} mass in a single PMA particle, which was then correlated with the particle stiffness. Interestingly, the stiffness of PMA particles was linearly dependent on the cross-linker amount (per particle) when the cross-linking concentration was in the range of 0.1–0.5 mg mL^{-1}. This correlation between particle stiffness and cross-linker amount in a single particle can be advantageous for predicting particle properties and suitability for various applications.

The loading amount of different polymers in the same type of MS particles is typically different due to their inherent properties, such as polymer conformation in solution, charge, molecular weight, and hydrophilicity (Wang, Angelatos, Dunstan, & Caruso, 2007). In other words, the density of polymer replica particles varies with different compositions, which can induce different particle stiffness. For example, PMA particles (0.04–2.53 MPa) exhibit different E_Y compared with PGA particles (2–25 kPa), even though similar concentrations of cross-linker were used for particle assembly (Best, Cui, et al., 2013; Cui et al., 2013). The MS templating method can also be used for preparing "supersoft" particles (0.2–3.3 kPa), when using PEG as the building block, in which the E_Y is 10 times lower than for PGA particles (Fig. 4) (Cui et al., 2014).

The size, shape, and nanostructure of polymer replica particles are dependent on those of the MS templates. Therefore, MS particles with fully porous or solid core/mesoporous shell structures result in polymer replica particles with a sponge structure, or polymer capsules with a hollow cavity, respectively (Cui, Faria, et al., 2016). Obviously, when nonporous silica particles are used as templates, hollow polymer capsules are obtained. Interestingly, polymer particles and capsules with different structures, but with similar nanoscale stiffness in the range of 2.4–21.3 mN m^{-1}, exhibit significantly different deformation behaviors at microscale dimension in a microfluidic blood capillary model (Sun, Björnmalm, et al., 2015), thus indicating the importance of particle nanostructure on the microscale stiffness.

3.2 LbL Assembly

LbL assembly is a prominent method for fabricating hollow polymer capsules, in which polymers are sequentially layered on the surface of a sacrificial template, followed by template removal (Richardson, Cui, et al., 2016; Richardson, Bjornmalm, & Caruso, 2015; Yan, Björnmalm, & Caruso, 2014). The stiffness of polymer capsules can be tuned by the shell composition, shell thickness, composition of the liquid core, morphology, and cross-linking density in the shell during the LbL assembly process. The polymer compositions have an important effect on the stiffness of the resulting capsules (Richardson, Tardy, et al., 2016). For example, capsules constructed by DNA/poly(allylamine hydrochloride) (PAH) multilayer films layered on melamine formaldehyde (MF) templates had a E_Y of ~100 MPa, which is twofold softer than poly(sodium 4-styrenesulfonate)/poly(allylamine hydrochloride) (PSS/PAH) capsules (Vinogradova et al., 2005). Similarly, PAH/PSS capsules with a stiffness of ~5 N m^{-1} are also twofold softer than dextran sulfate sodium salt/poly-L-arginine hydrochloride (DextS/PLArg) capsules (Hartmann, Weidenbach, Neubauer, Fery, & Parak, 2015). The introduction of inorganic nanoparticles into the capsule shell can result in pronounced reinforcement (e.g., by a factor of eight) of capsule stiffness compared with polymer capsules without nanoparticles (Dubreuil, Shchukin, Sukhorukov, & Fery, 2004).

LbL assembly is a versatile method for controlling the shell thickness of polymer capsules by increasing polymer layer numbers, which allows for convenient control over capsule stiffness. For $CaCO_3$-templated PAH/PSS capsules, an increase in layer numbers resulted in increased shell thicknesses from 150 to 800 nm, which increased particle stiffness (Palankar et al., 2013). Using a similar strategy by altering the layer numbers from 2 to 16, LbL assembly was used to increase the stiffness of DextS/PLArg capsules from ~0.2 to ~10 N m^{-1} (Hartmann et al., 2015).

Changing the composition of the liquid core (e.g., using polymers solutions) is another strategy to increase the stiffness of polymer capsules (Lebedeva, Kim, & Vinogradova, 2004; Lulevich, Radtchenko, Sukhorukov, & Vinogradova, 2003). In one example, hollow polyelectrolyte microcapsules of PAH/PSS were treated with acetone to make the multilayer shell permeable for solutions of strong polyelectrolyte PSS molecules. A combination of AFM and CLSM showed that the "filled" capsules prepared by this method were stiffer than "hollow" ones, which reflects the contribution of the excess osmotic pressure on the capsule stiffness (Lebedeva et al., 2004).

Because LbL assembly is independent of the shape of templates, nonspherical templates result in polymer capsules with different morphologies (Ariga et al., 2014; Ariga, Hill, & Ji, 2007). The stiffness or E_Y of LbL assembled capsules varies when morphology changes. It has been observed that cubic capsules (E_Y ~0.8 MPa) are softer than spherical capsules (E_Y ~4.3 MPa) when composition and layer numbers are the same for both capsules (Alexander et al., 2015; Shchepelina, Lisunova, Drachuk, & Tsukruk, 2012). However, dodecahedral capsules composed of PSS and poly(allylamine) (PA) were almost incompressible, while the E_Y of spherical PSS/PA capsules was about 89 MPa (Ejima, Yanai, et al., 2013).

Another strategy to tune the stiffness of polymer capsules involves controlling the cross-linking density, as described in the MS templating method section. Introducing disulfide bond formation is one of the most commonly used cross-linking techniques. In this approach, cross-linking is achieved using disulfide cross-linkers, or by cross-linking thiolated polymers based on oxidizing reagents or thiol-disulfide exchange (Zelikin, Quinn, & Caruso, 2006). For PMA_{SH} capsules assembled through the LbL assembly technique, the stiffness can be tuned by adjusting the percentage of thiol-modification of the polymers (Best, Neubauer, et al., 2013). The stiffness of the resulting polymer capsules, as measured by CP-AFM, ranged from 0.9 to 16.9 $mN\,m^{-1}$ over a thiol-modification range of 5–20 mol%. In addition, the degradation of disulfide cross-linked LbL assembled polymer capsules, either in the presence of reducing reagent or in intracellular microenvironment, becomes increasingly difficult when increasing the cross-linker concentration (Liang et al., 2012).

In a specific example related to conventional LbL assembly, atom transfer radical polymerization-mediated continuous assembly of polymers (CAP_{ATRP}) has been recently used to fabricate multilayered polymer capsules via repeated ATRP initiator adsorption and polymer film growth, followed by template removal (Fig. 5) (Sun, Wong, et al., 2015). Polymer capsules with different compositions, including hyaluronic acid (HA), poly(*N*-(2-hydroxypropyl)-methacrylamide) (PHPMA), poly(glutamic acid) (PGA), PEG, and PMA capsules, have been prepared via the CAP_{ATRP} method (Sun et al., 2017; Wong et al., 2013). The shell thickness (4.4–20 nm) and associated capsule stiffness (~7.5 to 28 $mN\,m^{-1}$) can be tuned by changing the number of CAP_{ATRP} steps from 1 to 4 (Fig. 5) (Sun, Wong, et al., 2015). Similarly, the stiffness of HA replica particles via CAP_{ATRP} on MS templates can be controlled in the range from ~2.4 to 21.3 $mN\,m^{-1}$ by varying the number of CAP_{ATRP} steps (Sun, Björnmalm, et al., 2015). By controlling the shell thickness and correspondingly the stiffness, the CAP_{ATRP} method is a facile approach to prepare particles for stiffness-related biological studies.

3.3 Particle Replication in Nonwetting Templates (PRINT)

PRINT is a high-resolution molding technology, which allows the design and synthesis of precisely defined micro- and nanoparticles (Xu et al., 2013). Several critical factors, including the size, shape, surface chemistry, and stiffness of the particles can be independently controlled by the PRINT technique (Palange et al., 2017). The PRINT technique provides a versatile route for varying particle shape and stiffness, although it is not easily applicable for spherical or hollow polymer particles (Fig. 6) (Palange et al., 2017; Xu et al., 2013). For example, deformable RBC-like microgel particles with defined size and stiffness have been fabricated via the PRINT technique (Chen et al., 2012; Merkel et al., 2011). One wt% poly(ethylene glycol) diacrylate (PEGDA) cross-linker resulted in a bulk modulus of less than 20 kPa, while a 0.05 wt% cross-linker resulted in polymer particles with a modulus of 6.5 kPa (Chen et al., 2012). By decreasing the amount of cross-linker from 10% to 1%, the modulus of the resultant polymer particle could be tuned from 63.9 to 7.8 kPa. Only the discoidal polymer particles with a low modulus (<17 kPa) could pass through a microfluidic channel

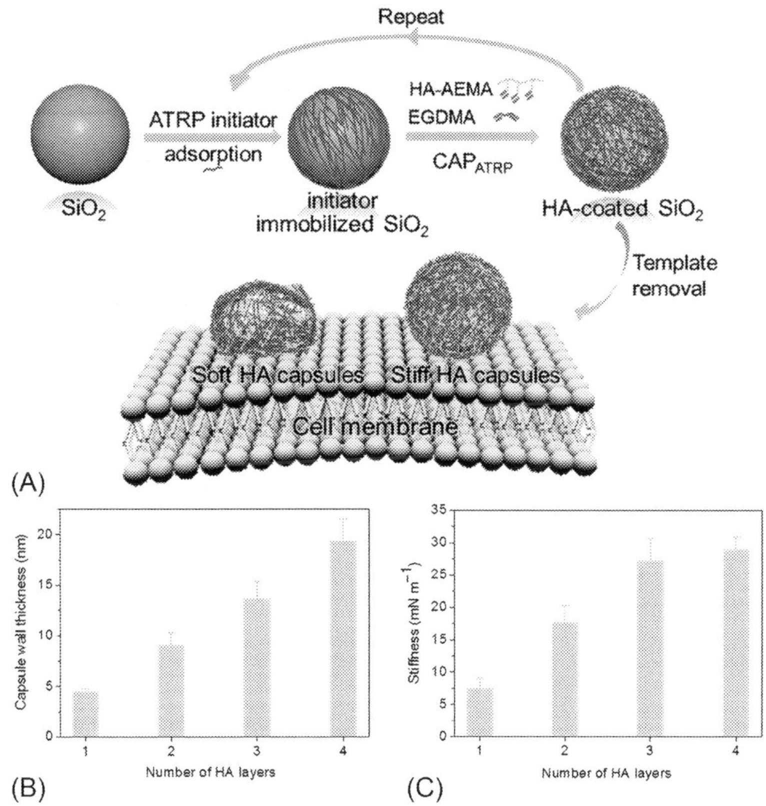

Fig. 5

(A) Illustration of HA capsule formation via CAP$_{ATRP}$ method including the ATRP initiator adsorption, ATRP promoted HA film growth, and SiO_2 template removal. (B) HA capsule shell thickness versus CAP layer numbers. (C) Stiffness of HA capsules versus the number of HA layers, as determined by CAP-AFM analysis. *Reproduced with permission from Sun, H., Wong, E. H. H., Yan, Y., Cui, J., Dai, Q., Guo, J., et al. (2015). The role of capsule stiffness on cellular processing. Chemical Science, 6, 3505.*

with a twofold smaller diameter compared with the particles, by stretching in length in the microchannel and recovering their discoidal shape after passing through the microchannel (Merkel et al., 2011). Compared with RBC-like polymer particles prepared via the PRINT method, the modulus (hundreds of MPa) of LbL assembled capsules with discoidal shape (about 6.7 μm in diameter and 2.8 μm thick) using $Ca(OH)_2$ particles as templates is more than three orders of magnitude higher (She et al., 2013). However, interestingly, these LbL assembled capsules could still pass through a capillary with an inner diameter of 5 μm, which is smaller than the capsule diameter of 6.7 μm (She et al., 2013). This can be attributed to the fact that the hollow capsule interior, like the RBC structure filled with fluid and wrapped by a flexible cell membrane (Palange et al., 2017), facilitates better deformability compared with PRINT hydrogel particles with a matrix interior structure.

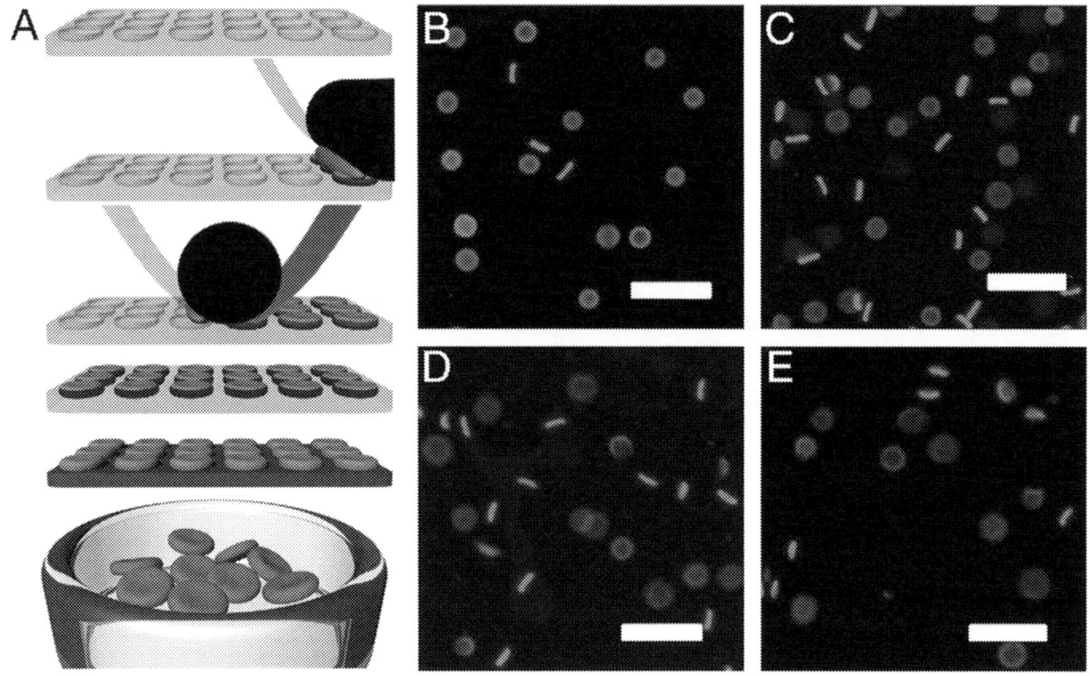

Fig. 6

(A) Schematic illustration of the fabrication of RBC-like polymer particles using the PRINT technique. (B–E) Fluorescence microscopy images of (B) 10% cross-linked, (C) 5% cross-linked, (D) 2% crosslinked, and (E) 1% cross-linked polymer particles. Scale bars are 20 μm. *Reproduced with permission from Merkel, T. J., Jones, S. W., Herlihy, K. P., Kersey, F. R., Shields, A. R., Napier, M., et al. (2011). Using mechanobiological mimicry of red blood cells (RBCs) to extend circulation times of hydrogel microparticles. Proceedings of the National Academy of Sciences of the United States of America, 108, 586.*

Except for the discoidal shape, the PRINT approach could also tune the stiffness of filamentous-shaped polymer particles with different aspect ratios by changing the cross-linking density within the particle matrix. The polymer particles possessed a constant width of 80 nm, and tunable lengths ranging from 180 to 5000 nm. The deformability was assessed by measuring the percentage of particles that could squeeze through a filter with 0.2 μm pores. The results showed that all the particles, regardless of aspect ratio or cross-linking density, had a high recovery efficiency of more than 70%, with the exception of 80 × 5000 nm particles containing 96 wt% PEGDA, which clogged the filtration membrane (Kersey, Merkel, Perry, Napier, & DeSimone, 2012).

3.4 Nanoemulsion Method

Nanoemulsions are typically prepared using two immiscible phases (i.e., water and oil), and commonly exhibit a diameter of up to 500 nm (Anton, Benoit, & Saulnier, 2008). Similar to PRINT particles, the stiffness of polymer particles based on nanoemulsion methods can be controlled by the content of cross-linkers. The Mitragotri group has recently reported a nanoemulsion method for the preparation of polymer particles with tunable modulus (Fig. 7) (Anselmo et al., 2015). The polymer particles composed of PEGDA were synthesized via a water/PEGDA-in-oil nanoemulsion templating method. By varying the volume fraction of PEGDA, the modulus of polymer particles could be controlled over a range of 0.255 to 3000 kPa (Anselmo et al., 2015). Similarly, zwitterionic polymer particles with tunable modulus were prepared by an inverse nanoemulsion polymerization method (Zhang et al., 2012). The modulus of the zwitterionic polymer particles (~250 nm in diameter) was

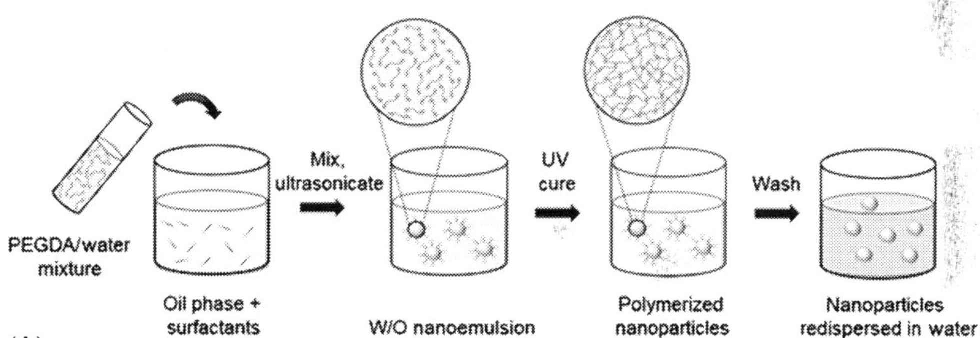

(A)

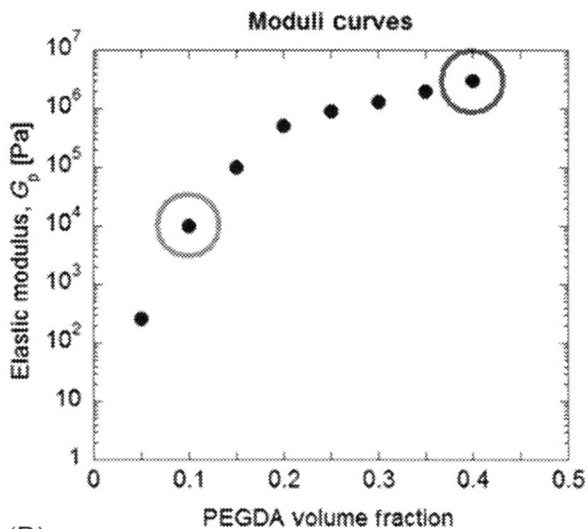

(B)

Parameter	Soft particles	Hard particles
Bulk modulus [kPa]	10	3000
Diameter in Z-average [nm]	209	217
PDI	0.23	0.27
ζ-potential [mV]	−36.8 ± 9.2	−35.3 ± 9.0
Anti-ICAM [μg protein per trillion particles]	3.8 ± 0.8	4.1 ± 0.6

(C)

Fig. 7
(A) Schematic representation of the assembly of polymer particles based on nanoemulsion technique. (B) The stiffness of polymer particles versus PEGDA volume fraction. (C) Comparison of soft and hard polymer particles. *Reproduced with permission from Anselmo, A. C., Zhang, M., Kumar, S., Vogus, D. R., Menegatti, S., Helgeson, M. E., et al. (2015). Elasticity of nanoparticles influences their blood circulation, phagocytosis, endocytosis, and targeting. ACS Nano, 9, 3169.*

controlled from 0.18 to 1.35 MPa by varying the cross-linker content in the range of 2%–15%. Softer particles with a low modulus could pass through filters with pore sizes around 220 nm more easily compared with hard particles. Given the advantage in particle stability and ease for scale up, nanoemulsion methods have been extensively used for fabrication of polymer particles and provide a facile route for adjusting particle stiffness/modulus for biological applications and bio-nano studies.

3.5 Metal-Phenolic Network (MPN) Formation

The formation of MPNs has proven to be an easy and fast way to engineer films and capsules based on the coordination of phenolic ligands and metal ions (Ejima, Richardson, et al., 2013; Guo et al., 2014). Different compositions (e.g., phenolic ligands) influence the stiffness/modulus of the resultant MPN capsules. For example, the modulus of tannic acid TA/Fe^{III} capsules is about 1 GPa (Ejima, Richardson, et al., 2013). When simple phenolic ligands were used for MPN formation, the modulus was decreased, with the modulus of gallic acid (GA)/Fe^{III} and pyrocatechol (PC)/Fe^{III} capsules found to be around 620 and 870 MPa, respectively (Rahim et al., 2015). The difference of modulus induced by changes in composition is probably due to the dynamic nature of the films and the different molecular structures and coordination modes of MPNs during assembly. For LbL assembled capsules, the modulus of capsules typically increases as layer numbers increase. However, the reverse has been observed for some MPN capsule systems. The modulus of $(TA/Fe^{III})_{4.5}$ multistep capsules (420 MPa), which is in the intermediate range for typical LbL assembled capsule systems (10–1000 MPa) (Vinogradova, Lebedeva, & Kim, 2006), was about 40% of the modulus reported for the one-step TA/Fe^{III} capsules, although both capsules had similar shell thicknesses (Rahim et al., 2014). This may be because tris-complex exists as dominant species in TA/Fe^{III} capsules, resulting in higher stiffness compared with multistep capsules composed of bis- and tris-complexes.

MPN films typically form within a finite time (~1 min) and the thickness remains largely constant, regardless of ligand or metal ion used (~10 nm per step). The Caruso group recently reported rust-mediated continuous assembly of MPNs for the formation of capsules (Rahim et al., 2017). Using this method, the shell thickness of GA/Fe^{III} capsules could be finely controlled from 5 to 70 nm by varying the incubation time during the continuous assembly process. Accordingly, the stiffness of GA/Fe^{III} capsules increased from 56 and 166 to 470 mN m^{-1} for the 10, 20, and 30 nm thick capsules, respectively. Although the composition and thickness (~10 nm) of GA/Fe^{III} capsules prepared by conventional one-step assembly and rust-mediated assembly are similar, the modulus of rust-mediated MPN capsules (~330 MPa) is only half compared with the capsules assembled using conventional one-step assembly.

3.6 Microfluidic Platform

Microfluidic platforms have been used to fabricate polymer hybrid particles via rapid reaction and precise manipulation of fluids inside microchannels (Fig. 8) (Sun et al., 2013; Yuan et al., 2012). Using a two-stage microfluidic platform, it can produce core-shell poly(lactic-*co*-glycolic acid) (PLGA)-lipid particles in a single-step method (Sun et al., 2013; Sun, Zhang, et al., 2015). Particle stiffness can be tuned by varying the amounts of interfacial water

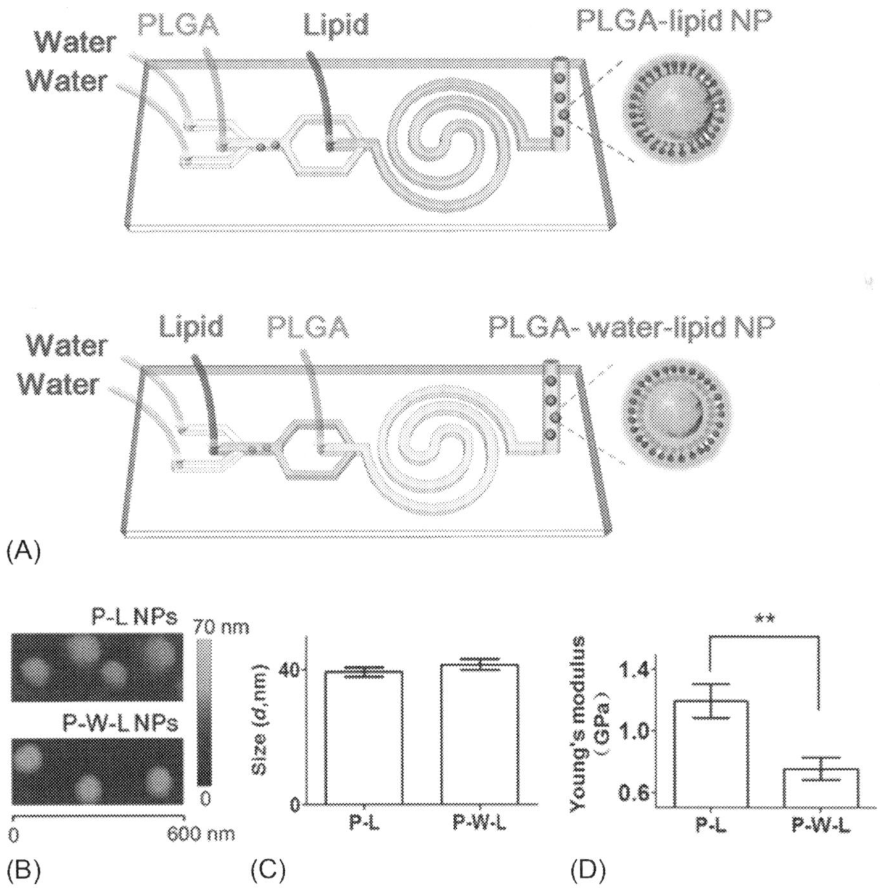

Fig. 8

(A) Scheme of microfluidic platforms for preparing polymer-lipid hybrid particles. (B) AFM images of P-L NPs and P-W-L particles. Size (C) and modulus (D) of P-L particles without interfacial water layer and P-W-L particles with interfacial water layer. *Reproduced with permission from Sun, J., Zhang, L., Wang, J., Feng, Q., Liu, D., Yin, Q., et al. (2015). Tunable rigidity of (polymeric core)-(lipid shell) nanoparticles for regulated cellular uptake. Advanced Materials, 27, 1402.*

between the PLGA core and lipid shell of the hybrid particles, which was achieved by altering the injection order of the PLGA and lipid-PEG organic solutions in the microfluidic device. The different order of injection would subsequently generate PLGA-lipid particles with an interfacial water layer (P-W-L), and ones without interfacial water (P-L). The introduction of interfacial water between the PLGA core and the lipid shell was observed to decrease the E_Y of P-L particles, from 1.20 to 0.76 GPa (P-W-L particles). This difference in modulus was induced by the different structure of the water layer and was found to influence cell association of the resultant particles.

A summary of strategies for the fabrication of polymer particles, the characterization methods for investigating the stiffness of polymer particles, and the influence of various factors on particles stiffness is presented in Table 1.

4 Bio-Nano Interactions of Soft Polymer Particles

Stiffness governs the behavior of many biological processes. For example, the stiffness of substrates and matrices can directly affect the behavior of tissues (Discher, Janmey, & Wang, 2005) and the function and gene activity of cells (Ning Wang & Ingber, 2009; Rehfeldt, Engler, Eckhardt, Ahmed, & Discher, 2010), which can direct cell phenotype and stem cell maturation (Discher, Mooney, & Zandstra, 2009; Engler, Sen, Sweeney, & Discher, 2006; Kloxin, Benton, & Anseth, 2010). Another example is the remarkable deformability of RBCs (Noguchi & Gompper, 2005; Skalak & Branemark, 1969), a property that changes as the cells age, as well as during pathological conditions such as malaria and some anemias (Mokken, Kedaria, Henny, Hardeman, & Gelb, 1992). Macrophages can sense these changes and use that information to target old or diseased RBCs for phagocytosis (Sosale et al., 2015). Stiffness is therefore a key factor to consider when designing synthetic materials for biological applications (Mädler Lutz et al., 2009; Mager, LaPointe, & Stevens, 2011), for example, during the nanoengineering of particles and drug delivery systems (Anselmo & Mitragotri, 2017; Best et al., 2012; Björnmalm et al., 2017; Cui, Richardson, et al., 2016). In this section, we provide an overview of recent results and insights into how stiffness affects bio-nano interactions of soft polymer particles, both in vitro and in vivo.

4.1 In Vitro Studies of Soft Polymer Particles

A range of different studies have been performed in vitro to investigate the "cell-like" behaviors of particles with different stiffness. Microfluidic systems have proven valuable for these efforts, as they can be used to recapitulate aspects of biological environments in a well-controlled setting (Björnmalm, Yan, & Caruso, 2014). For example, microfluidic systems have been used to investigate the deformability of LbL assembled polyelectrolyte capsules (Prevot et al., 2003; She et al., 2013), protein particles prepared using calcium carbonate

Table 1: Methodology for the preparation of polymer particles with tunable composition, morphology, and stiffness

Methods for Particle Engineering	Controllable Factors	Type of Particles	γ or E_Y	Stiffness Measurements	Reference
MS templating	Composition, cross-linking density, structure	PGA particles	E_Y: 2–25 kPa	AFM	(Cui et al., 2013)
		PEG particles	E_Y: 0.2–3.3 kPa γ: 0.3–3 mN m^{-1}	CP-AFM Microfluidic device	(Cui et al., 2014)
		PMA_{SH} particles	E_Y: 0.04–2.53 MPa γ: 1.6–28.4 mN m^{-1}	AFM	(Best, Cui, et al., 2013)
		PMA_{SH} capsules	γ: 0.9–16.9 mN m^{-1}	CP-AFM	(Best, Neubauer, et al., 2013)
LbL assembly	Composition, shell thickness, cross-linking density, morphology	DNA/PAH capsules	E_Y: ~100 MPa	CP-AFM	(Vinogradova et al., 2005)
		DextS/PLArg capsules PSS/PAH capsules	γ: ~10 N m^{-1} γ: ~ 5 N m^{-1}	CP-AFM	(Hartmann et al., 2015)
		PAH/PSS capsules	–	Microfluidic device	(She et al., 2012)
		HA capsules HA particles	γ: ~4.6–13.6 mN m^{-1} γ: ~2.4–21.3 mN m^{-1}	CP-AFM	(Sun, Björnmalm, et al., 2015)
		PSS/$PAH_{8-15kDa}$ capsules PSS/$PAH_{50-65kDa}$ capsules	E_Y: 500 MPa E_Y: 750 MPa	Osmotic Pressure Method	(Gao et al., 2001)
		PMA_{SH} capsules	E_Y: 22–27 MPa	Osmotic Pressure Method	(Chen et al., 2016)
		Spherical TA/PVPON capsules Cubical TA/PVPON capsules	E_Y: 4.30 $\pm$ 0.40 MPa E_Y: 0.61 $\pm$ 0.08 MPa	CP-AFM	(Shchepelina et al., 2012)
		Spherical PSS/PA capsules Dodecahedral PSS/PA capsules	E_Y: 89.1 $\pm$ 24.3 MPa Incompressible	AFM	(Ejima, Yanai, et al., 2013)

Continued

Table 1: Methodology for the preparation of polymer particles with tunable composition, morphology, and stiffness—cont'd

Methods for Particle Engineering	Controllable Factors	Type of Particles	γ or E_Y	Stiffness Measurements	Reference
PRINT	Cross-linking density	RBC-like particles	E_Y: 6.5 kPa, 20 kPa	Tension measurement	(Chen et al., 2012)
		RBC-like particles	E_Y: 7.8–63.9 kPa	Tension measurement	(Merkel et al., 2011)
Nanoemulsion	Cross-linking density	PEG-based particles	E_Y: 0.255 kPa (soft), E_Y: 3 MPa (hard)	Rheometer	(Anselmo et al., 2015)
		pCB particles	E_Y: 0.18–1.35 MPa	Tension measurement	(Zhang et al., 2012)
MPN	Composition, shell thickness, incubation time	TA/Fe^{III} capsules	E_Y: 1.0 ± 0.2 GPa	CP-AFM	(Ejima, Richardson, et al., 2013)
		GA/Fe^{III} capsules PC/Fe^{III} capsules	E_Y: 620 MPa E_Y: 870 MPa	CP-AFM	(Rahim et al., 2015)
		(TA/Fe^{III})$_{4.5}$ capsules	E_Y: 420 ± 30 MPa	CP-AFM	(Rahim et al., 2014)
		Rust-mediated GA/Fe^{III} capsules	γ: 56 ± 11 mN m^{-1}, γ: 166 ± 17 mN m^{-1}, γ: 470 ± 59 mN m^{-1}	CP-AFM	(Rahim et al., 2017)
Microfluidic platform	Interfacial water: without, with	Polymer-lipid NPs	E_Y: 1.20 ± 0.11 GPa E_Y: 0.76 ± 0.07 GPa		(Sun, Zhang, et al., 2015)

γ, stiffness; *EY*, Young's modulus; *PGA*, poly(L-glutamic acid); *PEG*, poly(ethylene glycol); *PMASH*, thiolated poly(methacrylic acid); *PAH*, poly(allylamine hydrochloride); *PDPA*, poly(2-diisopropylaminoethyl methacrylate); *DextS/PLArg*, dextran sulfate sodium salt/poly-L-arginine hydrochloride; *PSS*, poly(sodium 4-styrenesulfonate); *PAH*, poly(allylamine hydrochloride); *PA*, poly(allylamine); *RBCs*, red blood cells; *HA*, hyaluronic acid; *pCB*, poly(carboxybetaine); *TA*, tannic acid; *PVPON*, poly(*N*-vinylpyrrolidone); *GA*, gallic acid; *PC*, pyrocatechol.

templates (Schmidt et al., 2013), and particles of different shapes (prepared through stop-flow lithography) composed of PEGDA (Haghgooie et al., 2010). These types of microfluidic devices have also been used to investigate flow-based behaviors of microgel (PRINT) particles having tunable stiffness, and it was found that by adjusting their stiffness, the flow-based deformability of these types of particles could be made similar to that of human RBCs (Chen et al., 2012; Merkel et al., 2011). Similarly, a microfluidic blood capillary model was used to investigate the flow-based deformability of PEG particles prepared using the MS templating method (Cui et al., 2014), and HA particles prepared via CAP_{ATRP} (Sun, Björnmalm, et al., 2015). These studies showed that the flow-based deformability behavior of soft polymer particles can be tuned to be similar to human RBCs, by adjusting either the cross-linker concentration used during synthesis (Cui et al., 2014), or the internal structure of the resulting particles (Sun, Björnmalm, et al., 2015). While tunable stiffness of particles can be used to prepare soft particles with certain "cell-like" behaviors, it also affects the interactions between cells and particles.

Particles can be internalized by cells through several related, but distinct, pathways, including phagocytosis, clathrin-mediated endocytosis, caveolin-mediated endocytosis, clathrin/caveolae-independent endocytosis, and micropinocytosis (Behzadi et al., 2017). It is important to note that the terms of cellular internalization or cellular uptake are related to the term of cellular association, but they are not equivalent (Johnston, 2017). Several studies have investigated the effect of stiffness of different types of soft polymer particles on cellular internalization. For polymer microparticles (~2–4 μm), slower and less cellular internalization has been observed with increasing particle stiffness (Hartmann et al., 2015; Sun, Wong, et al., 2015), while for polymer nanoparticles (~40 nm) increasing stiffness was observed to increase cellular uptake (Fig. 9) (Sun, Zhang, et al., 2015). In a separate study, hydrogel nanoparticles (~200 nm) with intermediate stiffness were found to have the largest cellular uptake (Banquy et al., 2009). The softer particles were preferentially internalized via macropinocytosis, and more rigid particles were primarily internalized via clathrin-mediated routes, while the particles with intermediate stiffness were found to be internalized via multiple routes, thus providing an explanation for the observed increase in cellular uptake. These effects remain to be fully explored and are a multifaceted problem as particle design parameters such as size, shape, and stiffness (in addition to particle composition and cell type used) all affect biological outcomes through combinatorial effects (Garapaty & Champion, 2017). It can also be challenging to decouple related properties from stiffness. For example, stiffer particles can be prepared by increasing the amount of cross-linker used during synthesis. However, while an increased amount of cross-linker can lead to polymer particles with tunable stiffness, it can also lead to differences in intracellular stability (Liang et al., 2012), or biological effects (e.g., immune cell activation) (Cui et al., 2013), and the exact contribution of each component can therefore be challenging to elucidate. Emerging methods for probing cellular internalization mechanisms of soft polymer particles may aid in these efforts (Chen et al., 2016).

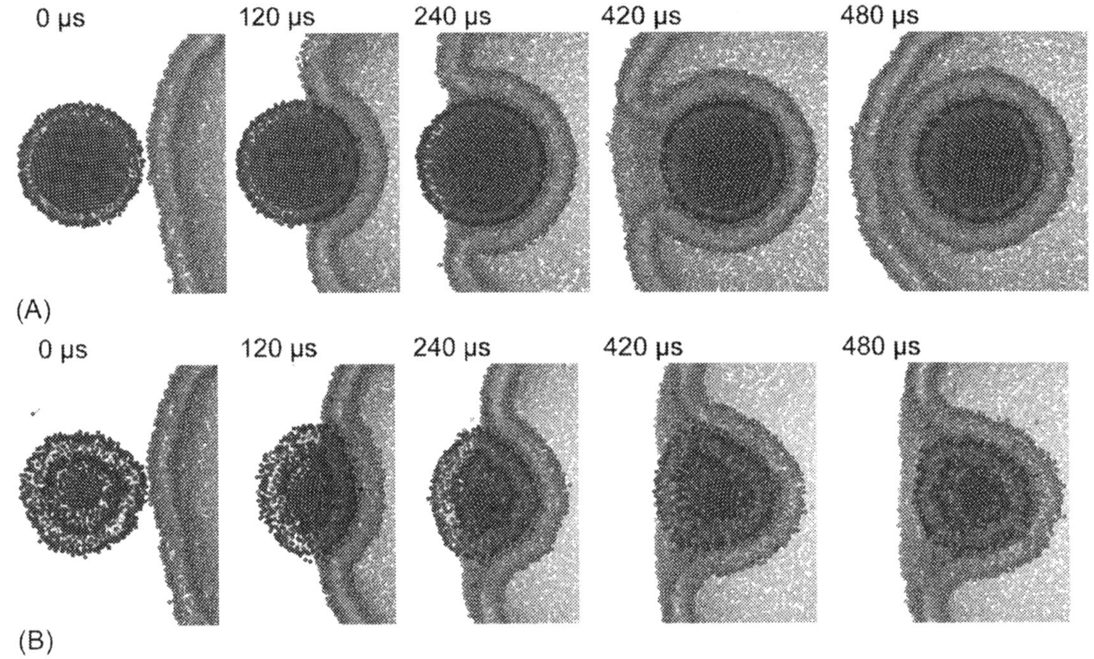

Fig. 9

Molecular dynamic simulation of the internalization of soft and hard lipid particles as a function of time. (A) The stiff particle is internalized via the wrapping of the cellular membrane, showing mild shape deformation. (B) The soft particle undergoes large deformation during the internalization and finally is trapped on the surface of the cell. *Reproduced with permission from Sun, J., Zhang, L., Wang, J., Feng, Q., Liu, D., Yin, Q., et al. (2015). Tunable rigidity of (polymeric core)-(lipid shell) nanoparticles for regulated cellular uptake. Advanced Materials, 27, 1402.*

4.2 In Vivo Studies of Soft Polymer Particles

Designing polymer particles with well-controlled blood circulation behaviors and biodistribution profiles is important for many applications in drug delivery and theranostics. In studies that compare blood circulation times (e.g., distribution and elimination half-lives) of polymer particles with tunable stiffness, softer particles can typically display longer circulation (Anselmo et al., 2015; Cui et al., 2015; Key et al., 2015; Merkel et al., 2011; Müllner et al., 2015; Zhang et al., 2012). This can be explained by the capability of softer particles to more easily avoid blood filtration mechanisms in organs such as the spleen (Mebius & Kraal, 2005). For example, soft PEG particles are more deformable and can pass through microcapillaries much more easily compared with MS@PEG particles in the presence of templates (Cui et al., 2014). It can be used to explain why PEG particles have longer circulation time and much less accumulation in the liver or spleen (Fig. 10) (Cui et al., 2015). Similarly, the incorporation of a crystalline core compartment (e.g., polycaprolactone) into cylindrical polymer brushes leads to more cylindrical and less

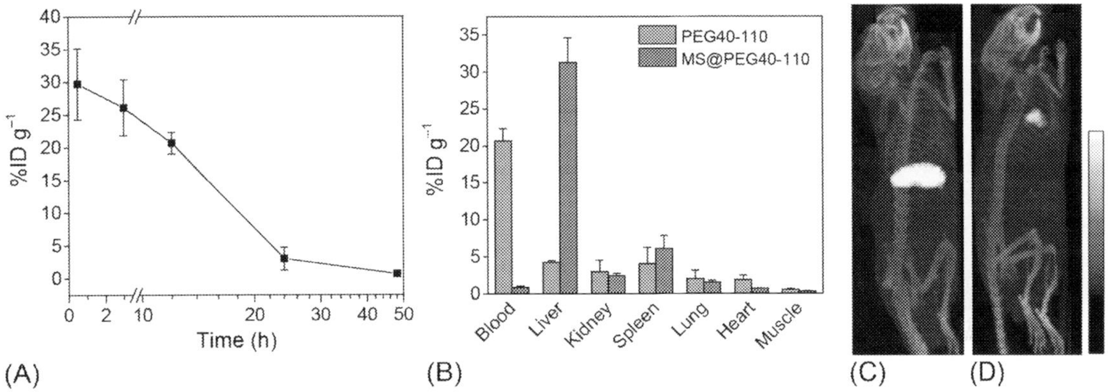

Fig. 10
(A) The presence of PEG particles in blood at different time points. (B) Biodistribution of stiff particles (MS@PEG) and soft particles (PEG) in various organs at 12 h after intravenous injection. PET/CT images of mice 12 h postinjection of (C) MS@PEG and (D) PEG particles. *Reproduced with permission from Cui, J., De Rose, R., Alt, K., Alcantara, S., Paterson, B. M., Liang, K., et al. (2015). Engineering poly(ethylene glycol) particles for improved biodistribution. ACS Nano, 9, 1571.*

randomly coiled structures with increased rigidity when comparing them with a hydrophilic core of poly(hydroxyethylmethacrylate). As a result, the more rigid brushes resulted in more rapid clearance from plasma and lower circulation time. These results suggest that the stiffness of polymer particles plays a significant role in their filtration and clearance (Müllner et al., 2015). Although softer particles exhibit improved biodistribution compared with stiff particles, they still display accumulation in organs such as the spleen and lungs (Zhang et al., 2012).

Taken together, these studies clearly illustrate the challenge in designing soft polymer particles with tailored bio-nano interactions, as predicting biological behavior from particle properties remains difficult. Much work in the field has therefore been of a more empirical nature, that is, preparing a range of particles with diverse properties, investigating bio-nano interactions, and then picking the top performer—instead of designing a specific particle system to achieve an explicit biological outcome. Important factors complicating these issues include the behavior of soft polymer particles under physiological flow conditions, which may change their behavior compared with results observed during conventional static cell culture conditions (Björnmalm, Faria, Chen, Cui, & Caruso, 2016; Braunger et al., 2017; Cui, Faria, et al., 2016; Howard et al., 2014), as well as the impact of biomolecular coronas (Hadjidemetriou & Kostarelos, 2017), which could strongly affect targeting outcomes of polymer particles (Dai et al., 2015). For inorganic particles, recent studies have also shown the importance of differential flow rates, cellular phenotypes, and types of tissue organization present in vivo for determining bio-nano interactions and biological outcomes (MacParland et al., 2017; Tsoi et al., 2016; Yang et al., 2017). These types of studies are yet to be conducted for soft polymer particles, and may prove useful for uncovering and understanding the complexities of bio-nano interactions governing the behavior of soft polymer particles, both in vitro and in vivo.

5 Conclusions and Outlook

In this chapter, we have outlined recent advances in (i) methods for the characterization of particle stiffness, (ii) strategies used for the engineering of soft polymer particles, and (iii) advances in our understanding of the biological behavior of soft polymer particles. For example, soft polymer particles have been shown to be capable of significantly reducing nonspecific interactions and avoiding filtration effects, and can exhibit improved biodistribution and enhanced circulation behavior. This is important for overcoming biological barriers; for example, to achieve enhanced tumor targeting and improved drug delivery and efficacy.

The rational design and engineering of soft polymer particles toward achieving specific biological outcomes is a challenge but important interdisciplinary research topic, which is being accelerated by the ongoing convergence of the chemical sciences, engineering, and biomedicine (Björnmalm, Faria, & Caruso, 2016). Although substantial efforts have gone into the engineering and study of soft polymer particles, and some of these particle systems have already shown promise for biomedical applications, unmet needs still exist. Different physicochemical properties (e.g., surface chemistry) of polymer particles should be taken in account with particle stiffness to improve the efficacy of theranostic delivery. This is important, as it would facilitate the establishment of particle design-biological performance relationships, which would accelerate the development of the next generation of soft polymer particles for biomedical applications.

Acknowledgments

This work was supported by the National Natural Science Foundation of China (21603120) and the Fundamental Research Funds of Shandong University (2017JC003). J.C. acknowledges the Thousand Talents Plan Program of China. M.B. acknowledges support from the European Commission's H2020 programme through a Marie Skłodowska-Curie Individual Fellowship (grant agreement no. 745676).

References

Albanese, A., Tang, P. S., & Chan, W. C. W. (2012). The effect of nanoparticle size, shape, and surface chemistry on biological systems. *Annual Review of Biomedical Engineering*, *14*, 1.

Alexander, J. F., Kozlovskaya, V., Chen, J., Kuncewicz, T., Kharlampieva, E., & Godin, B. (2015). Cubical shape enhances the interaction of layer-by-layer polymeric particles with breast cancer cells. *Advanced Healthcare Materials*, *4*, 2657.

Anselmo, A. C., & Mitragotri, S. (2017). Impact of particle elasticity on particle-based drug delivery systems. *Advanced Drug Delivery Reviews*, *108*, 51.

Anselmo, A. C., Zhang, M., Kumar, S., Vogus, D. R., Menegatti, S., Helgeson, M. E., et al. (2015). Elasticity of nanoparticles influences their blood circulation, phagocytosis, endocytosis, and targeting. *ACS Nano*, *9*, 3169.

Anton, N., Benoit, J.-P., & Saulnier, P. (2008). Design and production of nanoparticles formulated from nano-emulsion templates—a review. *Journal of Controlled Release*, *128*, 185.

Ariga, K., Hill, J. P., & Ji, Q. (2007). Layer-by-layer assembly as a versatile bottom-up nanofabrication technique for exploratory research and realistic application. *Physical Chemistry Chemical Physics*, *9*, 2319.

Ariga, K., Yamauchi, Y., Rydzek, G., Ji, Q., Yonamine, Y., Wu, K.C.-W., et al. (2014). Layer-by-layer nanoarchitectonics: invention, innovation, and evolution. *Chemistry Letters*, *43*, 36.

Banquy, X., Suarez, F., Argaw, A., Rabanel, J.-M., Grutter, P., Bouchard, J.-F., et al. (2009). Effect of mechanical properties of hydrogel nanoparticles on macrophage cell uptake. *Soft Matter*, *5*, 3984.

Behzadi, S., Serpooshan, V., Tao, W., Hamaly, M. A., Alkawareek, M. Y., Dreaden, E. C., et al. (2017). Cellular uptake of nanoparticles: journey inside the cell. *Chemical Society Reviews*, *46*, 4218.

Bermudez, H., Brannan, A. K., Hammer, D. A., Bates, F. S., & Discher, D. E. (2002). Molecular weight dependence of polymersome membrane structure, elasticity, and stability. *Macromolecules*, *35*, 8203.

Berry, J. D., Mettu, S., & Dagastine, R. R. (2017). Precise measurements of capsule mechanical properties using indentation. *Soft Matter*, *13*, 1943.

Best, J. P., Cui, J., Müllner, M., & Caruso, F. (2013). Tuning the mechanical properties of nanoporous hydrogel particles via polymer cross-linking. *Langmuir*, *29*, 9824.

Best, J. P., Javed, S., Richardson, J. J., Cho, K. L., Kamphuis, M. M. J., & Caruso, F. (2013). Stiffness-mediated adhesion of cervical cancer cells to soft hydrogel films. *Soft Matter*, *9*, 4580.

Best, J. P., Neubauer, M. P., Javed, S., Dam, H. H., Fery, A., & Caruso, F. (2013). Mechanics of pH-responsive hydrogel capsules. *Langmuir*, *29*, 9814.

Best, J. P., Yan, Y., & Caruso, F. (2012). The role of particle geometry and mechanics in the biological domain. *Advanced Healthcare Materials*, *1*, 35.

Björnmalm, M., Cui, J., Bertleff-Zieschang, N., Song, D., Faria, M., Rahim, M. A., et al. (2017). Nanoengineering particles through template assembly. *Chemistry of Materials*, *29*, 289.

Björnmalm, M., Faria, M., & Caruso, F. (2016). Increasing the impact of materials in and beyond bio-nano science. *Journal of the American Chemical Society*, *138*, 13449.

Björnmalm, M., Faria, M., Chen, X., Cui, J., & Caruso, F. (2016). Dynamic flow impacts cell-particle interactions: sedimentation and particle shape effects. *Langmuir*, *32*, 10995.

Björnmalm, M., Yan, Y., & Caruso, F. (2014). Engineering and evaluating drug delivery particles in microfluidic devices. *Journal of Controlled Release*, *190*, 139.

Blanco, E., Shen, H., & Ferrari, M. (2015). Principles of nanoparticle design for overcoming biological barriers to drug delivery. *Nature Biotechnology*, *33*, 941.

Bolshakova, A. V., Kiselyova, O. I., & Yaminsky, I. V. (2004). Microbial surfaces investigated using atomic force microscopy. *Biotechnology Progress*, *20*, 1615.

Braunger, J. A., Björnmalm, M., Isles, N. A., Cui, J., Henderson, T. M. A., O'Connor, A. J., et al. (2017). Interactions between circulating nanoengineered polymer particles and extracellular matrix components in vitro. *Biomaterials Science*, *5*, 267.

Chen, H., Zhang, W., Zhu, G., Xie, J., & Chen, X. (2017). Rethinking cancer nanotheranostics. *Nature Reviews Materials*, *2*, 17024.

Chen, K., Merkel, T. J., Pandya, A., Napier, M. E., Luft, J. C., Daniel, W., et al. (2012). Low modulus biomimetic microgel particles with high loading of hemoglobin. *Biomacromolecules*, *13*, 2748.

Chen, X., Cui, J., Ping, Y., Suma, T., Cavalieri, F., Besford, Q. A., et al. (2016). Probing cell internalisation mechanics with polymer capsules. *Nanoscale*, *8*, 17096.

Cui, J. (2017). Probing bio-nano interactions with templated polymer particles. *Chem*, *2*, 606.

Cui, J., Björnmalm, M., Liang, K., Xu, C., Best, J. P., Zhang, X., et al. (2014). Super-soft hydrogel particles with tunable elasticity in a microfluidic blood capillary model. *Advanced Materials*, *26*, 7295.

Cui, J., De Rose, R., Alt, K., Alcantara, S., Paterson, B. M., Liang, K., et al. (2015). Engineering poly(ethylene glycol) particles for improved biodistribution. *ACS Nano*, *9*, 1571.

Cui, J., De Rose, R., Best, J. P., Johnston, A. P. R., Alcantara, S., Liang, K., et al. (2013). Mechanically tunable, self-adjuvanting nanoengineered polypeptide particles. *Advanced Materials*, *25*, 3468.

Cui, J., Faria, M., Björnmalm, M., Ju, Y., Suma, T., Gunawan, S. T., et al. (2016). A framework to account for sedimentation and diffusion in particle-cell interactions. *Langmuir*, *32*, 12394.

Cui, J., Liu, Y., & Hao, J. (2009). Multiwalled carbon-nanotube-embedded microcapsules and their electrochemical behavior. *Journal of Physical Chemistry C*, *113*, 3967.

Cui, J., Richardson, J. J., Björnmalm, M., Faria, M., & Caruso, F. (2016). Nanoengineered templated polymer particles: navigating the biological realm. *Accounts of Chemical Research*, *49*, 1139.

Cui, J., Yan, Y., Wang, Y., & Caruso, F. (2012). Templated assembly of pH-labile polymer-drug particles for intracellular drug delivery. *Advanced Functional Materials*, *22*, 4718.

Dai, Q., Yan, Y., Guo, J., Björnmalm, M., Cui, J., Sun, H., et al. (2015). Targeting ability of affibody-functionalized particles is enhanced by albumin but inhibited by serum coronas. *ACS Macro Letters*, *4*, 1259.

Discher, D. E., Janmey, P., & Wang, Y.-l. (2005). Tissue cells feel and respond to the stiffness of their substrate. *Science*, *310*, 1139.

Discher, D. E., Mooney, D. J., & Zandstra, P. W. (2009). Growth factors, matrices, and forces combine and control stem cells. *Science*, *324*, 1673.

Donath, E., Sukhorukov, G. B., Caruso, F., Davis, S. A., & Möhwald, H. (1998). Novel hollow polymer shells by colloid-templated assembly of polyelectrolytes. *Angewandte Chemie, International Edition*, *37*, 2201.

Dubreuil, F., Shchukin, D. G., Sukhorukov, G. B., & Fery, A. (2004). Polyelectrolyte capsules modified with YF_3 nanoparticles: an AFM study. *Macromolecular Rapid Communications*, *25*, 1078.

Ducker, W. A., Senden, T. J., & Pashley, R. M. (1991). Direct measurement of colloidal forces using an atomic force microscope. *Nature*, *353*, 239.

Ejima, H., Richardson, J. J., Liang, K., Best, J. P., van Koeverden, M. P., Such, G. K., et al. (2013). One-step assembly of coordination complexes for versatile film and particle engineering. *Science*, *341*, 154.

Ejima, H., Yanai, N., Best, J. P., Sindoro, M., Granick, S., & Caruso, F. (2013). Near-incompressible faceted polymer microcapsules from metal-organic framework templates. *Advanced Materials*, *25*, 5767.

Engler, A. J., Sen, S., Sweeney, H. L., & Discher, D. E. (2006). Matrix elasticity directs stem cell lineage specification. *Cell*, *126*, 677.

Farokhzad, O. C., & Langer, R. (2009). Impact of nanotechnology on drug delivery. *ACS Nano*, *3*, 16.

Fox, M. E., Szoka, F. C., & Fréchet, J. M. J. (2009). Soluble polymer carriers for the treatment of cancer: the importance of molecular architecture. *Accounts of Chemical Research*, *42*, 1141.

Gao, C., Donath, E., Moya, S., Dudnik, V., & Möhwald, H. (2001). Elasticity of hollow polyelectrolyte capsules prepared by the layer-by-layer technique. *European Physical Journal E: Soft Matter and Biological Physics*, *5*, 21.

Garapaty, A., & Champion, J. A. (2017). Tunable particles alter macrophage uptake based on combinatorial effects of physical properties. *Bioengineering & Translational Medicine*, *2*, 92.

Gause, K. T., Yan, Y., O'Brien-Simpson, N. M., Cui, J., Lenzo, J. C., Reynolds, E. C., et al. (2016). Codelivery of nod2 and tlr9 ligands via nanoengineered protein antigen particles for improving and tuning immune responses. *Advanced Functional Materials*, *26*, 7526.

Gregory, R. D., Milac, T. I., & Wan, F. Y. M. (1998). A thick hollow sphere compressed by equal and opposite concentrated axial loads: an asymptotic solution. *SIAM Journal on Applied Mathematics*, *59*, 1080.

Guo, J., Ping, Y., Ejima, H., Alt, K., Meissner, M., Richardson, J. J., et al. (2014). Engineering multifunctional capsules through the assembly of metal-phenolic networks. *Angewandte Chemie, International Edition*, *53*, 5546.

Hadjidemetriou, M., & Kostarelos, K. (2017). Nanomedicine: evolution of the nanoparticle corona. *Nature Nanotechnology*, *12*, 288.

Haghgooie, R., Toner, M., & Doyle, P. S. (2010). Squishy non-spherical hydrogel microparticles. *Macromolecular Rapid Communications*, *31*, 128.

Hartmann, R., Weidenbach, M., Neubauer, M., Fery, A., & Parak, W. J. (2015). Stiffness-dependent in vitro uptake and lysosomal acidification of colloidal particles. *Angewandte Chemie, International Edition*, *54*, 1365.

Hertz, H. (1882). On the contact of elastic solids. *Journal fur die Reine und Angewandte Mathematik*, *92*, 156.

Howard, M., Zern, B. J., Anselmo, A. C., Shuvaev, V. V., Mitragotri, S., & Muzykantov, V. (2014). Vascular targeting of nanocarriers: perplexing aspects of the seemingly straightforward paradigm. *ACS Nano*, *8*, 4100.

Johnston, A. P. R. (2017). Life under the microscope: quantifying live cell interactions to improve nanoscale drug delivery. *ACS Sensors*, *2*, 4.

Kersey, F. R., Merkel, T. J., Perry, J. L., Napier, M. E., & DeSimone, J. M. (2012). Effect of aspect ratio and deformability on nanoparticle extravasation through nanopores. *Langmuir*, *28*, 8773.

Key, J., Palange, A. L., Gentile, F., Aryal, S., Stigliano, C., Di Mascolo, D., et al. (2015). Soft discoidal polymeric nanoconstructs resist macrophage uptake and enhance vascular targeting in tumors. *ACS Nano*, *9*, 11628.

Kloxin, A. M., Benton, J. A., & Anseth, K. S. (2010). In situ elasticity modulation with dynamic substrates to direct cell phenotype. *Biomaterials*, *31*, 1.

Lebedeva, O. V., Kim, B.-S., & Vinogradova, O. I. (2004). Mechanical properties of polyelectrolyte-filled multilayer microcapsules studied by atomic force and confocal microscopy. *Langmuir*, *20*, 10685.

Liang, K., Such, G. K., Zhu, Z., Dodds, S. J., Johnston, A. P. R., Cui, J., et al. (2012). Engineering cellular degradation of multilayered capsules through controlled cross-linking. *ACS Nano*, *6*, 10186.

Lulevich, V. V., Radtchenko, I. L., Sukhorukov, G. B., & Vinogradova, O. I. (2003). Deformation properties of nonadhesive polyelectrolyte microcapsules studied with the atomic force microscope. *The Journal of Physical Chemistry B*, *107*, 2735.

Lulevich, V. V., & Vinogradova, O. I. (2004). Effect of pH and salt on the stiffness of polyelectrolyte multilayer microcapsules. *Langmuir*, *20*, 2874.

MacParland, S. A., Tsoi, K. M., Ouyang, B., Ma, X.-Z., Manuel, J., Fawaz, A., et al. (2017). Phenotype determines nanoparticle uptake by human macrophages from liver and blood. *ACS Nano*, *11*, 2428.

Mädler Lutz, N. A. E., Velegol, D., Xia, T., Hoek, E. M. V., Somasundaran, P., Klaessig, F., et al. (2009). Understanding biophysicochemical interactions at the nano-bio interface. *Nature Materials*, *8*, 543.

Mager, M. D., LaPointe, V., & Stevens, M. M. (2011). Exploring and exploiting chemistry at the cell surface. *Nature Chemistry*, *3*, 582.

Markutsya, S., Jiang, C., Pikus, Y., & Tsukruk, V. V. (2005). Freely suspended layer by-layer nanomembranes: testing micromechanical properties. *Advanced Functional Materials*, *15*, 771.

Mebius, R. E., & Kraal, G. (2005). Structure and function of the spleen. *Nature Reviews. Immunology*, *5*, 606.

Merkel, T. J., Jones, S. W., Herlihy, K. P., Kersey, F. R., Shields, A. R., Napier, M., et al. (2011). Using mechanobiological mimicry of red blood cells to extend circulation times of hydrogel microparticles. *Proceedings of the National Academy of Sciences of the United States of America*, *108*, 586.

Mitragotri, S., Burke, P. A., & Langer, R. (2014). Overcoming the challenges in administering biopharmaceuticals: formulation and delivery strategies. *Nature Reviews Drug Discovery*, *13*, 655.

Mokken, E. C., Kedaria, M., Henny, C. P., Hardeman, M. R., & Gelb, A. W. (1992). The clinical importance of erythrocyte deformability, a hemorrheological parameter. *Annals of Hematology*, *64*, 113.

Müllner, M., Dodds, S. J., Nguyen, T.-H., Senyschyn, D., Porter, C. J. H., Boyd, B. J., et al. (2015). Size and rigidity of cylindrical polymer brushes dictate long circulating properties in vivo. *ACS Nano*, *9*, 1294.

Ning Wang, J. D. T., & Ingber, D. E. (2009). Mechanotransduction at a distance: mechanically coupling the extracellular matrix with the nucleus. *Nature Reviews. Molecular Cell Biology*, *10*, 75.

Noguchi, H., & Gompper, G. (2005). Shape transitions of fluid vesicles and red blood cells in capillary flows. *Proceedings of the National Academy of Sciences of the United States of America*, *102*, 14159.

Palange, A. L., Palomba, R., Rizzuti, I. F., Ferreira, M., & Decuzzi, P. (2017). Deformable discoidal polymeric nanoconstructs for the precise delivery of therapeutic and imaging agents. *Molecular Therapy*, *25*, 1514.

Palankar, R., Pinchasik, B.-E., Schmidt, S., De Geest, B. G., Fery, A., Möhwald, H., et al. (2013). Mechanical strength and intracellular uptake of CaCO3-templated LbL capsules composed of biodegradable polyelectrolytes: the influence of the number of layers. *Journal of Materials Chemistry B*, *1*, 1175.

Petros, R. A., & DeSimone, J. M. (2010). Strategies in the design of nanoparticles for therapeutic applications. *Nature Reviews Drug Discovery*, *9*, 615.

Prevot, M., Cordeiro, A. L., Sukhorukov, G. B., Lvov, Y., Besser, R. S., & Möhwald, H. (2003). Design of a microfluidic system to investigate the mechanical properties of layer-by-layer fabricated capsules. *Macromolecular Materials and Engineering*, *288*, 915.

Rahim, M. A., Björnmalm, M., Bertleff-Zieschang, N., Besford, Q., Mettu, S., Suma, T., et al. (2017). Rust-mediated continuous assembly of metal-phenolic networks. *Advanced Materials*, *29*, 1606717.

Rahim, M. A., Ejima, H., Cho, K. L., Kempe, K., Müllner, M., Best, J. P., et al. (2014). Coordination-driven multistep assembly of metal-polyphenol films and capsules. *Chemistry of Materials*, *26*, 1645.

Rahim, M. A., Kempe, K., Müllner, M., Ejima, H., Ju, Y., van Koeverden, M. P., et al. (2015). Surface-confined amorphous films from metal-coordinated simple phenolic ligands. *Chemistry of Materials*, *27*, 5825.

Rehfeldt, F., Engler, A. J., Eckhardt, A., Ahmed, F., & Discher, D. E. (2010). Cell responses to the mechanochemical microenvironment-implications for regenerative medicine and drug delivery. *Biomaterials*, *31*, 1.

Reissner, E. (1946). Stresses and small displacements of shallow spherical shells. *Studies in Applied Mathematics*, *25*, 279.

Richardson, J. J., Björnmalm, M., & Caruso, F. (2015). Technology-driven layer-by-layer assembly of nanofilms. *Science*, *348*, aaa2491.

Richardson, J. J., Cui, J., Björnmalm, M., Braunger, J. A., Ejima, H., & Caruso, F. (2016). Innovation in layer-by-layer assembly. *Chemical Reviews*, *116*, 14828.

Richardson, J. J., Tardy, B. L., Ejima, H., Guo, J., Cui, J., Liang, K., et al. (2016). Thermally induced charge reversal of layer-by-layer assembled single-component polymer films. *ACS Applied Materials & Interfaces*, *8*, 7449.

Schmidt, S., Behra, M., Uhlig, K., Madaboosi, N., Hartmann, L., Duschl, C., et al. (2013). Mesoporous protein particles through colloidal $CaCO_3$ templates. *Advanced Functional Materials*, *23*, 116.

Schmidt, S., Zeiser, M., Hellweg, T., Duschl, C., Fery, A., & Möhwald, H. (2010). Adhesion and mechanical properties of PNIPAM microgel films and their potential use as switchable cell culture substrates. *Advanced Functional Materials*, *20*, 3235.

Shchepelina, O., Lisunova, M. O., Drachuk, I., & Tsukruk, V. V. (2012). Morphology and properties of microcapsules with different core releases. *Chemistry of Materials*, *24*, 1245.

She, S., Li, Q., Shan, B., Tong, W., & Gao, C. (2013). Fabrication of red-blood-cell-like polyelectrolyte microcapsules and their deformation and recovery behavior through a microcapillary. *Advanced Materials*, *25*, 5814.

She, S., Xu, C., Yin, X., Tong, W., & Gao, C. (2012). Shape deformation and recovery of multilayer microcapsules after being squeezed through a microchannel. *Langmuir*, *28*, 5010.

Shi, J., Kantoff, P. W., Wooster, R., & Farokhzad, O. C. (2017). Cancer nanomedicine: progress, challenges and opportunities. *Nature Reviews. Cancer*, *17*, 20.

Skalak, R., & Branemark, P. I. (1969). Deformation of red blood cells in capillaries. *Science*, *164*, 717.

Sosale, N. G., Rouhiparkouhi, T., Bradshaw, A. M., Dimova, R., Lipowsky, R., & Discher, D. E. (2015). Cell rigidity and shape override CD47's "self"-signaling in phagocytosis by hyperactivating myosin-II. *Blood*, *125*, 542.

Suma, T., Cui, J., Müllner, M., Fu, S., Tran, J., Noi, K. F., et al. (2017). Modulated fragmentation of proapoptotic peptide nanoparticles regulates cytotoxicity. *Journal of the American Chemical Society*, *139*, 4009.

Sun, H., Björnmalm, M., Cui, J., Wong, E. H. H., Dai, Y., Dai, Q., et al. (2015). Structure governs the deformability of polymer particles in a microfluidic blood capillary model. *ACS Macro Letters*, *4*, 1205.

Sun, H., Cui, J., Ju, Y., Chen, X., Wong, E. H. H., Tran, J., et al. (2017). Tuning the properties of polymer capsules for cellular interactions. *Bioconjugate Chemistry*, *28*, 1859.

Sun, H., Wong, E. H. H., Yan, Y., Cui, J., Dai, Q., Guo, J., et al. (2015). The role of capsule stiffness on cellular processing. *Chemical Science*, *6*, 3505.

Sun, J., Xianyu, Y., Li, M., Liu, W., Zhang, L., Liu, D., et al. (2013). A microfluidic origami chip for synthesis of functionalized polymeric nanoparticles. *Nanoscale*, *5*, 5262.

Sun, J., Zhang, L., Wang, J., Feng, Q., Liu, D., Yin, Q., et al. (2015). Tunable rigidity of (polymeric core)-(lipid shell) nanoparticles for regulated cellular uptake. *Advanced Materials*, *27*, 1402.

Tsoi, K. M., MacParland, S. A., Ma, X.-Z., Spetzler, V. N., Echeverri, J., Ouyang, B., et al. (2016). Mechanism of hard-nanomaterial clearance by the liver. *Nature Materials*, *15*, 1212.

Vinogradova, O. I., Lebedeva, O. V., & Kim, B.-S. (2006). Mechanical behavior and characterization of microcapsules. *Annual Review of Materials Research*, *36*, 143.

Vinogradova, O. I., Lebedeva, O. V., Vasilev, K., Gong, H., Garcia-Turiel, J., & Kim, B.-S. (2005). Multilayer DNA/poly(allylamine hydrochloride) microcapsules: assembly and mechanical properties. *Biomacromolecules*, *6*, 1495.

Wang, Y., Angelatos, A. S., Dunstan, D. E., & Caruso, F. (2007). Infiltration of macromolecules into nanoporous silica particles. *Macromolecules*, *40*, 7594.

Wang, Y., Yu, A., & Caruso, F. (2005). Nanoporous polyelectrolyte spheres prepared by sequentially coating sacrificial mesoporous silica spheres. *Angewandte Chemie, International Edition*, *44*, 2888.

Wong, E. H. H., van Koeverden, M. P., Nam, E., Guntari, S. N., Wibowo, S. H., Blencowe, A., et al. (2013). Assembly of nanostructured films with hydrophobic subcompartments via continuous assembly of polymers. *Macromolecules*, *46*, 7789.

Xu, J., Wong, D. H. C., Byrne, J. D., Chen, K., Bowerman, C., & DeSimone, J. M. (2013). Future of the particle replication in nonwetting templates (print) technology. *Angewandte Chemie, International Edition*, *52*, 6580.

Yan, Y., Björnmalm, M., & Caruso, F. (2014). Assembly of layer-by-layer particles and their interactions with biological systems. *Chemistry of Materials*, *26*, 452.

Yang, Y.-S.S., Atukorale, P. U., Moynihan, K. D., Bekdemir, A., Rakhra, K., Tang, L., et al. (2017). High-throughput quantitation of inorganic nanoparticle biodistribution at the single-cell level using mass cytometry. *Nature Communications*, *8*, 14069.

Yuan, B., Jin, Y., Sun, Y., Wang, D., Sun, J., Wang, Z., et al. (2012). A strategy for depositing different types of cells in three dimensions to mimic tubular structures in tissues. *Advanced Materials*, *24*, 890.

Zelikin, A. N., Quinn, J. F., & Caruso, F. (2006). Disulfide cross-linked polymer capsules: en route to biodeconstructible systems. *Biomacromolecules*, *7*, 27.

Zhang, L., Cao, Z., Li, Y., Ella-Menye, J.-R., Bai, T., & Jiang, S. (2012). Softer zwitterionic nanogels for longer circulation and lower splenic accumulation. *ACS Nano*, *6*, 6681.

CHAPTER 14

Inorganic Nanotheranostic Platforms for Rapid and Reliable Molecular Profiling of Diseases

Davide A. Martella, Mary Okesola, Ciro Chiappini
Centre for Craniofacial and Regenerative Biology, King's College London, London, United Kingdom

1 Introduction

The prospect of integrating therapeutic, imaging, and molecular recognition functionalities promises to improve the efficacy of treatments by combining an understanding of the molecular characteristics of diseases with a targeted therapeutic selection that can be monitored in real-time. This vision can significantly advance the prospects of precision medicine approaches, but it cannot be realized using molecular agents, as building multiple functionality on them is often simply not possible, or occurs at the expense of other functionalities, and may require synthetic processes beyond what is economically feasible. Nanomaterials, on the other hand, can be more easily integrated with multiple molecular agents, functionalized with active moieties, and assembled into suprastructures, in order to bestow multiple concurrent functionalities across therapeutics and diagnostics. At the simplest level, nanomaterials can serve as carriers for multiple agents, each bringing an individual functionality, insuring that they are delivered as a single entity in known relative concentrations. Yet, in more advanced applications, they themselves incorporate functional elements that are expanded and extended by the incorporation of complementary bioactive agents. In this chapter we will review the progress made toward clinical theranostic application by four widely employed inorganic nanomaterials: iron oxide, gold, silica, and porous silicon (Fig. 1).

2 Iron Oxide Nanomaterials

2.1 Synthesis

Iron oxide nanoparticles can be synthesized through several physical, chemical, and even biological routes, making them a versatile nanomaterial to obtain, with the possibility to control their properties over broad ranges in order to optimize their theranostic

Handbook of Nanomaterials for Cancer Theranostics. https://doi.org/10.1016/B978-0-12-813339-2.00014-1

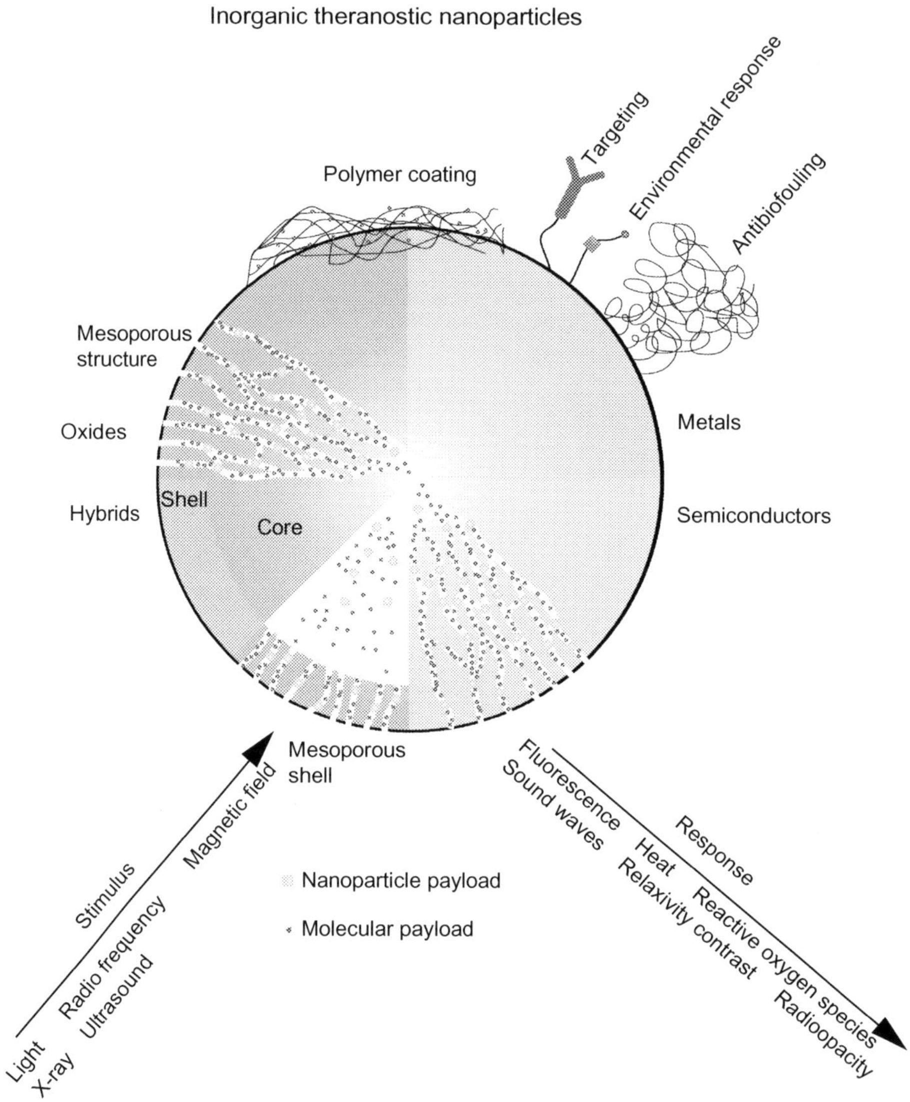

Fig. 1
Schematic representation of key characteristics and functionalities of inorganic theranostic platforms.

functionality (Fig. 2). Co-precipitation is a largely adopted chemical method to obtain superparamagnetic iron oxide nanoparticles (SPIONs) in aqueous solutions from mixtures of ferrous and ferric salts in alkaline conditions, and at mild temperatures (up to 90°C), in the absence of oxygen. Controlling the synthesis parameters, such as the type, concentration, and rate of addition of the base, the duration of the various phases of the

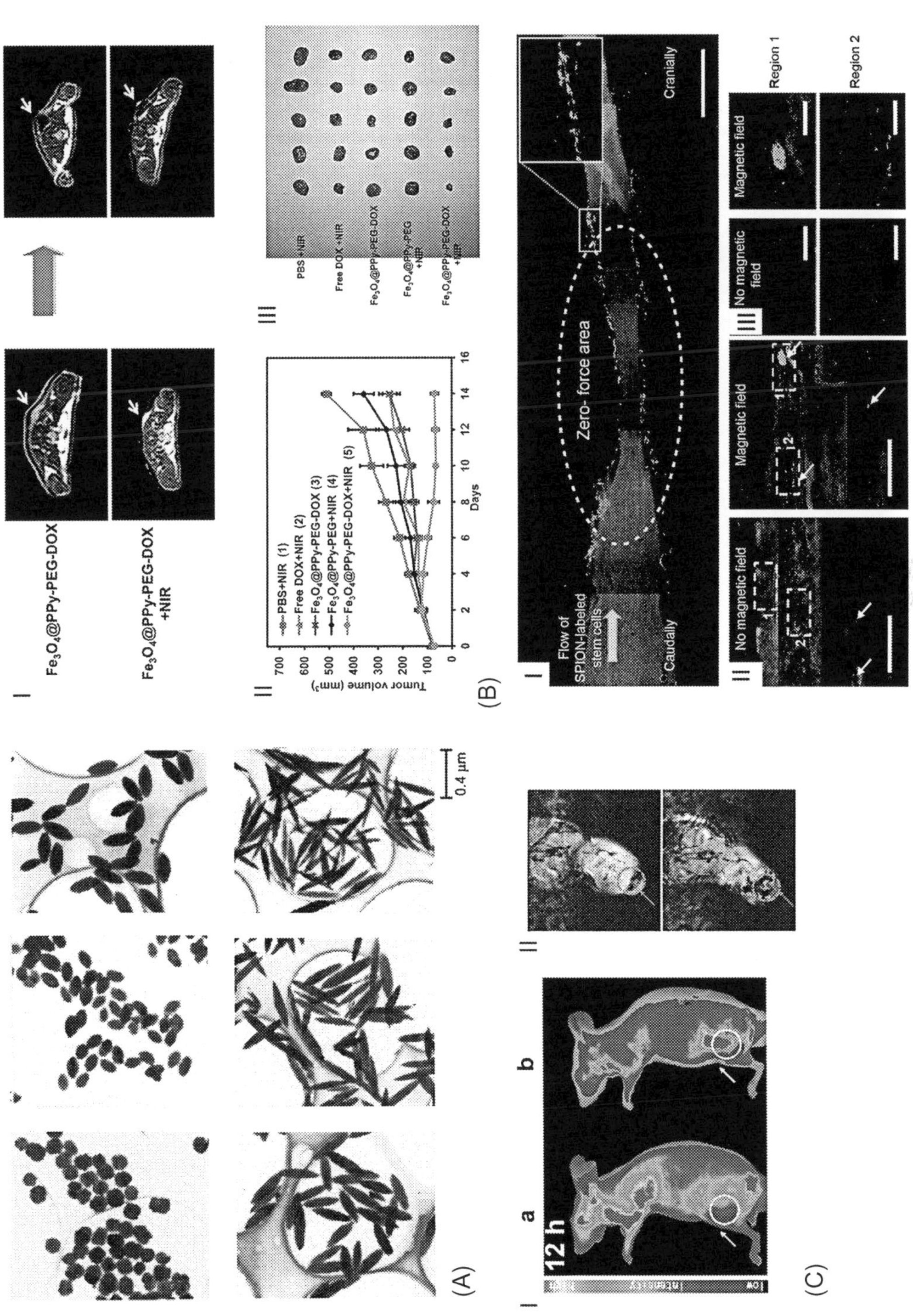

See the legend on next page.

reaction, the use and nature of surfactants, and the ratio of ferrous precursors influences the shape, and in turn, the magnetic properties of the resulting SPIONs (Thanh, Maclean, & Mahiddine, 2014). By controlling the rate of base addition, it is possible to determine whether lance-shaped or spherical particles are formed. Further, by tuning the ratio of ferrous precursors, it is possible to control the magnetic properties of the resulting material, between ferromagnetic and superparamagnetic (Vikram et al., 2015). Yet, this synthetic method yields a low crystallinity and a broad particle size distribution, which significantly limits their saturation magnetization.

Thermal decomposition in organic solvents at high temperatures with the use of surfactants can yield highly monodisperse and crystalline SPIONs, with a fine control over their size. Yet, this synthetic route employs harsh chemicals and synthetic conditions and yields highly hydrophobic particles that require further functionalization to be dispersed in aqueous solutions (Maity, Ding, & Xue, 2008). The strategies for aqueous dispersion involve either ligand substitution or the use of surfactants. These methods tends to be labor intensive, change the magnetic properties of the SPIONs, form aggregates, and reduce their dispersibility, as well as have low yield (Liu et al., 2014). An interesting approach is to exploit the high affinity of dopamine toward SPIONs to coat them with dopamine-PEG conjugates, making them simultaneously water-soluble and antifouling (Xu et al., 2008). By the hydrothermal method, it is possible to obtain

Fig. 2

Superparamagnetic iron oxide nanoparticles for theranostics. (A) Different morphologies of nanostructures ranging from spherical nanoparticle can be obtained by varying the synthetic parameters. (B) Nanoparticles with iron oxide core and polypyrrole shell, loaded with doxorubicin and functionalized with PEG for antibiofouling can visualize tumors by MRI (I) and mediate tumor burden reduction when exposed to near-infrared light that enhances cellular uptake and drug release (II and III). (C) SPIONs coated with chitosan and loaded with a NIR dye and the methotraxate chemotherapy drug are a theranostic nanoagent that can be tracked both by NIR live imaging (I) and MRI (II). (D) Magnetically-guided SPION-labeled stem cells injected in the caudal region of the spine migrate across spinal lesions (I) to localize at sites of intense magnetic field (II and III). *Modified from (A) Itoh, H., & Sugimoto, T. (2003). Systematic control of size, shape, structure, and magnetic properties of uniform magnetite and maghemite particles. Journal of Colloid and Interface Science, 265, 283–295. Copyright (2003) Elsevier; (B) Wang, C., Xu, H., Liang, C., Liu, Y., Li, Z., Yang, G., et al. (2013). Iron oxide @ polypyrrole nanoparticles as a multifunctional drug carrier for remotely controlled cancer therapy with synergistic antitumor effect. ACS Nano, 7, 6782–6795. Copyright (2013) American Chemical Society; (C) Lin, J., Li, Y., Li, Y., Wu, H., Yu, F., Zhou, S., et al. (2015). Drug/dye-loaded, multifunctional PEG–chitosan–iron oxide nanocomposites for methotraxate synergistically self-targeted cancer therapy and dual model imaging. ACS Applied Materials & Interfaces, 7, 11908–11920. Copyright (2015) American Chemical Society; (D) Tukmachev, D., Lunov, O., Zablotskii, V., Dejneka, A., Babic, M., Syková, E., et al. (2015). An effective strategy of magnetic stem cell delivery for spinal cord injury therapy. Nanoscale, 7, 3954–3958. Copyright (2015) Royal Society of Chemistry.*

particles with a defined size from an aqueous solution, in the presence of surfactants, by applying high temperatures and pressure to a polytetrafluorethylene sealed container. The synthesis can control the shape of the particles (rod, sphere, flower-like) and the value of their saturation magnetization (Ramesh, Rajalakshmi, Muthamizhchelvan, & Ponnusamy, 2012). Using microemulsion methods to synthesize SPIONs allows controlling size, and obtaining small-size SPIONs in mild environmental conditions (Okoli, Boutonnet, Mariey, Järås, & Rajarao, 2011). Yet, the removal of the unreacted reagents and surfactants is highly labor intensive. Synthetic routes using external energy sources such as ultrasound and microwaves also yield SPIONs, where particle properties can be controlled by the density of energy applied and its duration (Carenza et al., 2014; Dolores, Raquel, & Adianez, 2015).

The synthesis of SPIONs using biological systems promises to be more environmentally friendly and cost-effective than the traditional chemical routes. Using plant extracts to synthesize SPIONs yields amorphous particles, whose stability and monodispersity is limited by the propensity of reduced iron to re-oxidize in solution (Makarov et al., 2014). Bacteria can also produce iron oxide nanoparticles extracellularly, both in aerobic and anaerobic conditions, where the properties of the resulting materials can be controlled by the choice of precursors fed to the bacteria (Bharde et al., 2008). The sol-gel route is also viable to synthesize SPIONS with defined geometry and magnetic properties, including magnetite and meghemite (Itoh & Sugimoto, 2003).

2.2 Magnetic Properties

SPIONs possess magnetic moment due to the unpaired 3d electrons in the Fe cations present within their lattice. In the absence of an externally applied field, the singe phase magnetic domain established within the nanoparticles randomly orient due to thermal energy, while the presence of an external field allows them to align and exhibit a collective magnetic moment, giving rise to their superparamagnetic behavior. The size of the particle controls the magnetic moments of SPIONs with smaller size, reducing them. Defects at the particle surface reduce magnetic moments there, an effect known as surface spin canting, and particles with higher radii of curvature experience a more pronounced canting (Kim, Lee, et al., 2011). Similarly, defects within the particle bulk give rise to volume canting. Volume canting appears to rise in particles up to 11 nm in size and then decline rapidly (Baaziz et al., 2014). The nature of the capping agents and ligands present on the particle surface also influence their magnetic moment. Both the presence of catechols and the nature of the functional group used within the surface ligands can affect the magnetic moment (Darbandi et al., 2012; Yuen, Hutton, Masters, & Maschmeyer, 2012). Coating SPIONs with Au shells improves their magnetization by reducing magnetic disorder and improving crystallinity, while coating with Au nanostars reduces it (Quaresma et al., 2014). Shape also influences the magnetization of SPIONs, with elliptical nanoparticles displaying increased surface canting, and thus reduced magnetization compared with spherical counterparts (Choi, Cha, & Lee, 2013), while quasicubic and octapod-shape particles have an enhanced magnetization (Wortmann et al., 2014; Zhao et al., 2013).

2.3 Theranostic Applications

As with all nanomaterials, the toxicity of SPIONs is primarily regulated by their surface chemistry, and thus largely influenced by both the synthetic strategy and by the surface functionalizations employed for their biological applications. Yet, broadly discussing, SPIONs can be readily formulated as highly cytocompatible materials that can be well tolerated, even at high concentrations (Jeng & Swanson, 2007; Karlsson, Gustafsson, Cronholm, & Möller, 2009; Kim et al., 2006). Even studies in humans have identified that the side effects of SPION administration intravenously are uncommon and mild (Anzai et al., 2003; Thoeny et al., 2009). The limited toxicity is thought to arise from the biodegradability of the particles and their processing by the body where they are first metabolized in the liver and subsequently used for the production of red blood cells, or being excreted (Anzai et al., 2003).

The superparamagnetic characteristics of SPIONs make them a prime candidate as magnetic resonance imaging (MRI) contrast agents. Indeed, several variants of SPIONs are approved for MRI, although their practical use in the clinic remains limited (Wang, 2015). SPIONs accumulating within a tissue bestow it MRI contrast with respect to the surroundings. SPIONS can accumulate passively in the presence of leaky vasculature, leveraging the enhanced permeability and retention effect associated with tumors. Alternatively, SPIONs can actively target specific ligands, thus providing a strategy for biomarker-based molecular diagnostics. The possibility for active targeting and biomarker imaging is an advantage of nanoparticle-based contrast agents over molecular ones, where it would require complex chemical modification strategies to maintain their functionality while providing targeting specific to individual molecules. Indeed, iron oxide nanoparticles have been conjugated to target cancer cells using peptides such as RGD and Bombesin, antibodies against HER-2 and EGFR, the VEGF soluble factor, aptamers, and folate. Using folate coating, it was possible to observe focal MRI contrast within the tumor region in small animal models (Li et al., 2015; Scialabba et al., 2017).

In a similar strategy, SPIONS can be targeted to apoptotic cells within artherosclerotic lesions, by surface functionalization with Annexin V (Smith et al., 2007). MRI imaging of the lesions in a rabbit model shows negative contrast associated with SPION accumulation at the lesions, and provides the ability to differentiate between occlusive and mural plaques. Commercially available SPIONs with carbohydrate coatings can target atherosclerotic lesions and ischemia to provide MRI contrast for their morphological detection (Trivedi et al., 2004). With similar strategies, the ischemic regions of the heart following myocardial infarction can be better delineated using SPIONs and MRI (Alam et al., 2012).

The advanced applications and the potential of SPIONs to provide morphological and molecular contrast for diseases provides a platform to integrate either multiple diagnostics modalities or combine them with therapeutic strategies within a single agent, in a theranostics approach. As a minimum, SPIONs provide the ability to track the biodistribution of the therapeutic agent through MRI (Yun et al., 2012). Further, using magnetic guidance, it is possible to accumulate SPIONs and their associated therapy to a desired site (Tukmachev et al., 2015). SPIONs decorated with a combination of CREKA peptide and fluorescent agent can dual-image microthrombi in an ischemia-reperfusion rat model (Song, Huang, et al., 2014).

On the therapy side, many chemotherapies have been formulated with SPIONs, using different strategies that include covalent conjugation, layer-by-layer coating, and environmentally-responsive linkers (Jeon et al., 2016; Li et al., 2014; Zhang et al., 2014). A SPION decorated with pH-responsive polymers that coordinate cisplatin molecules can selectively release the cisplatin in low-pH environments, enhancing the potential release at the cancer site. Further decorating the particle with folate provides for active tumor targeting, so that this system has an enhanced antitumoral activity compared with the nontargeted one, in a mouse model (Zhang et al., 2014). SPIONs can be coated with PLGA shells that load drugs and targeting agents while being pH responsive. Such particles targeted with antibodies against Her-2 and loaded with tamoxifen selectively released the drug at pH 5.0, while no release was observed at physiological pH (Vivek et al., 2016). Temperature-responsive strategies leverage the transition from hydrophilic to hydrophobic of the PNIPAMM polymer when crossing the lower critical solution temperature (LCST). An elegant approach uses SPIONs covered in fluorescent carbon dots as the template for a PNIPAM coating loaded with curcumin (Wang, Yi, Mukherjee, Banerjee, & Zhou, 2014). This composite nanoparticle can simultaneously monitor temperature, image cancer cells, and trigger drug release in response to magnetic or near-infrared stimuli. The PNIPAM state controls drug release, and it is in turn controlled both by the environmental temperature (which can be elevated at cancer sites) and by the conversion of near-infrared light into heat by the carbon coating, or that of alternating magnetic fields by the SPIONs. Further, the change in temperature modulates the quenching efficiency of the PNIPAM toward the carbon dots, resulting in an ability to monitor temperature increase by the increased fluorescence.

Due to the increased ECM-remodelling activity of tumors, MMP-cleavable linkers can actively target the release drugs from SPIONs (Ansari et al., 2014). Using commercially available SPIONs with azademethylcolchicine conjugated through an MMP-14 cleavable peptide, it was possible to induce apoptosis in an MMP-14 expressing tumor model in-vivo, with negligible cell death for cells not expressing MMP-14. In a more advanced strategy where the pro-drugs also act as active targeting moieties, chitosan-coated SPIONs were the carriers for methotraxate conjugated through an amide bond to a PEG linker. The methotrexate both targeted cancer cells expressing the folate receptor, and acted as the drug, once cleaved by lysosomal proteases, upon cell internalization and endolysosomal trafficking (Lin, Li, et al., 2015).

SPIONs can also be delivery vectors for nucleic acids. SPIONs coated with chitosan grafted to polyethylene imine (PEI) were used to deliver siRNA in vivo, thanks to the electrostatic interaction between the PEI and the siRNA (Wang, Kievit, et al., 2016). The particle was targeted to the tumor using antibodies against the GPC3 receptor. When injected in the tail vein of the mouse, the particles could target the tumor and suppress the expression of the target reporter gene (luciferase). SPIONs can combine chemotherapy with immunomodulation by simultaneously delivering a cisplatin pro-drug and the immunostimulant Polyinosinic:polycytidylic acid (Poly I:C), a dsRNA analogue (Hernández Gil et al., 2015). This formulation enhances, by more than one order of magnitude, the toxicity of the pro-drug, and the Poly I:C activates dendridic cells, which are the key antigen-presenting cells leveraged in immunotherapy.

Particles obtained from layer-by-layer coating of SPIONs with polypyrrole (PPy) and PEG were used for photothermal therapy guided under combined MRI and photoacoustic imaging. The particles accumulated strongly at the tumor site at 24-h postinjection and were able to mediate photothermal therapy with near-infrared light, which led to complete tumor remission following treatment (Song, Gong, et al., 2014). Doxorubicin was incorporated within this system through π-stacking with the PPy (Wang, Xu, et al., 2013). The temperature increase during photothermal treatment enhanced the release of doxorucibin, but only in acidic pH conditions, providing an effective strategy to target both thermal and chemotherapy treatment to the cancer site and reduce side effects.

SPIONs can generate heat when placed in alternating magnetic fields, and the heat generated can be toxic for cancer cells, as in photothermal therapy. Clustering small SPIONs (10 nm) and targeting them against the folate receptor enables them to accumulate better at the tumor 24 h following intravenous injection. Further, this clustering increases their relaxivity and specific absorption rate (Hayashi et al., 2013). When subject to alternating magnetic fields for 20 min, these particles increase the temperature of the tumor tissue by 6°C with respect to the surrounding tissue, and lead to significant tumor burden reduction up to 35 days following the treatment.

This effect can be combined with thermotherapy by loading up to 30% DOX within SPIONs coated with phoshoplipid-PEG (Quinto, Mohindra, Tong, & Bao, 2015). The DOX released from these particles for up to 72 h and the SPIONs could generate temperatures up to 43°C. The combination of magnetic hyperthermia and chemotherapy enhanced cytotoxicity in vitro over hyperthermia or chemotherapy alone.

3 Gold Nanomaterials

3.1 Synthesis

The most common route for the synthesis of gold nanoparticles is the reduction of gold salts (typically $HAuCl_4$) in solution in the presence of stabilizers, which are usually necessary to prevent aggregation (Fig. 3). The Turkevich method and its variants are widely adopted due

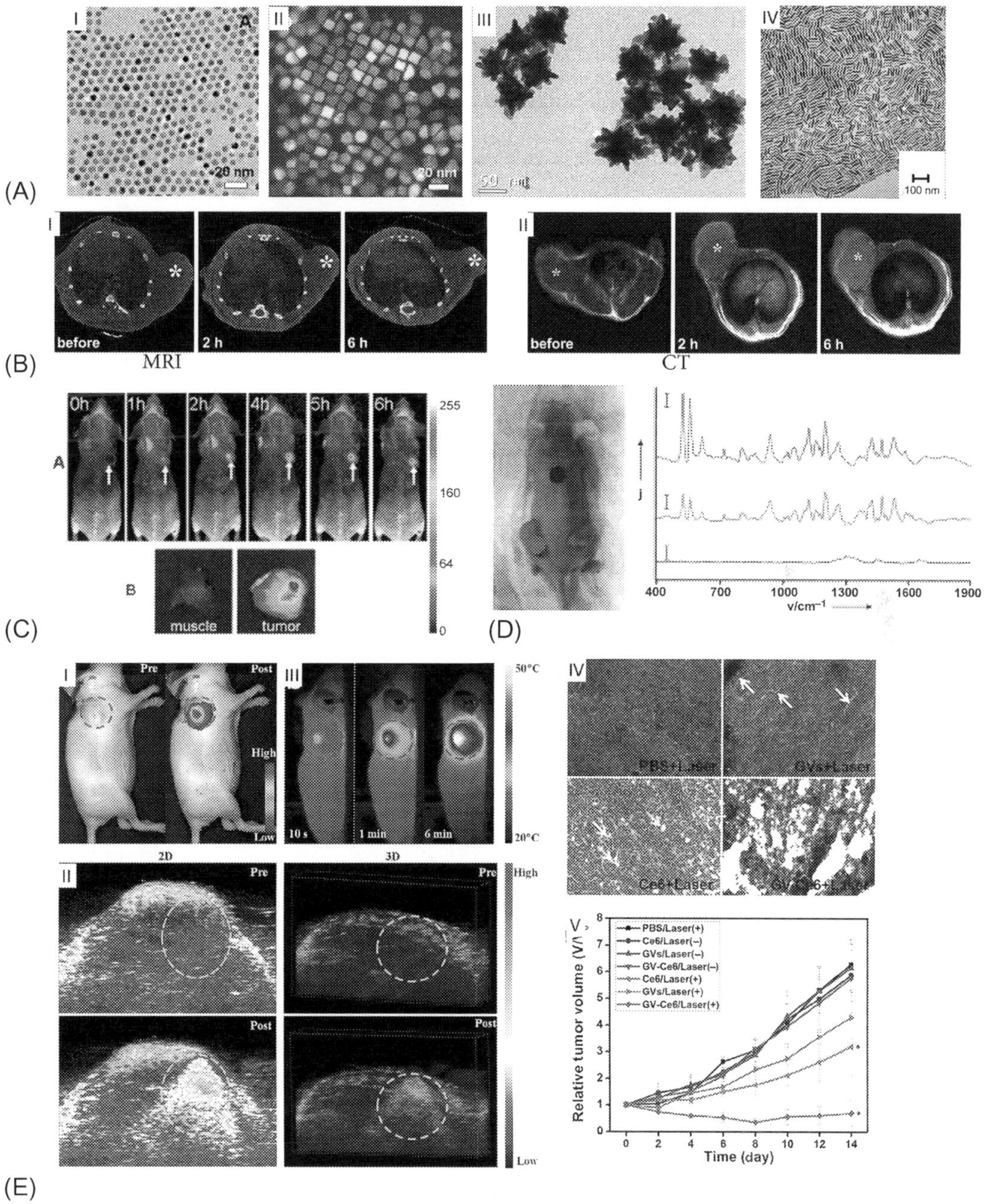

See the legend on next page.

to their simplicity, broad tunability, and aqueous process yielding water-soluble nanoparticles (Frens, 1972; Turkevich, Stevenson, & Hillier, 1951). The ratio of Au salt to citrate stabilizer, and the order of reagent addition alongside the pH of the reaction controls the kinetics and the resulting nanoparticle size (Ji et al., 2007; Sivaraman, Kumar, & Santhanam, 2011). Broadly, with this method, particles from 5 to 40 nm can be obtained. The particles obtained by the Turkevich method can then be used as seeds for further growth process in the presence of CTAB stabilizer, extending the size of nanoparticles to be formed beyond 100 nm (Jana, Gearheart, & Murphy, 2001a). The Brust method forms AuNPs in an organic phase, by reduction of the $HAuCl_4$ in sodium borohydrate, typically using octane- to dodecane-thiols as capping agents (Quinto et al., 2015). This method yields particles of sizes that can range between 1 and 20 nm, depending on the concentration of the Au salt and the capping agent (Leff, Ohara, Heath, & Gelbart, 2002). These particles are hydrophobic and require further processing before transferring to an aqueous solution.

Au nanoclusters are particles with defined numbers of Au atoms, arranged according to magic numbers that fill entire coordination shells. They are smaller than nanoparticles,

Fig. 3

Gold nanomaterials for theranostics. (a) A wide variety of nanostructures can be obtained from gold, including nanoparticles (I), nanocubes (II), nanostars (III), and nanorods (IV). (b) Dendrimer-entrapped gold nanoparticles loaded with gadolinium and targeted to tumors with folate serve as dual-modality contrast agents for MRI (I) and CT (II). (c) Gold nanoclusters exhibit near-infrared fluorescence that can be used to image tumors. (d) Au nanoparticles functionalized with SERS reporter CyNAMLA-381 act as SERS nanotags, exhibiting a characteristic signature spectrum (*top*). The SERS signal from the nanotags can be tracked in vivo, showing their accumulation within the tumor site (*mid*), but not in the upper dorsal area (*bottom*). (e) Gold-based nanovescicles, loaded with a photosensitizer can be visualized by NIR live imaging (I) and photoacoustic imaging (II), while mediating photothermal therapy (III) that yields tumor burden reduction (IV and V). *Modified from (a-I, a-II) Jin, R., Egusa, S., & Scherer, N. F. (2004). Thermally-induced formation of atomic Au clusters and conversion into nanocubes. Journal of the American Chemical Society, 126, 9900–9901. Copyright (2004) American Chemical Society, (a-III) Guerrero-Martínez, A., Barbosa, S., Pastoriza-Santos, I. & Liz-Marzán, L. M. (2011). Nanostars shine bright for you: colloidal synthesis, properties and applications of branched metallic nanoparticles. Current Opinion in Colloid & Interface Science, 16, 118–127. Copyright (2011) Elsevier, (a-IV) Gole, A. & Murphy, C. J. (2008). Azide-derivatized gold nanorods: functional materials for 'click' chemistry. Langmuir, 24, 266–272. Copyright (2008) American Chemical Society; (b) Chen, Q., Li, K., Wen, S., Liu, H., Peng, C., Cai, H., et al. (2013). Targeted CT/MR dual mode imaging of tumors using multifunctional dendrimer-entrapped gold nanoparticles. Biomaterials, 34, 5200–5209. Copyright (2013) Elsevier; (c) Wu, X., He, X., Wang, K., Xie, C., Zhou, B. & Qing, Z. (2010). Ultrasmall near-infrared gold nanoclusters for tumor fluorescence imaging in vivo. Nanoscale, 2, 2244–2249. Copyright (2010) Royal Society of Chemistry; (d) Samanta, A., Maiti, K. K., Soh, K. S., Liao, X., Vendrell, M., Dinish, U. S., et al. (2011). Ultrasensitive near-infrared Raman reporters for SERS-based in vivo cancer detection. Angewandte Chemie, International Edition, 50, 6089–6092. Copyright (2011) Wiley; (e) Lin, J., Wang, S., Huang, P., Wang, Z., Chen, S., Niu, G., et al. (2013). Photosensitizer-loaded gold vesicles with strong plasmonic coupling effect for imaging-guided photothermal/ photodynamic therapy. ACS Nano, 7, 5320–5329. Copyright (2013) American Chemical Society.*

with sizes <2 nm. Their properties arise at the molecular level, or from the overlapping of atomic properties, rather than from collective behaviors that characterize the larger Au nanoparticles. Indeed, they do not possess a plasmon resonance, but defined electronic levels that can be computed by the overlapping of atomic orbitals for the constituent gold and capping agent. By varying the number of gold atoms and the interaction with the capping agents, the levels can be tuned (Lu & Chen, 2012). Thanks to this electronic configuration, Au nanoclusters are both light absorbers and light emitters, which makes them suitable as optical imaging contrast agents. Their light emission is stable in comparison with that of fluorescent molecules, and it can be tuned during the synthesis process by the choice of capping ligand and overall particle size. Bottom-up approaches synthesize Au clusters starting from atomic species using phosphines or thiols as capping ligands. The thiols also act to partly reduce the Au, followed by the addition of sodium borohydrate to produce metallic Au (Zhu, Lanni, Garg, Bier, & Jin, 2008). The synthesis yields polydispersed clusters with different numbers of Au atoms, as well as nanoparticles. The overall distribution depends on the reaction conditions and the type of thiol capping ligand. By lowering the reaction temperature, it is possible to better control cluster formation and improve both uniformity and yield. Top-down generation of clusters starts from conventional Au nanoparticles, where thiols are substituted to the citrate as capping agents thanks to stronger affinity. The particles are then broken down into clusters by "loosening off" peripheral atoms through the etching ability of the ligands, helped by external energy provided, for example, through sonication (Jin, Egusa, & Scherer, 2004). These peripheral atoms re-arrange into clusters protected by the thiols, and go in solution. The composition of Au clusters can be determined by mass spectrometry, or by gel electrophoresis, distinguishing them by their overall mass (Jin, 2010; Negishi, Nobusada, & Tsukuda, 2005; Qian, Zhu, & Jin, 2012).

Au nanoshells are thin gold coatings deposited around dielectric nanoparticle cores. Polystyrene and silica are the most widely employed templates (Liu, Chen, et al., 2011). For the synthesis of Au nanoshells, small Au nanoparticles are seeded on thiols that cap the dielectric nanoparticles, and then grown until the desired shell thickness is achieved. Tuning the thickness of the shell controls the LSPR in the visible and NIR ranges. Less conventional cores for Au nanoshells include liposomes that provide a combined strategy for drug carrying and photothermal triggered release by bursting of the liposomes (Jin & Gao, 2009a); quantum dots to obtain particles with LSPR absorbance and fluorescence (Jin & Gao, 2009b), and silica-coated iron oxide nanoparticles to combine superparamagnetic and LSPR capabilities (Jin, Jia, Huang, O'Donnell, & Gao, 2010). Au shell structures with empty cores can carry drugs in their interior while maintaining a tunable LSPR. These structures are broadly classified as Au nanocages or nanoboxes, and are obtained by templating the Au shell over a sacrificial metal nanoparticle, which is then etched away or galvanically substituted (Skrabalak, Au, Li, & Xia, 2007; Sun & Xia, 2002; Wan, Xia, Wang, & Xia, 2013).

Anisotropic gold nanostructures can more readily tune their longitudinal local surface plasmon resonance (LSPR) by modulating the aspect ratio (Hu et al., 2006). In this way, they can readily achieve light absorption in the biologically relevant near-infrared window, while maintaining a small size that improves passive targeting and avoidance of the mononuclear phagocytic system. Nanorods are the most investigated anisotropic Au nanoparticles. They can be formed by templating within the nanopores of anodized aluminium oxide, but this strategy is fairly limited in tuning their geometry (Martin, 1996). On the flip side, it allows generating particles with multiple segments of different metals (Hurst, Payne, Qin, & Mirkin, 2006; Nicewarner-Peña et al., 2001). The most popular strategy for nanorod synthesis is seed-mediated growth, which yields products of high quality and uniformity with a simple way to control their aspect ratio (Jana, Gearheart, & Murphy, 2001b). This method uses citrate-stabilized Au nanoparticles as seeds for the growth of nanorods in the presence of a surfactant that preferentially caps one of the crystal surfaces. CTAB is the most commonly employed surfactant for this, which preferentially adsorbs on the (100) surfaces leading to nanorod growth in the {100} direction (Malikova, Pastoriza-Santos, Schierhorn, Kotov, & Liz-Marzán, 2002).

By slow reduction of Au it is possible to obtain nanoplateles, where thickness, sharpness of bends, and length of sides can be controlled. Changing these parameters tunes their LSPR from the visible to the NIR (Pastoriza-Santos, Alvarez Puebla, & Liz-Marzán, 2010). Other forms of anisotropic gold include nanostars/nanoflowers, where their anisotropic structure also provides for tuning of the LSPR without changing overall particle size (Chen, Wang, Ballato, Foulger, & Carroll, 2003; Cheng et al., 2014; Guerrero-Martínez, Barbosa, Pastoriza-Santos, & Liz-Marzán, 2011). Nanostars can be obtained at high purity and yield in a simple synthesis strategy (Kumar, Pastoriza-Santos, Rodríguez-González, de Abajo, & Liz-Marzán, 2008). Thanks to the simplicity and robustness of this approach, the LSPR and Raman properties of the resulting structures, nanostars are being broadly investigated for biological applications.

Typically, Au nanostructures have surface capping agents that do not provide sufficient stability in biologically relevant buffers. In some instances, the particles are outright nondispersible in aqueous buffers, while in others, they precipitate upon increase in ionic strength in the presence of serum. A broadly adopted approach for surface stabilization is ligand exchange. Ligand exchange employs thiols with high affinity to the nanostructure to substitute the original ligand. This strategy allows exchanging the hydrophobic thiols on the surface of Burst nanoparticles to make them dispersible in water (Shichibu, Negishi, Tsukuda, & Teranishi, 2005; Simard, Briggs, Boal, & Rotello, 2000), as well as to graft thiolated molecules with biological functionality such as peptides (Fabris et al., 2005), DNA (Taton, Mucic, Mirkin, & Letsinger, 2000) and PEG (Perrault, Walkey, Jennings, Fischer, & Chan, 2009). Using appropriate thiolated ligands, it is possible to conjugate molecules through click chemistry (Gole & Murphy, 2008), amide bonding (Wang et al., 2012), or by electrostatic interaction (Samanta et al., 2009). Au nanoclusters are usually coordinated with hydrophobic capping agents. Water-soluble nanoclusters can be formed by templating them within biomolecules, such as BSA or short helicoidal peptides that provide free thiols for nucleation and a limited volume for size confinement (de La Rica et al., 2014; Xie, Zheng, & Ying, 2009). Alternatively, the mesopores of silica shells can serve as templates to generate nanoclusters (Hembury et al., 2015).

3.2 Optical Properties

Gold can display photoemission when the dimension of Au nanocrystals within a structure decreases below 2 nm. Typically, a range of Au nanoclusters is photoluminescent, with efficiency and emission wavelength depending of the coordination number, the nature of the capping ligand, and the solvent (Wu & Jin, 2010). Larger Au nanostructures can also display fluorescence, provided that they host nanocrystals of suitable size. For example, polycrystalline nanoparticles with diameters >20 nm can exhibit fluorescence, being composed of several grains with cluster-like dimensions (Braun & von Delft, 1999). Equally, nanostars (Yuan et al., 2012), nanorods (Wang et al., 2005), and nanoshells (Gao, Vadakkan, & Nammalvar, 2011) can all exhibit fluorescence when their polycrystalline nature is adequately controlled.

The LSPR of Au nanoparticles sits squarely in the visible range. While the strength of absorption and scattering increases significantly with size, the absorption peak only moderately red-shifts, making it challenging to practically obtain nanoparticles with an NIR LSPR. An NIR LSPR is largely advantageous for in vivo application due to the relative optical transparency of tissue in the so-called "NIR biological window." Indeed, the LSPR is significantly more sensitive to changes in the shape of the Au nanostructure. In anisotropic particles, the LSPR peak splits into a transverse mode that remains associated with size and a longitudinal mode that varies with the geometry (Jain & El-Sayed, 2010). For nanorods, the longitudinal LSPR red-shifts linearly with their aspect ratio (Jun et al., 2008). Controlling the thickness of an Au nanoshell also tunes the LSPR peak across the NIR,with thinner shells inducing red-shift (Oldenburg, Averitt, Westcott, & Halas, 1998).

Metal nanostructures can enhance the Raman signal by several orders of magnitude, through a local enhancement of the electromagnetic field (Willets & Van Duyne, 2007). Raman is extremely valuable in the molecular characterization of biological samples, as it provides a label-free and rapid strategy for fingerprinting the composition of biological fluids (Xiao, Wang, Zhu, & Long, 2016), as well as mapping tissue composition (Lu et al., 2016). Gold nanostructures in particular are widely employed for surface enhanced Raman scattering (SERS), thanks to their strong enhancement, superior stability and biocompatibility with respect to the more SERS-active Ag (Le Ru & Etchegoin, 2009).

Particularly, Au nanorods and nanostars are broadly employed as SERS-active agents, because shape plays a fundamental role in determining the effect. Indeed, while nanorods display SERS enhancement factors of 4 to 5 orders of magnitude, comparable nanoparticles do not display any enhancement (Nikoobakht, Wang, & El-Sayed, 2002; Yu, Chang, Lee, & Chris Wang, 1997). The sharp tips of nanostars are deemed responsible for their ability to locally enhance the electromagnetic field (Indrasekara et al., 2014; Rodríguez-Lorenzo, Alvarez Puebla, de Abajo, & Liz-Marzán, 2010). Narrow gaps between metal surfaces, so-called "hot spots" can enhance SERS, while also providing ideal nanosized pockets to localize analyties (Le Ru, Etchegoin, & Meyer, 2006).

3.3 Theranostic Applications

The fluorescence of Au nanoclusters can be used to image cells following their uptake, in a potential strategy for live cell tracking (Zhao et al., 2013). Still, uptake of nanoclusters by cells is low, and the use of ligands that promote uptake can be beneficial. Conjugating the nuclear localization sequence by ligand exchange can enhance significantly the uptake of Au nanoclusters by cells (Lin, Chen, Sum, Lo, & Yang, 2008). Similarly, targeting agents such as folic acid and insulin improve uptake (Liu, Wu, et al., 2011; Retnakumari et al., 2010) as does the coating with proteins such as streptavidin or BSA (Lin et al., 2009). BSA- (Wu et al., 2010) and GSH-stabilized (Liu et al., 2013) nanoclusters are sufficiently stable to be injected in the vasculature, where they can be tracked by live NIR imaging. The clearance of these nanoclusters by the body is one to two orders of magnitude faster than equivalent functionalized nanoparticles.

Thanks to their light absorption in the NIR, certain Au nanomaterials can function as contrast agents for optical coherence tomography (OCT). Nanorods (Wei, Oldenburg, Zweifel, Hansen, & Boppart, 2006) and PEGylated nanoshells (Gobin et al., 2007) were used as contrast agents in vivo. The nanoshells accumulated at the tumor site within 20 h from tail vein injection. When imaging through the skin, the tumor could be detected by the improved contrast in the nanoshell injected group.

Iodine compounds are routinely used in the clinic as CT contrast agents, but Au has an almost three-fold greater attenuation coefficient, while it can be functionalized for active targeting, reside at the target for longer, and have a reduced renal toxicity (Dorsey et al., 2013; Jackson, Rahman, Wong, Ackerly, & Geso, 2010). Au nanoparticles as small as 2 nm can effectively act as CT contrast agents in vivo. They rapidly accumulate at tumors by passive targeting, where they can be observed as early as 15 min postinjection and reside there for up to 24 h (Copland et al., 2004). Dendrimer-entrapped Au nanoparticles loaded with gadolinium were targeted to tumors in vivo thanks to folic acid functionalization. Using these particles, it was possible to image tumors in vivo by a combination of CT and MRI (Chen, Li, et al., 2013).

Combined PET- or SPECT-CT is routinely employed in clinical practice to obtain coordinated molecular and morphological information. Incorporating a radionucleotide ^{64}Cu center within the lattice structure of Au nanoparticles generated a nanostructure with dual CT and PET capability that allowed imaging a tumor model in dual modality (Zhao, Sultan, Detering, Luehmann, & Liu, 2014).

SERS also enables cell imaging for diagnostics. The SERS signal from anti-EGFR targeted gold nanorods was enhanced upon detection of their target cells in vitro (Huang, El-Sayed, Qian, & El-Sayed, 2007). Au nanoparticles actively targeted with anti-VEGF or anti-HER2, and functionalized with Raman active molecules to carry a specific SERS signature, effectively accumulated at the tumor site in vivo and could be detected by Raman spectroscopy (Samanta et al., 2011). As a further development, using multiple SERS tags it was possible to discriminate the Raman signals originating from the injection of 10 multiplexed tags, and correlate them with the local concentration of the tag (Zavaleta et al., 2009).

A multifunctional nanoparticle with an Au core, surrounded by a Raman active layer within a silica shell coated with gadolinium was used as triple-modality contrast agent integrating Raman, photoacoustic, and MRI imaging. Using this strategy, it was possible to detect orthotropic brain tumors in a mouse model following tail vein injection of the contrast agent, with all three modalities. Intraoperative Raman was also used to guide tumor resection in the mouse model, contributing to remove infiltrating lesions and improve maximal resection (Kircher et al., 2012).

The efficient and tunable light absorption cross-section of Au nanostructures and its conversion into heat enables both photoacoustic imaging and photothermal therapy. Au nanoparticles functionalized with Herceptin could image Her2+ breast cancer cells in vitro as well as in vivo (Wang et al., 2004). Similarly, Au nanocages were able to image a melanoma animal model through photoacoustic imaging, thanks to targeting by either enhanced permeability and retention or by active targeting of the surface receptor, with the active targeting providing three times the signal as the passive (Kim, Cho, et al., 2010). An active area of investigation for the use of photoacoustic imaging is in the detection of sentinel lymph nodes. In this context, nanorods and nanocages were capable of improving the contrast for their detection (Pan et al., 2010; Song, Kim, Maslov, & Wang, 2009). Nanoshells have been extensively used for photothermal therapy since their initial application for the thermal ablation of tumor tissue in vivo, which resulted in the long-term survival of the animals (Hirsch et al., 2003). Photothermal treatment was coupled with dual MRI/CT imaging, by developing Au nanoparticles decorated with DOTA-chelated Gadolinium and a targeting agent for hepatocellular carcinoma. The particles were able to target hepatic cells and induce their death in vitro, while acting as efficient dual contrast agents (Zeng et al., 2014). DNA origami can encapsulate and stabilize Au nanorods, enhancing their passive targeting of tumors (Du et al., 2016). This system enabled photoacousting imaging and photothermal treatment of tumors that improved survival in an animal model.

Assembling block copolymer vesicles with Au nanoparticles embedded within the shell and photosensitizers loaded within the core enabled a theranostic approach that combined three imaging modalities, photothermal therapy, and photodynamic therapy. The particles could be imaged in vivo by fluorescence, thermal, and photoacoustic imaging. The full combined treatment showed a significant tumor burden reduction with respect to each treatment in isolation, that was sustained for 14 days (Lin et al., 2013).

Photothermal treatment can be used to release drugs also by cleaving Au-S bonds, for example, yielding to the releasse of thiolated DNA (Takahashi, Niidome, & Yamada, 2005). The same increase in temperature can release an siRNA strand by melting a double-stranded oligonucleotide where one strand is attached to the Au nanostructure (Poon et al., 2010).

Drugs can be loaded on Au nanoparticles either by direct conjugation of sulfhydryl groups to the surface (Cheng et al., 2008), through functional capping ligands (Brown et al., 2010) and by adsorption through electrostatic (Han et al., 2012) or hydrophobic interaction (Funkhouser, Arevalo, Glatzhofer, & O'Rear, 1995). Nanoshells and nanocages lend themselves for efficient drug loading within their inner hollow cores, and thus for combined chemotherapy and photothermal therapy (You, Zhang, & Li, 2010). Using the thermoresponsive polymer PNIPAAM to seal the opening of nanocages allowed controlling release of the cargo drugs by raising the temperature above the LCST (Yavuz et al., 2009). Further, the combination of Au mediated photothermal therapy with chemotherapy enhanced the cytotoxicity of each treatment alone, both in vitro and in animal models (You, Zhang, Xiong, et al., 2012; You, Zhang, Zhang, et al., 2012). In an integrated theranostic approach, by building an Au nanoshell around a pH- and reduction-sensitive polymeric vesicle, it was possible to deliver the nanoparticle to the tumor by passive targeting and trigger the release of the drug in association with the photothermal treatment, inducing a significant reduction of tumor size in comparison with each isolated treatment (Wang, Yuan, et al., 2016). The progress of the treatment could be followed by near infrared imaging.

4 Silica Nanoparticles

4.1 Synthesis

Unlike gold and iron oxide, silica nanoparticles, per se, do not have any intrinsic property arising from the nanoscale that make them suitable as theranostic agents. Yet the versatility and simplicity of their synthesis and functionalization, and the simple control over the generation and tuning of their mesoporosity, make them among the most promising platforms on which to build and integrate theranostic functionality (Fig. 4). The Stöber method is the classical strategy to synthesize silica particles with sizes ranging from nanometers to several microns (Stöber, Fink, & Bohn, 1968). The method relies on the hydrolysis of tetraalkyl silicates (typically tetraethyl orthosilicate i.e., TEOS) in an

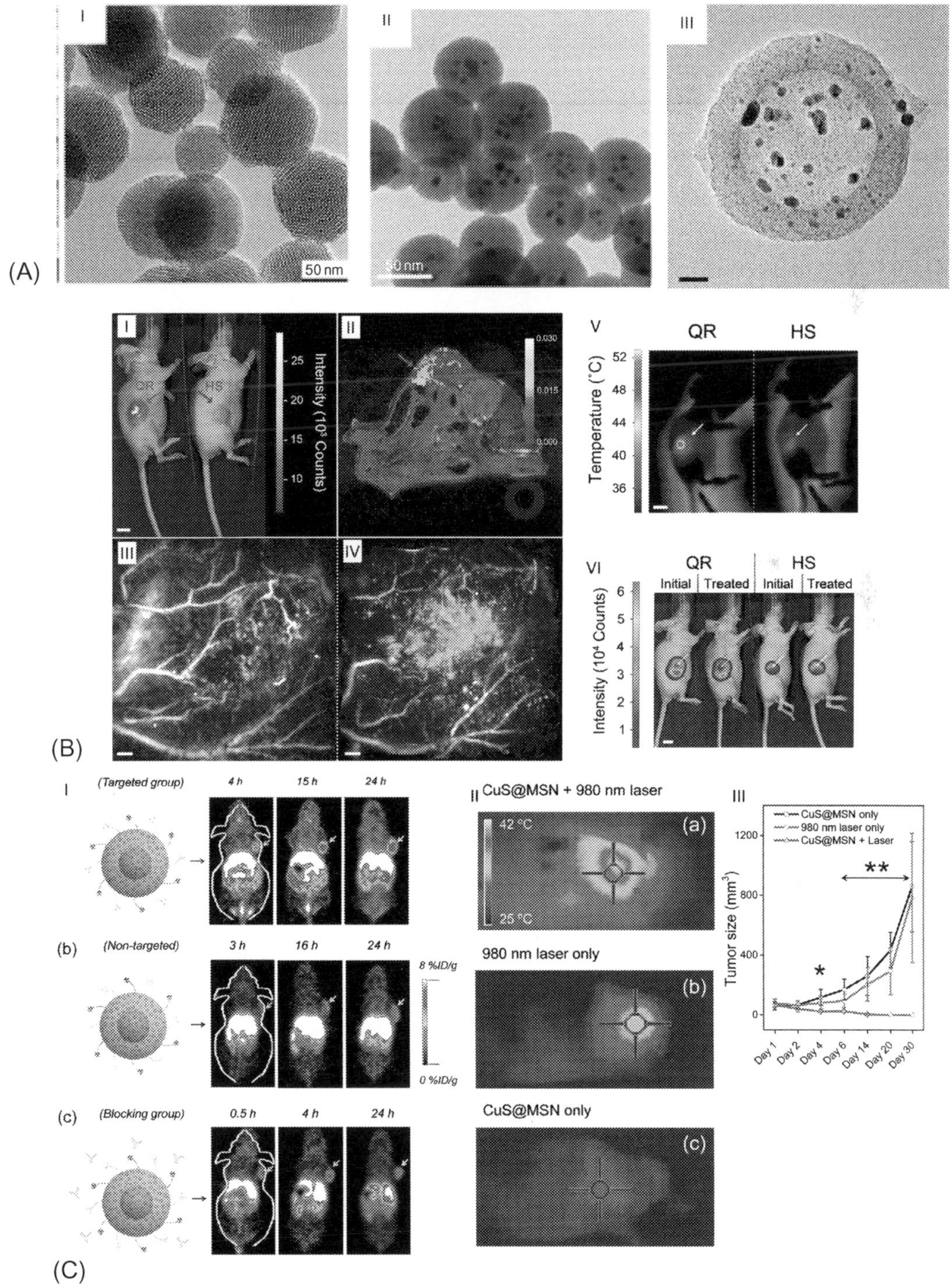

See the legend on next page.

aqueous solution containing ethanol, in the presence of ammonium hydroxide. Introducing aminopropyltriethoxysilane (APTES) during the synthesis, it is possible to provide the particles with amine functional groups at the surface for conjugation with dyes and bioactive agents. By addition of surfactants to the reaction, the Stöber method can be modified to synthesize mesoporous silica particles (Grün, Lauer, & Unger, 1997; Nooney, Thirunavukkarasu, Chen, Josephs, & Ostafin, 2002). Controlling the reaction condition, it is possible to tune particle size alongside the size of the pores (Nakamura, Mizutani, Nozaki, Suzuki, & Yano, 2007), their order, and orientation (Yano & Fukushima, 2004). The silica particles can be grown controllably by adding silica precursors that react preferentially with already formed particles, rather than forming new ones (Nakamura et al., 2007). The pH during synthesis is the most important parameter that controls the size of the resulting particles (Chiang et al., 2011). The addition of pluronic F127 can also regulate overall particle size (Kim, Chung, & Lin, 2010). Extending pore size beyond 2.3 nm is challenging and requires using surfactants other than CTAB (Fuertes, Valle-Vigón, & Sevilla, 2010; Han & Ying, 2005) and coupling them with strategies for pore enlargement (Kim, Na, et al., 2011). Alternatively, silica particles with large pores can be obtained by aerosol methods (Lu et al., 1999). The various surfactants and templates are then removed either by calcination or solvent extraction to yield the mesoporous particles.

It is possible to generate mesoporous silica shells, with highly ordered pores, by templating the silica condensation around another nanoparticle (Hembury et al., 2015). Templates can then either be removed by extraction or calcination, or, if they provide specific functionality, such as the case of metal and metal oxide nanoparticles, they can be retained to bestow multifunctionality to the particle. Nanorattles represent a peculiar class of silica shells, where the nanoparticle(s) within the shell have a size smaller than the inner core. Nanorattles can be generated by etching the template following the formation of the silica shell (Chen, Li, Tang, & Qi, 2009), or by using the silica shell as a nanoreactor for the synthesis of the particles

Fig. 4

Silica nanomaterials for theranostics. (A) Mesoporous silica nanoparticles (I) are a highly versatile template to generate hybrid nanomaterials. Core-shell Au-MSN (II) and Au@Si quantum rattles hosting AuNPs within the core and AuNCs within the shell (III) are silica-based multifunctional theranostics nanoagents. (B) The quantum rattles can be imaged by NIR live imaging (I), MRI (II) and photoacoustic imaging (III and IV) while they mediate photothermal therapy (V and VI). (C) CD105-targeted MSN nanoparticles decorated with Cus nanoparticles served as efficient PET contrast agents (I) while being able to mediate photothermal therapy (II) leading to a reduced tumor burden (III). *Modified from (A-I, A-II) Tang, F., Li, L., & Chen, D. (2012). Mesoporous silica nanoparticles: synthesis, biocompatibility and drug delivery. Advanced Materials, 24, 1504–1534. Copyright (2012) Royal Society for Chemistry; (C) Chen, F., Hong, H., Goel, S., Graves, S. A., Orbay, H., Ehlerding, E. B., et al. (2015). In vivo tumor vasculature targeting of CuS@MSN based theranostic nanomedicine. ACS Nano, 9, 3926–3934, under ACS AuthorChoice license.*

within the core (Hembury et al., 2015). This later strategy in particular was used to develop the quantum rattles, which host Au nanoparticles within the core and Au nanoclusters within the shell to bestow multimodal imaging and therapeutic functionality.

4.2 Surface Functionalization

A major capability of silica nanoparticles is the versatile surface functionalization through organosilanes. The organosilanes can either be incorporated during the synthesis process (Vaidya, Thaplyal, & Ganguli, 2011), or they can be grafted on the surface at a later time (Lee, Park, Chung, Kim, & Kim, 2011). This grafting can provide functional groups to be used for downstream grafting of functional agents. The available groups include amino, carboxyl, aldehyde, and thiol, for which established and selective bioconjugation strategies exist (Espiard, Mark, & Guyot, 1990; Oh et al., 2006). Using organosilanes as ligands, it is possible to conjugate enzymes (Cheng et al., 2015), peptides (Pan et al., 2012), antifouling agents (Otsuka, Nagasaki, & Kataoka, 2003), targeting moieties (Salvati et al., 2013), fluorophores (Lin et al., 2005) and other nanoparticles (Liu et al., 2010). The inner (pore) and outer surface of the particles can be selectively functionalized with different moieties (Kecht, Schlossbauer, & Bein, 2008). Alternatively, silica can be functionalized by adsorption, a strategy widely employed for nucleic acids and peptides (Puddu & Perry, 2012; Zhu et al., 2004).

4.3 Theranostic Applications

While solid silica particles are fundamentally stable, mesoporous silica nanoparticles (MSNs) are biodegradable, due to the hydrolysis of the silica network (He, Shi, Zhu, Chen, & Chen, 2010). Surface modification of MSNs can regulate their degradation profile (Ehlerding, Chen, & Cai, 2016; Shen et al., 2014). The toxicity of mesoporous silica nanoparticles depends on multiple parameters of the material, which include size, shape, surface chemistry porosity, and pore size (Lu et al., 2007; Zhu et al., 2013). Appropriately tailored MSNs are highly biocompatible both in vitro (Lin, Zhao, Yan, Hu, & Xu, 2015; Lu et al., 2007; Slowing, Trewyn, & Lin, 2007) and in vivo (Shang, Nienhaus, & Nienhaus, 2014). Both intravenous and subcutaneous injection of MSN is well tolerated in animal models at high concentrations (Dai et al., 2009; Lu, Liong, Li, Zink, & Tamanoi, 2010). Biodistribution is also affected by the same parameters as cytocompatibility.

Silica nanoparticles are usually not fluorescent, but they can be optically tracked by integrating fluorescent dyes or nanoparticles such as quantum dots and upconverting nanomaterials (Idris, Jayakumar, Bansal, & Zhang, 2015). These fluorescent trackers can be incorporated during synthesis, adsorbed, or covalently linked to the silica. By generating silicon quantum dots through oxygen deficiencies within the particles, it is possible to create photo-emitting centers that can be used to track the nanoparticle in vitro. Coupling this with delivery of doxorubicin provided a theranostic strategy to monitor chemotherapy (Chen, Zhen, et al., 2013)

Mesoporous silica shells are often employed as a more biocompatible interface for SPIONs and QDs, providing a platform for combined imaging and drug delivery (Singh, Kim, Patel, Knowles, & Kim, 2012; Yao et al., 2017). Multiple nanoparticles such as QDs, SPIONs, and Au nanomaterials can be simultaneously incorporated within a silica shell (Liu, Wang, et al., 2012; Wang & Gu, 2015). NPs can also be decorated over the silica surface. AuNPs were linked to silica particles through the interaction between an aptamer recognizing ATP grafted on the Au and the ATP grafted on the silica (Zhu, Lu, Song, Yang, & Wang, 2011). The AuNPs acted as pore caps, and in the presence of high ATP concentrations, they would be displaced to release the loaded drug. SPIONs were decorated on the surface of MSNs synthesized with encapsulated dye, to provide a dual NIR and MRI imaging modality (Lee et al., 2010). The MSN could be further loaded with DOX and passively targeted to the tumor, while tracking them with both MRI and fluorescence. Doxorubicin activity at the tumor was observed. With the aid of ultrasound, zinc sulphide nanoparticles can be decorated on MSN surfaces (Dolores et al., 2015). Layer by layer polymer coatings can be formed around the nanoparticles as well (Ariga, Hill, & Ji, 2007). All these particles can also be linked to the MSN surface through stimuli-responsive ligands (Chen et al., 2016; Liu, Qing, et al., 2015; Zhang, Wu, Li, & Feng, 2015), allowing them to serve as caps that control the release of drugs loaded within the mesoporous structure. More advanced coupling strategies involve encapsulating the silica particle within a lipid vescicle. This strategy could generate a multifunctional particle for dual MRI and optical imaging by encapsulating a quantum dot core within a silica shell coated by the lipid layer incorporating a gadolinium agent. The particle was further pegylated for antibiofouling and targeted with RGD peptides (Koole et al., 2008).

MSNs can effectively load and sustain the release of drugs, thanks to their large surface to volume ratio providing extended interaction for loading, and the well-defined porous structure enabling sustained release. The particles can then be capped to trigger drug release, and decorated with targeting moieties to optimize their biodistribution, overall enhancing the therapeutic index. Hollow MSN particles have a larger payload volume and a reduced mass, which enables them to load drugs and nucleic acids at higher mass (Singh, Kim, Mahapatra, Patel, & Kim, 2015; Wu et al., 2016). Particularly, rattle-type nanoparticles can leverage the interaction between the nanoparticles within the core and the drug, to improve both loading capacity and sustained release (Hembury et al., 2015). Using quantum rattles, it was possible to combine multimodal MRI, NIR, and photoacoustic imaging enabled by the Au nanocluster encased within the silica shell with photothermal therapy and an enhanced loading of the drug doxorubicin thanks to its interaction with hydrophobic Au nanoparticles within the hollow silica core. This allowed prolonging the sustained release of the drug, and monitoring nanoparticle localization with the three imaging modalities simultaneously, as well as mediating tumor burden reduction by photothermal therapy. Similar rattles containing SPIONs and functionalized with PEG and folate were able to actively target tumors while mediating MRI imaging and the release camptothecin to slow tumor growth (Lu, Li, Zink, & Tamanoi,

2012). The co-loading of multiple drugs and/or biologicals allow for their simultaneous delivery at known relative concentrations, with potential beneficial effects on combination therapy strategies. SiRNA targeting protein components of MDR pumps loaded alongside doxorubicin were capable of addressing multidrug resistance and improving doxorubicin toxicity by silencing their expression (Chen, Zhang, et al., 2009; Meng, Liong, et al., 2010).

Stimuli-responsive drug release allows optimizing the therapeutic index by increasing the ratio of drug delivered at the target. The pH is a broadly exploited stimulus because the interstitial space of tumors is known to be acidic (Zhang, Lin, & Gillies, 2010), and during intracellular trafficking of nanoparticles, the endocytotic vescicles becomes progressively more acidic (Repnik, Česen, & Turk, 2013). Both α- and β-cyclodextrins were able to trigger drug release upon acidification of the environment, showing the ability to deliver drug payload to a cancer target (Du, Song, & Liao, 2012; Meng, Xue, et al., 2010). PNIPAM was used to trigger the release of drugs upon temperature increase due to photothermal effects (Zhou, Zhu, & Di, 2007). The strongly reducing environment present in the intracellular space can also serve as a trigger stimulus. Systems based on cleavage of disulphide bonds, or redox-responsive polymer networks, can control drug release (Giménez et al., 2015; Lai et al., 2003). Drug release can also be stimulated externally, for example, using light to cleave photolabile linkers, and open the pores (He, He, Wang, Cao, & Zhao, 2012). Although the wavelength necessary is in the UV, which has potential toxic effects and negligible depth of penetration in tissue, recent two-photon strategies using NIR radiation are contributing to overcome these limitations. An alternative light-mediated strategy is to induce photothermal effects with NIR light, in turn leveraging temperature-mediated switching. Externally applied magnetic fields can also generate heat leading to drug release by releasing SPIONs that are used to cap the MSN (Rittikulsittichai et al., 2016). When coupling the SPIONS through a DNA hybridization, it is possible to melt the double-stranded DNA and release the payload with an alternating magnetic field generating heat (Ruiz-Hernández, Baeza, & Vallet-Regí, 2011).

By incorporating Au nanostructures with NIR LSPR within silica nanoparticles it is possible to combine the efficient drug loading and release of silica with efficient photothermal treatment (Liu, Detrembleur, et al., 2015; Wang, Wang, et al., 2016). Dual-timing PTT with an initial low-dose followed by step-up can first stimulate drug release and sensitize the tissue to the high-dose hyperthermia treatment (Zhang et al., 2012). Thanks to the incorporation of Au, these systems can be tracked by CT to bestow theranostic functionality, while further incorporating SPIONs allows dual CT/MRI tracking (Dong et al., 2011; Xia et al., 2014). Other multimodal strategies involve the use of indocyanine green as a NIR imaging agent and 5-fluorouracil in combination with silica coated gold nanorods (Fang et al., 2017). Silica nanocarriers can also effectively target agents for photodynamic therapy to the tumor site, where they can be activated through NIR radiation to induce tumor burden reduction (Gary-Bobo et al., 2011). Combinatorial treatment involving PTT, PDT and chemotherapy used MSN coated Au nanorods that were loaded with a photosensitizer, with β-CD capping agents that are held in place by a platinum pro-drug. The particle was further functionalized for antibiofouling and active targeting. In a reducing environment, the platinum pro-drug was converted to cisplatin, the β-CD freed releasing the photosensitizer. Upon irradiation with NIR light, the system induced an antitumoral effect superior to all its components in isolation. Incorporating perfluorohexane within the MSNs generated a sensitizer for ultrasound therapy with enhanced performance. The system could be combined with Au and SPION nanoparticles to incorporate PTT and CT/MRI guidance (Chen et al., 2011; Wang et al., 2013).

As MRI contrast agents, MSN can improve the stability and solubility of SPIONs and provide simpler strategies for targeting and antifouling surface functionalization, while preserving relaxivity (Kim et al., 2008). MSNs functionalized with gadolinium or loaded with gadolinium oxide nanoparticles also make for efficient MRI contrast agents that can be coupled with SPION cores to further enhance their efficacy (Shao et al., 2012; Stallmach, Kärger, Krause, Jeschke, & Oberhagemann, 2000; Taylor et al., 2008).

Near infrared imaging can be readily achieved by incorporating fluorescent dyes during particle synthesis, or functionalizing the particles afterward (Lee et al., 2009). Alternatively, upconversion nanoparticles that are excitable in the NIR and benefit from significantly reduced autofluorescence background have been loaded within MSNs, and could be tracked in vivo (Liu, Bu, et al., 2012). This approach was combined with the conjugation of Gd as MRI contrast agent alongside the delivery of a radio and photo-sensitizer and a chemotherapy drug. Under combined X-ray and light irradiation, the system yielded efficient antitumoral activity, coupled with the ability to track it by both MRI and live imaging (Fan et al., 2014). Integrating ^{64}Cu within MSN provided a strategy to introduce PET within theranostic approaches that included drug loading and targeted delivery (Chen et al., 2015; Huang et al., 2012). Trimodal imaging by CT, MRI, and upconversion imaging, coupled with trimodal therapy by photothermal, photodynamic, and chemotherapy, was possible thanks to a lanthanide yolk-silica shell particle. The lanthanides provided optical, MRI, and CT contrast, while carbon dots, ZnPc, and doxorubicin within the silica shell provided the three therapeutic modalities (Lv et al., 2015). The particles were successfully tested for multimodal tracking and tumor burden reduction by combinatorial therapy in vivo.

5 Porous Silicon Nanomaterials

5.1 Fabrication

Unlike the other nanomaterials considered in this chapter, porous silicon is largely fabricated using a top-down approach (Fig. 5). Crystalline silicon is made porous by etching in a solution containing hydrofluoric acid (HF). The conventional fabrication route applies an anodic potential to a silicon substrate in a galvanic cell in the presence of HF (Cullis & Canham, 1991).

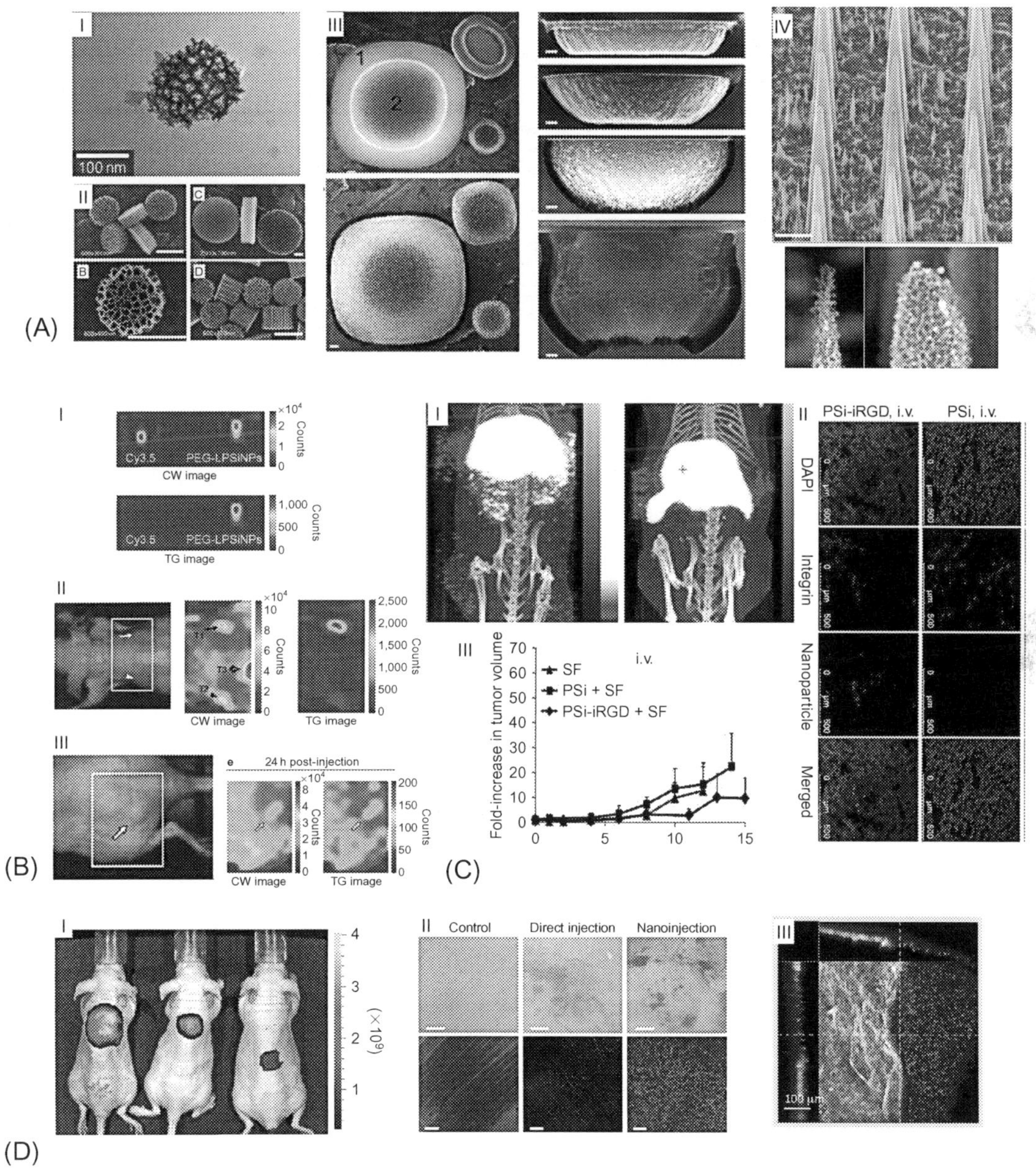

See the legend on next page.

The anodic potential oxidizes the silicon surface, leading to its dissolution in HF. Depending on the applied current, the resistivity of the silicon and the details of the solution composition, it is possible to tune the porosity between 20% to more than 90%, and pore size from less than a nanometre to the micron range, granting a much broader flexibility than mesoporous silica (Canham, 2014). By varying the applied current over time, it is possible to generate multilayered structures with variable pore size (Cunin et al., 2002; Lin, Motesharei, Dancil, Sailor, & Ghadiri, 1997). Using periodic variations of the current, bragg reflectors and photonic cavities can be generated, and used for optical investigation. The resulting material is intrinsically fluorescent,

Fig. 5

Porous silicon nanomaterials for theranostics. (a) Thanks to the top-down fabrication, the morphology of porous silicon nanostructures can be tightly controlled to generate (I) quasispherical particles, (II) discoidal particles, (III) quasihemispherical particles with controllable aspect ratio, and (IV) arrays of vertically aligned high aspect ratio nanostructures with tunable tip geometry. (b) The intrinsic photoluminescence of porous silicon enables high-contrast in vivo imaging by time-gated fluorescence, enabling to discriminate the nanoparticles from fluorophores in vitro (I) and in vivo (II) and allowing to highlight their accumulation within tumors following I.V. injection (III). (c) A theranostic system composed of RGD-targeted porous silicon loaded with sorafenib and tagged with ^{111}In could be used for combined PET imaging (I) and antiangiogenic therapy targeted against the tumor-associated vasculature (II) yielding a reduced tumor growth (III). (d) Porous silicon nanoneedles (a-IV) can deliver bioactive agents topically maintaining high localization and prolonged residence (I). Used to deliver a VEGF-expressing plasmid yield the local formation of new blood vessels (II). Functionalized with a Cathepsin B cleavable fluorescent peptide and applied to clinical samples of tumors, they can highlight Cahtepsin B activity across the margin region (III). *Modified from (a-I) Liu, D., Zhang, H., Mäkilä, E., Fan, J., Herranz-Blanco, B., Wang, C.-F., et al. (2015). Microfluidic assisted one-step fabrication of porous silicon@acetalated dextran nanocomposites for precisely controlled combination chemotherapy. Biomaterials, 39, 249–259. Copyright (2015) Elsevier, (a-II) Godin, B., Chiappini, C., Srinivasan, S., Alexander, J. F., Yokoi, K., Ferrari, M., et al. (2012). Discoidal porous silicon particles: fabrication and biodistribution in breast cancer bearing mice. Advanced Functional Materials, 22, 4225–4235. Copyright (2015) Wiley, (a-III) Chiappini, C., Tasciotti, E., Fakhoury, J. R., Fine, D., Pullan, L., Wang, Y.-C., et al., (2010). Tailored porous silicon microparticles: fabrication and properties. ChemPhysChem, 11, 1029–1035. Copyright (2012) Wiley, (a-IV) Chiappini, C., De Rosa, E., Martinez, J. O., Liu, X., Steele, J., Stevens, M. M., et al. (2015). Biodegradable silicon nanoneedles delivering nucleic acids intracellularly induce localized in vivo neovascularization. Nature Materials, 14, 532–539. Copyright (2015) Nature Publishing group; (b) Gu, L., Hall, D. J., Qin, Z., Anglin, E., Joo, J., Mooney, D. J., et al. (2013). In vivo time-gated fluorescence imaging with biodegradable luminescent porous silicon nanoparticles. Nature Communications, 4, under open access licensing; (c) Wang, C.-F., Sarparanta, M. P., Mäkilä, E. M., Hyvönen, M. L. K., Laakkonen, P. M., Salonen, J. J., et al. (2015). Multifunctional porous silicon nanoparticles for cancer theranostics. Biomaterials, 48, 108–118. Copyright 2015 Elsevier; (d-II) Chiappini, C., De Rosa, E., Martinez, J. O., Liu, X., Steele, J., Stevens, M. M., et al. (2015). Biodegradable silicon nanoneedles delivering nucleic acids intracellularly induce localized in vivo neovascularization. Nature Materials, 14, 532–539. Copyright 2015 Elsevier, (d-III) Chiappini, C., Campagnolo, P., Almeida, C. S., Abbassi-Ghadi, N., Chow, L. W., Hanna, G. B., et al. (2015). Mapping local cytosolic enzymatic activity in human esophageal mucosa with porous silicon nanoneedles. Advanced Materials, 27, 5147–5152, under creative commons license.*

due to the formation of silicon quantum dots within its skeleton (Cullis & Canham, 1991). The fluorescence intensity and wavelength is tunable within the visible and NIR range as a function of silicon resistivity, porosity, and surface functionalization (Peter et al., 1996). This etching process forms continuous porous silicon films that can be detached by the substrate and processed into particles by sonication, or ball milling (Bimbo et al., 2010; Dorvee, Derfus, Bhatia, & Sailor, 2004). Alternatively, through photolithography, it is possible to precisely define the geometry of the nanoparticles and their aspect ratio, either before the etching step, or directly on the porous layer (Chiappini, Tasciotti, et al., 2010; Cunin et al., 2002; Godin et al., 2012).

Alternative methods to generate porous silicon include stain etching, where an oxidizing compound in the HF solution induces the local oxidation of silicon, rather than the applied current (Kolasinski, 2005). This method was originally mired by slow etching processes, limited reproducibility, and lower flexibility than its electrochemical counterpart. Recent advancements with the introduction of a new class of metal-based oxidizing compounds, and their use in combination with secondary oxidants, significantly enhanced both the speed and the reliability of the process, suggesting a future broadening of its application (Dudley & Kolasinski, 2009; Kolasinski et al., 2017). Of particular value is the applicability of this strategy to the porosification of silicon nanoparticles, which can be readily obtained in large quantities at competitive pricing through the reduction process of silicates from plants, or accessing the metallurgical feedstock (Batchelor, Loni, Canham, Hasan, & Coffer, 2012; Loni et al., 2011). While isolated particles cannot be porosified by electrochemical etching, high porosity silicon particles were obtained by stain etching in suspension (Loni et al., 2011).

An alternative electroless method involves the use of metals as catalysts for the anodization. In the so-called metal-assisted chemical etching (MACE), silicon is efficiently oxidized in solutions containing hydrogen peroxide (or other oxidizing agent) thanks to the catalytic activity of metal deposited on the surface (Chartier, Bastide, & Lévy-Clément, 2008). MACE yields the directional etching of silicon under the metal deposits and, in appropriate conditions dependent on the solution composition and the resistivity of the substrate, the local porosification of the silicon structures adjacent to the metal deposits (Chiappini, Liu, Fakhoury, & Ferrari, 2010). By varying the concentration of oxidant over time, it is possible to generate multilayered structures, including periodic ones. Combining this strategy with lithographic techniques allows generating nanostructures with defined geometry, including nanoparticles, nanopillars, and nanocones. The particles can then be released from the substrate and used for theranostic purposes.

5.2 Surface Functionalization

Mesoporous silicon benefits from greater flexibility than silica in regard to surface functionalization. By oxidizing the surface following etching, it is possible to generate an oxide surface that is amenable to silane chemistry, analogously to porous silica (De Rosa et al., 2011).

Thermal carbonization is an alternative derivatization route that yields carbon-terminated surfaces with different grades of hydrogen and hydroxyl groups (Salonen, Laine, & Niinistö, 2002; Salonen & Mäkilä, 2018). By controlling the process, it is possible to modulate the hydrophilicity of the surface. Hydrosilylation of the surface of porous silicon, either by wet or thermal processes, provides an alternative strategy to obtain functional groups in the absence of a silicon oxide surface (Boukherroub, Wojtyk, Wayner, & Lockwood, 2002; Song & Sailor, 1998).

5.3 Theranostic Applications

Porous silicon particles have been largely tested for their safety in vitro and in vivo (Bimbo et al., 2010; Decuzzi et al., 2010; Godin et al., 2012), all the way to trials in humans (Goh et al., 2007). While specific surface chemistries have known toxicities, the vast literature highlights that the material can be functionalized to provide elevated biocompatibility at high dosages, which are well tolerated in multiple compartments, including the blood (Martinez et al., 2012), within tissue (Goh et al., 2007), the gastrointestinal tract (Bimbo et al., 2010), and the intraocular space (Low, Voelcker, Canham, & Williams, 2009). Intravenous injection and oral delivery of porous silicon is safe, and the biodistribution and clearance of the particle can be tracked by NIR imaging through the intrinsic fluorescence, showing that particles progressively biodegrade within the body (Park et al., 2009; Tzur-Balter, Shatsberg, Beckerman, Segal, & Artzi, 2015). The biodegradation kinetics are dependent on the porosity and pore size, as well as the surface chemistry (Godin et al., 2008, 2010; Martinez et al., 2013).

Porous silicon particles can effectively load drugs, biologicals, and nanoparticle payloads within their porous structure. Different from MSNs, the larger size of the pore allows for loading of quantum dots (Chiappini, Tasciotti, et al., 2010), SPIONs (Dorvee et al., 2004), carbon nanotubes (Tasciotti et al., 2008), and many others, providing simpler strategies to integrate functionality. The release of drugs is modulated by the interaction with the mesopores (Salonen et al., 2005), while nanoparticles can be trapped within the pores either by pore size reduction through oxidation (Dorvee et al., 2004), or by coating with polymer layers (De Rosa et al., 2011), or biological membranes (Parodi et al., 2012). All of these strategies contribute to modulate the release profile of payloads. Biologically labile payloads can be protected from the external environment through coatings that do not affect their release kinetics (De Rosa et al., 2011).

Drug release can be triggered with stimuli-responsive approaches. With a microfluidic system, it was possible to generate, by nanoprecipitation, porous silicon nanoparticles encapsulated within acetalated dextran, which is responsive to changes in pH. The dextran surface was further decorated with cell penetrating peptides, and the system used to deliver anticancer drug payloads was uptaken with high efficiency by cancer cells (Liu, Zhang, et al., 2015). Coating the particles with poly(beta-amino ester) also resulted in the pH responsive, sequential release of doxorubicin and paclitaxcel (Xu et al., 2015). Coating the nanoparticles with cell membranes derived from leukocytes enabled them to prolong evasion by the immune system by preventing opsonisation and through self-recognition. Also, with this strategy, the particles enhanced their permeation across endothelial cells to improve the delivery and toxicity toward cancer cells residing across artificial endothelia, and their accumulation at the tumor site (Parodi et al., 2012).

Grafting of PEG as an antifouling agent can prolong the circulation of porous silicon while also improving its stability and passive targeting potential (Bimbo et al., 2012; Näkki et al., 2015). Active targeting moieties such as antibodies, aptamers, peptides, and biopolymers increases cell recognition and uptake of porous silicon particles (Alhmoud et al., 2015; Almeida et al., 2014; Xia, Zhang, Shi, & Xiao, 2013). In a theranostic approach, particles targeted using mannose 6-phosphate and functionalized with porphyrin were effective at targeting cancer for photothermal therapy, while enabling NIR imaging (Chaix et al., 2016). RGD-targeted particles loaded with sulphorafane were able to target endothelial cells associated with tumors, and coupled with loading of the radiolabel ^{111}In enabled monitoring biodistribution by SPECT imaging while having antitumoral activity (Wang, Mäkilä, et al., 2014; Wang et al., 2015). Surface chemistry can also be leveraged to modulate the immune system response. The thermal carbonization of porous silicon could stimulate antigen-presenting cells to induce the proliferation of T-cells that were leveraged to modulate the immune response toward cancer cells (Shahbazi et al., 2014). Targeting the CD40 receptor through surface-conjugated antibodies also stimulated the activation of antigen presenting cells (Gu et al., 2012). Porous silicon itself is able to generate reactive oxygen species upon irradiation, and it was shown capable of mediating photodynamic therapy through singlet oxygen generation (Osminkina et al., 2012; Xiao, Gu, Howell, & Sailor, 2011). Grafting of porphyrins to porous silicon enabled their concentration within cells to improve the cytotoxicity of the PDT approach using both single and two-photon excitation (Secret et al., 2013). Porous silicon also mediated photothermal therapy that could induce tumour regression (Hong & Lee, 2013) and acted as active agent for sonodynamic and radiofrequency therapy, with the particles inducing hyperthermia upon application of ultrasounds or RF fields (Osminkina et al., 2014). RF therapy in vivo was able to induce tumor growth inhibition (Tamarov et al., 2014).

The long emission lifetime of porous silicon enables to distinguish it from background signal and other fluorophores through time-gated fluorescence imaging of tissues in vivo. Indeed, this approach allowed distinguishing the fluorescence originating from porous silicon from that of a fluorescent payload and to improve the signal to noise ratio more than 100-fold by suppressing autofluorescence (Gu et al., 2013; Joo et al., 2015).

The ability to load multiple nanoparticles within porous silicon enables developing multicomponent and multistage delivery systems, where functionality can be easily integrated to negotiate the multiple barriers toward the target, and achieve theranostic goals. In this context, a multistage system was developed capable of loading simultaneously multiple therapeutic and diagnostic agents, both in nanoparticle and molecular form (Tasciotti et al., 2008). The system enabled the sustained release of nanoparticles, including siRNA loaded liposomes that were able to target ovarian cancer and induce tumor growth inhibition by downregulation of the target gene (Tanaka et al., 2010). Loading gadolinium nanoparticles within the porous structure resulted in a marked enhancement of their relaxivity, improving their performance as MRI contrast agents (Ananta et al., 2010).

An altogether different class of porous silicon nanomaterials are the nanoneedles. They belong to the category of vertically aligned nanowire arrays that have recently emerged as a promising platform for combined intracellular delivery, sensing, and cell manipulation (Chiappini, 2017). Nanoneedles are arrays of nanocones with a tailored diameter, porosity, and density (Chiappini, De Rosa, et al., 2015). They can simultaneously mediate the intracellular delivery of payloads, and the sensing of the intracellular space, with minimal to no toxicity (Chiappini, Martinez, et al., 2015). They effectively deliver intracellularly nanoparticles, as well as nucleic acids and proteins, to regulate cell function accordingly (Chiappini, De Rosa, et al., 2015). When employed in vivo, they can locally deliver a payload within soft tissues, where it remains localized in the long term. Different from microneedles, they deliver to the cell cytosol, do not elicit an acute or chronic immune response, and the payload delivery is uniform across the delivery area. Nanoneedles grafted with a biosensing element can specifically sense intracellular enzymatic activity, to discriminate cancer cells from healthy ones (Chiappini, Campagnolo, et al., 2015). Tested on clinical samples of tumor margin regions, this strategy allowed discriminating pockets of tumor biomarker activity across the margin.

6 Conclusions

In the past two and a half decades, a plethora of nanomaterial systems have been designed that cleverly exploit the emergent properties of the nanoscale to provide functionality that matches or surpasses that of the molecular counterparts, and to integrate them within multifunctional systems that actively target diseases based on molecular composition, deliver predesigned amounts of therapeutics with regulated kinetics or as a reaction to the local environment, and that can be tracked with all the clinically relevant imaging modalities. Despite these enormous achievements, theranostics systems have not made their way into the clinic. The regulatory hurdles toward the approval of multifunctional systems with both drug and device components, alongside the technical complexity and costs associated with the fabrication of these technologies at pharmaceutical quality levels on the large scale, and in many instances, the performance gap when the systems are deployed in animal models or in humans, represent the major hurdles still to be faced for translating these technologies.

The newest generation of systems are becoming more efficient, and particularly more functional in clinically relevant settings, and are targeting niche applications where their functionality can clearly fill established unmet needs. Synthetic strategies are becoming more reliable, "green," and cost-effective, and regulatory agencies are becoming better equipped at dealing with nanoagents, highlighting a path forward for nanotheranostics translation.

References

Alam, S. R., Shah, A. S. V., Richards, J., Lang, N. N., Barnes, G., Joshi, N., et al. (2012). Ultrasmall superparamagnetic particles of iron oxide in patients with acute myocardial infarction: early clinical experience. *Circulation Cardiovascular Imaging*, *5*, 559–565.

Alhmoud, H., Delalat, B., Elnathan, R., Cifuentes Rius, A., Chaix, A., Rogers, M. L., et al. (2015). Porous silicon nanodiscs for targeted drug delivery. *Advanced Functional Materials*, *25*, 1137–1145.

Almeida, P. V., Shahbazi, M.-A., Mäkilä, E., Kaasalainen, M., Salonen, J., Hirvonen, J., et al. (2014). Amine-modified hyaluronic acid-functionalized porous silicon nanoparticles for targeting breast cancer tumors. *Nanoscale*, *6*, 10377–10387.

Ananta, J. S., Godin, B., Sethi, R., Moriggi, L., Liu, X., Serda, R. E., et al. (2010). Geometrical confinement of gadolinium-based contrast agents in nanoporous particles enhances T1 contrast. *Nature Nanotechnology*, *5*, 815–821.

Ansari, C., Tikhomirov, G. A., Hong, S. H., Falconer, R. A., Loadman, P. M., Gill, J. H., et al. (2014). Development of novel tumor-targeted theranostic nanoparticles activated by membrane-type matrix metalloproteinases for combined cancer magnetic resonance imaging and therapy. *Small*, *10*, 566–575.

Anzai, Y., Piccoli, C. W., Outwater, E. K., Stanford, W., Bluemke, D. A., Nurenberg, P., et al. (2003). Evaluation of neck and body metastases to nodes with ferumoxtran 10-enhanced MR imaging: phase III safety and efficacy study. *Radiology*, *228*, 777–788.

Ariga, K., Hill, J. P., & Ji, Q. (2007). Layer-by-layer assembly as a versatile bottom-up nanofabrication technique for exploratory research and realistic application. *Physical Chemistry Chemical Physics*, *9*, 2319–2340.

Baaziz, W., Pichon, B. P., Fleutot, S., Liu, Y., Lefevre, C., Greneche, J.-M., et al. (2014). Magnetic iron oxide nanoparticles: reproducible tuning of the size and nanosized-dependent composition, defects, and spin canting. *Journal of Physical Chemistry C*, *118*, 3795–3810.

Batchelor, L., Loni, A., Canham, L. T., Hasan, M., & Coffer, J. L. (2012). Manufacture of mesoporous silicon from living plants and agricultural waste: an environmentally friendly and scalable process. *Silicon*, *4*, 259–266.

Bharde, A. A., Parikh, R. Y., Baidakova, M., Jouen, S., Hannoyer, B., Enoki, T., et al. (2008). Bacteria-mediated precursor-dependent biosynthesis of superparamagnetic iron oxide and iron sulfide nanoparticles. *Langmuir*, *24*, 5787–5794.

Bimbo, L. M., Sarparanta, M., Mäkilä, E., Laaksonen, T., Laaksonen, P., Salonen, J., et al. (2012). Cellular interactions of surface modified nanoporous silicon particles. *Nanoscale*, *4*, 3184–3192.

Bimbo, L. M., Sarparanta, M., Santos, H. A., Airaksinen, A. J., Mäkilä, E., Laaksonen, T., et al. (2010). Biocompatibility of thermally hydrocarbonized porous silicon nanoparticles and their biodistribution in rats. *ACS Nano*, *4*(6), 3023–3032.

Boukherroub, R., Wojtyk, J. T. C., Wayner, D. D. M., & Lockwood, D. J. (2002). Thermal hydrosilylation of undecylenic acid with porous silicon. *Journal of the Electrochemical Society*, *149*, H59–H63.

Braun, F., & von Delft, J. (1999). Superconductivity in ultrasmall metallic grains. *Physical Review B*, *59*, 9527–9544.

Brown, S. D., Nativo, P., Smith, J.-A., Stirling, D., Edwards, P. R., Venugopal, B., et al. (2010). Gold nanoparticles for the improved anticancer drug delivery of the active component of oxaliplatin. *Journal of the American Chemical Society*, *132*, 4678–4684.

Canham, L. T. (2014). *Handbook of porous silicon*. Springer International Publishing.

Carenza, E., Barceló, V., Morancho, A., Montaner, J., Rosell, A., & Roig, A. (2014). Rapid synthesis of water-dispersible superparamagnetic iron oxide nanoparticles by a microwave-assisted route for safe labeling of endothelial progenitor cells. *Acta Biomaterialia*, *10*, 3775–3785.

Chaix, A., El Cheikh, K., Bouffard, E., Maynadier, M., Aggad, D., Stojanovic, V., et al. (2016). Mesoporous silicon nanoparticles for targeted two-photon theranostics of prostate cancer. *Journal of Materials Chemistry B*, *4*, 3639–3642.

Chartier, C., Bastide, S., & Lévy-Clément, C. (2008). Metal-assisted chemical etching of silicon in HF–H2O2. *Electrochimica Acta*, *53*, 5509–5516.

Chen, Y., Chen, H., Sun, Y., Zheng, Y., Zeng, D., Li, F., et al. (2011). Multifunctional mesoporous composite nanocapsules for highly efficient MRI-guided high-intensity focused ultrasound cancer surgery. *Angewandte Chemie*, *123*, 12713–12717.

Chen, F., Hong, H., Goel, S., Graves, S. A., Orbay, H., Ehlerding, E. B., et al. (2015). In vivotumor vasculature targeting of CuS@MSN based theranostic nanomedicine. *ACS Nano*, *9*, 3926–3934.

Chen, D., Li, L., Tang, F., & Qi, S. (2009). Facile and scalable synthesis of tailored silica 'nanorattle' structures. *Advanced Materials*, *21*, 3804–3807.

Chen, Q., Li, K., Wen, S., Liu, H., Peng, C., Cai, H., et al. (2013). Targeted CT/MR dual mode imaging of tumors using multifunctional dendrimer-entrapped gold nanoparticles. *Biomaterials*, *34*, 5200–5209.

Chen, S., Wang, Z. L., Ballato, J., Foulger, S. H., & Carroll, D. L. (2003). Monopod, bipod, tripod, and tetrapod gold nanocrystals. *Journal of the American Chemical Society*, *125*, 16186–16187.

Chen, G., Xie, Y., Peltier, R., Lei, H., Wang, P., Chen, J., et al. (2016). Peptide-decorated gold nanoparticles as functional nano-capping agent of mesoporous silica container for targeting drug delivery. *ACS Applied Materials & Interfaces*, *8*, 11204–11209.

Chen, A. M., Zhang, M., Wei, D., Stueber, D., Taratula, O., Minko, T., et al. (2009). Co-delivery of doxorubicin and Bcl-2 siRNA by mesoporous silica nanoparticles enhances the efficacy of chemotherapy in multidrug-resistant cancer cells. *Small*, *5*, 2673–2677.

Chen, H., Zhen, Z., Tang, W., Todd, T., Chuang, Y.-J., Wang, L., et al. (2013). Label-free luminescent mesoporous silica nanoparticles for imaging and drug delivery. *Theranostics*, *3*, 650–657.

Cheng, Y., C Samia, A., Meyers, J. D., Panagopoulos, I., Fei, B., & Burda, C. (2008). Highly efficient drug delivery with gold nanoparticle vectors for in vivo photodynamic therapy of cancer. *Journal of the American Chemical Society*, *130*, 10643–10647.

Cheng, K., Kothapalli, S.-R., Liu, H., Koh, A. L., Jokerst, J. V., Jiang, H., et al. (2014). Construction and validation of nano gold tripods for molecular imaging of living subjects. *Journal of the American Chemical Society*, *136*, 3560–3571.

Cheng, Y.-J., Luo, G.-F., Zhu, J.-Y., Xu, X.-D., Zeng, X., Cheng, D.-B., et al. (2015). Enzyme-induced and tumor-targeted drug delivery system based on multifunctional mesoporous silica nanoparticles. *ACS Applied Materials & Interfaces*, *7*, 9078–9087.

Chiang, Y.-D., Lian, H.-Y., Leo, S.-Y., Wang, S.-G., Yamauchi, Y., & Wu, K. C. W. (2011). Controlling particle size and structural properties of mesoporous silica nanoparticles using the taguchi method. *Journal of Physical Chemistry C*, *115*, 13158–13165.

Chiappini, C. (2017). Nanoneedle-based sensing in biological systems. *ACS Sensors*, *2*, 1086–1102.

Chiappini, C., Campagnolo, P., Almeida, C. S., Abbassi-Ghadi, N., Chow, L. W., Hanna, G. B., et al. (2015). Mapping local cytosolic enzymatic activity in human esophageal mucosa with porous silicon nanoneedles. *Advanced Materials*, *27*, 5147–5152.

Chiappini, C., De Rosa, E., Martinez, J. O., Liu, X., Steele, J., Stevens, M. M., et al. (2015). Biodegradable silicon nanoneedles delivering nucleic acids intracellularly induce localized in vivo neovascularization. *Nature Materials*, *14*, 532–539.

Chiappini, C., Liu, X., Fakhoury, J. R., & Ferrari, M. (2010). Biodegradable porous silicon barcode nanowires with defined geometry. *Advanced Functional Materials*, *20*, 2231–2239.

Chiappini, C., Martinez, J. O., De Rosa, E., Almeida, C. S., Tasciotti, E., & Stevens, M. M. (2015). Biodegradable nanoneedles for localized delivery of nanoparticles in vivo:exploring the biointerface. *ACS Nano*, *9*, 5500–5509.

Chiappini, C., Tasciotti, E., Fakhoury, J. R., Fine, D., Pullan, L., Wang, Y.-C., et al. (2010). Tailored porous silicon microparticles: fabrication and properties. *ChemPhysChem, 11*, 1029–1035.

Choi, J., Cha, J., & Lee, J.-K. (2013). Synthesis of various magnetite nanoparticles through simple phase transformation and their shape-dependent magnetic properties. *RSC Advances, 3*, 8365–8371.

Copland, J., Eghtedari, M., Popov, V., Kotov, N., Mamedova, N., Motamedi, M., et al. (2004). Bioconjugated gold nanoparticles as a molecular based contrast agent: implications for imaging of deep tumors using optoacoustic tomography. *Molecular Imaging & Biology, 6*, 341–349.

Cullis, A. G., & Canham, L. T. (1991). Visible light emission due to quantum size effects in highly porous crystalline silicon. *Nature, 353*, 335–338.

Cunin, F., Schmedake, T. A., Link, J. R., Li, Y. Y., Koh, J., Bhatia, S. N., et al. (2002). Biomolecular screening with encoded porous-silicon photonic crystals. *Nature Materials, 1*, 39–41.

Dai, C., Yuan, Y., Liu, C., Wei, J., Hong, H., Li, X., et al. (2009). Degradable, antibacterial silver exchanged mesoporous silica spheres for hemorrhage control. *Biomaterials, 30*, 5364–5375.

Darbandi, M., Stromberg, F., Landers, J., Reckers, N., Sanyal, B., Keune, W., et al. (2012). Nanoscale size effect on surface spin canting in iron oxide nanoparticles synthesized by the microemulsion method. *Journal of Physics D: Applied Physics, 45*, 195001.

de La Rica, R., Chow, L. W., Horejs, C.-M., Mazo, M., Chiappini, C., Pashuck, E. T., et al. (2014). A designer peptide as a template for growing Au nanoclusters. *Chemical Communications, 50*, 10648–10650.

De Rosa, E., Chiappini, C., Fan, D., Liu, X., Ferrari, M., & Tasciotti, E. (2011). Agarose surface coating influences intracellular accumulation and enhances payload stability of a nano-delivery system. *Pharmaceutical Research, 28*, 1520–1530.

Decuzzi, P., Godin, B., Tanaka, T., Lee, S. Y., Chiappini, C., Liu, X., et al. (2010). Size and shape effects in the biodistribution of intravascularly injected particles. *Journal of Controlled Release, 141*, 320–327.

Dolores, R., Raquel, S., & Adianez, G.-L. (2015). Sonochemical synthesis of iron oxide nanoparticles loaded with folate and cisplatin: effect of ultrasonic frequency. *Ultrasonics Sonochemistry, 23*, 391–398.

Dong, W., Li, Y., Niu, D., Ma, Z., Gu, J., Chen, Y., et al. (2011). Facile synthesis of monodisperse superparamagnetic Fe3O4 core@hybrid@Au shell nanocomposite for bimodal imaging and photothermal therapy. *Advanced Materials, 23*, 5392–5397.

Dorsey, J. F., Sun, L., Joh, D. Y., Witztum, A., Zaki, A. A., Kao, G. D., et al. (2013). Gold nanoparticles in radiation research: potential applications for imaging and radiosensitization. *Translational Cancer Research, 2*, 280–291.

Dorvee, J. R., Derfus, A. M., Bhatia, S. N., & Sailor, M. J. (2004). Manipulation of liquid droplets using amphiphilic, magnetic one-dimensional photonic crystal chaperones. *Nature Materials, 3*. nmat1253–899.

Du, Y., Jiang, Q., Beziere, N., Song, L., Zhang, Q., Peng, D., et al. (2016). DNA-nanostructure–gold-nanorod hybrids for enhanced in vivo optoacoustic imaging and photothermal therapy. *Advanced Materials, 28*, 10000–10007.

Du, L., Song, H., & Liao, S. (2012). A biocompatible drug delivery nanovalve system on the surface of mesoporous nanoparticles. *Microporous and Mesoporous Materials, 147*, 200–204.

Dudley, M. E., & Kolasinski, K. W. (2009). Stain etching with Fe(III), V(V), and Ce(IV) to form microporous silicon. *Electrochemical and Solid-State Letters, 12*, D22.

Ehlerding, E. B., Chen, F., & Cai, W. (2016). Biodegradable and renal clearable inorganic nanoparticles. *Advanced Science, 3*, 1500223.

Espiard, P., Mark, J. E., & Guyot, A. (1990). A novel technique for preparing organophilic silica by water-in-oil microemulsions. *Polymer Bulletin, 24*, 173–179.

Fabris, L., Antonello, S., Armelao, L., Donkers, R. L., Polo, F., Toniolo, C., et al. (2005). Gold nanoclusters protected by conformationally constrained peptides. *Journal of the American Chemical Society, 128*, 326–336.

Fan, W., Shen, B., Bu, W., Chen, F., He, Q., Zhao, K., et al. (2014). A smart upconversion-based mesoporous silica nanotheranostic system for synergetic chemo-/radio-/photodynamic therapy and simultaneous MR/UCL imaging. *Biomaterials, 35*, 8992–9002.

Fang, S., Lin, J., Li, C., Huang, P., Hou, W., Zhang, C., et al. (2017). Dual-stimuli responsive nanotheranostics for multimodal imaging guided trimodal synergistic therapy. *Small, 13*, 1602580.

Frens, G. (1972). Particle size and sol stability in metal colloids. *Kolloid-Zeitschrift und Zeitschrift für Polymere, 250*, 736–741.

Fuertes, A. B., Valle-Vigón, P., & Sevilla, M. (2010). Synthesis of colloidal silica nanoparticles of a tunable mesopore size and their application to the adsorption of biomolecules. *Journal of Colloid and Interface Science, 349*, 173–180.

Funkhouser, G. P., Arevalo, M. P., Glatzhofer, D. T., & O'Rear, E. A. (1995). Solubilization and adsolubilization of pyrrole by sodium dodecyl sulfate: polypyrrole formation on alumina surfaces. *Langmuir, 11*, 1443–1447.

Gao, L., Vadakkan, T. J., & Nammalvar, V. (2011). Nanoshells for in vivo imaging using two-photon excitation microscopy. *Nanotechnology, 22*, 365102.

Gary-Bobo, M., Mir, Y., Rouxel, C., Brevet, D., Basile, I., Maynadier, M., et al. (2011). Mannose-functionalized mesoporous silica nanoparticles for efficient two-photon photodynamic therapy of solid tumors. *Angewandte Chemie, International Edition, 50*, 11425–11429.

Giménez, C., de la Torre, C., Gorbe, M., Aznar, E., Sancenón, F., Murguía, J. R., et al. (2015). Gated mesoporous silica nanoparticles for the controlled delivery of drugs in cancer cells. *Langmuir, 31*, 3753–3762.

Gobin, A. M., Lee, M. H., Halas, N. J., James, W. D., Drezek, R. A., & West, J. L. (2007). Near-infrared resonant nanoshells for combined optical imaging and photothermal cancer therapy. *Nano Letters, 7*, 1929–1934.

Godin, B., Chiappini, C., Srinivasan, S., Alexander, J. F., Yokoi, K., Ferrari, M., et al. (2012). Discoidal porous silicon particles: fabrication and biodistribution in breast cancer bearing mice. *Advanced Functional Materials, 22*, 4225–4235.

Godin, B., Gu, J., Serda, R. E., Bhavane, R., Tasciotti, E., Chiappini, C., et al. (2010). Tailoring the degradation kinetics of mesoporous silicon structures through PEGylation. *Journal of Biomedical Materials Research A, 94*(4), 1236–1243.

Godin, B., Gu, J., Serda, R. E., Ferrati, S., Liu, X., Chiappini, C., et al. (2008). Multistage mesoporous silicon-based nanocarriers: biocompatibility with immune cells and controlled degradation in physiological fluids. *Controlled Release Newsletter/Controlled Release Society, 25*, 9–11.

Goh, A.S.-W., Chung, A.Y.-F., Lo, R.H.-G., Lau, T.-N., Yu, S.W.-K., Chng, M., et al. (2007). A novel approach to brachytherapy in hepatocellular carcinoma using a phosphorous32 (32P) brachytherapy delivery device—a first-in-man study. *International Journal of Radiation Oncology, Biology, Physics, 67*, 786–792.

Gole, A., & Murphy, C. J. (2008). Azide-derivatized gold nanorods: functional materials for 'click' chemistry. *Langmuir, 24*, 266–272.

Grün, M., Lauer, I., & Unger, K. K. (1997). The synthesis of micrometer- and submicrometer-size spheres of ordered mesoporous oxide MCM-41. *Advanced Materials, 9*, 254–257.

Gu, L., Hall, D. J., Qin, Z., Anglin, E., Joo, J., Mooney, D. J., et al. (2013). In vivo time-gated fluorescence imaging with biodegradable luminescent porous silicon nanoparticles. *Nature Communications, 4*, 2326.

Gu, L., Ruff, L. E., Qin, Z., Corr, M., Hedrick, S. M., & Sailor, M. J. (2012). Multivalent porous silicon nanoparticles enhance the immune activation potency of agonistic CD40 antibody. *Advanced Materials, 24*, 3981–3987.

Guerrero-Martínez, A., Barbosa, S., Pastoriza-Santos, I., & Liz-Marzán, L. M. (2011). Nanostars shine bright for you: colloidal synthesis, properties and applications of branched metallic nanoparticles. *Current Opinion in Colloid & Interface Science, 16*, 118–127.

Han, Y., & Ying, J. Y. (2005). Generalized fluorocarbon-surfactant-mediated synthesis of nanoparticles with various mesoporous structures. *Angewandte Chemie, International Edition, 44*, 288–292.

Han, L., Zhao, J., Zhang, X., Cao, W., Hu, X., Zou, G., et al. (2012). Enhanced siRNA delivery and silencing gold–chitosan nanosystem with surface charge-reversal polymer assembly and good biocompatibility. *ACS Nano, 6*, 7340–7351.

Hayashi, K., Nakamura, M., Sakamoto, W., Yogo, T., Miki, H., Ozaki, S., et al. (2013). Superparamagnetic nanoparticle clusters for cancer theranostics combining magnetic resonance imaging and hyperthermia treatment. *Theranostics, 3*, 366–376.

He, D., He, X., Wang, K., Cao, J., & Zhao, Y. (2012). A light-responsive reversible molecule-gated system using thymine-modified mesoporous silica nanoparticles. *Langmuir, 28*, 4003–4008.

He, Q., Shi, J., Zhu, M., Chen, Y., & Chen, F. (2010). The three-stage in vitro degradation behavior of mesoporous silica in simulated body fluid. *Microporous and Mesoporous Materials*, *131*, 314–320.

Hembury, M., Chiappini, C., Bertazzo, S., Kalber, T. L., Drisko, G. L., Ogunlade, O., et al. (2015). Gold-silica quantum rattles for multimodal imaging and therapy. *Proceedings of the National Academy of Sciences of the United States of America*, *112*, 1959–1964.

Hernández Gil, J., Cobaleda Siles, M., Zabaleta, A., Salassa, L., Calvo, J., & Mareque Rivas, J. C. (2015). An iron oxide nanocarrier loaded with a Pt(IV) prodrug and immunostimulatory dsRNA for combining complementary cancer killing effects. *Advanced Healthcare Materials*, *4*, 1034–1042.

Hirsch, L. R., Stafford, R. J., Bankson, J. A., Sershen, S. R., Rivera, B., Price, R. E., et al. (2003). Nanoshell-mediated near-infrared thermal therapy of tumors under magnetic resonance guidance. *Proceedings of the National Academy of Sciences of the United States of America*, *100*, 13549–13554.

Hong, C., & Lee, C. (2013). In vitro cell tests of pancreatic malignant tumor cells by photothermotherapy based on DMSO porous silicon colloids. *Lasers in Medical Science*, *29*, 221–223.

Hu, M., Chen, J., Li, Z.-Y., Au, L., Hartland, G. V., Li, X., et al. (2006). Gold nanostructures: engineering their plasmonic properties for biomedical applications. *Chemical Society Reviews*, *35*, 1084–1094.

Huang, X., El-Sayed, I. H., Qian, W., & El-Sayed, M. A. (2007). Cancer cells assemble and align gold nanorods conjugated to antibodies to produce highly enhanced, sharp, and polarized surface raman spectra: a potential cancer diagnostic marker. *Nano Letters*, *7*, 1591–1597.

Huang, X., Zhang, F., Lee, S., Swierczewska, M., Kiesewetter, D. O., Lang, L., et al. (2012). Long-term multimodal imaging of tumor draining sentinel lymph nodes using mesoporous silica-based nanoprobes. *Biomaterials*, *33*, 4370–4378.

Hurst, S. J., Payne, E. K., Qin, L., & Mirkin, C. A. (2006). Multisegmented one-dimensional nanorods prepared by hard-template synthetic methods. *Angewandte Chemie, International Edition*, *45*, 2672–2692.

Idris, N. M., Jayakumar, M. K. G., Bansal, A., & Zhang, Y. (2015). Upconversion nanoparticles as versatile light nanotransducers for photoactivation applications. *Chemical Society Reviews*, *44*, 1449–1478.

Indrasekara, A.S.D.S., Meyers, S., Shubeita, S., Feldman, L. C., Gustafsson, T., & Fabris, L. (2014). Gold nanostar substrates for SERS-based chemical sensing in the femtomolar regime. *Nanoscale*, *6*, 8891–8899.

Itoh, H., & Sugimoto, T. (2003). Systematic control of size, shape, structure, and magnetic properties of uniform magnetite and maghemite particles. *Journal of Colloid and Interface Science*, *265*, 283–295.

Jackson, P. A., Rahman, W.N.W.A., Wong, C. J., Ackerly, T., & Geso, M. (2010). Potential dependent superiority of gold nanoparticles in comparison to iodinated contrast agents. *European Journal of Radiology*, *75*, 104–109.

Jain, P. K., & El-Sayed, M. A. (2010). Plasmonic coupling in noble metal nanostructures. *Chemical Physics Letters*, *487*, 153–164.

Jana, N. R., Gearheart, L., & Murphy, C. J. (2001a). Seeding growth for size control of 5–40 nm diameter gold nanoparticles. *Langmuir*, *17*, 6782–6786.

Jana, N. R., Gearheart, L., & Murphy, C. J. (2001b). Seed-mediated growth approach for shape-controlled synthesis of spheroidal and rod-like gold nanoparticles using a surfactant template. *Advanced Materials*, *13*, 1389–1393.

Jeng, H. A., & Swanson, J. (2007). Toxicity of metal oxide nanoparticles in mammalian cells. *Journal of Environmental Science and Health Part A*, *41*, 2699–2711.

Jeon, H., Kim, J., Lee, Y. M., Kim, J., Choi, H. W., Lee, J., et al. (2016). Poly-paclitaxel/cyclodextrin-SPION nano-assembly for magnetically guided drug delivery system. *Journal of Controlled Release*, *231*, 68–76.

Ji, X., Song, X., Li, J., Bai, Y., Yang, W., & Peng, X. (2007). Size control of gold nanocrystals in citrate reduction: the third role of citrate. *Journal of the American Chemical Society*, *129*, 13939–13948.

Jin, R. (2010). Quantum sized, thiolate-protected gold nanoclusters. *Nanoscale*, *2*, 343–362.

Jin, R., Egusa, S., & Scherer, N. F. (2004). Thermally-induced formation of atomic Au clusters and conversion into nanocubes. *Journal of the American Chemical Society*, *126*, 9900–9901.

Jin, Y., & Gao, X. (2009a). Spectrally tunable leakage-free gold nanocontainers. *Journal of the American Chemical Society*, *131*, 17774–17776.

Jin, Y., & Gao, X. (2009b). Plasmonic fluorescent quantum dots. *Nature Nanotechnology*, *4*, 571–576.

Jin, Y., Jia, C., Huang, S.-W., O'Donnell, M., & Gao, X. (2010). Multifunctional nanoparticles as coupled contrast agents. *Nature Communications*, *1*. ncomms1042–8.

Joo, J., Liu, X., Kotamraju, V. R., Ruoslahti, E., Nam, Y., & Sailor, M. J. (2015). Gated luminescence imaging of silicon nanoparticles. *ACS Nano*, *9*, 6233–6241.

Jun, T., Yong-Hua, L., Rong-Sheng, Z., Kai-Qun, L., Zhi-Guo, X., Zhao-Feng, L., et al. (2008). Effect of aspect ratio distribution on localized surface plasmon resonance extinction spectrum of gold nanorods. *Chinese Physics Letters*, *25*, 4459–4462.

Karlsson, H. L., Gustafsson, J., Cronholm, P., & Möller, L. (2009). Size-dependent toxicity of metal oxide particles—a comparison between nano- and micrometer size. *Toxicology Letters*, *188*, 112–118.

Kecht, J., Schlossbauer, A., & Bein, T. (2008). Selective functionalization of the outer and inner surfaces in mesoporous silica nanoparticles. *Chemistry of Materials*, *20*, 7207–7214.

Kim, C., Cho, E. C., Chen, J., Song, K. H., Au, L., Favazza, C., et al. (2010). In vivo molecular photoacoustic tomography of melanomas targeted by bioconjugated gold nanocages. *ACS Nano*, *4*, 4559–4564.

Kim, T.-W., Chung, P.-W., & Lin, V. S. Y. (2010). Facile synthesis of monodisperse spherical MCM-48 mesoporous silica nanoparticles with controlled particle size. *Chemistry of Materials*, *22*, 5093–5104.

Kim, J., Kim, H. S., Lee, N., Kim, T., Kim, H., Yu, T., et al. (2008). Multifunctional uniform nanoparticles composed of a magnetite nanocrystal core and a mesoporous silica shell for magnetic resonance and fluorescence imaging and for drug delivery. *Angewandte Chemie, International Edition*, *47*, 8438–8441.

Kim, B. H., Lee, N., Kim, H., An, K., Park, Y. I., Choi, Y., et al. (2011). Large-scale synthesis of uniform and extremely small-sized iron oxide nanoparticles for high-resolution T1 magnetic resonance imaging contrast agents. *Journal of the American Chemical Society*, *133*, 12624–12631.

Kim, M.-H., Na, H.-K., Kim, Y.-K., Ryoo, S.-R., Cho, H. S., Lee, K. E., et al. (2011). Facile synthesis of monodispersed mesoporous silica nanoparticles with ultralarge pores and their application in gene delivery. *ACS Nano*, *5*, 3568–3576.

Kim, J. S., Yoon, T.-J., Yu, K. N., Kim, B. G., Park, S. J., Kim, H. W., et al. (2006). Toxicity and tissue distribution of magnetic nanoparticles in mice. *Toxicological Sciences*, *89*, 338–347.

Kircher, M. F., de la Zerda, A., Jokerst, J. V., Zavaleta, C. L., Kempen, P. J., Mittra, E., et al. (2012). A brain tumor molecular imaging strategy using a new triple-modality MRI-photoacoustic-Raman nanoparticle. *Nature Medicine*, *18*, 829–834.

Kolasinski, K. W. (2005). Silicon nanostructures from electroless electrochemical etching. *Current Opinion in Solid State and Materials Science*, *9*, 73–83.

Kolasinski, K. W., Gimbar, N. J., Yu, H., Aindow, M., Mäkilä, E., & Salonen, J. (2017). Regenerative electroless etching of silicon. *Angewandte Chemie, International Edition*, *56*, 624–627.

Koole, R., van Schooneveld, M. M., Hilhorst, J., Castermans, K., Cormode, D. P., Strijkers, G. J., et al. (2008). Paramagnetic lipid-coated silica nanoparticles with a fluorescent quantum dot core: a new contrast agent platform for multimodality imaging. *Bioconjugate Chemistry*, *19*, 2471–2479.

Kumar, P. S., Pastoriza-Santos, I., Rodríguez-González, B., de Abajo, F. J. G., & Liz-Marzán, L. M. (2008). High-yield synthesis and optical response of gold nanostars. *Nanotechnology*, *19*, 015606.

Lai, C.-Y., Trewyn, B. G., Jeftinija, D. M., Jeftinija, K., Shu, X., Jeftinija, S., et al. (2003). A mesoporous silica nanosphere-based carrier system with chemically removable CdS nanoparticle caps for stimuli-responsive controlled release of neurotransmitters and drug molecules. *Journal of the American Chemical Society*, *125*, 4451–4459.

Le Ru, E. C., & Etchegoin, P. G. (2009). *Principles of surface-enhanced Raman spectroscopy*. (pp. 1–27). Elsevier. https://doi.org/10.1016/B978-0-444-52779-0.00007-6.

Le Ru, E. C., Etchegoin, P. G., & Meyer, M. (2006). Enhancement factor distribution around a single surface-enhanced Raman scattering hot spot and its relation to single molecule detection. *The Journal of Chemical Physics*, *125*, 204701.

Lee, C. H., Cheng, S. H., Wang, Y. J., Chen, Y. C., Chen, N.-T., Souris, J., et al. (2009). Near-infrared mesoporous silica nanoparticles for optical imaging: characterization and in vivo biodistribution. *Advanced Functional Materials*, *19*, 215–222.

Lee, J. E., Lee, N., Kim, H., Kim, J., Choi, S. H., Kim, J. H., et al. (2010). Uniform mesoporous dye-doped silica nanoparticles decorated with multiple magnetite nanocrystals for simultaneous enhanced magnetic resonance imaging, fluorescence imaging, and drug delivery. *Journal of the American Chemical Society, 132*, 552–557.

Lee, C. H., Park, S. H., Chung, W., Kim, J. Y., & Kim, S. H. (2011). Preparation and characterization of surface modified silica nanoparticles with organo-silane compounds. *Colloids and Surfaces A: Physicochemical and Engineering Aspects, 384*, 318–322.

Leff, D. V., Ohara, P. C., Heath, J. R., & Gelbart, W. M. (2002). Thermodynamic control of gold nanocrystal size: experiment and theory. *The Journal of Physical Chemistry, 99*, 7036–7041.

Li, Z., Dong, K., Huang, S., Ju, E., Liu, Z., Yin, M., et al. (2014). A smart nanoassembly for multistage targeted drug delivery and magnetic resonance imaging. *Advanced Functional Materials, 24*, 3612–3620.

Li, L., Gao, F., Jiang, W., Wu, X., Cai, Y., Tang, J., et al. (2015). Folic acid-conjugated superparamagnetic iron oxide nanoparticles for tumor-targeting MR imaging. *Drug Delivery, 9*, 1–8.

Lin, S.-Y., Chen, N.-T., Sum, S.-P., Lo, L.-W., & Yang, C.-S. (2008). Ligand exchanged photoluminescent gold quantum dots functionalized with leading peptides for nuclear targeting and intracellular imaging. *Chemical Communications*, 4762–4764.

Lin, J., Li, Y., Li, Y., Wu, H., Yu, F., Zhou, S., et al. (2015). Drug/dye-loaded, multifunctional PEG–chitosan–iron oxide nanocomposites for methotraxate synergistically self-targeted cancer therapy and dual model imaging. *ACS Applied Materials & Interfaces, 7*, 11908–11920.

Lin, V. S. Y., Motesharei, K., Dancil, K.-P.S., Sailor, M. J., & Ghadiri, M. R. (1997). A porous silicon-based optical interferometric biosensor. *Science, 278*, 840–843.

Lin, Y.-S., Tsai, C.-P., Huang, H.-Y., Kuo, C.-T., Hung, Y., Huang, D.-M., et al. (2005). Well-ordered mesoporous silica nanoparticles as cell markers. *Chemistry of Materials, 17*, 4570–4573.

Lin, J., Wang, S., Huang, P., Wang, Z., Chen, S., Niu, G., et al. (2013). Photosensitizer-loaded gold vesicles with strong plasmonic coupling effect for imaging-guided photothermal/photodynamic therapy. *ACS Nano, 7*, 5320–5329.

Lin, C.-A.J., Yang, T.-Y., Lee, C.-H., Huang, S. H., Sperling, R. A., Zanella, M., et al. (2009). Synthesis, characterization, and bioconjugation of fluorescent gold nanoclusters toward biological labeling applications. *ACS Nano, 3*, 395–401.

Lin, X., Zhao, N., Yan, P., Hu, H., & Xu, F.-J. (2015). The shape and size effects of polycation functionalized silica nanoparticles on gene transfection. *Acta Biomaterialia, 11*, 381–392.

Liu, J., Bu, W., Zhang, S., Chen, F., Xing, H., Pan, L., et al. (2012). Controlled synthesis of uniform and monodisperse upconversion core/mesoporous silica shell nanocomposites for bimodal imaging. *Chemistry - A European Journal, 18*, 2335–2341.

Liu, H., Chen, D., Li, L., Liu, T., Tan, L., Wu, X., et al. (2011). Multifunctional gold nanoshells on silica nanorattles: a platform for the combination of photothermal therapy and chemotherapy with low systemic toxicity. *Angewandte Chemie, International Edition, 50*, 891–895.

Liu, Y., Chen, T., Wu, C., Qiu, L., Hu, R., Li, J., et al. (2014). Facile surface functionalization of hydrophobic magnetic nanoparticles. *Journal of the American Chemical Society, 136*, 12552–12555.

Liu, J., Detrembleur, C., De Pauw Gillet, M. C., Mornet, S., Jérôme, C., & Duguet, E. (2015). Gold nanorods coated with mesoporous silica shell as drug delivery system for remote near infrared light-activated release and potential phototherapy. *Small, 11*, 2323–2332.

Liu, C., Qing, Z., Zheng, J., Deng, L., Ma, C., Li, J., et al. (2015). DNA-templated in situ growth of silver nanoparticles on mesoporous silica nanospheres for smart intracellular GSH-controlled release. *Chemical Communications, 51*, 6544–6547.

Liu, H., Wang, T., Zhang, L., Li, L., Wang, Y. A., Wang, C., et al. (2012). Selected-control fabrication of multifunctional fluorescent–magnetic core–shell and yolk–shell hybrid nanostructures. *Chemistry - A European Journal, 18*, 3745–3752.

Liu, C. L., Wu, H. T., Hsiao, Y. H., Lai, C. W., Shih, C. W., Peng, Y. K., et al. (2011). Insulin-directed synthesis of fluorescent gold nanoclusters: preservation of insulin bioactivity and versatility in cell imaging. *Angewandte Chemie, International Edition, 50*, 7056–7060.

Liu, J., Yu, M., Zhou, C., Yang, S., Ning, X., & Zheng, J. (2013). Passive tumor targeting of renal-clearable luminescent gold nanoparticles: long tumor retention and fast normal tissue clearance. *Journal of the American Chemical Society, 135*, 4978–4981.

Liu, D., Zhang, H., Mäkilä, E., Fan, J., Herranz-Blanco, B., Wang, C.-F., et al. (2015). Microfluidic assisted one-step fabrication of porous silicon@acetalated dextran nanocomposites for precisely controlled combination chemotherapy. *Biomaterials*, *39*, 249–259.

Liu, R., Zhang, Y., Zhao, X., Agarwal, A., Mueller, L. J., & Feng, P. (2010). pH-responsive nanogated ensemble based on gold-capped mesoporous silica through an acid-labile acetal linker. *Journal of the American Chemical Society*, *132*, 1500–1501.

Loni, A., Barwick, D., Batchelor, L., Tunbridge, J., Han, Y., Li, Z. Y., et al. (2011). Extremely high surface area metallurgical-grade porous silicon powder prepared by metal-assisted etching. *Electrochemical and Solid-State Letters*, *14*, K25.

Low, S. P., Voelcker, N. H., Canham, L. T., & Williams, K. A. (2009). The biocompatibility of porous silicon in tissues of the eye. *Biomaterials*, *30*, 2873–2880.

Lu, F.-K., Calligaris, D., Olubiyi, O. I., Norton, I., Yang, W., Santagata, S., et al. (2016). Label-free neurosurgical pathology with stimulated raman imaging. *Cancer Research*, *76*, 3451–3462.

Lu, Y., & Chen, W. (2012). Sub-nanometre sized metal clusters: from synthetic challenges to the unique property discoveries. *Chemical Society Reviews*, *41*, 3594–3623.

Lu, Y., Fan, H., Stump, A., Ward, T. L., Rieker, T., & Brinker, C. J. (1999). Aerosol-assisted self-assembly of mesostructured spherical nanoparticles. *Nature*, *398*, 223–226.

Lu, C.-W., Hung, Y., Hsiao, J. K., Yao, M., Chung, T.-H., Lin, Y.-S., et al. (2007). Bifunctional magnetic silica nanoparticles for highly efficient human stem cell labeling. *Nano Letters*, *7*, 149–154.

Lu, J., Li, Z., Zink, J. I., & Tamanoi, F. (2012). In vivo tumor suppression efficacy of mesoporous silica nanoparticles-based drug-delivery system: enhanced efficacy by folate modification. *Nanomedicine: Nanotechnology, Biology and Medicine*, *8*, 212–220.

Lu, J., Liong, M., Li, Z., Zink, J. I., & Tamanoi, F. (2010). Biocompatibility, biodistribution, and drug-delivery efficiency of mesoporous silica nanoparticles for cancer therapy in animals. *Small*, *6*, 1794–1805.

Lv, R., Yang, P., He, F., Gai, S., Li, C., Dai, Y., et al. (2015). A yolk-like multifunctional platform for multimodal imaging and synergistic therapy triggered by a single near-infrared light. *ACS Nano*, *9*, 1630–1647.

Maity, D., Ding, J., & Xue, J.-M. (2008). Synthesis of magnetite nanoparticles by thermal decomposition: time, temperature surfactant and solvent effects. *Functional Materials Letters*, *01*, 189–193.

Makarov, V. V., Makarova, S. S., Love, A. J., Sinitsyna, O. V., Dudnik, A. O., Yaminsky, I. V., et al. (2014). Biosynthesis of stable iron oxide nanoparticles in aqueous extracts of Hordeum vulgare and Rumex acetosa plants. *Langmuir*, *30*, 5982–5988.

Malikova, N., Pastoriza-Santos, I., Schierhorn, M., Kotov, N. A., & Liz-Marzán, L. M. (2002). Layer-by-layer assembled mixed spherical and planar gold nanoparticles: control of interparticle interactions. *Langmuir*, *18*, 3694–3697.

Martin, C. R. (1996). Membrane-based synthesis of nanomaterials. *Chemistry of Materials*, *8*, 1739–1746.

Martinez, J. O., Boada, C., Yazdi, I. K., Evangelopoulos, M., Brown, B. S., Liu, X., et al. (2012). Short and long term, in vitro and in vivo correlations of cellular and tissue responses to mesoporous silicon nanovectors. *Small*, *9*, 1722–1733.

Martinez, J. O., Evangelopoulos, M., Chiappini, C., Liu, X., Ferrari, M., & Tasciotti, E. (2013). Degradation and biocompatibility of multistage nanovectors in physiological systems. *Journal of Biomedical Materials Research*, *102*, 3540–3549.

Meng, H., Liong, M., Xia, T., Li, Z., Ji, Z., Zink, J. I., et al. (2010). Engineered design of mesoporous silica nanoparticles to deliver doxorubicin and P-glycoprotein siRNA to overcome drug resistance in a cancer cell line. *ACS Nano*, *4*, 4539–4550.

Meng, H., Xue, M., Xia, T., Zhao, Y.-L., Tamanoi, F., Stoddart, J. F., et al. (2010). Autonomous in vitro anticancer drug release from mesoporous silica nanoparticles by pH-sensitive nanovalves. *Journal of the American Chemical Society*, *132*, 12690–12697.

Nakamura, T., Mizutani, M., Nozaki, H., Suzuki, N., & Yano, K. (2007). Formation mechanism for monodispersed mesoporous silica spheres and its application to the synthesis of core/shell particles. *Journal of Physical Chemistry C*, *111*, 1093–1100.

Näkki, S., Rytkönen, J., Nissinen, T., Florea, C., Riikonen, J., Ek, P., et al. (2015). Improved stability and biocompatibility of nanostructured silicon drug carrier for intravenous administration. *Acta Biomaterialia*, *13*, 207–215.

Negishi, Y., Nobusada, K., & Tsukuda, T. (2005). Glutathione-protected gold clusters revisited: bridging the gap between gold(I)-thiolate complexes and thiolate-protected gold nanocrystals. *Journal of the American Chemical Society*, *127*, 5261–5270.

Nicewarner-Peña, S. R., Freeman, R. G., Reiss, B. D., He, L., Peña, D. J., Walton, I. D., et al. (2001). Submicrometer metallic barcodes. *Science*, *294*, 137–141.

Nikoobakht, B., Wang, J., & El-Sayed, M. A. (2002). Surface-enhanced Raman scattering of molecules adsorbed on gold nanorods: off-surface plasmon resonance condition. *Chemical Physics Letters*, *366*, 17–23.

Nooney, R. I., Thirunavukkarasu, D., Chen, Y., Josephs, R., & Ostafin, A. E. (2002). Synthesis of nanoscale mesoporous silica spheres with controlled particle size. *Chemistry of Materials*, *14*, 4721–4728.

Oh, C., Lee, J.-H., Lee, Y.-G., Lee, Y.-H., Kim, J.-W., Kang, H.-H., et al. (2006). New approach to the immobilization of glucose oxidase on non-porous silica microspheres functionalized by (3-aminopropyl) trimethoxysilane (APTMS). *Colloids and Surfaces B: Biointerfaces*, *53*, 225–232.

Okoli, C., Boutonnet, M., Mariey, L., Järås, S., & Rajarao, G. (2011). Application of magnetic iron oxide nanoparticles prepared from microemulsions for protein purification. *Journal of Chemical Technology and Biotechnology*, *86*, 1386–1393.

Oldenburg, S. J., Averitt, R. D., Westcott, S. L., & Halas, N. J. (1998). Nanoengineering of optical resonances. *Chemical Physics Letters*, *288*, 243–247.

Osminkina, L. A., Sivakov, V. A., Mysov, G. A., Georgobiani, V. A., Natashina, U.A., Talkenberg, F., et al. (2014). Nanoparticles prepared from porous silicon nanowires for bio-imaging and sonodynamic therapy. *Nanoscale Research Letters*, *9*(1), 463.

Osminkina, L. A., Tamarov, K. P., Sviridov, A. P., Galkin, R. A., Gongalsky, M. B., Solovyev, V. V., et al. (2012). Photoluminescent biocompatible silicon nanoparticles for cancer theranostic applications. *Journal of Biophotonics*, *5*, 529–535.

Otsuka, H., Nagasaki, Y., & Kataoka, K. (2003). PEGylated nanoparticles for biological and pharmaceutical applications. *Advanced Drug Delivery Reviews*, *55*, 403–419.

Pan, L., He, Q., Liu, J., Chen, Y., Ma, M., Zhang, L., et al. (2012). Nuclear-targeted drug delivery of TAT peptide-conjugated monodisperse mesoporous silica nanoparticles. *Journal of the American Chemical Society*, *134*(13), 5722–5725.

Pan, D., Pramanik, M., Senpan, A., Ghosh, S., Wickline, S. A., Wang, L. V., et al. (2010). Near infrared photoacoustic detection of sentinel lymph nodes with gold nanobeacons. *Biomaterials*, *31*, 4088–4093.

Park, J.-H., Gu, L., von Maltzahn, G., Ruoslahti, E., Bhatia, S. N., & Sailor, M. J. (2009). Biodegradable luminescent porous silicon nanoparticles for in vivo applications. *Nature Materials*, *8*, 331–336.

Parodi, A., Quattrocchi, N., van de Ven, A. L., Chiappini, C., Evangelopoulos, M., Martinez, J. O., et al. (2012). Synthetic nanoparticles functionalized with biomimetic leukocyte membranes possess cell-like functions. *Nature Nanotechnology*, *8*, 61–68.

Pastoriza-Santos, I., Alvarez Puebla, R. A., & Liz-Marzán, L. M. (2010). Synthetic routes and plasmonic properties of noble metal nanoplates. *European Journal of Inorganic Chemistry*, *2010*, 4288–4297.

Perrault, S. D., Walkey, C., Jennings, T., Fischer, H. C., & Chan, W. C. W. (2009). Mediating tumor targeting efficiency of nanoparticles through design. *Nano Letters*, *9*, 1909–1915.

Peter, L. M., Riley, D. J., Wielgosz, R. I., Snow, P. A., Penty, R. V., White, I. H., et al. (1996). Mechanisms of luminescence tuning and quenching in porous silicon. *Thin Solid Films*, *276*, 123–129.

Poon, L., Zandberg, W., Hsiao, D., Erno, Z., Sen, D., Gates, B. D., et al. (2010). Photothermal release of single-stranded DNA from the surface of gold nanoparticles through controlled denaturating and Au-S bond breaking. *ACS Nano*, *4*, 6395–6403.

Puddu, V., & Perry, C. C. (2012). Peptide adsorption on silica nanoparticles: evidence of hydrophobic interactions. *ACS Nano*, *6*, 6356–6363.

Qian, H., Zhu, Y., & Jin, R. (2012). Atomically precise gold nanocrystal molecules with surface plasmon resonance. *Proceedings of the National Academy of Sciences of the United States of America*, *109*, 696–700.

Quaresma, P., Osório, I., Dória, G., Carvalho, P. A., Pereira, A., Langer, J., et al. (2014). Star-shaped magnetite@gold nanoparticles for protein magnetic separation and SERS detection. *RSC Advances*, *4*, 3659–3667.

Quinto, C. A., Mohindra, P., Tong, S., & Bao, G. (2015). Multifunctional superparamagnetic iron oxide nanoparticles for combined chemotherapy and hyperthermia cancer treatment. *Nanoscale*, *7*, 12728–12736.

Ramesh, R., Rajalakshmi, M., Muthamizhchelvan, C., & Ponnusamy, S. (2012). Synthesis of Fe_3O_4 nanoflowers by one pot surfactant assisted hydrothermal method and its properties. *Materials Letters*, *70*, 73–75.

Repnik, U., Česen, M. H., & Turk, B. (2013). The endolysosomal system in cell death and survival. *Cold Spring Harbor Perspectives in Biology*, *5*, a008755.

Retnakumari, A., Setua, S., Menon, D., Ravindran, P., Muhammed, H., Pradeep, T., et al. (2010). Molecular-receptor-specific, non-toxic, near-infrared-emitting Au cluster-protein nanoconjugates for targeted cancer imaging. *Nanotechnology*, *21*, 055103.

Rittikulsittichai, S., Kolhatkar, A. G., Sarangi, S., Vorontsova, M. A., Vekilov, P. G., Brazdeikis, A., et al. (2016). Multi-responsive hybrid particles: thermo-, pH-, photo-, and magneto-responsive magnetic hydrogel cores with gold nanorod optical triggers. *Nanoscale*, *8*, 11851–11861.

Rodríguez-Lorenzo, L., Alvarez Puebla, R. A., de Abajo, F. J. G., & Liz-Marzán, L. M. (2010). Surface enhanced Raman scattering using star-shaped gold colloidal nanoparticles. *Journal of Physical Chemistry C*, *114*, 7336–7340.

Ruiz-Hernández, E., Baeza, A., & Vallet-Regí, M. (2011). Smart drug delivery through DNA/magnetic nanoparticle gates. *ACS Nano*, *5*, 1259–1266.

Salonen, J., Laine, E., & Niinistö, L. (2002). Thermal carbonization of porous silicon surface by acetylene. *Journal of Applied Physics*, *91*, 456.

Salonen, J., Laitinen, L., Kaukonen, A., Tuura, J., Bjorkqvist, M., Heikkila, T., et al. (2005). Mesoporous silicon microparticles for oral drug delivery: loading and release of five model drugs. *Journal of Controlled Release*, *108*, 362–374.

Salonen, J., & Mäkilä, E. (2018). Thermally carbonized porous silicon and its recent applications. *Advanced Materials*, https://doi.org/10.1002/adma.201703819.

Salvati, A., Pitek, A. S., Monopoli, M. P., Prapainop, K., Bombelli, F. B., Hristov, D. R., et al. (2013). Transferrin-functionalized nanoparticles lose their targeting capabilities when a biomolecule corona adsorbs on the surface. *Nature Nanotechnology*, *8*, 137–143.

Samanta, A., Maiti, K. K., Soh, K. S., Liao, X., Vendrell, M., Dinish, U. S., et al. (2011). Ultrasensitive near-infrared Raman reporters for SERS-based in vivo cancer detection. *Angewandte Chemie, International Edition*, *50*, 6089–6092.

Samanta, B., Yang, X. C., Ofir, Y., Park, M. H., Patra, D., Agasti, S. S., et al. (2009). Catalytic microcapsules assembled from enzyme–nanoparticle conjugates at oil–water interfaces. *Angewandte Chemie, International Edition*, *48*, 5341–5344.

Scialabba, C., Puleio, R., Peddis, D., Varvaro, G., Calandra, P., Cassata, G., et al. (2017). Folate targeted coated SPIONs as efficient tool for MRI. *Nano Research*, *10*, 3212–3227.

Secret, E., Maynadier, M., Gallud, A., Gary-Bobo, M., Chaix, A., Belamie, E., et al. (2013). Anionic porphyrin-grafted porous silicon nanoparticles for photodynamic therapy. *Chemical Communications*, *49*, 4202–4204.

Shahbazi, M.-A., Fernández, T. D., Mäkilä, E. M., Le Guével, X., Mayorga, C., Kaasalainen, M. H., et al. (2014). Surface chemistry dependent immunostimulative potential of porous silicon nanoplatforms. *Biomaterials*, *35*, 9224–9235.

Shang, L., Nienhaus, K., & Nienhaus, G. U. (2014). Engineered nanoparticles interacting with cells: size matters. *Journal of Nanbiotechnology*, *12*, 5.

Shao, Y., Tian, X., Hu, W., Zhang, Y., Liu, H., He, H., et al. (2012). The properties of Gd2O3-assembled silica nanocomposite targeted nanoprobes and their application in MRI. *Biomaterials*, *33*, 6438–6446.

Shen, D., Yang, J., Li, X., Zhou, L., Zhang, R., Li, W., et al. (2014). Biphase stratification approach to three-dimensional dendritic biodegradable mesoporous silica nanospheres. *Nano Letters*, *14*, 923–932.

Shichibu, Y., Negishi, Y., Tsukuda, T., & Teranishi, T. (2005). Large-scale synthesis of thiolated Au 25 clusters via ligand exchange reactions of phosphine-stabilized Au 11 clusters. *Journal of the American Chemical Society*, *127*, 13464–13465.

Simard, J., Briggs, C., Boal, A. K., & Rotello, V. M. (2000). Formation and pH-controlled assembly of amphiphilic gold nanoparticles. *Chemical Communications*, 1943–1944.

Singh, R. K., Kim, T. H., Mahapatra, C., Patel, K. D., & Kim, H. W. (2015). Preparation of self-activated fluorescence mesoporous silica hollow nanoellipsoids for theranostics. *Langmuir*, *31*, 11344–11352.

Singh, R. K., Kim, T. H., Patel, K. D., Knowles, J. C., & Kim, H. W. (2012). Biocompatible magnetite nanoparticles with varying silica-coating layer for use in biomedicine: physicochemical and magnetic properties, and cellular compatibility. *Journal of Biomedical Materials Research Part A*, *100*, 1734–1742.

Sivaraman, S. K., Kumar, S., & Santhanam, V. (2011). Monodisperse sub-10 nm gold nanoparticles by reversing the order of addition in Turkevich method—the role of chloroauric acid. *Journal of Colloid and Interface Science*, *361*, 543–547.

Skrabalak, S. E., Au, L., Li, X., & Xia, Y. (2007). Facile synthesis of Ag nanocubes and Au nanocages. *Nature Protocols*, *2*, 2182–2190.

Slowing, I. I., Trewyn, B. G., & Lin, V. S. Y. (2007). Mesoporous silica nanoparticles for intracellular delivery of membrane-impermeable proteins. *Journal of the American Chemical Society*, *129*, 8845–8849.

Smith, B. R., Heverhagen, J., Knopp, M., Schmalbrock, P., Shapiro, J., Shiomi, M., et al. (2007). Localization to atherosclerotic plaque and biodistribution of biochemically derivatized superparamagnetic iron oxide nanoparticles (SPIONs) contrast particles for magnetic resonance imaging (MRI). *Biomedical Microdevices*, *9*, 719–727.

Song, X., Gong, H., Yin, S., Cheng, L., Wang, C., Li, Z., et al. (2014). Ultra-small iron oxide doped polypyrrole nanoparticles for in vivo multimodal imaging guided photothermal therapy. *Advanced Functional Materials*, *24*, 1194–1201.

Song, Y., Huang, Z., Xu, J., Ren, D., Wang, Y., Zheng, X., et al. (2014). Multimodal SPION-CREKA peptide based agents for molecular imaging of microthrombus in a rat myocardial ischemia-reperfusion model. *Biomaterials*, *35*, 2961–2970.

Song, K. H., Kim, C., Maslov, K., & Wang, L. V. (2009). Noninvasive in vivo spectroscopic nanorod-contrast photoacoustic mapping of sentinel lymph nodes. *European Journal of Radiology*, *70*, 227–231.

Song, J. H., & Sailor, M. J. (1998). Dimethyl sulfoxide as a mild oxidizing agent for porous silicon and its effect on photoluminescence. *Inorganic Chemistry*, *37*, 3355–3360.

Stallmach, F., Kärger, J., Krause, C., Jeschke, M., & Oberhagemann, U. (2000). Evidence of anisotropic self-diffusion of guest molecules in nanoporous materials of MCM-41 type. *Journal of the American Chemical Society*, *122*, 9237–9242.

Stöber, W., Fink, A., & Bohn, E. (1968). Controlled growth of monodisperse silica spheres in the micron size range. *Journal of Colloid and Interface Science*, *26*, 62–69.

Sun, Y., & Xia, Y. (2002). Shape-controlled synthesis of gold and silver nanoparticles. *Science*, *298*, 2176–2179.

Takahashi, H., Niidome, Y., & Yamada, S. (2005). Controlled release of plasmid DNA from gold nanorods induced by pulsed near-infrared light. *Chemical Communications*, 2247–2249.

Tamarov, K. P., Osminkina, L. A., Zinovyev, S. V., Maximova, K. A., Kargina, J. V., Gongalsky, M. B., et al. (2014). Radio frequency radiation-induced hyperthermia using Si nanoparticle-based sensitizers for mild cancer therapy. *Scientific Reports*, *4*. srep07034.

Tanaka, T., Mangala, L. S., Vivas-Mejia, P. E., Nieves-Alicea, R., Mann, A. P., Mora, E., et al. (2010). Sustained small interfering RNA delivery by mesoporous silicon particles. *Cancer Research*, *70*, 3687–3696.

Tasciotti, E., Liu, X., Bhavane, R., Plant, K., Leonard, A. D., Price, B. K., et al. (2008). Mesoporous silicon particles as a multistage delivery system for imaging and therapeutic applications. *Nature Nanotechnology*, *3*, 151–157.

Taton, T. A., Mucic, R. C., Mirkin, C. A., & Letsinger, R. L. (2000). The DNA-mediated formation of supramolecular mono- and multilayered nanoparticle structures. *Journal of the American Chemical Society*, *122*, 6305–6306.

Taylor, K. M. L., Kim, J. S., Rieter, W. J., An, H., Lin, W., & Lin, W. (2008). Mesoporous silica nanospheres as highly efficient MRI contrast agents. *Journal of the American Chemical Society*, *130*, 2154–2155.

Thanh, N. T. K., Maclean, N., & Mahiddine, S. (2014). Mechanisms of nucleation and growth of nanoparticles in solution. *Chemical Reviews*, *114*, 7610–7630.

Thoeny, H. C., Triantafyllou, M., Birkhaeuser, F. D., Froehlich, J. M., Tshering, D. W., Binser, T., et al. (2009). Combined ultrasmall superparamagnetic particles of iron oxide-enhanced and diffusion-weighted magnetic resonance imaging reliably detect pelvic lymph node metastases in normal-sized nodes of bladder and prostate cancer patients. *European Urology*, *55*, 761–769.

Trivedi, R. A., U-King-Im, J.-M., Graves, M. J., Cross, J. J., Horsley, J., Goddard, M. J., et al. (2004). In vivo detection of macrophages in human carotid atheroma: temporal dependence of ultrasmall superparamagnetic particles of iron oxide-enhanced MRI. *Stroke*, *35*, 1631–1635.

Tukmachev, D., Lunov, O., Zablotskii, V., Dejneka, A., Babic, M., Syková, E., et al. (2015). An effective strategy of magnetic stem cell delivery for spinal cord injury therapy. *Nanoscale*, *7*, 3954–3958.

Turkevich, J., Stevenson, P. C., & Hillier, J. (1951). A study of the nucleation and growth processes in the synthesis of colloidal gold. *Discussions of the Faraday Society*, *11*, 55–75.

Tzur-Balter, A., Shatsberg, Z., Beckerman, M., Segal, E., & Artzi, N. (2015). Mechanism of erosion of nanostructured porous silicon drug carriers in neoplastic tissues. *Nature Communications*, *6*, 6208.

Vaidya, S., Thaplyal, P., & Ganguli, A. K. (2011). Enhanced functionalization of Mn2 O3 @SiO2 core-shell nanostructures. *Nanoscale Research Letters*, *6*, 169.

Vikram, S., Dhakshnamoorthy, M., Vasanthakumari, R., Rajamani, A. R., Rangarajan, M., & Tsuzuki, T. (2015). Tuning the magnetic properties of iron oxide nanoparticles by a room-temperature air-atmosphere (RTAA) co-precipitation method. *Journal of Nanoscience and Nanotechnology*, *15*, 3870–3878.

Vivek, R., Thangam, R., Kumar, S. R., Rejeeth, C., Sivasubramanian, S., Vincent, S., et al. (2016). HER2 targeted breast cancer therapy with switchable "off/on" multifunctional "smart" magnetic polymer core–shell nanocomposites. *ACS Applied Materials & Interfaces*, *8*, 2262–2279.

Wan, D., Xia, X., Wang, Y., & Xia, Y. (2013). Robust synthesis of gold cubic nanoframes through a combination of galvanic replacement, gold deposition, and silver dealloying. *Small*, *9*, 3111–3117.

Wang, Y.-X.J. (2015). Current status of superparamagnetic iron oxide contrast agents for liver magnetic resonance imaging. *World Journal of Gastroenterology*, *21*, 13400.

Wang, X., Chen, H., Zheng, Y., Ma, M., Chen, Y., Zhang, K., et al. (2013). Au-nanoparticle coated mesoporous silica nanocapsule-based multifunctional platform for ultrasound mediated imaging, cytoclasis and tumor ablation. *Biomaterials*, *34*, 2057–2068.

Wang, Y., & Gu, H. (2015). Core–shell-type magnetic mesoporous silica nanocomposites for bioimaging and therapeutic agent delivery. *Advanced Materials*, *27*, 576–585.

Wang, H., Huff, T. B., Zweifel, D. A., He, W., Low, P. S., Wei, A., et al. (2005). In vitro and in vivo two-photon luminescence imaging of single gold nanorods. *Proceedings of the National Academy of Sciences of the United States of America*, *102*, 15752–15756.

Wang, K., Kievit, F. M., Sham, J. G., Jeon, M., Stephen, Z. R., Bakthavatsalam, A., et al. (2016). Iron-oxide-based nanovector for tumor targeted siRNA delivery in an orthotopic hepatocellular carcinoma xenograft mouse model. *Small*, *12*, 477–487.

Wang, C.-F., Mäkilä, E. M., Kaasalainen, M. H., Liu, D., Sarparanta, M. P., Airaksinen, A. J., et al. (2014). Copper-free azide–alkyne cycloaddition of targeting peptides to porous silicon nanoparticles for intracellular drug uptake. *Biomaterials*, *35*, 1257–1266.

Wang, C.-F., Sarparanta, M. P., Mäkilä, E. M., Hyvönen, M. L. K., Laakkonen, P. M., Salonen, J. J., et al. (2015). Multifunctional porous silicon nanoparticles for cancer theranostics. *Biomaterials*, *48*, 108–118.

Wang, Z., Wang, Y., Lu, M., Li, L., Zhang, Y., Zheng, X., et al. (2016). Janus Au–mesoporous silica nanocarriers for chemo-photothermal treatment of liver cancer cells. *RSC Advances*, *6*, 44498–44505.

Wang, Y., Xie, X., Wang, X., Geng, K., Gill, K. L., Patrick O'Neal, D., et al. (2004). Photoacoustic Tomography of a Nanoshell Contrast Agent in the in Vivo Rat Brain. *Nano Letters*, *4*, 1689–1692.

Wang, C., Xu, H., Liang, C., Liu, Y., Li, Z., Yang, G., et al. (2013). Iron oxide @ polypyrrole nanoparticles as a multifunctional drug carrier for remotely controlled cancer therapy with synergistic antitumor effect. *ACS Nano*, *7*, 6782–6795.

Wang, Y., Xu, J., Xia, X., Yang, M., Vangveravong, S., Chen, J., et al. (2012). SV119-gold nanocage conjugates: a new platform for targeting cancer cellsvia sigma-2 receptors. *Nanoscale*, *4*, 421–424.

Wang, H., Yi, J., Mukherjee, S., Banerjee, P., & Zhou, S. (2014). Magnetic/NIR-thermally responsive hybrid nanogels for optical temperature sensing, tumor cell imaging and triggered drug release. *Nanoscale*, *6*, 13001–13011.

Wang, L., Yuan, Y., Lin, S., Huang, J., Dai, J., Jiang, Q., et al. (2016). Photothermo-chemotherapy of cancer employing drug leakage-free gold nanoshells. *Biomaterials*, *78*, 40–49.

Wei, A., Oldenburg, A. L., Zweifel, D. A., Hansen, M. N., & Boppart, S. A. (2006). Plasmon-resonant gold nanorods as low backscattering albedo contrast agents. *Optics Express*, *14*, 6724–6738.

Willets, K. A., & Van Duyne, R. P. (2007). Localized surface plasmon resonance spectroscopy and sensing. *Annual Review of Physical Chemistry*, *58*, 267–297.

Wortmann, L., Ilyas, S., Niznansky, D., Valldor, M., Arroub, K., Berger, N., et al. (2014). Bioconjugated iron oxide nanocubes: synthesis, functionalization, and vectorization. *ACS Applied Materials & Interfaces*, *6*, 16631–16642.

Wu, X., He, X., Wang, K., Xie, C., Zhou, B., & Qing, Z. (2010). Ultrasmall near-infrared gold nanoclusters for tumor fluorescence imaging in vivo. *Nanoscale*, *2*, 2244–2249.

Wu, Z., & Jin, R. (2010). On the ligand's role in the fluorescence of gold nanoclusters. *Nano Letters*, *10*, 2568–2573.

Wu, M., Meng, Q., Chen, Y., Zhang, L., Li, M., Cai, X., et al. (2016). Large pore-sized hollow mesoporous organosilica for redox-responsive gene delivery and synergistic cancer chemotherapy. *Advanced Materials*, *28*, 1963–1969.

Xia, H.-X., Yang, X.-Q., Song, J.-T., Chen, J., Zhang, M.-Z., Yan, D.-M., et al. (2014). Folic acid-conjugated silica-coated gold nanorods and quantum dots for dual-modality CT and fluorescence imaging and photothermal therapy. *Journal of Materials Chemistry B*, *2*, 1945–1953.

Xia, B., Zhang, W., Shi, J., & Xiao, S.-J. (2013). Engineered stealth porous silicon nanoparticles via surface encapsulation of bovine serum albumin for prolonging blood circulation in vivo. *ACS Applied Materials & Interfaces*, *5*, 11718–11724.

Xiao, L., Gu, L., Howell, S. B., & Sailor, M. J. (2011). Porous silicon nanoparticle photosensitizers for singlet oxygen and their phototoxicity against cancer cells. *ACS Nano*, *5*, 3651–3659.

Xiao, R., Wang, C. W., Zhu, A. N., & Long, F. (2016). Single functional magnetic-bead as universal biosensing platform for trace analyte detection using SERS-nanobioprobe. *Biosensors & Bioelectronics*, *79*, 661–668.

Xie, J., Zheng, Y., & Ying, J. Y. (2009). Protein-directed synthesis of highly fluorescent gold nanoclusters. *Journal of the American Chemical Society*, *131*, 888–889.

Xu, W., Thapa, R., Liu, D., Nissinen, T., Granroth, S., Närvänen, A., et al. (2015). Smart porous silicon nanoparticles with polymeric coatings for sequential combination therapy. *Molecular Pharmaceutics*, *12*, 4038–4047.

Xu, C., Xie, J., Kohler, N., Walsh, E. G., Chin, Y. E., & Sun, S. (2008). Monodisperse magnetite nanoparticles coupled with nuclear localization signal peptide for cell-nucleus targeting. *Chemistry - An Asian Journal*, *3*, 548–552.

Yano, K., & Fukushima, Y. (2004). Synthesis of mono-dispersed mesoporous silica spheres with highly ordered hexagonal regularity using conventional alkyltrimethylammonium halide as a surfactant. *Journal of Materials Chemistry*, *14*, 1579–1584.

Yao, X., Niu, X., Ma, K., Huang, P., Grothe, J., Kaskel, S., et al. (2017). Graphene quantum dots-capped magnetic mesoporous silica nanoparticles as a multifunctional platform for controlled drug delivery, magnetic hyperthermia, and photothermal therapy. *Small*, *13*, 1602225.

Yavuz, M. S., Cheng, Y., Chen, J., Cobley, C. M., Zhang, Q., Rycenga, M., et al. (2009). Gold nanocages covered by smart polymers for controlled release with near-infrared light. *Nature Materials*, *8*. nmat2564–939.

You, J., Zhang, G., & Li, C. (2010). Exceptionally high payload of doxorubicin in hollow gold nanospheres for near-infrared light-triggered drug release. *ACS Nano*, *4*, 1033–1041.

You, J., Zhang, R., Xiong, C., Zhong, M., Melancon, M., Gupta, S., et al. (2012). Effective photothermal chemotherapy using doxorubicin-loaded gold nanospheres that target ephb4 receptors in tumors. *Cancer Research*, *72*, 4777–4786.

You, J., Zhang, R., Zhang, G., Zhong, M., Liu, Y., Van Pelt, C. S., et al. (2012). Photothermal-chemotherapy with doxorubicin-loaded hollow gold nanospheres: a platform for near-infrared light-trigged drug release. *Journal of Controlled Release*, *158*, 319–328.

Yu, Y. Y., Chang, S.-S., Lee, C.-L., & Chris Wang, C. R. (1997). Gold nanorods:electrochemical synthesis and optical properties. *The Journal of Physical Chemistry B*, *101*, 6661–6664.

Yuan, H., Khoury, C. G., Hwang, H., Wilson, C. M., Grant, G. A., & Vo-Dinh, T. (2012). Gold nanostars: surfactant-free synthesis, 3D modelling, and two-photon photoluminescence imaging. *Nanotechnology*, *23*, 075102.

Yuen, A. K. L., Hutton, G. A., Masters, A. F., & Maschmeyer, T. (2012). The interplay of catechol ligands with nanoparticulate iron oxides. *Dalton Transactions*, *41*, 2545–2559.

Yun, J., Sonabend, A. M., Ulasov, I. V., Kim, D.-H., Rozhkova, E. A., Novosad, V., et al. (2012). A novel adenoviral vector labeled with superparamagnetic iron oxide nanoparticles for real-time tracking of viral delivery. *Journal of Clinical Neuroscience: Official Journal of the Neurosurgical Society of Australasia*, *19*, 875–880.

Zavaleta, C. L., Smith, B. R., Walton, I., Doering, W., Davis, G., Shojaei, B., et al. (2009). Multiplexed imaging of surface enhanced Raman scattering nanotags in living mice using noninvasive Raman spectroscopy. *Proceedings of the National Academy of Sciences of the United States of America*, *106*, 13511–13516.

Zeng, Y., Zhang, D., Wu, M., Liu, Y., Zhang, X., Li, L., et al. (2014). Lipid-AuNPs@PDA nanohybrid for MRI/CT imaging and photothermal therapy of hepatocellular carcinoma. *ACS Applied Materials & Interfaces*, *6*, 14266–14277.

Zhang, X., Lin, Y., & Gillies, R. J. (2010). Tumor pH and its measurement. *Journal of Nuclear Medicine*, *51*, 1167–1170.

Zhang, Y., Wang, X.-J., Guo, M., Yan, H.-S., Wang, C.-H., & Liu, K.-L. (2014). Cisplatin-loaded polymer/magnetite composite nanoparticles as multifunctional therapeutic nanomedicine. *Chinese Journal of Polymer Science*, *32*, 1329–1337.

Zhang, Z., Wang, L., Wang, J., Jiang, X., Li, X., Hu, Z., et al. (2012). Mesoporous silica-coated gold nanorods as a light-mediated multifunctional theranostic platform for cancer treatment. *Advanced Materials*, *24*, 1418–1423.

Zhang, J., Wu, D., Li, M.-F., & Feng, J. (2015). Multifunctional mesoporous silica nanoparticles based on charge-reversal plug-gate nanovalves and acid-decomposable ZnO quantum dots for intracellular drug delivery. *ACS Applied Materials & Interfaces*, *7*, 26666–26673.

Zhao, Y., Sultan, D., Detering, L., Luehmann, H., & Liu, Y. (2014). Facile synthesis, pharmacokinetic and systemic clearance evaluation, and positron emission tomography cancer imaging of 64 Cu–Au alloy nanoclusters. *Nanoscale*, *6*, 13501–13509.

Zhao, Z., Zhou, Z., Bao, J., Wang, Z., Hu, J., Chi, X., et al. (2013). Octapod iron oxide nanoparticles as high-performance T_2 contrast agents for magnetic resonance imaging. *Nature Communications*, *4*. ncomms3266.

Zhou, Z., Zhu, S., & Di, Z. (2007). Grafting of thermo-responsive polymer inside mesoporous silica with large pore size using ATRP and investigation of its use in drug release. *Journal of Materials Chemistry*, *17*, 2428–2433.

Zhu, M., Lanni, E., Garg, N., Bier, M. E., & Jin, R. (2008). Kinetically controlled, high-yield synthesis of Au25 clusters. *Journal of the American Chemical Society*, *130*, 1138–1139.

Zhu, J., Liao, L., Zhu, L., Zhang, P., Guo, K., Kong, J., et al. (2013). Size-dependent cellular uptake efficiency, mechanism, and cytotoxicity of silica nanoparticles toward HeLa cells. *Talanta*, *107*, 408–415.

Zhu, C.-L., Lu, C.-H., Song, X.-Y., Yang, H.-H., & Wang, X.-R. (2011). Bioresponsive controlled release using mesoporous silica nanoparticles capped with aptamer-based molecular gate. *Journal of the American Chemical Society*, *133*, 1278–1281.

Zhu, S. G., Xiang, J. J., Li, X. L., Shen, S. R., Lu, H. B., Zhou, J., et al. (2004). Poly(l-lysine)-modified silica nanoparticles for the delivery of antisense oligonucleotides. *Biotechnology and Applied Biochemistry*, *39*, 179–187.

Further Reading

Tang, F., Li, L., & Chen, D. (2012). Mesoporous silica nanoparticles: synthesis, biocompatibility and drug delivery. *Advanced Materials*, *24*, 1504–1534.

CHAPTER 15

Hydrogels as a New Platform to Recapitulate the Tumor Microenvironment

Menekse Ermis*,†,a, Semih Calamak†,‡,a, Gizem Calibasi Kocal†,§, Sinan Guven†,§, Naside G. Durmus†, Imran Rizvi¶, Tayyaba Hasan¶,‖, Nesrin Hasirci*, Vasif Hasirci*, Utkan Demirci†

*Middle East Technical University, Ankara, Turkey †Stanford University, Palo Alto, CA, United States ‡Hacettepe University, Ankara, Turkey §Dokuz Eylul University, Izmir, Turkey ¶Harvard Medical School, Boston, MA, United States ‖Harvard University and Massachusetts Institute of Technology, Cambridge, MA, United States

1 Introduction

Well-engineered natural and synthetic hydrogels can mimic key features of ECMs while providing physical and biochemical characteristics of native tissues and the tumor microenvironment. In particular, biochemical and physical stimuli play critical roles in cell behavior, both in healthy and cancerous conditions (Cavo et al., 2016). The extremely heterogeneous tumor microenvironment is comprised of different elements, which can directly or indirectly affect the cancer cell phenotype via biological, physical, and chemical routes. Biological, physical, and chemical components of the tumor microenvironment, such as the extracellular matrix, cancerous, or noncancerous cells surrounding the cancer tissue, cytokines, hormones, degree of vascularization, pH, oxygen levels, flow, shear stress, and stiffness, contribute to the conditions required for tumor growth, and metastasis, as well as drug resistance. Mimicking and controlling the dynamic tumor microenvironment and its main characteristics by microfabricated three-dimensional cell cultures are crucial for understanding cancer biology, particularly with respect to limitless survival and metastasis. In other words, control of the tumor microenvironment is a basic step in cancer tissue engineering (Adjei & Blanka, 2015). This chapter provides an overview of hydrogels as a new platform to recapitulate the tumor microenvironment and their various applications in cancer therapy.

---

a These authors contributed equally to the work.

Handbook of Nanomaterials for Cancer Theranostics. https://doi.org/10.1016/B978-0-12-813339-2.00015-3

2 Hydrogels and Their Physical and Chemical Characteristics

Producing physiologically relevant cell culture models that mimic key biochemical and biophysical properties of the native extracellular matrix in vitro has tremendous promise for drug discovery, tissue engineering, and regenerative medicine (Brandenberg & Lutolf, 2016). During the past few decades, a great variety of research groups has been working in the field of hydrogel development to create different matrices to mimic cell media. Hydrogels are described as two-phase systems consisting of a hydrophilic polymer network and a solvent in the network (water or aqueous solution). Hydrogels are three-dimensional networks connected by physical intramolecular or covalent bonds. They can absorb from 10%, up to thousands of times their dry weight in an aqueous environment, and swell easily without dissolution (Hoffman, 2012). The hydrophilic nature of the hydrogels originates from the polar groups such as hydroxyl, amino, carboxyl, and amide groups, which form the backbone of the polymer chains. The physical properties of hydrogels, such as mechanical strength and swelling ratio, are controlled by the type of crosslinking agent or density of crosslinking (El-Sherbiny & Yacoub, 2013).

Various classification criteria can be used for hydrogels, depending on different parameters, including origin, structure, conformation, crosslinking type, physical form, charge, and ambient conditions. Depending on the source of the hydrogels, they can be distinguished as natural or synthetic. Also, hydrogels can be classified as homopolymeric, copolymeric, and multiple polymer interiors, based on polymer structure. Alternatively, classification can be made according to hydrogel configuration: amorphous, semicrystalline, and crystalline. Hydrogels are crosslinked either physically or chemically, and they have many different physical forms, including 3D bulk matrix, film (membrane), microsphere, nanosphere. Additionally, hydrogels can be classified as ionic, cationic, amphoteric, or neutral, depending on the charges of the constituent monomers. Finally, hydrogels can be classified as physical and chemical stimuli-responsive hydrogels (Table 1).

Hydrogels are promising candidates for a variety of applications in the pharmaceutical and medical industry. Because they have the ability to hold water in large quantities, and their soft and porous structures can be very similar to the extracellular matrix, they closely resemble living tissues (Fig. 1). In addition, such hydrogels better mimic the native extracellular microenvironment to help understand the role of microenvironmental cues on cellular functions and tissue formation (Shin, Olsen, & Khademhosseini, 2012). Hydrogels are chosen for cell encapsulation and 3D cell cultures because they promote uniform cell distributions and organized cell communities with adjustable pore sizes and diffusion of nutrients. 3D cell-laden hydrogels more closely mimic cell-cell and cell-matrix interactions in the body than two-dimensional (2D) cultures. Compared with conventional 2D scaffolds, hydrogels have pores large enough to accommodate living cells, and they may initiate the cells penetration and proliferation. One of the main disadvantages of hydrogels is the difficulty

Table 1: Classification of hydrogels

Hydrogel origin	Hydrogel Structure	Hydrogel Configuration	Crosslinking Type	Physical Form	Electrical Charge	Ambient Condition
Natural: Proteins: collagen, gelatin, and silk fibroin Polysaccharides alginate, chitosan, hyaluronic acid, and agarose	*Homopolymeric*: Hydrogel networks derived from a single type of monomers.	*Amorphous hydrogels*: Hydrogels, macromolecular chains are randomly located.	*Chemical crosslinking*: Networks have permanent and irreversible junctions as a consequence of conformational changes.	*3D matrix*: Tissue engineering and regenerative medicine, filtration and chemical applications.	*Neutral hydrogels*: Nonionic hydrogels consist of homopolymeric or copolymeric neutral hydrogels that do not contain charged groups in their structure. Swelling behavior is usually dependent on the change in ambient temperature.	*Responsive to physical stimuli*: Temperature, electricity, magnetic field, light, pressure, and sound.
Synthetic: Poly(ethylene glycol) (PEG), polymer oligo(poly(ethylene glycol) fumarate) polyvinyl alcohol (PVA)	*Copolymeric hydrogel*: The copolymeric hydrogels composed of two or more different monomer species arranged in block or alternate configuration.	*Semicrystalline hydrogels*: Hydrogels consisting of dense parts where the macromolecule chains are regularly organized in the structure. In addition, it is a complex structure of amorphous and crystalline structure.	*Physical crosslinking*: Crosslinking is through physical interactions, such as ionic interactions, crystallization, complexation, hydrogen bonds or hydrophobic interactions.	*Film*: Local drug delivery, biosensing, wound dressing.	*Ionic hydrogels*: These hydrogels contain cationic or anionic groups in their network.	*Responsive to chemical stimuli*: pH, solvent, ionic strength and molecular species.
	Multiple polymer interiors: Hydrogels consisting of two independent crosslinked synthetic and/or natural polymer components.	*Crystalline hydrogels*: The crystallization of the hydrogels is related to the alignment of the molecular chains in certain configurations.		*Microsphere*: Cell laden hydrogels, Systemic and sustained local drug delivery.	*Amphoteric hydrogels*: These hydrogels contain both acidic and basic groups simultaneously	
				Nanosphere: Systemic drug delivery imaging.	*Zwitterionic (polybetaines) hydrogels*: These hydrogels comprise of both anionic and cationic groups in each structural repeating unit.	

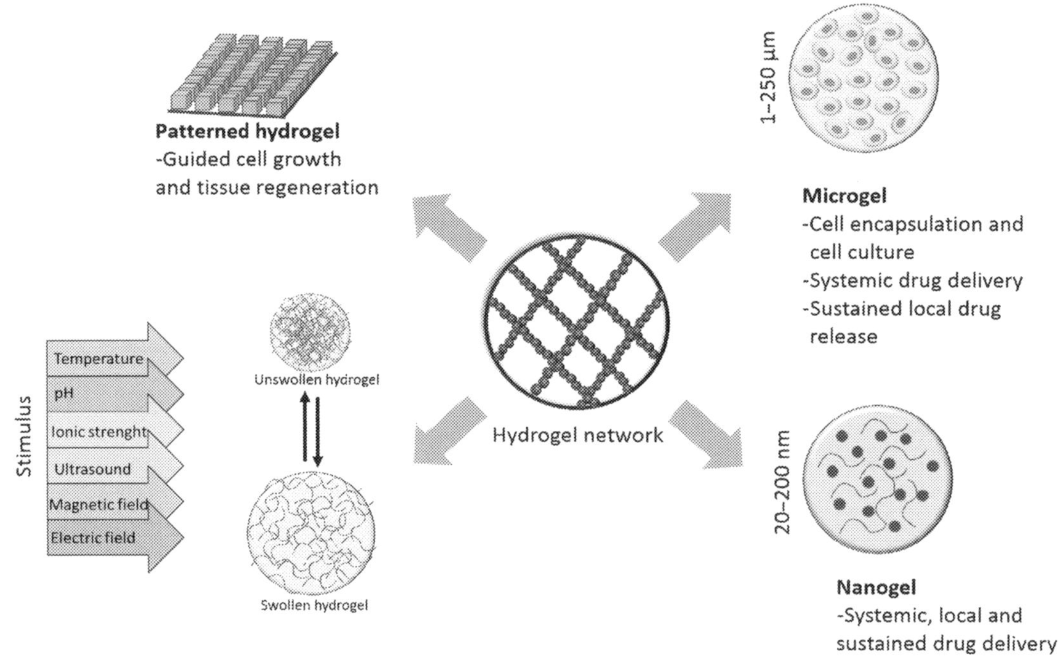

Fig. 1
Biomedical applications of hydrogels.

in assembling them in desired microgeometries and their sterilization. In recent years, 3D bioprinting technologies have eliminated these limitations with precise and aseptic production approaches (Park et al., 2017).

2.1 Hydrogel Stabilization Mechanisms

Hydrogels are soft and gelly structures, and because they absorb a large quantity of water, they can easily crumble. Stiffness and mechanical properties of hydrogels are often controlled by crosslinking strategies. There are various crosslinking strategies to adjust strength and pore morphology of the hydrogels. Hydrogel solutions generally expose either physical or chemical crosslinking mechanisms to create solidified hydrogels. The crosslinking method is chosen based on factors, such as the type of polymer, cell encapsulation strategy, drug release kinetics, and the target tissue.

2.1.1 Physical crosslinking

Crosslinking of hydrogels via physical interactions is one of the most common approaches for hydrogel stabilization. Physically crosslinked hydrogels are generally bound with weak forces such as hydrogen bonds, van der Waals forces, and ionic interactions, and are generally known as reversible systems. In recent years, there has been an increasing interest in physically crosslinked hydrogels for cell, protein, and drug encapsulation. The main reason is that the crosslinking agents do not affect the integrity of cells and biomolecules that are entrapped in the hydrogel (Hennink & Van Nostrum, 2012).

Physical crosslinking is carried out by an environmental trigger, such as changes in polyelectrolyte charge density, temperature, pH, crystallization, protein interactions, and secondary interactions (hydrophobic, electrostatic, and hydrogen bonding) (Chen et al., 2005). In addition, nanomaterials can be added into the hydrogel polymeric network to construct physically stabilized hydrogels (e.g., clay, minerals, polymeric, and metallic nanospheres) (Gaharwar, Peppas, & Khademhosseini, 2014). However, the addition of these materials changes the stiffness and diffusion properties of hydrogels, which impacts the phenotypes of the cells. Additionally, the incorporation of nanomaterials into the hydrogel networks could lead to nonspecific reactions with the cells (Cha et al., 2014). Hydrogel networks that are physically stabilized generally have a weak mechanical strength and show poor stability in in vitro and in vivo conditions.

2.1.2 Chemical crosslinking

Hydrogels formed by chemical crosslinking show greater stability, longer durability, and higher mechanical properties (tensile, shear, bending, etc.), compared with physically crosslinked hydrogels. Chemical crosslinked hydrogels are produced by grafting functional monomers into the polymer network or using a crosslinker to couple the two polymer chains. Chemically crosslinking reaction mechanisms rely on the crosslinking of functional groups of the polymers (-COOH, -OH, or -NH_2, etc.) with crosslinkers such as aldehydes (glutaraldehyde, adipic acid dihydrazide, etc.) and *N*,*N*-(3-dimethylaminopropyl)-*N*-ethyl carbodiimide (EDC) (Jiang et al., 2016). Additionally, radical reactions can be activated by application of high energy irradiation such as UV, gamma, electron beam, and microwave, and water-soluble polymers can be derivatized with the use of vinyl groups present in their structures or radicals created via the breaking of chemical bonds (Giammona et al., 1999).

Another chemical crosslinking method is enzymatic crosslinking of hydrogels. Enzymatic crosslinking strategies have received considerable attention for cell-laden and injectable hydrogels. Enzymatic reactions provide an appropriate chemical crosslinking strategy for in vivo applications, as the majority of the enzymes involved in the hydrogel crosslinking are responsible for catalyzing naturally occurring reactions in the human body (Teixeira et al., 2012). Tyrosinase and transglutaminase are mostly used for enzymatically crosslinking of hydrogels.

Photopolymerization systems usually consist of three main components: monomer, photo initiator, and additives. In the literature, cell-laden photocrosslinkable hydrogels are mainly used for tissue engineering applications. Especially, gelatin methacrylate (GelMA), alginate methacrylate (ALMA), poly(ethylene glycol) diacrylate, and PEG diacrylate are commonly used for cell encapsulation (Loessner et al., 2016; Meinert et al., 2017; Siltanen et al., 2016). In this technique, photoinitiator molecules in hydrogel solutions are triggered by UV light to initiate polymerization. The crosslinking process usually takes less than a second to a few minutes. In this method, reaction parameters such as photoinitiator concentration, UV light intensity, UV exposure time, and a distance between the hydrogel and the UV light source should be precisely controlled to ensure reproducibility and high cell viability. Otherwise, unreacted monomers and released radicals can damage cells and cellular components. Table 2 presents some examples of physically and chemically crosslinked hydrogels.

Table 2: Physically and chemically crosslinking strategies for hydrogels

Chemically Crosslinked Hydrogels	Physically Crosslinked Hydrogels
Crosslinking by radical polymerization: Methacrylic acid (pH-sensitive gels) (Bettini, Colombo, & Peppas, 1995) *N*-isopropylacrylamide (temperature-sensitive gels) (Cicek & Tuncel, 1998) Dextran (Brøndsted, Andersen, & Hovgaard, 1998) Poly(vinyl alcohol) (PVA) (Martens & Anseth, 2000) Hyaluronic acid (Jin et al., 2001)	*Ionic interactions*: Glycerol-phosphate disodium salt (Chenite et al., 2000) Calcium (Ca^{+2}) (Lin et al., 2005) Potassium (K^{+1}) (Watanabe et al., 1996)
Crosslinking by complementary groups: Aldehydes: Glutaraldehyde, adipic acid dihydrazide (Lee, Bouhadir, & Mooney, 2000; Mansur et al., 2008) Addition reactions: 1,6-Hexanedibromide, 1,6-hexamethylenediisocyanate, divinylsulfone (Brøndsted, Hovgaard, & Simonsen, 1995; Gehrke, Uhden, & McBride, 1998) Condensation reactions: *N,N*-(3-dimethylaminopropyl)-*N*-ethyl carbodiimide (EDC) (Kuijpers et al., 2000), isocyanide (de Nooy, Masci, & Crescenzi, 1999), hydrolysable, polyrotaxane (Ichi et al., 2001)	*Crystallization*: PVA (Peppas & Scott, 1992) Dextran (Stenekes, Talsma, & Hennink, 2001) Poly(lactic-L-acid) PLLA Poly(lactic-D-acid) PDLA (Ikada et al., 1987)
Crosslinking by high energy irradiation: Gamma, electron beam, microwave (Giammona et al., 1999)	*Crosslinking by hydrogen bonds*: Poly(acrylic acid) and poly(methacrylic acid with poly(ethylene glycol) (Bell & Peppas, 1996; Mathur et al., 1998) Poly(*N,N*-dimethylacrylamide-co-*N* acryolyloxysuccinimide) with oligodeoxyribonucleotides (Nagahara & Matsuda, 1996)
Crosslinking using enzymes: Tyrosinase, transglutaminase (Ca^{+2}), transferases, peroxidases (Teixeira et al., 2012)	*Crosslinking by protein interactions*: Genetically engineered proteins silk-like and elastin-like blocks human cardiac muscle (Chen, Kopeček, & Stewart, 2000)
Photoinitiation: UV and visible light (Kolesky et al., 2014; Nichol et al., 2010)	*Crosslinking by antigen-antibody interactions*: Polyacrylamide with rabbit IgG (Miyata, Asami, & Uragami, 1999)

3 3D Tissue Engineering Scaffolds as Hydrogels

In the past decade, hydrogel design and fabrication have been major areas of biomaterials research in the field of tissue engineering and regenerative medicine. Hydrogels provide an aqueous environment, which allows excellent transport of nutrient to cells and cell products. They are also easily modified with cell adhesion ligands and biomolecules (Hoffman, 2012). Hydrogels are being investigated as scaffolds for the encapsulation of cells, as well as the controlled release of drugs and bioactive molecules (Fig. 2).

Physical factors are highly effective in the fulfillment of cell functions and regulation of tissue morphogenesis. Hydrogel studies have indicated that differentiation and orientation of stem cells and neuronal cells can be regulated by changing the mechanical properties and surface stiffness of the hydrogel scaffolds (Li et al., 2014). Mechanical sensitivity depends on the ability of the cells to rearrange the adhesion ligands provided by the substrate and to apply tensile forces to the material. Despite tremendous advances in the field of hydrogel designs for tissue engineering, it may still remain challenging to design hydrogels with desirable mechanical properties (Yang et al., 2017). For hard tissues, especially cartilage and bone, hydrogels generally remain weak and soft enough to be applied for tissue engineering applications, which require load-bearing mechanical strength. Therefore, the development of

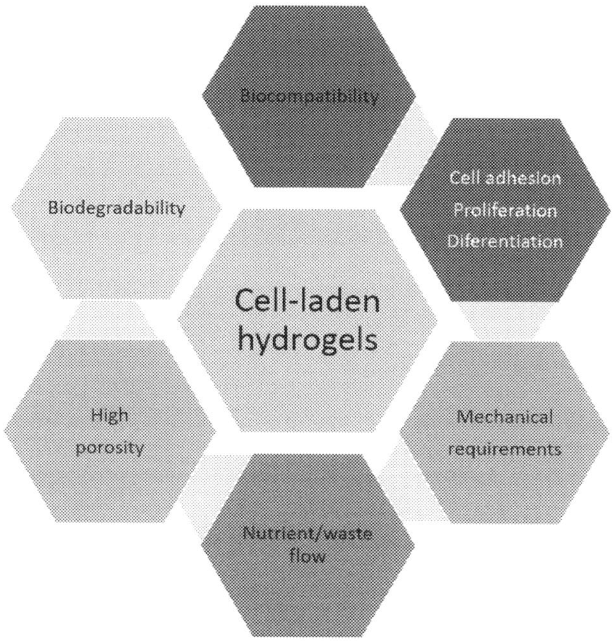

Fig. 2
The complex requirements of hydrogel-based ideal scaffold fabrication.

high mechanical-strength hydrogels with adjustable mechanical properties and the potential to mimic hard tissues are critical challenges to expanding the application areas of hydrogel-based tissue engineering scaffolds (Shin et al., 2012). In addition, the initial mechanical properties of biomaterials are not the only determining factor for the fate of cells encapsulated in the hydrogel matrix. Over time, the degradation of the matrix and the change in substrate mechanics can significantly alter the mechanical properties.

Additionally, recapitulating the cellular alignment and orientation is critical to organizing native tissues in 3D models. Many critical components of the cellular microarchitecture, such as cell differentiation, cell elongation, cell-cell interactions and signaling, can all be modulated through control of the hydrogel microenvironment. Significant research has been focused on controlling micro- and nanostructured hydrogel surfaces produced either by chemical or topographic patterning (Aubin et al., 2010). A variety of patterning methods have been previously developed to modify materials with precise geometry and morphology, including molding, soft lithography, and 3D printing (Wade et al., 2015). Cell alignment is known to play an important role for skeletal, cardiac muscles, tendons, and ligaments. Cell alignment also exists in nerve tissues, and it plays an important role during axonal regeneration (Li et al., 2014).

4 Biomimicking Tissue Structure and Composition With Cell-Laden Hydrogels

Cell-laden biomimetic hydrogels provide a novel and attractive approach for tissue regeneration. In the past two decades, hydrogels were frequently used as tissue engineering scaffolds (Hasirci et al., 2016). They served as tissue mimicking substrates for skin (Eke et al., 2017; Ng, Yeong, & Naing, 2016), corneas (Rizwan et al., 2017; Wang et al., 2016), cartilage (Ko et al., 2016; Naseri et al., 2016), nerves (Bouyer et al., 2016; Gnavi et al., 2017), and various other tissues. This increased interest for hydrogels in tissue engineering is due to their ability to immobilize and release cells, genes, proteins, and drugs (Hunt et al., 2014). Properties that are required for scaffold design, such as controlled release of growth factors, controllable mechanical properties, cytocompatibility, biomimicry, ease of tissue integration, and allowing injection in a liquid phase are possessed by the many varieties of hydrogels that are available for tissue engineering (Hasirci et al., 2016; Overstreet et al., 2012; Tan & Marra, 2010). In the literature, hydrogels are known as a promising class of biomaterials for both soft and hard 3D tissue engineering applications, and it is widely accepted that hydrogels combining both cells and growth factors have great potential to eliminate the challenge of 3D tissue regeneration. The biomimetic microenvironments created using cell-laden hydrogels enable improved representation of in vivo biology (Yang et al., 2017). Ideally, biomimetic scaffolds should meet basic requirements for living cells, including high porosity (relevant geometry and pore dimensions), optimal nutrient/waste flow, maintaining mechanical integrity during cell growth, and biodegradability with adjusted degradation time parallel to the tissue growth rate (Billiet et al., 2012). The use of hydrogel scaffolds is important, as they provide a support matrix for embedded cells with a highly hydrated microenvironment that is adjustable to nutrient transfer and oxygen diffusion (El-Sherbiny & Yacoub, 2013).

Hydrogel based on natural polymers, including proteins (collagen, gelatin, and silk fibroin), and polysaccharides (alginate, chitosan, hyaluronic acid, and agarose) have been widely used as a matrix due to their ease of gelation, as well as their adjustable mechanical and pore properties (Yang et al., 2017). In the past decade, 3D cell/organ printing techniques have been mainly used for developing 3D cell laden microtissues, with the desired geometry and internal morphology. In this regard, 3D printing technologies offer several advantages, including precise control of 3D constructs for patient-customized geometries, compositions, free-form fabrication, and containing the desired cells in the predetermined organization.

A great variety of specific tissue models has been designed, including cancers such as breast and lung tumors (Knowlton et al., 2016). However, the microsize of the encapsulated cells and the static culture condition lead to poor nutrient and oxygen exchange (Jiang et al., 2016). The use of microfluidics as a platform for cell-laden 3D hydrogels allows for the creation of laminar and turbulent flows within microfluidic channels, and it provides a significant advantage over traditional 2D static cultures (Knowlton et al., 2016). In addition, static cultures such as petri dishes or flasks do not recapitulate the dynamic physical forces, including hydrostatic pressure, compression, tension, and fluid shear stresses of interstitial flows. In recent years, in vitro microfluidic systems that contain cell-laden hydrogels have been explored to potentially produce functional microtissues and 3D cancer models. Today, these systems and organ-on-a-chip models are still an evolving field and require further research.

5 Characteristics of the Tumor Microenvironment

5.1 Biological Properties of the Tumor Microenvironment

5.1.1 Cells found in the tumor microenvironment

Except for cancerous cells, tumor tissue is surrounded by noncancerous cells, which can contribute to the tumor behavior in a positive or negative manner. Compared with cancerous cells, noncancerous cells are more likely to be genetically stable. However, cancerous cells tend to change their genetic material to evolve according to changing microenvironmental factors to survive. For example, acquiring or developing resistance to medication is a result of this genetic instability.

Fibroblast, immune cells, and endothelial cells that comprise blood vessels are among the main cell-based modulators of the tumor microenvironment. Cancer-associated fibroblasts reside within the tumor microenvironment, and they are key components of tumor stroma and have an important role in cancer development and progression. Cancer-associated fibroblasts function in the regulation of extracellular matrix turnover, due to both creating and degrading the extracellular matrix. Cancer-associated fibroblasts also have a role on immune reprogramming, regulation of inflammation, and wound healing through secretion of cytokines and chemokines (Kalluri, 2016). Immune cells such as macrophages, dendritic cells, natural killer cells, and regulatory T cells seek to kill the cancer cells. It has been shown that all of these cell types contribute to the tumor microenvironment via dysregulation of immune-mediated responses by the effect of cytokines produced within the tumor niche (Bussard et al., 2016). Cancer-associated endothelial cells are the other type of noncancerous cells in the tumor microenvironment. They boost metastasis and protect tumor cells from immune attacks and targeted chemotherapeutics via chaperoning cancer cells to distant locations (Yadav et al., 2013).

5.1.2 Angiogenesis and vasculature

An adequate supply of oxygen and nutrients, and removal of waste, are basic requirements for cancer cell survival. In the early stages of tumorigenesis, proximity to the host vascular system helps meet the demand for these crucial elements. As the tumor progresses and the demand for oxygen and nutrients to maintain tumor growth increases, the balance between proangiogenic and antiangiogenic factors (known as a angiogenic switch) shifts to trigger the formation of tumor neo-vasculature (Bergers & Benjamin, 2003). Unlike normal vasculature, tumor vasculature in a primitive shape shows disorganization and lack of the hierarchical branching during the formation of blood vessels. Angiogenesis under the condition of malignancy is diverse from normal tissues due to the slower proliferation rate of endothelial cells than tumor cells. Therefore, newly formed vessels have partial/missing basement membranes and endothelial lining, and deficiency of pericytes, smooth muscle, and receptors (Siemann & Horsman, 2015; Vaupel et al., 2004). The defective endothelial barrier leads to blood vessel leakiness that enables the transport of macromolecules from the blood stream to tumor cells and also contributes to the elevation of interstitial fluid flow and flow-associated forces in tumors, which can inhibit the accumulation of drugs in the tumor (Kobayashi, Watanabe, & Choyke, 2014). Additionally, there are plentiful functional deformities, such as inconsistent speed and direction of blood flow, elevated vascular slenderness, red blood cell sludging, leukocyte sticking, blockage of vessels by blood cell aggregates, or circulating tumor cells. All these factors lead to glucose and oxygen deficiency, production of excessive lactate, and acidosis in the tumor microenvironment (Siemann & Horsman, 2015; Vaupel et al., 2004).

5.1.3 Extracellular matrix and its properties

The extracellular matrix is composed of numerous components secreted by cells, such as proteins, polysaccharides, glycoproteins, and proteoglycans with diverse physical and biological properties; and supports cells by surrounding them in tissues (Pickup, Mouw, &

Weaver, 2014). The extracellular matrix has two major components: (i) the basement membrane, which separates epithelial and stromal layers of cells; and (ii) the interstitial matrix, which is composed of stromal cells. The basement membrane has a specific composition including laminins, collagen IV, fibronectin, and linker proteins, to make a less porous structure than the interstitial matrix. In contrast, the interstitial matrix has a highly charged, hydrated, and dense structure with fibrillar collagens and proteoglycans that provide tensile strength to normal and cancer tissues (Egeblad, Rasch, & Weaver, 2010). The extracellular matrix has a role in both biological and physical processes. In biological processes, it is associated with various direct and indirect signal transduction cascades from the cell surface to the nucleus, which allow the interaction of cells with their environments. In physical processes, it supports tissue architecture and integrity with its rigidity, porosity, insolubility, orientation, and other physical features that have a role in scaffolding (Lu, Weaver, & Werb, 2012).

5.2 Physical Properties of the Tumor Microenvironment

5.2.1 Stiffness

Disregulation of specific signaling cascades accompany tumor development and growth. However, changes in the mechanical characteristics of cancer cells and their neighboring microenvironment can also be seen. For example, cancer cells are less stiff than normal cells, and this alteration has been detected in various cancer types, including lung, breast, prostate, skin, and oral cancers (Cross et al., 2007; Fuhrmann et al., 2011; Plodinec et al., 2013; Remmerbach et al., 2009; Suresh, 2007). The decrease in cell stiffness is bigger in tumors with higher migration and invasion potential (Swaminathan et al., 2011). In addition to cell stiffness, tumor behavior is influenced by the tendency of cancer cells to bind to the other cells and extracellular components of their microenvironment. This tendency causes different expression patterns of specific adhesion proteins on the cell surface. Changing the mechanical characteristics of the extracellular, the microenvironment effects the proliferation rate, configuration, and invasiveness of tumors (Erler & Weaver, 2009). During the invasion by cancer cells, modification of the extracellular matrix involves degradation, repolymerization, and realignment of the matrix fibers (Katira, Bonnecaze, & Zaman, 2013). The reorganization and reorientation of these fibers determines the stiffness and elasticity of the modified extracellular matrix. Increased collagen deposition, fiber misalignment, and crosslinking during extracellular matrix reorganization leads to increased stiffness. Extracellular matrix stiffening provides essential signals to promote cellular plasticity and malignant progression by controlling the induction of epithelial-mesenchymal transition, therefore stiffness can be the key regulator of cellular behavior. However, stiffening remains a challenge to understanding how cells interact with each other and respond to their microenvironment (Bonnans, Chou, & Werb, 2014; Gilkes, Semenza, & Wirtz, 2014).

5.2.2 Fluid flow and shear stress

Mechanical features of the tumor microenvironment altered by biophysical cues have gained significant attention. Cancer cells located in the extracellular matrix are exposed to fluid flow driven by lymphatic drainage, increased pressure due to tumor growth or leaky microvessels, and muscle contraction for the transportation of molecules (Polacheck, Charest, & Kamm, 2011). This fluid flow, termed interstitial flow, influences the transport and formation of gradients of growth factors, chemokines, tumor antigens, and chemotherapeutic agents in the tumor microenvironment. Elevated interstitial flows can also play a key role in tumor progression (Harrell, Iritani, & Ruddell, 2007; Mitchell & King, 2013). Therefore, there is a critical need to understand the impact of fluid flow on transport, and physiology, as well as tissue architecture.

The content of interstitial flow depends on the location in the body. In soft tissues, its composition is identical to the blood plasma that leaks from capillaries. The velocity of flow can vary from 0.1 to 1.0 μm/s, but some studies have determined that flow can reach velocities as high as 4.0 μm/s (Mitchell & King, 2013; Polacheck et al., 2011). During increases in interstitial flow, cancer cells are exposed to two various fluid-associated forces: (i) stresses generated by elevated interstitial flows in the tumor microenvironment, and (ii) stresses generated by blood flow in the vascular microenvironment (Michor et al., 2011; Swartz & Lund, 2012). It is believed that flow-associated shear stress can be approximately 0.1 dyn cm^{-2} (Mitchell & King, 2013). During the metastatic process, increases in interstitial flow and fluid shear stress have emerged as possible stimuli for tumor cell invasion and circulating cancer cell adhesion to blood vessels (Chang et al., 2008; Hofmann et al., 2006).

5.3 Chemical Properties of the Tumor Microenvironment

5.3.1 Acidic extracellular pH

An acidic extracellular pH is a major chemical property of the tumor microenvironment due to the release of lactic acid from anaerobic glycolysis. According to the "Warburg effect," lactic acid production is a consequence of preferential energy production via the anaerobic glycolytic pathway, rather than oxidative phosphorylation, even in the presence of sufficient oxygen in the tumor microenvironment (Kato et al., 2013). This alteration in tumor metabolism results from an acidic extracellular (pH 6.5–6.9), compared with healthy tissue under physiologic conditions (pH 7.2–7.4) (Estrella et al., 2013).

Upregulation of glycolysis due to the glucose-lactate transformation drives microenvironmental acidosis, which is a requirement for the development of resistant cell populations to acid-induced cell toxicity and the promotion of cellular proliferation (Gatenby & Gillies, 2004). In tumor tissue, acidic pH is also an inevitable characteristic of inflammation, which is an important factor of tumor promotion and progression. Macrophages, key modulators of the link between inflammation and cancer, produce protons resulting from lactate production and glucose utilization, leading to increased extracellular acidity (Kato et al., 2013).

Studies have demonstrated that an acidic pH can induce proteinase activity, involved in invasion and tissue remodeling, both in vitro and in vivo. These findings support the "acid-mediated invasion" hypothesis, wherein hydronium ions (H_3O^+) move through concentration gradients from the tumor to neighboring normal tissue (Gatenby & Gawlinski, 1996; Moellering et al., 2008; Rofstad et al., 2006). The resulting ion-based movement induces extracellular matrix degradation by activated proteinases, angiogenesis via vascular endothelial growth factor secretion, inhibition of the immune response to tumor antigens, and the development of adaptive mechanisms (Estrella et al., 2013; Fukumura et al., 2001; Gatenby et al., 2006; Lardner, 2001; Xu, Fukumura, & Jain, 2002).

5.3.2 Hypoxia

In normal tissues, oxygen levels vary between 3% and 9%, and although rare, can fall below 2% (hypoxia). In tumor tissue, hypoxia is more predominant (<0.02) (Bertout, Patel, & Simon, 2008). Due to the uncontrolled and rapid proliferation of cancer cells, a tumor quickly consumes nutrients and oxygen, and a hypoxic condition arises in the tumor microenvironment. This decrease in oxygen levels triggers the upregulation of angiogenic factors for the generation of vessels in the hypoxic regions. However, newly formed tumor vessels do not organize like vasculature under normal physiological conditions. Tumor vessels are characterized by vascular leakiness, chaotic architecture with blunt ended, nonlaminar blood flow, and dysfunction. These characteristics result in mild or severe hypoxic regions, as well as necrosis (Eales, Hollinshead, & Tennant, 2016; Semenza, 2012). Mechanisms of hypoxia responses are managed by hypoxia inducible factors (HIFs) that are controlled by oxygen-dependent proteasomal degradation. The reduced oxygen levels in tumors stabilize HIFs and help tumors thrive under hypoxic conditions (Harris, 2002).

To expand and survive under hypoxic conditions, cancer cells develop adaptative mechanisms via increasing glycolytic metabolism, promoting proliferation with unlimited replication, escaping from apoptosis and immune checkpoints, inducing angiogenesis and moving to less hypoxic locations. Therefore, a hypoxic tumor microenvironment is a negative prognostic and predictive factor for an aggressive phenotype with supporting angiogenesis, invasiveness, metastasis, and resistance to chemo- and radiotherapy (Ruan, Song, & Ouyang, 2009).

6 Incorporation of 3D Hydrogel Tumor Models Into Microfluidic Devices for Downstream Testing and Administering Hydrodynamic, Shear, and Mechanical Forces to These Models

6.1 Tumor Models in Microfluidics Devices

After transformation of a single cell and initiation of cancer, the microenvironment surrounding the tumor also transforms to fit the needs of the growing cell mass. This tumor microenvironment consists of both biochemical and biophysical components. Although biochemical cues have been studied widely and in detail, the role of biophysical cues in the tumor microenvironment has only become a major area of focus in recent decades. During tumor growth, there are several factors influencing the invasive, migratory, and metastatic behavior of cancer cells. These include substrate stiffness, interstitial fluid pressure, shear stresses, topographic cues, and ligands available in the microenvironment (Fig. 3). Extracellular matrix proteins reorganize and accumulate, and stromal fibroblasts are activated around the tumor, which stiffens the microenvironment (Munson, Bellamkonda, & Swartz, 2013). There is also the hydrodynamic pressure of the interstitial flow and shear stresses introduced by fluid flow in the body cavities (Rizvi et al., 2013) and solid stresses generated from cells proliferating in a confined space (Shieh, 2011). Commonly used tissue engineering scaffolds and tumor models consist of 2D cell cultures or cell monolayers, but to

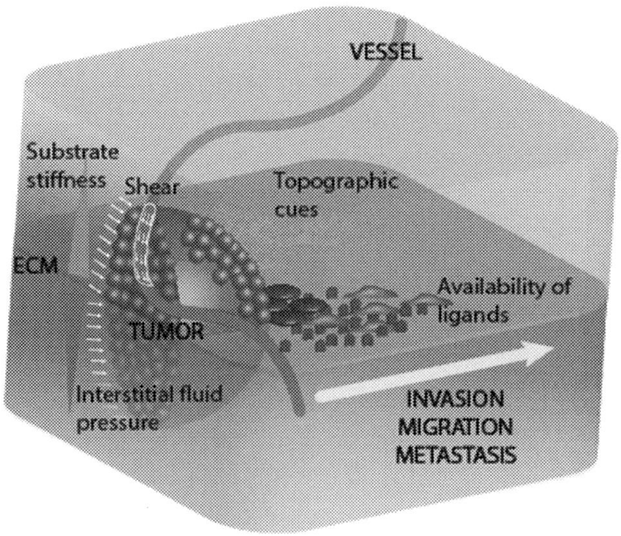

Fig. 3
Factors influencing the tumor growth, invasion, migration, and metastasis.

more closely mimic the tumor microenvironment, a matrix type culture platform, where the cancer cells and supporting tissue components are integrated, is required (Han et al., 2016). Complementary to the 3D culture, biophysical parameters such as solid stresses, stiffness, shear, and hydrodynamic pressure are required to mimic the mechanical microenvironment of the tumor. Every physical component of the microenvironment could be mimicked by the help of the microfluidics technology.

Bioengineered miniaturized in vitro physiological systems and technologies (lab-on-a-chip) are gaining the attention and interest of researchers in diverse disciplines, and have been widely applied in life sciences, particularly during the past decade. Such platforms are able to serve as systems to mimic both the biochemical and physical aspects of cellular microenvironments. Precise control of dynamic fluid flow and media composition over cell populations can be successfully achieved with microfluidic devices. Incorporation of cells in microfluidic devices was initiated to develop biosensors, diagnostic cell capture platforms, and cell sorting chips (Pappas, 2016; Ying & Wang, 2013). Further, with advances in 3D cell culture methods and tissue engineering approaches, more complex tissue/organ mimicries were developed in microphysiological systems. As in many biomedical and pharmaceutical applications, cancer research and development of chemotherapeutics are performed using conventional 2D cell culture platforms. Potential drug candidates identified with such platforms are further subjected to in vivo xenogenic cancer models. Outcomes of this approach are rarely in agreement with clinical trials. The inconsistency of in vitro models and patient results triggers the need for more relevant, predictive, and reliable settings for preclinical development of therapeutics. Bioengineered 3D cell culture systems can be successfully turned into natural tumor-like microenvironments in microfluidic devices that recapitulate the physical and biochemical cues of a tumor (Pappas, 2016).

6.2 Hydrodynamic, Shear, and Mechanical Forces

There are several approaches to mimic the tumor microenvironment together with its mechanical niche. We can separate the current literature into three groups: studies concentrating on (i) vascular networks, (i) tumor models, and (iii) tumor models with vascular networks (Table 3).

Solid stresses are somewhat easier to mimic. A growing tumor mass is commonly replicated in vitro in the form of spheroids, and it was shown that solid stresses of the surrounding semisolid matrix (agarose gel in this case) during the growth of the spheroids inhibits proliferation, but also decreases apoptotic rate. The solid stress resulting in arrested growth gives a survival advantage to the tumor (Helmlinger et al., 1997).

Microfluidic devices can be utilized for cancer detection, tumor cell isolation, and screening of therapeutics (Ozcelikkale et al., 2017; Warkiani et al., 2016; Zhang & Nagrath, 2013;

Table 3: Examples of microfluidics devices mimicking vascular networks, tumor models, and tumor models with vascular networks

Tissue	Material	Cells	Method	Force	
Vascular networks	Type I collagen gel	• Human umbilical vein endothelial cells (HUVECs) • Human brain vascular pericytes (HBVPCs) • Human umbilical arterial smooth muscle cells (HUASMCs)	• Microfluidics channels were produced in native Type I collagen gel by soft lithography • Bulk gel was loaded with perivascular cells • Microchannel network was seeded with endothelial cells	–	Zheng et al. (2012)
	Fibrin gel	• Normal human lung fibroblasts (NHLFs) • Endothelial cells (ECs)	• A polydimethylsiloxane microfluidic model with interconnected vascular networks • A long microchannel (550, 620, and 775 mm) for resistive circuit with a large hydraulic resistance • Microtissue chambers (0.12 μL) with an input impedance >50-fold larger than the long microchannel	Hydrostatic pressure	Hsu et al. (2013)
	Type I collagen gel	• Endothelial cells (EC)	• Two parallel microfluidic channels for delivering culture medium and a central gel cage with a micropillar array for housing cell-seeded scaffolds • A pressure differential (50 Pa) was imposed across the gel to demonstrate the capability to produce biophysical stimuli	Hydrostatic pressure	Vickerman et al. (2008)

Table 3: Examples of microfluidics devices mimicking vascular networks, tumor models, and tumor models with vascular networks—cont'd

Tissue	Material	Cells	Method	Force	
	Fibrinogen	• Human dermal lymphatic microvascular endothelial cells (LEC) • Normal human lung fibroblasts (NHLF)	• Two-sided channels were seeded with fibroblast-laden fibrin gel • Central channel was loaded with fibrin gel, and LECs were seeded to the surface of the fibrin were under paracrine interactions with stromal fibroblasts • Fluidics channels in between the fibroblasts and central channel had reservoirs for concentration and pressure gradients	Hydrostatic pressure	Kim, Chung, and Jeon (2016)
Tumor model	Type I collagen	• F98 rat glioblastoma cell line	• The microfluidic device design includes the cell culture chamber, the glass coverslip with etched pores, the flow channel, and the glass coverslip • The cells were seeded at a density of 0.3×10^6/mL in a 2 mg/mL Type I collagen gel • Controlled release of doxorubicin-TS-liposomes were studied under focused ultrasound trigger	-	Zervantonakis and Arvanitis (2016)
	0.1% hyaluronan with 0.12% rat tail Type I collagen	• RT2 malignant glioma cells	• Tumor cells (10^5) were seeded in 100 μL gel • An average velocity of 0.7 mm/s growth media through the cell/gel compartment or RT2 cells in hyaluronan/collagen matrix were added to radial flow chambers and exposed to pressure driven flow	Hydrostatic pressure and shear stress	Munson et al. (2013)

Continued

Table 3: Examples of microfluidics devices mimicking vascular networks, tumor models, and tumor models with vascular networks—cont'd

Tissue	Material	Cells	Method	Force	
Vascular networks with tumor model	Type I collagen and fibronectin	• Human breast cancer cells (MDA-MB-231)	• PDMS microfluidics device with 5 mm height ridges and 200 mm deep channels • PDMS devices were treated with oxygen plasma to confine collagen successfully • Interstitial flow was applied to the collagen gel with a syringe pump	Hydrostatic pressure	Huang et al. (2015)
	Type I collagen	• Human microvascular endothelial cells (hMVECs) • Human breast cancer cells (MDA-MB-231)	• Model consists of molded PDMS, through which access ports are bored and bonded to a cover glass to form a microfluidic channel and coated with a PDL solution to enhance matrix adhesion • Collagen Type I solution was injected into the gel regions of the device • Diluted Matrigel solution was introduced into the cell channel • Cancer cells were introduced so they adhere to and transmigrate across the endothelium into the gel region	–	Jeon et al. (2013)
	Type I collagen	• Oral squamous cell carcinoma (OSCC-3) • Human umbilical vein endothelial cells (HUVECs)	• PDMS wells of 4 mm diameter were produced and collagen gel was injected into them • In the collagen matrix, circular pores with a diameter of 50 mm were formed • Fluid flow with a speed of 1 mm/s was introduced • Micropatterned collagen gels were produced using a micropatterned PDMS stamp	Shear stress	Cross et al. (2010)

Table 3: Examples of microfluidics devices mimicking vascular networks, tumor models, and tumor models with vascular networks—cont'd

Tissue	Material	Cells	Method	Force	
	Matrigel	• Human umbilical vein endothelial cells (HUVECs) • Metastatic colon cancer (LOVO) • Nonmetastatic colon cancer (SW480)	• Cell-based microfluidic chip was fabricated using photolithography • To study cancer metastasis on a chip by combining intravasation and extravasation chambers • All infusions had a shear stress of 3 dyne/cm^2	Shear stress	Shin, Kim, and Jung (2011)
	Type I collagen with 10% Matrigel	• Human breast cancer cells (MDA-MB-231) • Human dermal lymphatic endothelial cells (LEC)	• The chamber is composed of a base with five microfluidic channels; a top, including the fluidic inlets and outlets, and five regions to house the 6.4 mm cell culture inserts • LECs were cultured to confluence on inserts pore diameters of 8 μm and MDA-MB-231 cells in Type I collagen with 10% Matrigel were added to the inserts • The transmural flow rate was set at either 0 or 2 mL/min	Shear stress	Pisano et al. (2015)
	Type I collagen	• Human breast cancer cells (MCF7) • Human microvascular endothelial cells (HMEC)	• The top layer has a 300 μm wide channel simulating the capillary of the tumor vasculature to simulate blood flow • Interstitial channel with a 1 mm by 2.7 mm central chamber for tumor cells • Two 300 μm wide side channels serving as lymphatics	Hydrostatic pressure	Kwak et al. (2014)

Continued

Table 3: Examples of microfluidics devices mimicking vascular networks, tumor models, and tumor models with vascular networks—cont'd

Tissue	Material	Cells	Method	Force	
	Type I collagen	• The human (Ph+) B lineage ALL cell line (SUP-B15) • Human bone marrow stromal cells (BMSC) • Human osteoblasts (HOB)	• The microchannels were fabricated via photolithography • 2D static, 3D static and 3D microfluidic • SUP-B15 cells, BMSC and HOB were fluorescently labeled using CellTracker Green CMFDA, Red CMTPX, and CellTrace Far Red DDAO-SE • Integrate interstitial flow into tri-culture where the velocity was $0.27 \pm 0.18\,\mu m/s$ (similar to bone marrow interstitial flow)	Hydrostatic pressure	Bruce et al. (2015)

Zhang, Zhang, & Zhang, 2017). Such in vitro biosystems can mimic and provide control over the dynamic interstitial fluid shear stress on the tumor. Recently, Calibasi Kocal et al. demonstrated the cancer cell plasticity under dynamic microfluidic culture compared with the conventional static cell culture (Calibasi Kocal et al., 2016). Esophageal cancer cells were cultured in a microfluidic chip with a flow rate simulating native interstitial fluid flow using cell culture media without any growth factors or stimulating agents. Results show that fluid shear stress induces cancer cells toward epithelial to mesenchymal character and makes them differentiate into a more cancer stem cell-like phenotype compared with the cells' culture in the static environment. Similarly, Rizvi et al. demonstrated that 3D ovarian cancer nodules grown under continuous flow showed increased epithelial to mesenchymal transition, increased expression and activation of the epidermal growth factor receptor, and induction of an aggressive and motile phenotype compared with static 3D cultures (Rizvi et al., 2013).

Cell physiology is dramatically affected by the culture environment as a response to substrate stiffness, cell polarity, reorganization of the cytoskeleton, and physical stimuli (Pineda, Nerem, & Ahsan, 2013; Ravi et al., 2015). Three-dimensional cell culture platforms, which recapitulate the biochemical and physical characteristics of the extracellular matrix of cells, have been of great interest to biomaterial scientists and bioengineers (Arslan-Yildiz et al., 2016; Green & Elisseeff, 2016; Guven et al., 2015). Many natural and synthetic biomaterials have been utilized as building units of 3D cell culture platforms, and find applications in tissue engineering, regenerative medicine, stem cell differentiation, drug delivery, and cancer research. Collagen, alginate, gelatin, hyaluronic acid, fibrin, silk, matrigel, polyethylene glycol (PEG), polycaprolactone (PCL), poly(lactide-co-glycolide) (PLGA), poly(vinyl alcohol) (PVA) are among the most common biomaterials already finding applications in the biomedical market and research. Hydrogels, which are highly versatile and modular in terms of chemical and physical properties, are among the most promising biomaterials to engineer a soft tissue and microenvironments similar to tumors (Asghar et al., 2013; El Assal et al., 2016). Hydrogel-based cell encapsulating 3D culture platforms can be generated through the crosslinking of hydrogel solutions via chemical, enzymatic, thermal, electrical, or photonic activation. Ultimately, the 3D cell encapsulating hydrogel generation method should preserve cell viability and functionality, as well as provide physical stability for a targeted culture period (Seliktar, 2012).

Hydrogels can be patterned in all types of shapes and broad scales, including submicrometer dimensions. Such processing versatility makes them excellent bioengineered cell microenvironments that can easily be incorporated into microfluidic systems for cancer research. Besides being more advantageous than monolayer cancer cell cultures, hydrogel-based microfluidic 3D tumor models can integrate human cells and provide a more realistic niche, thus potentially providing superior relevance compared with existing animal models (Zhang et al., 2017).

One of the components of the tumor microenvironment is new capillary growth, which is called the neovascularization (tumor cell-mediated neovascularization and lymphangiogenesis affected tumor progression and cancer metastasis). Having a vascular network in a tumor model is essential to understanding cancer physiology. By the use of microfluidics devices in vitro, vascular models could be produced (Fig. 4) (Helmlinger et al., 1997; Hsu et al., 2013; Zheng et al., 2012). Commonly, microchannels were manufactured, and the inside of the channels were epithelized to mimic vascular structures. The outside of the channels were filled with a hydrogel matrix to study neovascularization and sprout formation. Hydrogels mimic ECM and act as a medium to distribute hydrostatic pressure.

The hanging drop technique is one of the simplest methods to generate 3D cancer cell cultures. In this method, cells accumulate at the tip of a drop of medium, resulting in the formation of a spheroid due to gravitational forces. Limited direct manipulation and imaging difficulties are among the main disadvantages of this method. Similar 3D cancer cell clusters can be easily generated by arresting cancer cells in confined microfluidic systems. Elliott et al. densely seeded cancer cells, which rapidly formed 3D clusters in a chip, and easily monitored the extravascular transport and cellular uptake of drugs (Elliott & Yuan, 2012).

Tumor models also commonly include a hydrogel matrix (Fig. 5). Type 1 collagen is the most commonly used hydrogel due to being an ECM component, and its easy processability (Huang et al., 2015; Jeon et al., 2013; Kim et al., 2016; Zervantonakis & Arvanitis, 2016). To mimic the shear stress of interstitial flow in tissues, fluid flow is applied with the help of

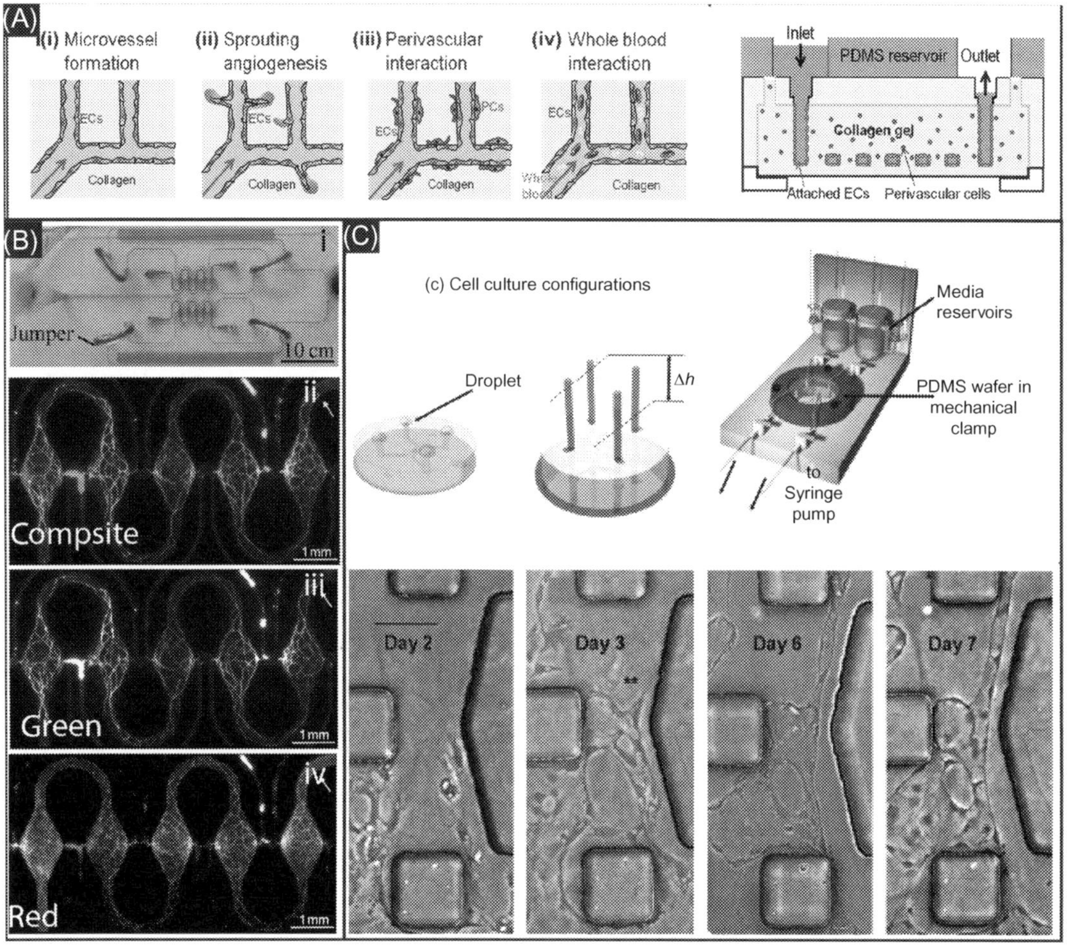

Fig. 4

Examples of vascular models in microfluidics chips. (A) Schematic cross-sectional view of a section of microvascular network illustrating (i) morphology and barrier function of endothelium, (ii) endothelial sprouting, (iii) perivascular association, and (iv) blood perfusion (*arrow*, flow direction) and schematic diagram of microfluidic collagen scaffolds after fabrication (Zheng et al., 2012). (B) The microtissue platform (i) and fluorescence image of developed human microtissues (ii). Microvascular network were labeled with CD31 (green, iii) and nuclei were labeled with DAPI (red, iv) (Hsu et al., 2013). (C) Cell culture flow configurations. Endothelial cells stimulated with enriched media. Results show a clear transition from cobblestone to multicellular, well-defined individual 2D tube structures with highly polarized tips (scale bar: 100 μm). The morphogenesis of EC can be tracked over several days. Cells migrate into gel region with a prominent increase in vacuolar structures that precede tube structure development (Vickerman et al., 2008).

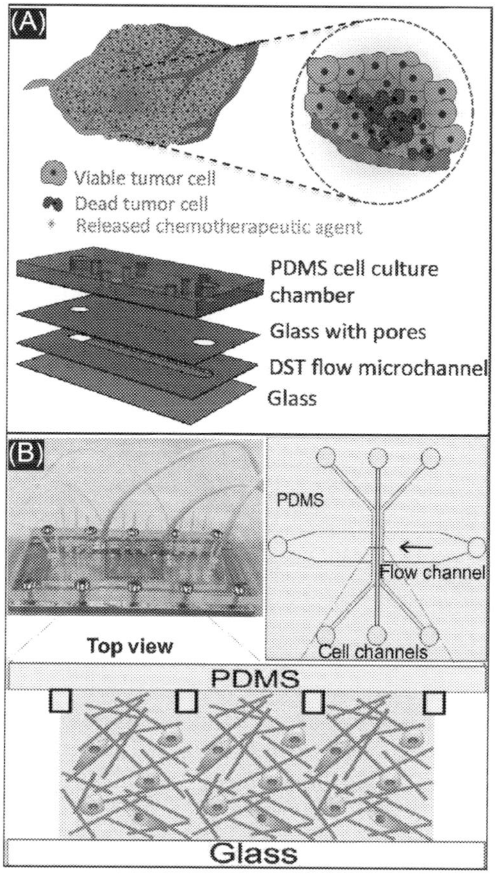

Fig. 5

Examples of vascular models in microfluidics chips. (A) Schematic showing localized drug release in a tumor. Assembly of the multilayer microfluidic device (Zervantonakis & Arvanitis, 2016). (B) Illustration of the microfluidic setup: three identical devices were patterned on a 1 mm thick PDMS membrane and sandwiched between a Plexiglas manifold. Within each device, cell embedded collagen is seeded in the three parallel cell channels. Fluid flows are introduced into horizontal flow channels. Collagen is confined within the three cell channels by the contact lines, marked by the *yellow rectangles* (Huang et al., 2015).

a pump (Huang et al., 2015; Zervantonakis & Arvanitis, 2016). Shear stress induced by fluid flow led to the alignment of the single cells (Munson et al., 2013).

Tumor models with vascular networks represent the anatomy true-to-life. There are several microfluidics designs that mimic the tumor microenvironment with shear and hydrostatic pressure components (Bruce et al., 2015; Kwak et al., 2014; Pisano et al., 2015; Shin et al., 2011). Examples of designs of such microfluidics devices are given in Fig. 6.

The delivery of chemotherapeutic drugs and nanoparticles have been widely investigated in 3D microfluidic tumor cultures (Ozcelikkale et al., 2017). Kwak et al. developed an integrated microfluidic system that recapitulates the components of solid breast cancer tumors to evaluate the transport of gold and polystyrene nanoparticles. Collagen-based 3D culture for human breast cancer cells and endothelial cells were compartmentalized in a microfluidic device, which contained both endothelial capillary and lymph channels (Fig. 6A) (Kwak et al., 2014).

The standard of care for most cancer patients typically involves surgery (when possible) along with chemotherapeutics and radiation therapy. Researchers continue to investigate

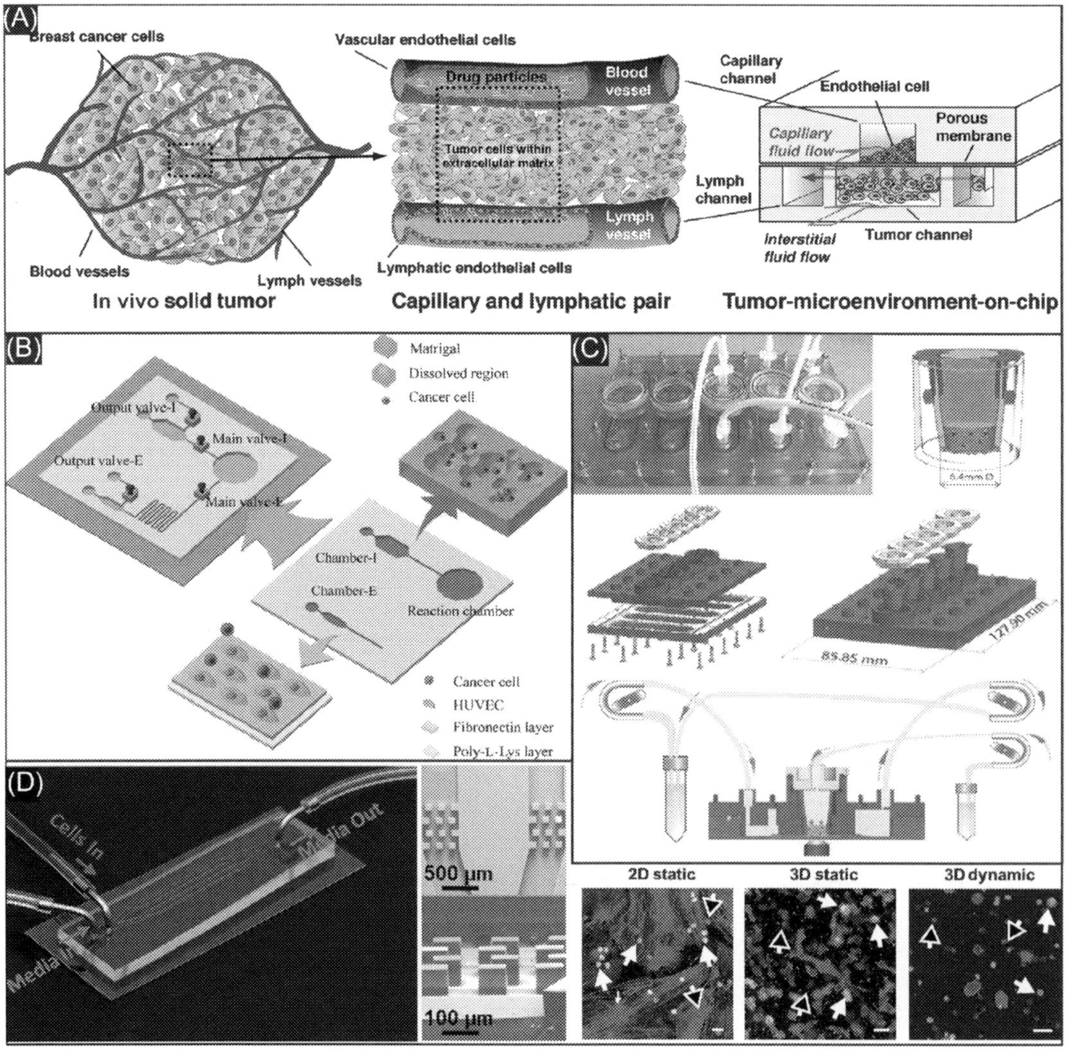

See the legend on opposite page.

the therapeutic effects of electromagnetic and radiofrequency fields in cancer treatment. Tumor-treating fields (TTFs), which refer to alternating electric fields with low-intensity and intermediate frequency (1–2 V/cm and 100–200 kHz, respectively), can actively disrupt mitosis and thereby inhibit cell division (Kirson et al., 2004). The availability of relevant model platforms is one of the bottlenecks in the development of these approaches. Pavesi et al. elegantly bioengineered a collagen-based 3D cancer coculture microenvironment in a microphysiological chip to address the effect of electromagnetic fields on cancer cells. Their research demonstrated how cancer signaling mechanisms and apoptotic pathways respond to electromagnetic fields (Pavesi et al., 2016).

Identifying the molecular mechanisms of cancer cell migration is critically important to better understand metastatic cascade in order to develop therapeutic approaches against this process (Kashaninejad et al., 2016). Many different 3D microfluidic cell culture platforms have been designed to evaluate cancer cell motility. Haessler et al. report how cancer cells move in the direction of simulated interstitial fluid flow and show a tendency to drain a lymphatic system in a dual 3D collagen hydrogel model (Haessler et al., 2012). Nguyen et al. investigated the effect of cell impedance on the migratory profile of tumors in a microfluidic chip fashioned with electrical sensors (Nguyen et al., 2013). Tumor microvasculature is as critical as 3D migration in determining the metastatic fate of cancer cells. In the past decade, advances in the design and fabrication of microfluidic systems enabled researchers to microfabricate real endothelial capillaries within 3D hydrogels embedded in chips. Vickerman et al. successfully demonstrated the formation of a perfusate artificial microvasculature by endothelial cells, which may be an ideal in vitro tumor metastasis model (Vickerman et al., 2008). This know-how opened a path to develop more sophisticated multitissues on chips and to

Fig. 6

Examples of tumor models with vascular networks. (A) A minimal functional unit of a solid tumor, that is, a pair of capillary and lymphatic vessels, is mimicked into a 3D microfluidic platform. The top channel *(red)* simulates the capillary with endothelium. The bottom layer has a central channel *(blue)* mimicking a 3D tumor microstructure and two-side channels simulating the lymphatics *(green)* (Kwak et al., 2014). (B) Schematic representation of the integrated cell-based microfluidic chip and the simulation of intra- and extra-vasation. It contains two main culture chambers; chamber-I is an intravasation chamber for cell migration, and chamber-E is an extra-vasation chamber for quantitative detection. The fluid flowing through each chamber can be controlled independently by screw valves (Shin et al., 2011). (C) Flow chamber, which hosts five independent channels. In a culture insert endothelial cells were cultured. Then, the solution containing fluorescently labeled tumor cells and extracellular matrix was placed in the upper chamber and allowed to polymerize. The chamber was mounted and primed with cell medium. Peristaltic pumps were connected to the device, allowing the fine control of the interstitial and luminal flow rates (Pisano et al., 2015). (D) Microfluidic platform and the SEM images of a top view and side view of the perfusion channels. 2D and 3D static and dynamic cultures of the cells, the actin filaments were stained with phalloidin in *red*, the nuclei were stained with DAPI in *blue (hollow arrows)*, and the nuclei of proliferating cells were stained with Ki67 in *green (white arrows)* (scale bars: 20 μm) (Bruce et al., 2015).

ask more specific questions in understanding extravasation (Bersini et al., 2014; Jeon et al., 2015), the molecular mechanisms of metastasis (Chen et al., 2016), and for use as high throughput drug screening platforms (Liu et al., 2016; Santo et al., 2017).

Besides culturing artificial 3D tumor microenvironments, microfluidic devices can also maintain the viability of real tumor tissue without digestion and primary cancer cell isolation. Indeed, maintaining native cellular composition and distribution, 3D ECM architecture and vascularization by culturing tumor tissues in vitro in microfluidic devices will overcome problems regarding the preservation and manipulation of tumors. Recently, Holton and coworkers developed a tumor biopsy trapping microfluidic device, which can also monitor the tumor microenvironment in real-time (Holton et al., 2017). This approach can also facilitate the utilization of primary cancer cells and eliminate the usage of cell lines, which do not necessarily represent the intrinsic cellular physiology.

7 Conclusion

Hydrogels are promising tissue engineering biomaterials. This property makes them just as suitable for cancer tissue modeling and creating controlled tumor microenvironments. Although there are several groups of biological and synthetic polymers available for hydrogel biomaterials, the most commonly used are collagen, fibrin, and gelatin, due to their ECM biomimicking properties. Advances in the field have proven that static cultures and encapsulation of cells in hydrogels alone are not enough to mimic the cell microenvironment. The tumor microenvironment is even more complicated and harder to imitate. Incorporation of hydrogels and microfluidics strategies showed promising results. This may fulfill the need for controllable and reproducible systems that can be used for cancer diagnosis and treatment. Cancer cell-laden hydrogels integrated into microfluidic model platforms will potentially further accelerate the progress of cancer metastasis and pharmaceutical applications. With the advances of the biomimicking microfluidic devices, personalized chemotherapeutic drug resistance and cancer therapy strategies will be enabled for future research in cancer.

References

Adjei, I. M., & Blanka, S. (2015). Modulation of the tumor microenvironment for cancer treatment: A biomaterials approach. *Journal of Functional Biomaterials*, *6*(1), 81–103.

Arslan-Yildiz, A., et al. (2016). Towards artificial tissue models: Past, present, and future of 3D bioprinting. *Biofabrication*, *8*(1), 014103.

Asghar, W., et al. (2013). In vitro three-dimensional cancer culture models. In Y. H. Bae, R. J. Mrsny, & K. Park (Eds.), *Cancer targeted drug delivery: An elusive dream* (pp. 635–665). New York, NY: Springer New York.

Aubin, H., et al. (2010). Directed 3D cell alignment and elongation in microengineered hydrogels. *Biomaterials*, *31*(27), 6941–6951.

Bell, C. L., & Peppas, N. A. (1996). Modulation of drug permeation through interpolymer complexed hydrogels for drug delivery applications. *Journal of Controlled Release*, *39*(2–3), 201–207.

Bergers, G., & Benjamin, L. E. (2003). Tumorigenesis and the angiogenic switch. *Nature Reviews Cancer*, *3*(6), 401–410.

Bersini, S., et al. (2014). A microfluidic 3D in vitro model for specificity of breast cancer metastasis to bone. *Biomaterials*, *35*(8), 2454–2461.

Bertout, J. A., Patel, S. A., & Simon, M. C. (2008). The impact of O_2 availability on human cancer. *Nature Reviews Cancer*, *8*(12), 967–975.

Bettini, R., Colombo, P., & Peppas, N. A. (1995). Solubility effects on drug transport through pH-sensitive, swelling-controlled release systems: Transport of theophylline and metoclopramide monohydrochloride. *Journal of Controlled Release*, *37*(1–2), 105–111.

Billiet, T., et al. (2012). A review of trends and limitations in hydrogel-rapid prototyping for tissue engineering. *Biomaterials*, *33*(26), 6020–6041.

Bonnans, C., Chou, J., & Werb, Z. (2014). Remodelling the extracellular matrix in development and disease. *Nature Reviews Molecular Cell Biology*, *15*(12), 786–801.

Bouyer, C., et al. (2016). A bio-acoustic levitational (BAL) assembly method for engineering of multilayered, 3D brain-like constructs, using human embryonic stem cell derived neuro-progenitors. *Advanced Materials*, *28*(1), 161–167.

Brandenberg, N., & Lutolf, M. P. (2016). In situ patterning of microfluidic networks in 3D cell-laden hydrogels. *Advanced Materials*, *28*(34), 7450–7456.

Brøndsted, H., Andersen, C., & Hovgaard, L. (1998). Crosslinked dextran—A new capsule material for colon targeting of drugs. *Journal of Controlled Release*, *53*(1), 7–13.

Brøndsted, H., Hovgaard, L., & Simonsen, L. (1995). Dextran hydrogels for colon-specific drug delivery. IV: Comparative release study of hydrocortisone and prednisolone sodium phosphate. *STP Pharma Sciences*, *5*(1), 65–69.

Bruce, A., et al. (2015). Three-dimensional microfluidic tri-culture model of the bone marrow microenvironment for study of acute lymphoblastic leukemia. *PLoS One*, *10*(10), e0140506.

Bussard, K. M., et al. (2016). Tumor-associated stromal cells as key contributors to the tumor microenvironment. *Breast Cancer Research*, *18*(1), 84.

Calibasi Kocal, G., et al. (2016). Dynamic microenvironment induces phenotypic plasticity of esophageal cancer cells under flow. *Scientific Reports*, *6*, 38221.

Cavo, M., et al. (2016). Microenvironment complexity and matrix stiffness regulate breast cancer cell activity in a 3D in vitro model. *Scientific Reports*, *6*, 35367.

Cha, C., et al. (2014). Structural reinforcement of cell-laden hydrogels with microfabricated three dimensional scaffolds. *Biomaterials Science*, *2*(5), 703–709.

Chang, S.-F., et al. (2008). Tumor cell cycle arrest induced by shear stress: Roles of integrins and Smad. *Proceedings of the National Academy of Sciences of the United States of America*, *105*(10), 3927–3932.

Chen, L., Kopeček, J., & Stewart, R. J. (2000). Responsive hybrid hydrogels with volume transitions modulated by a titin immunoglobulin module. *Bioconjugate Chemistry*, *11*(5), 734–740.

Chen, S., et al. (2005). Strong resistance of phosphorylcholine self-assembled monolayers to protein adsorption: Insights into nonfouling properties of zwitterionic materials. *Journal of the American Chemical Society*, *127*(41), 14473–14478.

Chen, M. B., et al. (2016). Elucidation of the roles of tumor integrin β1 in the extravasation stage of the metastasis cascade. *Cancer Research*, *76*(9), 2513–2524.

Chenite, A., et al. (2000). Novel injectable neutral solutions of chitosan form biodegradable gels in situ. *Biomaterials*, *21*(21), 2155–2161.

Cicek, H., & Tuncel, A. (1998). Immobilization of α-chymotrypsin in thermally reversible isopropylacrylamide-hydroxyethylmethacrylate copolymer gel. *Journal of Polymer Science Part A: Polymer Chemistry*, *36*(4), 543–552.

Cross, S. E., et al. (2007). Nanomechanical analysis of cells from cancer patients. *Nature Nanotechnology*, *2*(12), 780–783.

Cross, V. L., et al. (2010). Dense type I collagen matrices that support cellular remodeling and microfabrication for studies of tumor angiogenesis and vasculogenesis in vitro. *Biomaterials*, *31*(33), 8596–8607.

de Nooy, A. E., Masci, G., & Crescenzi, V. (1999). Versatile synthesis of polysaccharide hydrogels using the Passerini and Ugi multicomponent condensations. *Macromolecules*, *32*(4), 1318–1320.

Eales, K., Hollinshead, K., & Tennant, D. (2016). Hypoxia and metabolic adaptation of cancer cells. *Oncogenesis*, *5*(1), e190.

Egeblad, M., Rasch, M. G., & Weaver, V. M. (2010). Dynamic interplay between the collagen scaffold and tumor evolution. *Current Opinion in Cell Biology*, *22*(5), 697–706.

Eke, G., et al. (2017). Development of a UV crosslinked biodegradable hydrogel containing adipose derived stem cells to promote vascularization for skin wounds and tissue engineering. *Biomaterials*, *129*, 188–198.

El Assal, R., et al. (2016). 3-D microwell array system for culturing virus infected tumor cells. *Scientific Reports*, *6*, 39144.

Elliott, N. T., & Yuan, F. (2012). A microfluidic system for investigation of extravascular transport and cellular uptake of drugs in tumors. *Biotechnology and Bioengineering*, *109*(5), 1326–1335.

El-Sherbiny, I. M., & Yacoub, M. H. (2013). Hydrogel scaffolds for tissue engineering: Progress and challenges. *Global Cardiology Science and Practice*, *3*, 316–342.

Erler, J. T., & Weaver, V. M. (2009). Three-dimensional context regulation of metastasis. *Clinical & Experimental Metastasis*, *26*(1), 35–49.

Estrella, V., et al. (2013). Acidity generated by the tumor microenvironment drives local invasion. *Cancer Research*, *73*(5), 1524–1535.

Fuhrmann, A., et al. (2011). AFM stiffness nanotomography of normal, metaplastic and dysplastic human esophageal cells. *Physical Biology*, *8*(1), 015007.

Fukumura, D., et al. (2001). Hypoxia and acidosis independently up-regulate vascular endothelial growth factor transcription in brain tumors in vivo. *Cancer Research*, *61*(16), 6020–6024.

Gaharwar, A. K., Peppas, N. A., & Khademhosseini, A. (2014). Nanocomposite hydrogels for biomedical applications. *Biotechnology and Bioengineering*, *111*(3), 441–453.

Gatenby, R. A., & Gawlinski, E. T. (1996). A reaction-diffusion model of cancer invasion. *Cancer Research*, *56*(24), 5745–5753.

Gatenby, R. A., & Gillies, R. J. (2004). Why do cancers have high aerobic glycolysis? *Nature Reviews Cancer*, *4*(11), 891–899.

Gatenby, R. A., et al. (2006). Acid-mediated tumor invasion: A multidisciplinary study. *Cancer Research*, *66*(10), 5216–5223.

Gehrke, S. H., Uhden, L. H., & McBride, J. F. (1998). Enhanced loading and activity retention of bioactive proteins in hydrogel delivery systems. *Journal of Controlled Release*, *55*(1), 21–33.

Giammona, G., et al. (1999). New biodegradable hydrogels based on an acryloylated polyaspartamide cross-linked by gamma irradiation. *Journal of Biomaterials Science, Polymer Edition*, *10*(9), 969–987.

Gilkes, D. M., Semenza, G. L., & Wirtz, D. (2014). Hypoxia and the extracellular matrix: Drivers of tumour metastasis. *Nature Reviews Cancer*, *14*(6), 430–439.

Gnavi, S., et al. (2017). Gelatin-based hydrogel for vascular endothelial growth factor release in peripheral nerve tissue engineering. *Journal of Tissue Engineering and Regenerative Medicine*, *11*(2), 459–470.

Green, J. J., & Elisseeff, J. H. (2016). Mimicking biological functionality with polymers for biomedical applications. *Nature*, *540*(7633), 386–394.

Guven, S., et al. (2015). Multiscale assembly for tissue engineering and regenerative medicine. *Trends in Biotechnology*, *33*(5), 269–279.

Haessler, U., et al. (2012). Migration dynamics of breast cancer cells in a tunable 3D interstitial flow chamber. *Integrative Biology*, *4*(4), 401–409.

Han, B., et al. (2016). Recapitulation of complex transport and action of drugs at the tumor microenvironment using tumor-microenvironment-on-chip. *Cancer Letters*, *380*(1), 319–329.

Harrell, M. I., Iritani, B. M., & Ruddell, A. (2007). Tumor-induced sentinel lymph node lymphangiogenesis and increased lymph flow precede melanoma metastasis. *The American Journal of Pathology*, *170*(2), 774–786.

Harris, A. L. (2002). Hypoxia—A key regulatory factor in tumour growth. *Nature Reviews Cancer*, *2*(1), 38–47.

Hasirci, N., et al. (2016). *Hydrogels in regenerative medicine.* In *Gels handbook: Fundamentals, properties and applications volume 2: Applications of hydrogels in regenerative medicine*. Singapore: World Scientific.(pp. 1–52).

Helmlinger, G., et al. (1997). Solid stress inhibits the growth of multicellular tumor spheroids. *Nature Biotechnology*, *15*(8), 778–783.

Hennink, W., & Van Nostrum, C. F. (2012). Novel crosslinking methods to design hydrogels. *Advanced Drug Delivery Reviews*, *64*, 223–236.

Hoffman, A. S. (2012). Hydrogels for biomedical applications. *Advanced Drug Delivery Reviews*, *64*, 18–23.

Hofmann, M., et al. (2006). Lowering of tumor interstitial fluid pressure reduces tumor cell proliferation in a xenograft tumor model. *Neoplasia*, *8*(2), 89–95.

Holton, A. B., et al. (2017). Microfluidic biopsy trapping device for the real-time monitoring of tumor microenvironment. *PLoS One*, *12*(1), e0169797.

Hsu, Y.-H., et al. (2013). A microfluidic platform for generating large-scale nearly identical human microphysiological vascularized tissue arrays. *Lab on a Chip*, *13*(15), 2990–2998.

Huang, Y. L., et al. (2015). Interstitial flows promote amoeboid over mesenchymal motility of breast cancer cells revealed by a three dimensional microfluidic model. *Integrative Biology*, *7*(11), 1402–1411.

Hunt, J. A., et al. (2014). Hydrogels for tissue engineering and regenerative medicine. *Journal of Materials Chemistry B*, *2*(33), 5319–5338.

Ichi, T., et al. (2001). Controllable erosion time and profile in poly (ethylene glycol) hydrogels by supramolecular structure of hydrolyzable polyrotaxane. *Biomacromolecules*, *2*(1), 204–210.

Ikada, Y., et al. (1987). Stereocomplex formation between enantiomeric poly (lactides). *Macromolecules*, *20*(4), 904–906.

Jeon, J. S., et al. (2013). In vitro model of tumor cell extravasation. *PLoS One*, *8*(2), e56910.

Jeon, J. S., et al. (2015). Human 3D vascularized organotypic microfluidic assays to study breast cancer cell extravasation. *Proceedings of the National Academy of Sciences of the United States of America*, *112*(1), 214–219.

Jiang, W., et al. (2016). Cell-laden microfluidic microgels for tissue regeneration. *Lab on a Chip*, *16*(23), 4482–4506.

Jin, Y., et al. (2001). Recyclable characteristics of hyaluronate–polyhydroxyethyl acrylate blend hydrogel for controlled releases. *Journal of Controlled Release*, *73*(2), 173–181.

Kalluri, R. (2016). The biology and function of fibroblasts in cancer. *Nature Reviews Cancer*, *16*(9), 582–598.

Kashaninejad, N., et al. (2016). Organ-tumor-on-a-chip for chemosensitivity assay: A critical review. *Micromachines*, *7*(8), 130.

Katira, P., Bonnecaze, R. T., & Zaman, M. H. (2013). Modeling the mechanics of cancer: Effect of changes in cellular and extra-cellular mechanical properties. *Frontiers in Oncology*, *3*, 145.

Kato, Y., et al. (2013). Acidic extracellular microenvironment and cancer. *Cancer Cell International*, *13*(1), 89.

Kim, S., Chung, M., & Jeon, N. L. (2016). Three-dimensional biomimetic model to reconstitute sprouting lymphangiogenesis in vitro. *Biomaterials*, *78*, 115–128.

Kirson, E. D., et al. (2004). Disruption of cancer cell replication by alternating electric fields. *Cancer Research*, *64*(9), 3288–3295.

Knowlton, S., et al. (2016). 3D-printed microfluidic chips with patterned, cell-laden hydrogel constructs. *Biofabrication*, *8*(2), 025019.

Ko, C. Y., et al. (2016). In vitro and in vivo co-culture of chondrocytes and bone marrow stem cells in photocrosslinked PCL–PEG–PCL hydrogels enhances cartilage formation. *Journal of Tissue Engineering and Regenerative Medicine*, *10*(10), E485–E496.

Kobayashi, H., Watanabe, R., & Choyke, P. L. (2014). Improving conventional enhanced permeability and retention (EPR) effects; what is the appropriate target? *Theranostics*, *4*(1), 81.

Kolesky, D. B., et al. (2014). 3D bioprinting of vascularized, heterogeneous cell-laden tissue constructs. *Advanced Materials*, *26*(19), 3124–3130.

Kuijpers, A., et al. (2000). In vivo and in vitro release of lysozyme from cross-linked gelatin hydrogels: A model system for the delivery of antibacterial proteins from prosthetic heart valves. *Journal of Controlled Release*, *67*(2), 323–336.

Kwak, B., et al. (2014). Simulation of complex transport of nanoparticles around a tumor using tumor-microenvironment-on-chip. *Journal of Controlled Release*, *194*, 157–167.

Lardner, A. (2001). The effects of extracellular pH on immune function. *Journal of Leukocyte Biology*, *69*(4), 522–530.

Lee, K. Y., Bouhadir, K. H., & Mooney, D. J. (2000). Degradation behavior of covalently cross-linked poly (aldehyde guluronate) hydrogels. *Macromolecules*, *33*(1), 97–101.

Li, Y., et al. (2014). Engineering cell alignment in vitro. *Biotechnology Advances*, *32*(2), 347–365.

Lin, Y.-H., et al. (2005). Physically crosslinked alginate/N,O-carboxymethyl chitosan hydrogels with calcium for oral delivery of protein drugs. *Biomaterials*, *26*(14), 2105–2113.

Liu, W., et al. (2016). Heterotypic 3D tumor culture in a reusable platform using pneumatic microfluidics. *Lab on a Chip*, *16*(21), 4106–4120.

Loessner, D., et al. (2016). Functionalization, preparation and use of cell-laden gelatin methacryloyl-based hydrogels as modular tissue culture platforms. *Nature Protocols*, *11*(4), 727–746.

Lu, P., Weaver, V. M., & Werb, Z. (2012). The extracellular matrix: A dynamic niche in cancer progression. *The Journal of Cell Biology*, *196*(4), 395–406.

Mansur, H. S., et al. (2008). FTIR spectroscopy characterization of poly (vinyl alcohol) hydrogel with different hydrolysis degree and chemically crosslinked with glutaraldehyde. *Materials Science and Engineering: C*, *28*(4), 539–548.

Martens, P., & Anseth, K. (2000). Characterization of hydrogels formed from acrylate modified poly (vinyl alcohol) macromers. *Polymer*, *41*(21), 7715–7722.

Mathur, A. M., et al. (1998). Equilibrium swelling of poly (methacrylic acid-g-ethylene glycol) hydrogels: Effect of swelling medium and synthesis conditions. *Journal of Controlled Release*, *54*(2), 177–184.

Meinert, C., et al. (2017). Tailoring hydrogel surface properties to modulate cellular response to shear loading. *Acta Biomaterialia*, *52*, 105–117.

Michor, F., et al. (2011). What does physics have to do with cancer? *Nature Reviews Cancer*, *11*(9), 657–670.

Mitchell, M. J., & King, M. R. (2013). Computational and experimental models of cancer cell response to fluid shear stress. *Frontiers in Oncology*, *3*, 44.

Miyata, T., Asami, N., & Uragami, T. (1999). Preparation of an antigen-sensitive hydrogel using antigen–antibody bindings. *Macromolecules*, *32*(6), 2082–2084.

Moellering, R. E., et al. (2008). Acid treatment of melanoma cells selects for invasive phenotypes. *Clinical & Experimental Metastasis*, *25*(4), 411–425.

Munson, J. M., Bellamkonda, R. V., & Swartz, M. A. (2013). Interstitial flow in a 3D microenvironment increases glioma invasion by a CXCR4-dependent mechanism. *Cancer Research*, *73*(5), 1536–1546.

Nagahara, S., & Matsuda, T. (1996). Hydrogel formation via hybridization of oligonucleotides derivatized in water-soluble vinyl polymers. *Polymer Gels and Networks*, *4*(2), 111–127.

Naseri, N., et al. (2016). Nanocellulose-based interpenetrating polymer network (IPN) hydrogels for cartilage applications. *Biomacromolecules*, *17*(11), 3714–3723.

Ng, W. L., Yeong, W. Y., & Naing, M. W. (2016). Polyelectrolyte gelatin-chitosan hydrogel optimized for 3D bioprinting in skin tissue engineering. *International Journal of Bioprinting*, *2*(1), 53–62.

Nguyen, T. A., et al. (2013). Microfluidic chip with integrated electrical cell-impedance sensing for monitoring single cancer cell migration in three-dimensional matrixes. *Analytical Chemistry*, *85*(22), 11068–11076.

Nichol, J. W., et al. (2010). Cell-laden microengineered gelatin methacrylate hydrogels. *Biomaterials*, *31*(21), 5536–5544.

Overstreet, D. J., et al. (2012). Injectable hydrogels. *Journal of Polymer Science Part B: Polymer Physics*, *50*(13), 881–903.

Ozcelikkale, A., et al. (2017). In vitro microfluidic models of tumor microenvironment to screen transport of drugs and nanoparticles. *Wiley Interdisciplinary Reviews: Nanomedicine and Nanobiotechnology*, *9*(5), e1460.

Pappas, D. (2016). Microfluidics and cancer analysis: Cell separation, cell/tissue culture, cell mechanics, and integrated analysis systems. *Analyst*, *141*(2), 525–535.

Park, J., et al. (2017). Cell-laden 3D bioprinting hydrogel matrix depending on different compositions for soft tissue engineering: Characterization and evaluation. *Materials Science and Engineering: C*, *71*, 678–684.

Pavesi, A., et al. (2016). Engineering a 3D microfluidic culture platform for tumor-treating field application. *Scientific Reports*, *6*, 26584.

Peppas, N. A., & Scott, J. E. (1992). Controlled release from poly (vinyl alcohol) gels prepared by freezing-thawing processes. *Journal of Controlled Release*, *18*(2), 95–100.

Pickup, M. W., Mouw, J. K., & Weaver, V. M. (2014). The extracellular matrix modulates the hallmarks of cancer. *EMBO Reports*, *15*(2), 1243–1253.

Pineda, E. T., Nerem, R. M., & Ahsan, T. (2013). Differentiation patterns of embryonic stem cells in two- versus three-dimensional culture. *Cells Tissues Organs*, *197*(5), 399–410.

Pisano, M., et al. (2015). An in vitro model of the tumor–lymphatic microenvironment with simultaneous transendothelial and luminal flows reveals mechanisms of flow enhanced invasion. *Integrative Biology*, *7*(5), 525–533.

Plodinec, M., et al. (2013). The nanomechanical signature of breast cancer. *Biophysical Journal*, *104*(2), 321a.

Polacheck, W. J., Charest, J. L., & Kamm, R. D. (2011). Interstitial flow influences direction of tumor cell migration through competing mechanisms. *Proceedings of the National Academy of Sciences of the United States of America*, *108*(27), 11115–11120.

Ravi, M., et al. (2015). 3D cell culture systems: Advantages and applications. *Journal of Cellular Physiology*, *230*(1), 16–26.

Remmerbach, T. W., et al. (2009). Oral cancer diagnosis by mechanical phenotyping. *Cancer Research*, *69*(5), 1728–1732.

Rizvi, I., et al. (2013). Flow induces epithelial-mesenchymal transition, cellular heterogeneity and biomarker modulation in 3D ovarian cancer nodules. *Proceedings of the National Academy of Sciences of the United States of America*, *110*(22), E1974–E1983.

Rizwan, M., et al. (2017). Sequentially-crosslinked bioactive hydrogels as nano-patterned substrates with customizable stiffness and degradation for corneal tissue engineering applications. *Biomaterials*, *120*, 139–154.

Rofstad, E. K., et al. (2006). Acidic extracellular pH promotes experimental metastasis of human melanoma cells in athymic nude mice. *Cancer Research*, *66*(13), 6699–6707.

Ruan, K., Song, G., & Ouyang, G. (2009). Role of hypoxia in the hallmarks of human cancer. *Journal of Cellular Biochemistry*, *107*(6), 1053–1062.

Santo, V. E., et al. (2017). Drug screening in 3D in vitro tumor models: Overcoming current pitfalls of efficacy read-outs. *Biotechnology Journal*, *12*(1), 1600505.

Seliktar, D. (2012). Designing cell-compatible hydrogels for biomedical applications. *Science*, *336*(6085), 1124–1128.

Semenza, G. L. (2012). Hypoxia-inducible factors: Mediators of cancer progression and targets for cancer therapy. *Trends in Pharmacological Sciences*, *33*(4), 207–214.

Shieh, A. C. (2011). Biomechanical forces shape the tumor microenvironment. *Annals of Biomedical Engineering*, *39*(5), 1379–1389.

Shin, M. K., Kim, S. K., & Jung, H. (2011). Integration of intra-and extravasation in one cell-based microfluidic chip for the study of cancer metastasis. *Lab on a Chip*, *11*(22), 3880–3887.

Shin, H., Olsen, B. D., & Khademhosseini, A. (2012). The mechanical properties and cytotoxicity of cell-laden double-network hydrogels based on photocrosslinkable gelatin and gellan gum biomacromolecules. *Biomaterials*, *33*(11), 3143–3152.

Siemann, D. W., & Horsman, M. R. (2015). Modulation of the tumor vasculature and oxygenation to improve therapy. *Pharmacology & Therapeutics*, *153*, 107–124.

Siltanen, C., et al. (2016). Microfluidic fabrication of bioactive microgels for rapid formation and enhanced differentiation of stem cell spheroids. *Acta Biomaterialia*, *34*, 125–132.

Stenekes, R., Talsma, H., & Hennink, W. (2001). Formation of dextran hydrogels by crystallization. *Biomaterials*, *22*(13), 1891–1898.

Suresh, S. (2007). Biomechanics and biophysics of cancer cells. *Acta Materialia*, *55*(12), 3989–4014.

Swaminathan, V., et al. (2011). Mechanical stiffness grades metastatic potential in patient tumor cells and in cancer cell lines. *Cancer Research*, *71*(15), 5075–5080.

Swartz, M. A., & Lund, A. W. (2012). Lymphatic and interstitial flow in the tumour microenvironment: Linking mechanobiology with immunity. *Nature Reviews Cancer*, *12*(3), 210–219.

Tan, H., & Marra, K. G. (2010). Injectable, biodegradable hydrogels for tissue engineering applications. *Materials*, *3*(3), 1746–1767.

Teixeira, L. S. M., et al. (2012). Enzyme-catalyzed crosslinkable hydrogels: Emerging strategies for tissue engineering. *Biomaterials*, *33*(5), 1281–1290.

Vaupel, P., et al. (2004). Tumor microenvironmental physiology and its implications for radiation oncology. *Seminars in Radiation Oncology*, *14*(3), 198–206.

Vickerman, V., et al. (2008). Design, fabrication and implementation of a novel multi-parameter control microfluidic platform for three-dimensional cell culture and real-time imaging. *Lab on a Chip*, *8*(9), 1468–1477.

Wade, R. J., et al. (2015). Nanofibrous hydrogels with spatially patterned biochemical signals to control cell behavior. *Advanced Materials*, *27*(8), 1356–1362.

Wang, L., et al. (2016). A double network strategy to improve epithelization of a poly (2-hydroxyethyl methacrylate) hydrogel for corneal repair application. *RSC Advances*, *6*(2), 1194–1202.

Warkiani, M. E., et al. (2016). Ultra-fast, label-free isolation of circulating tumor cells from blood using spiral microfluidics. *Nature Protocols*, *11*(1), 134–148.

Watanabe, T., et al. (1996). NMR studies on water and polymer diffusion in dextran gels. Influence of potassium ions on microstructure formation and gelation mechanism. *Magnetic Resonance in Medicine*, *35*(5), 697–705.

Xu, L., Fukumura, D., & Jain, R. K. (2002). Acidic extracellular pH induces vascular endothelial growth factor (VEGF) in human glioblastoma cells via ERK1/2 MAPK signaling pathway mechanism of low pH-induced VEGF. *Journal of Biological Chemistry*, *277*(13), 11368–11374.

Yadav, A., et al. (2013). Tumor-associated endothelial cells promote tumor metastasis by chaperoning circulating tumor cells and protecting them from anoikis. *PLoS One*, *10*(10), e0141602.

Yang, J., et al. (2017). Cell-laden hydrogels for osteochondral and cartilage tissue engineering. *Acta Biomaterialia*, *15*(57), 1–25.

Ying, L., & Wang, Q. (2013). Microfluidic chip-based technologies: Emerging platforms for cancer diagnosis. *BMC Biotechnology*, *13*(1), 76.

Zervantonakis, I. K., & Arvanitis, C. D. (2016). Controlled drug release and chemotherapy response in a novel acoustofluidic 3D tumor platform. *Small*, *12*(19), 2616–2626.

Zhang, Z., & Nagrath, S. (2013). Microfluidics and cancer: Are we there yet? *Biomedical Microdevices*, *15*(4), 595–609.

Zhang, Y. S., Zhang, Y.-N., & Zhang, W. (2017). Cancer-on-a-chip systems at the frontier of nanomedicine. *Drug Discovery Today*, *22*(9), 1392–1399.

Zheng, Y., et al. (2012). In vitro microvessels for the study of angiogenesis and thrombosis. *Proceedings of the National Academy of Sciences of the United States of America*, *109*(24), 9342–9347.

CHAPTER 16

Lipid-Based Nanosystems for the Delivery of siRNA: Challenges and Trends

Nuno A. Fonseca[*,†], **Rui M. Lopes**[*], **Ana F. Cruz**[*], **Ana C. Gregório**[*], **Ângela Valério-Fernandes**[*], **Vera Moura**[*,†], **Sérgio Simões**[*], **João Nuno Moreira**[*]
[*]*University of Coimbra, Coimbra, Portugal* [†]*TREAT U, SA, Coimbra, Portugal*

1 Toward Therapeutic RNA Interference

The introduction of the RNA interference (RNAi) mechanism, including micro RNA (miRNA) and small-interference RNA (siRNA)-mediated gene silencing, while having an extensive impact on the study of cell biology, has also opened a wide range of novel therapeutic opportunities. Despite previous hints on the regulatory nature of small noncoding RNA, RNAi was first introduced upon the discovery that double-stranded RNA (dsRNA) enabled gene silencing in *Caenorhabditis elegans* (Fire et al., 1998; Ipsaro & Joshua-Tor, 2015), and later confirmed in mammalian cells (Elbashir et al., 2001). Meanwhile, in the past few decades, it has become clear that there are more than a thousand small noncoding RNAs that regulate more than half of all described human genes (Ipsaro & Joshua-Tor, 2015). This elicits a high complex regulation network, behind which rests a finely tuned mechanism of interference supporting cellular specialization, function, and signaling response.

1.1 RNA Interference

Despite the common features among miRNA and siRNA gene silencing, both pathways intrinsically differ in terms of dsRNA origin, as well as in the way that repression is achieved.

In the miRNA pathway (Fig. 1), long hairpinned primary miRNAs (pri-miRNAs) are cropped by a heterodimeric complex, termed a microprocessor (which includes Drosha), into the 70 nucleotides (nt) precursor miRNA (pre-miRNA) at the nucleus. Pre-miRNAs are then exported from the nucleus and processed by Dicer endonuclease into a 21–25 nt mature miRNA (dsRNA)—a small interfering RNA (siRNA)—which is then loaded into RISC (RNA-induced silencing complex), with concomitant discard of the passenger-strand (a.k.a nonsense strand). It is the guide strand-loaded Ago2, in the RISC complex

Handbook of Nanomaterials for Cancer Theranostics. https://doi.org/10.1016/B978-0-12-813339-2.00016-5

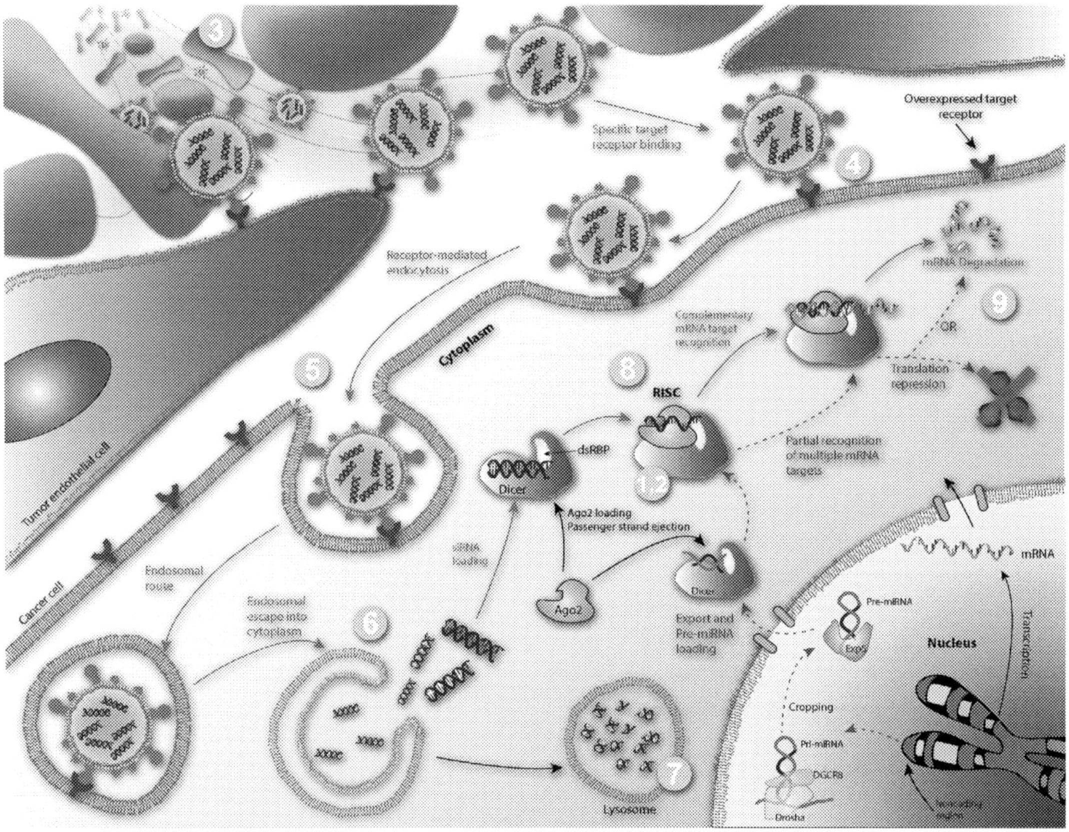

Fig. 1

Advancing RNA interference as a therapeutic strategy. (1) Endogenous RNAi. Protein synthesis regulation is paramount for cellular functions, representing a naturally occurring control phenomenon at the post-transcriptional level, mainly mediated by miRNA. Erratic gene expression, mediated by faulty RNAi, is one of the key checkpoints of carcinogenesis, and therefore passive of therapeutic intervention. (2) Exogenous siRNA-mediated RNAi. The exogenous delivery of processed double strand RNA—siRNA—represents the most prominent method to manipulate the expression of a given gene target in a human organism. However, its application is highly dependent on the development of delivery systems that can overcome different biological barriers upon systemic administration. (3) Prolonged circulation in the bloodstream. Upon systemic administration, siRNA-loaded nanoparticles, namely liposomes, should present prolonged blood circulation to enable their accumulation at the tumor site, mainly through the EPR effect, powered by aberrant angiogenesis activation and poor lymphatic drainage. (4) Beyond the liver by active targeting. Delivery technologies (mainly lipid-based) currently in clinical trials mainly target hepatic diseases. However, studies show that different lipid-based nanoparticles that explore overexpressed receptors in the tumor through ligand-mediated targeting to cancer cells, or to the tumor vasculature, enable a strategy with great potential to offer cancer targeting beyond the liver. (5) Liposomes internalization. Binding of ligand-mediated lipid-based nanoparticles should trigger receptor-mediated endocytosis. (6) Endosomal escape. Cell internalization does not guarantee full bioavailability of the siRNA at the intracellular level. The nanosystem should enable siRNA escape from the endocytic pathway, thus avoiding or limiting its degradation in the lysosomes (7). This will grant siRNA integrity and access to the RNAi machinery (8) for effective and specific gene silencing (9).

(which also includes the endonuclease Dicer), that finally mediates silencing through imperfect recognition of target messenger RNA (mRNA) at the 3′UTR (untranslated region), although not exclusively, thus impairing translation (Ipsaro & Joshua-Tor, 2015).

The processing of exogenous dsRNA (imported directly into the cytoplasm) is, in every aspect, similar to the processing of pre-miRNAs. The exogenous dsRNA is processed by Dicer into interfering dsRNAs—siRNAs—(as mature miRNA) which are subsequently loaded into RISC (Fig. 1). In the case of exogenous dsRNA-mediated generation of siRNAs, a perfect complementary of the guide strand does exist at the open reading frame (ORF) region, leading to target mRNA degradation (Elbashir et al., 2001; Gomes-da-Silva et al., 2012; Ipsaro & Joshua-Tor, 2015). Thus, whereas siRNAs generated endogenously may regulate several targets (owing to the absence of full complementarity); the siRNAs generated from exogenous dsRNA only regulate one target protein (Zhang, Wang, & Gemeinhart, 2013).

Regardless of the pathway, RNAi represents a highly specific post-transcriptional regulatory process by which cells modulate the expression of elementary protein effectors (Gomes-da-Silva, Fonseca, et al., 2012; Ipsaro & Joshua-Tor, 2015). While RNAi, physiologically, represents a pivotal process in cellular biology, its deregulation could be useful for treating several diseases, including cancer.

1.2 RNAi: A Driver in Cancer Waiting for Therapeutic Intervention

Cell signaling cascades play an irreplaceable role in the biological control and relay of information inside and between different cells, establishing a network that works seamlessly in normal conditions. However, a disease such as cancer sets its fundamental basis in the loss of the tight control over the information flow, led by the acquisition of, eventually, activating or silencing mutations that hinder the action of key regulators of a signaling cascade. This hindrance leads to complete impairment of the affected signaling pathway, culminating in the phenotypic expression of erratic cell behavior, which supports the disease (Hanahan & Weinberg, 2011). Ultimately, the erratic signaling leads to the involvement of a multitude of different cell types, including cancer cells, fibroblasts, endothelial cells, or even cells from the immune system, performing very dissimilar roles (Hanahan & Weinberg, 2011).

As a part of the gene expression regulatory machinery, and consequently influencing cell signaling and behavior, RNAi impairment may represent a key checkpoint in the carcinogenesis process, leading to deregulated gene expression, including oncogenes and tumor suppressors. In fact, the expression of several miRNAs, the oligonucleotide sequences targeting for translation blocking, have been found to be expressed at different levels in normal and breast cancer tissue (Iorio et al., 2005). This actually represents a common event in either benign or malignant breast tumors (Tahiri et al., 2014). Different miRNAs, either clusters or families, are aberrantly expressed in different solid or liquid tumors (Calin & Croce, 2006; Nana-Sinkam & Croce, 2014). For example, aberrantly expressed *let-7* family or the *miR-15a-miR-16-1* cluster target essential transcripts such as *Ras* and *BCL2* oncogenes, respectively (Calin & Croce, 2006). The altered expression of different miRNA clusters may be a result of their localization, often in chromosomal regions susceptible of deletion, amplification, or translocation; and many of those may be coincidently altered in different cancers (Nana-Sinkam & Croce, 2014). Further complicating the matter, regions coding for miRNA may exist in introns, exons, or even in the host gene, under further epigenetic control (Calin & Croce, 2006; Nana-Sinkam & Croce, 2014). Considering that each miRNA may regulate several protein-coding transcripts, and each of these may be regulated by different miRNAs (Calin & Croce, 2006), the arising exponential complexity exposes the impact that a single mutation may have on the cell biology, and consequently in the carcinogenesis process.

1.3 Beyond Virus-Mediated RNAi

Despite its critical importance in physiological cell biology, RNAi deregulation involvement in disease, namely cancer, presents itself as an opportunity for the development of innovative therapeutic strategies (Couto & High, 2010; Gomes-da-Silva, Fonseca, et al., 2012; Liao & Tang, 2015). Strictly at the cell level, viruses are the best known vehicle for the efficient transduction of DNA-based oligonucleotide sequences (Boudreau & Davidson, 2012). With such a strategy, stable suppression of tumorigenicity of pancreatic cancer cells was attained upon generating siRNAs against a mutant form of K-RAS, demonstrating the potential to reverse the oncogenic phenotype of cancer cells (Brummelkamp, Bernards, & Agami, 2002a, 2002b). Since then, different strategies using viral-mediated delivery of DNA-based short-hairpin RNAs (shRNA) or miRNA mimics have been developed and characterized in vitro and in vivo as potential RNAi therapeutic tools for different diseases, including cancer (Couto & High, 2010; Liao & Tang, 2015).

Although highly efficient delivery agents, viral vectors present several drawbacks, including elevated complexity in preparation and handling, as well as oncogenic activity to some extent, and are susceptible to eliciting extensive immune responses (Zhang et al., 2013). These aspects have limited their clinical use to certain specific situations as local treatment. This is the case for the recent Food and Drug Administration (FDA) approval of the first oncolytic virus-based cancer therapy, named talimogene laherparepvec, from Amgen (T-VEC, a.k.a. IMLYGIC™). IMLYGIC™ is a herpes simplex virus-based vector for delivery of GM-CSF indicated for the local treatment of unresectable cutaneous, subcutaneous, and nodal lesions in patients with recurrent melanoma, enabling cell death by lytic activity (Andtbacka et al., 2015; Ledford, 2015). Alternatively to viral vectors, the use of chemically modified (naked) siRNA or miRNA mimics has been attempted in order to modify their highly unfavorable pharmacokinetics, as result of elevated kidney elimination and blood nuclease degradation. Nevertheless, success has been limited (Wan, Allen, & Cullis, 2014; Zhang et al., 2013), notwithstanding the fact that naked RNAs can be immunogenic (Wan et al., 2014).

2 Aiming at In Vivo RNAi

To overcome the aforementioned hurdles within the context of intravenous administration, nanotechnology-based delivery systems, especially lipid-based, represent an important opportunity. This relies on the solid demonstration of pharmacokinetic modulation arising from the development of Caelyx/Doxil, where a significant increase of blood circulation half-life was achieved with liposomal doxorubicin, relative to the free drug, without immunogenicity (Barenholz, 2012).

Different lipid-based nanosystems have been developed for the delivery of mature interfering RNA toward direct inclusion in the RISC complex (Wan et al., 2014; Wang et al., 2015). Delivery of miRNAs has often been used to attempt to substitute faulty endogenous miRNA, aiming at multitarget regulation, or even modulation, of other miRNAs (beyond the scope of this chapter, but reviewed elsewhere) (Chen, Gao, & Huang, 2015; Wang et al., 2015). On the other hand, the delivery of siRNA, owing to its full complementarity to the target sequence, enables specific protein downregulation (Lam, Chow, Zhang, & Leung, 2015), a strategic feature for the shutdown of key signaling pathways involved in cancer, further explored as follows.

2.1 Lipid-Based Nanosystems: Evolving siRNA Delivery

2.1.1 Lipid complexes: The basics

The first iteration of an siRNA-Lipid complex (a.k.a. *lipoplex*) was based on the spontaneous electrostatic attraction between the anionic nucleic acids and cationic lipids, described for the development of a DNA-transfection vector (Felgner et al., 1987). The formation of a positively charged lipid-DNA complex, arising from the spontaneous interaction between liposomes containing a synthetic cationic lipid (DOTMA, N-[1-(2,3-dioleyloxy)propyl]-N,N,N-trime-thylammonium chloride) and catalase-encoding DNA, was reported. Upon interaction and uptake by negatively charged cell membranes, it resulted in successful protein expression by the transfected cells (Felgner et al., 1987).

The high therapeutic potential associated with the intracellular delivery of nucleic acids led to the translation of this strategy to siRNA. However, unlike DNA, siRNAs are smaller and present a lower charge density and a more rigid backbone structure. These render a lower binding affinity to cationic lipids, which results in low siRNA loading efficiency and less compact and stable lipoplexes (Kwok & Hart, 2011; Shu, Shu, Haque, Abdelmawla, & Guo, 2011). Attempts to improve their structural stability in vivo relied on the use of higher amounts of cationic lipids. Nonetheless, the extensive use of cationic lipids has been associated with severe on- and off-target toxicity (such as activation of complement system and immune or inflammatory responses) (Dokka, Toledo, Shi, Castranova, & Rojanasakul, 2000; Lv, Zhang, Wang, Cui, & Yan, 2006; Yew & Scheule, 2005). In addition, the highly positive surface charge of the complex induces strong interactions with negatively charged serum proteins, leading to changes in the complex surface charge and size, triggering a rapid blood clearance by the mononuclear phagocyte system (MPS) (Levchenko, Rammohan, Lukyanov, Whiteman, & Torchilin, 2002).

2.1.2 Cationics *redesign: Improving delivery efficiency and minimizing toxicity*

The elevated toxicity and poor pharmacokinetics provided by cationic complexes necessitated the development of new strategies to improve in vivo siRNA delivery. A key aspect resided in the development of novel cationic lipids providing enhanced complexation efficiency, while reducing the cationic lipid load *per* liposome, and thus toxicity. Bouxsein et al. observed that siRNA complexes containing a pentavalent lipid MVL5 exhibited superior silencing efficiency in vitro over a large range of lipid content and cationic lipid/nucleic acid molar ratios, relative to a similar formulation containing the monovalent DOTAP (Bouxsein, Mcallister, Ewert, Samuel, & Safinya, 2007). With the former lipid, siRNA complexes with five-fold less lipid molecules than DOTAP, with reduced toxicity, are generated (Bouxsein et al., 2007). Although the strategy could improve the lipoplex-cell interaction and endosomal escape, the predicted strong siRNA-MLV interaction may compromise the dissociation of the nucleic acid in the cytoplasm after endosomal escape (Ahmad et al., 2005).

In that context, further effort led to the development of several ionizable cationic lipids, in particular with a pK_a lower than 7, as a strategy to improve siRNA release from the liposome and endosome (Semple et al., 2001). At low pH values (pH ~4), these ionizable lipids display a positive charge that can interact with the negatively charged siRNA, enabling loading into the liposomes. However, at physiological pH values, the liposomes exhibit a low positive surface charge that enables (upon pegylation) longer circulation times and reduced toxicity. Once in the acidic environment of endosomes, the positively charged ionizable lipid promotes the subsequent destabilization of the endosomal membranes and siRNA release into the cytoplasm.

The first ionizable cationic lipid used for the encapsulation of nucleic acids was 1,2-dioleoyl-3-dimethylammonium propane (DODAP) (Semple et al., 2001). It was used in a new lipid-based nanoparticle—the SNALP (stabilized nucleic acid lipid particle). SNALP are formed by a mixture of an ionizable lipid (e.g., DODAP), cholesterol, a high-transition temperature lipid (as DSPC), and a poly(ethylene glycol) (PEG)-derivatized lipid that contributes to the neutral and hydrophilic exterior that shields the lipid bilayer in vivo, preventing rapid systemic clearance (Harris & Chess, 2003; Lasic, Martin, Gabizon, Huang, & Papahadjopoulos, 1991; Semple et al., 2001). Since then, several other new ionizable lipids have been synthesized, including the 1,2-dilinoleyloxy-3-dimethylaminopropane (DLinDMA). Intravenous delivery of Hepatitis B virus (HBV)-specific siRNA encapsulated in DLinDMA-incorporating SNALP resulted in persistent in vivo gene silencing activity against a model of HBV at low doses (Morrissey et al., 2005). Building upon this with a similar delivery strategy, following systemic administration, silencing of essential cell-cycle proteins, such as polo-like kinase

1 (PLK1) and kinesin spindle protein (KSP), displayed potent antitumor efficacy, in both orthotopic and subcutaneous liver cancer animal models. Protein downregulation persisted for up to 10 days, at a magnitude sufficient to induce the mitotic disruption and apoptosis of tumor cells (Judge et al., 2009). These were important steps in making siRNA a clinically viable therapeutic approach (Morrissey et al., 2005), representing the precursors of Alnylam's ALN-VSP02 and Tekmira's TKM-PLK1 candidates, addressed as follows.

Other important ionizable cationic lipids (such as DLin-KC2-DMA) have been identified through structure-activity relationship-guided synthesis, and using an in vivo screening model to silence Factor VII (FVII) in the liver (Akinc et al., 2008). SNALPs incorporating DLin-KC2-DMA ionizable lipid, and containing antiandrogen receptor siRNA, enabled effective in vitro silencing in prostate cancer cells. Upon systemic administration, androgen receptor silencing was further demonstrated in distal prostate xenograft tumors, with a consequent decrease in the serum levels of prostate specific antigens (Lee et al., 2012). Using DLin-KC2-DMA as a starting point, a novel ionizable cationic lipid, the DLin-MC3-DMA, with higher transfection activity was screened (Jayaraman et al., 2012). It is currently used in siRNA-containing lipid formulations under clinical development programs of Alnylam for noncancer indications (hypercholesterolemia (NCT02314442) and familial amyloidotic polyneuropathy (FAP) (NCT01960348)).

2.1.3 Cationic-free *liposomes: A strategy to overcome charge-associated limitations*

As an alternative strategy to overcome the limitations associated with the systemic administration of positively charged nanoparticles, there has been an increased interest in the use liposomes for nucleic acid delivery, without incorporating cationic lipids. Lacking surface charge, these liposomes present residual interaction with serum proteins after systemic administration, thereby achieving longer circulation times and presenting lower toxicity (Ozpolat, Sood, & Lopez-Berestein, 2014). This neutral liposomal formulation composed of 1,2-dioleoyl-sn-glycero-3-phosphatidylcholine (DOPC) for the systemic delivery of anti-EphA2 siRNA, presented an encapsulation efficiency of ~65% and in vivo, enabled a significant reduction of EphA2 expression after a single dose. This was translated in tumor growth inhibition in an orthotropic mouse model of ovarian cancer (Landen et al., 2005). The potential of this neutral formulation was further confirmed in an orthotropic mouse model of ovarian cancer using siRNAs against other proteins, such as FAK (Halder et al., 2006), IL-8 (Merritt et al., 2008), JNK-1 (Vivas-Mejia et al., 2010), PELP1 (Chakravarty et al., 2011), and Survivin 2B (Vivas-Mejia et al., 2011). In addition, this formulation has shown to be active in other orthotopic and subcutaneous xenograft models of cancer, including colorectal carcinoma (targeting of Id2), pancreatic (Shao et al., 2012), lung and colon (Pecot et al., 2014), prostatic (Shao et al., 2012), melanoma (Villares et al., 2008), and ER (−) and ER (+) breast cancer (Tekedereli et al., 2013), with limited toxicity or immune responses observed after a single or a 4-week multiadministration schedule

(Landen et al., 2005; Ozcan, Ozpolat, Coleman, Sood, & Lopez-Berestein, 2015; Shao et al., 2012). In addition, when combined with chemotherapy, such as paclitaxel (Landen et al., 2005; Pecot et al., 2014), docetaxel (Halder et al., 2006; Merritt et al., 2008; Vivas-Mejia et al., 2010, 2011), cisplatin (Halder et al., 2006), gemcitabine (Verma et al., 2008), or doxorubicin (Pecot et al., 2014), DOPC liposomes significantly enhanced the therapeutic efficacy relative to the single drug regimen.

2.2 Landscape of Clinical Development of siRNA-Based Therapeutics in Oncology

The discovery of the RNAi mechanism raised enormous interest from the scientific community, which is expected to explore the biology behind RNAi to conceive innovative therapeutic solutions against several diseases, including cancer (Conde & Artzi, 2015). In fact, based on the discoveries discussed in the previous section, in the past decade, different candidates have reached clinical trials. Nevertheless, there has been a decreased investment in this field, namely from large pharmaceutical companies (Conde & Artzi, 2015).

This reality is indeed reflected in the current number of clinical trials exploring RNAi as a therapeutic strategy against cancer (Table 1), which might be associated with the difficulties posed by the challenging siRNA pharmacokinetics. These issues have been addressed by relying on backbone chemical modifications or conjugation with biocompatible molecules, such as cholesterol, aiming at minimizing rapid renal and immune system-mediated clearance, as well as degradation (Whitehead, Langer, & Anderson, 2009).

Given the limited success of that strategy, the knowledge accumulated from lipid-based nanoparticles containing small drugs presented a rational step toward aiming at siRNA protection and stabilization (Gomes-da-Silva, Fonseca, et al., 2012). The use of refined cationic lipids, promoting effective siRNA complexation, combined with a PEG coating, to minimize MPS-mediated elimination, has enabled promising candidates to reach the clinical trials stage working against different cancers, targeting PLK-1 (TKM-080301) and KSP/VEGF (ALN-VSP02), respectively (Table 1) (Allen & Cullis, 2013; Bouchie, 2012; Judge et al., 2009; Zuckerman & Davis, 2015). From the phase I/II trials of TKM-080301 against different solid tumors, it was estimated that a maximum tolerated dose (MTD) was 0.75 mg/Kg, with transient increased release of cytokines, in which symptoms were manageable without sequelae (Northfelt et al., 2013; Ramanathan et al., 2013). The dose escalation indicated that the pharmacokinetic parameters AUC and C_{max} were proportional to the dose, with preliminary efficacy observed in 40% of the patients, for doses >0.6 mg/Kg, while confirming PLK-1 silencing (Northfelt et al., 2013; Ramanathan et al., 2013, 2014). Regarding the ALN-VSP02 candidate, it was well tolerated at doses as high as 1.25 mg/Kg, with pharmacokinetic parameters proportional to the dose (Cervantes et al., 2011). Demonstrating evidence of the anti-VEGF effect, 56% of the patients (with different solid tumors) presented a 40% decrease in the Ktrans signal at the tumor site, a measure of

Table 1: Clinical development of RNAi therapeutics in oncology

Start Date-End Date	Drug	Target	Vehicle/Formulation	Route	Condition/Indication	Phase/Status	Clinical Trial ID	Sponsors
2012–2015	siRNA-EphA2-DOPC	EphA2 gene	Neutral liposomal	IV	Advanced cancers	Phase I/Recruiting	NCT01591356	MD Anderson Cancer Center
2009–2013	Atu027	PKN3	AtuPLEX® and AtuRNAi®	IV	Advanced solid tumors	Phase I/Completed	NCT00938574	Silence Therapeutics GmbH
2013-	Atu027	PKN3	AtuPLEX® and AtuRNAi®	IV	Advanced or metastatic pancreatic adenocarcinoma	Phase Ib/IIa/Ongoing	NCT01808638	Silence Therapeutics GmbH
2011–2014	TKM 080301	PLK1 (polo-like kinase-1)	Stabilized nucleic acid lipid particle (SNALP)	Hepatic arterial infusion (HAI)	Colorectal cancer with hepatic metastases Pancreatic cancer with hepatic metastases Gastric cancer with hepatic metastases Breast cancer with hepatic metastases Ovarian cancer with hepatic metastases	Phase I/Completed	NCT01437007	National Cancer Institute (NCI), Tekmira Pharma
2014–2015	DCR-MYC	MYC	EnCore™ lipid nanoparticle	IV	Solid tumors Multiple myeloma Non-Hodgkin's lymphoma	Phase I/Recruiting	NCT02110563	Dicerna Pharmaceuticals, Inc.
2014–2015	DCR-MYC	MYC	EnCore™ lipid nanoparticle	IV	Hepatocellular carcinoma	Phase Ib/II/Recruiting	NCT02314052	Dicerna Pharmaceuticals, Inc.
2009–2012	ALN-VSP02	Kinesin spindle protein (KSP) and VEGF	SNALP	IV	Patients With Advanced Solid Tumors With Liver Involvement	Phase I/Completed	NCT01158079	Alnylam Pharmaceuticals

Continued

Table 1: Clinical development of RNAi therapeutics in oncology—cont'd

Start Date-End Date	Drug	Target	Vehicle/Formulation	Route	Condition/Indication	Phase/Status	Clinical Trial ID	Sponsors
2009–2011	ALN-VSP02	Kinesin spindle protein (KSP) and VEGF	SNALP	IV	Patients with advanced solid tumors with liver involvement	Phase I/ Completed	NCT00882180	Alnylam Pharmaceuticals
2011–2012	TKM 080301 (TKM-PLK1)	Polo-kinase-1	SNALP	IV	Colorectal cancer with hepatic metastases Pancreatic cancer with hepatic metastases Gastric cancer with hepatic metastases Breast cancer with hepatic metastases Ovarian cancer with hepatic metastases	Phase I Complete	NCT01437007	Tekmira Pharma
2010–2015	TKM 080301 (TKM-PLK1)	Polo-kinase-1	SNALP	IV	Neuroendocrine tumors (NET); Adrenocortical carcinoma (ACC)	Phase I/II Complete	NCT01262235	Tekmira Pharma
2014–	TKM 080301 (TKM-PLK1)	Polo-kinase-1	SNALP	IV	Advanced hepatocellular carcinoma	Phase I/II/ Recruiting	NCT02191878	Tekmira Pharma

capillary permeability obtained by dynamic contrast-enhanced magnetic resonance (DCE-MRI). Recently, Dicerna Pharmaceutics and Silence Therapeutics sponsored phase I clinical trials with lipid-based nanoparticles encapsulating siRNAs against MYC and PKN3 (a known protein controlling metastasis formation) (Strumberg et al., 2012), respectively (Table 1). Silence Therapeutics presented promising results with its Atu027 nanoparticle containing anti-PKN3 (protein kinase N3) siRNA, with 41% of the patients with advanced solid tumors demonstrating stable disease at the end of treatment. Atu027 was well tolerated up to dose levels of 336 μg siRNA/Kg with residual toxicity (Schultheis et al., 2014). Following a different trend, and supported by the data described in the previous section, neutral liposomes encapsulating an anti-EphA2 siRNA are under phase I clinical trials against advanced cancers at the MD Anderson Cancer Center (Table 1).

2.2.1 Major barriers to the delivery of nucleic acids

Despite those encouraging results, the current landscape of RNAi trials in cancer reveals two important aspects: major trials are in very early phases, evaluating mainly safety, and, in many of them, the main target organ is the liver (Table 1). These are, in part, a reflection of the early realization that the liposomal formulations of siRNAs accumulated preferentially in the liver, owing to their ability to bind ApoE and trigger LDL-like uptake into hepatocytes, therefore rendering the liver a target for proof-of-concept of in vivo RNAi, using nanoparticles (Wan et al., 2014). In addition, despite the molecular targeting specificity associated with siRNA technology, the systemic administration of liposomal siRNA is not devoid of side effects such as nausea, vomiting, or pyrexia, likely owing to off-target effects, in part arising from the lack of specific accumulation at the target organ (Gomes-da-Silva, Fonseca, et al., 2012; Northfelt et al., 2013; Ramanathan et al., 2013).

Additionally, the use of PEG is somehow hindered as it induces the production of IgM as a part of an immune response, which consequently leads to the accelerated blood clearance phenomenon of a second dose of PEGylated liposomes (Dams et al., 2000; Ishida et al., 2006). This affects siRNA accumulation at the tumor site, especially in organs other than the liver (Tagami, Nakamura, Shimizu, Ishida, & Kiwada, 2009). Furthermore, it is known that liposomes encapsulating nucleic acids present an enhanced PEG-related immune response, in part because of the presence of CpG or G-quartet motifs in the siRNAs, an effect that could be considerably inhibited by the use of nonimmunostimulatory siRNA (Tagami et al., 2009, 2010). Yet, at the cellular and intracellular level, PEG limits nanoparticle-membrane contact in a manner that is dependent on both the acyl chain length of the anchor lipid and the molecular weight of the PEG moiety. This results in a reduction of the cellular internalization of the liposomes, and besides that, depending on the PEG content, it might as well as limit the endosomal escape of the nucleic acid and subsequently, its stability and activity (Mishra, Webster, & Davis, 2004).

3 Innovate to Conquer: Transferring Tumor Biology into Technology

A foreseeable game changer of the current landscape of siRNA-mediated RNAi as a therapeutic strategy in cancer will be the further development of stimuli-responsive nanotechnology-based delivery systems, aiming at increasing siRNA bioavailability and cell targeting, in organs other than the liver.

3.1 Enhancing Systemic Exposure and Intracellular Bioavailability

Framed by the limitations described herein, the development of delivery systems that, upon systemic administration, target tissues other than the liver, enabling both spatial and temporal control of gene silencing, has proven to be a challenge. If the introduction of PEGylated liposomes has represented a significant leap forward in increasing the blood circulation half-life of the encapsulated drug, it did not come without a compromise of immunostimulatory effects, as discussed herein. Despite that, different strategies to enhance siRNA bioavailability are mandatory to provide better efficacy and limit off-target effects, and therefore safety.

Overcoming the immunostimulatory potential of the PEG/siRNA combination may translate into increased overall bioavailability of the effector siRNA, while minimizing severe side effects. In fact, this has been addressed by chemical modifying immunostimulatory siRNAs or by the selection of specific siRNA sequences designed without immunostimulatory activity. Chemical modification of siRNAs often involve the incorporation of 2′OMe nucleotides (2′OMe-guanine, 2′OMe-uracil, or 2′OMe-adenosine residues) throughout both strands of the siRNA duplex (Judge et al., 2009), whereas specific selection of sequences often foresees the minimization of CpG content (Tagami et al., 2009, 2010). Indeed, these strategies have been put in practice in Alnylam's and Tekmira's candidates (Table 1) (Wan et al., 2014).

One of the strategies devised to potentially overcome the PEG-limiting endosomal escape of the nucleic acid relies on the use of exchangeable PEG-derived lipids that present variable exchange rates from the particle's surface, depending on the size of the lipid anchors. Different examples include PEG coupled typically with ceramides (PEG-Cer) or succinoyl diacylglycerols (PEG-S-DAGs), but also with amphiphiles, based on *N*-methyl-4-alkylpyridinium chlorides (SAINT-PEGs). Generally, shorter chain lipid anchors show higher transfection efficiencies in vitro and dissociate more quickly from the bilayer. However, SNALPs containing longer chain lipid anchors are more stable, with longer circulation times, yielding an increase in tumor delivery (Ambegia et al., 2005). Another strategy uses PEG-disulphide lipids that incorporate a disulphide bond between the polymer and the lipid anchor that is cleaved by thiolytic agents in the reductive environment of the endosome (Takae et al., 2008). Similarly, vinyl ether PEG-lipids and orthoester PEG-lipids have been developed using chemical linkages that are sensitive to the reduced pH of the endosomal compartment (Heyes, Hall, Tailor, Lenz, & Maclachlan, 2006).

An alternative approach, and sometimes complementary to the ones aforementioned, to increase the intracellular bioavailability of encapsulated nucleic acids relies on engineering liposomes to destabilize and fuse with the endosomal membrane upon acidification of the endosome (Gomes-da-Silva et al., 2013; Gomes-da-Silva, Fonseca, et al., 2012; Ishida, Okada, Kobayashi, & Kiwada, 2006; Simoes, Slepushkin, Duzgunes, & Pedroso de Lima, 2001). This is achieved by the incorporation of pH-sensitive components, such as the fusogenic lipid 1,2-dioleoyl-sn-glycero-3-phosphoethanolamine (DOPE). At neutral or basic pH, and upon stabilization by amphiphilic molecules, such as cholesteryl hemisuccinate (CHEMS), it forms an organized and stable lipid bilayer. However, in an acidic environment, such as the endocytic pathway, protonation of the amphiphiles reduces their stabilizing effect, leading to liposomal disruption and fusion with the endocytic membrane. Gomes-da-Silva, et al. have demonstrated the feasibility of incorporating the DOPE/CHEMS pH-sensitive system developed for small drug delivery (Ishida, Okada, et al., 2006; Moura et al., 2012) into SNALPs, enabling effective gene silencing upon efficient endosomal escape. pH-sensitive SNALPs were able to retain >80% of the encapsulated siRNA over a 24 h period, while not eliciting an immunogenic response (Gomes-da-Silva, Fernandez, et al., 2013). In a similar approach, Sato et al., synthesized a new pH-dependent ionizable cationic lipid, YSK05 (pKa 6.6) (Sato et al., 2012), which formulated with DOPE or POPE, cholesterol, and PEG-derivatized lipids into liposomes, and enabled higher PLK1-silencing at the tumor site, upon local administration, as compared with liposomes containing cationic lipids, DOTAP and DODAP (nonionizable and a ionizable lipids, respectively) (Sato et al., 2012). It was reasoned that the observed therapeutic effect was subsequent to the intracellular pH-dependent endosomal escape (Sato et al., 2012).

3.2 Ligand-Mediated Targeted Delivery to Cancer Cells and Tumor Microenvironment

Notwithstanding bioavailability and off-target effects that have become major themes in siRNA delivery, there is an increasing awareness on the importance of targeting specific cell populations responsible for different tumor properties (Hanahan & Weinberg, 2011). Among the different cellular components within the tumor microenvironment, cancer cell populations and tumor endothelial cells of angiogenic vessels remain crucial targets by their role in cancer burden and accessibility (in the case of solid tumors), respectively (Carmeliet & Jain, 2011; Hanahan & Weinberg, 2011). Surface functionalization of nanoparticles with internalizing-targeting ligands that specifically recognize an overexpressed protein on the cell membrane is a strategy with enormous potential to significantly impact the therapeutic efficacy and safety of siRNA-based medicines (Fonseca, Gregorio, Valerio-Fernandes, Simoes, & Moreira, 2014; Gomes-da-Silva, Fonseca, et al., 2012). Of utmost importance, it represents an opportunity that would enable one to target other solid tumors beyond those from the liver.

A transferrin-targeted cyclodextrin-based formulation, containing a siRNA against the M2 subunit of ribonucleotide reductase (CALAA-01, Arrowhead Research Corp), represented the first attempt to achieve ligand-mediated targeted delivery of siRNAs in humans. It was withdrawn from a phase 1 clinical trial (Fonseca et al., 2014), and several new technologies are under preclinical study, presenting the potential to change the landscape for siRNA-based therapeutic strategies (Lorenzer, Dirin, Winkler, Baumann, & Winkler, 2015).

In this context, Meissner et al. have recently developed anti-CD20-targeted liposomes encapsulating anti-BCL2 siRNA (Meissner et al., 2015). CD20 is a marker of B-cells and is the target of rituximab, a monoclonal antibody approved for non-Hodgkin's lymphoma, for which resistance is often developed (Smith, 2003). CD20-targeted liposomes demonstrated significant antitumor activity in an in vivo lymphoma model, upon inducing apoptosis by downregulation of BCL2, apparently overcoming rituximab resistance (Meissner et al., 2015). Despite the fact that it was demonstrated that human peripheral blood mononuclear cells did not exhibit an immune response in vitro (Meissner et al., 2015), it is known that antibodies may present significant immune responses (Peer et al., 2007). An alternative strategy may rely on the use of small targeting moieties, less prone to development of immune responses, including aptamers or peptides. Exploring this concept, Hada and colleagues proposed an siRNA delivery system targeted by the RGD peptide, known to specifically bind to $\alpha_v\beta_3$ integrin expressed on tumor endothelial cells, and incorporating the novel pH-sensitive ionizable lipid, YSK05. The antiangiogenic potential of YSK05-incorporating liposomes containing anti-CD31 siRNA was proved upon enabling a 10-fold increase in gene silencing (ED_{50} below <0.75 mg/kg), relative to the nontargeted counterpart (Hada, Sakurai, & Harashima, 2015; Sato et al., 2012).

Gomes-da-Silva and colleagues proposed the nucleolin-binding F3 peptide as a targeting ligand of pH sensitive SNALPs containing anti-PLK1 siRNA. Nucleolin is a receptor overexpressed in different cell populations within the tumor microenvironment, including endothelial cells and cancer cells. Accordingly, the F3 peptide-targeted SNALPs effectively promoted intracellular delivery of the siRNA, increasing its bioavailability by enabling efficient endosomal escape (Gomes-da-Silva et al., 2012; Gomes-da-Silva, Fernandez, et al., 2013; Gomes-da-Silva, Ramalho, Pedroso de Lima, Simoes, & Moreira, 2013). This translated into an efficient PLK1 mRNA and protein downregulation, significantly impairing cell viability of a model of endothelial cells from angiogenic blood vessels and prostate cancer cells, by blocking cell cycle progression (Gomes-da-Silva, Ramalho, et al., 2013). Li et al. propose PEGylated liposomes targeted by the nucleolin-binding AS1411 aptamer (Li et al., 2014). The results demonstrated that AS1411-targeted liposomes effectively silence the BRAF oncogene and extensively accumulate at the tumor site in a melanoma mouse model, inhibiting tumor growth (Li, Hou, et al., 2014).

Overall, simultaneous targeted delivery of siRNAs to multiple cell populations within the tumor microenvironment may enable broad RNAi. Nevertheless, achieving an antitumor effect using siRNA assumes that every cell or cell population is highly dependent on the targeted gene for their cellular functions. Unfortunately, single gene or signaling pathway addiction is not often true in cancer (Hanahan & Weinberg, 2011). It thus represents an important (and additional) challenge to siRNA technologies.

3.3 Beyond Single Molecular Targeting: Combinatorial Strategies Involving RNAi

The dynamics of different cancer cell clones dictates that every other cell may respond differently to a given drug (Eirew et al., 2015; Hanahan & Weinberg, 2011; Ye & Weinberg, 2015). This is often behind the development of drug resistance, as well as the relative dependence of cells in a particular signaling pathway or molecular effector (Hanahan & Weinberg, 2011; Marjanovic, Weinberg, & Chaffer, 2013). To overcome this hindrance, it has been rationalized that successful therapeutic intervention may rely on the identification of drugs that, upon combination, generate a synergistic antitumor effect (Ramaswamy, 2007). Building on this concept, siRNA strategies may benefit from combinatorial design, either with other siRNAs or small drugs, to improve their therapeutic outcome.

Following the example of the phase I trial of Atu027, different combinatorial approaches with siRNA technologies are being tested pre-clinically, many of them exploring simultaneous delivery. As an example, Yang et al. developed dual peptide-targeted liposomes for the co-delivery of anti-VEGF siRNA and docetaxel to gliomas (Yang, Li, Wang, Dong, & Qi, 2014). The targeting ligands are Angiopep-2 and tLyP-1 peptides that specifically recognize LRP (low-density lipoprotein receptor-related protein) and neuropilin-1, respectively. The dual peptide-targeted liposomes demonstrated superior antitumor efficacy as the result of the simultaneous impairment of the angiogenic process (mediated by the anti-VEGF siRNA) and apoptosis induction by docetaxel, upon systemic administration (Yang et al., 2014). Similarly, Li and colleagues developed stimulus-sensitive liposomes for the *co*-encapsulation of a camptothecin prodrug (conjugated to pH sensitive zwitterionic poly[carboxybetaine] polymer) and anti-PLK1 siRNA (Li et al., 2014). These liposomes enabled extensive accumulation of camptothecin in the nucleus with enhanced synergistic cytotoxic effects. Upon in vivo systemic administration, those effects translated into enhanced antitumor activity against a HeLa cancer model (Li, Liu, et al., 2014).

Exploring both cell cycle blocking and mitotic spindle formation inhibition, mediated by anti-PLK1 siRNA encapsulated in F3 peptide-targeted liposomes and paclitaxel, respectively, the authors demonstrated that the combination decreased cell viability of prostate cancer model by 3-fold as compared to paclitaxel alone, as a result of significant cell cycle arrest (Gomes et al., 2013; Gomes-da-Silva, Ramalho, et al., 2013).

Overall, these examples illustrate that combining different strategies that impair multiple molecular targets or signaling pathways may translate into improved efficacy, representing a key aspect for successful cancer therapeutic intervention.

4 Conclusion and Future Prospects

Overcoming the highly unfavorable pharmacokinetics of siRNAs has been a demanding challenge that has limited the number of potential candidates in the clinical trial phase. Still, with the current generation of RNAi medicines currently under trial, significant progress has been made in achieving gene silencing, especially by exploring the nanotechnology-based delivery systems. Nevertheless, these achievements were mainly in liver-borne diseases. Cell targeting using lipid-based nanosystems, exploiting the overexpression of receptors by cancer cells, and particularly, the more accessible tumor endothelial cells, may enable one to explore RNAi to organs other than the liver. This achievement will ultimately enable the delivery of emerging gene editing tools, such TALENs (transcription activator-like effector nucleases) or the CRISP/Cas9 (clustered regularly interspaced short palindromic repeat-associated nuclease Cas9) endonuclease systems (Cox, Platt, & Zhang, 2015). These programmable nucleases have enabled gene editing in several disease models and, despite their simplicity, their delivery is currently limited to viral vectors, known to be highly immunogenic, and to be used systemically (Cox et al., 2015). Nevertheless, systemic delivery of nuclease systems with nonviral delivery systems may be developed, accounting for all precedent knowledge associated with oligonucleotide delivery, and exploring cell penetrating peptides as well as novel and more efficient nucleases (Hsu, Lander, & Zhang, 2014; Zetsche et al., 2015).

As discussed herein, RNAi as a therapeutic approach in cancer is still in its infancy. The design of innovative nanotechnology-based delivery systems may represent the turnover favoring gene silencing and gene editing as integrant parts of a truly personalized therapeutic intervention.

Acknowledgments

R. Lopes was a recipient of a post-doc fellowship from "Incentivo 2014—Estímulo à contratação e retenção de recursos humanos" from the Portuguese Foundation for Science and Technology (FCT). A.F. Cruz was a recipient of the fellowship SFRH/BD/85554/2012 from FCT. Ana C. Gregório and Ângela Valério-Fernandes were students of PhD Program in Experimental Biology and Biomedicine (PDBEB), Center for Neuroscience and Cell Biology, University of Coimbra, and recipients of the FCT fellowships SFRH/BD/51190/2010 and SFRH/BD/51191/2010, respectively. The work was supported by the European Regional Development Fund (ERDF) through the Centro 2020 Regional Operational Programme: project CENTRO-01-0145-FEDER-000012-HealthyAging2020, the COMPETE 2020 - Operational Program for Competitiveness and Internationalization, and the Portuguese national funds via FCT under project POCI-01-0145-FEDER-007440.

References

Ahmad, A., Evans, H. M., Ewert, K., George, C. X., Samuel, C. E., & Safinya, C. R. (2005). New multivalent cationic lipids reveal bell curve for transfection efficiency versus membrane charge density: Lipid-DNA complexes for gene delivery. *The Journal of Gene Medicine*, *7*, 739–748.

Akinc, A., Zumbuehl, A., Goldberg, M., Leshchiner, E. S., Busini, V., Hossain, N., et al. (2008). A combinatorial library of lipid-like materials for delivery of RNAi therapeutics. *Nature Biotechnology*, *26*, 561–569.

Allen, T. M., & Cullis, P. R. (2013). Liposomal drug delivery systems: From concept to clinical applications. *Advanced Drug Delivery Reviews*, *65*, 36–48.

Ambegia, E., Ansell, S., Cullis, P., Heyes, J., Palmer, L., & Maclachlan, I. (2005). Stabilized plasmid-lipid particles containing PEG-diacylglycerols exhibit extended circulation lifetimes and tumor selective gene expression. *Biochimica et Biophysica Acta*, *1669*, 155–163.

Andtbacka, R. H., Kaufman, H. L., Collichio, F., Amatruda, T., Senzer, N., Chesney, J., et al. (2015). Talimogene Laherparepvec improves durable response rate in patients with advanced melanoma. *Journal of Clinical Oncology*, *33*, 2780–2788.

Barenholz, Y. (2012). Doxil(R)—The first FDA-approved nano-drug: Lessons learned. *Journal of Controlled Release*, *160*, 117–134.

Bouchie, A. (2012). Companies in footrace to deliver RNAi. *Nature Biotechnology*, *30*, 1154–1157.

Boudreau, R. L., & Davidson, B. L. (2012). Generation of hairpin-based RNAi vectors for biological and therapeutic application. *Methods in Enzymology*, *507*, 275–296.

Bouxsein, N. F., Mcallister, C. S., Ewert, K. K., Samuel, C. E., & Safinya, C. R. (2007). Structure and gene silencing activities of monovalent and pentavalent cationic lipid vectors complexed with siRNA. *Biochemistry*, *46*, 4785–4792.

Brummelkamp, T. R., Bernards, R., & Agami, R. (2002a). Stable suppression of tumorigenicity by virus-mediated RNA interference. *Cancer Cell*, *2*, 243–247.

Brummelkamp, T. R., Bernards, R., & Agami, R. (2002b). A system for stable expression of short interfering RNAs in mammalian cells. *Science*, *296*, 550–553.

Calin, G. A., & Croce, C. M. (2006). MicroRNA signatures in human cancers. *Nature Reviews. Cancer*, *6*, 857–866.

Carmeliet, P., & Jain, R. K. (2011). Molecular mechanisms and clinical applications of angiogenesis. *Nature*, *473*, 298–307.

Cervantes, A., Alsina, M., Tabernero, J., Infante, J. R., Lorusso, P., Shapiro, G., et al. (2011). Phase I dose-escalation study of ALN-VSP02, a novel RNAi therapeutic for solid tumors with liver involvement. *ASCO Meeting Abstracts*, *29*, 3025.

Chakravarty, D., Roy, S. S., Babu, C. R., Dandamudi, R., Curiel, T. J., Vivas-Mejia, P., et al. (2011). Therapeutic targeting of PELP1 prevents ovarian cancer growth and metastasis. *Clinical Cancer Research*, *17*, 2250–2259.

Chen, Y., Gao, D. Y., & Huang, L. (2015). In vivo delivery of miRNAs for cancer therapy: Challenges and strategies. *Advanced Drug Delivery Reviews*, *81*, 128–141.

Conde, J., & Artzi, N. (2015). Are RNAi and miRNA therapeutics truly dead? *Trends in Biotechnology*, *33*, 141–144.

Couto, L. B., & High, K. A. (2010). Viral vector-mediated RNA interference. *Current Opinion in Pharmacology*, *10*, 534–542.

Cox, D. B., Platt, R. J., & Zhang, F. (2015). Therapeutic genome editing: Prospects and challenges. *Nature Medicine*, *21*, 121–131.

Dams, E. T., Laverman, P., Oyen, W. J., Storm, G., Scherphof, G. L., van der Meer, J. W., et al. (2000). Accelerated blood clearance and altered biodistribution of repeated injections of sterically stabilized liposomes. *The Journal of Pharmacology and Experimental Therapeutics*, *292*, 1071–1079.

Dokka, S., Toledo, D., Shi, X. G., Castranova, V., & Rojanasakul, Y. (2000). Oxygen radical-mediated pulmonary toxicity induced by some cationic liposomes. *Pharmaceutical Research*, *17*, 521–525.

Eirew, P., Steif, A., Khattra, J., Ha, G., Yap, D., Farahani, H., et al. (2015). Dynamics of genomic clones in breast cancer patient xenografts at single-cell resolution. *Nature*, *518*, 422–426.

Elbashir, S. M., Harborth, J., Lendeckel, W., Yalcin, A., Weber, K., & Tuschl, T. (2001). Duplexes of 21-nucleotide RNAs mediate RNA interference in cultured mammalian cells. *Nature*, *411*, 494–498.

Felgner, P. L., Gadek, T. R., Holm, M., Roman, R., Chan, H. W., Wenz, M., et al. (1987). Lipofection: A highly efficient, lipid-mediated DNA-transfection procedure. *Proceedings of the National Academy of Sciences of the United States of America*, *84*, 7413–7417.

Fire, A., Xu, S., Montgomery, M. K., Kostas, S. A., Driver, S. E., & Mello, C. C. (1998). Potent and specific genetic interference by double-stranded RNA in *Caenorhabditis elegans*. *Nature*, *391*, 806–811.

Fonseca, N. A., Gregorio, A. C., Valerio-Fernandes, A., Simoes, S., & Moreira, J. N. (2014). Bridging cancer biology and the patients' needs with nanotechnology-based approaches. *Cancer Treatment Reviews*, *40*, 626–635.

Gomes, C. P., Gomes-da-Silva, L. C., Ramalho, J. S., de Lima, M. C., Simoes, S., & Moreira, J. N. (2013). Impact of PLK-1 silencing on endothelial cells and cancer cells of diverse histological origin. *Current Gene Therapy*, *13*, 189–201.

Gomes-da-Silva, L. C., Fernandez, Y., Abasolo, I., Schwartz, S., Jr., Ramalho, J. S., Pedroso de Lima, M. C., et al. (2013). Efficient intracellular delivery of siRNA with a safe multitargeted lipid-based nanoplatform. *Nanomedicine (London, England)*, *8*, 1397–1413.

Gomes-da-Silva, L. C., Fonseca, N. A., Moura, V., Pedroso de Lima, M. C., Simoes, S., & Moreira, J. N. (2012). Lipid-based nanoparticles for siRNA delivery in cancer therapy: Paradigms and challenges. *Accounts of Chemical Research*, *45*, 1163–1171.

Gomes-da-Silva, L. C., Ramalho, J. S., Pedroso de Lima, M. C., Simoes, S., & Moreira, J. N. (2013). Impact of anti-PLK1 siRNA-containing F3-targeted liposomes on the viability of both cancer and endothelial cells. *European Journal of Pharmaceutics and Biopharmaceutics*, *85*, 356–364.

Gomes-da-Silva, L. C., Santos, A. O., Bimbo, L. M., Moura, V., Ramalho, J. S., Lima, M. C., et al. (2012). Towards a siRNA-containing nanoparticle targeted to breast cancer cells and the tumor microenvironment. *International Journal of Pharmaceutics*, *434*, 9–19.

Hada, T., Sakurai, Y., & Harashima, H. (2015). Optimization of a siRNA carrier modified with a pH-sensitive cationic lipid and a cyclic RGD peptide for efficiently targeting tumor endothelial cells. *Pharmaceutics*, *7*, 320–333.

Halder, J., Kamat, A. A., Landen, C. N., Jr., Han, L. Y., Lutgendorf, S. K., Lin, Y. G., et al. (2006). Focal adhesion kinase targeting using in vivo short interfering RNA delivery in neutral liposomes for ovarian carcinoma therapy. *Clinical Cancer Research*, *12*, 4916–4924.

Hanahan, D., & Weinberg, R. A. (2011). Hallmarks of cancer: The next generation. *Cell*, *144*, 646–674.

Harris, J. M., & Chess, R. B. (2003). Effect of pegylation on pharmaceuticals. *Nature Reviews Drug Discovery*, *2*, 214–221.

Heyes, J., Hall, K., Tailor, V., Lenz, R., & Maclachlan, I. (2006). Synthesis and characterization of novel poly(ethylene glycol)-lipid conjugates suitable for use in drug delivery. *Journal of Controlled Release*, *112*, 280–290.

Hsu, P. D., Lander, E. S., & Zhang, F. (2014). Development and applications of CRISPR-Cas9 for genome engineering. *Cell*, *157*, 1262–1278.

Iorio, M. V., Ferracin, M., Liu, C. G., Veronese, A., Spizzo, R., Sabbioni, S., et al. (2005). MicroRNA gene expression deregulation in human breast cancer. *Cancer Research*, *65*, 7065–7070.

Ipsaro, J. J., & Joshua-Tor, L. (2015). From guide to target: Molecular insights into eukaryotic RNA-interference machinery. *Nature Structural & Molecular Biology*, *22*, 20–28.

Ishida, T., Ichihara, M., Wang, X., Yamamoto, K., Kimura, J., Majima, E., et al. (2006). Injection of PEGylated liposomes in rats elicits PEG-specific IgM, which is responsible for rapid elimination of a second dose of PEGylated liposomes. *Journal of Controlled Release*, *112*, 15–25.

Ishida, T., Okada, Y., Kobayashi, T., & Kiwada, H. (2006). Development of pH-sensitive liposomes that efficiently retain encapsulated doxorubicin (DXR) in blood. *International Journal of Pharmaceutics*, *309*, 94–100.

Jayaraman, M., Ansell, S. M., Mui, B. L., Tam, Y. K., Chen, J., Du, X., et al. (2012). Maximizing the potency of siRNA lipid nanoparticles for hepatic gene silencing in vivo. *Angewandte Chemie (International Edition in English)*, *51*, 8529–8533.

Judge, A. D., Robbins, M., Tavakoli, I., Levi, J., Hu, L., Fronda, A., et al. (2009). Confirming the RNAi-mediated mechanism of action of siRNA-based cancer therapeutics in mice. *The Journal of Clinical Investigation*, *119*, 661–673.

Kwok, A., & Hart, S. L. (2011). Comparative structural and functional studies of nanoparticle formulations for DNA and siRNA delivery. *Nanomedicine*, *7*, 210–219.

Lam, J. K., Chow, M. Y., Zhang, Y., & Leung, S. W. (2015). siRNA Versus miRNA as therapeutics for gene silencing. *Molecular Therapy Nucleic Acids*, *4*, e252.

Landen, C. N., Chavez-Reyes, A., Bucana, C., Schmandt, R., Deavers, M. T., Lopez-Berestein, G., et al. (2005). Therapeutic EphA2 gene targeting in vivo using neutral liposomal small interfering RNA delivery. *Cancer Research*, *65*, 6910–6918.

Lasic, D. D., Martin, F. J., Gabizon, A., Huang, S. K., & Papahadjopoulos, D. (1991). Sterically stabilized liposomes: A hypothesis on the molecular origin of the extended circulation times. *Biochimica et Biophysica Acta*, *1070*, 187–192.

Ledford, H. (2015). Cancer-fighting viruses win approval. *Nature*, *526*, 622–623.

Lee, J. B., Zhang, K., Tam, Y. Y., Tam, Y. K., Belliveau, N. M., Sung, V. Y., et al. (2012). Lipid nanoparticle siRNA systems for silencing the androgen receptor in human prostate cancer in vivo. *International Journal of Cancer*, *131*, E781–90.

Levchenko, T. S., Rammohan, R., Lukyanov, A. N., Whiteman, K. R., & Torchilin, V. P. (2002). Liposome clearance in mice: The effect of a separate and combined presence of surface charge and polymer coating. *International Journal of Pharmaceutics*, *240*, 95–102.

Li, L., Hou, J., Liu, X., Guo, Y., Wu, Y., Zhang, L., et al. (2014). Nucleolin-targeting liposomes guided by aptamer AS1411 for the delivery of siRNA for the treatment of malignant melanomas. *Biomaterials*, *35*, 3840–3850.

Li, Y., Liu, R., Yang, J., Ma, G., Zhang, Z., & Zhang, X. (2014). Dual sensitive and temporally controlled camptothecin prodrug liposomes codelivery of siRNA for high efficiency tumor therapy. *Biomaterials*, *35*, 9731–9745.

Liao, Y., & Tang, L. (2015). Inducible RNAi system and its application in novel therapeutics. *Critical Reviews in Biotechnology*, 1–9.

Lorenzer, C., Dirin, M., Winkler, A. M., Baumann, V., & Winkler, J. (2015). Going beyond the liver: Progress and challenges of targeted delivery of siRNA therapeutics. *Journal of Controlled Release*, *203*, 1–15.

Lv, H., Zhang, S., Wang, B., Cui, S., & Yan, J. (2006). Toxicity of cationic lipids and cationic polymers in gene delivery. *Journal of Controlled Release*, *114*, 100–109.

Marjanovic, N. D., Weinberg, R. A., & Chaffer, C. L. (2013). Poised with purpose: Cell plasticity enhances tumorigenicity. *Cell Cycle*, *12*, .

Meissner, J. M., Toporkiewicz, M., Czogalla, A., Matusewicz, L., Kuliczkowski, K., & Sikorski, A. F. (2015). Novel antisense therapeutics delivery systems: In vitro and in vivo studies of liposomes targeted with anti-CD20 antibody. *Journal of Controlled Release*, *220*, 515–528.

Merritt, W. M., Lin, Y. G., Spannuth, W. A., Fletcher, M. S., Kamat, A. A., Han, L. Y., et al. (2008). Effect of interleukin-8 gene silencing with liposome-encapsulated small interfering RNA on ovarian cancer cell growth. *Journal of the National Cancer Institute*, *100*, 359–372.

Mishra, S., Webster, P., & Davis, M. E. (2004). PEGylation significantly affects cellular uptake and intracellular trafficking of non-viral gene delivery particles. *European Journal of Cell Biology*, *83*, 97–111.

Morrissey, D. V., Lockridge, J. A., Shaw, L., Blanchard, K., Jensen, K., Breen, W., et al. (2005). Potent and persistent in vivo anti-HBV activity of chemically modified siRNAs. *Nature Biotechnology*, *23*, 1002–1007.

Moura, V., Lacerda, M., Figueiredo, P., Corvo, M. L., Cruz, M. E., Soares, R., et al. (2012). Targeted and intracellular triggered delivery of therapeutics to cancer cells and the tumor microenvironment: Impact on the treatment of breast cancer. *Breast Cancer Research and Treatment*, *133*, 61–73.

Nana-Sinkam, S., & Croce, C. M. (2014). MicroRNA regulation of tumorigenesis, cancer progression and interpatient heterogeneity: Towards clinical use. *Genome Biology*, *15*, 445.

Northfelt, D. W., Hamburg, S. I., Borad, M. J., Seetharam, M., Curtis, K. K., Lee, P., et al. (2013). A phase I dose-escalation study of TKM-080301, a RNAi therapeutic directed against polo-like kinase 1 (PLK1), in patients with advanced solid tumors: Expansion cohort evaluation of biopsy samples for evidence of pharmacodynamic effects of PLK1 inhibition. *Journal of Clinical Oncology*, *31*. https://doi.org/10.1200/jco.2013.31.15_suppl.tps2621.

Ozcan, G., Ozpolat, B., Coleman, R. L., Sood, A. K., & Lopez-Berestein, G. (2015). Preclinical and clinical development of siRNA-based therapeutics. *Advanced Drug Delivery Reviews*, *87*, 108–119.

Ozpolat, B., Sood, A. K., & Lopez-Berestein, G. (2014). Liposomal siRNA nanocarriers for cancer therapy. *Advanced Drug Delivery Reviews*, *66*, 110–116.

Pecot, C. V., Wu, S. Y., Bellister, S., Filant, J., Rupaimoole, R., Hisamatsu, T., et al. (2014). Therapeutic silencing of KRAS using systemically delivered siRNAs. *Molecular Cancer Therapeutics*, *13*, 2876–2885.

Peer, D., Karp, J. M., Hong, S., Farokhzad, O. C., Margalit, R., & Langer, R. (2007). Nanocarriers as an emerging platform for cancer therapy. *Nature Nanotechnology*, *2*, 751–760.

Ramanathan, R. K., Hamburg, S. I., Borad, M. J., Seetharam, M., Kundranda, M. N., Lee, P., et al. (2013). A phase I dose escalation study of TKM-080301, a RNAi therapeutic directed against PLK1, in patients with advanced solid tumors. *Cancer Research*, *73*.

Ramanathan, R. K., Hamburg, S. I., Halfdanarson, T. R., Borad, M. J., Seetharam, M., Curtis, K. K., et al. (2014). A phase I/II dose escalation study of TKM-080301, a RNAi therapeutic directed against PLK1, in patients with advanced solid tumors, with an expansion cohort of patients with NET or ACC. *Pancreas*, *43*, 502.

Ramaswamy, S. (2007). Rational design of cancer-drug combinations. *The New England Journal of Medicine*, *357*, 299–300.

Sato, Y., Hatakeyama, H., Sakurai, Y., Hyodo, M., Akita, H., & Harashima, H. (2012). A pH-sensitive cationic lipid facilitates the delivery of liposomal siRNA and gene silencing activity in vitro and in vivo. *Journal of Controlled Release*, *163*, 267–276.

Schultheis, B., Strumberg, D., Santel, A., Vank, C., Gebhardt, F., Keil, O., et al. (2014). First-in-human phase I study of the liposomal RNA interference therapeutic Atu027 in patients with advanced solid tumors. *Journal of Clinical Oncology*, *32*, 4141–4148.

Semple, S. C., Klimuk, S. K., Harasym, T. O., dos Santos, N., Ansell, S. M., Wong, K. F., et al. (2001). Efficient encapsulation of antisense oligonucleotides in lipid vesicles using ionizable aminolipids: Formation of novel small multilamellar vesicle structures. *Biochimica et Biophysica Acta*, *1510*, 152–166.

Shao, L., Tekedereli, I., Wang, J., Yuca, E., Tsang, S., Sood, A., et al. (2012). Highly specific targeting of the TMPRSS2/ERG fusion gene using liposomal nanovectors. *Clinical Cancer Research*, *18*, 6648–6657.

Shu, D., Shu, Y., Haque, F., Abdelmawla, S., & Guo, P. (2011). Thermodynamically stable RNA three-way junction for constructing multifunctional nanoparticles for delivery of therapeutics. *Nature Nanotechnology*, *6*, 658–667.

Simoes, S., Slepushkin, V., Duzgunes, N., & Pedroso de Lima, M. C. (2001). On the mechanisms of internalization and intracellular delivery mediated by pH-sensitive liposomes. *Biochimica et Biophysica Acta*, *1515*, 23–37.

Smith, M. R. (2003). Rituximab (monoclonal anti-CD20 antibody): Mechanisms of action and resistance. *Oncogene*, *22*, 7359–7368.

Strumberg, D., Schultheis, B., Traugott, U., Vank, C., Santel, A., Keil, O., et al. (2012). Phase I clinical development of Atu027, a siRNA formulation targeting PKN3 in patients with advanced solid tumors. *International Journal of Clinical Pharmacology and Therapeutics*, *50*, 76–78.

Tagami, T., Nakamura, K., Shimizu, T., Ishida, T., & Kiwada, H. (2009). Effect of siRNA in PEG-coated siRNA-lipoplex on anti-PEG IgM production. *Journal of Controlled Release*, *137*, 234–240.

Tagami, T., Nakamura, K., Shimizu, T., Yamazaki, N., Ishida, T., & Kiwada, H. (2010). CpG motifs in pDNA-sequences increase anti-PEG IgM production induced by PEG-coated pDNA-lipoplexes. *Journal of Controlled Release*, *142*, 160–166.

Tahiri, A., Leivonen, S. K., Luders, T., Steinfeld, I., Ragle Aure, M., Geisler, J., et al. (2014). Deregulation of cancer-related miRNAs is a common event in both benign and malignant human breast tumors. *Carcinogenesis*, *35*, 76–85.

Takae, S., Miyata, K., Oba, M., Ishii, T., Nishiyama, N., Itaka, K., et al. (2008). PEG-detachable polyplex micelles based on disulfide-linked block catiomers as bioresponsive nonviral gene vectors. *Journal of the American Chemical Society*, *130*, 6001–6009.

Tekedereli, I., Alpay, S. N., Akar, U., Yuca, E., Ayugo-Rodriguez, C., Han, H. D., et al. (2013). Therapeutic silencing of Bcl-2 by systemically administered siRNA Nanotherapeutics inhibits tumor growth by autophagy and apoptosis and enhances the efficacy of chemotherapy in Orthotopic xenograft models of ER (−) and ER (+) breast cancer. *Molecular Therapy Nucleic Acids*, *2*, e121.

Verma, A., Guha, S., Diagaradjane, P., Kunnumakkara, A. B., Sanguino, A. M., Lopez-Berestein, G., et al. (2008). Therapeutic significance of elevated tissue transglutaminase expression in pancreatic cancer. *Clinical Cancer Research*, *14*, 2476–2483.

Villares, G. J., Zigler, M., Wang, H., Melnikova, V. O., Wu, H., Friedman, R., et al. (2008). Targeting melanoma growth and metastasis with systemic delivery of liposome-incorporated protease-activated receptor-1 small interfering RNA. *Cancer Research*, *68*, 9078–9086.

Vivas-Mejia, P., Benito, J. M., Fernandez, A., Han, H. D., Mangala, L., Rodriguez-Aguayo, C., et al. (2010). c-Jun-NH2-kinase-1 inhibition leads to antitumor activity in ovarian cancer. *Clinical Cancer Research*, *16*, 184–194.

Vivas-Mejia, P. E., Rodriguez-Aguayo, C., Han, H. D., Shahzad, M. M., Valiyeva, F., Shibayama, M., et al. (2011). Silencing survivin splice variant 2B leads to antitumor activity in taxane-resistant ovarian cancer. *Clinical Cancer Research*, *17*, 3716–3726.

Wan, C., Allen, T. M., & Cullis, P. R. (2014). Lipid nanoparticle delivery systems for siRNA-based therapeutics. *Drug Delivery and Translational Research*, *4*, 74–83.

Wang, H., Jiang, Y., Peng, H., Chen, Y., Zhu, P., & Huang, Y. (2015). Recent progress in microRNA delivery for cancer therapy by non-viral synthetic vectors. *Advanced Drug Delivery Reviews*, *81*, 142–160.

Whitehead, K. A., Langer, R., & Anderson, D. G. (2009). Knocking down barriers: Advances in siRNA delivery. *Nature Reviews Drug Discovery*, *8*, 129–138.

Yang, Z. Z., Li, J. Q., Wang, Z. Z., Dong, D. W., & Qi, X. R. (2014). Tumor-targeting dual peptides-modified cationic liposomes for delivery of siRNA and docetaxel to gliomas. *Biomaterials*, *35*, 5226–5239.

Ye, X., & Weinberg, R. A. (2015). Epithelial-mesenchymal plasticity: A central regulator of cancer progression. *Trends Cell Biol*, *25*, 675–686.

Yew, N. S., & Scheule, R. K. (2005). Toxicity of cationic lipid-DNA complexes. *Advances in Genetics*, *53*, 189–214.

Zetsche, B., Gootenberg, J. S., Abudayyeh, O. O., Slaymaker, I. M., Makarova, K. S., Essletzbichler, P., et al. (2015). Cpf1 is a single RNA-guided endonuclease of a class 2 CRISPR-Cas system. *Cell*, *163*, 759–771.

Zhang, Y., Wang, Z., & Gemeinhart, R. A. (2013). Progress in microRNA delivery. *Journal of Controlled Release*, *172*, 962–974.

Zuckerman, J. E., & Davis, M. E. (2015). Clinical experiences with systemically administered siRNA-based therapeutics in cancer. *Nature Review Drug Discovery*, *14*, 843–856.

CHAPTER 17

Multiparametric Preclinical Assessment of Theranostics Materials

Adriele Prina-Mello[*,†,‡]

[*]Department of Clinical Medicine, Trinity College Dublin, Dublin, Ireland [†]Centre for Research on Adaptive Nanostructures and Nanodevices (CRANN) Institute and AMBER Centre, Trinity College Dublin, Dublin, Ireland [‡]Trinity Translational Medicine Institute, Trinity College Dublin, Dublin, Ireland

1 Introduction

In the past 5 years, the field of nanomaterials applied to medicine developed a new translational model assessing, from bench to bedside, products containing "nano" properties (Bawa, Audette, & Rubinstein, 2016; Bobo, Robinson, Islam, Thurecht, & Corrie, 2016). However, the lack of specific protocols and characterization methodologies for the preclinical assessment has impacted the clinical translation and subsequent market approval.

Recently, several medicinal products and medical devices containing nanomaterials have been clinically approved and promoted as first line treatment for cancer (e.g., Doxil) or, more recently, as adjuvant in combination treatment (e.g., nanoAbraxane, Onivyde) (Björnmalm, Thurecht, Michael, Scott, & Caruso, 2017). The approval of theranostics products has been slower than for each single diagnostic or treatment component due to the complexity in addressing their primary mode of action (PMoA), safety, efficacy, and quality attributes. The current preclinical assessment approaches are no longer a prerequisite for translation. This creates the need for multilayered and/or multiparametric approaches (Bawa et al., 2016; Gobbo, Sjaastad, Radomski, Volkov, & Prina-Mello, 2015) to understand and account for the physical, chemical, and biological properties and their clinical manifestations. There are several challenges to these approaches because the more complex the theranostic product, the harder the PMoA assessment can be. Nonetheless, regulators have agreed on some common basic parameters linked to the physicochemical properties (e.g., particle size distribution, surface charge, chemical composition, drug loading and release kinetics, or heating capabilities or surface enhanced properties) as the critical factors for the assessment of the quality, as well as for the in vivo efficacy and safety of nanomaterial-based theranostic products. In the past 60 years, a wealth of knowledge and research has been developed around

Handbook of Nanomaterials for Cancer Theranostics. https://doi.org/10.1016/B978-0-12-813339-2.00017-7

iron oxide materials (Commission, 2011; Southern & Pankhurst, 2017). This has led to the fact that to date, multiple clinical trials are ongoing to explore the treatment capabilities that theranostics products can offer. Iron oxide nanoparticles (IONP), such as gold (GNP), silver (AgNP), hafnium dioxide, and other nanomaterials, display unique properties due to their ability to be manufactured at the nanoscale level, which is below 100 nm, as defined by the European Commission (Commission, 2011), which differentiates them from larger particles and bulk material. IONP in the 1–100 nm range have a large surface to volume ratio, can have metallic and/or magnetic properties depending on their composition, and can act as heat, or vibrational, core sources. The core can also act as the scaffold for the functionalization of a wide range of therapeutic compounds. Multifunctionality is conveyed by specially designed architectures of the core responding to several types of external stimuli such as magnetic fields, light, or particle irradiation (Chen, Zhang, Yu, Zhang, & He, 2016; Dou et al., 2016; Kim et al., 2010; Li et al., 2016; Pankhurst, 2013). Furthermore, IONP can also be tailored on a case by case basis by functionalizing them with targeting moieties and/or therapeutic agents (Corr et al., 2006; Gultepe et al., 2010; Mejias et al., 2011; Roca et al., 2009; Tartaj, Morales, Gonzalez-Carreno, Veintemillas-Verdaguer, & Serna, 2011). Such properties make IONP suitable candidates for theranostic applications and their role as magnetic resonance imaging (MRI) contrast agents in drug and gene delivery platforms (de Jong & Borm, 2008; Gupta & Gupta, 2005; Veiseh et al., 2009) and magnetic hyperthermia (Hirsch et al., 2003; Sadeghi-Aliabadi, Mozaffari, Behdadfar, Raesizadeh, & Zarkesh-Esfahani, 2013). These are being intensively researched.

First proposed in 1957 by Gilchrist et al., magnetic hyperthermia has had >3000 peer-reviewed articles related to its theoretical application published (Périgo et al., 2015), and the interest in this field continues to grow. Considerable effort has been put forth by the several communities involved; for example, in Europe, the largest funding is provided by the European Union Research Programme in an effort to bring the first theranostic product to the first human clinical trial. Achieving clinical grade product therefore requires compliance with good manufacturing practice (GMP), good laboratory practice (GLP), and international standards (ISO, ASTM, CEN) to achieve or minimize contaminants, mix ups, and errors.

A rigorous preclinical sterility, including toxin-contaminations and physico-chemical characterization (PCC) of IONP, prior to the in vivo experimentation, is necessary to ensure adherence to the necessary predefined safety standard levels for any type of parenteral administration, with intravenous being the most common route. Efficacy and quality have to be achieved by minimizing variation in the product's critical quality attributes (CQAs); where the most important of all is the batch-to-batch variability.

In a global effort to provide standardized means for assessment, the U.S. National Cancer Institute Nanotechnology Characterization Laboratory (NCI-NCL) and the European

Nanomedicine Characterization Laboratory (EUNCL) have been created for supporting the preclinical assessment for medical translation of new nanotechnology-enabled products. At first, primarily for cancer, but also for other life-threatening conditions such as cardiovascular, neurodegenerative, and autoimmune conditions (European Nanomedicine Characterisation Laboratory, 2017b). Under the theranostics products where only some "nano"-related standardized methods can be transferred over, effort has been put into developing crucial PCC in biological systems to screen for multimodality or multifunctionalities. There is, therefore, a need to establish a prioritized approach for assays and protocols development for the measurements of theranostic products sterility and PCC such as bacterial contamination, endotoxins, core particle size, hydrodynamic diameter, surface charge, and colloidal stability, but also for the assessment of the multifunctionalities, for example, multimodal imaging and photothermal activation for gold nanoparticles based on their plasmonic responses; and magnetic hyperthermia or thermal ablation for IONP based on their specific absorption rate (SAR).

This is of great importance because, based on the NCI-NCL and EUNCL experience, more than half of all nanomedicine screened failed in achieving industrially acceptable quality standards, and was therefore deemed to be unsuitable for in vivo translation in humans. This thus constitutes the first preclinical hurdle as "go/no go" criteria for translation. An example is the evaluation of size distribution and stability for theranostic-NP in both water and complete culture medium (e.g., containing serum). This can provide a thorough understanding of the NP properties, especially if designed for systemic administration, based on their PCC behavior, and therefore provide the grounds, or not, for relevant subsequent in vitro and in vivo testing (Murdock, Braydich-Stolle, Schrand, Schlager, & Hussain, 2008). Although this is neither exhaustive nor comprehensive, it provides a good PCC base when planning the biological interaction or kinetics experiment for the specific NP. Such an approach, which goes under the safe-by-design industrial product development, is introducing multiple tiers for the screening of the promising theranostic NP to then allow the investigation of the advanced biological interactions and systemic response (e.g., toxicity, immunotoxicity, efficacy, pharmacokinetics, and biodistribution). For example, an aggregating IONP or GNP in serum or plasma would therefore not meet the basic quality criteria for any systemic administration, and therefore not be suitable for any further in vitro biocompatibility or efficacy study, as presented in previous works (Movia et al., 2014).

Adopting a safe-by-design and a quality-by-design approach at the design phase for theranostic and nanomedicine products has been proven to accelerate the translational path (Bawa et al., 2016; Bobo et al., 2016), enable compliance with the 3Rs directives (Flecknell, 2002; Musazzi, Marini, Casiraghi, & Minghetti, 2017; Wells, 2011) (replacement, reduction, and refinement), and therefore reduce in vivo preclinical costs and accelerate the translation to clinical trials. This is schematically shown in Fig. 1.

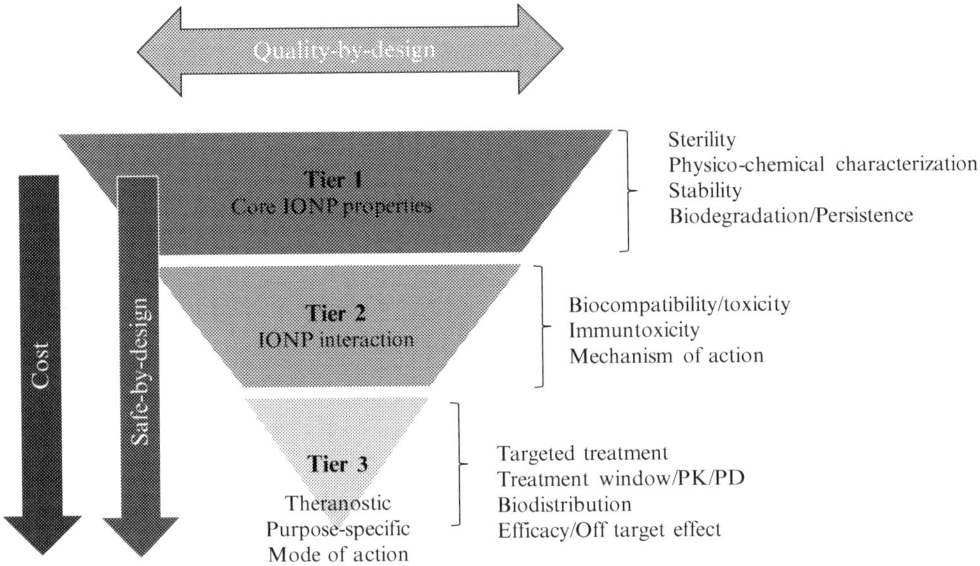

Fig. 1
Safe-by-design and quality-by-design considerations introduced into a preclinical assessment tiered approach for theranostics product translation.

Another important aspect for fast and effective clinical translation is to secure a continuous dialogue between the active research innovators, the industrial developers, business investors, and regulatory agencies. To this aim, the Community of Researchers (CoR) is a flexible framework for EU-USA cooperation that includes regulatory agencies, scientific bodies, and metrology institutes with the goal of working together to anticipate information needs and monitor the progress in technological development to guide harmonization of standards, protocols, and assays for the approval of new nanomedicine products. The CoR in nanomedicine has highlighted as a key priority the need for harmonization of assessment protocols across the community to achieve a unified approach for all nanotechnology-enabled products.

When combining the knowledge acquired through the engagement with NCI-NCL, EUNCL, CoR, and the national and international regulatory and metrology agencies, with the expertise developed through the exploitation of the safe- and quality-by-design approach, the following preclinical decision-making process can be drawn for theranostic product selection (see Fig. 2).

Testing and verification of such processes was carried out with a large scale, industrially driven Pan-European project, or flagship initiative; in this instance, for the case of iron oxide nanoparticles for theranostic use as a multifunctional tool for magnetic hyperthermia, imaging, and drug delivery. The key enabling properties of this decision-making process are expanded in the next sections.

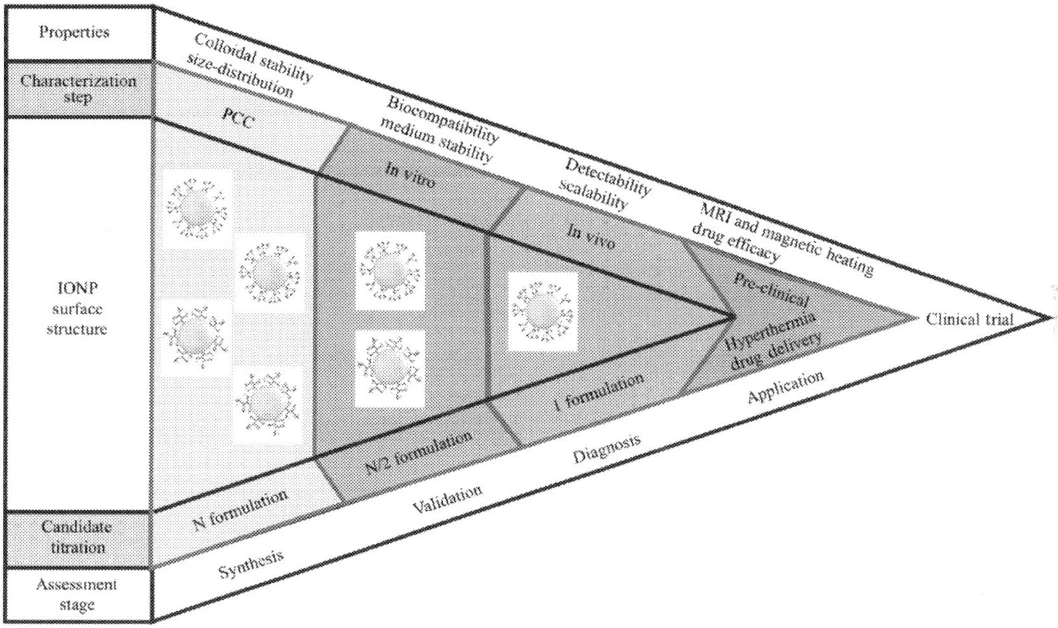

Fig. 2
Preclinical decision-making process toward theranostic product selection. A three-step decision making process is employed in order to select the most suitable candidate for future clinical translation. (1) Physico-chemical characterization (PCC) decision making based on IONP colloidal stability in native water suspension and nanoparticle hydrodynamic size distribution below 100 nm as defined by the European Commission. (2) IONP selected for in vitro testing were categorized as biocompatible if they did not induce an acute cytotoxic effect as determined through a multiparametric approach and were not destabilized in solution through interaction with biological media. (3) Suitable IONP can then undertake scale-up assessment, based on the capability to produce industrial quantities while maintaining strict physico-chemical specifications and the absence of acute toxicity in a small animal model.

2 Physico-Chemical Characterization: Importance of Size, Coating, Charge, and Stability

Nanomaterials, such as IONP for use in theranostics, ultimately come into contact with protein-rich physiological fluids; especially when administered systematically. Therefore, the importance of understanding the PCC properties of IONP should not be overlooked. Knowing how the properties are affected by dispersion in a physiological medium is necessary for the success of the IONP in its specific application.

2.1 Size

Slight changes in PCC properties have significant implications on the behavior of IONP in the biological environment, such as the mechanism of uptake (Alexis, Pridgen, Molnar, &

Farokhzad, 2008). Particle size plays a decisive role in nanoparticle properties, and therefore particle sizing is an essential task in nanoparticle characterization. The size of the core material can dictate the applications for which they are suitable, for example, nanometer sized particles can be used for targeted drug delivery, and are small enough to enter cells to carry out their intended function; whereas particles in the micron range have shown great promise in protein and cell separation applications. The degree of reactivity of IONP is size dependent and is related to their surface area, with smaller nanomaterials (greater surface area) having higher dissolution reactivity compared with their larger analogues (Fig. 3). Enhancement in IONP reactivity can be particularly high at sizes less than or equal to 10 nm (Auffan et al., 2009).

2.2 Surface Coating

A key component to ensure long circulation time with increased presentation at a tumor site or organ of interest is the avoidance of the ReticuloEndothelial system (RES) and the organs responsible for nanoparticle clearance, namely the liver, spleen, and kidney (Brigger, Dubernet, & Couvreur, 2002; Laurent et al., 2008; Müller & Wallis, 1993). While IONP size has been shown to be a major predictor for nanoparticle clearance from the blood (Longmire, Choyke, & Kobayashi, 2008), the presence of hydrophobic residues on the nanoparticle surface have also been shown to reduce circulation time because of increased protein opsonization (Chouly, Pouliquen, Lucet, Jeune, & Jallet, 1996). Furthermore, when entire

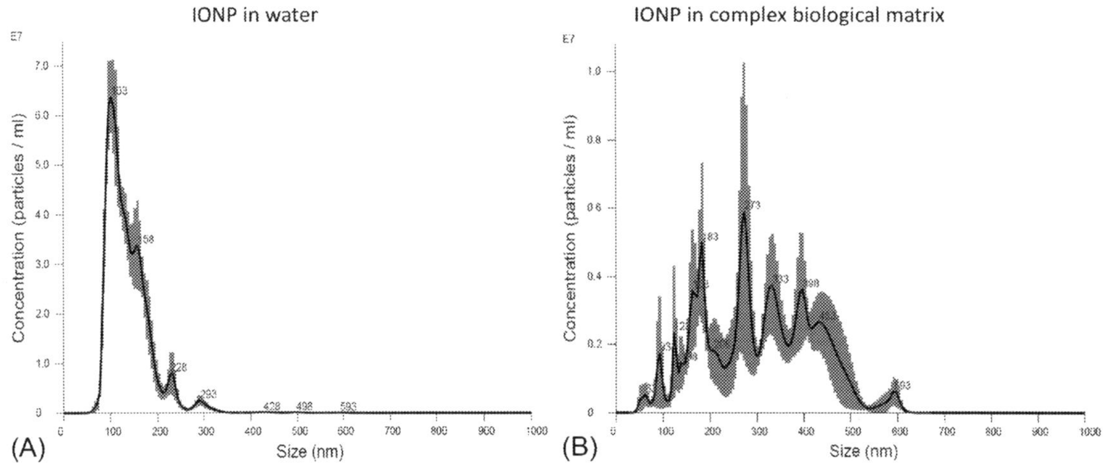

Fig. 3

Nanomaterials characterization by NTA. IONP were incubated with DI water (A) or complete culture medium (B) for 24 h and were evaluated in their size distribution for hydrodynamic changes. Colored red lines show overlapping of individual measurements of IONP (average of 6 consecutive measurements). *Courtesy of Dr. C.M. Maguire, unpublished data.*

protein layers are adsorbed to nanoparticles' surface, forming their biological identity in the blood, the distinction in time and binding kinetics can determine the difference between soft and hard protein adsorption, or corona. This is a dynamic coating that can alter the PCC of the nanoparticle (e.g., size, surface charge, and colloidal stability), and also immunological response/recognition (e.g., immune suppression, stimulation, or avoidance) and therefore impacting the results from in vivo studies (Pino et al., 2014). Despite the limited literature on the characterization of the protein corona with IONP, concerning the in vivo safety and efficacy, these have been extensively presented in a previous publication (Hannon, Tutty, & Prina-Mello, 2018).

2.3 Surface Charge

Surface charge is an important characteristic of nanomaterials for theranostics, and determines cellular uptake and biodistribution. Positively charged nanoparticles have a higher rate of cell uptake against the negatively charged plasma membrane at nonspecific sites, when compared with neutral and negatively charged nanoparticles (Rejman, Oberle, Zuhorn, & Hoekstra, 2004; Villanueva et al., 2009; Yue et al., 2011). The increase in cellular association generally confers that positively charged nanomaterials are more toxic compared with negative variants of the same size and composition (Bhattacharjee et al., 2010; Goodman, Mccusker, Yilmaz, & Rotello, 2004). Thus, neutral and negative nanomaterials have been shown to limit the extent of protein opsonization, have an increased circulation time, and have a lower rate of nonspecific cellular uptake (Alexis et al., 2008). This can be also applicable in the case of protein corona where the surface charge plays a major role on the protein adsorption (Sakulkhu, Mahmoudi, Maurizi, Salaklang, & Hofmann, 2014).

2.4 Physico-Chemical Characterization Techniques

The basic characterization techniques requested by the regulatory agencies for nanomaterials used for theranostics is transmission electron microscopy (TEM). Providing information on the core nanoparticle structure, and dynamic light scattering (DLS) provides information on the hydrodynamic diameter surrounding the core nanoparticle. Nanoparticle tracking analysis (NTA) has also been introduced complementarily to DLS (Hole et al., 2013; Maguire et al., 2017) to provide sizing, multimodal distribution, and concentration data that are CQAs. Thus, it is clear that the combination of TEM, DLS, and NTA techniques provide powerful PCC tools to enable continued nanoparticle and IONP assessment and to facilitate better understanding of their behavior upon exposure to the biological environment. This has been recently emphasized in the latest work of Gioria et al. (Gioria et al., 2018) in which solutions have also been provided in the case of polydisperse samples.

3 Chemical Characterization

3.1 Multimodal Theranostics With Targeted Drug Delivery Capabilities

Chemical characterization, drug loading, and drug delivery for theranostics nanomaterials are being investigated in a multiplicity of options and models, which include active compound detection and targeting to cell models (e.g., tumor), active targeting to angiogenic cell surface markers of endothelial cells, magnetic-field assisted delivery, and triggered drug delivery. Thus, there are several analytical techniques that are at the basis of the characterization, such as high performance liquid chromatography, mass spectrometry and nuclear magnetic resonance. These come into play when the chemical composition, total amount, and structural identity have to be measured for the active product indication (API) or the chemical moieties coupled to these NPS (Forum, 2017).

Nonetheless, with the additional surface functionalization comes a further layer of complexity, whereby the nanoparticles must remain colloidally stable, but must also have a high drug-loading capacity. Steric hindrance, availability of sufficient functional groups, and employing spacer molecules play a role in the amount of drug incorporated. Functionalization strategies for IONP include electrostatic and covalent attachment of the molecule of interest to the surface of the IONP. Electrostatic attachment involves the interaction of an oppositely charged moiety to the surface of the IONP, and attractive force or bond strength depends on the charge of both the moiety to be attached and the nanoparticle surface, and can easily be disrupted in the presence of a high salt solution. In such an environment, drug release occurs as a passive process (Kossatz et al., 2015). Taking this limitation into account, the covalent functionalization of drugs and targeting moieties onto the surface of the nanomaterials seems to be the best approach.

4 Theranostics Characterization

4.1 Key physical Properties for Therapeutic Hyperthermia

The basic idea behind hyperthermia in theranostic applications is to exploit this heating mechanism, turning the unwanted increase of the body temperature into a controlled and localized process to target cell killing, drug release, or enhanced contrast for imaging. Thus, the evaluation of the physical properties around magnetic hyperthermia are critical, especially because, for the induction of localized cell-apoptosis, there is the need to deliver stable and localized temperatures around 43°C.

4.1.1 Magnetic hyperthermia

The application of magnetic hyperthermia was first demonstrated by Gilchrist et al. in 1957 using iron oxide particles with a size range of 10 nm to 1 μm, to heat lymph nodes using a 1.2–2.0 MHz magnetic field (Gilchrist et al., 1957). It has been shown that sustaining a temperature above 43°C is sufficient for clinical hyperthermia and that sustaining the temperature for more than 1 h would cause cancer cell death (Field & Morris, 1983). This requires a sufficient quantity of nanomaterials in the tumor, which need to be quantified to secure an average temperature dose delivered according to the CEM43T90 (cumulative equivalent minutes of T90 above 43°C), whereby 90% of the tumor area has reached the therapeutic dose of 43°C for a defined time period. The need to physically quantify the total amount of IONP accumulated into a tumor, and also the SAR of the injected IONP, are fundamental to achieving the CEM43T90. Despite the extensive scientific research published on this matter, there is limited literature on the standardization on the characterization for the physical properties around magnetic hyperthermia (Périgo et al., 2015). In past decades, several works have focused on the identification of the key properties and parameters associated with multilayered IONP. Among the many magnetic properties identified, as functions of the IONP's final nature, the SAR is the main parameter that has been used to quantify the heating potential of IONP (Hilger et al., 2002). This is found to depend on the IONP intrinsic magnetic properties (e.g., total iron content, ferromagnetic domains, coercivity, magnetic saturation, and others), the liquid or suspension where they are dispersed, and heating temperature and time (Hilger et al., 2002). This allowed for IONP classification and assessment based on their magnetic hyperthermia potential, and has enabled the in silico modeling for heat dissipation that has enabled the prediction of the IONP efficiency in line with the "safe and quality by design" approaches. More recently, in the fundamental work from Teran and collaborators, they have been able to determine the IONP magnetic hyperthemic efficacy based on the heat dissipation, and heat transfer models. Furthermore, the ability to also account for the interparticles' interactions within a tissue-like model based on Monte Carlo simulations has supported the experimental data based on SAR-values obtained by many research groups (Périgo et al., 2015). It is known that the SAR-value is dependent on the measurement technique adopted: calorimetric or magnetometric.

In parallel, the work of Pankhurst and co-workers has also introduced new methodologies on the practicalities of experimental measurement and characterization for magnetic hyperthermia IONP. Their work pushes the characterization toward the assessment of the PMoA for the local heat generation into phantom samples, agarose and ex vivo tissue, to account for the unique IONP physical transduction mechanism (Southern & Pankhurst, 2017). Through this groundbreaking work, it is possible to determine, at the preclinical level, the extent of adverse effects in a clinical trial. This approach allows for a cost-effective identification of indicative IONP dose limits to be used in vivo, under a set administration route, without the burden of undertaking any scale up or in vivo study. This work effectively develops a guidance model for clinical scenario projections for magnetic therapeutic efficacy starting from acquired preclinical PCC characterization, and thus reducing in vitro and in vivo experimental work to the assessment of the cell-IONP interaction.

5 Cell-NP Interaction in Biological Environment: Importance of Preclinical Assessment

Nanomaterials, such as IONP, for use in theranostics, have to comply with safety standards when in contact with physiological fluids or cells. Sterility and endotoxicity are therefore the first two critical parameters that should be evaluated before any in vitro or in vivo experimental work can be carried out. Sterility, in particular, accounts for the highest failure rate in medical devices, whereas endotoxicity is the limiting parameter for in vivo screening of medicinal products. These two main assays are recommended to be carried out as preclinical entry assays for product translation, according to the EUNCL (European Nanomedicine Characterisation Laboratory, 2017a). The adoption of validated assays will also ensure that high quality data is obtained. This is of importance for nanomaterials for theranostics where the assay interference is very common due to the nature of the core material to be used, and insight on the technical aspects associated with these assessments falls under the safety considerations and immunotoxicity for IONP (Hannon et al., 2018).

5.1 Mechanism of Interaction

Nanoparticle uptake can occur by energy independent (passive) and dependent (active) mechanisms, and these provide a means for targeted drug delivery and understanding the mechanism's potential toxicity. Passive uptake involves the movement across the plasma membrane without any energy driven processes, as shown to occur for nanoparticles <5 nm in size (Samuel et al., 2012; Wang, Bai, Jiang, & Nienhaus, 2012), whereas larger nanomaterials can cross the cell membrane through lipid bilayer defects that can facilitate the uptake of large nanoparticles. Active uptake involves a size-specific response with a given contact or interaction with the cell membrane for mechanical ligand-receptor uptake, or the endocytosis process, which includes clathrin mediated, caveolin mediated, macropinocytosis, or phagocytosis (Calero et al., 2015). Cell specific interactions should be recorded to distinguish between the different cell phenotypes.

6 In Vitro Screening in Biological Environments: Experimental Design

It is important that reliable and reproducible in vitro based safety tests are applied to each new nanoparticle formulation (Nel et al., 2013; Prina-Mello, Crosbie-Staunton, Salas, del Puerto Morales, & Volkov, 2013). Biocompatibility testing of nanomaterials is typically conducted using assays for determining toxicity of chemicals and drugs (Kroll, Pillukat, Hahn, & Schnekenburger, 2009). However, it has been widely demonstrated that nanoparticle interference can occur with colorimetric assays employing dyes such as Comassie blue, alamar blue™, neutral red, propidium iodide, dichlorofluorescein, MTT, and WST-1; and the need exists for an alternative approach (Casey et al., 2007; Guadagnini et al., 2013). The move toward automated microscopy platforms utilizing fluorescence-based assays dramatically reduces IONP interference through direct visualization of subcellular organelles and disregarding extracellular fields of view. High content screening (HCS) provides the flexibility required in image analysis to enable subcellular data mining without the interference that free iron from IONP may otherwise induce in a colorimetric test. Previous studies from my laboratory have demonstrated the suitability of HCS for monitoring the biocompatibility of a number of nanomaterials in a range of cell types (Collins et al., 2017; Prina-Mello et al., 2013). Several other studies have adopted HCS to investigate the effects of transition metal oxide nanoparticles in zebrafish (Lin et al., 2011), metal, and metal oxide (Ag, Au, Pt, Al_2O_3, SiO_2, ZnO and QD1) in bronchial epithelial (BEAS-2B) and macrophage (RAW 264.7) cell lines (George et al., 2011). This raises HCSA to become a gold standard method for nanomaterial testing, providing rapid toxicological assessment and multiparametric output not affected by assay interference, as presented in previously published works and book chapters (Collins et al., 2017; Prina-Mello, Mohamed, Verma, Jain, & Volkov, 2014).

6.1 In Vitro Screening: Experimental Model

Screening of each formulation to determine the biocompatibility and capacity to induce acute toxicity, as part of the intracellular degradation and resulting appearance of the constitutive parts of the IONP, would represent an essential checkpoint to be carried out prior to the in vivo study. In line with the presented three-tiered approach (Fig. 1), the testing of nanomaterials constitutes Tier 2, whereby in vitro assays are employed to screen the nanomaterials for biocompatibility, toxicity, immunotoxicity, and mechanism of action. Nanomaterials can enter the body and induce a wide range of effects, which are dependent on the mode of entry and whether nanomaterials reside outside or inside the cell. Many nanomaterials may not induce toxic effects when localized to the extracellular environment, but may induce toxicity upon entry into the acidic environment of the lysosome. Two approaches can be taken to determine nanomaterials toxicity in vitro, first, the utilization of viability tests that rely on a color change to provide a measure of cellular viability, such as those used to determine the suitability of drugs in the pharmaceutical setting; or second, the use of fluorescent viability tests against specific cellular markers or organelles. Both approaches suffer from interference to some degree, with conflicting reports as to the extent of nanoparticle-induced toxicity, using the same nanomaterials. High throughput fluorescent techniques provide multiple parameter readouts compared with the single output obtained using colorimetric assays only.

Testing of nanomaterials has been typically conducted using assays originally developed to determine the toxicity of chemicals and drugs (Kroll et al., 2009). However, it has been demonstrated that nanomaterials interference can occur with assays employing dyes such as those found in the lactate dehydrogenase (LDH), 3-(4,5-dimethylthiazol-2-yl)-2, 5-diphenyltetrazolium bromide (MTT) and WST-1 assays (Casey et al., 2007; Guadagnini et al., 2013). Thus, this establishes the importance of introducing suitable controls for the assay reagents or for identifying its limit of detections, such as the upper and lower response limit. This can be applied for any parameter adopted for the multiparametric screening (e.g., pH, kinetics, tracking, and specific organelle labeling). In some instances, the quenching of fluorescence by IONP can also be quite advantageous, such as in the production of smart nanoparticles when closer to the limit of detection of the instrumentation. This is the case of IONP in magnetic resonance imaging, where to compensate for the lack of image resolution, small particles were identified by both contrast and fluorescence using near-infrared fluorescence spectroscopy (Josephson, Kircher, Mahmood, Tang, & Weissleder, 2002).

6.1.1 High-throughput screening: Multiparametric assessment

High-throughput screening (HTS) also provides the flexibility required in image analysis to enable subcellular data mining without the interference that IONP may induce in a colorimetric test. HCS, as a specific type of HTS, is a convergence between high resolution fluorescence-based imaging and cell-based assays. Thus, the move toward an automated microscopy platform utilizing a fluorescent based assay has proven to dramatically reduce extracellular nanoparticle interference through direct visualization of the subcellular organelles (Prina-Mello et al., 2014). Simultaneous detection of many biological pathways allows us to screen thousands of compounds in a fully automated robotic platform. Following the decision making step of Fig. 2, the IONP most suitable for identification for in vivo study is then determined by the HCSA multiparametric assessment.

Intracellular degradation or extracellular matrix corrosion and liberation of IONP constitutive chemicals can result in the excessive formation of reactive oxygen species (ROS). ROS are highly reactive molecules and present in the cell as by-products of metabolism. Under normal cellular conditions, the ROS low level is due to the superoxide dismutase family of enzymes catalyzing the dismutation of superoxide into hydrogen peroxide and water (Winterbourn, 1995). Cellular compartmentalization or damage is manifested by ROS increase, as oxidative stress, and subsequent DNA, RNA, lipids, and proteins damage (Winterbourn, 1995). The release of Fe^{2+}, as electrochemical reduction, can directly increase formation of ROS through the production of hydroxide (OH^-) and reactive hydroxyl radicle (OH) (Huang et al., 2013; Lemire, Harrison, & Turner, 2013; Winterbourn, 1995; Yan, Gu, & Zhao, 2013). Thus, ROS is one of the parameters to always test when screening IONP for toxicity in vitro.

7 In Vitro Screening for Immunotoxicity

As stated herein, ROS plays a major role in the immunotoxicity associated with IONP. Upon administration, IONP interacts and internalizes into cells to induce the upregulation of ROS, which may cause undesired immune dysfunction, and thus reduce the safety of the IONP. Thus, the unintentional interactions with many components of the immune system, including the array of leukocytes, macrophages, red blood cells, and serum proteins are the causative of immunotoxicity responses. Hypersensitivity and other immune modulating effects with IONP are commonly observed during clinical application (Bircher & Auerbach, 2014; Szebeni et al., 2015).

PCCs are usually responsible for immune-responses such as endotoxin contamination and sterility. Capitalizing on past extensive physico-chemical and biological characterization knowledge and expertise developed internationally (Giannakou et al., 2016), the key aspects in immunotoxicity and safety associated with the translational process of IONP are associated with the immunotoxic profiling, assessment, and remediation for the manufacturing preparation of theranostics nanomaterials. This represents a huge concern for the clinical approval of IONP and those in the preclinical pipeline. It is therefore essential that immunotoxicity is evaluated early and efficiently during the development and testing of IONP nanoparticles. This has been extensively presented in a previous work where consideration for improvement in the quality and safety preparation of clinically relevant IONP are presented in details, such as coagulation, haemolysis, monocyte phagocytic system, protein corona formation, and immune system reaction (Hannon et al., 2018).

Overcoming the immunotoxicity hurdle will represent a major step forward in the 60-year scientific research on iron oxide nanoparticles for clinical applications. There are many reasons for this poor translation to the clinic; nonetheless, oxidative stress and immunotoxicity caused by IONPs are certainly contributing factors to be considered and to overcome in future translational research.

8 In Vivo Testing for Theranostic Nanomaterials and Clinical Translation Consideration

The preclinical evaluation of the safety and efficacy of nanomaterials for theranostics in vivo has to be carefully planned because there is not a simplistic "one size fits all" solution for the extremely diverse nature of the theranostics nanomaterials, in which there are numerous different clinical indications requiring multiple endpoints, specific administration routes, formulations delivery, and technical details for the experimental testing of the associated theranostics properties. An example can span from enhanced MRI imaging, image guided surgery, magnetic and environmental hyperthermia, and photodynamic therapies to multifunctional or multimodal applications.

It is extremely difficult to try to draw any generally applicable in vivo experimental design or conclusions on the matter because theranostics nanomaterials fall between medicinal products and the medical devices classification. Thus, there is the need, as indicated herein, for the theranostic section, to first identify the PMoA of the nanomaterials, and then couple it with the external medical device that is delivering the multifunctionality and multimodality. Once this is identified and classified by the national regulatory body, adherence to the specific guidelines on the in vivo data for preclinical assessment has to be followed. For the medicinal products, this follows the Pharmacopeia guidelines for the assessment of at least maximal tolerated toxicity, biodistribution, pharmacokinetics, and pharmacodynamics and efficacy. Whereas for medical devices, Europe has recently published the new medical device directives (Commission, 2017), in which nanomaterials for medical devices are classified at the highest level of risk, and thus are in need of a full preclinical assessment dossier on toxicity, biocompatibility, and safety based on the mode of action.

Independently from the nature of the applications, consideration should be given to the fact the preclinical testing in vivo should be as close as possible to the clinical one in which every study protocol has to be evaluated on its own experimental and usefulness merit by an ethical committee. This guarantees compliance with the 2010 European Commission directives (2010/63) (Commission, 2010) on animal testing in which an alternative, or replacement, has to be found where possible (Doke & Dhawale, 2015; Flecknell, 2002). This has opened a large investigation on several alternatives, ranging from 3D multilayers, embedded or scaffoldless spheroids, up to organ-on-chip. In past years, the attention of the scientific community has also been focused in this direction, where a multistakeholders' discussion has been brought forward internationally (Nel et al., 2013). Furthermore, very recently the Scientific Committee on Health Environmental and Emerging Risks (SCHEER) has published the final opinion regarding the need for nonhuman primates in biomedical research, production, and testing of products and devices (Risks, 2017), where questions are raised on the complex and multidisciplinary issues where nanotechnology is used.

Nonetheless, for the specific case of IONP adopted in the clinic, nanoparticles are employed to enhance the signal as advanced imaging for guided surgery or as a contrast agent, as a non-PMoA. It was the use as contrast agents that brought the first iron oxide nanomaterials into the clinic, with a range of approved compounds for intravenous delivery over the past few years. Currently, IONP are the most clinically tested inorganic nanoparticle (Anselmo & Mitragotri, 2016). However, despite this huge interest, and high biocompatibility, the therapeutic potential of IONP has yet to be successfully met in the clinic, while many FDA (U. S. Food and Drug Administration) and EMA (European Medicines Agency) approved diagnostic IONP-formulations have been discontinued. Examples include Endorem® (Dextran coated MNP, 80–150 nm), Resovist® (Carboxydextran-coated ultra-small MNP, 20 nm), Sinerem® (Dextran-coated ultra-small MNP 15–30 nm), Feruglose (pegylated starch-coated ultra-small MNP 20 nm), and iron oxide-based agents for gastrointestinal contrast, which are given orally such as Ferumoxsil® (silicon-coated MNP, 300 nm) and Abdoscan®

(sulphonated styrene-divinylbenzene-coated MNP, >300 nm) (Li et al., 2013; Wang, 2011). There are many proposed reasons for this poor translation to the clinic; nonetheless, oxidative stress and immunotoxicity cause an off target accumulation, especially in the liver, spleen, and kidney, which is certainly a contributing factor that must be considered and overcome in future preclinical assessment research. This leaves a large gap for the production of clinically viable IONP, with the potential for providing novel nanoparticles for molecular imaging, targeted drug delivery, and thermal therapy.

9 Conclusions and Future Perspective

Iron oxide nanoparticles (IONP) are a hugely promising manufacturing nanomaterial and are already used in a wide array of clinical applications. This chapter presented an overall preclinical decision making process approach focused toward the clinical translation of IONP as a theranostic nanomaterial. Considerations based on safety and quality by design are brought into a tiered approach for the successful screening and assessment of the nanomaterial. A three step tiered approach is presented to systematically identify suitable IONP that could act as basic scaffolds for theranostic applications. Through comprehensive PCC, the evaluation of the main properties and parameters for theranostics applications were presented with several examples of critical issues related to safety and quality of the nanomaterials. The understanding of the cell-IONP biological interaction and its investigation in vitro presented by the opportunity for fast multiparametric screening provided the basis for cost-effective early stage fail / pass assessment for further in vivo preclinical experimental investigation. Combined considerations of immunotoxicity response and in vivo testing were presented with a view to show how IONP have progressed in the clinical translation in the past 60 years, and their current potential as theranostic tools.

Financial and Competing Interest Disclosure

This work was partially supported by the MULTIFUN (European Commission 7th Framework grant ref.# 262943). NOCANTHER project (European Commission H2020, grant ref.# 685795). EUNCL project (EC, H2020, grant ref.# 654190).

Conflict of Interest

The author has no financial conflict with the materials disclosed.

Acknowledgments

The author would like to thank Prof. I. Hilger, M.d.P. Morales, Prof. Q. Pankhurst, Prof. Y. Volkov, Dr. D. Movia, Dr. O. Gobbo, Dr. C.M. Maguire, Dr. K. Crosbie-Staunton, Dr. G. Salas, Dr. N. Liptrott, Mr. G Hannon, Ms. M.A. Tutty, Ms. A. Bogdanska for the many opportunities for technical discussions under the following EC funded projects.

References

Alexis, F., Pridgen, E., Molnar, L. K., & Farokhzad, O. C. (2008). Factors affecting the clearance and biodistribution of polymeric nanoparticles. *Molecular Pharmaceutics*, *5*, 505–515.

Anselmo, A. C., & Mitragotri, S. (2016). Nanoparticles in the clinic. *Bioengineering and Translational Medicine*, *1*, 10–29.

Auffan, M., Rose, J., Bottero, J. Y., Lowry, G. V., Jolivet, J. P., & Wiesner, M. R. (2009). Towards a definition of inorganic nanoparticles from an environmental, health and safety perspective. *Nature Nanotechnology*, *4*, 634–641.

Bawa, R., Audette, G. F., & Rubinstein, I. (2016). *Handbook of clinical nanomedicine: Nanoparticles, imaging, therapy and clinical applications*. CRC Press.

Bhattacharjee, S., de Haan, L. H., Evers, N. M., Jiang, X., Marcelis, A. T., Zuilhof, H., et al. (2010). Role of surface charge and oxidative stress in cytotoxicity of organic monolayer-coated silicon nanoparticles towards macrophage NR8383 cells. *Particle and Fibre Toxicology*, *7*, 25.

Bircher, A. J., & Auerbach, M. (2014). Hypersensitivity from intravenous iron products. *Immunology and Allergy Clinics of North America*, *34*, 707–723.

Björnmalm, M., Thurecht, K. J., Michael, M., Scott, A. M., & Caruso, F. (2017). Bridging bio-nano science and cancer nanomedicine. *ACS Nano*, *11*, 9594–9613.

Bobo, D., Robinson, K. J., Islam, J., Thurecht, K. J., & Corrie, S. R. (2016). Nanoparticle-based medicines: A review of FDA-approved materials and clinical trials to date. *Pharmaceutical Research*, *33*, 2373–2387.

Brigger, I., Dubernet, C., & Couvreur, P. (2002). Nanoparticles in cancer therapy and diagnosis. *Advanced Drug Delivery Reviews*, *54*, 631–651.

Calero, M., Chiappi, M., Lazaro-Carrillo, A., Rodriguez, M. J., Chichon, F. J., Crosbie-Staunton, K., et al. (2015). Characterization of interaction of magnetic nanoparticles with breast cancer cells. *Journal of Nanobiotechnology*, *13*, 16.

Casey, A., Herzog, E., Davoren, M., Lyng, F. M., Byrne, H. J., & Chambers, G. (2007). Spectroscopic analysis confirms the interactions between single walled carbon nanotubes and various dyes commonly used to assess cytotoxicity. *Carbon*, *45*, 1425–1432.

Chen, W. W., Zhang, S. H., Yu, Y. Y., Zhang, H. S., & He, Q. J. (2016). Structural-engineering rationales of gold nanoparticles for cancer theranostics. *Advanced Materials*, *28*, 8567–8585.

Chouly, C., Pouliquen, D., Lucet, I., Jeune, J. J., & Jallet, P. (1996). Development of superparamagnetic nanoparticles for MRI: Effect of particle size, charge and surface nature on biodistribution. *Journal of Microencapsulation*, *13*, 245–255.

Collins, A. R., Annangi, B., Rubio, L., Marcos, R., Dorn, M., Merker, C., et al. (2017). High throughput toxicity screening and intracellular detection of nanomaterials. *Wiley Interdisciplinary Reviews: Nanomedicine and Nanobiotechnology*, *9*.

Commission. (2010). *Directive 2010/63/EU of the European Parliament and of the Council of 22 September 2010 on the protection of animals used for scientific purposes*. EUR-LEX European Union.

Commission. (2011). *Commission Recommendation of 18 October 2011 on the definition of nanomaterial text with EEA relevance*. Official Journal of the European Union. EUR-LEX.

Commission. (2017). *Regulation (EU) 2017/745 of the European Parliament and of the Council of 5 April 2017 on medical devices, amending Directive 2001/83/EC, Regulation (EC) No 178/2002 and Regulation (EC) No 1223/2009 and repealing Council Directives 90/385/EEC and 93/42/EEC*. EUR-LEX.

Corr, S. A., O'Byrne, A., Gun'ko, Y. K., Ghosh, S., Brougham, D. F., Mitchell, S., et al. (2006). Magnetic-fluorescent nanocomposites for biomedical multitasking. *Chemical Communications (Cambridge)*, 4474–4476.

Doke, S. K., & Dhawale, S. C. (2015). Alternatives to animal testing: A review. *Saudi Pharmaceutical Journal*, *23*, 223–229.

Dou, Y., Guo, Y. Y., Li, X. D., Li, X., Wang, S., Wang, L., et al. (2016). Size-tuning ionization to optimize gold nanoparticles for simultaneous enhanced CT imaging and radiotherapy. *ACS Nano*, *10*, 2536–2548.

European Nanomedicine Characterisation Laboratory. (2017a). *EU-NCL Assay Cascade [online]* Available from:http://www.euncl.eu/about-us/assay-cascade/.

European Nanomedicine Characterisation Laboratory. (2017b). *EU-NCL: European Nanomedicine Characterisation Laborator [online]*. Available from: http://www.euncl.eu/.

Field, S. B., & Morris, C. C. (1983). The relationship between heating time and temperature: Its relevance to clinical hyperthermia. *Radiotherapy and Oncology*, *1*, 179–186.

Flecknell, P. (2002). Replacement, reduction and refinement. *ALTEX*, *19*, 73–78.

Forum, I. (2017). *International Pharmaceutical Regulators Forum [online]*. Available from: https://www.i-p-r-f.org/index.php/en/.

George, S., Xia, T., Rallo, R., Zhao, Y., Ji, Z., Lin, S., et al. (2011). Use of a high-throughput screening approach coupled with in vivo zebrafish embryo screening to develop hazard ranking for engineered nanomaterials. *ACS Nano*, *5*, 1805–1817.

Giannakou, C., Park, M.V.D.Z., de Jong, W. H., van Loveren, H., Vandebriel, R. J., & Geertsma, R. E. (2016). A comparison of immunotoxic effects of nanomedicinal products with regulatory immunotoxicity testing requirements. *International Journal of Nanomedicine*, *11*, 2935–2952.

Gilchrist, R. K., Medal, R., Shorey, W. D., Hanselman, R. C., Parrott, J. C., & Taylor, C. B. (1957). Selective inductive heating of lymph nodes. *Annals of Surgery*, *146*, 596–606.

Gioria, S. C. F., Urbán, P., Maguire, C. M., Bremer-Hoffmann, S., Prina-Mello, A., Calzolai, L., et al. (2018). Are existing standard methods suitable for the evaluation of nanomedicines: some case studies. *Nanomedicine (London)*, https://doi.org/10.2217/nnm-2017-0338 [in press].

Gobbo, O. L., Sjaastad, K., Radomski, M. W., Volkov, Y., & Prina-Mello, A. (2015). Magnetic nanoparticles in cancer theranostics. *Theranostics*, *5*, 1249–1263.

Goodman, C. M., Mccusker, C. D., Yilmaz, T., & Rotello, V. M. (2004). Toxicity of gold nanoparticles functionalized with cationic and anionic side chains. *Bioconjugate Chemistry*, *15*, 897–900.

Guadagnini, R., Halamoda Kenzaoui, B., Cartwright, L., Pojana, G., Magdolenova, Z., Bilanicova, D., et al. (2013). Toxicity screenings of nanomaterials: Challenges due to interference with assay processes and components of classic in vitro tests. *Nanotoxicology*, *9*, 13–24.

Gultepe, E., Reynoso, F. J., Jhaveri, A., Kulkarni, P., Nagesha, D., Ferris, C., et al. (2010). Monitoring of magnetic targeting to tumor vasculature through MRI and biodistribution. *Nanomedicine (London, England)*, *5*, 1173–1182.

Gupta, A. K., & Gupta, M. (2005). Synthesis and surface engineering of iron oxide nanoparticles for biomedical applications. *Biomaterials*, *26*, 3995–4021.

Hannon, G., Tutty, M. A., & Prina-Mello, A. (2018). Immunotoxicity and safety considerations for iron oxide nanoparticles. In N. T. Thanh (Ed.), *Clinical applications of magnetic nanoparticles*. CRC Press/Taylor and Francis.

Hilger, I., Frühauf, K., Andrä, W., Hiergeist, R., Hergt, R., & Kaiser, W. A. (2002). Heating potential of iron oxides for therapeutic purposes in interventional radiology. *Academic Radiology*, *9*, 198–202.

Hirsch, L. R., Stafford, R. J., Bankson, J. A., Sershen, S. R., Rivera, B., Price, R. E., et al. (2003). Nanoshell-mediated near-infrared thermal therapy of tumors under magnetic resonance guidance. *Proceedings of the National Academy of Sciences of the United States of America*, *100*, 13549–13554.

Hole, P., Sillence, K., Hannell, C., Maguire, C. M., Roesslein, M., Suarez, G., et al. (2013). Interlaboratory comparison of size measurements on nanoparticles using nanoparticle tracking analysis (NTA). *Journal of Nanoparticle Research*, *15*, 2101.

Huang, G., Chen, H., Dong, Y., Luo, X., Yu, H., Moore, Z., et al. (2013). Superparamagnetic iron oxide nanoparticles: Amplifying ROS stress to improve anticancer drug efficacy. *Theranostics*, *3*, 116–126.

de Jong, W. H., & Borm, P. J. (2008). Drug delivery and nanoparticles: Applications and hazards. *International Journal of Nanomedicine*, *3*, 133–149.

Josephson, L., Kircher, M. F., Mahmood, U., Tang, Y., & Weissleder, R. (2002). Near-infrared fluorescent nanoparticles as combined MR/optical imaging probes. *Bioconjugate Chemistry*, *13*, 554–560.

Kim, D.-H., Rozhkova, E. A., Ulasov, I. V., Bader, S. D., Rajh, T., Lesniak, M. S., et al. (2010). Biofunctionalized magnetic-vortex microdiscs for targeted cancer-cell destruction. *Nature Materials*, *9*, 165–171.

Kossatz, S., Grandke, J., Couleaud, P., Latorre, A., Aires, A., Crosbie-Staunton, K., et al. (2015). Efficient treatment of breast cancer xenografts with multifunctionalized iron oxide nanoparticles combining magnetic hyperthermia and anti-cancer drug delivery. *Breast Cancer Research*, *17*, 66.

Kroll, A., Pillukat, M. H., Hahn, D., & Schnekenburger, J. (2009). Current in vitro methods in nanoparticle risk assessment: Limitations and challenges. *European Journal of Pharmaceutics and Biopharmaceutics*, *72*, 370–377.

Laurent, S., Forge, D., Port, M., Roch, A., Robic, C., Vander Elst, L., et al. (2008). Magnetic iron oxide nanoparticles: Synthesis, stabilization, vectorization, physicochemical characterizations, and biological applications. *Chemical Reviews*, *108*, 2064–2110.

Lemire, J. A., Harrison, J. J., & Turner, R. J. (2013). Antimicrobial activity of metals: Mechanisms, molecular targets and applications. *Nature Reviews Microbiology*, *11*, 371–384.

Li, L., Jiang, W., Luo, K., Song, H., Lan, F., Wu, Y., et al. (2013). Superparamagnetic iron oxide nanoparticles as MRI contrast agents for non-invasive stem cell labeling and tracking. *Theranostics*, *3*, 595–615.

Li, L. L., Fu, S. Y., Chen, C. F., Wang, X. D., Fu, C. H., Wang, S., et al. (2016). Microenvironment-driven bioelimination of magnetoplasmonic nanoassemblies and their multimodal imaging-guided tumor photothermal therapy. *ACS Nano*, *10*, 7094–7105.

Lin, S., Zhao, Y., Xia, T., Meng, H., Ji, Z., Liu, R., et al. (2011). High content screening in zebrafish speeds up hazard ranking of transition metal oxide nanoparticles. *ACS Nano*, *5*, 7284–7295.

Longmire, M., Choyke, P. L., & Kobayashi, H. (2008). Clearance properties of nano-sized particles and molecules as imaging agents: Considerations and caveats. *Nanomedicine (London, England)*, *3*, 703–717.

Maguire, C. M., Sillence, K., Roesslein, M., Hannell, C., Suarez, G., Sauvain, J.-J., et al. (2017). Benchmark of nanoparticle tracking analysis on measuring nanoparticle sizing and concentration. *Journal of Micro and Nano-Manufacturing*, *5*, 041002–041012.

Mejias, R., Perez-Yague, S., Gutierrez, L., Cabrera, L. I., Spada, R., Acedo, P., et al. (2011). Dimercaptosuccinic acid-coated magnetite nanoparticles for magnetically guided in vivo delivery of interferon gamma for cancer immunotherapy. *Biomaterials*, *32*, 2938–2952.

Movia, D., Gerard, V., Maguire, C. M., Jain, N., Bell, A. P., Nicolosi, V., et al. (2014). A safe-by-design approach to the development of gold nanoboxes as carriers for internalization into cancer cells. *Biomaterials*, *35*, 2543–2557. https://doi.org/10.1016/j.biomaterials.2013.12.057.

Müller, R. H., & Wallis, K. H. (1993). Surface modification of i.v. injectable biodegradable nanoparticles with poloxamer polymers and poloxamine 908. *International Journal of Pharmaceutics*, *89*, 25–31.

Murdock, R. C., Braydich-Stolle, L., Schrand, A. M., Schlager, J. J., & Hussain, S. M. (2008). Characterization of nanomaterial dispersion in solution prior to in vitro exposure using dynamic light scattering technique. *Toxicological Sciences*, *101*, 239–253.

Musazzi, U. M., Marini, V., Casiraghi, A., & Minghetti, P. (2017). Is the European regulatory framework sufficient to assure the safety of citizens using health products containing nanomaterials? *Drug Discovery Today*, *22*(6), 870–882. https://doi.org/10.1016/j.drudis.2017.01.016.

Nel, A. E., Nasser, E., Godwin, H., Avery, D., Bahadori, T., Bergeson, L., et al. (2013). A multi-stakeholder perspective on the use of alternative test strategies for nanomaterial safety assessment. *ACS Nano*, *7*, 6422–6433.

Pankhurst, D.O.A.Q.A. (2013). Magnetic hyperthermia. In P. O'Brien (Ed.), *Nanoscience: Nanostructures through chemistry*. Cambridge: Royal Society of Chemistry.

Périgo, E. A., Hemery, G., Sandre, O., Ortega, D., Garaio, E., Plazaola, F., et al. (2015). Fundamentals and advances in magnetic hyperthermia. *Applied Physics Reviews*, *2*, 041302. https://doi.org/10.1063/1.4935688.

Pino, P. D., Pelaz, B., Zhang, Q., Maffre, P., Nienhaus, G. U., & Parak, W. J. (2014). Protein corona formation around nanoparticles—From the past to the future. *Materials Horizons*, *1*, 301–313.

Prina-Mello, A., Crosbie-Staunton, K., Salas, G., del Puerto Morales, M., & Volkov, Y. (2013). Multiparametric toxicity evaluation of SPIONs by high content screening technique: Identification of biocompatible multifunctional nanoparticles for nanomedicine. *IEEE Transactions on Magnetics*, *49*, 377–382.

Prina-Mello, A., Mohamed, B. M., Verma, N. K., Jain, N., & Volkov, Y. (2014). Advanced methodologies and techniques for assessing nanomaterial toxicity. In N. A. Monteiro-Riviere & C. Lang Tran (Eds.), *Nanotoxicology: Progress toward nanomedicine*. (2nd ed.). Boca Raton, FL, USA: CRC Press Taylor and Francis.

Rejman, J., Oberle, V., Zuhorn, I. S., & Hoekstra, D. (2004). Size-dependent internalization of particles via the pathways of clathrin- and caveolae-mediated endocytosis. *The Biochemical Journal*, *377*, 159–169.

Risks, S.C.O.H.E.A.E. (2017). *The need for non-human primates in biomedical research, production and testing of products and devices (update 2017)*. European Commission.

Roca, A. G., Veintemillas-Verdaguer, S., Port, M., Robic, C., Serna, C. J., & Morales, M. P. (2009). Effect of nanoparticle and aggregate size on the relaxometric properties of MR contrast agents based on high quality magnetite nanoparticles. *The Journal of Physical Chemistry. B*, *113*, 7033–7039.

Sadeghi-Aliabadi, H., Mozaffari, M., Behdadfar, B., Raesizadeh, M., & Zarkesh-Esfahani, H. (2013). Preparation and cytotoxic evaluation of magnetite (Fe_3O_4) nanoparticles on breast cancer cells and its combinatory effects with doxorubicin used in hyperthermia. *Avicenna Journal of Medical Biotechnology*, *5*, 96–103.

Sakulkhu, U., Mahmoudi, M., Maurizi, L., Salaklang, J., & Hofmann, H. (2014). Protein corona composition of superparamagnetic iron oxide nanoparticles with various physico-chemical properties and coatings. *Scientific Reports*, *4*, 5020.

Samuel, S. P., Jain, N., O'Dowd, F., Paul, T., Kashanin, D., Gerard, V. A., et al. (2012). Multifactorial determinants that govern nanoparticle uptake by human endothelial cells under flow. *International Journal of Nanomedicine*, *7*, 2943–2956.

Southern, P., & Pankhurst, Q. A. (2017). Commentary on the clinical and preclinical dosage limits of interstitially administered magnetic fluids for therapeutic hyperthermia based on current practice and efficacy models. *International Journal of Hyperthermia*, 1–16.

Szebeni, J., Fishbane, S., Hedenus, M., Howaldt, S., Locatelli, F., Patni, S., et al. (2015). Hypersensitivity to intravenous iron: Classification, terminology, mechanisms and management. *British Journal of Pharmacology*, *172*, 5025–5036.

Tartaj, P., Morales, M. P., Gonzalez-Carreno, T., Veintemillas-Verdaguer, S., & Serna, C. J. (2011). The iron oxides strike back: From biomedical applications to energy storage devices and photoelectrochemical water splitting. *Advanced Materials*, *23*, 5243–5249.

Veiseh, O., Gunn, J. W., Kievit, F. M., Sun, C., Fang, C., Lee, J. S., et al. (2009). Inhibition of tumor-cell invasion with chlorotoxin-bound superparamagnetic nanoparticles. *Small*, *5*, 256–264.

Villanueva, A., Canete, M., Roca, A. G., Calero, M., Veintemillas-Verdaguer, S., Serna, C. J., et al. (2009). The influence of surface functionalization on the enhanced internalization of magnetic nanoparticles in cancer cells. *Nanotechnology*, *20*, 115103.

Wang, T., Bai, J., Jiang, X., & Nienhaus, G. U. (2012). Cellular uptake of nanoparticles by membrane penetration: A study combining confocal microscopy with FTIR spectroelectrochemistry. *ACS Nano*, *6*, 1251–1259.

Wang, Y. X. (2011). Superparamagnetic iron oxide based MRI contrast agents: Current status of clinical application. *Quantitative Imaging in Medicine and Surgery*, *1*, 35–40.

Wells, D. J. (2011). Animal welfare and the 3Rs in European biomedical research. *Annals of the New York Academy of Sciences*, *1245*, 14–16.

Winterbourn, C. C. (1995). Toxicity of iron and hydrogen peroxide: The Fenton reaction. *Toxicology Letters*, *82–83*, 969–974.

Yan, L., Gu, Z., & Zhao, Y. (2013). Chemical mechanisms of the toxicological properties of nanomaterials: Generation of intracellular reactive oxygen species. *Chemistry, an Asian Journal*, *8*, 2342–2353.

Yue, Z. G., Wei, W., Lv, P. P., Yue, H., Wang, L. Y., Su, Z. G., et al. (2011). Surface charge affects cellular uptake and intracellular trafficking of chitosan-based nanoparticles. *Biomacromolecules*, *12*, 2440–2446.

CHAPTER 18

RNAi Nanomaterials and Nanovehicles: Inorganic Nanoparticles for RNAi

Joseph Hardie*, Shuaidong Huo†, Ying Jiang*, Xing-Jie Liang†, Michelle E. Farkas*, Vincent M. Rotello*

**University of Massachusetts Amherst, Amherst, MA, United States* †*National Center for Nanoscience and Technology, Chinese Academy of Sciences (CAS), Beijing, China*

1 Introduction

RNA interference (RNAi) is an endogenous cellular pathway controlling protein expression, which uses specific short RNA sequences that cooperate with endonucleases to induce the destruction of complementary mRNAs (Fire et al., 1998). This Noble Prize-winning finding has, in recent years, blossomed into an effective approach for selective control of protein expression in vitro and in vivo. The utility of this technology for disease treatment has been demonstrated in previous research (Aagaard & Rossi, 2007; Leung & Whittaker, 2005; Whitehead, Langer, & Anderson, 2009). Because siRNA is unstable and cannot independently cross the cell membrane, a suitable vehicle is required (Yin et al., 2014; Yuan, Naguib, & Wu, 2011). Currently available methods have successfully delivered siRNA to induce RNAi in many different biological settings (Barton & Medzhitov, 2002). However, major challenges remain for its therapeutic use, including controlled release, targeting, uptake, immunogenicity, and biocompatibility (Whitehead et al., 2009).

In recent years, the number of nanomaterials and corresponding nanovehicles in existence has increased dramatically (Aagaard & Rossi, 2007; Leung & Whittaker, 2005; Whitehead et al., 2009). Due to the numerous and difficult criteria required of any siRNA delivery system, a large number of agents have been developed and investigated, among them: cationic polymers, liposomes, virus-like particles, cell-penetrating peptides, inorganic nanoparticles, and others (Finlay et al., 2015; Han et al., 2015; Jing et al., 2016; Li, Liu, Ding, Chen, & Tang, 2015; Shao et al., 2012). Of these entities, nanoparticles are a notable class of carriers for RNA-based therapeutics in vitro and in vivo. Owing to their unique physical and optical properties, as well as their easily modifiable surfaces, inorganic nanoparticles with specific functionalizations can overcome obstacles associated with cellular, local, and systemic

Handbook of Nanomaterials for Cancer Theranostics. https://doi.org/10.1016/B978-0-12-813339-2.00018-9

delivery of siRNA (Zhang, Li, & Yu, 2014). This chapter will discuss recent progress on the design and fabrication of inorganic nanoparticles for siRNA delivery.

2 Inorganic Nanoparticles for RNAi

Inorganic nanoparticles (NPs) have shown great potential as nanocarriers for gene delivery (Sokolova & Epple, 2008). The development of versatile chemical syntheses has enabled a wide range of inorganic materials to be used as vehicles for RNA cargo in in vitro and in vivo applications. (Sokolova & Epple, 2008; Yuan et al., 2011; Zhang et al., 2014). In this section, we summarize recent progress in the surface functionalization of inorganic NPs that confer unique physicochemical properties required for siRNA delivery. Specifically, we will discuss methods of loading and encapsulation of siRNA for RNAi, including covalent conjugation, noncovalent conjugation, and charge-reversal strategies. We will also discuss alternative methods that enable cytosolic delivery and codelivery with other therapeutics.

2.1 Covalent Conjugation

The most straightforward means to associate siRNA with its inorganic delivery vehicle is via covalent conjugation to the nanoparticle surface. Gold nanoparticles (AuNPs), one of the most widely studied inorganic materials for siRNA delivery, are easily conjugated to siRNA via thiol-gold covalent chemistry. One of the first applications of this strategy appeared in 2006 by Nagasaki et al. Here, a poly(ethylene glycol)-*block*-poly(2-(*N*,*N*-dimethylamino) ethyl methacrylate) copolymer (PEG-PAMA) was complexed with AuNPs, followed by covalent immobilization of SH-siRNA (Oishi, Nakaogami, Ishii, & Nagasaki, 2006). These siRNA-functionalized AuNPs were successfully delivered into human liver cancer cells and induced 65% knockdown of the targeted gene. While the siRNA remained associated with the gold in this instance, Li et al. generated a folate-targeted gold nanoshell vehicle covalently attached to thiolated siRNA, which was designed to be liberated upon near-infrared (NIR) irradiation (Lu et al., 2010). The release of siRNA from the vehicle was efficiently controlled by laser irradiation, owing to the gold nanoshell's surface plasmon resonance (SPR) properties. This platform was shown to be efficacious in vivo, reducing HeLa cell growth in nude mice.

In order to enhance siRNA delivery efficiency, covalent siRNA/AuNP conjugates can be coated with cell penetrating peptides or cationic polymers. Reich et al. developed siRNA/hollow gold nanoshell conjugates, whose shell coating consisted of a protective streptavidin layer to which cell penetrating peptides were attached (Fig. 1) (Huang et al., 2015). Using NIR laser irradiation, the siRNA was released from the nanoparticle surface, resulting in controlled RNAi action that induced differentiation of the hESC. Park et al. conjugated thiolated sense and antisense siRNA onto different samples of AuNP and hybridized them to create multimeric networks (Kong et al., 2011). These groupings were coated with

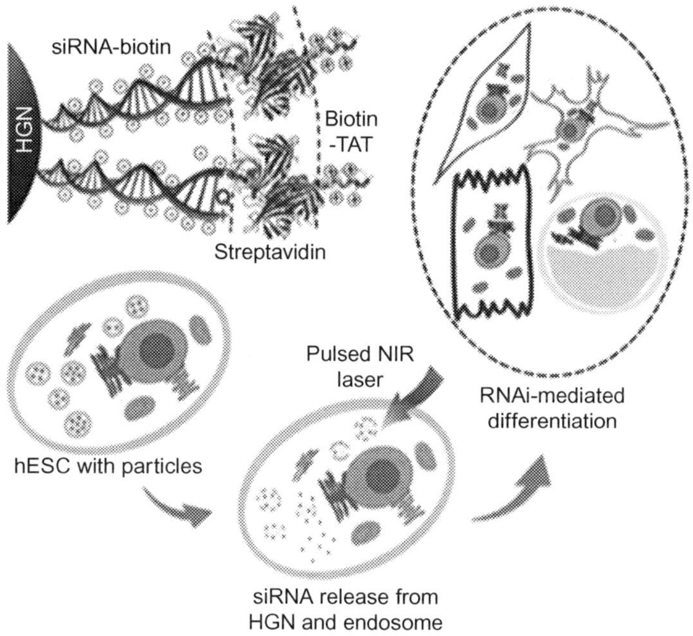

Fig. 1

Schematic of the HGN-siRNA-TAT architecture, NIR laser-activated siRNA delivery and RNAi-mediated differentiation in hESC. *Reprinted with permission Huang, X., Hu, Q., Braun, G. B., Palleoro, A., Morales, D. P., Zasadsinski, J., et al. (2015). Light-activated RNA interference in human embryonic stem cells. Biomaterials, 63, 70–79. Copyright 2015 Elsevier.*

polyethylenamine (PEI) to improve siRNA delivery and gene silencing efficiency. This multimeric covalent AuNP-siRNA vehicle was shown to successfully knock down vascular endothelial growth factor (VEGF) transcription, with no off-target toxicity. This network was also imaged in cells using a microcomputed tomography (CT) instrument, useful for in vivo studies and translational work. One of the most advanced applications of gold nanoparticles and covalently attached siRNA involves the work of the Mirkin group. Mirkin et al. created spherical nucleic acid (SNA) nanoparticles using gold nanoparticle templates (Giljohann, Seferos, Prigodich, Patel, & Mirkin, 2009). They conjugated thiolated antisense-siRNA and oligoethylglycol thiol to 13 nm AuNPs, then allowed the sense strand of the siRNA to hybridize to the bound antisense strand. Using this construct to target luciferase in luciferase-expressing HeLa cells resulted in a 70% knockdown in luciferase activity after 4 days.

Magnetic nanoparticles (MNPs), which exhibit superparamagnetism and respond to magnetic fields, have many unique properties that make them an attractive platform for siRNA delivery. MNPs can be remotely controlled by an external magnetic field to facilitate siRNA release, as well as applied in magnetic resonance imaging (MRI) to track delivery in vivo. In an example, Medarova et al. designed a magnetic tumor-targeted nanodrug (MN-EPPT-siBIRC5) for siRNA delivery and tumor imaging (Kumar, Yigit, Dai, Moore, & Medarova, 2010). Thiolated siRNA targeting the antiapoptosis gene survivin (BIRC5), peptide targeting tumor-specific antigen mucin-1(MUC-1), and the NIR fluorophore, Cyanine 5.5 (Cy5.5), were simultaneously bound to iron oxide MNPs. This construct demonstrated tumor-specific delivery of siRNA, and enabled simultaneous imaging of the process via fluorescent imaging. The activity of siRNA targeting the tumor-specific antiapoptotic gene BIRC5 resulted in the induction of necrosis and apoptosis in the tumors, translating into a significant decrease in tumor growth rate. In addition, MRI could be used to quantitatively monitor nanodrug bioavailability in the tumor tissue throughout the course of treatment.

Quantum dots (QDs) are colloidal semiconductor crystals, typically a few nanometers in size. QDs exhibit unique optical properties, and can be used as photostable probes to track siRNA delivery. Furthermore, QD surfaces can be chemically modified to improve biocompatibility and enhance the transfection efficiency of the siRNA being delivered. Bhatia et al. conjugated thiol-siRNA and a tumor-homing peptide (F3) onto QDs for the targeted silencing of enhanced green fluorescent protein (EGFP) in an EGFP-expressing HeLa cell line. In vivo delivery efficacy of the construct has not yet been determined (Derfus, Chen, Min, Ruoslahti, & Bhatia, 2007). Lee et al. reported the development of multifunctional QDs/siRNA nanoparticles conjugated with cell-penetrating (TAT) and tumor-targeting (RGD) peptides for targeted delivery into human U87 glioblastoma cells (Jung et al., 2010). The silencing of epidermal growth factor receptor variant III (EGFRvIII) using the QDs/siRNA conjugates significantly prohibited tumor cell growth. Despite the achievements in using QDs for siRNA delivery, potential problems for in vivo use arise from their inherent cytotoxicities. A careful assessment of the delivery efficiency and long-term cytotoxicity of QDs should be evaluated before they can be used translationally (Fig. 2).

2.2 Noncovalent Conjugation

Association of unmodified siRNA with an inorganic vehicle through noncovalent binding provides another attractive strategy for delivery. Due to the anionic nature of siRNA, inorganic nanoparticles modified to display positive charges on their surfaces can interact with siRNA via electrostatic binding. Rotello et al. generated a noncovalent siRNA delivery vehicle using AuNPs functionalized with dendrimer units that terminated in positively charged triethylenetetramine (TETA) units (Kim et al., 2012). These dendrimer-AuNPs successfully loaded and delivered β-galactosidase targeted siRNA into murine pancreatic cells with over 50% knockdown efficiency.

The extent of electrostatic binding of siRNA to inorganic nanoparticles can also be mediated by the decoration of surfaces with densely positively charged, polycationic polymers. For example, Zhang et al. reported a layer-by-layer (LBL) synthesis of polyallylamine hydrochloride (PAH)-coated AuNPs for siRNA delivery (Elbakry et al., 2009). After the

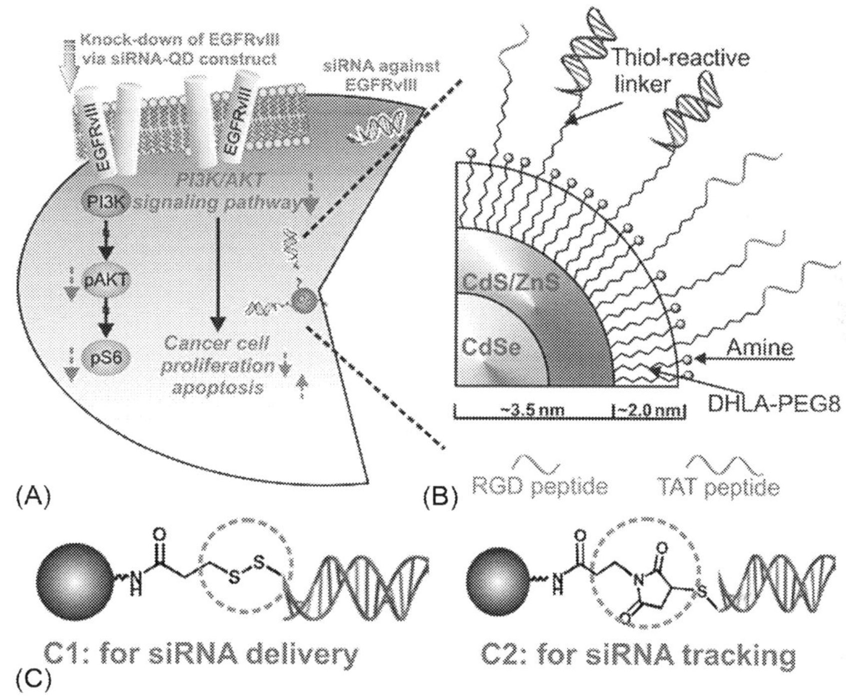

Fig. 2

(A) Quantum dots as a multifunctional nanoplatform to deliver siRNA and to elucidate the effects of EGFRvIII-knockdown within the PI3K signaling pathway in U87-EGFRvIII cells. (B) Schematic of multifunctional siRNA-QDs structure. (C) Two strategies for the siRNA-QD conjugation. L1 utilizes a disulfide moiety, which is easily reduced within cells to release the siRNA. L2 possesses a noncleavable maleimide linker for covalently conjugating siRNA to QDs, which enables the tracking of siRNA-QDs within the cells. *Reprinted with permission from Jung, J., Solanki, A., Memoli, K. A., Kamei, K., Kim, H., Drahl, M. A., et al. (2010). Selective inhibition of human brain tumor cells through multifunctional quantum-dot-based siRNA delivery. Angewandte Chemie International Edition in English, 49(1), 103–107. Copyright 2009 John Wiley and Sons.*

synthesis of the gold core, alternating layers of PAH and siRNA were deposited onto the nanoparticle surfaces using a centrifugal synthesis method. The surface potential of the nanoparticles was monitored and used as the benchmark to determine the optimal amount of deposition for maximum loading, which was determined to be approximately 1170 siRNA molecules per nanoparticle. The PAH-AuNP showed little cytotoxicity and was used to knock down EGFR with 40% efficiency.

Mesoporous silica nanoparticles possess increased surface areas as a result of the materials' porous nature, and can also make effective electrostatic carriers for siRNA. Shen et al. produced mesoporous silica nanoparticles pore-functionalized with PEI that have been grafted to cyclodextrin to reduce charge-induced toxicity associated with unmodified PEI (Shen et al., 2014). Using fluorescently labeled siRNA with in vivo microscopy, the ability of

CD-PEI-MSNPs to deliver siRNA in nude mice with MDA-MB-231 breast cancer xenografts was demonstrated. Six hours after treatment, diffuse fluorescence through the targeted cells was evident, which was most likely a result of endosomal escape of siRNA by the "proton sponge effect" (Shen et al., 2014). The "proton sponge effect" results from the buffering capability of the polyvalent cationic surface functionality, leading to a salt influx into the endosomes trying to maintain the correct pH, eventually rupturing the endosomal membrane and releasing the siRNA into the cytosol (Shen et al., 2014). Additionally, Li et al. developed a PEI-functionalized iron oxide-mesoporous silica yolk-shell nanocapsule (NC) that electrostatically bound siRNA on the surface via interaction with PEI (Fig. 3) (Zhang et al., 2012). The "yolk-shell" aspect of this vehicle uses multiple materials, and exploits the properties of each. In this case, the fluorescence of the mesoporous silica shell and the paramagnetism of the iron oxide yolk enable simultaneous fluorescent imaging and magnetic guidance. The PEI-Fe_3O_4@fmSiO_2 yolk-shell NCs, with external magnetic guidance, effectively knocked down β-actin in HeLa cells without displaying the toxicity commonly associated with PEI.

Carbon-based nanomaterials, including carbon nanotubes (CNT) and graphene, also possess large surface areas that can be functionalized in various ways for siRNA delivery.

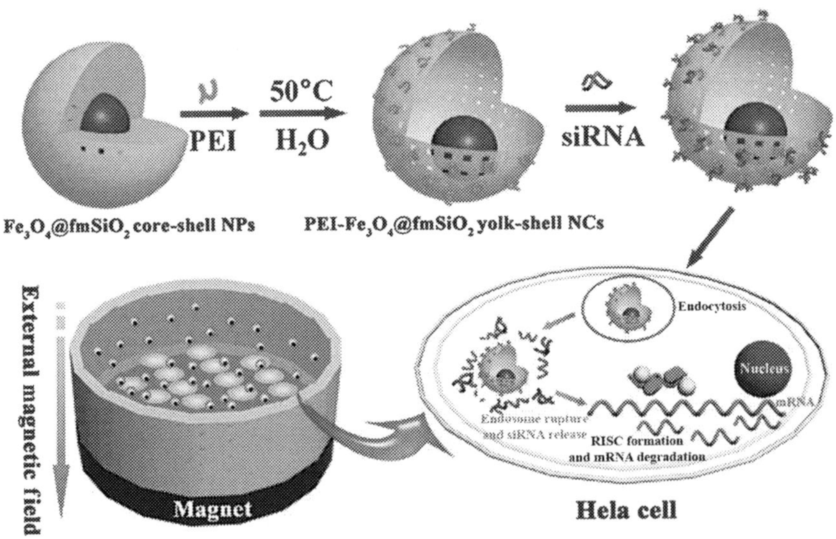

Fig. 3

Schematic illustration of the synthetic procedure for the PEI-Fe_3O_4@fmSiO_2 yolk-shell NCs and their application in simultaneous fluorescent imaging and magnetically guided siRNA delivery into Hela cells. *Reprinted with permission from Zhang, Y., Wang, T., Li, L., Wang, C., Su, Z., Li, J. (2012). Multifunctional fluorescent-magnetic polyethyleneimine functionalized Fe_3O_4-mesoporous silica yolk-shell nanocapsules for siRNA delivery. Chemical Communications, 48, 8706–8708. Copyright 2012 Royal Society of Chemistry.*

For example, ammonium functionalized CNT delivered siRNA against telomerase reverse transcriptase (TERT) and knocked down TERT expression in a Lewis lung carcinoma mouse tumor model, showing feasibility for use as a cancer therapeutic (Zhang et al., 2006). Kostarelos et al. used similarly positively charged multiwalled carbon nanotubes (MWNTs) to deliver a proprietary toxic siRNA (siTOX) to tumor xenografts, which arrested tumor growth (Podesta et al., 2009). The MWNTs also delivered siRNA to the central nervous system (CNS) to knock down the expression of Caspase 3 in an attempt to reduce neurodegeneration as a therapeutic against strokes.

Graphene has a two-dimensional carbon nanosheet structure; it also has high water solubility and good colloidal stability, which are ideal for siRNA delivery (Zhang et al., 2011). Zhang et al. grafted graphene oxide (GO) independently with conjugated PEI and noncovalent 1-pyrenemethylamine adsorption, both of which resulted in efficient siRNA loading and intracellular delivery. The sequential delivery of siRNA against Bcl-2 and followed by doxorubicin treatment using PEI grafted GO nanoparticles led to significantly enhanced antitumor efficacy.

By functionalizing the surfaces of inorganic nanoparticles with polymers, the agents' stabilities, circulation times, and levels of siRNAs complexed can be fine-tuned. Pierre et al. generated antifouling magnetic iron oxide particles functionalized with an inner shell of cationic poly(dimethylaminoethyl acrylate) (P(DMAEA)) polymer and covered with an external poly(ethylene glycol) (PEG) shell (Boyer et al., 2010). With the external PEG coating, these particles showed enhanced stability in the presence of serum proteins and low levels of nonspecific association with proteins. Delivery of eGFP targeted siRNA using these particles resulted in an approximately 80% decrease in fluorescence in eGRP-expressing human neuroblastoma SHEP cells. In an unusual twist on traditional metal oxide nanoparticles, Park et al. produced hollow manganese oxide nanoparticles (HMON) that were functionalized with PEI- 3,4-dihydroxy-l-phenylalanine (DOPA) and the anti-HER2 drug Herceptin (Fig. 4) (Bae, Lee, Kim, & Park, 2011). The manganese oxide core of this vehicle facilitated higher levels of MRI contrast, leading to better imaging of the agent in vivo, while Herceptin conjugation enabled targeting of HER2-overexpressing breast cancer cells. Using this vehicle to deliver VEGF-targeted siRNA into various breast cancer cell lines resulted in decreased cell viability (50% on average).

Aside from via electrostatic attraction with the inorganic nanoparticle surface, siRNAs can be noncovalently associated through their encapsulation on the particle interior. Calcium phosphate (CaP), a natural mineral with excellent biocompatibility, is commonly used to encapsulate siRNA due to ease of hollow CaP nanoparticle synthesis (Li, Chen, Tseng, Mozumdar, & Huang, 2010). The chemical nature of CaP NPs also provides a convenient siRNA release mechanism due to their dissolution under acidic conditions, including those present in the intracellular environment (Li et al., 2010). Taking advantage of this

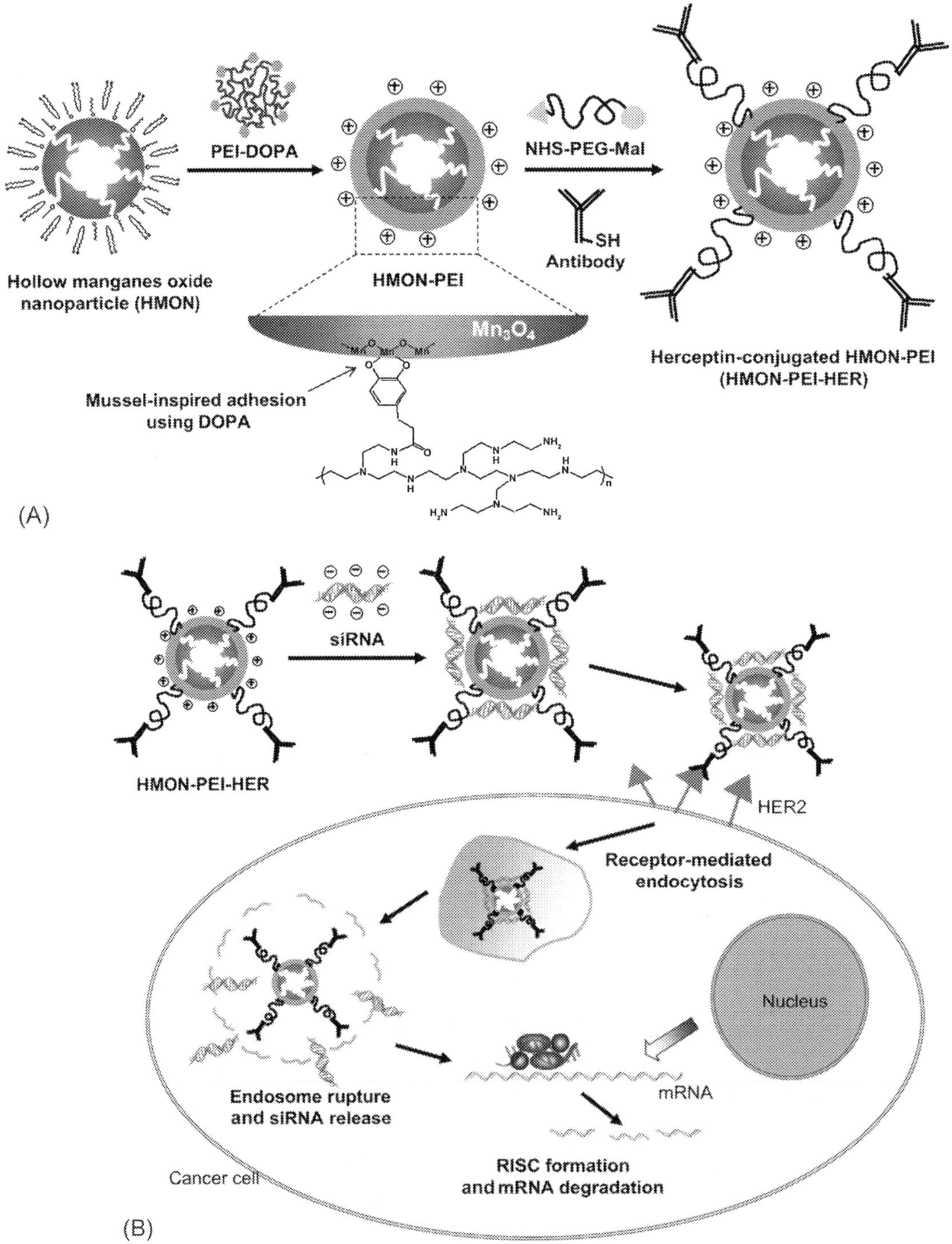

See the legend on opposite page.

inherent biodegradability, Huang et al. presented an siRNA-encapsulating 1,2-dioleoyl-3-trimethylammonium-propane chloride salt (DOTAP)-liposome CaP NP, designed to take advantage of the acid-triggered degradation of CaP, and the prolonged circulation time of cholesterol-based liposomes (Li et al., 2010). The CaP NP core was synthesized and loaded with siRNA through a water-in-oil emulsion method, then purified, resuspended in water, and decorated with DOTAP-based liposomes. Delivery of these liposome-functionalized CaP NPs loaded with antiluciferase siRNA yielded a 50% decrease in luminescence in nude mice bearing subcutaneous luciferase-expressing lung cancer tumors. In similar work by Kwong et al., triple-shelled CaP NPs surface-coated with PEG were generated to study the mechanism of nanoparticle entry and uptake (Tobin et al., 2013). The triple-shelled feature of these particles enabled various and increased loadings of the NPs through differential packing of each "shell." In vivo delivery of the X-linked inhibitor of apoptosis protein (XIAP)-targeted siRNA-loaded triple shelled CaP NPs into nude mice bearing subcutaneous lung tumors revealed tumor localization and accumulation of the NPs through quantitative fluorescent microscopy. CaP-based vehicles have been shown to be effective in encapsulating siRNA; however, the inability to control the size dispersity of the nanoparticles make CaP NPs difficult to use in therapeutic applications.

Charge-reversal polyelectrolytes have also been recently employed as a method for controlling the degradation of NPs in response to the intracellular environment, improving siRNA delivery (Mok, Park, & Park, 2008). In an acidic environment, the net charge of the polymers changes from positive to negative, resulting in intracellular siRNA release (Mok et al., 2008). To use this strategy, Park et al. developed QD NPs conjugated with citraconylated PEG-poly-L-lysine (PLL) copolymers via biotinylation-avidin labeling (Mok et al., 2008). The vehicle was designed so that upon entering the acidic intracellular environment, the citraconylated amides would be cleaved, yielding an anionic particle surface, thus facilitating its degradation and cargo release. These modified QD NPs were delivered into HeLa cells and demonstrated enhanced uptake as determined by fluorescent confocal microscopy.

Fig. 4

(A) Synthetic scheme for the production of PEI-DOPA-coated hollow manganese oxide nanoparticles (HMON-PEI) followed by Herceptin-conjugation to generate HMON-PEI (HMON-PEI-HER), a multifunctional nanocarrier for cancer-targeted siRNA delivery and magnetic resonance imaging. (B) Schematic diagram illustrating the formation of polyelectrolyte complexes of HMON-PEI-HER with siRNA, receptor-mediated cellular uptake, endosomal escape, loading of siRNA to RNA-induced silencing complexes (RISC), and target mRNA degradation in cancer cells. *Reprinted with permission from Bae, K. H., Lee, K., Kim, C., Park, T. G. (2011). Surface functionalized hollow manganese oxide nanoparticles for cancer targeted siRNA delivery and magnetic resonance imaging. Biomaterials, 32(1), 176–184. Copyright 2010 Elsevier.*

2.3 Direct Cytosolic Delivery Strategy

One of the continuing issues in developing siRNA delivery vehicles is the fate of the carrier and cargo within the cell. The majority of vehicles enter cells through an endocytosis pathway, resulting in endosomal entrapment (Dominska & Dykxhoorn, 2010). In order for RNAi to occur, the siRNA must leave the endosome and enter the cytosol. The method traditionally used to overcome this problem is chemical modification of the vehicle to facilitate endosomal rupture, which also releases the siRNA (Dominska & Dykxhoorn, 2010). Decoration of inorganic nanoparticles with highly positively charged polyamine functionalities to utilize the "proton sponge effect" has been used considerably over the years, unfortunately the release efficacy for these methods are generally low and greatly reduce the resulting knockdown efficiency of the delivered siRNA (Dominska & Dykxhoorn, 2010; Shen et al., 2014). However, there are new inorganic vehicles that employ nonendocytotic pathways for cellular uptake, which can avoid the problem of endosomal entrapment. Rotello et al. recently developed a gold nanoparticle-based capsule platform that undergoes direct cytosolic delivery and avoids endosomal entrapment (Jiang et al., 2015). This vehicle consists of a linoleic acid emulsion coated on the exterior with gold nanoparticles, suspended in aqueous solution (Fig. 5). The gold NP surfaces are decorated with organic ligands that terminate in arginine, which are able to bind electrostatically to siRNA. The vehicle's fatty acid core causes the whole complex to associate with hydrophobic portions of the cellular lipid bilayer, enabling the siRNA to directly enter the cytosol through a process known as membrane fusion. This effect was prevented by cell treatment with nystatin, an inhibitor of the lipid raft-mediated membrane fusion process, further confirming our hypothesis for uptake mechanism (Jiang et al., 2015). This capsule vehicle also demonstrated highly effective siRNA delivery of PLK-targeted siRNA, resulting in a 70% decrease in cell viability.

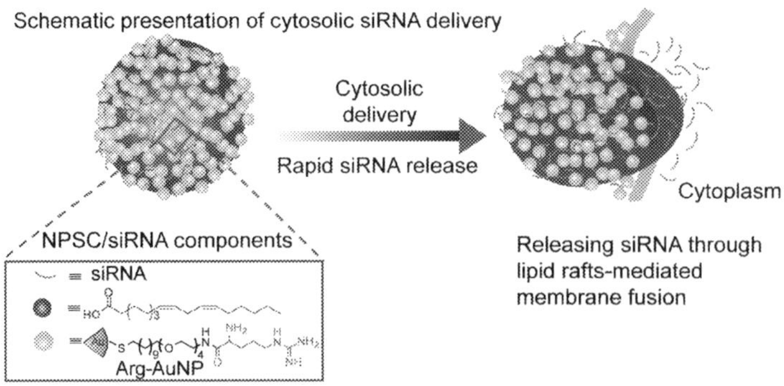

Fig. 5
NPSC/siRNA components and schematic presentation of NPSC-mediated cytosolic siRNA delivery. *Reprinted with permission from Jiang, Y., Tang, R., Duncan, B., Jiang, Z., Yan, B., Mout, R., et al. (2015). Direct cytosolic delivery of siRNA using nanoparticle-stabilized nanocapsules. Angewandte Chemie, International Edition 54 (2), 2015, 506–510.*

2.4 Codelivery

Many inorganic siRNA vehicles allow the simultaneous accommodation of small molecule cargo, thus facilitating codelivery of anticancer drugs and siRNA combination therapies. By loading different layers sequentially, Kwong et al. used triple shelled CAP NPs to codeliver doxorubicin with XIAP-targeted siRNA into subcutaneous lung tumors in vivo, resulting in enhanced reduction of tumor size (Fig. 6) (Tobin et al., 2013). Similarly, Zhao et al. fabricated a pH-responsive hollow mesoporous silica nanoparticle (HMSNP)-based drug/siRNA codelivery system, taking advantage of the controlled NP pore size to load both cargoes. They demonstrated that the codelivery of doxorubicin and siRNA targeted to the antiapoptotic protein Bcl-2 successfully suppressed Bcl-2 expression and led to enhanced cytotoxicity and apoptosis than either individual treatment alone (Ma, Zhao, Ng, & Zhao, 2013). Mao et al. recently synthesized QD NPs surface-functionalized with β-cyclodextrin (β-CD) polymers that electrostatically associate with siRNA and hydrophobically encapsulated doxorubicin (Li et al., 2012). Delivering these QD NPs, loaded with both doxorubicin and multidrug

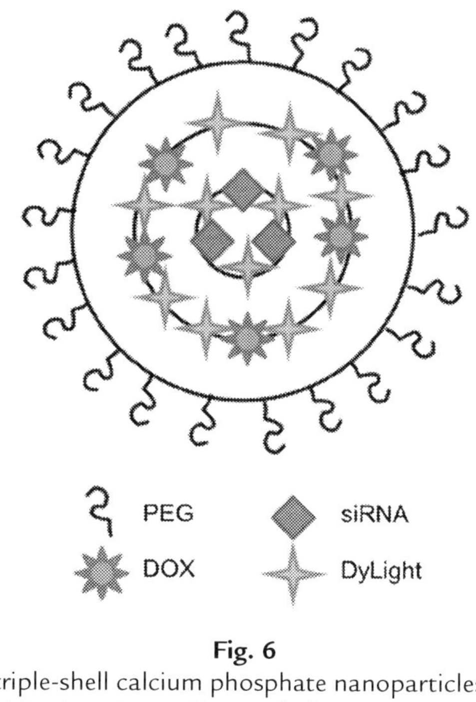

Fig. 6

Composition of pegylated triple-shell calcium phosphate nanoparticles with DOX in the secondary (middle) shell and siRNA in the primary (inner) shell. DyLight 550 (Dy550) or 755 (Dy755) fluorescent dyes were incorporated in the primary and secondary shells. *Reprinted with permission from Tobin, L. A., Xie, Y., Tsokos, M., Chung, S. I., Merz, A. A., Arnold, M. A. et al. (2013). Pegylated siRNA-loaded calcium phosphate nanoparticle-driven amplification of cancer cell internalization in vivo. Biomaterials, 34(12), 2980–2990. Copyright 2013 Elsevier.*

resistance gene-1 (MDR-1) targeted siRNA, into doxorubicin-resistant HeLa cells resulted in increased cell death due to RNAi-induced knockdown of the cellular broad-spectrum drug resistance mechanisms.

3 Conclusion

The inorganic nanoparticle-based siRNA delivery approach has evolved over the past decade into an array of multifaceted strategies to achieve efficient RNAi. Advancements in the understanding of inorganic nanoparticle synthesis have led to increased degrees of control in generating new vehicles. NPs can now be synthesized in a range of sizes, from single digit to hundreds of nanometers, enabling the optimization of NPs for delivery. Additionally, there are now a plethora of chemical approaches for NP surface functionalization that have led to the incorporation and utilization of new types of physicochemical properties. These include the modification of NPs with PEG for tuning in vivo circulation times and conjugation of ligands for targeted delivery to diseased cells or tissues. The unique physical and optical properties of inorganic NPs can be combined to design multifunctional inorganic NPs that display a diverse set of functionalities to facilitate efficient and effective siRNA delivery.

Though the efficiency of inorganic nanoparticles for siRNA delivery has been well proven, there are still barriers to clinical translation. Determination of toxicology, biodistribution, and pharmacokinetic profiles of these NPs still requires much effort and is critical for ensuring reproducible effects and for finding the best siRNA carrier. There are additional considerations of the cost and time associated with inorganic nanoparticle synthesis, including its ability to be scaled up, which must also be considered before inorganic nanomaterials can be effectively used in a therapeutic setting.

Acknowledgments

This work was supported by the grants from the NIH (EB014277), National Distinguished Young Scholars grant from NSFC (31225009) and financial support from China Scholarship Council.

Conflict of Interest

The authors declare no competing financial interests.

References

Aagaard, L., & Rossi, J. J. (2007). RNAi therapeutics: Principles, prospects and challenges. *Advanced Drug Delivery Reviews*, *59*(2–3), 75–86.

Bae, K. H., Lee, K., Kim, C., & Park, T. G. (2011). Surface functionalized hollow manganese oxide nanoparticles for cancer targeted siRNA delivery and magnetic resonance imaging. *Biomaterials*, *32*(1), 176–184.

Barton, G. M., & Medzhitov, R. (2002). Retroviral delivery of small interfering RNA into primary cells. *Proceedings of the National Academy of Sciences of the United States of America, 99*(23), 14943–14945.

Boyer, C., Priyanto, P., Davis, T. P., Pissuwan, D., Bulmus, V., Kallavaris, M., et al. (2010). Anti-fouling magnetic nanoparticles for siRNA delivery. *Journal of Materials Chemistry, 20*, 255–265.

Derfus, A. M., Chen, A. A., Min, D., Ruoslahti, E., & Bhatia, S. N. (2007). Targeted quantum dot conjugates for siRNA delivery. *Bioconjugate Chemistry, 18*(5), 1391–1396.

Dominska, M., & Dykxhoorn, D. M. (2010). Breaking down the barriers: siRNA delivery and endosome escape. *Journal of Cell Science, 123*(8), 1183–1189.

Elbakry, A., Zaky, A., Liebl, R., Rachel, R., Goepferich, A., & Breunig, M. (2009). Layer-by-layer assembled gold nanoparticles for siRNA delivery. *Nano Letters, 9*(5), 2059–2064.

Finlay, J., Roberts, C. M., Dong, J., Zink, J. I., Tamanoi, F., & Glackin, C. A. (2015). Mesoporous silica nanoparticle delivery of chemically modified siRNA against TWIST1 leads to reduced tumor burden. *Nanomedicine: Nanotechnology, Biology, and Medicine, 11*(17), 1657–1666.

Fire, A., Xu, S., Montgomery, M. K., Kostas, S. A., Driver, S. E., & Mello, C. C. (1998). Potent and specific genetic interference by double-stranded RNA in *Caenorhabditis elegans*. *Nature, 391*(6669), 806–811.

Giljohann, D. A., Seferos, D. S., Prigodich, A. E., Patel, P. C., & Mirkin, C. A. (2009). Gene regulation with polyvalent siRNA-nanoparticle conjugates. *Journal of the American Chemical Society, 131*(6), 2072–2073.

Han, J., Cai, J., Borjihan, W., Ganbold, T., Rana, T. M., & Baigude, H. (2015). Preparation of novel curdlan nanoparticles for intracellular siRNA delivery. *Carbohydrate Polymers, 117*, 324–330.

Huang, X., Hu, Q., Braun, G. B., Palleoro, A., Morales, D. P., Zasadsinski, J., et al. (2015). Light-activated RNA interference in human embryonic stem cells. *Biomaterials, 63*, 70–79.

Jiang, Y., Tang, R., Duncan, B., Jiang, Z., Yan, B., Mout, R., et al. (2015). Direct cytosolic delivery of siRNA using nanoparticle-stabilized nanocapsules. *Angewandte Chemie, International Edition, 54*(2), 506–510.

Jing, X., Foged, C., Martin-Bertelsen, B., Yaghmur, A., Knapp, K. M., Malmsten, M., et al. (2016). Delivery of siRNA complexed with palmitoylated α-peptide/β-peptoid cell-penetrating peptidomimetics: Membrane interaction and structural characterization of a lipid-based nanocarrier system. *Molecular Pharmaceutics, 13*(6), 1739–1749.

Jung, J., Solanki, A., Memoli, K. A., Kamei, K., Kim, H., Drahl, M. A., et al. (2010). Selective inhibition of human brain tumor cells through multifunctional quantum-dot-based siRNA delivery. *Angewandte Chemie International Edition in English, 49*(1), 103–107.

Kim, S. T., Chompoosor, A., Yeh, Y. C., Agasti, S. S., Solfiell, D. J., & Rotello, V. M. (2012). Dendronized gold nanoparticles for siRNA delivery. *Small, 8*(21), 3253–3256.

Kong, W. H., Bae, K. H., Hong, C. A., Lee, Y., Hahn, S. K., & Park, T. G. (2011). Multimerized siRNA cross-linked by gold nanoparticles. *Bioconjugate Chemistry, 22*(10), 1962–1969.

Kumar, M., Yigit, M., Dai, G., Moore, A., & Medarova, Z. (2010). Image-guided breast tumor therapy using a small interfering RNA nanodrug. *Cancer Research, 70*, 7553–7561.

Leung, R. K., & Whittaker, P. A. (2005). RNA interference: From gene silencing to gene-specific therapeutics. *Pharmacology & Therapeutics, 107*(2), 222–239.

Li, J., Chen, Y., Tseng, Y., Mozumdar, S., & Huang, L. (2010). Biodegradable calcium phosphate nanoparticle with lipid coating for systemic siRNA delivery. *Journal of Controlled Release, 142*, 416–421.

Li, Y., Liu, F., Ding, F., Chen, P., & Tang, M. (2015). Inhibition of liver fibrosis using vitamin A-coupled liposomes to deliver matrix metalloproteinase-2 siRNA in vitro. *Molecular Medicine Reports, 12*(3), 3453–3461.

Li, J., Wang, Y., Zhao, M., Tan, C., Li, Y., Le, X., et al. (2012). Multifunctional QD-based co-delivery of siRNA and doxorubicin to HeLa cells for reversal of multidrug resistance and real-time tracking. *Biomaterials, 33*(9), 2780–2790.

Lu, W., Zhang, G., Zhang, R., Flores, L. G., II, Huang, Q., Gelovani, J. G., et al. (2010). Tumor site-specific silencing of *NF*-κB p65 by targeted hollow gold nanosphere-mediated photothermal transfection. *Cancer Research, 70*, 3177–3188.

Ma, X., Zhao, Y., Ng, K. G., & Zhao, Y. (2013). Integrated hollow mesoporous silica nanoparticles for target drug/siRNA co-delivery. *Chemistry A European Journal, 19*, 15593–15603.

Mok, H., Park, J. W., & Park, T. G. (2008). Enhanced intracellular delivery of quantum dot and adenovirus nanoparticles triggered by acidic pH via surface charge reversal. *Bioconjugate Chemistry*, *19*(4), 797–801.

Oishi, M., Nakaogami, J., Ishii, T., & Nagasaki, Y. (2006). Smart PEGylated gold nanoparticles for the cytoplasmic delivery of siRNA to induce enhanced gene silencing. *Chemistry Letters*, *35*(9), 1046–1047.

Podesta, J. E., Al-Jamal, K. T., Herrero, M. A., Tian, B., Ali-Boucetta, H., Hedge, V., et al. (2009). Antitumor activity and prolonged survival by carbon-nanotube-mediated therapeutic siRNA silencing in a human lung xenograft model. *Small*, *5*(10), 1176–1185.

Shao, W., Paul, A., Abbasi, S., Chahal, P. S., Mena, J. A., Montes, J., et al. (2012). A novel polyethyleneimine-coated adeno-associated virus-like particle formulation for efficient siRNA delivery in breast cancer therapy: Preparation and in vitro analysis. *International Journal of Nanomedicine*, *7*, 1575–1586.

Shen, J., Kim, H., Su, H., Wang, F., Wolfram, J., Kirui, D., et al. (2014). Cyclodextrin and polyethylenimine functionalized mesoporous silica nanoparticles for delivery of siRNA. *Cancer Therapeutics Theranostics*, *4*(5), 487–497.

Sokolova, V., & Epple, M. (2008). Inorganic nanoparticles as carriers of nucleic acids into cells. *Angewandte Chemie, International Edition*, *47*(8), 1382–1395.

Tobin, L. A., Xie, Y., Tsokos, M., Chung, S. I., Merz, A. A., Arnold, M. A., et al. (2013). Pegylated siRNA-loaded calcium phosphate nanoparticle-driven amplification of cancer cell internalization *in vivo*. *Biomaterials*, *34*(12), 2980–2990.

Whitehead, K. A., Langer, R., & Anderson, D. G. (2009). Knocking down barriers: Advances in siRNA delivery. *Nature Reviews Drug Discovery*, *8*(2), 129–138.

Yin, H., Kanasty, R. L., Eltoukhy, A. A., Vegas, A. J., Dorkin, J. R., & Anderson, D. G. (2014). Non-viral vectors for gene-based therapy. *Nature Reviews Genetics*, *15*(8), 541–555.

Yuan, X., Naguib, S., & Wu, Z. (2011). Recent advances of siRNA delivery by nanoparticles. *Expert Opinion on Drug Delivery*, *8*(4), 521–536.

Zhang, L., Li, Y., & Yu, J. C. (2014). Chemical modification of inorganic nanostructures for targeted and controlled drug delivery in cancer treatment. *Journal of Materials Chemistry B*, *2*(5), 452–470.

Zhang, L., Lu, Z., Zhao, Q., Huang, J., Shen, H., & Zhang, Z. (2011). Enhanced chemotherapy efficacy by sequential delivery of siRNA and anticancer drugs using PEI-grafted graphene oxide. *Small*, *7*(4), 460–464.

Zhang, Y., Wang, T., Li, L., Wang, C., Su, Z., & Li, J. (2012). Multifunctional fluorescent-magnetic polyethyleneimine functionalized Fe_3O_4-mesoporous silica yolk-shell nanocapsules for siRNA delivery. *Chemical Communications*, *48*, 8706–8708.

Zhang, Z., Yang, X., Zhang, Y., Zeng, B., Wang, S., Zhu, T., et al. (2006). Delivery of telomerase reverse transcriptase small interfering RNA in complex with positively charged single-walled carbon nanotubes suppresses tumor growth. *Clinical Cancer Research*, *12*, 4933–4939.

CHAPTER 19

Theranostic Magnetic Nanoparticles as Molecular Imaging Agents for siRNA Delivery

Zhe Wang*, **Junqing Wang***,†, **Gang Liu**†
*The Chinese University of Hong Kong, Hong Kong SAR, China †Xiamen University, Xiamen, China

1 Introduction and Background

Since the discovery of RNA-mediated inhibition of protein expression in plants and fungi in the early 1900s (Napoli, Lemieux, & Jorgensen, 1990; Romano & Macino, 1992), and later the important study of RNA interference (RNAi) in *Caenorhabditis elegans* in 1998 (Fire et al., 1998), small interfering RNAs (siRNAs) have been exploited as a powerful research tool, and the promising therapeutic agents for post-transcriptional protein-coding gene knockdown in diseased cells (Ryther, Flynt, Phillips, & Patton, 2005). Their intriguing features include: (1) the unique specificity (sequence-specific binding of base pairs between siRNA and target messenger RNAs); (2) high efficiency with a catalytic nature (RNAi affects amplification via an RNA-dependent RNA polymerase mediated secondary siRNA replication process) (Lipardi, Wei, & Paterson, 2001; Sijen et al., 2001); and (3) wide applicability (in principle, siRNAs can be designed and synthesized to knock down any target gene with a complementary sequence). These features are highly beneficial for the personalized novel treatment of various diseases, especially in use against previously undruggable targets, for example, the messenger RNAs of oncogene proteins (Wu, Lopez-Berestein, Calin, & Sood, 2014).

To date, siRNA-based therapeutics have moved into clinical trials since the first-in-human study of bevasiranib (Cand5) for the treatment of wet age-related macular degeneration in 2004 (OPKO Health, Inc., n.d.; Ozcan, Ozpolat, Coleman, Sood, & Lopez-Berestein, 2015). Despite some encouraging success of these clinical trials, a number of challenges are still faced by RNAi-based therapeutics in systemic use. For example, delivery efficiency of siRNAs or other small therapeutic RNAs to target tissue is impeded by many factors, including nuclease degradation in plasma, and rapid systemic elimination of naked siRNA (Alexis, Pridgen, Molnar, & Farokhzad, 2008); entrapment of nano-sized carriers in the reticuloendothelial system and phagocytized by phagocytic cells (Wang, Lu, Wientjes, & Au, 2010); the vascular

Handbook of Nanomaterials for Cancer Theranostics. https://doi.org/10.1016/B978-0-12-813339-2.00019-0

endothelial barrier in many tissues, and the resistance of extracellular matrix and intracellular release via endosome escape (Haussecker, 2014). Furthermore, the siRNA-based therapeutics have raised safety concerns related to adverse side effects, such as liver toxicity, which can be induced by the saturation effect that occurs when introduced foreign siRNAs compete with endogenous RNA through the miRNA regulatory pathway (Grimm et al., 2006); the induction of innate immune responses caused by irrational design of the siRNA sequence (Hornung et al., 2005; Reynolds et al., 2006); and the unpredictable phenotypic toxicity of off-target effects induced by siRNA (Fedorov et al., 2006). Therefore, it is necessary to overcome these major obstacles of unwanted biodistribution, insufficient therapeutic efficiency, and undesired toxicity for the clinical translation of siRNA-based therapeutics.

Molecular imaging with theranostic nanoplatforms can serve as powerful tools for siRNA-based therapeutic product development and optimization of disease management (Fig. 1), through the in vivo monitoring of the siRNA delivery process and therapeutic efficacy (Liu, Swierczewska, Lee, & Chen, 2010; Moore & Medarova, 2009). Generally, a variety of theranostic nanosystems have been used for multiple molecular imaging modalities guided in vivo siRNA delivery and silencing, including positron emission tomography (PET) and single-photon emission computed tomography (SPECT) imaging (radioisotope integrated organic/inorganic nanoparticles); optical fluorescence imaging (semiconductor/carbon/silica/ Ag_2S quantum dots, single-walled carbon nanotubes, colloidal gold nanoparticles) (Wang et al., 2015); magnetic resonance imaging (MRI contrast agents, mainly iron oxide-based

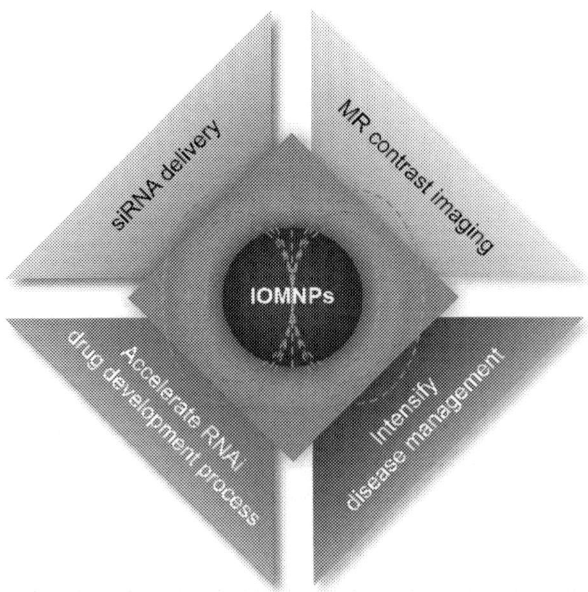

Fig. 1
Key utilizations of IOMNPs in RNAi theranostic application.

magnetic nanoparticles (IOMNPs)); ultrasound (US, microbubbles) and computed tomography (CT contrast agents, e.g., gold nanoparticles) (James & Gambhir, 2012; Liu et al., 2010). In fact, most imaging modalities have limited use due to their significant shortcomings: PET and SPECT lack anatomic information that needs to be compensated by the combination of CT or MRI (Pichler, Judenhofer, & Pfannenberg, 2008). Besides, using radionuclide/X-ray imaging techniques poses safety risks when exposed to ionizing radiation, which limits the number of scans for the same patient in a period of time (Pichler et al., 2008). Optical fluorescence imaging is limited by its superficial penetration (<1 cm) because of light scattering and autofluorescence in tissues, which needs to be improved by using near-infrared (NIR) fluorescence imaging (Lee, Park, Hong, Choi, & Choi, 2012; Wang et al., 2015). US is also limited by its depth of penetration. By contrast, MRI shows a number of superiorities compared with other imaging modalities including: (1) the whole body imaging without depth limitation; (2) high spatial resolution: ~1 mm; (3) no ionizing radiation exposure; (4) excellent soft tissue contrast; and (5) high clinical utility (James & Gambhir, 2012). However, the poor signal sensitivity of MRI can result in prolonged acquisition times (James & Gambhir, 2012). To overcome this major limitation and simultaneously achieve concurrent siRNA-based therapy for theranostic purposes, multifunctional magnetic nanoparticles have been developed for high sensitivity targeted MRI-guided siRNA delivery (Xing, Liu, Zhu, Hou, & Chen, 2014). In particular, IOMNPs have generated a sustained interest because of their versatility (not only their unique magnetic properties, but also a well-developed surface functionalization for active targeting, loading, and controlled release of siRNA) and a reasonable safety profile (biodegradability and low toxicity) (Jain, Reddy, Morales, Leslie-Pelecky, & Labhasetwar, 2008; Medarova, Pham, Farrar, Petkova, & Moore, 2007; Xie, Liu, Eden, Ai, & Chen, 2011; Yoo, Lee, Shin, & Cheon, 2011). Presently, a variety of IOMNP formulations have been clinically approved by the U.S. Food and Drug Administration (FDA) and the European Medicines Agency (EMA) for MR imaging and therapeutic applications (Table 1; Sun, Lee, & Zhang, 2008; Wang, 2011; Singh & Sahoo, 2014).

Table 1: Overview of the various clinically approved IOMNP formulations (Singh & Sahoo, 2014; Wang, 2011)

Generic Name/Brand Name	Approved Year	Surface Coating	Target Site	Contrast Effect
Ferumoxide/Feridex®, Endorem™	1996	Dextran	Liver lesions	Predominantly $T2$ enhancement
Ferucarbotran/Resovist®	2001	Carboxydextran	Liver lesions	Predominantly $T2/T1$ enhancement
Ferumoxsil/GastroMARK®, Lumirem®	1996	Silicone	Gastrointestinal tract	$T2^*$ enhancement
Ferristene/Abdoscan®	2000	Sulfonated styrene-divinylbenzene copolymer	Gastrointestinal tract	Predominantly $T2^*$ enhancement

In this account, IOMNPs can be rendered as an ideal theranostic nanoplatform for real time in vivo monitoring of siRNA delivery, following the treatment responses to evaluate and address the translational challenges, thereby accelerating the progress of RNAi therapeutics development. This chapter focuses on recent conspectus in the synthesis and surface functionalization of IOMNPs for targeted imaging and magnetic-guided siRNA delivery. This chapter also discusses the pharmacokinetics, targeting approaches, cellular internalization, intracellular trafficking, and toxicity of IOMNPs-mediated siRNA delivery. In addition, some prominent reports of applications for siRNA imaging based on the IOMNPs delivery platform are outlined.

2 Magnetism of IOMNPs

The application of magnetic nanoparticles in a variety of medical practices has been witnessed in past decades. The novelty of using these nanoparticles in vitro/in vivo primarily depends on their nanoscale superparamagnetic behavior being easily manipulated by using an applied magnetic field. Among various magnetic nanomaterials that consist of magnetic elements such as iron, nickel and cobalt; IOMNPs, including magnetite (Fe_3O_4) and its oxidized form maghemite (γ-Fe_2O_3), have been mostly exploited in the biomedical field due to their better biocompatibility and safety profile than other metallic-based magnetic nanomaterials. In regard to the biomedical application field, several important parameters are generally considered for shaping the nanoscale magnetism of magnetic nanoparticles, including the transverse relaxation rate ($R2$) or relaxivity coefficient ($r2$), coercivity (H_c), specific loss powers (SLPs), and translational attractive force (F_m).

Before the discussion of the above-mentioned parameters of IOMNPs, we recall some fundamental concepts of magnetism. The magnetism in materials is a class of physical phenomena that is medicated by the magnetic field, which arises from the electron magnetic moment in atoms (Fig. 2). Virtually all materials exhibit diamagnetism that can generate an opposed magnetic field to repeal the applied magnetic field (H), but they are usually very weak, and often shielded by other, stronger intrinsic magnetic effects, such as paramagnetism and ferromagnetism. The atoms in diamagnetic materials have no net magnetic moment when there is no external H applied. In contrast with diamagnetism, paramagnetism is a type of magnetic state in which paramagnetic materials can be attracted by the H due to the partially induced magnetic moment of atoms, which is linearly aligned in same direction as the H. These materials do not maintain their induced magnetic properties when the H is absent, because the direction of the magnetic moment is randomly oriented under the influence of the Néel temperature (T_N). Comparably, superparamagnetism shows similar magnetic behavior to paramagnetism, which is raised from small, single-domain ferromagnetic/ferrimagnetic nanoparticles. Such nanoscale behavior can be retained, even when the temperature drops below the T_N, because the energy required to undermine the magnetic ordering of a nanoparticle is close to ambient temperature.

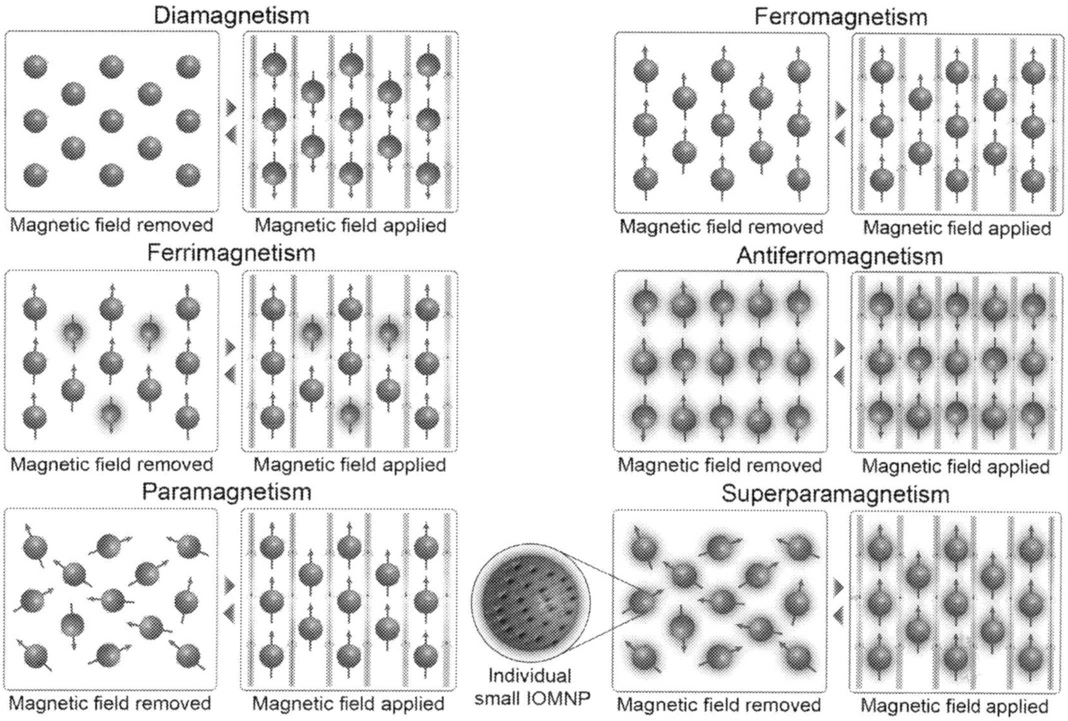

Fig. 2
An overview of different types of magnetic behavior.

One common magnetic behavior exhibited on permanent magnets is called ferromagnetism. Ferromagnetic materials stem from a parallel alignment of the atomic magnetic moments, and are maintained at a lowered energy state with a favorable tendency. As for ferrimagnetism, it is found in ionic compounds, and possesses a more complex crystal structure. Ferrimagnetism exhibits similar behavior to that of ferromagnetism, but holds different magnetic ordering within the structure. Such magnetic structure can be divided into two types of sub-lattices with unequal magnetic moments, leading to a spontaneous net magnetization. The populations of atoms in these materials are grouped together into magnetic domains. In each domain, all atomic moments are aligned in the same direction. However, some domains with opposing magnetic moments weaken the overall magnetization of the material. As for antiferromagnetism, the magnetic moments of neighboring atoms in the materials are aligned in an orderly manner in the opposite direction. In the absence of external H, they have similar magnetic properties to those of paramagnetism, as the atomic moments cancel out the total magnetization.

IOMNPs with enhanced magnetic parameters are favorable theranostic agents for high sensitivity MRIs, efficient magnetofection, magnetothermal-based hyperthermia therapy, and controlled therapeutics release. At first, IOMNPs are specifically designed as the *T2/negative* contrast agents for MRI applications. Because a secondary magnetic field of IOMNPs is generated by applying an external magnetic field (EMF), the transverse (spin-spin) relaxation process of protons in water molecules surrounding IOMNPs is agitated by local inhomogeneous magnetic moments, which results in a reduced transverse relaxation time (*T*2) of those surrounding protons. This effect leads to a negative enhancement by decreasing MR signal intensity, which is displayed as a darker *T*2-weighted image. The contrast enhancement ability of IOMNPs is defined in terms of its transverse relaxation rates, $R2$ (s^{-1}) (Eq. 1) and the relaxivity coefficient ($r2$), a gradient of line given by $R2$ against concentrations of IOMNP (Fig. 3C). The $R2$ is measured as $1/T2$, and the equation predicts that a high saturation magnetization (M_S) and large particle diameter (r) are essential for a higher $R2$, and thus an enhanced $T2$ contrast image (Fig. 3C). And yet, the IOMNP coercivity (H_c) is worthy of attention when particles' size is increased for contrast enhancement. Once IOMNPs are under an external magnetic field (H), their induced magnetic moments are

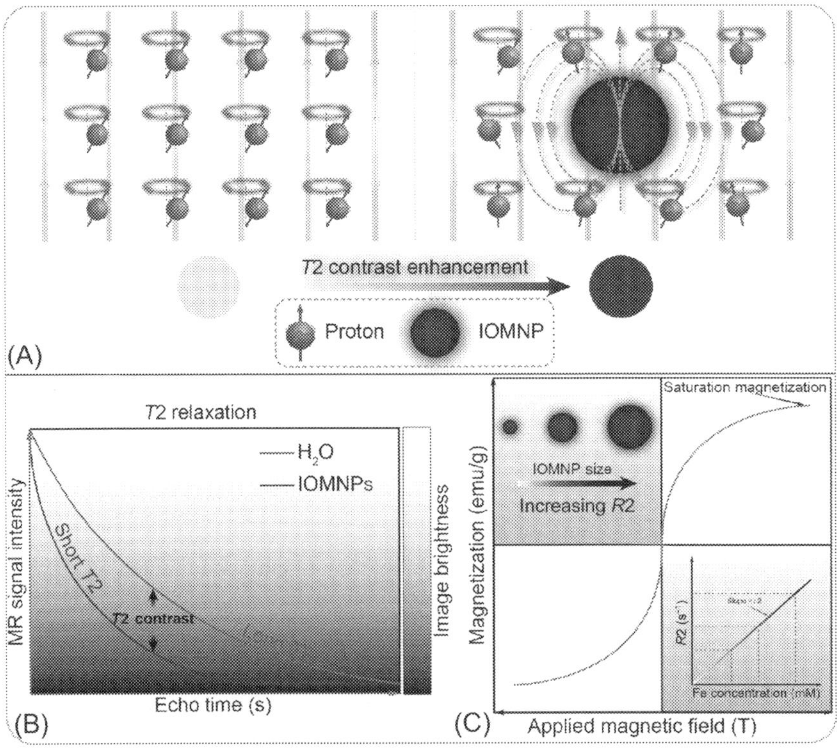

Fig. 3
(A) Schematic representation of the IOMNPs medicated *T*2 contrast enhancement. (B) The MR signals intensity of water and IOMNPs against time in *T*2 relaxation. (C) The key factors and parameters in measurement of *R*2.

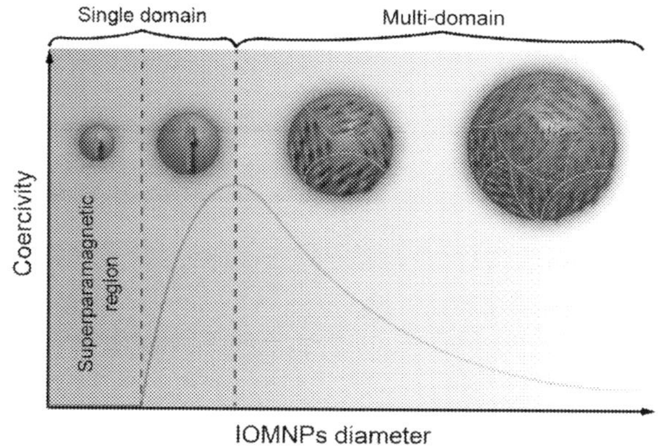

Fig. 4
Schematic illustration of the coercivity and size relations of IOMNPs.

aligned along the H direction, and reach the M_S. The coercivity is the strength of the applied magnetic field that required demagnetizing these ferromagnetic/ferrimagnetic particles, which is size-dependent (Fig. 4). Because the units of magnetic moments in particles are grouped in magnetic domains, this determines their magnetic behavior. The multiple magnetic domain particles can convert to single domain when their size decreases to a certain value, and further reduction of the particle size below a critical diameter can exhibit superparamagnetic characteristics, that is, $H_c = 0$. Therefore, after the removal of the EMF, the remnant magnetization of IOMNPs can be readily demagnetized through thermal agitation at an ambient temperature. This favorable property avoids undesired magnetic dipole-induced aggregation of IOMNPs in liquid media, which provides them a unique advantage for biological application.

$$R2 = \frac{1}{T2} = \frac{256\pi^2\gamma^2}{405} \cdot M_s^2 \varnothing \cdot \frac{r^2}{D\left(1 + \frac{L}{r}\right)} \tag{1}$$

Where $T2$ is transverse relaxation time, γ is the proton gyromagnetic ratio, M_s is saturation magnetization, $\varnothing$ is the nanoparticle volume fraction, r is the nanoparticle core radius, D is the diffusion coefficient of water molecules, and L is the thickness of the nanoparticle surface coating (Koenig & Kellar, 1995).

In addition to the MR contrast imaging and drug release, IOMNPs-based therapeutic carriers can also be directed (to push or attract) by a designed magnet system for magnetofection and magnetic fields controlled in vivo therapeutic delivery (Fig. 5). The magnetic force experienced by a single IOMNP is defined by magnetic fields (F_{mag}) (Eq. 2), which depends on the magnetic intensity (H), field gradient in space, particle size, saturation magnetization, and the position of the particle. It should be noted that particle size should remain small enough to retain the superparamagnetism.

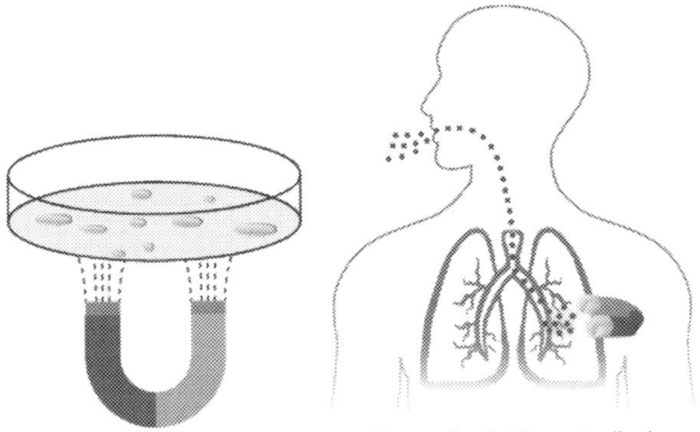

Fig. 5
Schematic illustration of the IOMNP system for magnetofection and magnetic fields controlled in vivo therapeutic delivery.

$$F_{\text{mag}} = \frac{4\pi a^3 \mu_0 \chi}{3\left(1+\frac{\chi}{3}\right)} \cdot \left[\frac{\partial H}{\partial x}\right]^T \tag{2}$$

Where a is the radius of the IOMNP, μ_0 is the vacuum magnetic permeability, χ is the magnetic susceptibility, H is the magnetic intensity, $x=(x, y, z)$ denotes the position of the IOMNP.

3 Synthesis and Surface Modification of IOMNPs

3.1 Synthetic Approaches

In the biomedical filed, IOMNPs typically refer to spinel-structured magnetite (Fe_3O_4), maghemite (γ-Fe_2O_3) and their analogous ferrites ($MO \cdot Fe_2O_3$ or MFe_2O_4, where the M is the divalent metal iron such as Mn^{2+}, Zn^{2+}, Fe^{2+}, Co^{2+}, Ni^{2+}). During the past five decades, the synthetic development of monodispersed and uniform sized IOMNPs has been intensively investigated (Schwertmann & Cornell, 2000). A variety of synthetic routes, including co-precipitation, thermal decomposition, hydrothermal/solvothermal synthesis, microwave irradiation, the sol-gel process, the microemulsion method, the electrochemical method, spray/laser pyrolysis, and sonolysis have been developed. Among these, to meet the requirements of the biomedical application; the uniform size distribution, the monodispersity, the particle size control, and the purity are crucial for in vivo biocompatibility, toxicity, and performance of the IOMNPs. Currently, the synthesis of IOMNPs via the alkaline co-precipitation in an aqueous solution, in microemulsions, and the thermal decomposition of iron organometallic precursors in non-aqueous media are mostly studied; and these are

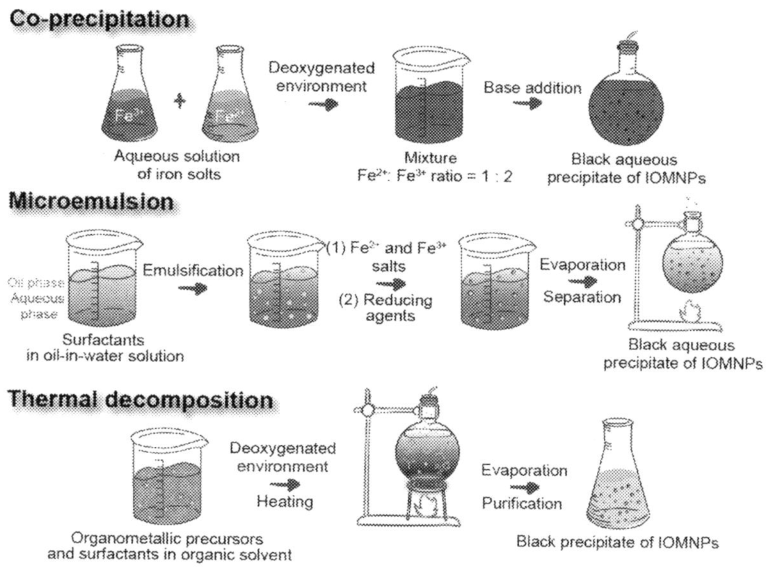

Fig. 6
A summary of major synthetic approaches of IOMNPs.

discussed in this section (Fig. 6). Other synthetic methods are less popular in respect to biomedical application, and have reviewed elsewhere (Laurent et al., 2008; Sun et al., 2014; Tartaj, Morales, Veintemillas-Verdaguer, Gonzalez-Carreno, & Serna, 2003).

The classical synthesis by alkaline co-precipitation in aqueous solution is a facile and inexpensive chemical pathway for mass production of IOMNPs. Fe_3O_4 and γ-Fe_2O_3 are generally obtained by co-precipitation of a specific amount of ferrous (Fe^{2+}) and ferric (Fe^{3+}) salts in a basic aqueous solution. For complete precipitation of Fe_3O_4, the reaction conditions, including the Fe^{2+}: Fe^{3+} ratio, which is fixed at 1: 2, the pH should be controlled between 8 and 14, and the reaction needs to be processed in a non-oxidizing oxygen environment (Jolivet, Chaneac, & Tronc, 2004; Massart & Cabuil, 1987). However, the as-prepared Fe_3O_4 is not thermodynamically stable, and can be readily oxidized to form γ-Fe_2O_3 in air (Jolivet et al., 2004). In order to obtain the relatively stable IOMNPs, the intended oxidation of IOMNPs is usually performed after the initial formation of Fe_3O_4 (Massart, 1981). In addition, it is possible to tailor the particle size with a mean diameter ranging from 2 to 20 nm, by modifying the reaction parameters, such as pH, ion concentration, reaction temperature, Fe^{2+}: Fe^{3+} ratio, and the type of alkali salts (Babes, Denizot, Tanguy, Le Jeune, & Jallet, 1999; Jean-Pierre, 2000; Lopez-Lopez, Duran, Delgado, & Gonzalez-Caballero, 2005; Massart, 1981). Nonetheless, the size distribution of IOMNPs can hardly be controlled within a narrow range through co-precipitation in an aqueous system (Lu, Salabas, & Schuth, 2007), because two critical steps are involved in the formation of the IOMNPs: (1) nucleation and (2) crystal growth on the nuclei

(Schwertmann & Cornell, 2000). Nucleation occurs when the critical supersaturation value is reached; then a growth of the nuclei spontaneously takes place. A number of kinetic factors govern this process (Boistelle & Astier, 1988; Etienne & Stéphane, 2012; LaMer & Dinegar, 1950). According to the LaMer mechanism and the Ostwald ripening phenomenon (Boistelle & Astier, 1988; LaMer & Dinegar, 1950), the periods of nucleation and crystal growth should be separated, and nucleation needs to be prevented during the process of crystal growth. A slowly controlled crystal growth, as well as other optimizations, are desirable for improving size distribution control and producing monodispersed IOMNPs (Etienne & Stéphane, 2012; Schwarzer & Peukert, 2004; Sugimoto, 2003).

Alternatively, the alkaline co-precipitation in microemulsion has been developed to overcome the shortcomings of the conventional co-precipitation method by achieving controlled homogeneity of size and morphology (Chin & Yaacob, 2007). This method is also applied to fabricate analogous ferrite nanoparticles (MFe_2O_4). A microemulsion system is a chemically stable isotropic dispersion of two immiscible liquids; for example, nanosized water droplets disperse in the oil phase (water-in-oil microemulsions), or vice versa, that is, oil-in-water microemulsions (Drmota, Koseli, & Žnidaršič, 2012). The interface between the two phases is stabilized by surfactants and co-surfactants, which provide a restriction effect that confines particle growth and agglomeration. In this method, IOMNPs can be simply prepared by combining two water-in-oil microemulsions, with microemulsion 1 containing metal salts and microemulsion 2 containing reducing agents (Boutonnet, Kizling, Stenius, & Maire, 1982; Boutonnet, Kizling, Touroude, Maire, & Stenius, 1991). The co-precipitation occurs in aqueous droplet-dimmers/aggregates due to the random collision between the mixtures of water-in-oil droplets. The microemulsion method takes advantage of using size-adjustable nano-droplets as nanoreactors to synthesize size-controlled IOMNPs with an average diameter from a few, to several hundred nanometers. The size of nano-droplets can also be controlled by modifying the water/oil/surfactant ratio (Feltin & Pileni, 1997; Muller & Muller, 1984). However, this method suffers from the drawbacks of poor crystalline particles (usually reduces the superparamagnetic properties) obtained from low-temperature synthesis, which could be improved by post-processing with thermal annealing (Mahmoudi, Sant, Wang, Laurent, & Sen, 2011). Moreover, large-scale production is also limited due to the massive quantities of oil required, and complicated post-processing for surface functionalization (Lin, Kim, El Haj, & Dobson, 2008).

In addition to co-precipitation methods, the thermal decomposition of organometallic precursors in high-boiling organic solvents has been specifically developed for the synthesis of monodispersed IOMNPs and Fe-*M* alloy nanoparticles that offer enhanced magnetism and advanced homogeneous control over the crystallinity and morphology. Various organometallic precursors, including metal acetylacetonates: $[M(acac)_n]$, cupferronate: $[M(C_6H_5N_2O_2)_n]$, pentacarbonyl: $[M_x(CO)_n]$, and acetate: $[M(C_2H_3O_2)_n]$, (where the M=Mn, Zn, Fe, Co, Ni; n depends on the oxidation number of *M*) are often used for thermal decomposition in non-aqueous solvents containing stabilizing surfactants (Jeong, Teng, Wang, Yang, & Xia, 2007). Thereby, the small hydrophobic IOMNPs are generated from the nucleation of monomers, and larger particles can be subsequently obtained through the seeded growth process (Jun, Choi, & Cheon, 2006). Moreover, the average diameter of IOMNPs can also be tuned by altering the ratios between organometallic precursors, solvents, stabilizing surfactants, and other reaction parameters, such as the heating rate and reaction time. However, thermal decomposition is confronted with some potential challenges, such as the additional surface modifications of particles, which are often required to afford hydrophilicity, and a strict product purification process, which is also essential prior to the in vivo application, because toxic solvents are sometimes used.

3.2 Primary Surface Coating and Functionalization With siRNA

There are several potential problems, including immunogenicity (Bessis, GarciaCozar, & Boissier, 2004), carcinogenesis (Baum, Kustikova, Modlich, Li, & Fehse, 2006), insufficient siRNA loading capacity (Thomas, Ehrhardt, & Kay, 2003), toxin generation (Nayerossadat, Maedeh, & Ali, 2012), and complicated production processes, that limit the use of viral vectors for gene therapy (Bouard, Alazard-Dany, & Cosset, 2009). Like other inorganic-based gene carriers, IOMNP-based siRNA carriers have been increasingly studied in the past decade, and they are expected to become one of the non-viral carriers for replacing viral mediated siRNA delivery (Xing et al., 2014). Despite the fact that the monodispersed and uniformly-sized IOMNPs can be synthesized in organic media, they must be coated to achieve hydrophilicity, long-term stability, additional functionality (i.e., targeting and siRNA packaging), and modulation of $r2$ relaxivity for in vivo application. The surface coating of hydrophobic IOMNPs is often performed through the ligand exchange, using high-affinity hydrophilic ligands, or encapsulation with organic or inorganic materials (Fig. 7).

In ligand exchange methods, the ideal coating ligands should have a nucleophilicmoietyon on one end that binds the iron oxide surface (such as hydroxyl, amino, carboxyl, thiol, and phosphate groups), and a hydrophilic functional group on the other end that is exposed to the biological environment. Due to the simple reaction approach and desirable thickness of the coating layer, many exchange ligands have been developed to meet the requirements of certain biomedical applications (Xie et al., 2011). For example, carboxylic acid, dopamine, and their analogues (e.g., dimercaptosuccinic acid and zwitterionic dopamine sulfonate (Xie et al., 2011)) are commonly used for compact surface coating and functionalization. In contrast, various kinds of polymeric ligands with a relatively large molecular weight are more widely used due to better stability and multi-functionality. Among these polymeric ligands, the FDA-approved poly (ethylene glycol) (PEG), dextran, poly (vinylpyrrolidone) (PVP), poly (vinyl alcohol) (PVA) and their derivatives are favorable for in vivo application because of excellent biocompatibility. In addition, poly (ethylenimine) (PEI), an efficient gene delivery vector, is specifically utilized for metal surface coating and gene transfection.

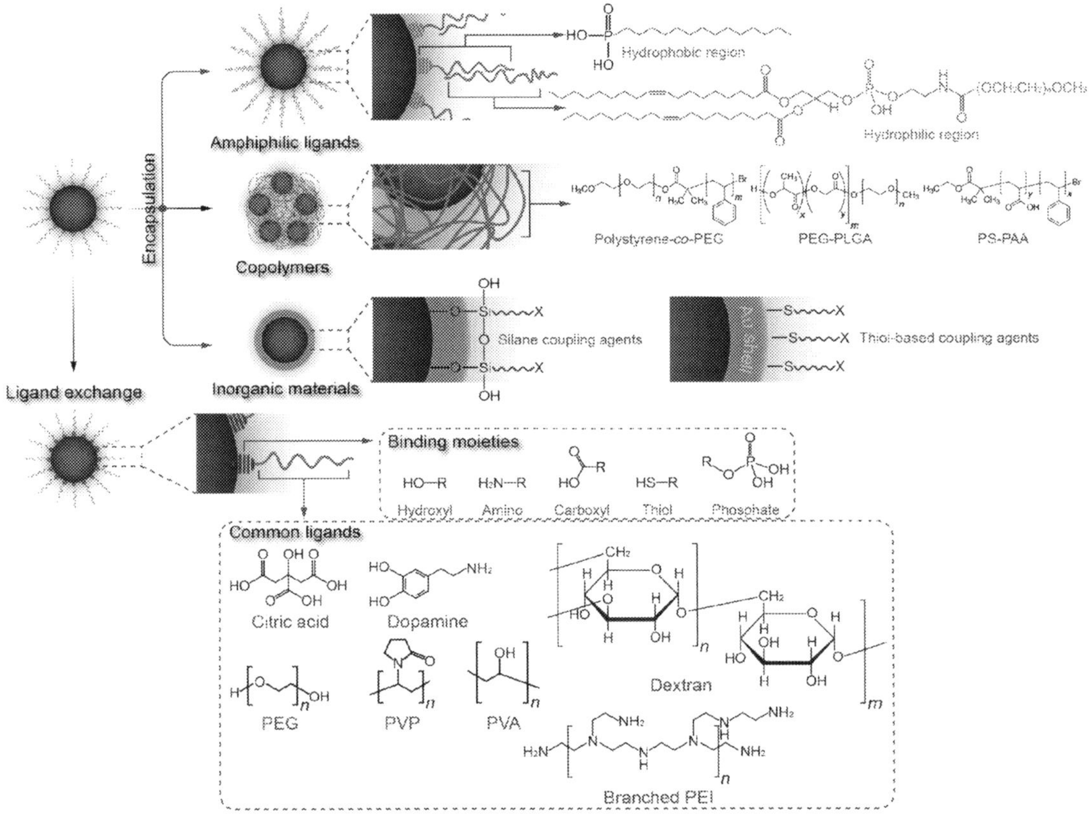

Fig. 7
Schematic illustration of the general methods for primary surface coating of IOMNPs.

The other strategy for surface coating is the encapsulation of particles within amphiphilic ligands, the hydrophilic polymer matrix, and inorganic materials. The self-assembling of amphiphilic ligands on the hydrophobic surface of the particles is based on the intercalations between the hydrophobic surfactants of particles and hydrophobic segments of amphiphilic ligands, namely hydrophobic interactions. Many amphiphilic copolymers have been used for coating the IOMNPs, such as PEG-phospholipid, tetradecylphosphonate, maleimide/methoxy-PEG-poly (lactide) (Nasongkla et al., 2006), polyoxyethylene 2 alkyl ether (Robinson et al., 2005), and so on. However, surface coating through the addition of amphiphilic copolymers can significantly increase the hydrodynamic diameter (HD), thus using amphiphilic oligomers (e.g., *alpha*-cyclodextrin) could be a desirable method of providing a thinner surface layer for the physiological environment (Wang, Wong, Teng, Lin, & Yang, 2003). In addition to the single-core encapsulation, multiple IOMNPs can also be embedded by copolymers to form a hydrophilic IOMNP-polymer complex particle. Numerous copolymers, including polystyrene-*co*-PEG, poly(lactic-*co*-glycolic acid)-*co*-PEG, and polystyrene-*co*-poly-(acrylic acid) have been used as polymer matrixes. Such hybridizations can produce large particle size, as well as increasing the *T*2 signal intensity (Neely et al., 2013).

In comparison with the organic ligands and polymers, silica (Si) and gold (Au)-based inorganic materials provide several advantages, including tunable shell thickness, better stability, and ease of surface functionalization for the surface coating of IOMNPs. Tetraethyl orthosilicate and tetramethyl orthosilicate are common silane coupling agents that are often used for silica coating of metal surfaces; the hydroxyl groups on the IOMNP surface are essential to initiate the oxane (Si-O-Metal) bridging via the Stöber process. For hydrophobic IOMNPs that lack hydroxyl groups on the surface, coating can be achieved by reverse microemulsion techniques (Yi, Lee, Papaefthymiou, & Ying, 2006). The silica-coated surface can be further modified by coupling with organo-functional silanes (amino-silane, epoxy-silane, vinyl-silane, etc.). Moreover, the coating of IOMNPs by a thin layer of gold is another attractive approach to enabling even more functionality toward sensing, drug delivery, and hyperthermia therapy. The preparation of a surface coating with gold can be adopted by the reverse micelle method (Lin et al., 2001), and the additional gold layer can be facilely functionalized using gold-thiol chemistry.

Based on the primary surface coating, siRNA molecules can be conjugated with IOMNPs through various strategies (Fig. 8). In general, two basic considerations need to be taken into account, (1) the siRNA loaded IOMNPs must remain physiologically stable until entering the cytoplasm; (2) the siRNA needs to be unpacked from the IOMNPs, and released from the endosome for the initiation of RNA-induced silencing complex activation. One of the attractive attributes of the siRNA delivery is that it relies on the electrostatic interactions between water-soluble cationic polymer-based IOMNPs and the siRNA anionic phosphate backbone. The linear/branched polyethylenimine PEI and their variations are widely-used, water-soluble cationic polymers that consist of repeating amine units, which can efficiently attach to the siRNA under physiological pH. In addition, high siRNA loading efficacy can be achieved by layer-by-layer (LbL) assembly of siRNA and cationic polymers on the surface of a single IOMNP based on its multilayered, core-shell structure. Furthermore, after the intracellular uptake of siRNA-cationic polymer-IOMNP complexes by endosomes, those amine groups are protonated by absorb protons, referred to as the proton-sponge effect, which leads to the spontaneous influx of water and Cl^- ions to balance the proton flux inside the endosomal vesicles (Sonawane, Szoka, & Verkman, 2003). This results in osmotic swelling, and the membrane rupture of endosomes with the subsequent leakage of its contents (siRNA with cationic polymer-based IOMNPs) into the cytosol (Akinc, Thomas, Klibanov, & Langer, 2005; Sonawane et al., 2003). Other alternative cationic polymers and lipids, including polyamidoamine (PAMAM) dendrimers (Taratula et al., 2011), poly-L-lysine (Cui et al., 2014), and Metafectene™ (Mykhaylyk et al., 2007), have also been reported to functionalize IOMNPs for electrostatic-based siRNA delivery.

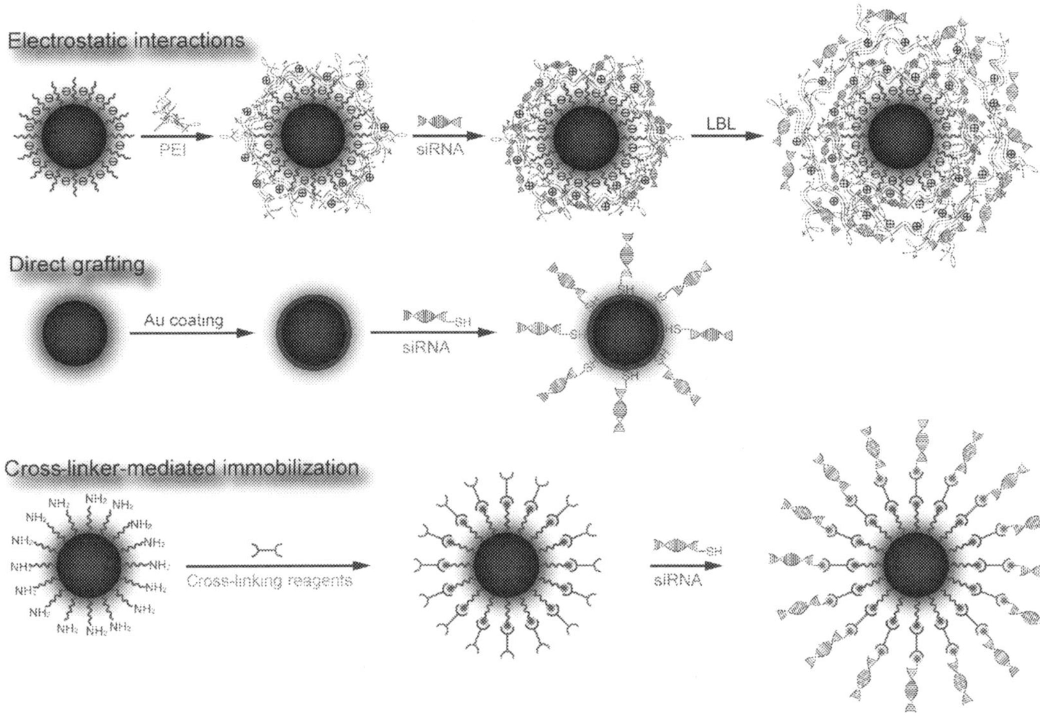

Fig. 8
Different strategies for siRNA molecules binding with functionalized IOMNPs.

In addition to using attractive electrostatic interaction, the siRNA molecules can also be attached on the surface of IOMNPs by either direct grafting or cross-linker-mediated immobilization, depending on the primary surface coating. In fact, the terminal site of siRNA molecules is often functionalized with the thiol group (SH-siRNA) for further conjugation. Thereby, direct grafting can be achieved by forming the thiolate bond between thiolated siRNA and gold-coated IOMNPs. Besides, it has been proposed that the intracellular SH-siRNAs can be released from the surface of gold-coated IOMNPs by exchanging SH-siRNA with glutathione (GSH, which is an important thiolcontaining antioxidant and highly abundant in most cells) in the cytosol (Kim et al., 2014; Oishi, Nakaogami, Ishii, & Nagasaki, 2006). Another popular strategy for siRNA attachment on surface functionalized IOMNPs is to establish a cross-linkage by using heterobifunctional cross-linking reagents. For example, *N*-succinimidyl 3-(2-pyridyldithio)propionate (SPDP) is one of the most commonly used cross-linkers for amine-to-thiol conjugation via amide (the activated *N*-hydroxysuccinimideester end of SPDP reacts with primary amine groups) and disulfide (the pyridyldithiol group at the other end of SPDP reacts with thiolresidues) linkage (Hermanson, 2013). These newly formed disulfide bonds can be readily broken by intrinsic redox molecules, such as the GSH and thioredoxinin cell, either with or without the enzymes' assistance (Jere et al., 2009). In addition to SPDP, many of its analogs, including long-chain (LC)-SPDP, Sulfo-LC-SPDP, SMPT (4-succinimidyloxycarbonyl-methyl-α-(2-pyridyldithio) toluene), Sulfo-LC-SMPT, *m*-maleimidobenzoyl-*n*-hydroxysuccinimide ester (MBS), Sulfo-MBS, and etc. so forth, have been use for siRNA conjugation applications (Hermanson, 2013). The LC cross-linking agents are designed for greater spatial accessibility, and the sulfonic group modified cross-linking agents may increase the solubility in an aqueous solution.

4 IOMNPs for Systemic siRNA Delivery

4.1 Pharmacokinetics of IOMNP-Based siRNA Carriers

For therapeutic delivery purposes, the fastest delivery method is via the intravenous route, which maximizes the bioavailability compared with other administration routes. Typically, the IOMNPs-assisted siRNA delivery offers several advantages over the small naked-siRNA molecules on their own, such as: (1) the IOMNPs-assisted delivery approach can protect the siRNAs from RNase degradation in the bloodstream; (2) functionalized siRNA-IOMNPs can increase the blood circulation half-life by reducing the rate of renal clearance; (3) nanoparticle-conjugated siRNAs can ensure the distribution of siRNA-IOMNPs throughout the body; (4) access to the diseased area via passive/active targeting and (5) promoting the cellular uptake and intracellular trafficking; as well as (6) MR imaging-guided siRNA delivery for therapeutic evaluation and disease management. Like other metallic nanoparticles, the pharmacokinetics of siRNA-IOMNPs are determined by their morphology, final HD, surface charge, surface functionalization, protein binding, dose, and animal species (Lin, Monteiro-Riviere, & Riviere, 2015). Based on available evidence, siRNA-loaded metallic nanoparticles without surface engineering tend to accumulate in mononuclear phagocyte system (MPS) organs, including the liver, spleen, blood, bone marrow, and lymph nodes for long periods of time, due to the non-specific uptake of serum protein-bound siRNA-IOMNPs by phagocytic cells, primarily monocytes and macrophages (Song et al., 2014; Wang et al., 2010). Therefore, proper surface functionalization with bio-inert and hydrophilic polymers, including PEG, *N*-(2-hydroxypropyl)methacrylamide, PEI, PVP, and PVA can prolong circulation time by escaping from the MPS-mediated nonspecific recognition (Kanasty, Dorkin, Vegas, & Anderson, 2013).

The HD of siRNA-IOMNPs is also an important parameter that influences their biodistribution, and it has been consistently found in the size-dependent (~4–100 nm) distribution for metallic nanoparticles, with the smaller nanoparticles exhibiting the wider area of distribution, including the liver, spleen, lymph nodes, kidneys, bone marrow, brain, and heart. In addition, the vascular endothelium is negatively charged due to the heparan sulfate proteoglycans that are present on the cellular surface and in the extracellular matrix (Sarrazin, Lamanna, & Esko, 2011), thus the modification of siRNA-IOMNPs with a tunable surface charge can achieve adjustable absorption by tissues and organs. The metabolism of metallic nanoparticles greatly depends on surface coating. Besides, the elimination of metallic nanoparticles is primarily through the biliary and urinary pathways, but a considerable amount of nanoparticles are retained in the liver and spleen.

4.2 Targeting Strategies

The specificity and affinity of the siRNA-IOMNPs to targeted diseased organs, tissues, and cells is crucial for effective siRNA delivery (Fig. 9). One widely used strategy is the active targeting approach that involves targeting components such as small molecules (folic acid, triphenylphosphonium, and (*S*)-2-(3-((*S*)-5-amino-1-carboxypentyl)ureido)pentanedioic acid), peptides, and proteins (arginine-glycine-aspartic acid (RGD), antibodies, aptides, transferrin andaptamers), fatty acids, DNA-targeting sequences, polysaccharides(hyaluronic acid), and so forth (Bertrand, Wu, Xu, Kamaly, & Farokhzad, 2014). It can selectively interact with the substrate binding site of targets (including overexpressed receptors and enzymes in

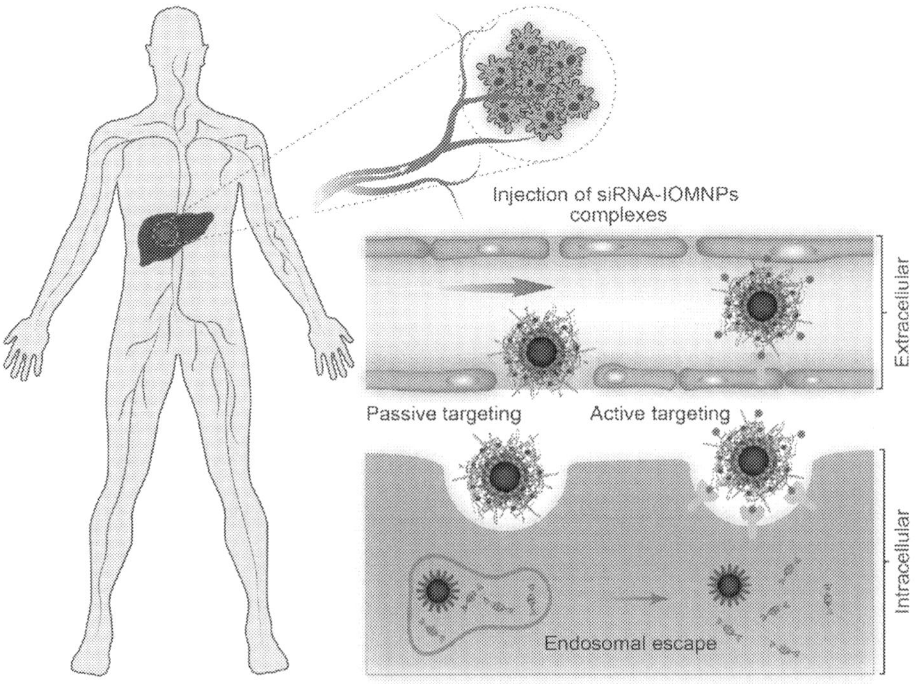

Fig. 9
The systemic delivery of IOMNP-siRNA complexes and targeting strategies for disease site approach.

lesion domains) by decorating with an appropriate type of ligand. Therefore, the purpose of using this strategy is to enhance the specific cellular uptake of the therapeutic nanocarriers, and many considerations need to be addressed for the desired delivery efficacy, especially the non-specific binding between nanocarriers and endogenous protein, physicochemical properties (net surface charge, morphology, final HD, and aqueous solubility), and the ligand density.

In addition to active targeting approaches, the passive targeting of nanocarriers to diseased areas (typically tumors) is another potential strategy that is associated with organ-specificity, and it is also governed by certain pathophysiological characteristics of irregular tumor vessels (angiogenesis-promoted fenestrations of capillaries with the pore size ranging from a few hundred nm to several μm), which is known as the effect of MPS, and was previously called the EPR effect (Hume, 2006). Although the concept of the MPS phenomenon can be exploited for passive, targeted nanoparticle-based therapeutic delivery, numerous factors, including the physicochemical properties (morphology, final HD, surface charge, ligand density, protein binding, etc.) of nanocarriers, tumor location, size, and type, as well as the specific microenvironment, can influence the distribution and accumulation of nanocarriers in tumors. For these reasons, a full understanding of the biological behavior of nanomaterials in tumor physiological environments is thus essential to the development of efficient site-specific delivery.

Another attractive targeting method is the EMF-controlled delivery. As mentioned earlier, in this method, a constant EMF is used to transport the therapeutics-loaded IOMNPs to a specific target site within the body. This concept has also been successfully applied to enhance the in vitro gene transfection (magnetofection) for various cell types. Compared with active and passive targeting methods, the major advantage of EMF-controlled delivery is that it reduces both of the undesired systemic distribution methods of the therapeutic IOMNPs and the effective dosage, and enhances the accumulation in target sites by overcoming the biological hindrances such as limited penetration from blood vessels to the tumor site (Lee et al., 2015; Prijic & Sersa, 2011). Based on its physical properties of delivery, this strategy can be applied in various tissues and organs, and the ultimate success in clinical application is counting on the development of high-gradient magnetic field facilities, and high magnetic performance IOMNPs with optimal biocompatibility and safety.

4.3 Cellular Uptake and Intracellular Trafficking

Once the siRNA-loaded IOMNPs reach the target tissue/organ, they have to enter into the target cells for successful delivery of siRNAs and IOMNPs. siRNA-IOMNPs, combined with the aforementioned targeting strategies (i.e., active, passive, and EMF-controlled delivery) can facilitate the cellular uptake through the endocytosis (Fig. 9). In most cases, the positively charged siRNA-IOMNPs can be increasingly taken up by cells because of the negatively charged cell membrane. While for the targeting therapy and imaging, the receptor-mediated cellular internalization using active targeting approaches is often favorable. Usually, the endosomes uptake the siRNA-loaded nanoparticles after the cellular internalization, followed by the fusion with lysosomes, and this entrapment undermines the therapeutic function of siRNAs, which must take place in cytoplasm. One potential strategy designed to enhance the endosomal escape is using cationic polymers (discussed in previous section), which elevates the osmotic pressure and leads to the swelling and rupture of the endosomal and lysosomal membranes.

4.4 Toxicity of IOMNP-Based RNAi Therapy

The in vivo applications of IOMNPs for MR imaging, drug delivery, and combination therapy have been extensively reported (Gao et al., 2013; Liu, Wang, Lee, Ai, & Chen, 2012; Song et al., 2014; Wan et al., 2013; Xing et al., 2013). In the context of IOMNP-based siRNA delivery, the toxicological effects of both IOMNPs and siRNA molecules can no longer be overlooked because the safety issue is one of the major obstacles for clinical translation (Wan et al., 2013). For IOMNPs in general, the nanotoxicity is subject to the nanoparticle surface area-to-volume ratio, chemical stability, composition of the nanomaterial, and amount of dose. In contrast to other metallic nanoparticles, IOMNPs can be degraded into iron ions in liver and spleen tissues as an iron pool, and subsequently transformed to antiferromagnetic ferritin for the needs of various biological processes (Levy et al., 2011). However, a certain high concentration of iron ions in cells can induce oxidative stress owing to free-radical formation, which leads to cell death (van den Bos et al., 2003). It has been demonstrated that possible toxicity from released iron ions can be prevented by adding an effective organic coating, such as polysaccharide, PEG derivatives, and antioxidants to protect the IOMNP surface and improve biocompatibility. For example, dextran-coated IOMNPs are a well-established nanoplatform for imaging and therapeutic delivery; several relevant IOMNPs-based imaging agents such as Resovist® and Feridex® have been approved by the FDA for clinical use (Ros et al., 1995). Although some success has been achieved in clinical translation, the complex physicochemical interaction between IOMNPs and endogenous carbohydrates, lipids, proteins, and nucleic acids still need to be elucidated.

The siRNA-based therapeutic safety is another important issue for siRNA-IOMNPs reaching the clinical stage. In some cases, adverse effects may be induced by unsuspected "off-target" silencing through the undesired mRNA knockdown in normal tissues and organs, because of the irrational siRNA design and chemical modification, accompanied by a lack of specificity of the siRNA sequence (Jackson et al., 2006). Besides, excessive length of the siRNA sequence (>30 bp) is likely to induce the interferon reaction based on the innate immune system (Hornung et al., 2005; Judge et al., 2005). Thus, a shorter siRNA sequence ranging between 21 and 23 bp is highly favorable to avoid sequence-dependent immunostimulatory effects (Hornung et al., 2005). Moreover, the formulation and quality control of siRNA-IOMNPs are also important to govern the nanotoxicity and immunogenicity for systemic delivery. Additionally, the animal model plays a significant role in nanotoxicity investigations. Different animal models may exhibit resultant side effects inconsistently due to the differences at the physiological and anatomical level between species. Therefore, in the approach to clinical translation, the design for good human-like animal models should be necessary. In order to accelerate the in vivo preclinical therapeutic and safety evaluation, the IOMNP-based siRNA delivery can afford real-time MRI monitoring with non-invasive quantitative assessments, including the biophysicochemical processes, and spatial and temporal distribution of siRNA-IOMNPs in vivo (Willmann, van Bruggen, Dinkelborg, & Gambhir, 2008).

5 Molecular Imaging of IOMNPs-Mediated siRNA Delivery

The IOMNPs-based siRNA delivery platform has emerged as a promising tool for non-invasive in vivo monitoring of siRNA delivery and silencing by evaluating whole-body biological events on the molecular level in a real-time manner. One of the most important studies has been reported by Medarova et al., their strategy permitted in vivo siRNA delivery with simultaneous MR and NIR imaging of its uptake process in tumors by using a Cy5.5 dye, myristoylated polyarginine peptides, and siRNA labeled IOMNPs (MN-NIRF-siGFP) (Medarova et al., 2007). In this study, the MN-NIRF-siGFP was intravenously injected into two separate mouse models, one with bilaterally implanted 9L-GFP and 9L-RFP tumors, the other bearing subcutaneous LS174T human colorectal adenocarcinoma. A significant drop of $T2$ relaxivity in the tumors was observed after 24 h post-injection, which reflects the accumulation of MN-NIRF-siGFP in tumor tissue and describes the feasibility of in vivo tracking by using an IOMNPs-based siRNA delivery system. In addition, the authors followed the silencing process and corroborated the RNAi efficiency by employing optical imaging (Medarova et al., 2007). Their results indicate that the combination of MR and NIR-guided siRNA delivery in mice can provide sufficient imaging quality and biological information for the development and optimization of an RNAi-based therapeutic product.

Another study demonstrated the use of IOMNPs as a theranostic tool for combined imaging and delivery of siRNA to breast tumors in mice (Kumar, Yigit, Dai, Moore, & Medarova, 2010). In this report, dextran-coated IOMNP cores are triply conjugated to peptide sequences (which specifically interact with tumor-specific antigen uMUC-1), the Cy5.5 dye (for NIR imaging), and siRNA molecules (that binds the tumor-specific antiapoptotic gene BIRC5). This tumor-targeted nanodrug (MN-EPPT-siBIRC5) was then intravenously injected into subcutaneous mouse models bearing human breast tumors, and the in vivo quantitative MRI showed a significant shortening of tumor $T2$ relaxation times after 24 h of injection (Kumar et al., 2010). Furthermore, the NIR imaging showed a high fluorescence signal from tumor site, indicating the preferential tumor uptake of MN-EPPT-siBIRC5, and a significant decrease of the tumor growth rate was observed (Kumar et al., 2010). All data revealed that

IOMNPs-based siRNA delivery system illustrates the potential feasibility for clinical studies, because there are several IOMNPs that have already been approved.

In addition to the preceding studies, several relevant investigations have also been carried out to demonstrate the theranostic function of IOMNPs. For example, Lee et al. developed a multifunctional manganese-doped IOMNPs (MnMEIO) coated with bovine serum albumin; its surface is functionalized by a PEG decorated cell-specific targeting ligand (cyclicRGD peptide), and a fluorescent dye (Cy5) labeled siRNA (siGFP) (Lee et al., 2009). In this study, the breast-cancer cells (over express $\alpha_v\beta_3$ integrin) and lung-carcinoma cells ($\alpha_v\beta_3$ integrin deficient) were used as two separate cell lines and treated with MnMEIO-siGFP-Cy5/PEG-RGD nanocomplexes. The *T*2 contrast MR imaging showed an increased 1/*T*2 signal intensity in breast-cancer cell lines with rising amounts of the nanoparticles (Lee et al., 2009). No contrast change was observed in lung-carcinoma cell lines, which indicated that these RGD-labeled nanocomplexes can selectively uptake by $\alpha_v\beta_3$ integrin expressed cell lines (Lee et al., 2009). Furthermore, the subcellular distribution and knockdown efficiency of MnMEIO-siGFP-Cy5/PEG-RGD nanocomplexes were monitored by fluorescence imaging ofCy5 dye and GFP expression level (Lee et al., 2009). All results demonstrated that such an imaging-guided siRNA delivery strategy can offer potential benefits, including simultaneous therapeutic treatment and visualization for in vivo and future clinical studies.

One recent attempt has been made to establish an in vivo EMF-controlled and multimodality imaging-guided siRNA delivery for the treatment of hepatocellular carcinoma (HCC). In this effort, Chen et al. prepared a theranostic IOMNPs-based siRNA delivery strategy by functionalizing a commercial IOMNPs (SilenceMag) with ^{131}I labeled human siVEGF, and following their (^{131}I-hsiVEGF/SilenceMag) subcutaneous injection into HCC tumor-bearing mice, and the injected ^{131}I-hsiVEGF/SilenceMag was magnetically guided to the tumor site by using EMF devices. The biodistribution of ^{131}I-hsiVEGF/SilenceMag was observed by SPECT and MR imaging studies. Both of the imaging modalities indicated a successful attraction and retention of the ^{131}I-hsiVEGF/SilenceMag in HCC tumors. The results demonstrated that both the growth rate of tumor mass and VEGF protein level were deceased in ^{131}I-hsiVEGF/SilenceMag injected mice compared with that of the control group.

6 Conclusion and Future Perspectives

The RNAi-based therapeutics remain one of the most revolutionary approaches for personalized treatment of diseases, and currently many siRNA therapeutics are under clinical studies. Although some success has been achieved, there are still many obstacles to systemic delivery of siRNA due to the intrinsic nature of siRNA molecules. IOMNP-based carriers are available for multifunction and effective delivery of siRNAs to the cytoplasm of target cells within the body. In contrast to other nanoplatforms, IOMNPs provide significant advantages, including low toxicity, biodegradability, MRI contrast enhancement, and magnetic force-controlled therapeutic delivery, which could contribute to the applicable theranostic functions that not only affect the development process of siRNA therapeutics, but also intensify disease management.

In order to translate IOMNPs from the laboratory bench to the clinical bedside, additional research is needed to further investigate synthesis of high magnetic performance IOMNPs with low toxicity. The major challenges in pharmacological investigations are the incomplete and inconsistent experimental criteria for pharmacokinetics and the in vivo study of the IOMNPs-based delivery platform, which has resulted in difficulties in integrating and generating a conclusive review from various published experimental data. It has been suggested that the PBPK models could be the best option for pharmacokinetic studies of a variety of metallic nanoparticles (Lin et al., 2015), which can integrate and compare all associated data among different studies to generate more insights into many parameters' dependent pharmacokinetic outcomes.

Acknowledgments

Address correspondence to junqing@me.com and gangliu.cmitm@xmu.edu.cn. This study was supported by the Major State Basic Research Development Program of China (Grant Nos. 2017YFA0205201, 2014CB744503 and 2013CB733802), the National Natural Science Foundation of China (NSFC) (Grant Nos. 81422023, 81371596, 51273165, U1705281 and U1505221), the Key Project of Chinese Ministry of Education (Grant No. 212149), the Fundamental Research Funds for the Central Universities, China (Grant No. 2013121039 and 20720150141), and the Program for New Century Excellent Talents in University (Grant No. NCET-13-0502).

References

Akinc, A., Thomas, M., Klibanov, A. M., & Langer, R. (2005). Exploring polyethylenimine-mediated DNA transfection and the proton sponge hypothesis. *Journal of Gene Medicine*, *7*(5), 657–663. https://doi.org/10.1002/jgm.696.

Alexis, F., Pridgen, E., Molnar, L. K., & Farokhzad, O. C. (2008). Factors affecting the clearance and biodistribution of polymeric nanoparticles. *Molecular Pharmaceutics*, *5*(4), 505–515. https://doi.org/10.1021/Mp800051m.

Babes, L., Denizot, B., Tanguy, G., Le Jeune, J. J., & Jallet, P. (1999). Synthesis of iron oxide nanoparticles used as MRI contrast agents: a parametric study. *Journal of Colloid and Interface Science*, *212*(2), 474–482. https://doi.org/10.1006/jcis.1998.6053.

Baum, C., Kustikova, O., Modlich, U., Li, Z. X., & Fehse, B. (2006). Mutagenesis and oncogenesis by chromosomal insertion of gene transfer vectors. *Human Gene Therapy*, *17*(3), 253–263. https://doi.org/10.1089/hum.2006.17.253.

Bertrand, N., Wu, J., Xu, X. Y., Kamaly, N., & Farokhzad, O. C. (2014). Cancer nanotechnology: the impact of passive and active targeting in the era of modern cancer biology. *Advanced Drug Delivery Reviews*, *66*, 2–25. https://doi.org/10.1016/j.addr.2013.11.009.

Bessis, N., GarciaCozar, F. J., & Boissier, M. C. (2004). Immune responses to gene therapy vectors: influence on vector function and effector mechanisms. *Gene Therapy*, *11*, S10–S17. https://doi.org/10.1038/sj.gt.3302364.

Boistelle, R., & Astier, J. P. (1988). Crystallization mechanisms in solution. *Journal of Crystal Growth*, *90*(1–3), 14–30. https://doi.org/10.1016/0022-0248(88)90294-1.

Bouard, D., Alazard-Dany, D., & Cosset, F. L. (2009). Viral vectors: from virology to transgene expression. *British Journal of Pharmacology*, *157*(2), 153–165. https://doi.org/10.1038/bjp.2008.349.

Boutonnet, M., Kizling, J., Stenius, P., & Maire, G. (1982). The preparation of monodisperse colloidal metal particles from microemulsions. *Colloids and Surfaces*, *5*(3), 209–225. https://doi.org/10.1016/0166-6622(82)80079-6.

Boutonnet, M., Kizling, J., Touroude, R., Maire, G., & Stenius, P. (1991). Monodispersed colloidal metal particles from nonaqueous solutions—catalytic behavior in hydrogenolysis and isomerization of hydrocarbons of supported platinum particles. *Catalysis Letters*, *9*(5–6), 347–354. https://doi.org/10.1007/Bf00764825.

Chin, A. B., & Yaacob, I. I. (2007). Synthesis and characterization of magnetic iron oxide nanoparticles via w/o microemulsion and Massart's procedure. *Journal of Materials Processing Technology*, *191*(1–3), 235–237. https://doi.org/10.1016/j.jmatprotec.2007.03.011.

Cui, X. W., Liu, R. J., Liu, Z. W., Shen, X. Q., Wang, Q. J., & Tan, X. L. (2014). Cationic poly-L-lysine Fe_2O_3/SiO_2 nanoparticles loaded with small interference RNA: application to silencing gene expression in primary rat neurons. *Journal of Nanoscience and Nanotechnology*, *14*(4), 2810–2815. https://doi.org/10.1166/jnn.2014.8554.

Drmota, A., MD, J. K., & Žnidaršič, A. (2012). Microemulsion method for synthesis of magnetic oxide nanoparticles. In R. Najjar (Ed.), *Microemulsions—an introduction to properties and applications* (pp. 250). London, UK: InTech.

Etienne, D., Marie-Hélène, D., & Stéphane, M. (2012). Synthesis and characterisation of iron oxide ferrite nanoparticles and ferrite-based aqueous fluids. In *Magnetic nanoparticles* (pp. 47–72). Boca Raton, FL: CRC Press. https://doi.org/10.1201/b11760-5.

Fedorov, Y., Anderson, E. M., Birmingham, A., Reynolds, A., Karpilow, J., Robinson, K., et al. (2006). Off-target effects by siRNA can induce toxic phenotype. *RNA*, *12*(7), 1188–1196. https://doi.org/10.1261/Rna.28106.

Feltin, N., & Pileni, M. P. (1997). New technique for synthesizing iron ferrite magnetic nanosized particles. *Langmuir*, *13*(15), 3927–3933. https://doi.org/10.1021/La960854q.

Fire, A., Xu, S. Q., Montgomery, M. K., Kostas, S. A., Driver, S. E., & Mello, C. C. (1998). Potent and specific genetic interference by double-stranded RNA in caenorhabditis elegans. *Nature*, *391*(6669), 806–811. https://doi.org/10.1038/35888.

Gao, L., Xie, L., Long, X., Wang, Z., He, C.-Y., Chen, Z.-Y., et al. (2013). Efficacy of MRI visible iron oxide nanoparticles in delivering minicircle DNA into liver via intrabiliary infusion. *Biomaterials*, *34*(14), 3688–3696. https://doi.org/10.1016/j.biomaterials.2013.01.094.

Grimm, D., Streetz, K. L., Jopling, C. L., Storm, T. A., Pandey, K., Davis, C. R., et al. (2006). Fatality in mice due to oversaturation of cellular microRNA/short hairpin RNA pathways. *Nature*, *441*(7092), 537–541. https://doi.org/10.1038/Nature04791.

Haussecker, D. (2014). Current issues of RNAi therapeutics delivery and development. *Journal of Controlled Release*, *195*, 49–54. https://doi.org/10.1016/j.jconrel.2014.07.056.

Hermanson, G. T. (2013). *Heterobifunctional crosslinkers. Bioconjugate techniques* (3rd ed., pp. 299–339). https://doi.org/10.1016/B978-0-12-382239-0.00006-6.

Hornung, V., Guenthner-Biller, M., Bourquin, C., Ablasser, A., Schlee, M., Uematsu, S., et al. (2005). Sequence-specific potent induction of IFN-alpha by short interfering RNA in plasmacytoid dendritic cells through TLR7. *Nature Medicine*, *11*(3), 263–270. https://doi.org/10.1038/Nm1191.

Hume, D. A. (2006). The mononuclear phagocyte system. *Current Opinion in Immunology*, *18*(1), 49–53. https://doi.org/10.1016/j.coi.2005.11.008.

Jackson, A. L., Burchard, J., Schelter, J., Chau, B. N., Cleary, M., Lim, L., et al. (2006). Widespread siRNA "off-target" transcript silencing mediated by seed region sequence complementarity. *RNA*, *12*(7), 1179–1187. https://doi.org/10.1261/rna.25706.

Jain, T. K., Reddy, M. K., Morales, M. A., Leslie-Pelecky, D. L., & Labhasetwar, V. (2008). Biodistribution, clearance, and biocompatibility of iron oxide magnetic nanoparticles in rats. *Molecular Pharmaceutics*, *5*(2), 316–327.

James, M. L., & Gambhir, S. S. (2012). A molecular imaging primer: modalities, imaging agents, and applications. *Physiological Reviews*, *92*(2), 897–965. https://doi.org/10.1152/physrev.00049.2010.

Jean-Pierre, J. (2000). *Metal oxide chemistry and synthesis: From solution to solid state*. New York: Wiley.

Jeong, U., Teng, X. W., Wang, Y., Yang, H., & Xia, Y. N. (2007). Superparamagnetic colloids: controlled synthesis and niche applications. *Advanced Materials*, *19*(1), 33–60. https://doi.org/10.1002/adma.200600674.

Jere, D., Arote, R., Jiang, H. L., Kim, Y. K., Cho, M. H., & Cho, C. S. (2009). Bioreducible polymers for efficient gene and siRNA delivery. *Biomedical Materials*, *4*(2), 025020https://doi.org/10.1088/1748-6041/4/2/025020.

Jolivet, J. P., Chaneac, C., & Tronc, E. (2004). Iron oxide chemistry. From molecular clusters to extended solid networks. *Chemical Communications*, (5), 481–487. https://doi.org/10.1039/b304532n.

Judge, A. D., Sood, V., Shaw, J. R., Fang, D., McClintock, K., & MacLachlan, I. (2005). Sequence-dependent stimulation of the mammalian innate immune response by synthetic siRNA. *Nature Biotechnology*, *23*(4), 457–462. https://doi.org/10.1038/nbt1081.

Jun, Y. W., Choi, J. S., & Cheon, J. (2006). Shape control of semiconductor and metal oxide nanocrystals through nonhydrolytic colloidal routes. *Angewandte Chemie, International Edition*, *45*(21), 3414–3439. https://doi.org/10.1002/anie.200503821.

Kanasty, R., Dorkin, J. R., Vegas, A., & Anderson, D. (2013). Delivery materials for siRNA therapeutics. *Nature Materials*, *12*(11), 967–977. https://doi.org/10.1038/NMAT3765.

Kim, H. J., Takemoto, H., Yi, Y., Zheng, M., Maeda, Y., Chaya, H., et al. (2014). Precise engineering of siRNA delivery vehicles to tumors using Polyion complexes and gold nanoparticles. *ACS Nano*, *8*(9), 8979–8991. https://doi.org/10.1021/nn502125h.

Koenig, S. H., & Kellar, K. E. (1995). Theory of 1/T1 and 1/T2 NMRD profiles of solutions of magnetic nanoparticles. *Magnetic Resonance in Medicine*, *34*(2), 227–233.

Kumar, M., Yigit, M., Dai, G. P., Moore, A., & Medarova, Z. (2010). Image-guided breast tumor therapy using a small interfering RNA nanodrug. *Cancer Research*, *70*(19), 7553–7561. https://doi.org/10.1158/0008-5472.CAN-10-2070.

LaMer, V. K., & Dinegar, R. H. (1950). Theory, production and mechanism of formation of monodispersed hydrosols. *Journal of the American Chemical Society*, *72*(11), 4847–4854. https://doi.org/10.1021/ja01167a001.

Laurent, S., Forge, D., Port, M., Roch, A., Robic, C., Elst, L. V., et al. (2008). Magnetic iron oxide nanoparticles: synthesis, stabilization, vectorization, physicochemical characterizations, and biological applications. *Chemical Reviews*, *108*(6), 2064–2110. https://doi.org/10.1021/cr068445e.

Lee, J. H., Lee, K., Moon, S. H., Lee, Y., Park, T. G., & Cheon, J. (2009). All-in-one target-cell-specific magnetic nanoparticles for simultaneous molecular imaging and siRNA delivery. *Angewandte Chemie, International Edition*, *48*(23), 4174–4179. https://doi.org/10.1002/anie.200805998.

Lee, J. H., Park, G., Hong, G. H., Choi, J., & Choi, H. S. (2012). Design considerations for targeted optical contrast agents. *Quantitative Imaging in Medicine and Surgery*, *2*(4), 266–273. https://doi.org/10.3978/j.issn.2223-4292.2012.12.04.

Lee, N., Yoo, D., Ling, D., Cho, M. H., Hyeon, T., & Cheon, J. (2015). Iron oxide based nanoparticles for multimodal imaging and magnetoresponsive therapy. *Chemical Reviews*, https://doi.org/10.1021/acs.chemrev.5b00112.

Levy, M., Luciani, N., Alloyeau, D., Elgrabli, D., Deveaux, V., Pechoux, C., et al. (2011). Long term in vivo biotransformation of iron oxide nanoparticles. *Biomaterials*, *32*(16), 3988–3999. https://doi.org/10.1016/j.biomaterials.2011.02.031.

Lin, M. M., Kim, D. K., El Haj, A. J., & Dobson, J. (2008). Development of superparamagnetic iron oxide nanoparticles (SPIONS) for translation to clinical applications. *IEEE Transactions on Nanobioscience*, *7*(4), 298–305. https://doi.org/10.1109/Tnb.2008.2011864.

Lin, Z. M., Monteiro-Riviere, N. A., & Riviere, J. E. (2015). Pharmacokinetics of metallic nanoparticles. *Wiley Interdisciplinary Reviews Nanomedicine and Nanobiotechnology*, *7*(2), 189–217. https://doi.org/10.1002/wnan.1304.

Lin, J., Zhou, W. L., Kumbhar, A., Wiemann, J., Fang, J. Y., Carpenter, E. E., et al. (2001). Gold-coated iron (Fe@Au) nanoparticles: synthesis, characterization, and magnetic field-induced self-assembly. *Journal of Solid State Chemistry*, *159*(1), 26–31. https://doi.org/10.1006/jssc.2001.9117.

Lipardi, C., Wei, Q., & Paterson, B. M. (2001). RNAi as random degradative PCR: siRNA primers convert mRNA into dsRNAs that are degraded to generate new siRNAs. *Cell*, *107*(3), 297–307.

Liu, G., Swierczewska, M., Lee, S., & Chen, X. (2010). Functional nanoparticles for molecular imaging guided gene delivery. *Nano Today*, *5*(6), 524–539. https://doi.org/10.1016/j.nantod.2010.10.005.

Liu, G., Wang, Z., Lee, S., Ai, H., & Chen, X. (2012). Chapter thirteen—design and fabrication of N-alkyl-polyethylenimine-stabilized iron oxide nanoclusters for gene delivery. In D. Nejat (Ed.), *Vol. 509. Methods in enzymology*. (pp. 263–276). Academic Press. https://doi.org/10.1016/B978-0-12-391858-1.00009-5.

Lopez-Lopez, M. T., Duran, J. D. G., Delgado, A., & Gonzalez-Caballero, F. (2005). Stability and magnetic characterization of oleate-covered magnetite ferrofluids in different nonpolar carriers. *Journal of Colloid and Interface Science*, *291*(1), 144–151. https://doi.org/10.1016/j.jcis.2005.04.099.

Lu, A. H., Salabas, E. L., & Schuth, F. (2007). Magnetic nanoparticles: synthesis, protection, functionalization, and application. *Angewandte Chemie, International Edition*, *46*(8), 1222–1244. https://doi.org/10.1002/anie.200602866.

Mahmoudi, M., Sant, S., Wang, B., Laurent, S., & Sen, T. (2011). Superparamagnetic iron oxide nanoparticles (SPIONs): development, surface modification and applications in chemotherapy. *Advanced Drug Delivery Reviews*, *63*(1–2), 24–46. https://doi.org/10.1016/j.addr.2010.05.006.

Massart, R. (1981). Preparation of aqueous magnetic liquids in alkaline and acidic media. *IEEE Transactions on Magnetics*, *17*(2), 1247–1248. https://doi.org/10.1109/Tmag.1981.1061188.

Massart, R., & Cabuil, V. (1987). Effect of some parameters on the formation of colloidal magnetite in alkaline-medium—yield and particle-size control. *Journal de Chimie Physique*, *84*(7–8), 967–973.

Medarova, Z., Pham, W., Farrar, C., Petkova, V., & Moore, A. (2007). In vivo imaging of siRNA delivery and silencing in tumors. *Nature Medicine*, *13*(3), 372–377.

Moore, A., & Medarova, Z. (2009). Imaging of siRNA delivery and silencing. *Methods in Molecular Biology*, *487*, 93–110. https://doi.org/10.1007/978-1-60327-547-7_5.

Muller, B. W., & Muller, R. H. (1984). Particle-size distributions and particle-size alterations in microemulsions. *Journal of Pharmaceutical Sciences*, *73*(7), 919–922. https://doi.org/10.1002/jps.2600730714.

Mykhaylyk, O., Vlaskou, D., Tresilwised, N., Pithayanukul, P., Moller, W., & Plank, C. (2007). Magnetic nanoparticle formulations for DNA and siRNA delivery. *Journal of Magnetism and Magnetic Materials*, *311*(1), 275–281. https://doi.org/10.1016/j.jmmm.2006.10.1178.

Napoli, C., Lemieux, C., & Jorgensen, R. (1990). Introduction of a chimeric chalcone synthase gene into petunia results in reversible co-suppression of homologous genes in trans. *Plant Cell*, *2*(4), 279–289. https://doi.org/10.1105/Tpc.2.4.279.

Nasongkla, N., Bey, E., Ren, J. M., Ai, H., Khemtong, C., Guthi, J. S., et al. (2006). Multifunctional polymeric micelles as cancer-targeted, MRI-ultrasensitive drug delivery systems. *Nano Letters*, *6*(11), 2427–2430. https://doi.org/10.1021/nl061412u.

Nayerossadat, N., Maedeh, T., & Ali, P. A. (2012). Viral and nonviral delivery systems for gene delivery. *Advanced Biomedical Research*, *1*, 27. https://doi.org/10.4103/2277-9175.98152.

Neely, L. A., Audeh, M., Phung, N. A., Min, M., Suchocki, A., Plourde, D., et al. (2013). T2 magnetic resonance enables nanoparticle-mediated rapid detection of candidemia in whole blood. *Science Translational Medicine*, *5*(182).

Oishi, M., Nakaogami, J., Ishii, T., & Nagasaki, Y. (2006). Smart PEGylated gold nanoparticles for the cytoplasmic delivery of siRNA to induce enhanced gene silencing. *Chemistry Letters*, *35*(9), 1046–1047. https://doi.org/10.1246/Cl.2006.1046.

OPKO Health, Inc. (n.d.). Open Label Study for the Evaluation of Tolerability of Five Dose Levels of Cand5 (NCT00722384). (Study Start Date: August 2004) OPKO Health, Inc. https://clinicaltrials.gov/ct2/show/study/NCT00722384?term=NCT00722384&rank=1 (accessed 24 July 2015).

Ozcan, G., Ozpolat, B., Coleman, R. L., Sood, A. K., & Lopez-Berestein, G. (2015). Preclinical and clinical development of siRNA-based therapeutics. *Advanced Drug Delivery Reviews*, *87*, 108–119. https://doi.org/10.1016/j.addr.2015.01.007.

Pichler, B. J., Judenhofer, M. S., & Pfannenberg, C. (2008). Multimodal imaging approaches: PET/CT and PET/MRI. *Handbook of Experimental Pharmacology*, *185*(Pt 1), 109–132. https://doi.org/10.1007/978-3-540-72718-7_6.

Prijic, S., & Sersa, G. (2011). Magnetic nanoparticles as targeted delivery systems in oncology. *Radiology and Oncology*, *45*(1), 1–16. https://doi.org/10.2478/v10019-011-0001-z.

Reynolds, A., Anderson, E. M., Vermeulen, A., Fedorov, Y., Robinson, K., Leake, D., et al. (2006). Induction of the interferon response by siRNA is cell type- and duplex length-dependent. *RNA*, *12*(6), 988–993. https://doi.org/10.1261/Rna.2340906.

Robinson, D. B., Persson, H. H. J., Zeng, H., Li, G. X., Pourmand, N., Sun, S. H., et al. (2005). DNA-functionalized MFe_2O_4 (M = Fe, Co, or Mn) nanoparticles and their hybridization to DNA-functionalized surfaces. *Langmuir*, *21*(7), 3096–3103. https://doi.org/10.1021/La047206o.

Romano, N., & Macino, G. (1992). Quelling—transient inactivation of gene-expression in neurospora-crassa by transformation with homologous sequences. *Molecular Microbiology*, *6*(22), 3343–3353. https://doi.org/10.1111/j.1365-2958.1992.tb02202.x.

Ros, P. R., Freeny, P. C., Harms, S. E., Seltzer, S. E., Davis, P. L., Chan, T. W., et al. (1995). Hepatic Mr-imaging with ferumoxides—a multicenter clinical-trial of the safety and efficacy in the detection of focal hepatic-lesions. *Radiology*, *196*(2), 481–488.

Ryther, R. C. C., Flynt, A. S., Phillips, J. A., & Patton, J. G. (2005). siRNA therapeutics: big potential from small RNAs. *Gene Therapy*, *12*(1), 5–11.

Sarrazin, S., Lamanna, W. C., & Esko, J. D. (2011). Heparan sulfate proteoglycans. *Cold Spring Harbor Perspectives in Biology*, *3*(7), a004952https://doi.org/10.1101/cshperspect.a004952.

Schwarzer, H. C., & Peukert, W. (2004). Tailoring particle size through nanoparticle precipitation. *Chemical Engineering Communications*, *191*(4), 580–606. https://doi.org/10.1080/00986440490270106.

Schwertmann, U., & Cornell, R. M. (2000). *Iron oxides in the laboratory: Preparation and characterization* (2nd ed.). Weinheim: Wiley-VCH.https://doi.org/10.1002/9783527613229.

Sijen, T., Fleenor, J., Simmer, F., Thijssen, K. L., Parrish, S., Timmons, L., et al. (2001). On the role of RNA amplification in dsRNA-triggered gene silencing. *Cell*, *107*(4), 465–476.

Singh, A., & Sahoo, S. K. (2014). Magnetic nanoparticles: a novel platform for cancer theranostics. *Drug Discovery Today*, *19*(4), 474–481. https://doi.org/10.1016/j.drudis.2013.10.005.

Sonawane, N. D., Szoka, F. C., Jr., & Verkman, A. S. (2003). Chloride accumulation and swelling in endosomes enhances DNA transfer by polyamine-DNA polyplexes. *Journal of Biological Chemistry*, *278*(45), 44826–44831. https://doi.org/10.1074/jbc.M308643200.

Song, X., Gong, H., Yin, S., Cheng, L., Wang, C., Li, Z., et al. (2014). Ultra-small iron oxide doped polypyrrole nanoparticles for in vivo multimodal imaging guided photothermal therapy. *Advanced Functional Materials*, *24*(9), 1194–1201. https://doi.org/10.1002/adfm.201302463.

Sugimoto, T. (2003). Formation of monodispersed nano- and micro-particles controlled in size, shape, and internal structure. *Chemical Engineering & Technology*, *26*(3), 313–321. https://doi.org/10.1002/ceat.200390048.

Sun, C., Lee, J. S. H., & Zhang, M. Q. (2008). Magnetic nanoparticles in MR imaging and drug delivery. *Advanced Drug Delivery Reviews*, *60*(11), 1252–1265. https://doi.org/10.1016/j.addr.2008.03.018.

Sun, S. N., Wei, C., Zhu, Z. Z., Hou, Y. L., Venkatraman, S. S., & Xu, Z. C. (2014). Magnetic iron oxide nanoparticles: synthesis and surface coating techniques for biomedical applications. *Chinese Physics B*, *23*(3), 037503https://doi.org/10.1088/1674-1056/23/3/037503.

Taratula, O., Garbuzenko, O., Savla, R., Wang, Y. A., He, H. X., & Minko, T. (2011). Multifunctional nanomedicine platform for cancer specific delivery of siRNA by superparamagnetic iron oxide nanoparticles-dendrimer complexes. *Current Drug Delivery*, *8*(1), 59–69. https://doi.org/10.2174/156720111793663642.

Tartaj, P., Morales, M. D., Veintemillas-Verdaguer, S., Gonzalez-Carreno, T., & Serna, C. J. (2003). The preparation of magnetic nanoparticles for applications in biomedicine. *Journal of Physics D: Applied Physics*, *36*(13), R182–R197. https://doi.org/10.1088/0022-3727/36/13/202.

Thomas, C. E., Ehrhardt, A., & Kay, M. A. (2003). Progress and problems with the use of viral vectors for gene therapy. *Nature Reviews Genetics*, *4*(5), 346–358. https://doi.org/10.1038/nrg1066.

van den Bos, E. J., Wagner, A., Mahrholdt, H., Thompson, R. B., Morimoto, Y., Sutton, B. S., et al. (2003). Improved efficacy of stem cell labeling for magnetic resonance imaging studies by the use of cationic liposomes. *Cell Transplantation*, *12*(7), 743–756.

Wan, Q., Xie, L., Gao, L., Wang, Z., Nan, X., Lei, H., et al. (2013). Self-assembled magnetic theranostic nanoparticles for highly sensitive MRI of minicircle DNA delivery. *Nanoscale*, *5*(2), 744–752. https://doi.org/10.1039/C2NR32438E.

Wang, Y. X. (2011). Superparamagnetic iron oxide based MRI contrast agents: current status of clinical application. *Quantitative Imaging in Medicine and Surgery*, *1*(1), 35–40. https://doi.org/10.3978/j.issn.2223-4292.2011.08.03.

Wang, J., Liu, G., Leung, K. C. F., Loffroy, R., Lu, P. X., & Wang, Y. X. (2015). Opportunities and challenges of fluorescent carbon dots in translational optical imaging. *Current Pharmaceutical Design*, *21*(42). https://doi.org/10.2174/1381612821666150917093232.

Wang, J., Lu, Z., Wientjes, M. G., & Au, J. L. S. (2010). Delivery of siRNA therapeutics: barriers and carriers. *AAPS Journal*, *12*(4), 492–503. https://doi.org/10.1208/s12248-010-9210-4.

Wang, Y., Wong, J. F., Teng, X. W., Lin, X. Z., & Yang, H. (2003). "Pulling" nanoparticles into water: phase transfer of oleic acid stabilized monodisperse nanoparticles into aqueous solutions of alpha-cyclodextrin. *Nano Letters*, *3*(11), 1555–1559. https://doi.org/10.1021/nl034731j.

Wang, J., Zhang, P., Huang, C., Liu, G., Leung, K.C.-F., & Wáng, Y. X. J. (2015). High performance photoluminescent carbon dots for in vitro and in vivo bioimaging: effect of nitrogen doping ratios. *Langmuir*, *31*(29), 8063–8073. https://doi.org/10.1021/acs.langmuir.5b01875.

Willmann, J. K., van Bruggen, N., Dinkelborg, L. M., & Gambhir, S. S. (2008). Molecular imaging in drug development. *Nature Reviews Drug Discovery*, *7*(7), 591–607. https://doi.org/10.1038/nrd2290.

Wu, S. Y., Lopez-Berestein, G., Calin, G. A., & Sood, A. K. (2014). RNAi therapies: drugging the Undruggable. *Science Translational Medicine*, *6*(240).

Xie, J., Liu, G., Eden, H. S., Ai, H., & Chen, X. Y. (2011). Surface-engineered magnetic nanoparticle platforms for cancer imaging and therapy. *Accounts of Chemical Research*, *44*(10), 883–892.

Xing, R., Bhirde, A., Wang, S., Sun, X., Liu, G., Hou, Y., et al. (2013). Hollow iron oxide nanoparticles as multidrug resistant drug delivery and imaging vehicles. *Nano Research*, *6*(1), 1–9. https://doi.org/10.1007/s12274-012-0275-5.

Xing, R. J., Liu, G., Zhu, J. H., Hou, Y. L., & Chen, X. Y. (2014). Functional magnetic nanoparticles for non-viral gene delivery and MR imaging. *Pharmaceutical Research*, *31*(6), 1377–1389. https://doi.org/10.1007/s11095-013-1205-2.

Yi, D. K., Lee, S. S., Papaefthymiou, G. C., & Ying, J. Y. (2006). Nanoparticle architectures templated by SiO_2/Fe_2O_3 nanocomposites. *Chemistry of Materials*, *18*(3), 614–619. https://doi.org/10.1021/cm0512979.

Yoo, D., Lee, J. H., Shin, T. H., & Cheon, J. (2011). Theranostic magnetic nanoparticles. *Accounts of Chemical Research*, *44*(10), 863–874.

Index

Note: Page numbers followed by *f* indicate figures and *t* indicate tables.

F

G

K

L

M

N

Q

R

S